Specialty Polymers

This comprehensive volume provides current, state-of-the-art information on specialty polymers that can be used for many advanced applications. The book covers the fundamentals of specialty polymers, synthetic approaches, and chemistries to modify their properties to meet the requirements for special applications, along with current challenges and prospects. Chapters are written by global experts, making this a suitable textbook for students and a one-stop resource for researchers and industry professionals.

Key Features:

- Presents synthesis, characterization, and applications of specialty polymers for advanced applications.
- Provides fundamentals and requirements for polymers to be used in many advanced and emerging areas.
- Details novel methods and advanced technologies used in polymer industries.
- Covers the state-of-the-art progress on specialty polymers for a range of advanced applications.

Specialty Polymers

Fundamentals, Properties, Applications and Advances

Edited by
Ram K. Gupta

CRC Press
Taylor & Francis Group
Boca Raton London New York

CRC Press is an imprint of the
Taylor & Francis Group, an **informa** business

Cover image © Shutterstock

First edition published 2023
by CRC Press
6000 Broken Sound Parkway NW, Suite 300, Boca Raton, FL 33487-2742

and by CRC Press
4 Park Square, Milton Park, Abingdon, Oxon, OX14 4RN

CRC Press is an imprint of Taylor & Francis Group, LLC

© 2023 selection and editorial matter, Ram K. Gupta; individual chapters, the contributors

Library of Congress Cataloging-in-Publication Data
Names: Gupta, Ram K., editor.
Title: Specialty polymers : fundamentals, properties, applications and
advances / edited by Ram K. Gupta.
Other titles: Specialty polymers (CRC Press)
Description: First edition. | Boca Raton : CRC Press, 2023. |
Includes bibliographical references and index. |
Identifiers: LCCN 2022037366 (print) | LCCN 2022037367 (ebook) |
Subjects: LCSH: Polymers. | Polymers–Industrial applications.
Classification: LCC TP1087 .S64 2023 (print) |
LCC TP1087 (ebook) | DDC 668.9–dc23/eng/20221108
LC record available at https://lccn.loc.gov/2022037366
LC ebook record available at https://lccn.loc.gov/2022037367

ISBN: 9781032243726 (hbk)
ISBN: 9781032243740 (pbk)
ISBN: 9781003278269 (ebk)

DOI: 10.1201/9781003278269

Typeset in Times
by Newgen Publishing UK

Contents

Contributors

Nazlı Albayrak
School of Medicine, Acibadem M. A. Aydınlar
    University, Istanbul, 34752, Turkey

Cassia A. Allison
National Institute for Materials Advancement,
    Pittsburg State University, Pittsburg, KS
    66762, USA
Pittsburg High School, Pittsburg, KS
    66762, USA

Khairunnisa Amreen
MEMS, Microfluidics and Nanoelectronics
    Lab, Department of Electrical and
    Electronics Engineering, Birla Institute of
    Technology and Science Pilani, Hyderabad
    Campus, Hyderabad 500078, India

Ana Arnaiz
Departamento de Química, Facultad de
    Ciencias, Universidad de Burgos, Plaza de
    Misael Bañuelos s/n, 09001 Burgos, Spain
Universidad Politécnica de Madrid, Calle
    Ramiro de Maeztu 7, 28040 Madrid, Spain

Rossella Arrigo
Department of Applied Science and
    Technology, Politecnico di Torino, and Local
    INSTM Unit, Viale Teresa Michel 5, 15121,
    Alessandria, Italy

Magdalene A. Asare
National Institute for Materials Advancement,
    Pittsburg State University, Pittsburg, KS
    66762, USA
Department of Chemistry, Pittsburg State
    University, Pittsburg, KS 66762, USA

Larissa Bach-Toledo
Grupo de Pesquisas em Macromoléculas e
    Interfaces (GPMIn), Universidade Federal do
    Paraná, CP 19032, CEP 81531-980 Curitiba,
    PR, Brazil

Mattia Bartoli
Center for Sustainable Future, Italian Institute
    of Technology, Via Livorno 60, 10144
    Turin, Italy
Consorzio Interuniversitario Nazionale per la
    Scienza e Tecnologia dei Materiali (INSTM),
    Via G. Giusti 9, 50121 Florence, Italy

Saeed Bastani
Surface Coatings and Corrosion Dept.,
    Institute for Color Science and Technology,
    Vafamanesh St., Tehran, 1654-16765, Iran
Printing Ink Science and Technology Dept.,
    Institute for Color Science and Technology,
    Vafamanesh St., Tehran, 1654-16765, Iran

Dipankar Chattopadhyay
Department of Polymer Science and
    Technology, University of Calcutta, 92
    A.P.C. Road, Kolkata – 700 009, West
    Bengal, India
Center for Research in Nanoscience and
    Nanotechnology, Acharya Prafulla Chandra
    Roy Sikhsha Prangan, University of Calcutta,
    JD-2, Sector-III, Saltlake City, Kolkata-
    700098, West Bengal, India

Mahesh Chaudhari
National Institute for Materials Advancement,
    Pittsburg State University, Pittsburg, KS
    66762, USA
Department of Chemistry, Pittsburg State
    University, Pittsburg, KS 66762, USA

Pragati Chauhan
Department of Chemistry, Banasthali
    Vidyapith, Rajasthan 304022, India

Karan W. Chugh
Department of Polymer and Surface
    Engineering, Institute of Chemical
    Technology, Nathalal Parekh Marg, Matunga
    (E), Mumbai-400019, India

Farnaz Dabbagh Moghaddam
Institute for Photonics and Nanotechnologies,
  National Research Council, Via Fosso del
  Cavaliere, 100, 00133, Rome, Italy

Tim Dawsey
National Institute for Materials Advancement,
  Pittsburg State University, Pittsburg, KS
  66762, USA

Gabriela De Alvarenga
Grupo de Pesquisas em Macromoléculas e
  Interfaces (GPMIn), Universidade Federal do
  Paraná, CP 19032, CEP 81531-980 Curitiba,
  PR, Brazil

Felipe M. de Souza
National Institute for Materials Advancement,
  Pittsburg State University, Pittsburg, KS
  66762, USA

Andrei Deller
Grupo de Pesquisas em Macromoléculas e
  Interfaces (GPMIn), Universidade Federal do
  Paraná, CP 19032, CEP 81531-980 Curitiba,
  PR, Brazil

Yash N. Desai
National Institute for Materials Advancement,
  Pittsburg State University, Pittsburg, KS
  66762, USA
Department of Chemistry, Pittsburg State
  University, Pittsburg, KS 66762, USA

Emine Dilara Kocak
Marmara University Faculty of Technology,
  Recep Tayyip Erdogan Kulliye Aydınevler
  Mah. Idealtepe Yolu no:15, 34854, Maltepe/
  İstanbul, Turkey

Alireza Fatahi
Department of Chemistry, University of
  Isfahan, Isfahan 81746-73441, I.R. Iran

Morteza Ganjaee Sari
Nano Technology Dept., Institute for Color
  Science and Technology, Vafamanesh St.,
  Tehran, 1654-16765, Iran

José M. García
Departamento de Química, Facultad de
  Ciencias, Universidad de Burgos, Plaza de
  Misael Bañuelos s/n, 09001 Burgos, Spain

Neena George
Post Graduate and Research Department of
  Chemistry, Maharaja's College, Eranakulam-
  682 011, Kerala, India

Matineh Ghomi
School of Chemistry, Damghan University,
  Damghan, 36716-41167, Iran

Adrija Ghosh
Department of Polymer Science and
  Technology, University of Calcutta, 92
  A.P.C. Road, Kolkata – 700 009, West
  Bengal, India

Sanket Goel
MEMS, Microfluidics and Nanoelectronics
  Lab, Department of Electrical and
  Electronics Engineering, Birla Institute of
  Technology and Science Pilani, Hyderabad
  Campus, Hyderabad 500078, India

Çiğdem Gül
Marmara University, Institute of Pure and
  Applied Sciences, Goztepe Campus, The
  Buildings of Institutes, Floor: 2, 34722,
  Kadıköy/İstanbul, Turkey

Anjali Gupta
National Institute for Materials Advancement,
  Pittsburg State University, Pittsburg, KS
  66762, USA
Pittsburg High School, Pittsburg, KS
  66762, USA

Najihah Binti Mohd Hashim
Department of Pharmaceutical Chemistry,
  Universiti Malaya, 50603 Kuala Lumpur,
  Malaysia

Md. Saddam Hossain
Department of Chemistry, Khulna University
  of Engineering & Technology, Khulna-9203,
  Bangladesh

Bruna M. Hryniewicz
Grupo de Pesquisas em Macromoléculas e
    Interfaces (GPMIn), Universidade Federal do
    Paraná, CP 19032, CEP 81531-980 Curitiba,
    PR, Brazil

Md. Mahinur Islam
Department of Chemistry, University of Dhaka,
    Dhaka-1000, Bangladesh

Md. Mominul Islam
Department of Chemistry, University of Dhaka,
    Dhaka-1000, Bangladesh

Samane Jafarifard
Polymer Engineering and Color Technology
    Dept., Amirkabir University of Technology,
    Hafez St., Tehran, 15875-4413, Iran

Rani Joseph
Department of Polymer Science and Rubber
    Technology, Cochin University of Science
    and Technology, Cochin-22, Kerala, India

Jingjing Kang
The Key Laboratory of Environmental
    Pollution Monitoring and Disease Control,
    Ministry of Education, Guizhou Medical
    University, Guiyang 550025, China
Engineering Research Center of Higher
    Education Institutions of Guizhou Province,
    School of Public Health, Guizhou Medical
    University, Guiyang 550025, PR China

Selcan Karakuş
Department of Chemistry, Faculty of
    Engineering, Istanbul University-Cerrahpasa,
    Avcılar, 34320, Istanbul, Turkey

Vanessa Klobukoski
Grupo de Pesquisas em Macromoléculas e
    Interfaces (GPMIn), Universidade Federal do
    Paraná, CP 19032, CEP 81531-980 Curitiba,
    PR, Brazil

Emine D. Kocak
Department of Textile Engineering, Faculty
    of Technology, Marmara University, RTE
    Campus, Aydinevler District, Uyanik Street,
    34840, Istanbul, Turkey

Panagiota Koralli
National Hellenic Research Foundation, 48
    Vassileos Constantinou Avenue, 11635,
    Athens, Greece

Dinesh Kumar
School of Chemical Sciences, Central
    University of Gujarat, Gandhinagar
    382030, India

K.A.U. Madhushani
Department of Chemistry, Pittsburg State
    University, Pittsburg, Kansas 66762, USA
National Institute for Materials Advancement,
    Pittsburg State University, Pittsburg, KS
    66762, USA

Umesh R. Mahajan
Department of Polymer and Surface
    Engineering, Institute of Chemical
    Technology, Nathalal Parekh Marg, Matunga
    (E), Mumbai-400019, India

Syed Mahmood
Department of Pharmaceutical Technology,
    Universiti Malaya, 50603 Kuala Lumpur,
    Malaysia

Giulio Malucelli
Department of Applied Science and Technology
    and Local INSTM Unit, Politecnico di
    Torino, Viale Teresa Michel 5, 15121,
    Alessandria, Italy

S.T. Mhaske
Department of Polymer and Surface
    Engineering, Institute of Chemical
    Technology, Nathalal Parekh Marg, Matunga
    (E), Mumbai-400019, India

Álvaro Miguel
Departamento de Química, Facultad de
    Ciencias, Universidad de Burgos, Plaza de
    Misael Bañuelos s/n, 09001 Burgos, Spain
Universidad Autónoma de Madrid, Calle
    Einstein 3, 28049 Madrid, Spain

Abbas Mohammadi
Department of Chemistry, University of
    Isfahan, Isfahan 81746-73441, I.R. Iran

Jyotidarsan Mohanty
Department of Polymer and Surface
  Engineering, Institute of Chemical
  Technology, Nathalal Parekh Marg, Matunga
  (E), Mumbai-400019, India

Dionysios E. Mouzakis
Department of Military Studies, Division of
  Mathematics and Engineering Sciences,
  Hellenic Army Academy, Vari, 16673,
  Attica, Greece

Ajalesh B. Nair
Post Graduate and Research Department of
  Chemistry, Union Christian College, Aluva-
  683 102, Kerala, India

Nektarios K. Nasikas
Department of Military Studies, Division of
  Mathematics and Engineering Sciences,
  Hellenic Army Academy, Vari, 16673,
  Attica, Greece

Suresh Nayar
Institute for Reconstructive Sciences in
  Medicine, 16940, 87 Avenue, Edmonton,
  Alberta, Canada
Department of Surgery, Faculty of Medicine
  and Dentistry, University of Alberta, Canada

Ehsan Nazarzadeh Zare
School of Chemistry, Damghan University,
  Damghan, 36716-41167, Iran

Sinem Özlem Enginler
Department of Obstetrics and Gynecology,
  Faculty of Veterinary Medicine, Istanbul
  University-Cerrahpasa, Avcılar, 34320,
  Istanbul, Turkey

Styliani Papatzani
University of West Attica, 28 Agiou Spiridonos,
  12243, Egaleo, Greece
Hellenic Army Academy, Leoforos Eyelpidon
  (Varis – Koropiou) Avenue, 16673,
  Vari, Greece

A.A.P.R. Perera
Department of Chemistry, Pittsburg State
  University, Pittsburg, Kansas 66762, USA

National Institute for Materials Advancement,
  Pittsburg State University, Pittsburg, KS
  66762, USA

Buwanila T. Punchihewa
Department of Chemistry, University of
  Missouri-Kansas City, Missouri 64110, USA

Simi Pushpan K.
Post Graduate and Research Department of
  Chemistry, Union Christian College, Aluva-
  683 102, Kerala, India

Mohamad Reza Sarfjoo
Department of Chemistry, Isfahan University of
  Technology, Isfahan 415683111, I.R. Iran

Dinesh Rokaya
Department of Clinical Dentistry, Walailak
  University International College of Dentistry,
  Bangkok 10400, Thailand

Jean Gustavo de A. Ruthes
Grupo de Pesquisas em Macromoléculas e
  Interfaces (GPMIn), Universidade Federal do
  Paraná, CP 19032, CEP 81531-980 Curitiba,
  PR, Brazil

Sasiwimol Sanohkan
Department of Prosthetic Dentistry, Faculty of
  Dentistry, Prince of Songkla University, Hat
  Yai, Songkhla 90110, Thailand

Ashutosh K. Singh
Department of Oral and Maxillofacial
  Surgery, Tribhuvan University Teaching
  Hospital, Institute of Medicine, Kathmandu
  44600, Nepal

Mansi Sharma
Department of Chemistry, Banasthali
  Vidyapith, Rajasthan 304022, India

Rekha Sharma
Department of Chemistry, Banasthali
  Vidyapith, Rajasthan 304022, India

Md. Sadiqul Islam Sheikh
Department of Chemistry, University of Dhaka,
  Dhaka-1000, Bangladesh

Department of Chemistry, Khulna University
  of Engineering & Technology, Khulna-9203,
  Bangladesh

Rafael J. Silva
Grupo de Pesquisas em Macromoléculas e
  Interfaces (GPMIn), Universidade Federal do
  Paraná, CP 19032, CEP 81531-980 Curitiba,
  PR, Brazil

Suprakash Sinha Ray
Department of Textile Technology, Kaduna
  Polytechnic, P. M. B. 2021, Tudun-Wada,
  Kaduna, Nigeria
DST-CSIR National Centre for Nanostructured
  Materials, Council for Scientific and
  Industrial Research, Pretoria 0001,
  South Africa

Zarif Mohamed Sofian
Department of Pharmaceutical Technology,
  Universiti Malaya, 50603 Kuala Lumpur,
  Malaysia

Melbha Starlin Chellathurai
Department of Pharmaceutical Technology,
  Universiti Malaya, 50603 Kuala Lumpur,
  Malaysia

Vishwa Suthar
Department of Chemistry, Pittsburg State
  University, Pittsburg, KS 66762, USA
National Institute for Materials Advancement,
  Pittsburg State University, Pittsburg, KS
  66762, USA

Alberto Tagliaferro
Consorzio Interuniversitario Nazionale per la
  Scienza e Tecnologia dei Materiali (INSTM),
  Via G. Giusti 9, 50121 Florence, Italy
Department of Applied Science and
  Technology, Politecnico di Torino, C.so Duca
  degli Abruzzi 24, 10129 Turin, Italy
Faculty of Science, Ontario Tech University,
  2000 Simcoe Street North, Oshawa, Canada

Jonathan Tersur Orasugh
Department of Polymer Science and
  Technology, University of Calcutta, 92

A.P.C. Road, Kolkata – 700 009, West
  Bengal, India
Department of Chemical Sciences, University
  of Johannesburg, Doorfontein, Johannesburg
  2028, South Africa
Department of Textile Technology, Kaduna
  Polytechnic, P. M. B. 2021, Tudun-Wada,
  Kaduna, Nigeria
DST-CSIR National Centre for Nanostructured
  Materials, Council for Scientific and
  Industrial Research, Pretoria 0001,
  South Africa

Miriam Trigo-López
Departamento de Química, Facultad de
  Ciencias, Universidad de Burgos, Plaza de
  Misael Bañuelos s/n, 09001 Burgos, Spain

Tatiana L. Valerio
Grupo de Pesquisas em Macromoléculas e
  Interfaces (GPMIn), Universidade Federal do
  Paraná, CP 19032, CEP 81531-980 Curitiba,
  PR, Brazil

Saúl Vallejos
Departamento de Química, Facultad de
  Ciencias, Universidad de Burgos, Plaza de
  Misael Bañuelos s/n, 09001 Burgos, Spain

Marcio Vidotti
Grupo de Pesquisas em Macromoléculas e
  Interfaces (GPMIn), Universidade Federal do
  Paraná, CP 19032, CEP 81531-980 Curitiba,
  PR, Brazil

Franciele Wolfart
Instituto Federal de Educação, Ciência e
  Tecnologia Farroupilha – Campus São Borja,
  Rua Otaviano Castilho Mendes, 355, Betim,
  CEP 97670-000, São Borja – RS, Brazil

Jiao Xie
The Key Laboratory of Environmental
  Pollution Monitoring and Disease Control,
  Ministry of Education, Guizhou Medical
  University, Guiyang 550025, China
Engineering Research Center of Higher
  Education Institutions of Guizhou Province,
  School of Public Health, Guizhou Medical
  University, Guiyang 550025, PR China

Mahtab Yadolahi
School of Chemistry, Damghan University,
  Damghan, 36716-41167, Iran

Zehra Yildiz
Department of Textile Engineering, Faculty
  of Technology, Marmara University, RTE

Campus, Aydinevler District, Uyanik Street,
  no:6, T1/319, 34840, Istanbul, Turkey

Priyesh Zalavadiya
National Institute for Materials Advancement,
  Pittsburg State University, Pittsburg, KS
  66762, USA

About the Author

Ram K. Gupta is Associate Professor at Pittsburg State University. Dr. Gupta's research focuses on green energy production, storage using 2D materials, optoelectronics and photovoltaics devices, bio-based polymers, flame-retardant polyurethanes, conducting polymers and composites, organic–inorganic hetero-junctions for sensors, bio-compatible nanofibers for tissue regeneration, scaffold and antibacterial applications, and bio-degradable metallic implants. Dr. Gupta has published around 250 peer-reviewed articles, made over 350 national/international/regional presentations, chaired many sessions at national/international meetings, and edited/wrote several books/chapters for leading publishers such as the American Chemical Society, Royal Society of Chemistry, CRC, Elsevier, Springer, and Wiley. He has published many research articles related to energy materials in high-impact journals such as *Nature Communications*, *Solar Energy*, *Scientific Reports*, *Journal of Materials Chemistry A*, and *Chemistry of Materials*, to name but a few. He has received several million dollars for research and educational activities from external agencies. He is serving as Editor-in-Chief, Associate Editor, and editorial board member for several journals.

1 Specialty Polymers
An Introduction

Yash N. Desai,[1,2] Magdalene A. Asare,[1,2]
Felipe M. de Souza,[1] and Ram K. Gupta[1,2]

[1] National Institute for Materials Advancement,
Pittsburg State University, Pittsburg, KS 66762, USA
[2] Department of Chemistry, Pittsburg State University, Pittsburg,
KS 66762, USA

1 INTRODUCTION

Polymers are generally composed of organic compounds mostly based on C, H, O, and N, which form a long chain of small repeating units known as monomers [1]. The human body and other organisms also consist of polymers in the forms of proteins, carbohydrates, and glucose. Subsequently, polymers can be classified into several categories according to their behavior, morphology, process-induced polymers, the source from which they are obtained, and based on their application and properties [2]. The first category is presented as natural and synthetic polymers which heavily depends on the source from which the polymer is obtained. For example, plant-based polymers such as lignin and cellulose are natural polymers and biodegradable substances. Synthetic polymers are artificially produced by humans and have varying biodegradability. For instance, polyolefin is obtained through the cracking of hydrocarbons which is prominently known as naphtha cracking, and these polymers are non-biodegradable, however, the first polyethylene was produced through sugarcane cellulose with an old-school conventional method.

The second type of classification is based on the behavior of the polymer when they encounter heat or high temperature. Thermo-plastics are the type of polymers that can be reheated, and they tend to flow so they can be reprocessed or remolded into the desired shape. Polystyrene is an example of a thermoplastic polymer that is generally manufactured by the tower process. However, thermoset plastics, as inferred from the name, reflect the property that once they attain curing and settle down, they cannot be reheated and remolded as they degrade if exposed to a temperature higher than their service temperature. Formaldehyde resins are considered thermoset polymers, they don't soften on the re-application of heat mainly because of their cross-linked structure and the bonds don't tend to stretch or gain energy from the external source.

Polymers are also categorized based on their structure: linear, branched, and cross-linked. Linear polymers have no branches and the monomers are joined to each other from end to end of the monomers. A typical example of a linear polymer is polytetrafluoroethylene (Teflon). Branched polymers have some bonds attached from the second polymer chain to the primary backbone chain, with starch as an example of a branched biopolymer. The cross-linked structure of the polymer is formed when too many secondary polymeric chains are attached to the primary polymer chain and form a 3D web-like structure. This structure decides the properties of the polymers. Polyurethanes, which are a versatile example of polymers, can be cross-linked.

Another polymer classification is defined by the polymerization process reaction by which the polymer is manufactured, which involves addition and condensation polymerization. In addition polymerization, all the monomers turn out into polymeric chains, and there is no liberation of any

DOI: 10.1201/9781003278269-1

molecules, an example is the synthesis of polyethylene. Whereas condensation polymerization primarily liberates by-products such as water molecules with the formation of the polymer. For instance, acrylonitrile butadiene styrene is the yield of a condensation polymerization with the liberation of water. Also, the polymer formed through the condensation is a heteropolymer where two or more monomers join together to yield a polymer.

Polymers are also classified according to their applications, based on their properties which can be presented as a commodity, engineering, and the category mainly discussed in this chapter, i.e. specialty polymers. Besides the type of application and properties, the manufacturing cost is also considered before selecting the particular polymer. Commodity polymers, as reflected in the name, are utilized because of their availability, manufacturing cost, and most importantly, the phase of their application. Polyethylene and its derivatives are commodity polymers due to the low properties they display. And to be more specific, high-density polyethylene (HDPE) which is the most superior in polyethylenes, displays a low melt and service temperature of 135 °C and 80 °C, respectively. In engineering polymers, properties in terms of heat deflection, and mechanical and chemical strength are higher than commodity polymers. For example, Nylon 6,6, which is an engineering polymer, displays a high service temperature of 220 °C, and robust physical and chemical features that outnumber the properties of commodity polymers such as HDPE, as discussed above.

Considering the differences among the various categories of polymers, specialty polymers reflect the top-notch properties for every aspect of consideration, such as mechanical properties and chemical resistance, and subsequently, they also sustain high service temperatures. Specialty polymers can also be defined as polymers that display specific properties and will be procured into a specific application. In the applications where specialty polymers are used, the manufacturing cost of the parts is a secondary concern, so the product's price is not considered to obtain the desired properties. When requiring a wide range of properties such as optimum service temperature, mechanical and chemical resistance, and biocompatibility of a polymer, specialty polymers excel in such areas wherever they are utilized. Specialty polymers are emerging daily and gaining popularity over other polymers because of the qualities they offer in the application. They have multifunctional properties and that's the reason they are available in several applications such as textiles, motion-sensing, electronics such as semiconductors, biomedical applications for drug-delivery systems, and energy supply solutions such as batteries, cells, and supercapacitors. In addition, an innovative application of specialty polymers is their use in the coating industry for corrosion protection. Luxury automobiles, as well as prime car parts, are nowadays being manufactured with specialty polymers due to their identical strength to metals that were conventionally utilized in the automobile industries. Additionally, specialty polymers are more attractive than metals due to their improved wear-and-tear properties as well as better abrasion resistance. They also exhibit excellent service temperatures, making them perfect for automobile and textile applications.

2 SPECIALTY POLYMERS AND THEIR APPLICATIONS

2.1 POLYMERS FOR CORROSION PROTECTION

Polymers have always been considered as insulators of electricity since their invention. However, in the 1960s, many scientists synthesized and researchers derived some polymers that conducted electric charges and were called semiconductors [3]. However, metals were only considered as conductors in addition to their high mechanical strength and better serviceability. On the other hand, metals tend to oxidize in the presence of normal air and moisture leading to corrosion. This widely known phenomenon of corrosion on metals causes huge losses as the product becomes unfit for further application. To overcome this issue, electrically active or semiconducting polymers have been introduced. The main idea behind using the polymers is they inhibit the reaction on the surface of the metals which eventually prevents corrosion. The use of intrinsically conducting polymers is one of the major aspects of considering specialty polymers for corrosion. Polyaniline and polypyrrole are applied on the metal surface in their oxidative state, which improves the oxidation reaction on the polymer/metal surface [3]. This can also be said as the application of this coating shifts the surface of corrosion reaction from the metal to the polymeric coating and protects the metal. In addition, the coating overcomes some defects on the metal surface in the passive domain. In the past, metals were compounded or alloyed with each other, which could protect the prime metal from corrosion up to some extent but could not provide the desired service. However, specialty polymers such as polyphenols and polythiophene can better serve this area of need. Polyphenols, especially natural occurring one's, comprise several hydroxyl groups and display excellent adhesive and good anti-oxidant properties when adopted for corrosion-resistant coatings [4]. Polythiophene also possesses electro-conductivity, resulting in the delocalization of electrons along its backbone chain that eventually inhibits corrosion on the metal surface [3].

Prior adsorption of the coatings on the metal surface acts as an anticorrosion coating, and the hydrophobicity of polymers serves as a great advantage for corrosion inhibition. Other polymers made up of epoxies have been found to display good water barrier properties and can be chosen for applications in the corrosion-resistant coating. To add to this, epoxy has better strength in terms of mechanical properties and does not easily wear off the metal surface once it is cured. However, the application of epoxy coatings compromises the aesthetic quality of the surface. Corrosion is also a major issue in the marine industry, due to high levels of salt in the water that escalates the rate of corrosion. Inhibitors in the form of azoles are incorporated with different heterocyclic polymers, polythiopropinate and polymaleic acid have been developed to inhibit iron and steel corrosion while in seawater and other acidic environments. Every metal is affected by corrosion, however coating of poly-N-vinyl imidazole (PVI) can reduce the corrosion. Subsequently, azole compounds such as benzotriazole act as corrosion-prevention agents on the copper surface [5]. Initially, corrosion pigments were added to paints before applying them to the surfaces of metals to prevent corrosion. Despite that, this method lacked serviceability, and pigments such as chromates that were used are toxic and carcinogenic. Their usage has therefore decreased and electronically conducting polymer coatings which have greater advantages are more popular these days.

The automobile manufacturing industry also faces issues arising from corrosion, especially during the primary stages of cutting coiled stainless steel parts for different automobile applications. Here, the exposed surfaces from the different cut sizes are inevitably susceptible to faster corrosion, hence, coating them with corrosion-resistant polymeric materials has proven a suitable solution. Offshore facilities such as oil and gas pipelines are exposed to high-saline water, and to protect them from serious corrosion and maintain their structural integrity, graphene-based polyurethane coatings are applied, as shown in Figure 1.1, which inhibit corrosion and demonstrate self-healing properties which itself is an excellent advantage to save the coating from general wear and tear. Here, the polymeric coating demonstrates better barrier properties, which act as the prime working

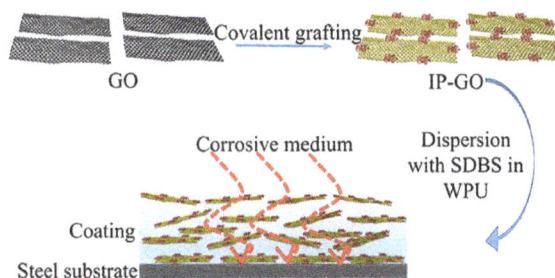

FIGURE 1.1 Image explaining covalent grafting of graphene-composed polyurethane for corrosion prevention on steel. Adapted with permission [6]. Copyright (2019) American Chemical Society. Further permission related to the material excerpted should be directed to the ACS.

mechanism behind its application on the metal surface [6]. Successful experiments have also been conducted where covalent grafting of graphene oxide with isophorone diisocyanate and N,N-dimethyl-ethanolamine have been used to manufacture waterborne polyurethane. Overall, graphene serves as the filler or additive utilized in the procured PU polymer matrix in the effort to protect the steel substrate from corrosion.

As ships in the marine industry corrode due to the saltwater they move in, so will electrodes of batteries due to their electrolytes. This affects the life cycle, efficiency, and proper output of the batteries. To deal with this, some scientists have synthesized a current collector as well as corrosion protector composite coating, incorporating carbon black into polyethylene. Based on their results, the redox reaction that involves the transfer of charge from the cathode to the anode, which eventually erodes the metal layer and promotes corrosion, was reduced in the batteries [7]. In effect, Li-ion batteries demonstrated superior corrosion resistance and had improved overall performance [8].

2.2 Polymers for Aerospace and Automobile Industries

Polymeric materials constitute a high fraction of structures for aerospace and automobile application. And, with the high demand for manufacturing lighter and more efficient machines, it has become crucial for industries to introduce polymers and their composites into the production system. Polymers can now be seen in both interior and exterior parts of machines and are gradually replacing metallic parts. In addition to their light weight, which is comparably 20% lesser than aluminum, for example, they require minimal maintenance, they have high performance, and they are the best fit for these industries [9]. Once the material weight is decreased, the total weight of the machine also decreases, and the high consumption of fuel is eventually reduced, bringing increased sustainability to the environment.

The automobile industry has revolutionized thus far and it will continue to do so at a very high speed. Gears that were initially manufactured from metals have now been replaced by nylons. Nylons are therefore specialty polymers due to their self-lubricating properties that result in less wear and tear caused by heavy friction while rotating. They also require low maintenance and are corrosion resistant. Silicones, which are also specialty polymers, can be used in gears and also as synthetic engine oils to lubricate and regulate the temperature of engines. However, if they are compared with conventional engine oils made from petroleum, then silicon-based oils have a longer service life and better performance. Silicon lubrication is also necessary for hydraulic cylinders and gas turbines in aircraft. The advantage of using silicone-based lubricants is they can be operated at very high temperatures.

On the other hand, fuel storage tanks and insulating materials for both automobiles and aircraft are being manufactured with specialty polymers like polyurethane. Polyurethane is also used in the manufacturing of seats and cushions. Due to the need for high impact resistance and excellent flexibility, body bumpers of automobiles are made of specialty polymers. Automobiles when used for some special purposes need to be bulletproof and for this purpose, Kevlar, polycarbonate, Taron, and other specialty materials are used [10]. In addition, polymers add an aesthetic look and offer outstanding properties identical to those of metallic parts, and these polymeric parts also have longer service lives.

Some examples should be taken into consideration, such as several types of aircrafts including the Airbus 380 have been made of plastic and consume around 60% less aluminum and other metals as compared to other models. Carbon fibers were once used for most parts as carbon fiber provided high tolerance to damage, and the application of rivets was reduced, which improved the part integrity [9]. Engine parts such as turbines were made of Kevlar shells which could sustain high service temperatures. Apart from carbon fiber composites, glass fiber-reinforced plastic was used to mold critical designs such as rear fin edges. Parts made from composites were very easy to install. According to reports from the manufacturing company, utilization of composites even increased the service life of the aircraft from six to twelve years.

2.3 Polymers for Shape-Memory Applications

The shape-memory polymers (SMPs) can be also classified as stimuli-sensitive polymers and they can be defined as polymers that tend to temporarily change their shape and retain their original form whenever specific stimuli such as heat, light, pH, moisture, and a magnetic and electrical field are applied [11]. The unique feature of SMPs is that they can recover a predefined shape by up to 100%, allowing them to be utilized in several applications such as self-healing plastics, actuators, biocompatible applications such as self-expandable stents, and many orthopedic applications. SMPs work in a combination of programming and recovery processes. In programming, the application of an external force results in a temporary shape change to the polymer, while the temperature is elevated above the phase transition temperature and constant stress. In the recovery phase, there is no stress involved but the temperature is elevated to the transition temperature of the polymer.

Moreover, SMPs are used to produce parts that can sustain high strain conditions, so to deal with this, acrylates are compounded into the shape-memory polymer which displays a high strain-bearing capacity, however, the characteristics of SMPs are dependent on temperature. Shape-memory properties are also delivered by natural specialty polymers. One good example of a bio-specialty polymer is poly-L-lactic acid (PLLA) which is used in the manufacturing of self-expandable stents for repairing blocked arteries and veins. Surgical stitches are also being made using the same PLLA due to its promising water response property that results in a 97% recovery rate of the stitches post-surgery [12].

SMPs also demonstrate self-healing properties, and so, damage that occurs due to mechanical or certain scratches in normal wear and tear is solved through self-healing properties. This can be attained with a blended system consisting of cross-linked and linear poly-caprolactone. For this reason, self-healing SMPs are in great demand for surface properties in automotive applications such as the coating on the outer body of parts to prevent minor scratches. These coatings can also be incorporated with paints when applied on the walls where cracks formed due to abrasion can be taken care of [13].

Ethylene-vinyl acetate (EVA) is a specialty polymeric material that naturally possesses shape-memory properties. EVA is compounded with polyurethane foams and both polymers combine to produce good-quality footwear, which is used in sports applications. When textiles and shape-memory polymers are combined, they produce new products which can even replace some body parts. Figure 1.2 reflects a prototype of a foot and ankle that was produced using a combination of acrylic, nylon, and other specialty polymers [14]. Athletes who practice running especially tend to use shoes manufactured from memory foams, as these materials are very lightweight and have good functionality which provides excellent comfort while running. However, footwear manufactured by compounding EVA and polyurethane is difficult to recycle, which creates a biohazard. [15].

Recently, greener technology has been adopted using polyvinyl alcohol (PVA), which is a biodegradable specialty polymer that exhibits the mechanism of shape memory as it reacts to the heat produced through walking and provides comfort. This is known as the thermo-responsive shape-memory effect. Ethylene-vinyl acetate has also been used in electronic applications such as sensors that monitor human movement through the advancement of technology in artificial intelligence, and this polymer is usually filled with carbon nanotubes. For instance, a EVA/CNT fiber strain sensor responded with significant results such as a fast response speed of just 312 ms and astonishing durability of 5000 cycles. Viewing these results, they were brought into service to measure human health with the help of wearable electronics such as smart-watches which measure pulse rates and distances traveled by the individuals using them [16]. This product is also used to measure and evaluate several movements in the human body such as the functioning of the finger joint or the movement of the wrist. EVA/CNT-incorporated materials can also enable precise human motion monitoring.

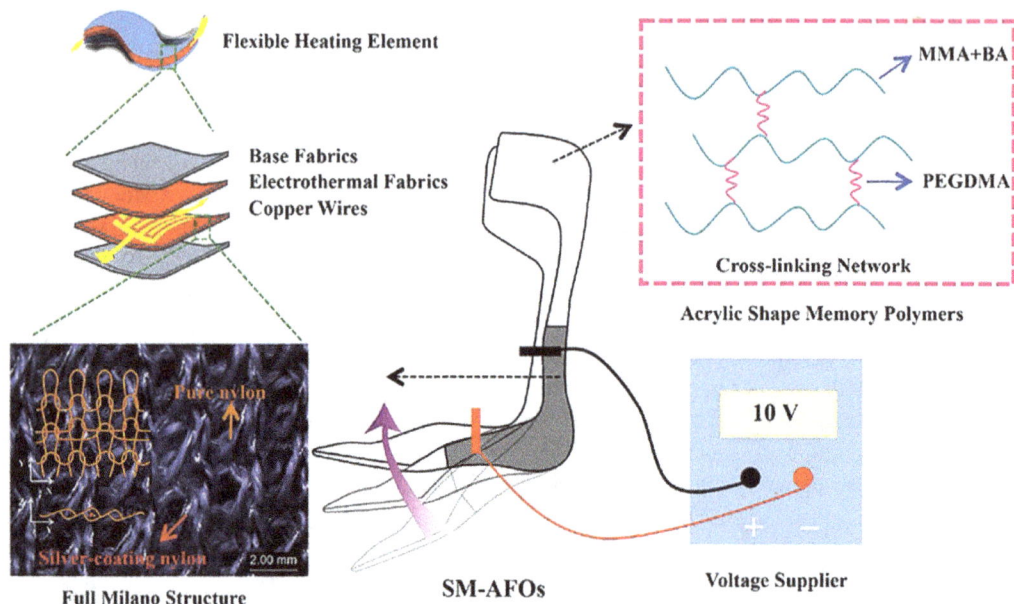

FIGURE 1.2   This image illustrates a prototype of an ankle and foot manufactured of shape-memory textile composites. Reproduced with permission [14]. Copyright (2018) American Chemical Society.

2.4   POLYMERS FOR ENERGY APPLICATIONS

Specialty polymers have many potential applications in solar technologies, energy storage devices, and fuel cells. When discussing solar applications, polymers are used as optical components in solar systems. Mirrors on parabolic troughs are made of specialty polymers such as fluoropolymers and acrylics. Polyesters are engaged with ultraviolet stabilizers and glass fiber composites to manufacture flat plate collectors of solar panels [17]. Photovoltaic cell arrays are manufactured and incorporated into silicones and acrylics for protection from the weather. Polyvinyl butyral is coated on the uppermost layer as a safety laminate. Solar energy generates an electric charge, but this charge must be stored someplace and, for this, energy storage methods need to be implemented such as using photocatalytic materials. Nano-sized ceramic fillers are intercalated into a polymer matrix to achieve high energy density with a combination of hybrids. For example, $BaTiO_3$ is a ceramic filler used in the polymer matrix poly {5-bis [(4 trifluoro-methoxyphenyl) oxycarbonyl] styrene}, a kind of fluoric-liquid-crystalline specialty polymer [18]. Some research has provided an innovative direction to manufacture nanostructured materials for applications related to energy storage such as Li-ion batteries and other renewable energy-storage devices with specialty polymers such as polyaniline and polythiophene [19]. Therefore, modifying the conducting polymers with redox-active groups to improve the electrochemical properties makes them a new emerging application to produce biodegradable batteries. In addition, the usage of polymeric materials in storage batteries also prevents corrosion at the electrodes [8], as depicted in Figure 1.3. Here, the carbon black mixed polyethylene forms a protective wall against corrosion in electrodes and still allows the efficient flow of charge.

Hydrogen fuel cells display a large-scale energy conversion rate because of renewable H_2. These fuel cells can be made of microporous cell membranes composed of specialty polymers such as perfluorosulfonic acid (Nafion). These polymers are very attractive in energy devices as proton exchange membranes as they exhibit high ionic conductivity, excellent mechanical and chemical strength suitable for hydrogen fuel cells, and very rapid ion transfer [20]. Similarly, as researchers

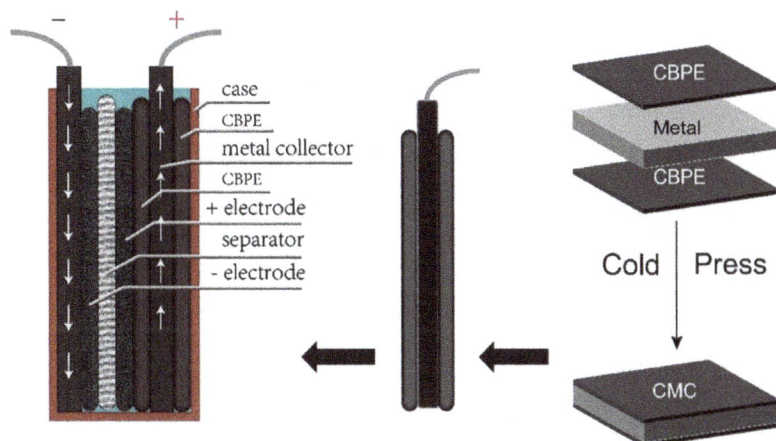

FIGURE 1.3 This figure illustrates the carbon black and polyethylene composite in polymer batteries to inhibit corrosion. Adapted with permission [8]. Copyright (2021) American Chemical Society.

have found the use of electronically active polymers in fuel cells, they have seen other advantages when used in supercapacitors for improved performance. Supercapacitors are used to gather charges and then deliver them to a circuit when needed. Certain advancements have been made using supercapacitors such that they can be charged for 40 s and have a power delivery time that can be extended for about 30 minutes. Graphene-based polydimethylsiloxane, a specialty material, is used for such supercapacitors. These graphene-based devices also have super flexibility, which provides a broad area of application [21].

2.5 POLYMERS FOR BIOMEDICAL APPLICATIONS

Polymers used in biomedical applications have made many advancements that have revolutionized the healthcare industry. These advancements in technology have by default improved the patient survival rate and quality of life. It has also supported the manufacturing of new and better medical devices. Improvisation in the field has improved components such as tissue engineering, nanotechnology, and delivery of drugs which replenish tissues, and several other functions such as repairing and treating a living body. Of all the materials being utilized for biomedical applications, polymers have been the greatest and most promising biomaterials. This is corroborated by their widespread use in medical applications. The success of polymeric materials in the biomedical field is solely due to their unique properties and functionalities [22].

Polymers that are suitable for biomedical applications have properties such as low toxicity, longer shelf-life, high chemical resistance, biocompatibility, drag-reducing, healing timeline management, and high molecular weight. Biomedical applications of polymers are associated with living organisms and are actuated to ease some processes in these organisms. The term drag-reducing used above is defined as decreasing the resistance to the flow of liquids without affecting the density. Specialty polymers such as polyethylene oxides and polyacrylamides demonstrate such qualities and ease blood circulation through the body. Such polymers are very useful but lack mechanical degradability. However, specialty polymers such as poly-N-vinyl formamide (PNVF) display low toxicity as well as excellent mechanical degradation [23]. Therefore, PNVF acts as a flow-increasing agent and provides turbulent flow through the veins. To facilitate the immediate start of drag-reducing polymers in the body, specialty polymers like poly(vinyl-caprolactam) microgels manufactured through methyl acrylate and methyl ether methacrylate can be adopted [24]. The polymer shows good stability and temperature sensitivity, which improves the drug release process whenever the polymer surpasses a certain temperature, and the above-discussed specialty polymer was also demonstrated to be non-toxic through tests conducted by blank microgels being introduced into a living body. Nevertheless, these microgels presented antitumor activities in the body. Therefore, a sufficient predefined quantity of drug can be transported by the application of this polymer.

Polymers, as discussed, reflect many qualities, with a unique one being their healing property. This advantage has been widely used in skincare and cosmetic applications. An example is silicon because of the elastomeric and biocompatibility effect it displays. This specialty material can be used to heal a deep wound by grafting the affected area with the help of binding material such as poly(dimethylsiloxane) [25]. This application has demonstrated promising results. Silicone particles are also capable of absorbing several nutrients from the human body, such as oils and deposited sebum which retains the natural glow of the skin. Sebum is a type of natural oil secreted through the skin pores that can be secreted from scalp hair [25].

Devices and apparatuses used for the medical operations of drug delivery and extraction of fluids from bodies often face challenges of contamination, which can increase the risk of infections. It is a tough and crucial task to prevent such contamination. The accumulation of antibiotics and pathogens on the surface in the form of biofilms results in the body experiencing infections [26]. Alternative solutions to this problem are discarding the medical device or separating the body part which is infected before the infection spreads to other organs. To address this, scientists have researched this issue and discovered that a polymeric coating on these devices will act as a sterilizing agent. This coating inhibits the process of accumulation of biofilms upon the devices and protects them from contamination. Consequently, these coatings can also be applied on devices that are inserted into the body and will remain for some time until their biodegradation [26]. However, to prevent infections due to the contamination of the devices being used, possible methods are discussed in Figure 1.4 which provides a broader idea of dealing with biological infections. Surface-engineered specialty polymer polyetheretherketone (PEEK) showed anti-microbial qualities when coated upon devices and the polymer demonstrated sterilizing properties when

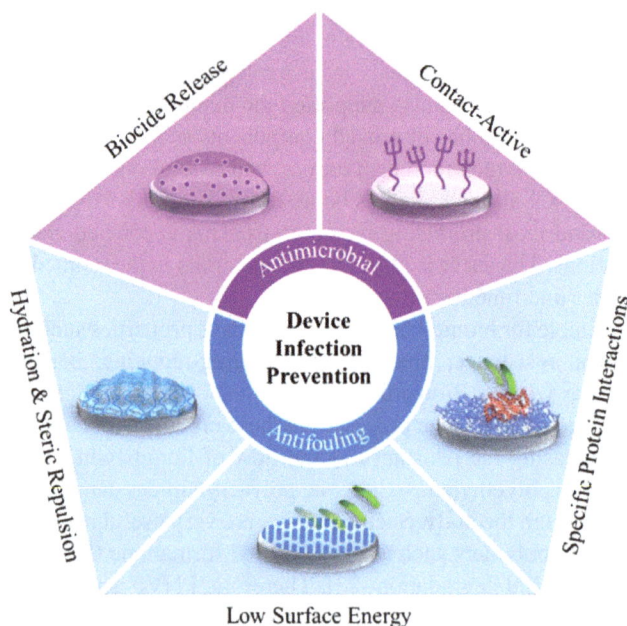

FIGURE 1.4   This image classifies the strategies applied for antimicrobial and antifouling polymeric coating to prevent infections. Adapted with permission [26]. Copyright (2017) American Chemical Society.

implanted into bodies. The surface engineering of PEEK is carried out by immersing the polymer in sulfuric acid, which improves the bacterial resistance of the polymer coating [27].

2.6 POLYMERS FOR SENSING APPLICATIONS

Sensors and their applications in various fields have gained serious attention in recent years. More specifically, sensors made of polymeric material are in high demand currently. In product manufacturing industries, quality control is considered a crucial aspect, and so sensors are of great importance to detect and separate defective products from satisfactory ones. Sensors are also being utilized in food industries to check the quality and quantity of the material being manufactured in a production line. While discussing sensors produced through polymers, the properties of the polymer, such as conductivity, require consideration. Hence, polymers that possess intrinsic conductivity are the most suitable for the sensing application [28].

In addition to the intrinsic conductivity of specialty polymers in sensing applications, they can also function as detectors for smoke and other chemicals. A typical example is a polyaniline that has both intrinsic conductivity and good sensing properties. Polyaniline can be made into films used for smoke detection, and these films can provide real-time smoke exposure data by measuring the quantity of smoke emitted [29]. Another useful application of such a gas sensor is to detect methanol in a solution of spirit made from sugarcane [30]. Due to the hazardous nature of the consumption of methanol to human health, the effective detection of even low concentrations of polyaniline during sugar production is an important task to achieve. Poly(2-dodecanoylsulfanyl-*p*-phenylenevinylene) (12COS-PPV) is another specialty conductive polymer that is applied in the form of a coating on tin electrodes. The unique ability of this polymer as compared to polyaniline which can detect methanol, ethanol, and other chemicals is that 12COS-PPV is engineered toward the specific detection of only methanol in the sample solution [30].

In addition to sensing films, molecularly imprinted polymers (MIPs) have gained attraction due to their ease of manufacture and excellent functioning durability. The simple process of making MIPs is casting the templates of parts that are to be detected on the film or the detector. Their relationship with many properties and receptors attests to their wide range of applications and excellent serviceability. MIP functioning can be illustrated as a lock and key and so, they only detect the material and parts that fix into the imprints. Enantioselective L-phenylalanine anilide (L-PAA) is a specialty polymer that is used as a template on the film to enable imprinting and sensing application.

2.7 POLYMERS FOR TEXTILE INDUSTRIES

Textile industry integration with specialty polymers brings novelty and they can both serve to fulfill several functionalities through their properties. Smart textile polymers are now stimuli sensitive and respond according to external changes which depend on several stimuli such as being thermal-, humidity-, light-, and pH-responsive. Each stimulus resembles its functionalities, such as temperature management, comfort, aesthetic appeal, and antimicrobial properties, respectively. In addition, smart textile polymers offer advantageous opportunities in wound monitoring and assisted drug release.

Furthermore, in exploring the temperature-responsive stimulus, the phase-changing materials are considered very important. These materials absorb and emit thermal energy, but this specialty function does not temper their internal temperature. Polymeric specialty waxes such as eicosane, octadecane, and heptadecane provide such properties under extremely low temperatures that they regulate and supply thermal energy, which stabilizes the temperature at 30–34 °C, which is a comfortable range for the human body under harsh weather conditions. The waxes mentioned above are transformed into microcapsules and incorporated into textiles. Due to their light and flexible characteristics, it becomes very easy to utilize them while taking advantage of their desired properties. The use of such materials for thermochromic applications as discussed in the above paragraph reduces the total weight of the product, such as thermal jackets and other thermal wear produced with lighter weights [31].

Temperature-maintaining textile materials are also of great importance in the military and other conditions where people are exposed to extremely harsh temperatures. Specialty polymers, therefore, play a huge role in textiles because they have been found to display high tensile strength even in elevated temperature conditions. In addition, test results of such materials show that they can withstand temperatures of around 300 °C and still provide the same performance for a longer period of time [32]. An example of such a material which is commercially available from Du-Pont Industries is "Nomex." This material has linear orientation and some crystallization effects. Other specialty polymeric materials that can provide high-temperature stability include polybenzimidazole and poly-oxadiazoles [32].

Apart from temperature, humidity also plays an important role in determining the shelf life of textile materials. This is because textiles are prone to attack by microorganisms under humidity and harsh temperatures. Consequently, these microorganisms aid in the propagation of infectious diseases such as skin rashes and unpleasant odors [33]. Self-sterilizing polymeric materials could have the potential to reduce the spread of such microorganisms. Polyaminopropyl biguanide is compounded with textile materials because of its intrinsic disinfectant properties which provide sufficient sterilization on textile materials and prevent further infection of the human body [34]. Textile materials have other uses as biomedical applications and sports applications where infections and contamination are typical due to microorganisms. Polyhexamethylene biguanide is also one of the specialty biocompatible materials that possess antimicrobial properties, and this material is also used to produce textile and clothing products [35].

Antimicrobial fabrics have also been applied on wounds as dressing. Chitosan and its derivatives demonstrate excellent antimicrobial and good wound-healing properties. Chitosan hydrogel is composed of cotton and is used on wounds to help re-establish the skin architecture. However, chitosan, a specialty biopolymer when enabled with a pH-responsive stimulus system, eases the drug-delivery system. Temperature and pH stimuli when acting together work as a system, so when the temperature is elevated, a pH shift is observed and the drug is released. In conclusion, the drug-delivery system works on a pH-responsive or thermal-responsive system [36].

2.8 POLYMERS FOR HIGH-STRENGTH APPLICATIONS

Polymeric materials that demonstrate high tenacity and high modulus are called high-strength specialty polymers. Along with high strength, these materials are rigid and comprise very good flexibility. The properties of high flexibility and rigidity are due to strong axial chemical bonding, other than that they illustrate a very low weight to volume ratio. Polymers of this type are generally obtained in fibers and when compounded with other polymers, they produce high-strength material and also enhance the properties of the binding material. Adding up, there are many specialty polymeric fibers such as polyhydroquinone-imidazopyridine (M5), other aramid polymeric fibers like Kevlar, Twaron, and fibers from polyethylene derivatives namely Spectra and Dyneema. M5 is synthesized through a condensation polymerization technique comprising of three monomers called poly{2,6-diimidazo[4,5-b:40,50-e], pyridinylene-1,4(2,5-dihydroxy), and phenylene}[37]. M5 is scientifically known as polyhydroquinone-diimidazopyridine (PIPD). Along with its outstanding rigidity and flexibility, this polymeric fiber also reflects exceptional fire-resistant properties. In the past, Dupont manufactured Nomex, a derivative of Nylon considered to be a fire-resistant material, however, with the gradual advancements in science and technology, M5, another specialty polymer, displays a 20 times greater fire performance index than Nomex [37]. Other than fire-resistant properties, M5 fiber also shows a high tensile strength and modulus of 2.5 GPa and 150 Pa, respectively.

Para-aramid fibers, generally known as Kevlar, were put into application in the 1970s and became very popular due to the properties they offered. Kevlar is used for many applications such as sporting gear, with the most novel application of Kevlar being the armored body vest. Kevlar displays an impressive tensile strength to weight ratio which makes the material five times stronger than steel and the most interesting part is that Kevlar is not susceptible to metallic corrosion. Kevlar was initially introduced as an alternative for tire threads which were made of Nylon, because Nylon wasn't capable of performing as well as desired and didn't provide strength to automobile tires. However, Kevlar fulfilled the requirements in tires and also replaced many other materials such as fiberglass and polyester. Kevlar, as discussed, nowadays is used as a protective material such as insulation for jet engines and gloves to deal with sharp instruments. In sports applications, kayaks and boats are made of Kevlar to provide the best impact resistance and do not add significant weight to the product. Kevlar has also been employed to hold large sea vessels at ports in the form of ropes. These ropes are even used in the assisted landing of aircraft on aircraft carriers.

Ultra-high-molecular-weight polyethylene, commonly known as Dyneema, is obtained in the form of a fiber which shows a strong nature in terms of mechanical properties and its main specialty is in waterproofing. In some instances, Dyneema is a more suitable material than Kevlar. For comparison, different fibers make up Kevlar and Dyneema which are polyamide and polyethylene derivatives, respectively. In UV exposure, Kevlar loses 20% of its strength as compared to Dyneema which is only affected by 5%. Kevlar absorbs around 3.5 times total weight in water, but Dyneema absorbs no water at all. Dyneema is considered the world's strongest material but lacks only one property, which is heat and fire resistance. Meanwhile, Kevlar material offers the property of fire resistance and can work at extremely elevated temperatures.

3   CONCLUDING REMARKS

As thoroughly covered in the chapter, specialty polymers are encountered in our day-to-day activities and can serve specific functions. Even though it might be assumed that specialty polymers have only one specific ability for particular functions, they rather have a versatile nature and can be used for more than one application. However, some specialty polymers may have drawbacks in certain applications. For example, in high-strength polymer applications, Dyneema displays excellent properties, however, it is not suitable for applications at high temperatures and is not fire-resistant. Kevlar, on the other hand, which has many identical properties to Dyneema, can be used in high-temperature applications as well as for its fire-resistant behaviors.

Specialty polymers are being employed in many industries and, nowadays, they are also well known as smart polymers, whereby they are used in memory applications with the help of intrinsic properties such as shape memory, where polymers can change their shape up to an extent through the effects of temperature and stress. After the return to normal stress and temperature they are able to return to their original shape and size. Ordinarily, polymers were primarily considered as insulators, such as polyurethane, but current discoveries have found polymers that have other applications in electrochemistry. There are intrinsically conductive polymers that can prevent corrosion by shifting the site of the corrosion reaction from the metal to themselves. This intrinsic conductivity also is useful in different areas such as electronic products, and sensing applications where health-monitoring devices analyze the body's health and can be in the form of wearables such as shoes and watches. Examples of specialty polymers like polyanilines possess this particular property. Aside from conductivity, polyanilines are also flexible and propose good mechanical strength which shows that they have all-round properties that place them in the specialty polymer category. The transportation industry also requires conductive polymers to supply data for the smooth functioning of the machines being manufactured and utilized. These industries require high-strength materials in the form of composites to reduce the total load. As a result, the consumption of fuel to run these transportation machines is reduced. Undoubtedly, specialty polymers have been a breakthrough and a promising route for progress for science and technology, although some challenges remain. Integrating polymers into manufacturing adds more value to products and improves the quality of parts. Additionally, we live in the plastic age today and the utilization of polymers or plastics sometimes may appear irresistible, making life occasionally feel impossible without them. Besides the mass application of polymeric materials, they have become an issue due to their improper disposal or recycling. A greater evolution of this field is needed by developing novel biodegradable specialty polymers which could be a step ahead in the direction of sustainability. Imagine if new biodegradable materials were to be developed that possess high-strength properties such as Kevlar, intrinsic conductivity as with polythiophenes, and shape-memory property as with ethyl vinyl acetate polymer. This would be a breakthrough and increase the scope of opportunities for specialty polymers in the near future.

REFERENCES

1. Carraher CE, Seymour RB (1985) Introduction To Polymer Science and Technology. ACS Symp Ser 13–47
2. Manas Chanda (2000) Advanced Polymer Chemistry. Marcel Dekker: New York
3. Zarras P, Stenger-Smith JD (2003) An introduction to corrosion protection using electroactive polymers. In: Peter Zarras, John D Stenger-Smith, Yen Wei (eds) Electroactive Polymers for Corrosion Control. American Chemical Society Volume 843: 2–17
4. Hlushko H, Cubides Y, Hlushko R, Kelly TM, Castaneda H, Sukhishvili SA (2018) Hydrophobic antioxidant polymers for corrosion protection of an aluminum alloy. ACS Sustain Chem Eng 6:14302–14313
5. Eng FP, Ishida H (1986) Corrosion protection on copper by polyvinylimidazole. ACS Symp Ser 4:268–282
6. Wen JG, Geng W, Geng HZ, Zhao H, Jing LC, Yuan XT, Tian Y, Wang T, Ning YJ, Wu L (2019) Improvement of corrosion resistance of waterborne polyurethane coatings by covalent and noncovalent grafted graphene oxide nanosheets. ACS Omega 4:20265–20274
7. Lin D, Liu Y, Li Y, Li Y, Pei A, Xie J, Huang W, Cui Y (2019) Fast galvanic lithium corrosion involving a Kirkendall-type mechanism. Nat Chem 11:382–389
8. Liu B, Yue J, Lv T, Wang S, Zhou A, Xiong X, Suo L (2021) Sandwich structure corrosion-resistant current collector for aqueous batteries. ACS Appl Energy Mater 4:4928–4934
9. Maria M (2013) Advanced composite materials of the future in aerospace industry. Incas Bull 5:139–150
10. Mittal V (2011) High Performance Polymers and Engineering Plastics. John Wiley & Sons.
11. Ahn SK, Deshmukh P, Kasi RM (2011) Exploiting architecture and composition of side-chain liquid crystalline polymers for shape memory applications. ACS Symp Ser 1066:39–51
12. Xie Y, Lei D, Wang S, Liu Z, Sun L, Zhang J, Qing FL, He C, You Z (2019) A biocompatible, biodegradable, and functionalizable copolyester and its application in water-responsive shape memory scaffold. ACS Biomater Sci Eng 5:1668–1676
13. Rodriguez ED, Luo X, Mather PT (2011) Linear/network poly(ε-caprolactone) blends exhibiting shape memory assisted self-healing (SMASH). ACS Appl Mater Interfaces 3:152–161
14. Chen J, Hu J, Leung AKL, Chen C, Zhang J, Zhang Y, Zhu Y, Han J (2018) Shape memory ankle-foot orthoses. ACS Appl Mater Interfaces 10:32935–32941
15. Bonadies I, Izzo Renzi A, Cocca M, Avella M, Carfagna C, Persico P (2015) Heat storage and dimensional stability of poly(vinyl alcohol) based foams containing microencapsulated phase change materials. Ind Eng Chem Res 54:9342–9350
16. Li Z, Qi X, Xu L, Lu H, Wang W, Jin X, Md ZI, Zhu Y, Fu Y, Ni Q, Dong Y (2020) Self-repairing, large linear working range shape memory carbon nanotubes/ethylene vinyl acetate fiber strain sensor for human movement monitoring. ACS Appl Mater Interfaces 12:42179–42192
17. Carroll WF, Schissel P (1982) Polymer in solar energy: Applications and opportunities. In: American Chemical Society, Polymer Preprints, Division of Polymer Chemistry. pp 195–196
18. Luo H, Chen S, Liu L, Zhou X, Ma C, Liu W, Zhang D (2019) Core-shell nanostructure design in polymer nanocomposite capacitors for energy storage applications. ACS Sustain Chem Eng 7:3145–3153

19. Jia X, Ge Y, Shao L, Wang C, Wallace GG (2019) Tunable conducting polymers: toward sustainable and versatile batteries. ACS Sustain Chem Eng 7:14321–14340

20. Du X, Yuan Y, Dong T, Chi X, Wang Z (2021) Polymer electrolyte membranes from microporous Troger's base polymers for fuel cells. ACS Appl Energy Mater 4:13327–13334

21. Zequine C, Bhoyate S, de Souza F, Arukula R, Kahol PK, Gupta RK (2020) Recent advancements and key challenges of graphene for flexible supercapacitors. In: Singh L, Mahapatra DM (eds) Adapting 2D Nanomaterials for Advanced Applications. American Chemical Society, pp 3–49

22. Tonzani S (2013) Polymers for biomedical applications. J Appl Polym Sci 129:527

23. Marhefka JN, Marascalco PJ, Chapman TM, Russell AJ, Kameneva M V. (2006) Poly(N-vinylformamide) – A drag-reducing polymer for biomedical applications. Biomacromolecules 7:1597–1603

24. Wang Y, Nie J, Chang B, Sun Y, Yang W (2013) Poly(vinylcaprolactam)-based biodegradable multiresponsive microgels for drug delivery. Biomacromolecules 14:3034–3046

25. Liles DT, Lin F (2010) Silicone elastomeric particles in skin care applications. ACS Symp Ser 1053:207–219

26. Zander ZK, Becker ML (2018) Antimicrobial and antifouling strategies for polymeric medical devices. ACS Macro Lett 7:16–25

27. Mo S, Mehrjou B, Tang K, Wang H, Huo K, Qasim AM, Wang G, Chu PK (2020) Dimensional-dependent antibacterial behavior on bioactive micro/nano polyetheretherketone (PEEK) arrays. Chem Eng J 392:123736

28. Partridge AC, Jansen ML, Arnold WM (2000) Conducting polymer-based sensors. Mater Sci Eng C 12:37–42

29. Liu Y, Antwi-Boampong S, BelBruno JJ, Crane MA, Tanski SE (2013) Detection of secondhand cigarette smoke via nicotine using conductive polymer films. Nicotine Tob Res 15:1511–1518

30. Péres LO, Li RWC, Yamauchi EY, Lippi R, Gruber J (2012) Conductive polymer gas sensor for quantitative detection of methanol in Brazilian sugar-cane spirit. Food Chem 130:1105–1107

31. Hu J, Lu J (2014) Smart polymers for textile applications. In: María Rosa Aguilar, Román JS (eds) Smart Polymers and their Applications. pp 437–475

32. Polymers RO, Pont D (1968) Symposium focuses on heat-resistant fibers. Chem Eng News 46:40–41

33. Varesano A, Vineis C, Aluigi A, Rombaldoni F (2011) Antimicrobial polymers for textile products. Sci Against Microb Pathog 1:99–110

34. Morais DS, Guedes RM, Lopes MA (2016) Antimicrobial approaches for textiles: From research to market. Materials (Basel) 9:1–21

35. Allen MJ, White GF, Morby AP (2006) The response of *Escherichia coli* to exposure to the biocide polyhexamethylene biguanide. Microbiology 152:989–1000

36. Hu J, Meng H, Li G, Ibekwe SI (2012) A review of stimuli-responsive polymers for smart textile applications. Smart Mater Struct 21:53001

37. Northolt MG, Sikkema DJ, Zegers HC, Klop EA (2002) PIPD, a new high-modulus and high-strength polymer fibre with exceptional fire protection properties. Fire Mater 26:169–172

2 Materials and Chemistries of Polymers

Mansi Sharma,[1] Pragati Chauhan,[1] Rekha Sharma,[1] and Dinesh Kumar[2]

[1] Department of Chemistry, Banasthali Vidyapith, Rajasthan 304022, India
[2] School of Chemical Sciences, Central University of Gujarat, Gandhinagar 382030, India

1 INTRODUCTION

The term "polymers" is derived from the Greek words "poly," which means "many," and "meres," which means "parts." They are macromolecules composed of numerous small molecules. These macromolecules are either straight, somewhat branching, or highly linked. Monomers are the tiny molecules that serve as the building blocks for these larger compounds. In the most basic polymers, monomers are identical. For high intensity, intense polyvinyl chloride (PVC) is produced from monomer vinyl chloride (VC). A polymer's monomers can be the same, in which case it is referred to as a homopolymer, or they can comprise multiple monomers, in which case it is referred to as a copolymer. Polymers can also incorporate additional monomers in random copolymers or block copolymers. Alternatively, monomers can be copolymerized in alternating blocks of identical monomers [1,2].

-H-I-H-I-H-I-H	-H-H-H-I-I-I-	-H-H-H-H-H-
Random copolymers	**Block copolymers**	**Homopolymers**

It is possible to find natural polymer molecules in flora and fauna (natural polymeric molecules) and synthesize them artificially (synthetic polymeric molecules). The physiochemical properties of polymers permit them to be used in a broad range of everyday scenarios. Chemicals and cross-links between long chains are present in some polymers. The mass and number of reiterating units in the molecules determine the size of polymers. The degree of polymerization is represented by this letter on the indicator. A polymer's relative molar mass is determined via the product of the repeating units and the degree of polymerization (DOP). It is composed of many structural units that are usually connected by covalent bonds that possess a large molecular mass (between 10,000 and 1,000,000 g/mol). Its organic nature (presense of carbon and hydrogen) is of value in the great majority of commercially available products. Significantly, they are carbon covalent composites. Hydrogen, chlorine, sulfur, oxygen, fluorine, and phosphorus are the most common elements used in polymer chemistry. These elements can form covalent bonds with carbon [3,4].

Besides the fundamental valence force, polymer molecules exhibit secondary intermolecular forces. An example of dipole force is the force generated between opposingly charged ends of polar bonds and the dispersion force caused by electron clouds around specific atoms within the polymer molecule. There are many polymers in which hydrogen bonds play a crucial role, particularly proteins, since hydrogen bonds are extremely powerful dipoles that emit electronegative charges such as oxygen or nitrogen. Hydrogen bonds bind molecules together and keep them in a

DOI: 10.1201/9781003278269-2

specific position. These stable structures are required for proteins to perform their unique activities in biological processes [5].

2 CLASSIFICATION

Due to their complex structures, diverse behaviors, and wide applications, polymers are divided into many categories. Figure 2.1 presents the classification of polymers. Polymers are classified on the following basis:

- Based on the available sources
- Based on polymerization
- Based on the chain structure of monomers
- Based on monomers
- Based on molecular forces [6].

2.1 BASED ON THE AVAILABLE SOURCES

The availability of polymers determines their classification.

Natural: They can be found in plants as well as mammals. Proteins, starch, cellulose, and rubber are examples. Finally, biopolymers are ecological polymers.

Semisynthetic: Semisynthetic polymers were developed by humans. The most common man-made polymer is plastic and this is the one that is used the most. It is utilized in a wide range of dairy products or industrial segments. Examples include polyether, nylon-6, and so on.

Synthetic: These are manufactured from chemically modified naturally existing polymers. Examples include cellulose acetate as well as nitrate [7].

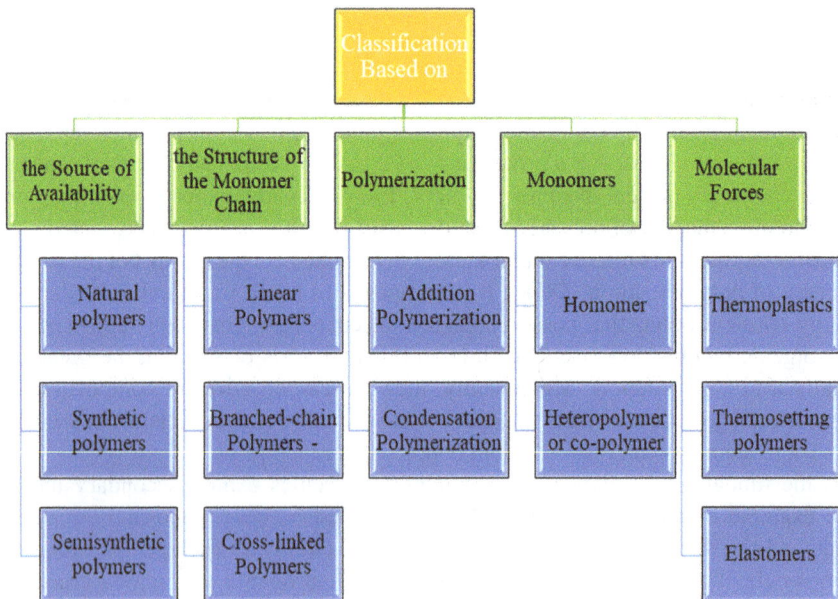

FIGURE 2.1 A schematic flow chart representing the classification of polymers on several bases.

2.2 BASED ON THE CHAIN STRUCTURE OF THE MONOMER

Linear: This category includes polymers with long, straight chains. PVC is a linear polymer often used to produce pipes and electrical cables.

Branched-chain: The linear chain of a branched-off polymer results in a branched-chain polymer. An example is polythene with low density.

Cross-linked: Monomers with bifunctional and trifunctional functions make up these molecules. Linear polymers have a weak covalent bond compared to cross-linked polymers, for example, Bakelite and Melamine [8].

2.3 BASED ON POLYMERIZATION

Addition: Teflon, PVC, and poly-ethane are a few examples.

Condensation: Perylene, polyesters, and Nylon-66 are a few examples [9].

2.4 BASED ON MONOMERS

Homomer: A solo monomeric unit is seen in this class. Polyethene is one such material.

Copolymer/heteropolymers: The various monomer units are seen in this class. A good example is Nylon-66.

2.5 BASED ON MOLECULAR FORCES

The three main categories of polymers are thermoplastics, thermosets, and elastomers. Looking at how they react to heat is the most effective approach to tell them apart.

Thermoplastics: Thermoplastic polymers come in two forms: amorphous and crystalline. They have a somewhat malleable temperament, yet they are typically weak. Intermediate forces attract these segments. PVC is an example of long-chain molecules in thermoplastics that are bound together by secondary van der Waals forces and can be in the form of linear bonding (secondary bonds). The binding force is overcome by activating the molecular chains at a high temperature, allowing them to traverse freely in a viscous liquid over another. The secondary bonds may have broken down. Glass transition occurs when secondary bonds melt at a certain temperature (Tg). Once the polymer cools, secondary forces begin to hold, causing the molecular chains to return to a limited form. The ability to melt and remelt thermoplastics makes them easily recyclable [10].

Thermosetting polymers: These are frequently formless, vigorous, and rigid overall, nevertheless they are hard in specific circumstances. These polymers considerably improve the mechanical characteristics of the material. Epoxies, silicones, and phenolics are examples of these types of materials. In thermosetting polymers, long-chain monomers are organized in a formless system with cross-linked bonding. It demonstrates that covalent bonds connect long molecule chains. Cross-linking occurs through a process known as curing. A thermosetting plastic cannot be remelted because cross-linking holds the molecular chains; instead, it decomposes when heated above its melting point. Thermosetting polymers can only be amorphous because cross-linking prevents molecular arrangement into an ordered crystalline structure.

Elastomers: Amorphous elastomers are employed in applications where the Tg is higher than the melting point. The deformation can be extremely large since they are elastically deformable while maintaining their shape. Weak interaction forces exist in these rubber-like substances. Elastomers have amorphous linear bonds with some cross-linking of long molecular chains. At room temperature, the chains' general degree of excitement has already reached the secondary van der Waals bonds; however, the cross-links in the structure work to restore the elastomer's inventive shape after alteration [11,12].

3 TYPES OF POLYMERS

3.1 NATURAL POLYMERS

Natural polymers are a subcategory of polymers derived from normal sources (plants or animals). They mostly consist of proteins and carbohydrates found in both flora and fauna, and provide stable strength. Thermosets, thermoplastics, rubbers, and elastomers are examples of other polymers. Polymers can be been extracted from their original bulk state, such as lignin and cellulose from wood. The polymers created via biotic processes like bacterial fermentation and production are also included. Polymers like synthetic and natural ones are classified as addition or condensation polymers depending on how they are formed. Condensation polymers make up most natural polymers when monomer units combine to generate a tiny fragment (typically water) as a siding material. Further polymers are created by combining the monomer units that make up the polymer directly, with no side products. Polymers found in nature can be classified into six categories: polysaccharides, proteins, polynucleotides, polyesters, lignin, and polyisoprenes [13,14].

3.1.1 Polysaccharides

These are acetic-bond-linked copolymers of amino sugars and glucose. Polysaccharides are widely considered the most abundant renewable resource on the planet. Polysaccharides are classified into numerous categories based on their structure or function. The three primary categories are storage polysaccharides like glycogen and starch, gel-forming polysaccharides like mucopolysaccharides and alginic acid, and structural polysaccharides like chitin and cellulose. Polymers can be straight-chained/branched and non-ionic/ionic (cationic and anionic). Amylopectin, amylase, and glycogen are the most prevalent storage polysaccharides, while chitin and cellulose are the most common types of structure. Chitin is a structural polysaccharide found in crustaceans, whereas cellulose, along with hemicelluloses, pectin, and lignin, is the major basic constituent of plants. An additional basic polysaccharide that originates in human cells is hyaluronan. The mono-polymers determine each polysaccharide's physiological property and their arrangement within the polymer structure [15,16].

In contrast to synthetic polymers, many polysaccharides exhibit irregular properties, making classification and characterization difficult. Table 2.1 presents the monomer and its applications. For instance, because of its highly branched nature, amylopectin should ideally be nanocrystalline, according to polymer science laws. Amylopectin, on the other hand, is semicrystalline. For many years, this has encouraged many researchers to prefer researching synthetic polymers to natural polymers, particularly polysaccharides, which exhibit numerous structural and functional abnormalities. Because of their biodegradability and biocompatibility, polysaccharides have become increasingly desirable [17].

3.1.2 Cellulose

Cellulose is the principal constituent of plant cell walls and is an extremely prevalent chemical. Along with additional constituents like hemicelluloses, lignin, and pectin, it helps distinguish animal and plant cells by forming the cell wall. The use of cellulose can be traced back to Egyptian pharaohs and Chinese dynasties when it was used in lingerie and writing constituents. For hundreds of years, cellulose has been profusely derived from flax, cotton, wood, linen, hemp, jute, and kenaf and has been extensively used in various commercial processes. Cellulose has a wide range of uses, from paper to food and clothing pulp, and it has a significant economic impact [18].

Bacteria like acetobacter, algae like microdicyon and valonia, and aquatic faunae of the ascites family are all sources of cellulose. The human body cannot digest cellulose, but animals, particularly

TABLE 2.1
Monomers and Their Applications

S. No.	Monomer	Polymer	Uses of Polymer
1.	1,3-Butadiene styrene	BUNA – S	Synthetic rubber
2.	Tetraflouroethane	Teflon	Non-stick cookware plastics
3.	Ethylene glycol and phthalic acid	Glyptal	Fabric
4.	Vinyl cyanide	PVC	Tubes, pipes
5.	Melamine and formaldehyde	Melamine formaldehyde resin	Ceramic plastic material
6.	Phenol and formaldehyde	Bakelite	Plastic switches, mugs, buckets
7.	Caprolactum	Nylon-6	Fabric
8.	Ethylene glycol and terephthalic acid	Terylene	Fabric
9.	Vinyl cyanide and 1,3-butadiene	BUNA-N	Synthetic rubber
10.	Isoprene (1,2-methyl-1,3-butadiene)	Rubber	Making elastic materials, tires

ruminants, can. It is also not soluble in water. A plant's inflexibility and dynamic nature are ensured via a combination of protein and cellulose inside its cell walls, and cellulose microfibrils provide tensile strength resistance while allowing the cells to grow and extend. Cellulose microfibers are formed and extended by complexes involving 5c-diphosphate (UDP), glucose molecules, cellulose synthase, and glucose [19].

3.1.3 Lignin

The most plentiful aromatic polymer is lignin, which is measured via cellulose. The free-radical polymerization of para-hydroxy cinnamic acid alcohols results in lignin formation. It has a part in the structure of cells of most vascular plants, similar to cellulose. It is a complex heteropolymer having a complicated structure. The following hydrophobic polymers have been discovered in plant cell walls or serve as a matrix for binding cellulose microfibrils and other cell wall components, resulting in biomechanical strength and stiffness. They control plants' upright development. Lignin in various nonvascular floras has also been discovered in several investigations. The creation of lignified cell walls has been a fundamental structural evolution after plants transitioned from marine to earthly settings around 475 million years ago [20].

It is generated in the gaps around the cellulose microfibrils in plant cell walls during the last stage of cell differentiation, resulting in a lignocellulose matrix that contributes to the plant's strength. Lignin is commonly thought of as a by-product of manufacturing methods such as ethanol synthesis and paper as well as pulp creation by lignocellulose biomass. It accounts for 20–30% of all cellulosic biomass, despite the fact that it is not fermentable. The pulp and paper industry generates 40–50 million tonnes of lignin as an unwanted side-product annually [21].

Lignin forms approximately 18–25% of wood, with the rest consisting of hemicelluloses and cellulose, which create a matrix inside the xylem. Plant cell walls include lignin, which is found in the cell's extracellular matrix and gives it stiffness and support. The major constituents include 10–15% carbonyl, 15–20% benzyl alcohol, 15–30% phenolic hydroxyl, and 92–96%, although there are some thermal aldehyde groups also. Apart from its basic purpose, lignin is important for nutrient and water transport throughout the plant and it stops harmful enzymes from penetrating, hence avoiding breakdown [22].

3.1.4 Pectin

Pectin is a complicated set of molecules made up primarily of -d-(1-4 galacturonan with a few units of -l-(1-2) rhamnose thrown in for good measure. Agar and mucopolysaccharides are examples of gel-forming polysaccharides. Pectin, along with other components including cellulose, hemicelluloses, and lignin, is discovered in the main cell wall of plants. Cell walls of dicotyledon higher plants contain up to 35% of this compound. Monocotyledon plants have a lower proportion of them, and it comes in various shapes. Besides having structural and developmental functions, pectin has also been shown to contribute to plants' ion exchange capacity, ensuring that ions move freely and pH is maintained within their walls [23].

3.1.5 Starch

Starch is an odorless and tasteless white powder. There are two types of molecules in this polysaccharide: amylose and amylopectin. Each has a different percentage depending on the kind of starch and its source. However, it is approximately 20–25% amylose and 70–75% amylopectin. Water can permanently dissolve starch, whereas alcohol and water are normally insoluble due to the existence of water and heat through "gelatinization" [24].

It is utilized in commercial processes, the most common being adhesives, paper, and clothes. Starch has been used for beautifying cream, foodstuff thickening, and paper manufacturing. In a wobbled helix configuration, the polysaccharide amylose is derived from poly(1-4)-α-d-linked polyglucans. The structure of amylopectin is branched, with a branching each 28–30 glucose units. Because of its branched structure, amylopectin is more vulnerable to degradation and hydrolysis than amylase due to the extra bare areas. Despite differences in amylase and amylopectin compositions, the microstructure of starch granules is almost indistinguishable for all kinds of starch [25].

3.1.6 Glycogen

Another storage polysaccharide is glycogen. The structure and size of amylopectin are similar to those of other storage polysaccharides, except for their greater branching and compactness. Glycogen is a glucose storage form found in the cytoplasm of animal cells. Although glycogen plays an important role in body metabolism, it is not used in industry [26].

3.1.7 Chitin

The chitin polysaccharide of animal origin is highly hydrophobic, and its units contain amino and acetyl groups. However, it can also be dissolved in particular diluters, including chloro-alcohols fused with aqueous solutions of acetamide dimethyl acetate, hexafluoro-isopropanol, and mineral acids, comprising 5% lithium chloride. It is abundant in nature, being common in the exoskeletons of crustaceans/insects, and in the mycelia and spores of fungi. The only difference between chitin and cellulose is that the hydroxyl groups are substituted by acetamido groups on the cellulose chain [27].

Chitin and chitosan are natural polysaccharide polymers. These properties allow it to be used as a film- and gel-forming material, and as a chelating agent for metallic ions to form polyoxysalts. Chitin is extensively used in biotech, especially in the improved form of chitosan found via deacetylation of chitin. It is generally used in transdermal drug delivery (TDD) due to its mechanical, mucoadhesive, responsive properties, insoluble nature in alkaline/neutral environments, and soluble nature in acid environments [28].

3.1.8 Alginate

Alginates are water-soluble polymers with a long chain found in the cell walls of seaweeds, where it provides flexibility and strength. As early as 600 BC, it was used as a food source. The decontaminated aspects of alginate were not removed from seaweed until 1896, when this was done by Akrefting. It was a commercially available invention in 1929, with Kelco being the first to use it as an ice cream stabilizer. It is frequently located in the presence of further positive ions, primarily calcium and sodium, as calcium and sodium alginate, respectively. The properties of alginate are affected by the positive ions attached to it. The characteristics of alginate are also influenced by the algae species, including *Macrocystis pyrifera*, *Ascophyllum nodosum*, *Laminaria digitata*, and *Laminaria hyperborea*. *Azotobacter* as well as *Pseudomonas* bacteria generate polymeric constituents like alginate. Alginic acid has several biotic as well as manufacturing functions, including stabilization, drug delivery, viscosity, and binding. It is also combined with other polymers like chitosan and hyaluronic acid to perform wider tasks [29,30].

3.1.9 Proteins

Amino acid groups and, in some cases, additional groups are merged into amide bonds, and peptide bonds in proteins. The form, size, solubility, content, and function of proteins can all be categorized. Proteins are divided into two categories depending on form and size: globular and fibrous proteins. Globular proteins are delicate, water-soluble proteins. Globular proteins include things like antibodies, enzymes, and hormones. Fibrous proteins are water-insoluble proteins that are more difficult to break down. Hair, nails, and skin are structural tissues where these proteins can be found. Proteins are classed as simple, complex, or derivative proteins, depending on how soluble they are. Simple proteins are those that yield simply amino acids when they are hydrolyzed. Albuminoids, albumins, glutelins, histones, globins, and prolamins are some of the subcategories of these proteins. Prosthetic groups and simple proteins are combined to form compound proteins and conjugate proteins.

Conjugate proteins come in various shapes and sizes, dependent on the prosthetic group connected to them. These could be chromoproteins, nucleoproteins, glycoproteins, lipoproteins, metalloproteins, mucoproteins, or phosphoproteins. Proteins obtained from incomplete, complete, enzymatic, alkali, and acidic hydrolysis of conjugated/simple proteins are referred to as derived proteins. These proteins can be primary or secondary. Proteins obtained from the main sources, including coagulated, metaproteins, and proteins, are formed during partial hydrolysis of peptide bonds inside proteins. Hydrolysis of peptide bonds results in secondary proteins. Storage, regulatory, exotic, transport, secretary, noxious, catalytic, contractile, as well as structural proteins are also categorized by their behavior. A controlling, catalytic, and protective protein is an enzyme, hormone, or antibody. A few of the most common proteins used in the industrial sector are zein, silk sericin, gelatin, collagen, wheat gluten, casein, soy, and silk fibroin protein [31,32]

3.1.10 Polyhydroxyalkanoates

Polyhydroxyalkanoates are bacterial complexes, although plants are used to manufacture them, specifically their leaves [33].

3.1.11 Polynucleotides

The building blocks of life, deoxyribonucleic acid (DNA) and ribonucleic acid (RNA) are polynucleotides, instructions that a cell must follow to perform its function. They are used in biomedical applications like DNA sequencing and gene therapy. Polynucleotides are 13 or more nucleotide monomers linked together in a chain. For example, the DNA molecule itself consists of polynucleotide chains that fold into a dual helix. Simultaneously, the nucleotide sequence determines thew instructions for a specific cell [34].

3.1.12 Polyisoprenes

Natural rubbers with thermosetting capabilities are known as polyisoprene. Trans and cis polyisoprenes, commonly recognized as E and Z polyisoprene, are the two forms of polyisoprenes. The isoprene unit present determines the type of polyisoprene produced, and both forms have dissimilar characteristics. The main commercial source of polyisoprene is the *Hevea brasiliensis* tree, which produces large amounts of polyisoprene and has the required mechanical characteristics. The most commonly used plants that produce trans polyisoprene are Gutta-percha (*Eucommia ulmoides* and *Palaquium gutta*) and Balata (*Minusops balata*). Rigidity, insulation, a very low thermal extension/reduction coefficient, and alkali and acid resistance are all desirable qualities of trans polyisoprene [35,36].

3.2 SYNTHETIC POLYMERS

Different materials are available to create synthetic polymers. As monomers are added to the chain one at a time, hydrocarbons such as ethylene and propylene become polymers (ethylene).

3.2.1 Poly(ethylene)

This polymer, which was originally commercially made in 1939 for use in electrical insulation, has one of the most basic molecular configurations ($[CH_2\text{-}CH_2\text{-}]_n$). The naming of this polymer is a challenge, ethene, rather than the earlier ethylene, is the International Union of Pure and Applied Chemistry (IUPAC)'s ideal term for the monomer. As a result, the polymer's IUPAC designation is poly(ethene).

Poly(ethylene) can be made in four different ways in the industrial setting, each with slightly different qualities. The following are the four options:

1. Conventional oil techniques
2. Philips method
3. Processes by Ziegler
4. Processes involving high pressure.

The first group uses a pressure of 1000–3000 atm and a temperature in the 80–300 °C range. The initiators of free radicals like oxygen and benzoyl peroxide are commonly utilized, and circumstances must be wisely regulated to avoid a runaway reaction that produces graphite, methane, and hydrogen instead of polymer. High-pressure methods produce lower-density poly(ethylene) with low molar masses, typically 0.915–0.945 g cm^{-3}.

The coordination reactions catalyzed via alkyl systems are the basis of the Ziegler processes. Karl Ziegler discovered these reactions in Germany, and G. Natta developed them in Milan in the early 1950s. The combination of titanium tetrachloride and triethyl aluminum is a classic Ziegler–Natta catalyst. It is supplied into the reaction vessel first, followed by the addition of ethylene. The reaction occurs at low temperatures, usually below 70 °C, with the presence of air and moisture strictly avoided, as they would destroy the catalyst. These procedures yield poly(ethylene) with 0.945 g cm^{-3} density, which is an intermediate value. By adjusting the proportion of catalyst constituents and injecting an insignificant quantity of H into the reaction vessel, a wide range of relative molar weights for such polymers can be achieved. Table 2.2 represents the complete method for formulating poly(ethylene).

Polymers are a common material in today's society. They are a waxy solid that is inexpensive, easy to work with, and chemically resistant. Low relative molar mass grades have the drawback of "environmental stress cracking." They break disastrously for no specific reason, subsequently being exposed to sunshine or humidity. Despite this disadvantage, poly(ethylene) comes in various grades with numerous applications. Pipes, packing, chemical plant components, crates, and insulation items fall under this category [37–39].

TABLE 2.2
Complete Methods for Formulating Poly(ethylene)

S. No.	Catalyst	Temperature (°C)	Pressure (atm)	Process	Density of Product
1.	Generally supported MoO_3 with promoters of Na, Ca metal, or hydride	230–270	40–80	Indiana (standard oil)	9.6 g cm^{-3}
2.	Five percent CrO_3 in alumina/silica	130–160	15–35	Philips	9.6 g cm^{-3}

3.2.2   Poly(propylene)

Following Natta's research into synthesizer catalysts of high relative molar mass polymers for alkenes, this polymer with the structure [-CH$_2$CH(CH$_3$)-]$_n$ emerged as a marketable substance. Natta demonstrated in 1954 that comparable polymers of propylene could be made as a result of his work on ethylene polymerization (propylene), which was first introduced in 1957, after quick commercial exploitation. There are formal issues with the naming of this polymer and with poly(ethylene) because its IUPAC name, poly(propene), is rarely, if ever, used by polymer scientists.

Poly(propylene) was discovered to exist in two forms when it was initially created. One resembled poly(ethylene), nevertheless having higher inflexibility and solidity; the other was discovered to be amorphous and weak. Carbon's isotactic nature means that it has a consistent stereochemistry for each alternating atom. There is also a second type of methyl-bearing carbon molecule, which is atactic, with a distribution of different stereochemical arrangements at each carbon atom.

It is slightly different from poly(ethylene) and approximately 90–95% isotactic in profitable poly(propylene):

1. The density is lower (0.90 g cm^{-3}).
2. Because of its advanced softening point, it is suitable for use at higher temperatures. It is commonly utilized in manufacturing jug-style containers and has proven to be very resistant to the effects of hot liquids.
3. It is not prone to snapping due to exposure to the elements.
4. The tertiary C–H bond in the molecule is further easily fragmented. The additional intriguing feature of isotactic poly(propylene) is that it can be flexed repetitively without becoming brittle. Hence it is used for one-piece moldings, such as boxes for card indexes, which can be used over many years without damage [40,41].

3.2.3   Poly(methyl methacrylate) (PMMA)

PMMA is an artificial polymer made from the monomer called (methyl methacrylate). Figure 2.2 represents the chemical structure of PMMA. It has the IUPAC names poly[1-(methoxycarbonyl)-1-methyl ethylene] for the hydrocarbon and poly(methyl 2-methyl propanoate) for the ester. Rowland Hill and John Crawford, a British chemist, revealed PMMA in the early 1930s. A German chemist, Otto Rohm, applied it for the first time in 1934. PMMA is an optically transparent thermoplastic that is commonly used as a substitute for inorganic glass. Its light weight, greater impact strength, shatter-resistant properties, easy dispensation, scratch resistance, and weather resistance are outstanding features.

The polymer structure is prevented from stuffing closely in a crystalline manner and easily revolving adjacent to the C–C bonds by a nearby methyl group (CH$_3$). PMMA was identified as a noncrystalline thermoplastic due to this discovery.

PMMA was first used in World War II in aircraft windows and bubble canopies for gun turrets. PMMA is a promising polymer with advantages in pneumatic actuation, optical, sensor, conductive

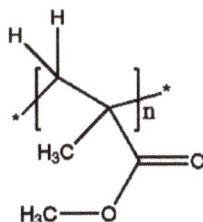

FIGURE 2.2   The chemical structure of PMMA.

devices, and analytical separation. Further advantages comprise the application of PMMA in biomedical areas. Polymer viscosity, electrolytes, and drug distribution utilizing electro-osmotic flow and electro-diffusion are some of the applications of PMMA [42,43].

3.2.4 Poly(glycolic acid) (PGA)

Poly(glycolic acid) is a hydroxy acetic acid and glycolic acid linear polymer with high crystallinity. Because of its high crystallinity, it has relatively high strength and lower solubility in water/organic solvents. Because of its low solubility in water, PGA degrades in vivo via a bulk mechanism. The first artificial resorbable suture was created with the help of PGA because of its fast degradation as well as outstanding fiber-forming capacity. DEXON, a biodegradable commercial suture, and Biofix, a bone interior fixation device, were the primary decomposable profitable sutures permitted via the FDA. Patients with fractures or osteotomies receive PGA self-reinforced composite inserts, which are harder than any other degradable polymer system. PGA has been used as a filler material in short-term tissue engineering scaffolds for regenerating bone, cartilage, tendons, intestinal tissues, and spinal cord tissue, and is frequently formed into a complicated network as well as the wider applications such as a scaffold for bone, cartilage, tendons, intestinal tissues, and lymphatic tissues. Glycolic acid produced by high degradation rates of PGA has been accompanied by a strong provocative response, even though it is bioresorbable; for this reason, other polymers, such as poly(lactide-co-glycolide) copolymers, have been united with PGA.

3.2.5 Poly(lactic) acid (PLA)

Poly(D,L-lactic acid) PDLLA and PLLA are the binary polymers from the polymerization of lactose monomers that show promise in biomedical applications. Lactide is a cyclic dimer of lactic acid with two enantiomers. PLLA is considered one of the most ideal biomaterials for load-bearing due to its ductile strength, low extension, and high modulus characteristics, as well as its slow degradation rate. On the other hand, PDLLA and PLLA contain the casual distribution of both isomeric forms of lactic acid, which confers a low ductile strength and a rapid deprivation rate. To reduce the degradation time of high-molecular-weight PLLA, blends or copolymers with other biodegradable polymers have been designed to reduce the degradation time. Combining PLLA and PDLLA with other degradable polymers like chitosan, PEG, or PLGA, has resulted in complexes with the required PLLA and PDLLA properties. Cartilage, ligament, biocomposite, and scaffolds for bone and tendon regeneration, as well as neural/vascular regeneration, have all used PLLA-based biomaterials. Orthopedic fixation devices like Sysorb interference screws, Phantom Suture Anchors (DePuy), Phantom Soft Thread Soft Tissue Fixation Screws, Full Thread Bio Interference Screws (Arthrex), Orthopaedics Phusiline, BioScrews, BIOFIX, and PL-FIX pins are PLLA-based commercially available products.

3.2.6 Polyanhydrides (PA)

Polyanhydrides are made up of two carbonyl groups that are joined together via an ether bond. Polyanhydrides have wide applications in the biomedical field. They were finally permitted as drug-delivery vehicles by the FDA because of their outstanding biocompatibility as well as hydrolytic variability. Polyanhydrides break down into non-hazardous biocompatible diacid monomers, which can be broken down or excreted. The hydrophobicity of monomers has a big impact on how fast they degrade. Polyanhydride is frequently inserted into nanoparticles/microparticles for short-range controlled transport of bioactive compounds via injection, oral administration, or aerosol. Likewise, aliphatic diacids and hydrophobic aromatic diacids can be copolymerized into aromatic dimers or dimers of aliphatic fatty acids to retard polymer degradation. PA's mechanical properties are limited due to their rapid and uniform degradation, so methacrylate polyanhydrides have been investigated as cross-linkable, as well as injectable biomaterials to increase their strength for tissue-engineering applications. Drug transport or structural support for bone tissue manufacturing have both been investigated using cross-linked polyanhydrides.

3.2.7 Polycarbonates (PC)

Polycarbonate is a linear polymer with two germinal ether bonds and a carbonyl bond in its backbone. Even though this bond is highly hydrolytically stable, it degrades much more quickly in vivo. PCs have significantly more applications than other polymers because of their unique mixture of dimensional stability, heat resistance, and optical clarity. PC is used in a comprehensive range of therapeutic apparatus such as critical medical devices, because of its minimum liquid absorption, biocompatibility, as well as ease of purification. It has excessive elasticity but is deprived of mechanical strength. Poly(trimethylene carbonate) (PTMC) is one of the most extensively studied polycarbonates. It is an ideal drug-delivery candidate because of its degradation into carbonic acid, and 1,3-propanediol is biocompatible. Microparticles, discs, and gels made from PTMC have been used to deliver angiogenic agents and antibiotics. In the fabrication of polymersomes, micelles, and sutures, it is frequently copolymerized with poly(L-glutamic acid), PLA, polyether, and PCL. Potential applications of degradable polycarbonates in tissue engineering are being explored through new polycarbonates [44].

3.2.8 Poly(tetrafluoroethylene) (PTFE)

Poly(tetrafluoroethylene), Teflon, is an extremely crystalline, high-molecular-weight perfluoro polymer made of PTFE. Fluorine replaces hydrogen in PE, resulting in dramatic changes to its chemical as well as physical properties. Due to the huge size as well as common repulsion between fluorine atoms, the intermolecular forces are relatively weak and PTFE macromolecule chains show a twisting helix. PTFE has the lowest coefficient of friction compared to other polymers because of its chain–chain slip and helical packing. PTFE is chemically resistant and hemocompatible, making it ideal for biomedical tubing and a key component of advanced multi-lumen small-gauge medical-grade tubing used in many new minimally invasive catheters. Clinical interventional (catheters in various forms) and permanent implants are among the biomedical applications of PTFE solids, fluorocarbon coatings, perfluorinated fluids, and gels (cardiovascular, dental, ocular, craniofacial, urological, and abdominal applications). Prolonged ePTFE is a fibrillated form of PTFE that is created through a series of extrusion, stretching, and heating processes to create a microporous material with pore sizes ranging from 30 to about 100 micrometers that has a higher strength-to-weight ratio and creep resistance than fully dense PTFE. Meshes, advanced catheters, medical tubing, sutures, vascular grafts, as well as further therapeutic implants have all benefited from the use of ePTFE fiber meshes (Gore-Tex). Patches made of ePTFE are also used for soft-tissue regeneration, such as hernia repair and surgical sutures. Graft thrombosis, on the other hand, is a common property of fluoropolymer meshes and weaves in blood, which quickly passivates surfaces for acute short-term use but limits long-term blood-contact applications.

3.2.9 Polyether Ether Ketone

In polyether ether ketones, there are ether-and-ketone functional groups associated with the aryl rings, and they are semicrystalline. Due to its special chemical structure, polyether ether ketone has chemical and physical characteristics that have remained the gold standard in interbody implants because of its versatility and elasticity. Its biocompatibility in vivo and in vitro is also excellent. However, its mechanical characteristics are very similar to human cortical bone. In orthopedic and traumatic applications, polyether ether ketone is a suitable replacement component for metal implants due to these characteristics.

In contrast, polyether ketone is biochemically inert due to its stable chemical structure, preventing its efficient bonding with adjacent bone tissue when implanted in vivo. Polyether ether ketone improves its bioactivity through composite preparation and surface modification. Due to advancements in processing, the ability to interact mechanically with polyether ether ketone with shape memory has opened up new orthopedic surgery applications.

3.2.10   Polyurethanes

Polyurethane is a chain of carbon-based components with carbamate (urethane) links, characteristically formed by reacting a diisocyanate with a polyol (polyether or polyester). It is commonly united through a hydrocarbon chain extender to provide the solid section of the polymer. It is a sequence of polyol block copolymers that contain a soft segment and a hard segment. A microphase separation can occur in polyurethanes with segmented polar and nonpolar segments due to their different polarities. Chemical incompatibility between soft and hard segments accounts for the exceptional mechanical properties and biocompatibility. Changing the structure and ratio of the hard and soft segments can change the mechanical properties, including processing and biodegradation. Biodegradable polyurethanes are formulated from amino-alkylated diisocyanates rather than aromatic diisocyanates due to their potential carcinogenic properties. These polyurethanes do not possess good hydrolytic stability, whilst polyurethanes with polyether segments do not have good oxidative stability. A variety of medical devices have been made with polyurethane, including pacemaker leads and ventricular assist devices. Polycarbonate and polyether macrodiols containing larger hydrocarbon segments between ether groups have been employed to design bio-resistant polyurethanes. The biodegradability of polyurethane has recently been enhanced. Due to their soft and biodegradable characteristics, tissue engineering scaffolds are commonly constructed with polylactide or polyglycolide, polycaprolactone, and polyethylene oxide. Most commonly, medical devices that contact blood will be constructed from polyurethanes. Recently, a cardiology stent product was released on the market, which is placed inside the blocked coronary arteries after dilation with the help of a balloon made from smart shape-memory polyurethane [45,46].

4   CONCLUSION

Recently, we have been able to accomplish better control over molecular weight distribution and arrangement in covalent polymers. Advances in technology have contributed to improving biopolymer applications, degradation potential, and environmental safety. Bioplastics have also benefited from natural polymer modifications that have enhanced their mechanical and physical properties. Bioplastics are becoming increasingly popular in industrial and packaging applications, and they have great potential. By expanding bioplastic use, we can strengthen green economies and reduce the effect of greenhouse gases on the environment.

REFERENCES

1. Young R J, Lovell P A (2011) Introduction to Polymers. CRC Press, Boca Raton, US.
2. Nicholson J (2017) The chemistry of Polymers. RSC Adv. Burlington House, Piccadilly, London.
3. Koltzenburg S, Maskos M, Nuyken O (eds) (2017) Elastomers. In: Polymer Chemistry, 477–491. Springer, Berlin, Heidelberg.
4. Hall C (2017) Polymer materials: an introduction for technologists and scientists. Macmillan International Higher Education. London.
5. Hong M, Chen E Y X (2019) Future directions for sustainable polymers. Trends Analyt. Chem. 1: 148–151.
6. Tan L J, Zhu W, Zhou K (2020) Recent progress on polymer materials for additive manufacturing. Adv. Funct. Mater. 30: 2003062.
7. Rajeswari S, Prasanthi T, Sudha N, Swain R P, Panda S, Goka V (2017) Natural polymers: a recent review. World J. Pharm. Pharm. Sci. 6: 472–494.
8. Dyson R W (ed) (1987) Polymer structures and general properties. In: Specialty Polymers, 3–19. Springer, Boston, MA.

9. Fakirov S (2019) Condensation polymers: Their chemical peculiarities offer great opportunities. Prog. Polym. Sci. 89 1–18.

10. Whelan D (2017) Thermoplastic elastomers. In: Gilbert M (ed) Brydson's Plastics Materials, 653–703. Elsevier, Butterworth-Heinemann.

11. Kulkarni G S (2018) Introduction to polymers and their recycling techniques. In: Sabu T, Ajay V R, Krishnan K, Abitha V K, Martin G T (eds) Recycling of Polyurethane Foams. William Andrew Publishing, 1–16. Elsevier, US.

12. Prajapati R, Kohli K, Maity S K, Sharma B K (2021) Recovery and recycling of polymeric and plastic materials. In: Jyotishkumar P, Sanjay M R, Arpitha GR, Suchart S (eds) Recent Developments in Plastic Recycling, 15–41. Springer, Singapore.

13. Visakh P M, Mathew A P, Thomas S (2013) Natural polymers: their blends, composites and nanocomposites: state of art, new challenges and opportunities, Adv. Natural Polym. 1–20. Springer, Germany.

14. Bhatia S (ed) (2016) Natural polymers vs synthetic polymer. In: Natural Polymer Drug Delivery Systems, 95–118. Springer, Cham. Heidelberg, Germany.

15. Linhardt R J, Galliher P M, Cooney C L (1987) Polysaccharide lyases. Appl. Biochem. Biotechnol. 12: 135–176.

16. Yui T, Ogawa K (2005) X-ray diffraction study of polysaccharides. In: Severian D (ed) Polysaccharides: Structural Diversity and Functional Versatility, 99–122. Dekker. New York.

17. Olatunji O (ed) (2016) Classification of natural polymers. In: Natural Polymers, 1–17. Springer, Cham. Heidelberg, Germany.

18. Pérez S, Mazeau K (2005) Conformations, structures, and morphologies of celluloses, Polysaccharides: Structural diversity and functional versatility, 2: 1–99. Marcel Dekker, New York.

19. Heinze T, El Seoud O A, Koschella A (eds) (2018) Structure and properties of cellulose and its derivatives. In: Cellulose Derivatives, 39–172. Springer, Cham. Heidelberg, Germany.

20. Tobimatsu Y, Schuetz M (2019) Lignin polymerization: how do plants manage the chemistry so well?. Curr. Opin. Biotechnol. 56: 75–81.

21. Cotana F, Cavalaglio G, Nicolini A, Gelosia M, Coccia V, Petrozzi A, Brinchi L (2014) Lignin as co-product of second-generation bioethanol production from ligno-cellulosic biomass. Energy Procedia. 45: 52–60.

22. Adler E (1977) Lignin chemistry—past, present and future. Wood Sci. Technol. 11: 169–218.

23. Thakur B R, Singh R K, Handa A K, Rao M A (1997) Chemistry and uses of pectin—a review. Crit Rev Food Sci Nutr. 37: 47–73.

24. Parker R, Ring S G (2001) Aspects of the physical chemistry of starch. J. Cereal Sci. 34: 1–17.

25. Rutenberg M W, Solarek D (1984) Starch derivatives: Production and uses. In: Whistler RL, Bemiller J N, Paschall (eds) Starch: Chemistry and Technology, 311–388. Academic Press, Cambridge, Massachusetts, US.

26. Besford Q A, Cavalieri F, Caruso F (2020) Glycogen as a building block for advanced biological materials. Adv. Mater. 32: 1904625.

27. Crini G (2019) Historical review on chitin and chitosan biopolymers. Environ Chem Lett. 17: 1623–1643.

28. Sharma K, Singh V, Arora A (2011) Natural biodegradable polymers as matrices in transdermal drug delivery. Int J Drug Dev Res. 3: 85–103.

29. Kim H S, Lee C G, Lee E Y (2011) Alginate lyase: structure, property, and application. Biotechnol. Bioprocess Eng. 16: 843–851.

30. Tvaroska I, Rochas C, Taravel F R, Turquois T (1992) Computer modeling of polysaccharide–polysaccharide interactions: An approach to the κ-carrageenan–mannan case. Biopolymers 32: 551–560.

31. Mangaraj S, Yadav A, Bal L M, Dash S K, Mahanti N K (2019) Application of biodegradable polymers in food packaging industry: a comprehensive review. Packag. Technol. Sci. 3: 77–96.

32. Bhatia S (ed) (2016) Natural polymers vs synthetic polymer. In: Natural Polymer Drug Delivery Systems, 95–118. Springer, Cham. Heidelberg, Germany.

33. Nawrath C, Poirier Y (2008) Pathways for the synthesis of polyesters in plants: cutin, suberin, and polyhydroxyalkanoates. Plant Mol. Biol. 1: 201–239.

34. Manavbasi Y, Süleymanoglu E (2007) Nucleic acid-phospholipid recognition: Fourier transform infrared spectrometric characterization of ternary phospho-lipid-inorganic cation-DNA complex and its

relevance to chemicopharmaceutical design of nanometric liposome-based gene delivery formulations. Arch. Pharm. Res. 30: 1027–1040.

35. Burford R (2019) Polymers: A historical perspective. J. Proc. – R. Soc. N. S.W., 152: 242–250.

36. Asawatreratanakul K, Zhang Y W, Wititsuwannakul D, Wititsuwannakul R, Takahashi S, Rattanapittayaporn A, Koyama T (2003) Molecular cloning, expression and characterization of cDNA encoding cis-prenyltransferases from Hevea brasiliensis: a key factor participating in natural rubber biosynthesis. Eur. J. Biochem. 270: 4671–4680.

37. Hacker M C, Krieghoff J, Mikos A G (2019) Synthetic polymers. Regen. Med., Academic Press, US. 559–590.

38. Gunatillake P, Mayadunne R, Adhikari R (2006) Recent developments in biodegradable synthetic polymers. Biotechnol. Annu. Rev. 12: 301–347.

39. Eisch J J (2012) Fifty years of Ziegler–Natta polymerization: from serendipity to science, A personal account. Organometallics 31: 4917–4932.

40. Cowie J M G, Arrighi V (2007) Polymers: chemistry and physics of modern materials, 41–69. CRC Press, Boca Raton, Florida, US.

41. Hasegawa M, Sudo A, Shikinami Y, Uchida A (1999) Biological performance of a three-dimensional fabric as artificial cartilage in the repair of large osteochondral defects in rabbit. Biomaterials 20: 1969–1975.

42. Ali U, Karim K J B A, Buang N A (2015) A review of the properties and applications of poly (methyl methacrylate) (PMMA). Polym Rev (Phila Pa) 55: 678–705.

43. Stroupe J D, Hughes R E (1958) The structure of crystalline poly-(methyl methacrylate). J. Am. Chem. Soc. 80: 2341–2342.

44. Simionescu B C, Ivanov D (2016) Natural and synthetic polymers for designing composite materials. In: Iulian Vasile Antoniac (ed) Handbook of Bioceramics and Biocomposites, 233–286. Springer, Cham.

45. Das R, Karumbaiah K M (2015) Biodegradable polyester-based blends and composites: manufacturing, properties, and applications. Biodegradable Polyesters 13: 321–340.

46. Zia K M, Bhatti H N, Bhatti I A (2007) Methods for polyurethane and polyurethane composites, recycling and recovery: A review. React. Funct. Polym. 67: 675–692.

3 Natural Resources for Polyurethanes Industries

Magdalene A. Asare,[1,2] Felipe M. de Souza,[2] and Ram K. Gupta[1,2]

[1] Department of Chemistry, Pittsburg State University, Pittsburg, KS 66762, USA

[2] National Institute for Materials Advancement, Pittsburg State University, Pittsburg, KS 66762, USA

1 INTRODUCTION

The discovery and use of polymers began during the ancient times and knowledge about them was advanced into more practical uses around 1600 BC by the Mesoamerican peoples, when natural rubber was made into balls, figurines, rubber bands, and other artifacts. In the production of final polymer products, they extracted latex from an indigenous tree usually found in the southernmost parts of South America, *Castilla elastica*, and mixed it with a liquid from *Ipomoea alba*, another ancient plant. History tells us that the rubber liquid generated was used for medicine, painting, rituals, and also was solidified into rubber balls used for religious and sporting events [1]. With the course of industrialization, numerous scientists have discovered different types of plastics that have served essential uses in the delivery of goods of services and other purposes. For example, in 1839, Charles Goodyear discovered vulcanized rubber that could be used in the making of tires for automobiles, erasers, life jackets, and soles of shoes, among many other uses. Other inventors included Eduard Simon who accidentally discovered polystyrene in 1839, Eugene Baumann who discovered polyvinyl chloride in 1872, and Reginald Gibson and Eric Fawcett who formulated polyethylene in 1933 [2].

Polymers are macromolecules that are formed by the chemical combination of smaller repeating units of molecules called monomers. There are numerous ways in which different types of monomers can be chemically bonded and oriented in a polymerization reaction, giving rise to a wide range of types and applications of polymers. Polymers can appear in diverse forms and shapes due to the different types of configurational sequences of chains they can form, known as tacticity, which affects the physical, mechanical, and other properties of polymers. Polymers can be natural or synthetic, depending on their building blocks. Natural ones include DNA, starch, and collagen, while nylon and polyester, among others, constitute synthetic polymers. Natural polymers have existed throughout history and have provided essential aspects needs for survival such as food and renewable resources for papermaking and wood processing, textiles, and vegetable oils for coating and adhesives. The advancements in science and technology, industrialization, and the creation of novel polymers have led to the production of synthetic polymers using both simple and sophisticated approaches.

Polyurethanes (PUs) are among the most widely produced and used polymers in the world, because it is possible to manipulate their building blocks in the synthesis of different types of materials, and they have therefore found their applications in numerous industries. Also, as a result of their tunability and broad quality characteristics, they are ubiquitous in many homes, industries, electronics, energy, healthcare, construction, transportation, sports, and textiles, among innumerable

DOI: 10.1201/9781003278269-3

29

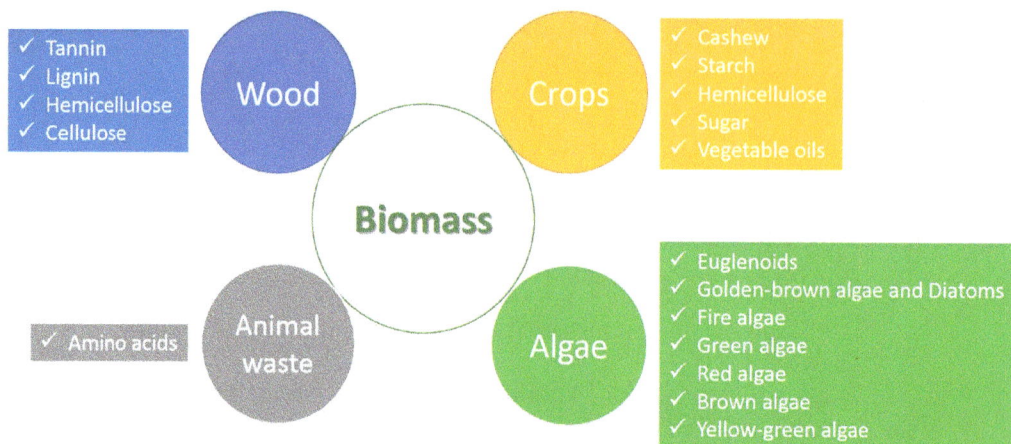

FIGURE 3.1 Examples of biomass. Reproduced with permission from Reference [4]. Copyright 2021, American Chemical Society.

others. Polymers are very advantageous as compared to the typically used metals or wood due to their tunability, easy production, low cost, high strength to density ratios, and excellent resistance to chemical and physical degradation. However, due to their high dependence on petrochemical sources for production, their non-recyclable nature, and non-biodegradability, they pose challenging ecological problems [3]. This has pushed for tremendous research in the effort to deal with global warming and the exhaustion of non-renewable natural resources, and has provided for strategic measures to be implemented. Scientists have seen the importance of immediately replacing fossil fuels with possible eco-friendly alternatives and technologies that use green energy in chemical reactions and polymer synthesis. Promising replacements that scientists have resorted to include renewable natural sources such as biomass which are sustainable materials from plants and animals (Figure 3.1) [4]. The use of biomass as an alternative is essential in polyurethanes because most PUs are not fully recyclable and when they are at the end of their life use, they either end up in landfills or are burnt.

2   CHEMISTRY AND TYPES OF POLYURETHANES

In the quest to find competitors for nylon, Professor Otto Bayer and coworkers discovered polyurethane in 1937. This discovery was a remarkable breakthrough in science even though it was disregarded from the start. And the initiation of this discovery produced beneficial products that were early adopted and used in World War II [5,6]. Currently, there is an abundant demand and production of polyurethanes due to their enticing morphology and a broad spectrum of desired mechanical properties, making them one of the most widely used polymers. Polyurethane can be used in the soles of shoes, furniture, medical devices, construction, packing, automobiles, and in countless other ways. Polyurethanes can therefore be found in many places due to their adaptability to be manipulated into different products using a diverse range of synthesis routes. The basic and general synthesis of polyurethane requires the presence of a polyol (a compound with two or more OH groups) and an isocyanate group, as shown in Figure 3.2. The physicochemical properties of PU are dependent on the nature, stoichiometry, and functionality of the present monomers, while their thermomechanical characteristics are associated with the probability of chains to phase segregate as a result of strong hydrogen bonds in the urethane linkage [7].

The functionality and type of polyols play a huge role in the characteristic properties of the final polyurethane product formed. For instance, an increase in the functionality of a polyol with a

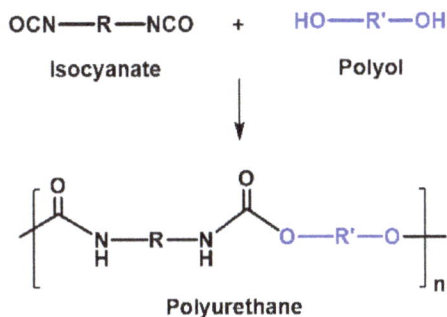

FIGURE 3.2 Chemical reaction showing the formation of polyurethane.

constant molecular weight can increase the hardness of foam and cause a small reduction in the tensile strength, tear strength, and elongation of the product [8]. Additionally, a high-molecular-weight polyol produces a flexible polyurethane and a low-molecular-weight polyol leads to the synthesis of rigid polyurethane material. There are different types of polyols such as polyether, polyester, polycarbonate, and acrylic polyols. It is worth noting that different polyols have distinct synthetic routes with different advantages and disadvantages. For example, polyether polyols are prepared through the copolymerization of propylene oxide and ethylene oxide in the presence of a suitable polyol precursor, while polyester polyols are synthesized similarly to polyester polymers. Factors that influence the type of polyol to use in a chemical reaction include the cost, hydrolytic and oxidative stability, viscosity, and other thermomechanical factors such as modulus strength and thermal flexibility [6].

In as much as polyols play a big role in PU synthesis, isocyanates are crucial in the formation and type of desired polyurethane manufactured. Isocyanates are very reactive and delicate when exposed, which requires precautions to be taken. Isocyanates can be categorized into aromatic or aliphatic, where aromatic ones are more reactive than aliphatic ones. Examples of aromatic isocyanates are toluene diisocyanate (TDI) and diphenyl-methane diisocyanate (MDI), which are more commonly used as compared to isophorone diisocyanate (IPDI) and hexamethylene diisocyanate (HDI), which are examples of aliphatic isocyanates. The use of isocyanates such as MDI and TDI has drawn a lot of questions and concerns due to some health complications that have been researched in people who have been subjected to prolonged exposure. Scientists have found that even with a low concentration of isocyanate, sensitive individuals could present with asthma [8].

Even though isocyanate is very reactive, its reaction with polyols, especially at room temperature, is relatively slow. This could be attributed to a lower affinity between the polar and less dense phase of the polyol with the non-polar and denser phase of the isocyanate. This drawback is usually attended to by scientists with the addition of appropriate concentrations of catalysts and/or surfactants to speed up the rate of the reaction [6]. Catalysts primarily speed up the chemical reaction between the polyol and isocyanate as well as other added chemicals. The most commonly used types are made of amines and tin. The proportion of catalysts used in a reaction is very important and needs to be specific for the particular reaction because it plays a major role in the polymerization and blowing of the polyurethane formed [8]. Another vital chemical, although only used in small amounts, is surfactants. Surfactants reduce the surface tension, stabilize the cell formation, and control the air permeability in the polyurethane. A common surfactant is one mainly composed of silicon [9]. In addition, blowing agents aid in the formation of the cellular structure and morphology of polyurethanes. They exist as either physical or chemical blowing agents. The physical ones are usually solvents with low boiling points, such as hexane and pentane. Chlorofluorocarbons (CFCs) are popular physical blowing agents due to their affordability and excellent thermal and chemical stability, but they are banned due to their destructive actions on the ozone layer. Chemical blowing agents expand the polymers when added with water as a typical example [8]. Other miscellaneous chemicals that could be added in the synthesis of polyurethane include plasticizers to reduce hardness, pigments for aesthetics, cross-linkers for structural modifications, fillers for improved mechanical properties, and flame retardants to minimize the flammability of the products manufactured [6]. Polyurethane can be grouped into different categories such as rigid and flexible foams, elastomers, sealants, adhesives, for paints and coatings, as well as composite materials. The different types of polyurethanes serve different functions and, in other instances, they have some overlapping characteristics but are used in different applications. Rigid polyurethanes can be used for building and appliance insulation, construction, and the packaging and storage of materials. Flexible polyurethanes, on the other hand, are used as cushions in furniture and bedding, in carpet underlays, and safety pads among many other uses. Elastomers that are more stretchy can be used in footwear, medical devices, pipe linings, and other industrial applications. Overall, polyurethane-based products are very convenient for the delivery of services to consumers [5].

3   CHEMICAL REQUIREMENTS IN BIOMATERIALS FOR POLYURETHANES

In today's news and scientific research, the replacement of raw materials from petroleum sources with biobased materials is a major highlight and topic of discussion. Promising renewable materials include the use of biobased masses such as cellulose, lignin, proteins, and vegetable oils among many others. These sources are under intensive investigation, with findings showing their advantages and disadvantages. However, vegetable-based oils have been one of the most popular substitutes for petroleum sources due to their easy availability, low cost, eco-friendly nature, and the potential modification of their active site for a diverse range of chemical syntheses and applications. Vegetable oils have a broad classification, with some consisting of a wide range of triglycerides and fatty acid residues. Some can have fatty acid chain lengths containing about 14–22 carbons and double bonds from 0–5 or more located at different points. Their unique compositions in addition to their degree of unsaturation (containing double or triple bonds) distinguish them. The naturally occurring carbon–carbon double bonds in most vegetable oils can be found between their 9th and 16th carbons, however, these bonds are usually not conjugated and are less reactive as a result of the trapping of radicals by allyl hydrogens of the methylene groups between the double bonds. On the other hand, some oils, like tung oil, have conjugated double bonds, rendering them highly reactive. Other vegetable oils have esters, hydroxyl, and epoxy groups which are reactive sites, with examples including castor and vernonia oil. Also, even though some oils have reactive centers and can directly be used in chemical synthesis, they can be modified in addition to the other unreactive oils through different methods to enhance their reactivity as starting monomers for polymerization reactions.

The variety of ways in which vegetable oils can be made more reactive include epoxidation, transesterification, hydroformylation, ozonolysis, metathesis, and thiol-ene method. The formation of an epoxide group from the conversion of a carbon–carbon double bond followed by a ring-opening reaction using alcohols, amines, or carboxylic acids leads to the generation of biobased polyols that can take part in the chemical synthesis of polyurethanes. The chemical representation of this reaction is shown in Figure 3.3. Here, there is an oxygen transfer from the peroxide to the carbon–carbon double bond and a ring-opening reaction with the acetic anhydride to form the polyol [10,11]. One of the most commonly used methods for the introduction of an epoxide into a double bond is the Prileshjev reaction which utilizes peracids [12]. Consequently, a number of researchers have tried different approaches and reactants with a diverse range of results. For instance, He et al. used hydrogen peroxide and formic acid in the formation of a polyol from soybean oil, and their results proved a safer and more efficient method with the production of rigid polyurethane foams that had better results as compared to commercial ones [13]. The mole ratios of the peroxide and formic acid are vital, and other researchers have tried different catalysts in conjunction with hydrogen peroxide such as Ti/SiO_2, tungsten-based chemicals, and enzymes [14,15].

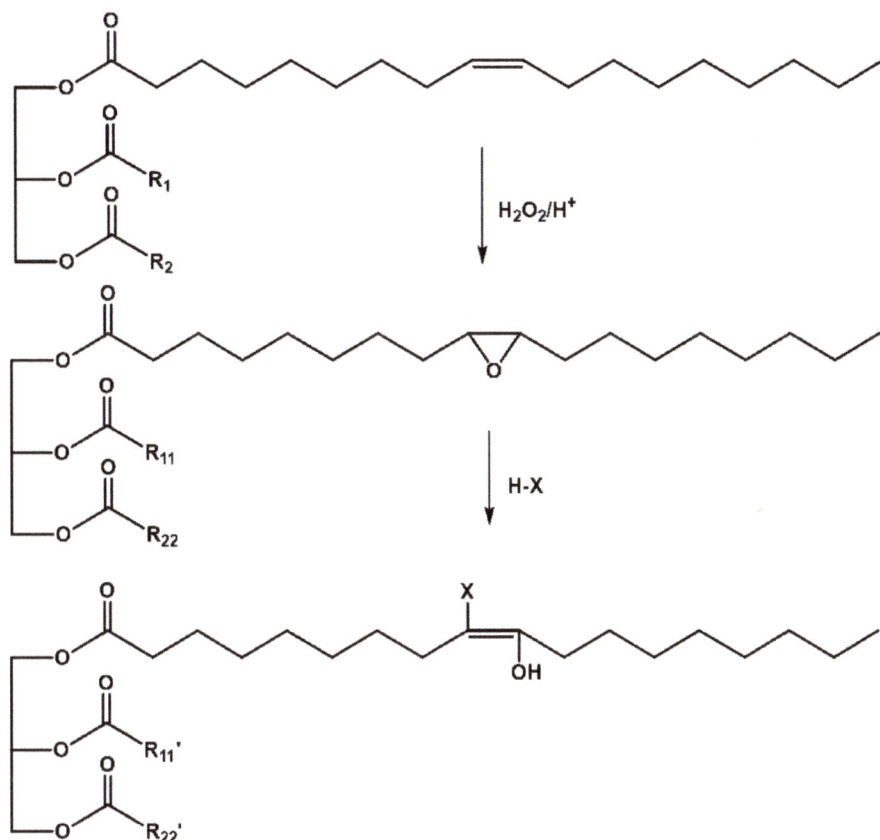

FIGURE 3.3 Formation of biobased polyol through epoxidation followed by ring-opening reaction. Reproduced with permission from Reference [11]. Copyright 2018, Springer Nature.

Hydroformylation is another important synthetic route for the conversion of some vegetable oils, such as soybean oil, into polyols, as shown in Figure 3.4. It incorporates a syngas (hydrogen and carbon monoxide) in a 1:1 mole ratio in the presence of either rhodium or cobalt as catalysts. This is usually performed around 70–130 °C and results in the formation of an aldehyde which is eventually turned into a polyol with a primary OH group following a hydrogenation step [16]. An experiment performed at the Kansas Polymer Research Center uncovered that there is a huge difference in the polymer formed when rhodium or cobalt was adopted as a catalyst for soybean oil. It was found that rhodium facilitated a higher conversion of the double bonds to polyol of 95% under mild conditions as compared to cobalt that required harsher conditions but led to a lower conversion rate of 67%. Also, since the rate of conversion affects the final product, it was observed that rhodium formed a rigid plastic, while cobalt provided a hard rubber. This was confirmed from the different results of the tensile strength, Young's modulus, and elongation break tests conducted. For example, rhodium had a high Young's modulus of 363 MPa, and cobalt recorded one as low as 13 MPa [17].

There is a difference in the type of polyol formed when epoxidation or hydroformylation is chosen. Usually, an epoxidation followed by ring-opening forms a polyol with a secondary alcohol, and a primary OH group is seen in a hydroformylation reaction which affects the further reactivity of the polyol and product formed. Guo and research team, for instance, recorded that a soybean polyol formed from hydroformylation had a shorter gel time and an improved curing ability when reacted with isocyanate as compared with the product from the epoxidation reaction [19]. Ozonolysis, which

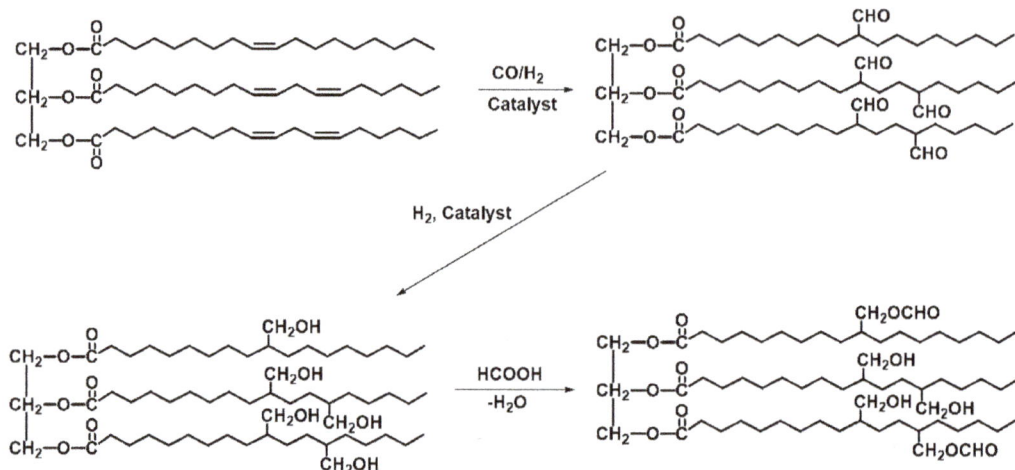

FIGURE 3.4 Hydroformylation process for polyol formation. Reproduced with permission from Reference [18]. Copyright 2007, John Wiley and Sons.

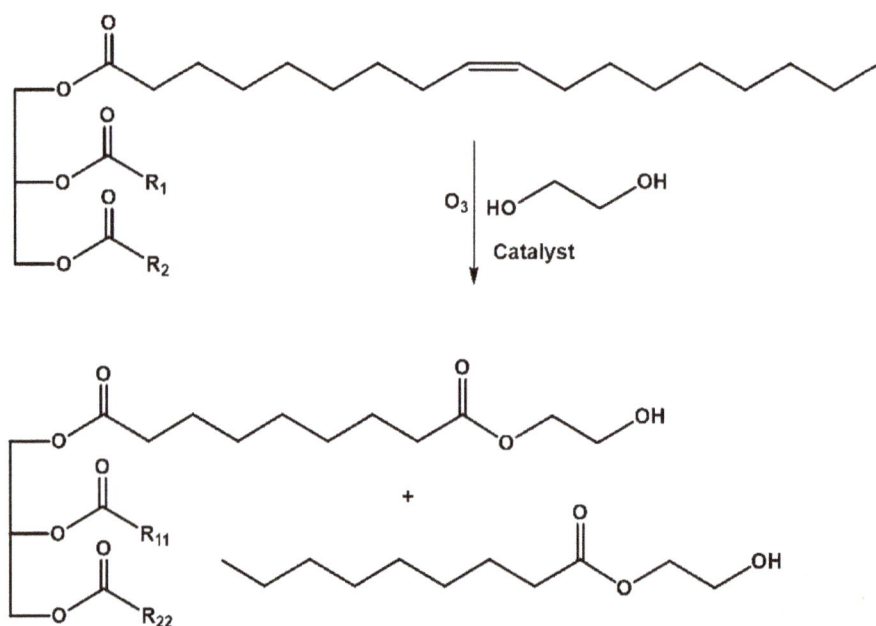

FIGURE 3.5 Ozonolysis of vegetable oils into polyol. Reproduced with permission from Reference [11]. Copyright 2018, Springer Nature.

is another method, involves the conversion of double bonds in some vegetable oils into ozonides which are then reduced into aldehydes and finally polyols. Polyols generated usually have primary OH groups which are more reactive with isocyanate than secondary OH groups. Also, the hydroxyl groups of the polyols are found at the end of the fatty acid chains which constitute aliphatic chains and eventually form rigid polyurethane foams when reacted with isocyanate [11]. As illustrated in Figure 3.5, ozone is used to breakdown and oxidize the double bonds of the oils, which are further reduced to ozonides and then polyols by different reducing catalysts.

$$R_1—CH{=}CH-R_2 \ + \ R'—SH \ \xrightarrow{\text{UV light}} \ R_1—CH_2—\underset{\underset{SR'}{|}}{CH}-R_2$$

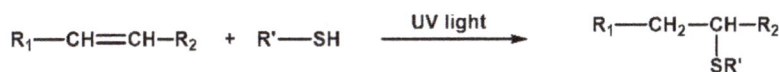

FIGURE 3.6   A representative thiol-ene reaction.

Transesterification and transamidation are similar chemical reactions that have been successful in the synthesis of polyols from vegetable oils. Transesterification uses an ester and polyols usually made with glycerol which has a mixture of monoglycerides, diglycerides, and triglycerides. However, the monoglycerides (containing two hydroxyl groups per molecule) are more crucial for the synthesis of polyurethane. In transamidation, amines such as diethanolamine can be used in making polyols. The fatty acid diethanolamides are bifunctional and have been found to enhance the compatibility of some polyols with improved physicomechanical properties when used in polyurethane. Successful experiments have been performed with rapeseed, soybean, linseed, and sunflower oil [11].

Metathesis is another chemical route for the production of polyols from vegetable oils. Here, a two-step process involving the metathesis and reduction in the presence of a catalyst takes place. The double bonds can be changed into polyols through an epoxidation–alcoholysis reaction. For example, in a metathesis reaction, scientists used a trioleine in reaction with ethylene in the presence of ruthenium and it formed a triglyceride with terminal bonds that were then converted into a polyol with an epoxidation followed by an alcoholysis with methanol.

The above functionalization reactions usually have at least two steps which may require solvents and catalysts, and they take a lot of time. The thiol-ene reaction, on the other hand, has gained lots of attention as a result of its simple synthesis, short reaction time, and high yield, while it falls under click chemistry. A general reaction scheme of thiol-ene is shown in Figure 3.6 [20]. Thiol-ene reactions occur in the following four steps: (1) free radical formation from thermal or photochemical generation; (2) production of thiyl radicals at sulfur atoms due to transfer reactions with thiol groups; (3) formation of new radicals through anti-Markovnikov addition; and (4) creation of final products and new thiyl radicals. These reactions are usually initiated with UV light of 200–400 nm wavelengths in the presence of photoinitiators.

4   BIO-DERIVED POLYOLS

Polymers, especially polyurethanes, that are in high demand depend extensively on raw materials that can be obtained from petroleum sources. However, due to price fluctuations, exhaustion, or the impending extinction of these non-renewable sources, biobased alternatives are being sought. In a much deeper perspective, researchers and manufacturers of polyurethane are concerned about finding ways of using different biobased materials for polyols and isocyanates rather than from petroleum sources. In addition, they are also mindful about how comparable the processing and quality of the biobased products will be to the already commercialized petroleum-based ones. Polyol has the largest percent by weight as compared to the other components involved in the synthesis of polyurethane. Given this, pressing concerns about finding sustainable sources in place of petrochemicals are increasing. Replacements that have been found for polyols (a vital starting material of polyurethane) include a wide variety of biomasses such as materials from plants and animals. Biomasses and other natural substitutes are becoming popular because they are acquired from non-polluting reserves and they also have a low carbon imprint. Biomass, in particular, comes in a large quantity and variety, it has zero sulfur and nitrogen, and it could potentially ease the current global energy crises that we face. This and many other reasons have convinced some companies such as Dow Chemical, Shell Chemicals Limited, and Bayer Material Science to commercialize biobased polyols.

In the synthesis of polyols from biomass, chemical processes such as oxypropylation and acid liquefaction have been attempted which have produced successful outcomes. Starting materials generated from the above were used to fully or partially replace petroleum sources in polyurethane foams which showed relatively close kinetics, cellular morphology, and thermal conductivity as compared to their commercialized petroleum counterparts [21,22]. The use of oxypropylation in the synthesis of polyols from biomass is a distinct reaction that not only increases the functionality of the polyols but also the number of hydroxyl groups present. Oxypropylation has been adopted and used by various scientists in the production of polyols from natural resources and waste materials such as chitosan, cork, lignin, sugar beet, cornstarch, and olive stone. For example, Pavier et al. synthesized polyols used for polyurethane products from sugar beet pulp (SBP). Based on these results, oxypropylation enhanced the hydroxyl groups present as well as the viscosity of the polyols, leading to more efficient use when adopted in the synthesis of polyurethane [23]. The oxypropylation of cork to polyol was performed to use in the preparation of polyurethane [24]. Cork, which is a natural composite and raw material, was converted under high pressure and temperature in the presence of KOH as a catalyst to make the polyol. The synthesized polyol had a low viscosity for easier manipulation in addition to an improved OH functionality. To mention another, one research team adopted oxypropylation in the synthesis of polyol from tree bark for polyurethane foams, as shown in Figure 3.7. When compared to commercial and other foams, the bark-derived foam had a comparable hydroxyl value and concentration in addition to a higher elastic modulus and compressive strength [25].

Aside from the great results generated from the adoption of oxypropylation, acid liquefaction is another interesting approach used, which does not require high temperatures, pressures, or the use of propylene oxide. In addition, it has an easy method of operation, high production and energy conversion, and an effective treatment for organic materials that are used. With this method, researchers have tried different techniques in the pretreatment of biomass such as the use of ultrasound and microwave. After that, a diverse range of liquefaction procedures, which include the use of solvents and catalysts, liquefaction with chemical modification and new technologies, or a combination of several techniques has been researched and used [26]. Even though the reaction mechanism of this method can be complex, it is very effective in the transformation of biomass into polyols. Liquefaction has been successfully used in making polyols and different materials from natural sources such as bamboo, wheat straw, ground coffee, corn, and lignin. Scientists have also used other methods to produce biobased polyols, as elaborated in the examples below.

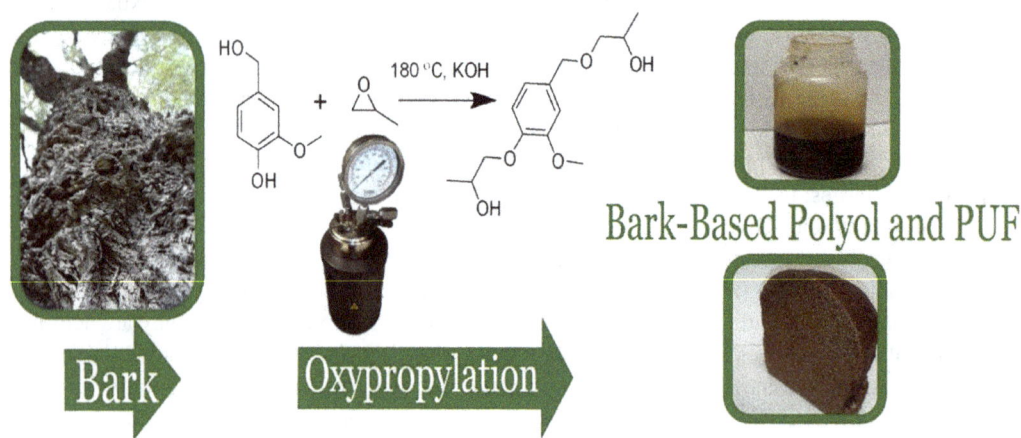

FIGURE 3.7 Biobased polyol from the oxypropylation of bark. Reproduced with permission from Reference [25]. Copyright 2015, Elsevier.

4.1 Lignin

Lignin is a popularly researched and abundant natural polymer that can be substituted for some petrochemical materials. It is one of the major components in the cell walls of plants and is the most abundant aromatic biopolymer which constitutes about 30% of the organic carbon biosphere [27]. Lignin has several unique properties such as carbon neutrality, low weight, high antioxidant and antimicrobial activities, biodegradability, and it does not compete with food crops, which makes it very suitable for different applications. Depending on the level of manufacturing, lignin can either be considered a raw material or waste, nevertheless, it serves a significant purpose. For instance, researchers found that about 50 million tons of lignin are generated annually as waste from chemical pulp industries. However, this serves as a promising material in different applications because they end up with functional groups, they have better stability, good rheology, and improved mechanical properties. Cellulose, hemicellulose, lignin, and small components of waxes and water-soluble compounds usually constitute lignocellulosic materials. The compositional fraction of each influences the properties of the biobased polymers that are synthesized.

Lignin has a distinct cross-linking pattern, depending on its source of extraction. For example, lignin from softwood is more branched and cross-linked, while that from hardwood is linear, which can impact the end product's properties and applications [28]. Apart from that, the method through which lignin is extracted can affect its physicochemical properties since it generally results in a breakdown to lower molecular weights. Different methods of extraction include sulfite, Kraft, soda, and other fractionation processes [29]. Due to the unique advantages of lignin, it has extensively been studied and is currently in use for polyurethane applications. In the aspect of polyols derived from lignin, diverse approaches have been used due to the rich hydroxyl content it contains. For example, to improve the property of lignin, Ahvazi et al. modified wheat straw soda lignin to make polyols [30]. These researchers observed that using acidic conditions produced higher aliphatic hydroxyl groups, hence there was higher reactivity as compared to the basic medium. However, scientists have found some discrepancies in the application of lignin polyols in polyurethanes and have performed different experiments for improved properties. For instance, Haridevan and research team investigated the dispersion and mechanical effects of Kraft lignin biobased polyols in rigid polyurethane foams. With different modifications in terms of weight percentages of the blends, there were some minimal improvements in the flame retardancy of the foams [31].

4.2 Vegetable Oils

Vegetable oils are among the most researched natural resources in the production of polymers, especially polyurethane. They have a wide variety and molecular weight, a versatile and easily manipulative chemical structure, in addition to their low cost, biodegradability, and comparable characteristics when used as substitutes for petroleum-based polyurethane products. Many vegetable oils, such as soybean oil, palm oil, castor oil, rapeseed oil, canola oil, and tung oil, are used in polyol production. The following sections report some of them in detail.

4.2.1 Soybean Oil

Soybean is a legume that is cultivated in subtropical and tropical zones. It is one of the most valuable crops produced, and serves huge functions such as food for livestock and aquaculture, protein for humans, oils for skincare, biofuels, and raw materials for some polymer-based products. Soybean in itself is made up of 82% soybean meal and 18% soybean oil. In the USA, for example, about 68% of total soybean oil produced is consumed, while the remainder is channeled into biofuels and raw materials for polymers [32]. An extensive amount of research has been conducted in the use of soybean oil as polyols for polyurethane.

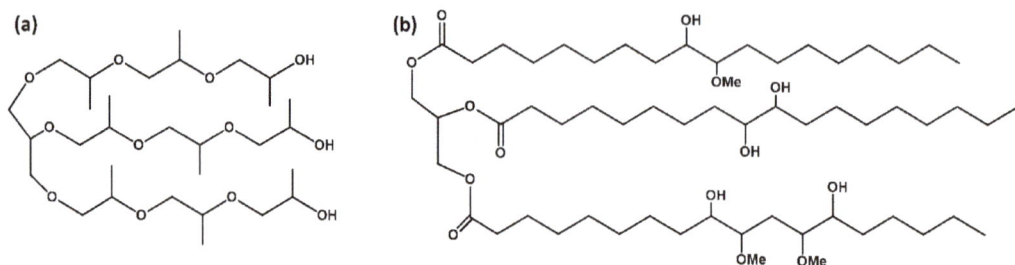

FIGURE 3.8 Comparison of petroleum-based polyol (a) and soybean polyol (b). Reproduced with permission from Reference [35]. Copyright 2011, Elsevier.

Gu et al. synthesized and characterized polyurethane foams made from soybean oils [33]. In their experiment, crude soybean oil was extracted and converted into a polyol through epoxidation with hydrogen peroxide and hydroxylation with a controlled amount of acid. It is interesting to note that these scientists made higher and lower hydroxyl value soybean polyols, used them in polyurethane foams, and compared them to three petroleum-based polyurethane foams. They found some differences in the foams based on the hydroxyl number, such that biobased foams with higher OH numbers had a better cell size and distribution as compared to those with a lower hydroxyl number. However, except for some poor cryogenics in the case of soybean foams, they could be excellent substitutes for petroleum-derived foams. Furthermore, epoxidation of the sucrose esters and ethers of soybean oil was conducted using base-catalyzed acid-epoxy reactions and acid-catalyzed alcohol-epoxy reactions, respectively. The biobased polyols generated were used for polyurethane coating on steel, which was then tested for different properties. The results showed that the biobased polyols had greater hardness and other appreciable mechanical properties due to their well-defined chemical structures and high hydroxyl functionality [34].

In addition, Tan et al. carried out an investigation to compare the properties of polyurethane foams made from a commercial petroleum polypropylene-based polyol and a commercially available soybean polyol, with their chemical structures illustrated in Figure 3.8 [35]. Comparatively, the soybean polyol had higher molecular weight, functionality, and viscosity than the petroleum-based polyol. Even though the very high viscosity could have some drawbacks in terms of processing, it was successfully used in the synthesis of polyurethane foams. From the various tests, it was observed that the biobased soybean foams had comparable foam kinetics, cellular morphology, density, and an even higher compressive strength than the petroleum source. Soybean oil is a promising alternative, and more research needs to be undertaken for its full optimization in the polymer industry.

4.2.2   Castor Oil

Castor oil is obtained from the seeds of a plant called *Ricinus communis*. In general, castor oil is used for medicinal purposes and, since it can be nauseating, it is usually not edible. Castor oil is composed of a high amount of unsaturated fatty acid called ricinoleic acid, which constitutes about 90% of its components. The presence of a high ricinoleic acid content makes castor oil very unique since it is uncommon to find it in other oils. Ricinoleic acid has hydroxyl groups that are not regularly seen in other oils. This facilitates the easy conversion and chemical synthesis of castor oil into other products. Castor oil can be hydrogenated, oxidized, halogenated, and polymerized due to the presence of a double bond. In terms of polyol production, castor oil can directly be used as a raw material in the synthesis of polyurethane due to the presence of hydroxyl groups. However, to maximize the functionality and full potential of castor oil, it can be passed through different reactions such as epoxidation followed by ring-opening. Scientists have used castor oil-based polyols in different polyurethanes and have found comparable and even higher characteristics compared to polyurethane made from petrochemical sources.

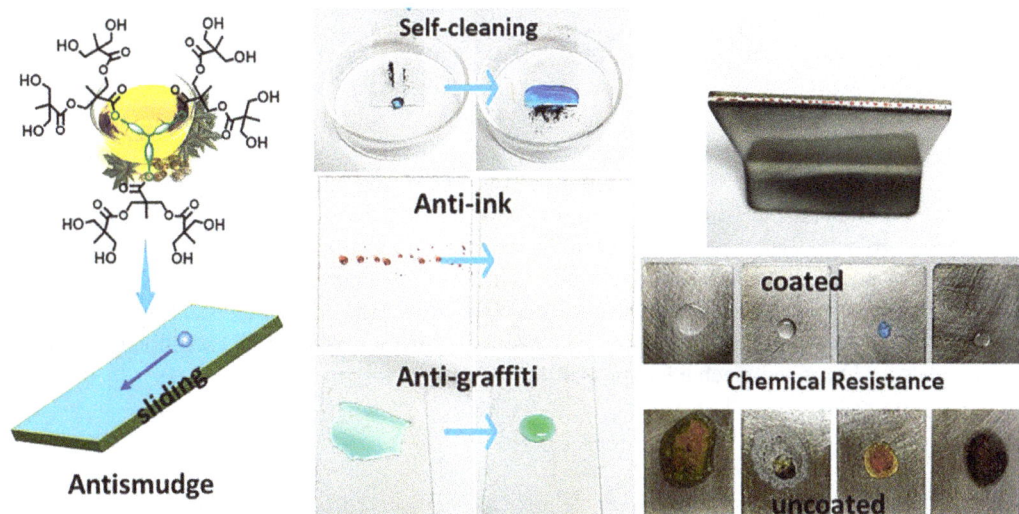

FIGURE 3.9 Castor oil-based polyol for antismudge application. Reproduced with permission from Reference [36]. Copyright 2021, American Chemical Society.

For instance, Wei et al. employed biobased castor oil as a green alternative for a polyurethane antismudge coating application [36]. A hyperbranched polyol made from castor oil was used as a precursor in a chemical reaction with other components such as hexamethylene diisocyanate and a mono-hydroxyl-terminated poly(dimethylsiloxane) to make a coating material. From the physical appearance (Figure 3.9) and other test results, it was observed that the biobased polyurethane coating had excellent liquid repellency and good self-cleaning efficiency and transparency, in addition to its mechanically robust nature. Diversified research was conducted by Ionescu and team using three different castor oil polyols synthesized with varying approaches and each was reacted with a different isocyanate to study the effects of the hydroxyl numbers and functionality on cast polyurethane and rigid polyurethane foams [37]. The thiol-ene reaction with the use of mercaptoethanol was used to make one of the polyols. After reacting the different polyols with distinct isocyanates, the scientists observed that the polyurethane showed good mechanical properties, which suggests that castor oil could serve as a promising replacement for petrochemical materials for use in polyurethanes [37].

4.2.3 Jatropha Oil

Jatropha oil is obtained from the seeds of the jatropha fruit. It contains about 78.9% unsaturated fatty acids, composed of oleic and linoleic acids, that are useful for chemical reactions. This is a promising biobased material because it has a high degree of unsaturation that gives room for a wide range of chemical modifications, syntheses, and applications. In addition, it is not edible and would be a better substitute for edible oils in the production of polymeric materials from renewable sources. Scientists are intrigued by this oil and have used it in many applications, including polyurethane coatings, adhesives, and elastomers. A research team used jatropha oil-based polyol in a waterborne polyurethane dispersion application [38]. The hydroxyl number of the polyol was varied by using an epoxidation reaction followed by oxirane ring-opening with methanol. Their findings showed that the biobased polyurethane dispersions had high thermal stability, good water repellency, quality pendulum hardness, and other characteristics that can make it suitable for other uses such as wood binding and as a decorative coating.

4.2.4   Palm Oil

Palm oil is one of the most widely used, produced, and consumed oils. It has recently gained a lot of attention for non-food applications. Scientists have used several methods in the production of polyols from palm oil which include epoxidation followed by ring-opening, alcoholysis, and transesterification. Ng et al. synthesized a palm oil-based polyester polyol (PPP) through a polycondensation reaction with deacidified glycerol monostearate and glutaric acid [39]. The PPP was then used in making a water-blown porous polyurethane utilizing a one-shot foaming method. The palm oil-based polyol was successful in this specific application and gave rise to products that had improved pore size and compactness, and high tensile strength and elastic modulus. In addition, the products showed tunable degradation characteristics. Other researchers also used palm-oil-based polyols in the production of semi-rigid bio-polyurethane foams with the addition of cellulose nanocrystals for improved mechanical properties [40].

4.3   Cornstarch

Cornstarch, a natural material, also finds its application as a starting material for different chemical reactions for various applications such as for the preparation of polymers. Starch has abundant polymer chains that are composed of hydroxyl groups which are important for chemical reactions. In an experiment, a specific ratio of cornstarch, polyethylene glycol, glycerol, and 1,4-butanediol was combined and used with other important chemicals to make polyurethane foams [41]. Successful foam formations ensued and the effects of the cornstarch on the foam were studied and the results showed good properties of the foam.

4.4   Wood

Another interesting source of polyol used for polyurethane is wood. In an experiment performed by Ertaş et al., eucalyptus and pine woods were combined with polyethylene glycol for the synthesis of rigid polyurethane foams [42]. The wood-based polyols were made through liquefaction with sulfuric acid at around 140–160 °C for 120 minutes. Before using them in the rigid foams, confirmatory tests were conducted. It was found that the foams had comparable compressive strength, density, elastic modulus, and thermal conductivity with petroleum-based foams. And even better, the wood-based foams were biodegradable making them a step toward solving some of the environmental crises the world faces.

5   BIO-DERIVED ISOCYANATES

Isocyanate (NCO), which is the second major component used in the synthesis of polyurethane, is very dangerous and can have detrimental health effects in humans and animals. Researchers have found that prolonged exposure to isocyanate can lead to serious respiratory complications such as asthma, acute poisoning, and skin irritations [43,44]. In addition, the process of making isocyanate requires phosgenation of amines, and phosgene is a noxious chemical. A lot of diverse research and time have been invested into discovering more human-friendly approaches to the synthesis of isocyanate and novel substitutes of isocyanate in the synthesis of polyurethane. In discovering different syntheses other than phosgenation, the Curtius, Hofmann, and Lossen reactions among others have been found. Despite that, the mechanisms involved in the above form carbonyl nitrene intermediates that release an exothermic peak during its rearrangement in the formation of isocyanate. This is an alarming safety concern, especially for industrial-scale production of isocyanate. In addition, most raw materials used for isocyanate are petroleum-based. Classical examples of petroleum-based isocyanates include toluene diisocyanate (TDI), methylene diphenyl diisocyanate (MDI), 1,6-hexamethylene diisocyanate (HMDI), and isophorone diisocyanate (IPDI) [45]. In the exploration for a better alternative, biomass as raw material has risen to the occasion and has been proven to be a potential solution to this issue because it is unlimited and natural, it has tunable chemical advantages that make it possible for different transformation, and it has shown great results when used in recent polymeric materials. Researchers have categorized some of the main natural sources from which isocyanates have been successfully made such as amino acids, sugars, furan, lignin, cashew nut shell liquid, vegetable oils, and algae [4].

5.1 Amino Acid-Based Isocyanate

Lysine is a classic example of an amino acid that has been researched in the production of biobased isocyanate. Lysine can be used to either synthesize methyl L-lysine diisocyanate (MELDI) or ethyl ester L-lysine diisocyanate (EELDI) following different reaction paths. L-leucine is a branched α-amino acid which varies a little from the aliphatic α-amino acid in L-lysine. However, either one can be used for biobased isocyanates. L-leucine in particular has an isobutyl group that affords it an advantage of a strong hydrophobicity which enables it to form an α-helix that stabilizes the protein. Due to its nature, it has been found to be a biocompatible material for use as artificial skin or fibers. In an anionic polymerization reaction, L-leucine isocyanate was made with L-leucine methyl ester hydrochloride with phosgene. Other scientists have used L-lysine in biodegradable injectable polyurethane for orthopedics and tissue repair and regeneration. They found that although L-lysine has some drawbacks such as a microphase-mixed behavior, they are biocompatible and have a lower vapor pressure, which makes them less toxic in comparison to petroleum-based hexamethylene diisocyanate, for example [46, 47].

5.2 Sugar-Based Isocyanate

Natural sugars can be found in different forms and sources. Carbohydrates, as an example, are a great source of green materials that can be used to make sustainable products that can replace petrochemical ones. They are receiving substantial attention because of the possibility of transforming their monosaccharides or disaccharides into different precursors or reactors for biobased products [48]. In addition, starch can be passed through hydrogenolysis to form dianhydrohexitols from the present monosaccharides. Researchers have confirmed that dianhydrohexitols have three diverse types with two hydroxyl functional groups that can be transformed into isocyanate through phosgenation. In another work, isosorbitol, which can be synthesized from sorbitol (a type of sugar alcohol), was used to make isocyanate. The chemical reaction involved double esterification of isosorbide with other chemicals like succinic anhydride and isomannide. It is worth noting that Zenner et al., in their research work, aimed to use alternative measures and reactions to petroleum-based reagents in the production of biobased isocyanate [49]. Another appreciable discovery is the first commercially available sugar-based isocyanate named pentamethylene diisocyanate. It has about 70% renewable content made from biomass and, as compared to its petroleum counterpart, it has a lower carbon footprint and better energy efficiency.

5.3 Vegetable Oil-Based Isocyanate

In the quest to find biobased alternatives to petroleum-sourced isocyanate, Çaylı et al. synthesized a soybean oil isocyanate [50]. The beginning of the reaction constituted the bromination of the triglyceride chains of soybean oil at the allylic position with N-bromosuccinimide. This was followed by the reaction with AgNCO, which resulted in the actual conversion into the isocyanate, as shown in Figure 3.10. It was interesting to find from this research that there was a high yield of conversion and, with an increasing amount of AgNCO, there was a greater yield in the amount of biobased isocyanate formed.

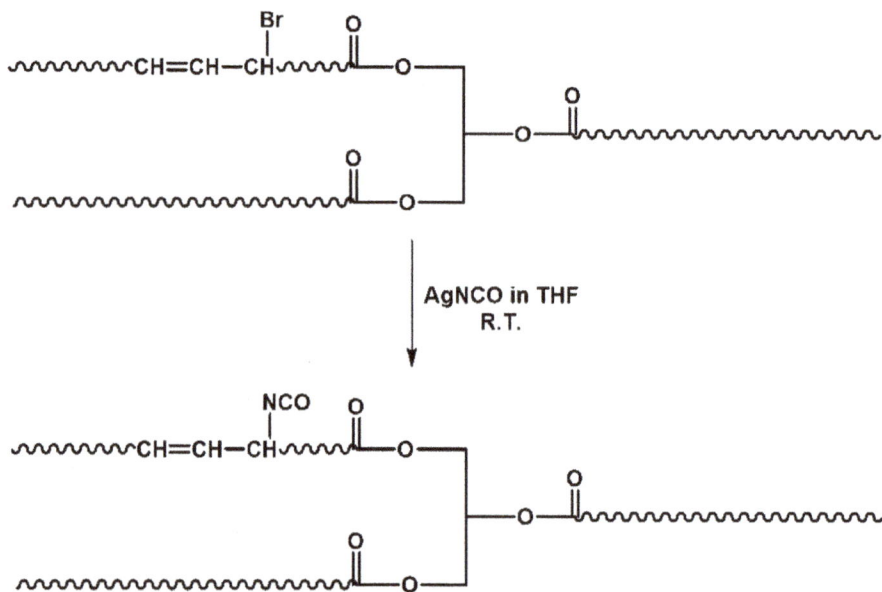

FIGURE 3.10 Synthesis for biobased isocyanate from soybean oil. Reproduced with permission from Reference [50]. Copyright 2008, John Wiley and Sons.

5.4 FURAN-BASED ISOCYANATE

Furan-based compounds are a group of biomasses that have numerous advantages for the development of sustainable raw materials for polymers. Examples include furfural, 5-hydroxymethylfurfural, and 2,5-furandicarboxylic acid. These have been discovered to be easily converted into derivatives and precursors for biobased polymers such as polyamides and polyurethane. Furfural in particular has been highly researched as a result of coming across its high reactivity due to several functional groups and its versatility. The first-ever furan-based isocyanate was patented in 1962 by Garber and, thus far, many other modifications and investigations for better sustainability have been done and are still ongoing [51]. Besides the advantages of finding biobased isocyanate from furan, they have some drawbacks also. For example, 2,5-furandiisocyanate is very reactive to moisture and oxygen, which makes its long-term storage more difficult. This factor is attributed to furan's electron-rich and planar ring structure, which increases its electrophilicity and speeds up nucleophilic additions to chemicals like polyols. Hence, scientists have found blocking agents as a solution. Here, the blocking agents in simple terms will protect the isocyanate and further expose it when it is necessary for its reaction with a polyol or another chemical [52].

5.5 LIGNIN-BASED ISOCYANATE

Lignin is the second most abundant natural polymer in the world next to cellulose [53]. It contains a tremendous amount of hydrogen and carbons and, as a result, it is usually classified as a promising natural source for gas and for replacing petroleum-based raw materials. The extraction of this resource has different methods that either require sulfur compounds or do not. Examples include Kraft, lignosulfonate, and organosolv reactions [54]. Using some transformation reactions, various biobased polymers can be obtained from lignin. A notable contribution to the utilization of natural resources is the synthesis of biobased isocyanates using vanillic acid (chemically derived from lignin). With intensive research conducted by Kuhire et al., biobased isocyanates discovered include bis(4-isocyanato-2-methoxyphenoxy) and bis(4-isocyanato-2,6-dimethoxyphenoxy)alkanes [55].

5.6 Cashew Nut Shell-Based Isocyanate

The cashew tree is an evergreen plant usually found in coastal areas which bears a cashew nut that is used as snacks and for other purposes. However, the cashew shell in particular contains a viscous dark brown liquid known as cashew nut shell liquid (CNSL), and it is of great interest for scientific research due to its composition of non-isoprenoid phenolic lipids [56]. Some scientists have been successful in the production of biobased isocyanate from cashew nut shell liquid, however, more research needs to be conducted with this natural raw material [4].

5.7 Algae-Based Isocyanate

Algae, photosynthetic organisms, have also gained a lot of attention as a raw material in the synthesis of biobased products because they are easy to cultivate and do not require arable land as is the case for some vegetable oils. In addition, they can also produce more unsaturated fatty acid chains that are very useful for different synthetic reactions. Scientists are currently moving deeper into this resource to find new alternatives and they have derived some biobased isocyanates also [4]. Notwithstanding, other scientists are focused on non-isocyanate polyurethane (NIPU) and ongoing research information can be found. Different synthetic routes formed include polycondensation, rearrangement, ring-opening polymerization, and polyaddition. Researchers have found that these reactions also have their advantages and disadvantages, with some dealing with industrial applications. However, this could be a promising alternative to using lethal isocyanate for polyurethane applications and therefore more research is in progress for an all-around method for NIPUs [57].

6 CONCLUSION

The fast-growing world and the emergence of new technologies in addition to other important factors have placed huge demand for more plastics and polymers. This has also led to consequences in finding renewable sources for their products due to the pressing need to protect the environment. Biomass in the form of vegetable oils, lignin, algae, and others has proven to be suitable for transformation into raw materials and precursors in the production of polyurethane. Scientists have taken advantage of the versatile and tunable nature of vegetable oils, for example, and have used different synthetic routes such as the epoxidation reaction followed by ring-opening, thiol-ene reaction, metathesis, and many others in the production of biobased polyols. Subsequently, these bio-derived polyols have found comparable and even greater advantages over petroleum-based sources when used in different polyurethane applications. However, some have disadvantages that are being investigated for corrective measures and solutions. In addition, researchers have discovered less toxic materials and synthetic methods for the production of isocyanate. Biomasses such as lignin, vegetable oils, and furans have been successful in the synthesis of biobased isocyanate. Also, other researchers have delved into the total eradication of isocyanate in polyurethane due to the detrimental health effects it presents after prolonged exposure. The replacement of petrochemical sources with biobased materials has also thrown a challenge on how efficient and comparable the use of natural sources can be to the already-existing petrochemical ones. Researchers are therefore investigating this area and coming up with unquestionable solutions and alternatives.

REFERENCES

1. Hosler Dorothy, Burkett Sandra L., Tarkanian Michael J. (1999) Prehistoric polymers: Rubber processing in ancient Mesoamerica. Science (80) 284:1988–1991
2. Andrady Anthony L., Neal Mike A. (2009) Applications and societal benefits of plastics. Philos. Trans. R. Soc. B Biol. Sci. 364:1977–1984

3. Siracusa Valentina, Rocculi Pietro, Romani Santina, Rosa Marco Dalla (2008) Biodegradable polymers for food packaging: a review. Trends Food Sci Technol 19:634–643

4. Hai Thien An Phung, Tessman Marissa, Neelakantan Nitin, Samoylov Anton A., Ito Yuri, Rajput Bhausaheb S., Pourahmady Naser, Burkart Michael D. (2021) Renewable polyurethanes from sustainable biological precursors. Biomacromolecules 22:1770–1794

5. Michae Szycher (1938) Szycher's Handbook of Polyurethanes. CRC Press

6. Akindoyo John O., Beg M. D.H., Ghazali Suriati, Islam M. R., Jeyaratnam Nitthiyah, Yuvaraj A. R. (2016) Polyurethane types, synthesis and applications-a review. RSC Adv 6:114453–114482

7. Gomez-Lopez Alvaro, Panchireddy Satyannarayana, Grignard Bruno, Calvo Inigo, Jerome Christine, Detrembleur Christophe, Sardon Haritz (2021) Poly(hydroxyurethane) adhesives and coatings: State-of-the-art and future directions. ACS Sustain Chem Eng 9:9541–9562

8. Gama Nuno V, Ferreira Artur, Barros-Timmons Ana (2018) Polyurethane foams: Past, present, and future. Materials (Basel) 11:1841

9. S.T. Lee, Ramesh N.S. (2004) Polymeric foams: mechanisms and materials [Book review]. CRC Press.

10. Somidi Asish K.R., Sharma Rajesh V., Dalai Ajay K. (2014) Synthesis of epoxidized canola oil using a sulfated-SnO$_2$ catalyst. Ind Eng Chem Res 53:18668–18677

11. Sawpan Moyeenuddin Ahmad (2018) Polyurethanes from vegetable oils and applications: a review. J Polym Res 25:184

12. Zhang Chaoqun, Garrison Thomas F., Madbouly Samy A., Kessler Michael R. (2017) Recent Advances in Vegetable Oil-Based Polymers and Their Composites. Elsevier B.V.

13. He Wei, Kang Peng, Fang Zheng, Hao Jingying, Wu Hao, Zhu Yuchen, Guo Kai (2020) Flow reactor synthesis of bio-based polyol from soybean oil for the production of rigid polyurethane foam. Ind Eng Chem Res 59:17513–17519

14. Campanella A., Baltanás M. A., Capel-Sánchez M. C., Campos-Martín J. M., Fierro J. L.G. (2004) Soybean oil epoxidation with hydrogen peroxide using an amorphous Ti/SiO$_2$ catalyst. Green Chem 6:330–334

15. Orellana-Coca Cecilia, Billakanti Jagan M., Mattiasson Bo, Hatti-Kaul Rajni (2007) Lipase mediated simultaneous esterification and epoxidation of oleic acid for the production of alkylepoxystearates. J Mol Catal B Enzym 44:133–137

16. Lligadas Gerard, Ronda Juan C., Galiá Marina, Cádiz Virginia (2010) Plant oils as platform chemicals for polyurethane synthesis: Current state-of-the-art. Biomacromolecules 11:2825–2835

17. Guo Andrew, Demydov Dima, Zhang Wei, Petrovic Zoran S. (2002) Polyols and polyurethanes from hydroformylation of soybean oil. J Polym Environ 10:49–52

18. Zoran S Petrovíc,* Andrew Guo, Ivan Javni Ivana Cvetkovíc and Doo Pyo Hong (2007) Polyurethane networks from polyols obtained by hydroformylation of soybean oil. Polym Int 55:275–281

19. Guo Andrew, Zhang Wei, Petrovic Zoran S. (2006) Structure–property relationships in polyurethanes derived from soybean oil. J Mater Sci 41:4914–4920

20. Ionescu Mihail, Radojčić Dragana, Wan Xianmei, Petrović Zoran S., Upshaw Thomas A. (2015) Functionalized vegetable oils as precursors for polymers by thiol-ene reaction. Eur Polym J 67:439–448

21. Aniceto José P.S., Portugal Inês, Silva Carlos M. (2012) Biomass-based polyols through oxypropylation reaction. ChemSusChem 5:1358–1368

22. Niu Min, Zhao Guang jie, Alma Mehmet Hakki (2011) Polycondensation reaction and its mechanism during lignocellulosic liquefaction by an acid catalyst: A review. For Stud China 13:71–79

23. Pavier Claire, Gandini Alessandro (2000) Oxypropylation of sugar beet pulp. 1. Optimisation of the reaction. Ind Crops Prod 12:1–8

24. Evtiouguina M., Barros-Timmons A., Cruz-Pinto J. J., Neto C. Pascoal, Belgacem M. N., Gandini A. (2002) Oxypropylation of cork and the use of the ensuing polyols in polyurethane formulations. Biomacromolecules 3:57–62

25. D'Souza Jason, George Ben, Camargo Rafael, Yan Ning (2015) Synthesis and characterization of bio-polyols through the oxypropylation of bark and alkaline extracts of bark. Ind Crops Prod 76:1–11

26. Ye Liyi, Zhang Jingmiao, Zhao Jie, Tu Song (2014) Liquefaction of bamboo shoot shell for the production of polyols. Bioresour Technol 153:147–153

27. Lv Zilu, Xu Jikun, Li Chenyu, Dai Lin, Li Huihu, Zhong Yongda, Si Chuanling (2021) PH-responsive lignin hydrogel for lignin fractionation. ACS Sustain Chem Eng 9:13972–13978

28. Thakur Vijay Kumar, Thakur Manju Kumari, Raghavan Prasanth, Kessler Michael R. (2014) Progress in green polymer composites from lignin for multifunctional applications: A review. ACS Sustain Chem Eng 2:1072–1092
29. Doherty William O.S., Mousavioun Payam, Fellows Christopher M. (2011) Value-adding to cellulosic ethanol: Lignin polymers. Ind Crops Prod 33:259–276
30. Soares Belinda, Gama Nuno, Freire Carmen S.R., Barros-Timmons Ana, Brandão Inês, Silva Rui, Neto Carlos Pascoal, Ferreira Artur (2015) Spent coffee grounds as a renewable source for ecopolyols production. J Chem Technol Biotechnol 90:1480–1488
31. Haridevan Hima, McLaggan Martyn S., Evans David A.C., Martin Darren J., Seaby Trent, Zhang Zhanying, Annamalai Pratheep K. (2021) Dispersion methodology for technical lignin into polyester polyol for high-performance polyurethane insulation foam. ACS Appl Polym Mater 3:3528–3537
32. Bote Sayli Devdas, Narayan Ramani (2021) Synthesis of biobased polyols from soybean meal for application in rigid polyurethane foams. Ind Eng Chem Res 60:5733–5743
33. Gu Ruijun, Konar Samir, Sain Mohini (2012) Preparation and characterization of sustainable polyurethane foams from soybean oils. JAOCS, J Am Oil Chem Soc 89:2103–2111
34. Pan Xiao, Webster Dean C. (2012) New biobased high functionality polyols and their use in polyurethane coatings. ChemSusChem 5:419–429
35. Tan Suqin, Abraham Tim, Ference Don, MacOsko Christopher W. (2011) Rigid polyurethane foams from a soybean oil-based Polyol. Polymer (Guildf) 52:2840–2846
36. Wei Daidong, Zeng Juanjuan, Yong Qiwen (2021) High-performance bio-based polyurethane antismudge coatings using castor oil-based hyperbranched polyol as superior cross-linkers. ACS Appl Polym Mater 3:3612–3622
37. Ionescu Mihail, Radojčić Dragana, Wan Xianmei, Shrestha Maha Laxmi, Petrović Zoran S., Upshaw Thomas A. (2016) Highly functional polyols from castor oil for rigid polyurethanes. Eur Polym J 84:736–749
38. Saalah Sariah, Abdullah Luqman Chuah, Aung Min Min, Salleh Mek Zah, Awang Biak Dayang Radiah, Basri Mahiran, Jusoh Emiliana Rose (2015) Waterborne polyurethane dispersions synthesized from jatropha oil. Ind Crops Prod 64:194–200
39. Ng Wei Seng, Lee Choy Sin, Chuah Cheng Hock, Cheng Sit Foon (2017) Preparation and modification of water-blown porous biodegradable polyurethane foams with palm oil-based polyester polyol. Ind Crops Prod 97:65–78
40. Zhou Xiaojian, Sain Mohini M., Oksman Kristiina (2016) Semi-rigid biopolyurethane foams based on palm-oil polyol and reinforced with cellulose nanocrystals. Compos Part A Appl Sci Manuf 83:56–62
41. Oh-Jin Kwon, Seong-Ryul Yang, Dae-Hyun Kim Jong-Shin Park (2006) Characterization of polyurethane foam prepared by using starch as polyol. J Appl Polym Sci 103:1544–1553
42. Ertaş Murat, Fidan M. Said, Alma Mehmet Hakki (2014) Preparation and characterization of biodegradable rigid polyurethane foams from the liquefied eucalyptus and pine woods. Wood Res 59:97–108
43. Gupta Swati, Upadhyaya Ravi (2014) Cancer and P21 protein expression: in relation with isocyanate toxicity in cultured mammalian cells. Int J Pharm Res Bio-Sciences 3:20–28
44. Karol Meryl H., Kramarik Jean A. (1996) Phenyl isocyanate is a potent chemical sensitizer. Toxicol Lett 89:139–146
45. Tawade Bhausaheb V, Shingte Rahul D, Kuhire Sachin S, Sadavarte Nilakshi V, Garg Kavita, Maher Deepak M, Ichake Amol B, More Arvind S, Wadgaonkar Prakash P (2017) Bio-based di-/polyisocyanates for polyurethanes: An overview. PU Today 41–46
46. Sanda Fumio, Takata Toshikazu, Endo Takeshi (1995) Synthesis of a novel optically active nylon-1 polymer: Anionic polymerization of L-leucine methyl ester isocyanate. J Polym Sci Part A Polym Chem 33:2353–2358
47. Hafeman Andrea E., Li Bing, Yoshii Toshitaka, Zienkiewicz Katarzyna, Davidson Jeffrey M., Guelcher Scott A. (2008) Injectable biodegradable polyurethane scaffolds with release of platelet-derived growth factor for tissue repair and regeneration. Pharm Res 25:2387–2399
48. Lichtenthaler Frieder W., Peters Siegfried (2004) Carbohydrates as green raw materials for the chemical industry. Comptes Rendus Chim 7:65–90
49. Zenner Michael D., Xia Ying, Chen Jason S., Kessler Michael R. (2013) Polyurethanes from isosorbide-based diisocyanates. ChemSusChem 6:1182–1185

50. Gökhan Çaylı Selim Küsefoğlu (2008) Biobased polyisocyanates from plant oil triglycerides: Synthesis, polymerization, and characterization. J Appl Polym Sci 109:2948–2955
51. Garber John D. (1962) Furfuryl Isocyanates. US Pat 1960–1962
52. Wicks Douglas A, Wicks Zeno W (1999) Blocked isocyanates III: Part A. Mechanisms and chemistry. Prog Org Coatings 36:148–172
53. Najarro Marleny Cáceres, Nikolic Miroslav, Iruthayaraj Joseph, Johannsen Ib (2020) Tuning the lignin-caprolactone copolymer for coating metal surfaces. ACS Appl Polym Mater 2:5767–5778
54. Arapova O. V., Chistyakov A. V., Tsodikov M. V., Moiseev I. I. (2020) Lignin as a renewable resource of hydrocarbon products and energy carriers (a review). Pet Chem 60:227–243
55. Sachin S. Kuhire Samadhan S. Nagane and Prakash P. Wadgaonkar (2017) Poly(ether urethane)s from aromatic diisocyanates based on lignin-derived phenolic acids. Polymer International 66:892–899
56. Kumar Shiva, Dinesha P., Rosen Marc A. (2018) Cashew nut shell liquid as a fuel for compression ignition engines: A comprehensive review. Energy and Fuels 32:7237–7244
57. Cornille Adrien, Auvergne Rémi, Figovsky Oleg, Boutevin Bernard, Caillol Sylvain (2017) A perspective approach to sustainable routes for non-isocyanate polyurethanes. Eur. Polym. J. 87:535–552

4 Recent Advances in Edible Polymers

Jiao Xie[1,2] and Jingjing Kang[1,2]

[1] The Key Laboratory of Environmental Pollution Monitoring and Disease Control, Ministry of Education, Guizhou Medical University, Guiyang 550025, China
[2] Engineering Research Center of Higher Education Institutions of Guizhou Province, School of Public Health, Guizhou Medical University, Guiyang 550025, PR China

1 INTRODUCTION

1.1 BACKGROUND

Worldwide, packaging polymers are common and play an important role in daily life. Among all packaging polymers, nonbiodegradable materials derived from petroleum-based products, such as polypropylene, polyethylene, polystyrene, etc., comprise a high proportion owing to their low cost, excellent stability, aesthetic appearance, barrier properties, and resistance to oil, chemical, microbial, and fungal interference [1]. However, their nonbiodegradable nature causes plastic waste to accumulate all around the world. According to a paper by Jia L. et al., the amount of plastic waste in 2015 alone reached five billion tons in natural sites [2]. The United States is reported to produce more than 32 million tons of plastic waste yearly. With the discovery of microplastics in a large number of animals (including fish, birds, and humans), people have begun to pay a great deal of attention to the safety of traditional packaging plastics, which will become a threat to human health and the environment [1,3]. Moreover, petroleum shortages are becoming increasingly serious and have resulted in a worldwide problem. To overcome this deteriorating situation, nondegradable polymers are restricted to use in our daily life, and various incentives are also offered by governments to appeal to people to reduce their usage of nonbiodegradable plastics.

In this context, the biodegradable polymer industry welcomes new development opportunities, and many biodegradable polymers have been newly applied in the packaging industry. To promote the development of biodegradable materials, the European Commission (EC) directive EU2019/904 puts forward higher requirements for the application of biodegradable materials in food packaging [4]. With the increasing application of new, safe, high-quality, and suitable food-grade products in functional food industries, the biodegradable packaging materials used in the food industry are labelled as safe, high-quality, biodegradable, sustainable, biocompatible, and easily consumed by lower animals and humans without any harmful effect on health. The concept of edible polymers thus comes into focus.

1.2 HISTORY OF EDIBLE POLYMERS

There is a long history of the use of food film as food packaging, and edible packaging for food preservation is considered an ancient technology. A strong representative example is the sausage (a food prepared by stuffing meat into animal intestines), which was developed by the Mesopotamian Sumerians at approximately 3000 BCE and Chinese settlers at 580 BCE [5]. Then, the bean curd sheet (the skin of boiled soy milk) was used to cover different foods by the Chinese and Japanese

DOI: 10.1201/9781003278269-4

during the 15th century, followed by applications of animal grease (lard) and wax to coat fruits and other foods in China and Europe during the 16th century. The USA then produced a patented coating made from gelatine for meat addition in the 19th century. However, advancements in the use of food film for packaging slowed as synthetic oil-based films began to dominate the packaging market after World War II [6]. With the awareness of problems related to traditional plastic materials that has developed in recent years, edible polymers have come back into focus.

Unlike synthetic polymer packaging, edible polymers are mainly produced from natural food-grade resources containing starch, cellulose, polylactic acid (PLA), and chitin and are degradable and recyclable; they possess good barrier properties for moisture and gases [7]. After edible polymers create a physical barrier between external spoilage factors and food surfaces through coating, immersion, and spraying, the shelf life of the wrapped food is markedly extended. In addition, edible polymers could also be applied in other applications, including cancer diagnostics, drug delivery, nanomedicine intervention, and other biomedical fields. Based on these facts, edible polymers as substitutes for synthetic plastics have received incredible attention.

1.3 MAIN PURPOSES OF EDIBLE POLYMERS

Edible films can be used as barriers by coating and placement onto food surfaces, in which edible films can be eaten like food without any health risk [8]. Composed of only food-grade components, edible polymer films are generally recognized as safe. Figure 4.1 lists the main qualities of edible food packaging. In addition to common qualities (such as being environmentally friendly, controlling gases and humidity, inhibiting surface contamination, and extending food shelf life and quality), edible food packaging has taken on new roles, including the promotion of mechanical protection, improvement of sensorial and functional properties, and addition of nutritional value.

In this chapter, the seven sections include an introduction, categories and properties, hydrophilic edible polymers, hydrophobic edible polymers, major techniques for making edible polymers in

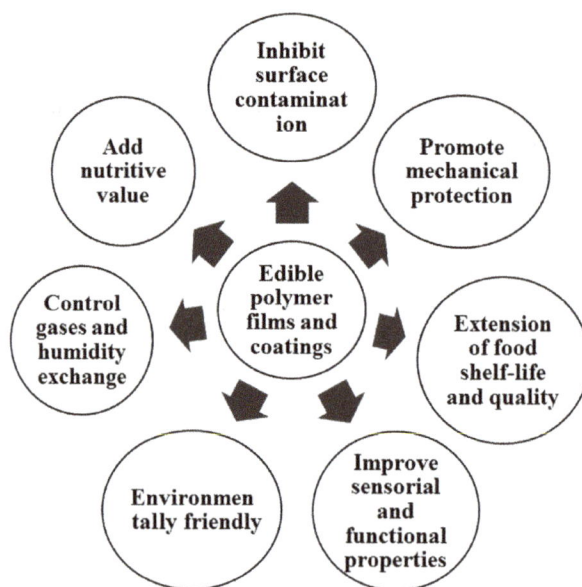

FIGURE 4.1 Main qualities of edible food packaging. Adapted with permission from Reference [6]. Copyright (2021), MDPI, with some modifications.

different forms, special applications, and conclusions, which are introduced in detail, in order to facilitate new insights for future research.

2 CATEGORIES AND PROPERTIES OF EDIBLE POLYMERS

Based on solubility, edible polymers are generally classified as hydrophilic or hydrophobic substances. Hydrophobic edible polymers are mainly derived from lipids, such as oils and fat, wax, and resins, and they show provide water vapour barriers because of their hydrophobic nature [9]. While hydrophilic edible polymers are mainly applied in the form of hydrocolloids, which consist of long-chain hydrophilic polymers (such as polysaccharides and proteins), their abundance of hydroxyl groups speeds up the formation of viscous dispersions or gels after dispersion into water by increasing the affinity between hydrocolloids to bind water molecules. Every hydrophilic edible polymer can thicken water solvents, but this situation changes when all different hydrocolloids are used to form a gel, as only a small percentage of hydrocolloids are capable of forming gels [10]. In a gel system, a three-dimensional network is built to trap or immobilize water molecules on hydrophilic molecules, and the properties of this network are different from the textural properties and sensory properties of two gels made by different hydrocolloids. These two polymers both have strong advantages and weaknesses. Compared to hydrophobic edible polymers, hydrocolloid-based polymers are more neutral for maintaining colour, sweetener, flavour, salt concentrations, etc. Therefore, food packaged with hydrophilic edible polymers shows superior sensory properties (appearance and smell) over food packaged with hydrophobic edible polymers, however the latter polymers are better water vapour barriers. By adding hydrophobic materials into hydrocolloids, both enhanced water vapour permeability and excellent sensory properties would appear in their composites [10].

The properties of edible polymers can be described by indicators including barrier properties, mechanical characteristics, optical parameters, and biodegradability. Therein, barrier properties and mechanical characteristics are considered the most important determinant factors of packed food material, optical parameters only are evaluated when edible polymers act as edible coating materials [11], while biodegradability is a factor that distinguishes edible polymers from traditional plastics. High moisture and oxygen barrier properties are conducive to extending the shelf life of packed food material, as they help to maintain the levels of moisture and oxygen in food less than what is required for microbial growth. Apart from the addition of hydrophobic material to hydrophilic polymers mentioned above, surfactants that are able to reduce surface tension are also applied to enhance the moisture barrier level of the hydrophilic polymers. The mechanical characteristics, as another important determinant factor of packaging polymers, include the tensile strength (TS), elongation at breakage (EAB), Young's storage modulus, and loss factor (tanδ), which are crucial for evaluating the mechanical potential of an edible polymer as a food packaging material [12]. With the resistance to breakage and integrity evaluated by TS and EAB, these two indicators play an important role in the mechanical characteristics of packing polymers [13]. Biodegradability indicates the ability of a material to degrade into monomers after its interaction with biological elements, as nonbiodegradable plastic polymers harm the environment and climate, and consumers' interest in eco-friendly biodegradable materials is increasing.

3 HYDROPHILIC EDIBLE POLYMERS

At present, the most common hydrophilic edible polymers on the market are mainly categorized into protein-based and polysaccharide-based edible films, and the performances of these two edible polymers are distinct from each other. For instance, owing to their strong hydrogen bonding, polysaccharide-based edible films could be easily bonded with moisture as well as some hydrophilic additives, such as pigments, flavours, and micronutrients; therefore, polysaccharide-based edible films are characterized as good oxygen barriers despite their poor moisture resistance. In contrast to polysaccharide-based polymers, protein-based edible films have better mechanical strength, which promotes the application of protein-based polymers to protect vegetables and fruits in circulation from damage.

3.1 POLYSACCHARIDE-BASED EDIBLE POLYMERS

Polysaccharides, including pectin, alginate, starch, carrageenan, and xanthan gum, are often introduced to produce edible polymers as sustainable materials in coatings and edible film formulations. As natural polysaccharides are safe, polysaccharide-based edible polymers could be used to permeate carbon dioxide and oxygen selectively without any side effects. These characteristics promote their use as a coating and edible film for extending the shelf life of fruits and vegetables.

3.1.1 Plant-Origin Polysaccharides

3.1.1.1 Starch

Starch, one of the most promising polymers of natural origin, is considered the perfect base material for producing edible polymers because of its easy availability, low price, and annual renewable nature [14]. A unique feature of starch is its film-forming property, with characteristics of being insoluble in cold water as well as degradation temperatures lower than the melting point. Starch does not melt like conventional plastics, which is conducive to the formation of edible films. Employing conventional techniques, such as solution casting and extrusion processing, a starch solution is capable of transforming into a polymer film [15].

In the process of solution casting, the starch aqueous solution (3–12%) reacts to gelatinization by heating above T_{gel}, the gelatinized starch solution is evaporated to obtain a polymer film through a drying process after being poured into a mould, and plasticizers (such as sorbitol, glycerol, urea, and simple sugars) are essential to enhance the flexibility and mechanical characteristics of starch films, and an antimicrobial starch-based film is obtained by activating-solution casting with lauroyl arginate (a cationic surfactant). Despite its attractive appearance, the starch-based film prepared by this method cannot not be produced on a commercial scale. However, the second masking technique, extrusion processing, can produce polymer films commercially. Accompanied by the application of polyethylene glycol or sorbitol (10–60%) plasticizers, the structural properties and physiochemical characteristics of extruded starch are altered by changing the screw speed, feed humidity, barrel temperature, mould pressure, energy input, etc. [16]. Compared with the first masking technique, extrusion processing could improve the ultimate elongation and transparency of polymers at high efficiency and low energy consumption; however, this process is only suitable for raw material blends at low moisture and temperature tolerance.

Polymers (edible) and coatings prepared using carbohydrates show significantly low barriers towards moisture because of their hydrophilic nature, and this situation could be improved by the addition of hydrophobic materials (such as plant oil and hydrophobic gelatine-based plasticizers) [17] and hydrophilic plasticizers (sodium trimetaphosphate, montmorillonite glycerol and sorbitol) [18]. Furthermore, starch is modified to block steam infiltration and enhance the mechanical strength of the polymer, and this change is directly proportional to the level of modification [19]. As the ratio of amylose to amylopectin could change the orientation of the molecular structure, the amylose percentage in starch somewhat determines the tensile strength of the polymer [20]; the lower the percentage of amylose is, the lower the elastic stretch of the starch-based polymer. The most common plasticizers used for the improvement of polymer mechanical properties are glycerol, sorbitol, and essential oil, and it has been suggested that nanoparticle incorporation into food packaging materials could heighten the mechanical properties of starch films significantly. In addition, the film thickness is reported to be more related to the mechanical resistance of the

TABLE 4.1

Physical and Mechanical Properties of Starch-Based Films

Type of Starch Used	Solubility	Film Thickness (μm)	Moisture Content (%)	Tensile Strength (MPa)	Elongation at Break (%)	Young Modulus (MPa)
Maize starch[27]	–	266	22.26	1.49	51	14.2
Potato starch[27]	–	332	9.74	3.05	70	14.5
Oat starch[27]	–	266	21.77	0.36	27	1.8
Rice starch[27]	–	145	18.72	1.8	49	9.6
Topaca starch[27]	–	136	17.22	0.78	137	0.8
Wheat starch[28]	14.49–19.67	35.4–80.8	2.01–3.24	2.03–2.10	13.18–14.16	0.08–0.1

Source: Adapted with permission from Reference [15], Copyright (2021), MDPI.

polymer. The commercial status of starch-based polymers is mainly determined by the mechanical parameters described in Table 4.1, where tensile strength is the most significant indicator, while film thickness, moisture level, WVP, elongation at break, and solubility follow in turn. The existence of ordered zones in starch-based polymers is considered to contribute to their optical properties, as more ordered zones could reduce film absorbance, which is involved in transparency [21]. To protect high-fat foods from light-mediated oxidation, polymers tend to be used with high light absorbance in food packaging. Despite the fact that the transparency of starch-based films varies with the starch type and their original source, it could also be decreased by adding nanoparticles [22] (such as calcium carbonate nanoparticles, nano-SiO_2, and cellulose nanocomposites), organic molecules (glycerol at 15–25%, coloured compounds, oil, and protein inducing Maillard reactions) [23, 24] and inorganic molecules, such as talc powder [25]. Furthermore, the drying temperature and concentration of the starch solution are two key factors that change the film transparency. The starch-based polymer has high biodegradability; according to a paper from Rachmawati [26], cassava and gold potato starch-based films exhibited weight losses of 99.35% and 90.03%, respectively, after 31 days. To shorten the biodegradation time of starch-based polymers, plasticizers (such as acetic acid and formic acid) and sustainable nanofillers (cellulose nanocrystals and nanoclay) are applied to improve the biodegradability.

3.1.1.2 Cellulose, Pectin, and Gum Arabic

Other important polysaccharides abundant in plants are cellulose, pectin, and gum arabic. These three polysaccharides have different molecular structures, which are bound to affect their film properties.

Cellulose, mainly from the cell wall of plants, is formed by D-glucose units through β-1,4 glycosidic bonds. With supramolecular structures consisting of crystalline regions and amorphous regions, cellulose shows an approximate crystallinity degree of 40–60%, as the hydroxyl groups in the amorphous parts more easily react compared to those in crystalline regions. Cellulose cannot be utilized in its native form [29] owing to its weak film-forming capability and water insolubility. To overcome these defects, chemical treatment is applied to produce derivatives containing methylcellulose, hydroxypropyl methylcellulose, carboxymethyl cellulose, and so on. Since glycosidic alpha linkages are easier to break than glycosidic beta linkages, cellulose derivatives rich in glycosidic beta linkages show a lower biodegradation rate than starch, as well as stronger resistance to microbial growth and enzymatic cleavages. Edible polymer films produced by cellulose derivatives have been proven to be odourless, tasteless, and biodegradable but with poor water vapour barriers [30].

Pectin, mainly found in vegetables and fruits, is an anionic polysaccharide, where α-d-galacturonic acid units are linked together through 1,4 glycosidic bonds. Apart from the use of pectin as stabilizers, thickeners, and gelling agents in yoghurts, jams, and ice creams, the pectin-based film is an excellent barrier to oil and aroma, oxygen, and has a high initial modulus. Plasticizer addition and cross-linking reactions with polyvalent cations [31] are facilitated to enhance the flexibility of pectin-based films because of their poor resistance to moisture, low elongation and frangibility.

Relatively, sources of **gum arabic** are spread over a limited area (stems of various *Acacia* species), consisting of rhamnose, galactose, arabinose, and glucuronic acid. Gum arabic is the most used heteropolysaccharide in the food, cosmetics, and pharmaceutical industries, owing to its low cost, pH stability, tolerance to temperature and ions, and features of film-forming, emulsification, and encapsulation. After fruits are coated with gum arabic, their respiration rate and ethylene production decline significantly. Gum arabic-based films have low extensibility, which can be improved by semirefined carrageenan [32].

3.1.2 Polysaccharides of Marine Origin: Alginate, Carrageenan, and Agar

As is well known, the common polysaccharide products in the seaweed industry are alginates, agars, and carrageenans, which have low price, bioavailability, biodegradability, and biocompatibility. These polysaccharides are widely applied in the food and pharmaceutical industries.

Alginate, a natural polysaccharide mainly isolated from different brown seaweeds, is often composed of linking units of a-L-guluronate and R-D-mannuronate at different ratios by 1,4 glycosidic bonds. Although alginate-based films exhibit poor water vapour barriers, they can be improved by calcium addition. When alginate reacts with calcium cations or polyvalent metals, it can produce a strong insoluble gel or polymer, and their gel formation is easier than that of alginate-based films [33].

Carrageenan is a linear sulphated water-soluble polysaccharide extracted from the Rhodophyceae family, and 3-linked-β-D-galactopyranose units and 4-linked-α-D-galactopyranose units alternate in the carrageenan molecule [7]. Based on its molecular structure, carrageenan can be classified into κ- (kappa), ι- (iota), and λ- (lambda) types [34], which are mainly used as emulgator, gel, and stabilizing ingredients in the pharmaceutical and food industries. Benefiting from its water-soluble behaviour and the capacity of its chemical structure to form a helical assembly, carrageenan has great potential for preparing films, conducive to its application in smearing and wrapping. In industry, carrageenan is often mixed with other components (such as glycerol, sodium alginate, and essential oil) to prepare films with good mechanical and physical properties.

Agar is obtained as a gel from red algae, and its unique solubility in hot water makes it a gel inducer generally used in desserts and candy.

3.1.3 Microbial Extracellular Polysaccharides: Pullulan, gellan, and xanthan gum

Pullulan is a highly water-soluble exopolysaccharide produced by *Aureobasidium pullulans*, and its molecular structure is composed of maltotriose units connected through α-(1,6)-glycosidic bonds [4]. With the high cost of pullulan extraction, it is considered an expensive neutral polysaccharide that is unsuitable for application in the food industry. However, it is nontoxic, odourless, and tasteless; films produced by pullulan appear odourless, colourless, and tasteless and are heat-sealable, water-permeable, transparent, and weakly oxygen- and oil-permeable. Therefore, it is necessary to produce pullulan economically and effectively [35].

Gellan is an exopolysaccharide possessing unique colloidal abilities and good abilities to form coatings and gelling properties [36], and the bacterium *Sphingomonas elodea* is the main microorganism that synthesizes gellan with a molecular structure of repeated tetrasaccharide units containing two glucose residues, rhamnose and glucuronic acid. Because of its solubility in water, gellan is usually applied as a gelling agent, for texturizing, and as a carrier of food additives in the food industry, while it has been proven that gellan-based edible coatings can enhance the quality and shelf life of freshly cut vegetables.

Xanthan gum is generally produced by the bacterium *Xanthomonas campestris* and is recognized as an exopolysaccharide that plays a role as a thickener, stabilizer, and emulsifier in the food industry. Based on a structure of 1,4-linked β-d-glucose residues and trisaccharide dehydrated to form new d-glucose residues, xanthan gum is capable of increasing solution viscosity without enzymatic degradation in any case [37], which has drawn wide attention. An edible coating of xanthan gum has been proven to improve the shelf life and quality of fresh-cut fruits as well as drug loads.

3.2 PROTEIN-BASED EDIBLE POLYMERS

Proteins are biological macromolecules that mainly occur as either fibrous proteins or globular proteins, where peptide bonds link all amino acid residues together. Accompanied by interesting optical activity and foam-forming and stabilizing abilities, as well as gel-forming ability, protein is mainly utilized as an emulsifier to change the interfacial tension of a solution [38]. It has gained remarkable attention from consumers and food manufacturers, as the protein-based film presents better mechanical and barrier properties (especially to carbon dioxide gases and oxygen) than polysaccharides. Currently, various proteins from soybean, milk, and animal skins have been applied in the food industry and packaging.

Soy protein is obtained from soybeans and exists as a protein product in various forms, such as soy flour, soy concentrate, and soy isolates. Because of their larger protein percentage (90% protein and 2% carbohydrate) compared to others, soy isolates are the main material used to form films, which are usually obtained through baked film methods or as Yuba films. Soy protein-based films are smoother, more flexible, and clearer than those formed by other plant proteins, and a superior gas barrier still occurs for soy protein films compared to that of lipid and polysaccharide films.

Milk, another main source of protein, consists of 80% casein protein and 20% whey protein, which can be separated by alkaline precipitation after a degreasing process (pH 4.6 at 20 °C). Casein-based films are mainly formed by drying either sodium or calcium caseinate solution, where the film of sodium caseinate appears to have excellent optical and tensile properties, while the barrier properties of calcium caseinate films are better. Despite having poor water vapour permeability, whey protein-based edible films exhibit a better barrier to oxygen at low or intermediate RH.

Gelatine is obtained from the hydrolysis of collagen, which mainly exists in animal skins, muscles, bones, and connective tissues. Pure and dry gelatine is a glass-like solid with a faint yellow colour, presenting transparent, tasteless, brittle, and odourless characteristics. Collagen edible films are formed after drying their hot solution on a plate in an oven [39]. Protein content is the main factor that affects the thickness and mechanical properties of collagen-based edible films. The food industry prefers to use collagen edible films for preserving the humidity of meat products and assisting in meat product cooking.

4 HYDROPHOBIC EDIBLE POLYMERS

Lipids are a large and diverse group of hydrophobic organic compounds, and according to their exact molecular structures, they can be further classified into glycerides, phospholipids, cerebrosides, fatty alcohols, and fatty acids. Because of their nonsolubility in water, lipids can block moisture (water) penetration, inducing great attention from the food industry with regard to lipid applications in coatings and edible films. Lipid-based edible films are often used to provide gloss and retain moisture in food packages, and this ability of lipids to block water vapour mainly depends on their polarity, which is codetermined by the chemical group distribution, carbon chain length, and unsaturated number of lipid molecules. However, it is difficult for most lipids to form films, which need to be turned over by the combination of lipids with hydrophilic macromolecules, such as polysaccharides and proteins; lipid-based edible polymeric materials can often be produced by waxes, phospholipids, and resins.

4.1 Waxes

Waxes, common hydrophobic macromolecules consisting of long-chain fatty acid alcohols or esters, can be separately obtained from vegetables and animals. As waxes can protect the covering tissues, they are often applied to coat or cover vegetables and fruits to reduce moisture permeability between food and the atmosphere. Once a fruit is coated with waxes, it exhibits lower weight loss, better firmness, better lightness, and better gloss than fruits without coating. To improve the film-forming capacity of wax, hydrophilic materials (such as sorbitol, Tween 40, gel, and glycerol) are added to prepare films with good mechanical, barrier, and stable physical properties for food packaging.

4.2 Resins

Although some insects (such as *Laccifer lacca*) can produce resin, most resins are translucent solid or semisolid plant secretions with yellowish-brown tones, which are produced only when plants respond to external injuries. According to the research, resin-based films exhibit a quick drying nature, transparency, glossiness, and sound emulsion stability. Because they provide a good barrier to water vapour and gases, resin-based films are often used to extend the shelf life of food.

4.3 Phospholipids

As important lipoids, phospholipids exist as mixtures owing to their emulsifying property. They have various origins (such as soybean, vegetable, and egg) and occupy an important position among all food emulsifiers. Soy lecithin is an important food emulsifier that is usually applied to prevent phase separation and maintain a hydrophilic–lipophilic balance of two immiscible phases; therefore, the existence of soy lecithin affects the colour, solubility, opacity, and microstructure of edible films or coatings [7].

5   MAJOR TECHNIQUES FOR MAKING EDIBLE POLYMERS IN DIFFERENT FORMS

As shown in Figure 4.2, edible polymers can be further categorized into particle, layer, and textile structures, which are respectively applied as emulsifiers, films, and edible fibres.

5.1 Major Techniques for Making Edible Particles

Based on the advantages of excellent biocompatibility, biodegradability, nontoxicity, harmlessness, and ease of control, edible particles (such as modified starch granules, cellulose microparticles, and starch nanoparticles) have been widely considered long-term stabilizers during the production of Pickering emulsions. Apart from Pickering emulsions, edible particles are also applied as self-adhesives to improve the external appearance of food.

There are many strategies for making edible particles, and the techniques used are different when the raw materials vary. For example, solid edible particles are mainly prepared by mechanical methods and chemical breakdown methods, with the assistance of mechanical methods (such as ball, wet, freezer milling, and high-pressure homogenizer); turbulence and cavitation shear at high speed decrease the size of particles and crystals, while during chemical methods, submicron cellulose or chitin particles are obtained by acid treatment without disrupting the crystalline regions [41]. Water-insoluble edible particles are prepared by precipitation, while protein-based edible particles

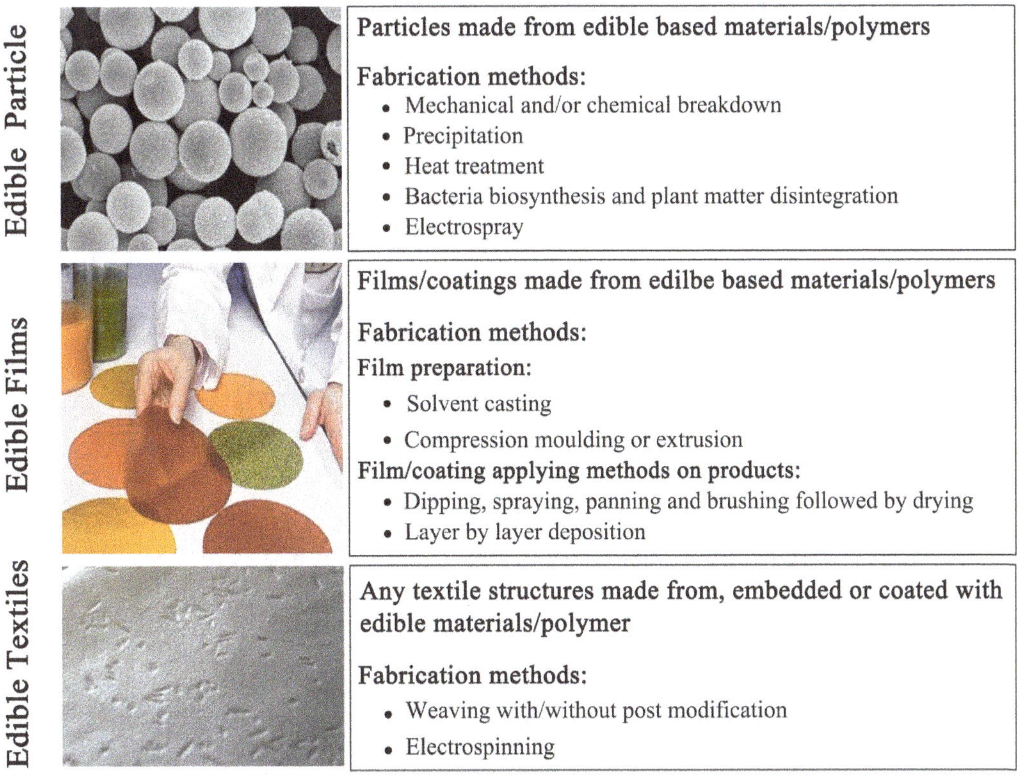

	Particles made from edible based materials/polymers
Edible Particle	**Fabrication methods:** • Mechanical and/or chemical breakdown • Precipitation • Heat treatment • Bacteria biosynthesis and plant matter disintegration • Electrospray
Edible Films	**Films/coatings made from edilbe based materials/polymers** **Fabrication methods:** Film preparation: • Solvent casting • Compression moulding or extrusion Film/coating applying methods on products: • Dipping, spraying, panning and brushing followed by drying • Layer by layer deposition
Edible Textiles	**Any textile structures made from, embedded or coated with edible materials/polymer** **Fabrication methods:** • Weaving with/without post modification • Electrospinning

FIGURE 4.2 Different preparative forms/structures of edible polymers. Adapted with permission from Reference [40]. Copyright (2020), Elsevier Ltd.

are prepared by heat treatment. A proper solvent should be selected for fully dissolving water-insoluble materials, and edible particles of micro-/nanosizes are collected after eliminating solvent [42]. Protein-based edible particles are often prepared by changing their natural polypeptide structure and aggregating into microparticles through heat treatment or solvent modification [43]. Protein/polysaccharide composite nanoparticles are formed by the complexation method, which is mainly performed by electrostatic interactions, hydrophobic interactions, and hydrogen bonding. Cellulose nanoparticles can also be obtained by bacterial biosynthesis as well as plant matter disintegration.

5.2 MAJOR TECHNIQUES FOR MAKING EDIBLE FILMS/COATINGS

According to the distinction definition of films and coatings from McHugh [44], films should be formed separately and then applied onto the food surface, while coatings can be smeared directly on the food surface; therefore, major techniques for making edible films are introduced here in detail. The main techniques used to produce edible films include extrusion, injection moulding, thermo-forming, and solvent casting. During the fabrication process, all biopolymers are able to form films. After being dissolved in solvent (water or alcohol), biopolymer solutions are added by various additive ingredients according to the final requirements of the edible film, which can also be regulated by changing the drying temperature, humidity, and solution pH [45]. In general, polysaccharides and proteins have good abilities to form films, which block the permeation of oxygen, aroma, and lipids at low and intermediate relative humidities, while lipids have weak abilities to form films but present excellent water vapour barriers. In actuality, polysaccharide, protein, and lipid components are usually combined to produce composite films with desired functionality. Edible films are applied by several methods (such as dipping, spraying, panning, and brushing) to form single-layer and layer-by-layer coatings.

5.3 MAJOR TECHNIQUES FOR MAKING EDIBLE TEXTILES

Defined as textile structures that are embedded or coated with edible polymers, edible textiles can be consumed by humans, animals, or microorganisms without any side effects. They provide a great opportunity for textile engineers and designers to develop novel materials and design methods. Edible fibres are woven into fabrics, and these woven structures are suitable for reinforcing hard foods, such as biscuits and chocolates. The electrospinning technique is applied to produce edible nanofibres and textiles under electrical forces, with defects of a low yield rate and chemical solvent usage, and the electrospinning technique has a limited scope of application [46].

6 SPECIAL APPLICATIONS OF EDIBLE POLYMERS

Various edible polymers are widely applied in the food and biomedical industries; moreover, the cosmetics, energy, and water treatment industries also require the application of edible polymers.

6.1 APPLICATION OF EDIBLE POLYMERS IN THE FOOD INDUSTRY

Edible polymers have been used in great applications in the food industry mainly in the forms of edible particles and edible films, which can protect food from physical, chemical, and biological deterioration. In the functional food industry, additives such as flavouring, pigments, antimicrobials, antioxidant agents, sweeteners, and nutraceuticals are incorporated into edible films to prolong food shelf life as well as enhance food safety and quality, which further results in a popular research topic regarding the release performance of edible polymers.

With the emergence of smart food packaging, the potential of edible polymers for protecting and monitoring the surrounding environmental conditions of packaged foods is being reinvestigated. For example, according to Zhai et al. [47], a composite edible film containing gellan gum, gelatine, and red radish anthocyanin (RRA) extract is used as a sensor for monitoring food spoilage, as it can change colour with the metamorphic processes of milk and fish. To reduce the environmental pollution problem of biomass residues (such as rice straw, pineapple peel, red algae waste, sago seed shells, and sugarcane bagasse), there is also great enthusiasm for the technology of edible polymers produced from biomass residues. Cellulose nanocrystals, as a successful product, are usually applied to improve the mechanical strength and toughness of other edible films [48].

Edible films prepared with nonbovine ingredients and biopolymers by cold casting are biocompatible with muscle cells [49], promoting the development of 3D printing technology in the food industry to some extent. Finally, the edible polymer can be applied as slippery coatings in food packaging material or fried food, and edible superhydrophobic coatings decrease liquid food residues because of their incompatibility with water. During deep frying, edible coatings with desirable mechanical and barrier properties can be applied as a barrier to lipids, which protects food from fat permeation [50].

6.2 APPLICATION OF EDIBLE POLYMERS IN BIOMEDICINE

The application scope of edible polymers in the biomedical field is very broad, ranging from drug carriers (such as drug delivery and protein release) to self-efficacy (tissue engineering, regenerative medicine, wound dressing, and biomedical devices). Edible polymers can be applied in various forms (such as particles, films, and textiles), where edible particles are employed to increase lipid/

oil oxidative stability after acting as a protective layer around emulsion droplets. Because of their smaller sizes, drug-delivery nanodevices fabricated with edible particles (such as liposomes, nanoparticles, dendrimers, micelles, and nanorods) [51] have developed rapidly, offering great possibilities for specific targeting. Owing to the biodegradability and biocompatibility of edible particles, their Pickering emulsion can prepare porous scaffolds for tissue engineering, making it the most attractive template for tissue engineering.

Edible polymer films for drug delivery have gained lots of attention, where oral administration is used for drugs that require speedy absorption, topical delivery is used for wound healing, drug vaginal delivery is used for curing gynaecological diseases, and drug gastroretentive delivery is used for transporting water or poorly soluble molecules [52]. In addition, because edible polymers in hydrogel forms can be manufactured to produce 3D structures with interconnected pores [53], they have been extensively researched in the tissue engineering field, promoting the development of regenerative medicine.

In addition to the applications mentioned above, edible polymers can be used in applications for stopping bleeding and as edible electronic devices. Because gel formations are conducive to solution absorption in a low adherent manner, dressing removal without much trauma, pain reduction, and fast granulation and re-epithelialization by providing a humid environment, edible polymer-based hydrogels are applied to treat wounds. In capsule size, edible electronic devices containing sensors and batteries can be orally administered through the digestive system to monitor changes in temperature, heartbeat, and pH level, and for wound or disease treatment as well as drug delivery, whereby they could be applied as power pacemakers, neurostimulators, drug-delivery devices, ingestible cameras, and glucose monitors [54].

6.3 OTHER APPLICATIONS

As edible polymers have the abilities to enhance dispersion stability and improve product efficacy, texture, biodegradability and biocompatibility with skin and aesthetic properties, they are often used in cosmetic products as rheological modifiers, water-soluble binders, thickeners, film-forming agents, conditioners, sensory ingredients, texturing agents, moisturizers, and hydrating agents (Figure 4.3). Unlike their applications in food industries, polysaccharides and proteins are applied in the cosmetic field to stabilize emulsions and act as active ingredients separately. Compared to

Application of edible polymer based structures

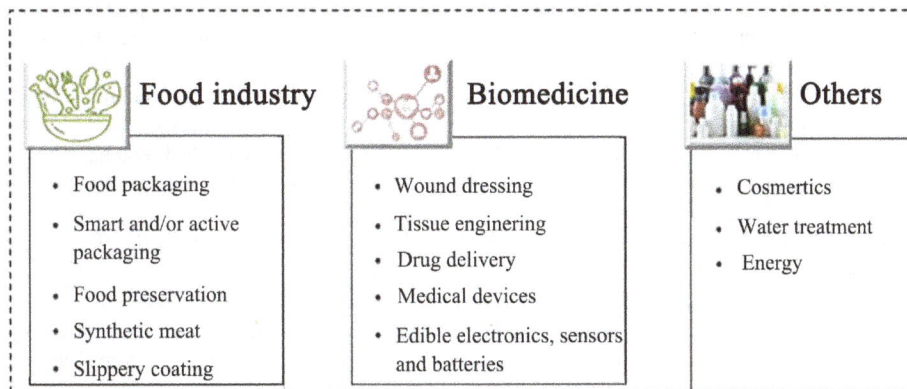

Food industry	Biomedicine	Others
• Food packaging	• Wound dressing	• Cosmertics
• Smart and/or active packaging	• Tissue enginering	• Water treatment
• Food preservation	• Drug delivery	• Energy
• Synthetic meat	• Medical devices	
• Slippery coating	• Edible electronics, sensors and batteries	

FIGURE 4.3 Applications of edible polymers. Adapted with permission from Reference [40]. Copyright (2020), Elsevier Ltd.

synthetic polymers, emulsions made by natural polymers spread better on the skin, which provides a stringiness feeling.

With the increased demand for energy storage devices over the past few decades, edible polymers capable of ion dissociation with metal salts have been used to produce solid polymer electrolytes. The corresponding fabrication process overcomes the disadvantage of high cost, and the resulting products all have properties of nontoxicity, eco-efficiency, flexibility, and good electrochemical stability [55]. In addition, carbohydrates containing cellulose, chitosan, and gums have been widely studied to remove contaminants (such as suspended solids, dyes, pesticides, toxicants, and heavy metals) from water by forming hydrogen bonds.

7 CONCLUSION

Edible polymers are nontoxic when consumed by living beings such as humans and animals due to their excellent digestibility and biocompatibility, which has promoted research on edible polymers as substitutes for synthetic plastics. Natural polymers are applied in the food industry due to their capacities to deliver bioactive components, emulsify, and thicken. Moreover, edible polymers play an important role in pharmaceutical sciences and biomedicine by improving patient outcomes. Although there remain some difficulties in the industrial applications of edible polymer films or textiles, edible films or coatings can always be used as sustainable alternatives to plastic packages in food packaging or biomedical systems.

REFERENCES

1. George, G., B.S. Kumar, and A. Srinivasan, Polymer nanocomposites for food packaging applications, Polymer Nanocomposites for Food Packaging Applications (2015) 743–768.
2. Jia, L., S. Evans, and S.V. Linden, Motivating actions to mitigate plastic pollution, Nat Commun. **10**(2019) 4582.
3. Gundogdu, S., N. Rathod, A. Hassoun, E. Jamroz, P. Kulawik, C. Gokbulut, A. Ait-Kaddour, and F. Ozogul, The impact of nano/micro-plastics toxicity on seafood quality and human health: facts and gaps, Crit Rev Food Sci Nutr. (2022) 1–19.
4. Zikmanis, P., K. Juhņeviča-Radenkova, V. Radenkovs, D. Segliņa, I. Krasnova, S. Kolesovs, Z. Orlovskis, A. Šilaks, and P. Semjonovs, Microbial polymers in edible films and coatings of garden berry and grape: Current and prospective use, Food and Bioprocess Technology **14**(2021) 1432–1445.
5. Mkandawire, M. and A.N.A. Aryee, Resurfacing and modernization of edible packaging material technology, Current Opinion in Food Science **19**(2018) 104–112.
6. Teixeira-Costa, B.E. and C.T. Andrade, Natural polymers used in edible food packaging—History, function and application trends as a sustainable alternative to synthetic plastic, Polysaccharides **3**(2021) 32–58.
7. Mohamed, S.A.A., M. El-Sakhawy, and M.A. El-Sakhawy, Polysaccharides, Protein and lipid-based natural edible films in food packaging: A Review, Carbohydr Polym. **238**(2020) 116178.
8. Liang, J., H. Yan, J. Zhang, W. Dai, X. Gao, Y. Zhou, X. Wan, and P. Puligundla, Preparation and characterization of antioxidant edible chitosan films incorporated with epigallocatechin gallate nanocapsules, Carbohydr Polym. **171**(2017) 300–306.
9. Galus, S. and J. Kadzińska, Food applications of emulsion-based edible films and coatings, Trends in Food Science & Technology **45**(2015) 273–283.
10. Jeevahan, J., M. Chandrasekaran, R.B. Durairaj, G. Mageshwaran, and G.B. Joseph, A brief review on edible food packing materials, Journal of Global Engineering Problems & Solutions **1**(2017) 9–19.
11. Sun, Q., T. Xi, Y. Li, and L. Xiong, Characterization of corn starch films reinforced with $CaCO_3$ nanoparticles, PLS ONE **9**(2014) e106727.
12. Salmieri, S., F. Islam, R.A. Khan, F.M. Hossain, H.M.M. Ibrahim, C. Miao, W.Y. Hamad, and M. Lacroix, Antimicrobial nanocomposite films made of poly(lactic acid)–cellulose nanocrystals

(PLA–CNC) in food applications—part B: effect of oregano essential oil release on the inactivation of *Listeria monocytogenes* in mixed vegetables, Cellulose **21**(2014) 4271–4285.

13. Santana, J.S., É.K.d.C. Costa, P.R. Rodrigues, P.R.C. Correia, R.S. Cruz, and J.I. Druzian, Morphological, barrier, and mechanical properties of cassava starch films reinforced with cellulose and starch nanoparticles, Journal of Applied Polymer Science **136**(2018) 47001.

14. Punia, S., K.S. Sandhu, S.B. Dhull, A.K. Siroha, S.S. Purewal, M. Kaur, and M.K. Kidwai, Oat starch: Physico-chemical, morphological, rheological characteristics and its applications – A review, International Journal of Biological Macromolecules **154**(2020) 493–498.

15. Bangar, S.P., S.S. Purewal, M. Trif, S. Maqsood, M. Kumar, V. Manjunatha, and A.V. Rusu, Functionality and applicability of starch-based films: An eco-friendly approach, Foods **10**(2021) 1–24.

16. Calderon-Castro, A., M.O. Vega-Garcia, J. de Jesus Zazueta-Morales, P.R. Fitch-Vargas, A. Carrillo-Lopez, R. Gutierrez-Dorado, V. Limon-Valenzuela, and E. Aguilar-Palazuelos, Effect of extrusion process on the functional properties of high amylose corn starch edible films and its application in mango (*Mangifera indica* L.) cv. Tommy Atkins, J Food Sci Technol. **55**(2018) 905–914.

17. Yang, J. and T. Warnow, Effect of different concentrations of olive oil and oleic acid on the mechanical properties of albumen (egg white) edible films, BMC Bioinformatics **12**(2011) 1–12.

18. Rompothi, O., P. Pradipasena, K. Tananuwong, A. Somwangthanaroj, and T. Janjarasskul, Development of non-water soluble, ductile mung bean starch based edible film with oxygen barrier and heat sealability, Carbohydrate Polymers **157**(2017) 748–756.

19. Colussi, R., V.Z. Pinto, S.L.M.E. Halal, B. Biduski, LucianaPrietto, D.D. Castilhos, E.d.R. Zavareze, and A.R. GuerraDias, Acetylated rice starches films with different levels of amylose: Mechanical, water vapor barrier, thermal, and biodegradability properties, Food Chemistry **221**(2017) 1614–1620.

20. Jha, P., Functional properties of starch-chitosan blend bionanocomposite films for food packaging: the influence of amylose-amylopectin ratios, Journal of Food Science and Technology, Mysore (2020) 1–11.

21. Geleta, T.T., S.A. Habtegebreil, and G.N. Tolesa, Physical, mechanical, and optical properties of enset starch from bulla films influenced by different glycerol concentrations and temperatures, Journal of Food Processing and Preservation **44**(2020) e14586.

22. Balakrishnan, P., S. Gopi, S. M S, and S. Thomas, UV resistant transparent bionanocomposite films based on potato starch/cellulose for sustainable packaging, Starch – Stärke **70**(2018) 1700139.

23. Geleta, T.T., S.A. Habtegebreil, and G.N. Tolesa, Physical, mechanical, and optical properties of enset starch from bulla films influenced by different glycerol concentrations and temperatures, Journal of Food Processing and Preservation **44**(2020) e14586.

24. Saberi, B., R. Thakur, Q.V. Vuong, S. Chockchaisawasdee, J.B. Golding, C.J. Scarlett, and C.E. Stathopoulos, Optimization of physical and optical properties of biodegradable edible films based on pea starch and guar gum, Industrial Crops and Products **86**(2016) 342–352.

25. Shi, A.M., L.J. Wang, D. Li, and B. Adhikari, Characterization of starch films containing starch nanoparticles: part 1: physical and mechanical properties, Carbohydr Polym. **96**(2013) 593–601.

26. Rachmawati, N., R. Triwibowo, and R. Widianto, Mechanical properties and biodegradability of acid-soluble chitosan-starch based film, Squalen Bulletin of Marine & Fisheries Postharvest & Biotechnology. **10**(2015) 1–7.

27. Nguyen Vu, H.P. and N. Lumdubwong, Starch behaviors and mechanical properties of starch blend films with different plasticizers, Carbohydr Polym. **154**(2016) 112–120.

28. Jiménez, A., M.J. Fabra, P. Talens, and A. Chiralt, Phase transitions in starch based films containing fatty acids. Effect on water sorption and mechanical behaviour, Food Hydrocolloids **30**(2013) 408–418.

29. Nechita, P. and M. Roman, Review on polysaccharides used in coatings for food packaging papers, Coatings **10**(2020) 566.

30. Tabari, M., Investigation of carboxymethyl cellulose (CMC) on mechanical properties of cold water fish gelatin biodegradable edible films, Foods **6**(2017) 1–7.

31. Sucheta, K. Chaturvedi, N. Sharma, and S.K. Yadav, Composite edible coatings from commercial pectin, corn flour and beetroot powder minimize post-harvest decay, reduces ripening and improves sensory liking of tomatoes, Int J Biol Macromol. **133**(2019) 284–293.

32. Setyorini, D., and Nurcahyani, P.R., *Effect of Addition of Semi Refined Carrageenan on Mechanical Characteristics of Gum Arabic Edible Film*, in IOP Conference Series: Materials Science and Engineering: Bandung, Indonesia (2016) 12011.

33. Martelli, M.R., T.T. Barros, M.R.D. Moura, L. Mattoso, and O. Assis, Effect of chitosan nanoparticles and pectin content on mechanical properties and water vapor permeability of banana puree films, Journal of Food Science **78**(2013) N98–N104.

34. Hassan, R.A., L.Y. Heng, and L.L. Tan, Novel DNA biosensor for direct determination of carrageenan, Sci Rep. **9**(2019) 6379.

35. Zikmanis, P., S. Kolesovs, and P. Semjonovs, Production of biodegradable microbial polymers from whey, Bioresources and Bioprocessing **7**(2020) 1–15.

36. Moreira, M.R., L. Cassani, O. Martin-Belloso, and R. Soliva-Fortuny, Effects of polysaccharide-based edible coatings enriched with dietary fiber on quality attributes of fresh-cut apples, J Food Sci Technol. **52**(2015) 7795–7805.

37. Sonu Sharma, and V. T. Ramana Rao, Xanthan gum based edible coating enriched with cinnamic acid prevents browning and extends the shelf-life of fresh-cut pears, Lwt Food Science & Technology (2015) 191–800.

38. Burger, T.G. and Y. Zhang, Recent progress in the utilization of pea protein as an emulsifier for food applications, Trends in Food Science & Technology **86**(2019) 25–33.

39. Mohamed, S., M. El-Sakhawy, E. Nashy, and A.M. Othman, Novel natural composite films as packaging materials with enhanced properties, International Journal of Biological Macromolecules. **136**(2019) 774–784.

40. Kouhi, M., M.P. Prabhakaran, and S. Ramakrishna, Edible polymers: An insight into its application in food, biomedicine and cosmetics, Trends in Food Science & Technology **103**(2020) 248–263.

41. Tzoumaki, M.V., T. Moschakis, E. Scholten, and C.G. Biliaderis, In vitro lipid digestion of chitin nanocrystal stabilized o/w emulsions, Food Funct. **4**(2013) 121–129.

42. Zou, Y., J. Guo, S.W. Yin, J.M. Wang, and X.Q. Yang, Pickering emulsion gels prepared by hydrogen-bonded zein/tannic acid complex colloidal particles, J Agric Food Chem. **63**(2015) 7405–7414.

43. Xiao, J., Y. Li, and Q. Huang, Recent advances on food-grade particles stabilized Pickering emulsions: Fabrication, characterization and research trends, Trends in Food Science & Technology **55**(2016) 48–60.

44. McHugh, T.H., Protein–lipid interactions in edible films and coatings, Nahrung. **44**(2000) 148–151.

45. Jolanta, Wróblewska-Krepsztul Tomasz, Rydzkowski Gabriel, Borowski Mieczysław, Szczypiński Tomasz, Klepka Vijay, Kumar, and Thakur, Recent progress in biodegradable polymers and nanocomposite-based packaging materials for sustainable environment, International Journal of Polymer Analysis & Characterization **23**(2018) 383–395.

46. Kouhi, M., V. Jayarama Reddy, and S. Ramakrishna, GPTMS-modified bredigite/PHBV nanofibrous bone scaffolds with enhanced mechanical and biological properties, Appl Biochem Biotechnol. **188**(2019) 357–368.

47. Zhai, X., Z. Li, J. Zhang, J. Shi, X. Zou, X. Huang, D. Zhang, Y. Sun, Z. Yang, M. Holmes, Y. Gong, and M. Povey, Natural biomaterial-based edible and pH-sensitive films combined with electrochemical writing for intelligent food packaging, J Agric Food Chem. **66**(2018) 12836–12846.

48. Kassab, Z., F. Aziz, H. Hannache, H. Ben Youcef, and M. El Achaby, Improved mechanical properties of k-carrageenan-based nanocomposite films reinforced with cellulose nanocrystals, Int J Biol Macromol. **123**(2019) 1248–1256.

49. Acevedo, C.A., N. Orellana, K. Avarias, R. Ortiz, and P. Prieto, Micropatterning technology to design an edible film for in vitro meat production, Food and Bioprocess Technology **11**(2018) 1267–1273.

50. Ananey-Obiri, D., Matthews, L., Azahrani, M.H., Ibrahim, S.A., Galanakis, C.M., Tahergorabi, R., Application of protein-based edible coatings for fat uptake reduction in deep-fat fried foods with an emphasis on muscle food proteins – ScienceDirect, Trends in Food Science & Technology **80**(2018) 167–174.

51. Felice, B., M.P. Prabhakaran, A. Rodríguez, and S. Ramakrishna, Drug delivery vehicles on a nano-engineering perspective, Materials Science & Engineering C. **41**(2014) 178–195.

52. Tiwari, R.R., M.S. Umashankar, and N. Damodharan, Recent update on oral films: A bench to market potential, International Journal of Applied Pharmaceutics **10**(2018) 920–927.

53. Zhou, Y., T. Xu, Y. Zhang, C. Zhang, Z. Lu, F. Lu, and H. Zhao, Effect of tea polyphenols on curdlan/chitosan blending film properties and its application to chilled meat preservation, Coatings **9**(2019) e9040262.

54. Zhang, S., K.J. Lee, M. Goudie, H.J. Kim, and A. Khademhosseini, Minimally invasive technologies for biosensing, Interfacing Bioelectronics and Biomedical Sensing (2020) 193–224.

55. Singh, R., A.R. Polu, B. Bhattacharya, H.W. Rhee, C. Varlikli, and P.K. Singh, Perspectives for solid biopolymer electrolytes in dye sensitized solar cell and battery application, Renewable and Sustainable Energy Reviews **65**(2016) 1098–1117.

5 Chicken Fat
A Promising and Sustainable Raw Material for Polyurethane Industries

A.A.P.R. Perera,[1,2] K.A.U. Madhushani,[1,2] Felipe M. de Souza,[2] Tim Dawsey,[2] and Ram K. Gupta[1,2]

[1] Department of Chemistry, Pittsburg State University, Pittsburg, Kansas 66762, USA
[2] National Institute for Materials Advancement, Pittsburg State University, Pittsburg, KS 66762, USA

1 INTRODUCTION

Polymers quickly made their mark in human history as they have been used in many products which have changed the living standards of society and played a crucial role in advancing science and technology. The application of polymers can be witnessed in nearly every sector, including automobiles, aerospace, electronics, agricultural, biomedical, construction, furniture, and packaging. This vast range of applications is related to the diversity of polymers and the tunability of their properties which can make the same polymer useful in different applications. Polyethylene (PE), which is one of the most common polymers, is a thermoplastic obtained through the polymerization of ethylene. PE is easily processed and provides versatile applications as it can be obtained in either a crystalline or amorphous form with a range of densities, and mechanical and optical properties (transparent or translucent). Low-density PE is mainly used for low-strength applications, while high-density PE is used in applications that require high strength. Polyethylene terephthalate is another widely used polymer that is employed in the manufacture of containers such as bottles, as well as some uses in buildings, clothes, and so on. Polyisoprene, polystyrene, polyvinylchloride, polypropylene, polyamides, and PUs are some of the other commercially important polymers.

PUs are perhaps one of the most versatile materials employed in the industry as they have a diverse range of properties. For example, flexible PUs can be used for the manufacture of furniture, while rigid PU foams can be used as a building material in construction due to their low weight, and appreciable mechanical and thermal properties. The high thermal insulting nature of PU foams makes them suitable for appliances also. For the same reasons, PUs are also employed in the automotive industry as their mechanical properties and light weight can improve fuel efficiency [1]. Aside from good thermal insulation, PUs can also be good sound insulators when they present an open cell structure that allows them to absorb and dampen the sound [2]. Medical and biomedical applications are another important area where PUs find wide use. PU ionomers (PUIs) are commonly used in biomedical applications. A PUI presents ionic groups along its polymeric chain which aids its dispersibility in polar solvents, can introduce self-healing properties, and prevents the adhesion of biomolecules such as proteins. A PUI can be used to produce tubes for body transporting fluids such as hemodialysis tubes, artificial hearts, and pacemaker components, among others. Waterborne

DOI: 10.1201/9781003278269-5

PUs are another class of PUs widely used for coatings, adhesives, and sealants with the advantage of low volatile organic components [3].

Despite the promising future of PUs, one of the main challenges lies in finding alternative chemicals for the manufacturing of polyols and isocyanates as they are mostly derived from petrochemical resources. There has been considerable demand for the development of bio-based PUs, with properties compared to petrochemical-derived PUs, as a means to reduce the dependence on petrochemical-based materials [4–6]. Vegetable oils from soybean, sunflower, olives, castor plant, palm, linseed, canola, and rapeseed, among others, have been shown to be viable starting materials that can be easily converted into polyols for PUs. Another seldomly explored possibility lies in the use of waste from the poultry industry. In this sense, some of the discarded materials can be suitable for the synthesis of polyols while offering a remarkably high performance in the product. Hence, a promising, and sustainable cycle can be established by converting waste material into value-added products. This is one of the core aspects of sustainability as it can introduce a novel material with satisfactory performance along with providing a proper new life for materials that would otherwise be discarded. One of the remaining challenges is scaling the use of these renewable starting materials, as their introduction into mainstream production can lead to more sustainable credentials in the industry. It is also necessary to employ methodologies that lead to better performance of PUs compared to status-quo petrochemical-based PUs. Through these concepts, this chapter provides a study focused on making PUs using chicken fat, along with some other examples derived from other renewable sources. In addition, the RPUFs obtained are incorporated with FR compounds that further improve their properties.

2   INDUSTRIAL NEED FOR RENEWABLE MATERIALS

PUs are an extremely versatile class of polymers that are commonly obtained through a step-growth polyaddition reaction between a polyol and an isocyanate. A polyol is a monomeric unit that presents several hydroxyl functions (–OH), whereas polyisocyanate consists of monomers presenting at least two isocyanate functions (–N=C=O). The reaction between these two organic functionalities leads to the urethane linkage (–NH(C=O)–O–), which is the rigid segment along the polymer chain, whereas the soft segments are usually defined by the alkyl chain derived from the polyol. This polyaddition reaction can be performed under mild conditions which usually include reactions at room temperature, without the formation of byproducts, along with straightforward procedures. Such factors, along with the attractive properties of PUs, make these polymers a highly attractive investment for industries, which results in a strong market. Isocyanates, which are one of their core components, can be produced through a nitration reaction of aromatic compounds such as toluene, or benzene, followed by hydrogenation to convert the nitro groups into amines. Then, carbonyl chloride ($COCl_2$), also known as phosgene gas, is used to react with the amine group to convert it into an isocyanate. Another synthetic path lies in the use of dimethyl carbonate. In this approach, different types of isocyanates (Figure 5.1) can be manufactured, among which the mostly used in the industry are toluene diisocyanate (TDI), methylene diphenyl diisocyanate (MDI), hexamethylene diisocyanate (HMDI), 1,6-hexamethylene diisocyanate (HDI), and isophorone diisocyanate (IPDI).

These isocyanates can be introduced individually, as blends, or in the form of oligomers containing diols having a low molecular weight with isocyanate end-groups. Also, isocyanates derived from aliphatic amines are usually employed for the synthesis of coatings as they tend to present better stability toward UV radiation and moisture. However, these aliphatic isocyanates tend to be considerably less reactive than aromatic ones. They tend to be employed in specialty applications rather than the manufacture of rigid or flexible PU foams, which are widely consumed and require a faster production rate. Under this perspective, the isocyanate structures found commercially usually present an aromatic ring and require multi-step conversions to yield the desired

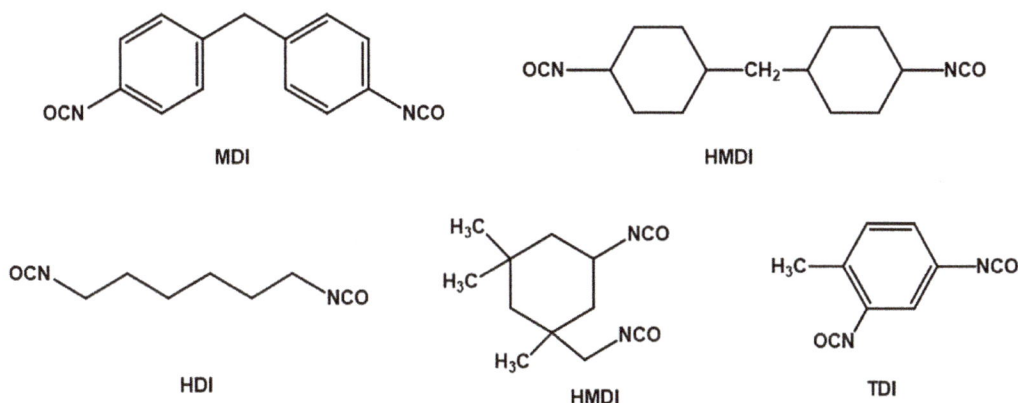

FIGURE 5.1 Structures of commonly used isocyanates for the manufacture of PU.

function. However, it is rare to find aromatic structures derived from renewable sources that could be converted into isocyanates in the same manner as petrochemical-based raw materials. Because of this, there is a limitation in the use of renewable sources for the synthesis of isocyanates with aromatic structures. On top of that, isocyanates are toxic materials that require careful handling during the processing of PU, which is another concern for the industry and community. There has been a trend in developing non-isocyanate PUs as a means to decrease the dependence upon non-renewable sources and introduce materials with less toxicity into the manufacturing process. One of the approaches employed for that consists of the reaction of cyclic carbonate compounds with amines to yield a hydroxyurethane group. One example of this process has been reported by Lee *et al.* [7], who performed an epoxidation reaction in soybean oil, followed by a carbonation reaction, to introduce the cyclic carbonate groups. Then, 3-aminopropyltriethoxysilane was reacted with the cyclic carbonated vegetable oil to form the hydroxyurethane linkages. After that, lignin was used as the polyol to react with the siloxane groups forming the polymeric structure.

Following that line, the polyols are the other key component for the synthesis of PUs. Unlike isocyanates, the polyols do not require such extensive reaction processes, which allows for a much broader number of starting materials that can be converted into polyols. Usually compounds with some degree of unsaturation can provide a reactive site for the chemical introduction of hydroxyl groups. Polyols began being manufactured mostly through petrochemical-based raw materials such as ethylene oxide (EO) and propylene oxide (PO) which yielded polyether diols. The processing allowed a considerable degree of versatility since the molecular weight of the polyether polyols could be controlled to obtain a PU with more elastomeric properties when the molecular weight was high, or a more rigid polymer when molecular weights were lower. On top of that, the functionality can be defined by introducing a starter which is usually small molecules with several hydroxyl groups such as glycerin, sorbitol, or sucrose-based polyols. Other types of commercially available polyols include polyesters, polycarbonates, acrylics, and polybutadiene diols. The same perspective of finding renewable and safer sources for the manufacture of isocyanates is applied to polyols. In this sense, finding novel bio-renewable materials that are suitable for polyol production can ease the burden on the excessive demand for petrochemicals aside from introducing products that can be competitive with the currently established ones.

There are several types of raw materials that can be employed for the synthesis of polyols. Vegetable oils are a practical example of that case, as they are composed of unsaturated fatty acids that are esterified with glycerin. These fatty acid segments can be found in different proportions depending on the vegetable oil. Based on this, there are generally three unsaturated fatty acids that contain 18 C atoms that can be functionalized with hydroxyl groups: oleic, linoleic, and linolenic acids, as they present one, two, and three unsaturated bonds, respectively. These fatty acids can be found in several vegetable oils derived from corn, soybean, castor plant, sunflower, rapeseed, linseed, coconut, cottonseed, olive, peanut, palm, sesame, tung, and jatropha, among others. These bio-renewable materials can potentially be converted into polyols to produce PUs [6,8–10]. Another seldomly explored resource that could potentially be used as a starting material for the synthesis of polyols is animal fat, commonly known to offer a considerable degree of unsaturation allowing for chemical functionalization on the double bonds. In this respect, the following sections discuss, in depth, the use of chicken fat as a promising material for the synthesis of polyols to produce rigid PUs that can be further incorporated with flame-retardant components.

3 METHODS TO SYNTHESIZE AND CHARACTERIZE POLYOLS

The polyurethane industries cover a wide share of the polymer sector due to the versatility of the starting materials, range of synthetic approaches, tunability, and applications of PUs. In this section, we provide some synthetic approaches used for the preparation of starting materials for polyurethanes and methods used for their characterization. Several synthetic approaches such as hydroformylation followed by reduction, ozonolysis, thiol-ene, epoxidation followed by ring-opening, and transesterification can be performed for the synthesis of polyols. Based on these approaches, hydroformylation is a process that consists of reacting a double bond with H_2 and CO gases to yield aldehyde groups, which are then converted into alcohols through a reduction reaction with H_2. Petrovic *et al.* [11] performed the hydroformylation process on soybean oil. One of the advantages of this method is that it allows almost full conversion of double bonds into hydroxyl groups. Another important factor regarding hydroformylation is that, after the hydrogenation process, primary hydroxyl groups are formed which are more reactive toward isocyanate than secondary or tertiary ones. Through that, the hydroxyl content can be controlled according to the desired application. It can be observed that viscosity follows a linear relation with the hydroxyl number, which is an important factor in terms of processability. Based on that, the authors performed esterification of the primary hydroxyl groups with formic acid, converting them into ester groups and therefore allowing precise control of the hydroxyl number. The reaction process can be seen in Figure 5.2.

Based on this discussion it was observed that the hydroformylation/reduction process leads to the formation of primary hydroxyl groups much like epoxidation/ring-opening [9,12,13]. In this

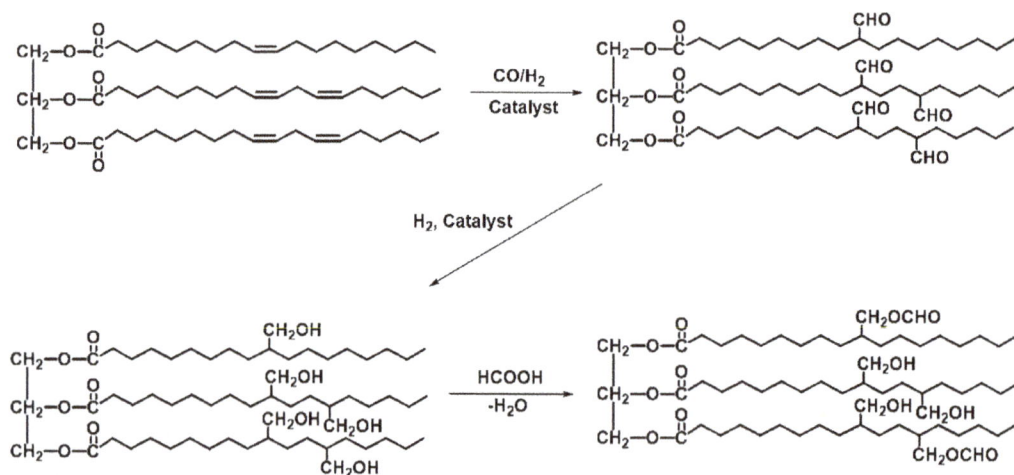

FIGURE 5.2 Hydroformylation approach for the synthesis of a polyol derived from soybean oil. Reproduced with permission [11]. Copyright (2007), Wiley.

sense, taking into consideration the structure of unsaturated vegetable oils, it is noted that their double bonds are in the middle of the fatty acid chain. Hence, the primary hydroxyl groups formed through either hydroformylation/reduction or epoxidation/ring-opening, are inherently located in the middle of the chain. Because of that, the rest of the alkyl segment remains inactive after the synthesis of the polyol. These dangling segments can prevent the proper settlement of the polymeric chain, by acting as plasticizers, which could lead to a deterioration of mechanical properties or make the PU more flexible. Ozonolysis is a chemical approach for the synthesis of polyols that can lead to structures with primary and terminal hydroxyl groups. The chemical process for this technique promotes cleavage of the carbon double bonds that can be converted into hydroxyl, aldehyde, or carboxylic acid groups according to the reaction parameters.

Petrovic et al. [14] synthesized polyols based on triolein, soybean oil, and canola oil through the ozonolysis process. One observation was that soybean oil-based polyols, synthesized through hydroformylation/reduction, or epoxidation/ring-opening, presented a functionality of around 4.5, whereas, when ozonolysis was performed the functionality was around 2.5, and for canola oil it was around 2.8. This can be attributed to the fact that vegetable oils contain a variety of fatty acids which can be both saturated and unsaturated. In this sense, saturated fatty acids such as stearic acid are left unreacted due to the absence of double bonds. Hence, the maximum functionality that can be achieved while performing ozonolysis would be around 3.0, as was observed for the case of triolein. Yet, structures without inert alkyl segments can be realized through this methodology which can lead to PUs with stronger mechanical properties. The structure of the triolein- and soybean-based polyols is depicted in Figure 5.3. The inherent challenge of this technique is that it employs ozone which is a toxic gas, along with requirements for temperature control and the use of organic solvents during the synthesis process. Because of that, such factors hinder the broader application of this approach.

Another approach that has been attracting the attention of the scientific community is thiol-ene coupling. This technique consists of the disruption of a double bond through a photocatalyzed radical addition of S• and H• radicals derived from a mercaptan. A compound that contains an SH group, along with hydroxyl groups on the other end of the molecule, can be used to convert compounds with unsaturation into polyols containing hydroxyl groups intermediated with a C–S bond at the carbon containing the former double bond [15]. The conveniences of this technique are usually related to procedures that can be performed at room temperature, without the need for solvent, no formation of byproducts, and facile photocatalysis. On the other hand, it is worth noting that mercaptans are toxic, which requires some safety precautions during the handling of such compounds. Gupta et al. [16] performed the thiol-ene approach to convert limonene, an unsaturated bio-waste compound found in orange peels, into a polyol. Through that, a rigid PU with

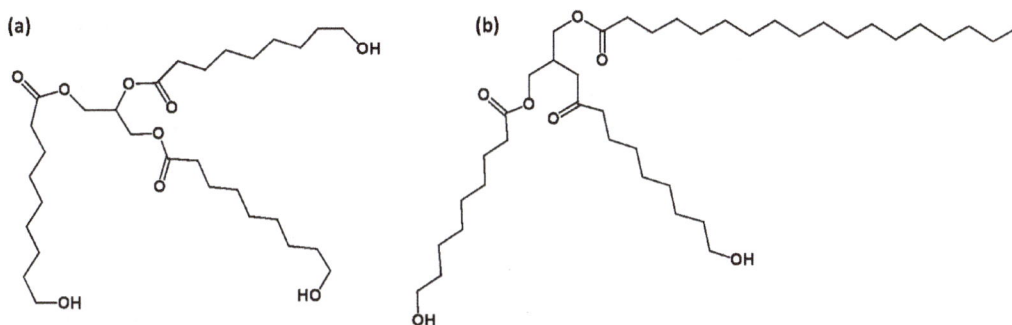

FIGURE 5.3 Structures of (a) triolein-based triol and (b) soybean-based diol were obtained through the ozonolysis approach. Adapted with permission [14]. Copyright (2005), American Chemical Society.

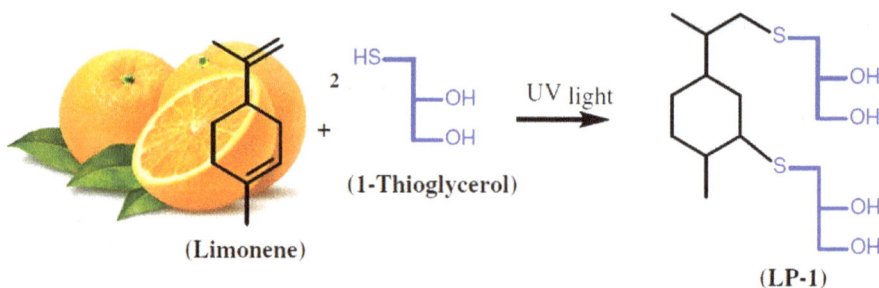

FIGURE 5.4 Use of the thiol-ene process to convert limonene into a polyol through its reaction with 1-thioglycerol. Adapted with permission [16]. Copyright (2014), Springer Nature.

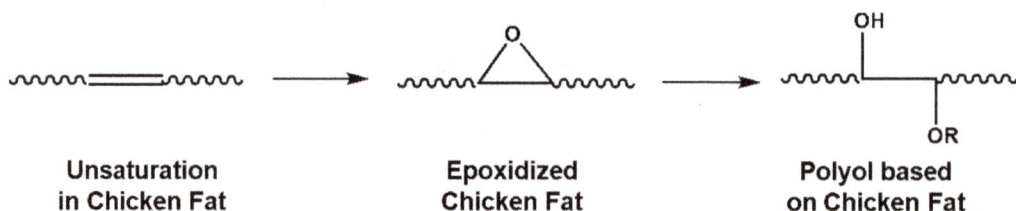

| Unsaturation in Chicken Fat | Epoxidized Chicken Fat | Polyol based on Chicken Fat |

FIGURE 5.5 Schematics of the reaction showing conversion of CF into polyol for polyurethanes.

satisfactory mechanical properties could be obtained. The thiol-ene reaction performed for the synthesis of a limonene-based polyol (LP-1) is presented in Figure 5.4. As can be seen through this example, thiol-ene coupling is an attractive technology as it can be performed through a facile procedure that occurs under mild conditions along with the possibility of being performed in different types of bio-based materials [1,6,17].

Another widely used technique for the synthesis of polyols consists of the two-step process of conversion of a double bond into an epoxy ring followed by a ring-opening reaction. Such a process was adopted to convert an unsaturated chicken fat (CF) into a polyol, as demonstrated through the reaction depicted in Figure 5.5. The unsaturated CF reacted with peracetic acid formed *in situ* through the reaction between hydrogen peroxide with acetic acid leading to the formation of the epoxy groups. The ring-opening reaction was then performed with methanol catalyzed by tetrafluorboric acid. The products obtained from each step were characterized through Fourier transformed infrared (FT-IR) provided in Figure 5.6. CF presented a small peak at around 3002 cm^{-1} related to the hydrogen stretch from unsaturated carbon (H–C=). After the epoxidation reaction, the resulting epoxidized chicken fat (ECF) was characterized by a weak peak at around 800 cm^{-1} related to the C–O–C bond which confirmed the presence of epoxy groups. Following that, after the ring-opening reaction, the CF polyol (CFP) product was confirmed based on the disappearance of the weak peak related to the C–O–C peak along with the presence of the characteristic –OH stretch at around 3500 cm^{-1} [18–20].

4 IMPORTANCE OF FLAME-RETARDANT POLYURETHANE FOAMS AND THEIR PREPARATION

The properties of the PU are greatly influenced by the formulation. Table 5.1 shows the amount (in grams) used for each of the components to make the rigid PU.

Each component plays a role in the foaming process. The CFP, SG-522 (commercial polyol), and isocyanate are the main reagents related to the formation of rigid RPUF. NIAX A-1, referred to

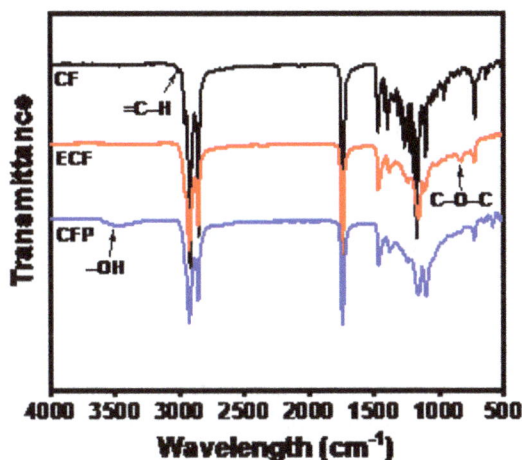

FIGURE 5.6 ATR-FTIR plots of chicken fat, epoxidized chicken fat, and polyol based on chicken fat.

TABLE 5.1
Formulation of the Polyurethane Foams

Ingredient	F-1	F-2	F-3	F-4	F-5	F-6	F-7
CFP	5	5	5	5	5	5	5
SG-522	15	15	15	15	15	15	15
A-1	0.18	0.18	0.18	0.18	0.18	0.18	0.18
Water	0.8	0.8	0.8	0.8	0.8	0.8	0.8
T-12	0.04	0.04	0.04	0.04	0.04	0.04	0.04
B8404	0.4	0.4	0.4	0.4	0.4	0.4	0.4
Isocyanate	32	32	32	32	32	32	32
DMMP	0	1	2	3	4	5	6

Note: All numbers are in grams.

as A-1, is a tertiary amine-based catalyst for the reaction between hydroxyl and isocyanate groups. In this case, the catalysts polarize either the –OH group by interacting with the H or the –N=C=O by interacting with the C atom, which further enhances the electrophilic nature of these species. Alongside that, DABCO T-12 is another catalyst based on dibutyltin dilaurate that also facilitates the reaction between hydroxyl and isocyanate groups through polarization of the bonds based on the presence of the complex Sn metallic center. The mechanism for this process is presented in Figure 5.7.

Water is another important foaming agent as it reacts with isocyanate forming an unstable carbamic acid that promptly decomposes into an amine followed by the release of CO_2. Through that, the evolution of gases promotes foam rise. The controlled addition of water is an important factor to regulate the foam's density, as well as maintaining its cellular structure. Hence, properties can be easily tuned based on the amount of water added into the formulation. This approach tends to be more eco-friendly as it avoids the use of chlorofluorocarbons (CFCs) and hydrochlorofluorocarbons (HCFCs), which are known for being harmful to the environment, due to their participation in the deterioration of the ozone layer. On top of that, the use of CFCs and HCFCs can also increase PU's flammability.

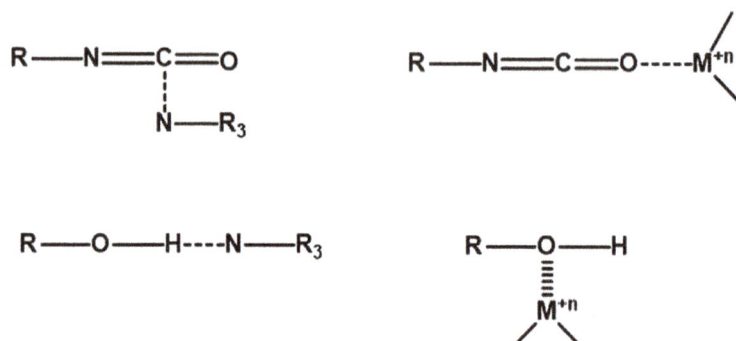

FIGURE 5.7 Bond polarization of hydroxyl and isocyanate groups through tertiary amine and transition metal-based catalysts. Adapted with permission [22]. Copyright (2011), Royal Society of Chemistry.

B8404, a silicone-based surfactant, was also used. Even though hydroxyls and isocyanates are reactive, their miscibility is relatively poor. Hence, the surfactant promotes a more homogeneous system, which allows them to react without leading to microphase separation. Lastly, dimethyl methylphosphonate (DMMP) was added as the component responsible for introducing flame retardancy. It was incorporated as a non-reactive FR which allows it to be simply blended along with the other components. Such an approach allows a facile introduction of FR properties into the foam, which is usually adopted in the industry, as it is relatively low cost and effective. Another way of introducing such properties can be realized by using reactive FR species that can be chemically attached to the polyol's structure. This has the advantage of promoting even distribution of the FR species throughout the foam along with preventing migration, or evaporation, of the FR compound over time. However, it also introduces higher cost as it requires an extra synthetic step, along with additional chemical groups that can act as plasticizers and influence the foam's mechanical properties in an undesirable manner. The shape of the obtained free-rise foams is presented in Figure 5.8, demonstrating that they maintained their structure regardless of the addition of DMMP, suggesting that there is acceptable compatibility between the active FR compound and the PU matrix.

5   CHARACTERISTICS OF POLYURETHANE FOAMS

The density of a material influences its application, as it is one of the key physical properties related to its mechanical behavior, morphology, and insulation, among other factors [23–25]. The concentration of water used as the chemical blowing agent was kept constant and the effect of increasing concentrations of DMMP was studied. It was observed that the addition of DMMP promoted a mild decrease in density. The control foam (F-1) presented the highest density of 32.5 kg/m^3, whereas the F-7 presented the lowest density of around 29 kg/m^3 (Figure 5.9). Materials with lower density values may be desirable in construction, thermal insulation, and packing. The CF-based RPUF presented values of density somewhat lower compared to other works that employed DMMP as non-reactive FR. For example, Feng *et al.* [26] made a series of RPUFs with increasing concentrations of DMMP and expandable graphite (EG). These authors observed that their density ranged from 36 to 46.3 kg/m^3 for 0 to 16 wt.%, respectively, of the total amount of FR in the foams showing a relatively predictable increase in density.

Closed-cell content (CCC) describes the percentage of a PU's cellular structure that is closed, meaning that it can block the passage of air and water, for instance. The CCC for the set of CF-based RPUFs is presented in Figure 5.10. This measurement provides important information that aids in the elucidation of the foam's properties related to the adsorption of air moisture, thermal insulation, as well as flame retardancy. In the case of the latter, it can be assumed that cellular structures that

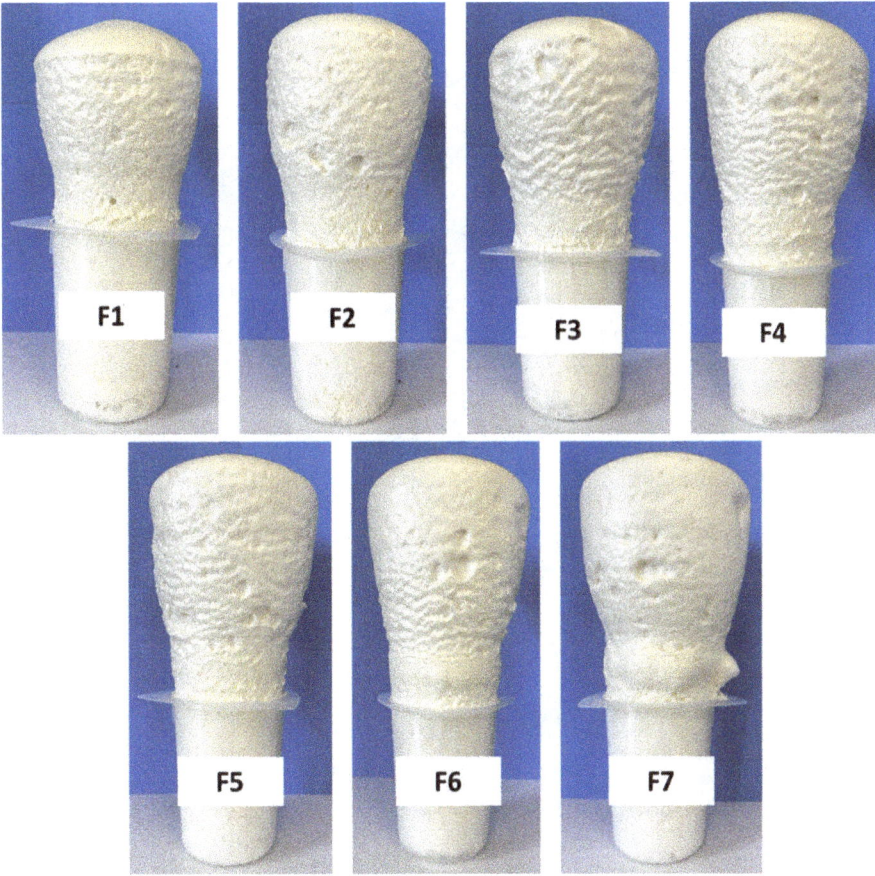

FIGURE 5.8   Digital photos of the prepared foams with varying amounts of DMMP.

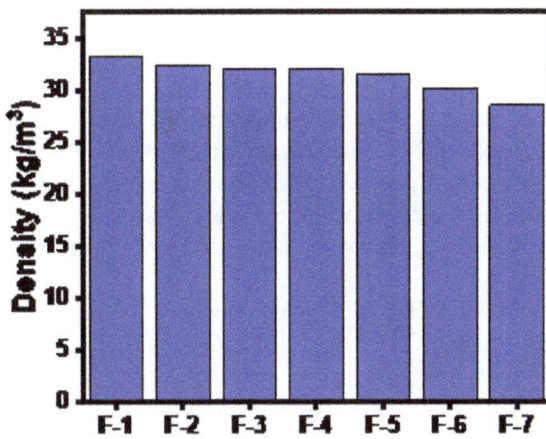

FIGURE 5.9   The density of polyurethane foams has different amounts of DMMP.

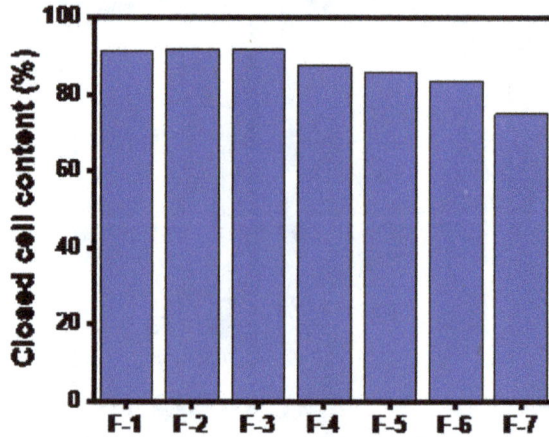

FIGURE 5.10   Closed-cell content of polyurethane foams with varying amounts of DMMP.

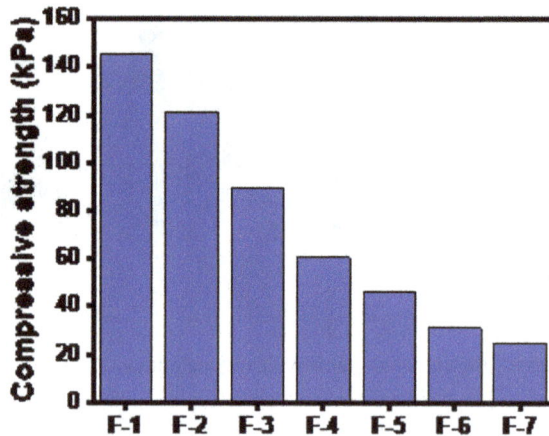

FIGURE 5.11   Compressive strengths of polyurethane foams having varying amounts of DMMP.

are closed can prevent the diffusion of oxygen within it which could catalyze the combustion of the foam. It was observed that the CF-based RPUFs presented around 90% of CCC, which tended to decrease with an increasing concentration of DMMP, reaching the lowest at around 75%. Hence, it can be noted that DMMP promoted an opening of the cell structure, suggesting that the optimal properties were more likely to be reached without the need for high concentrations of FR in the PU matrix.

The mechanical behavior of RPUF is another important measurement used to determine its usability and performance. Figure 5.11 presents the values of compressive strength at the yield for the CF-based RPUF under the influence of increasing concentrations of DMMP. From this perspective, it was noted that there was an exponential decrease in compressive strength with the addition of DMMP, as the highest value of yield at break was reached by the F-1 control sample which displayed around 145 kPa, whereas the lowest value was obtained for the F-7 (~10 wt.% DMMP) sample which presented 25 kPa. This decrease in compressive strength might be attributed to the plasticizing effect of the P=O groups which has been observed in other studies that employed DMMP as an FR in RPUF [8,20,27]. A similar trend was observed when comparing the CF-based RPUF results in terms of density and closed-cell content, as both faced a decrease when higher

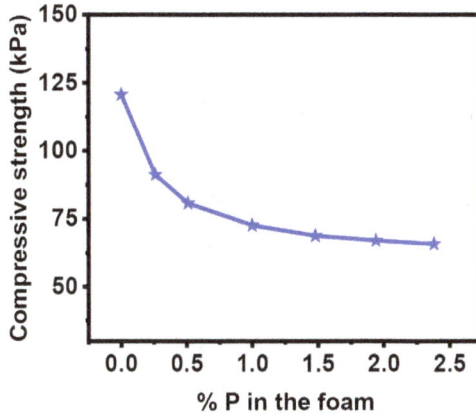

FIGURE 5.12 Compression strength for the corn oil-based RPUF. Figure adapted from [8]. Copyright (2019), Authors, Licensee MDPI, Basel, Switzerland. This article is an open access article distributed under the terms and conditions of the Creative Commons Attribution (CC BY) license.

concentrations of DMMP were added. Ramanujan *et al.* [8] presented the synthesis of a corn oil-based polyol synthesized through the thiol-ene approach which was used to make RPUF. The foams were mixed with varying amounts of DMMP, and their properties were evaluated. As in the case of CF-based RPUF, the corn oil-based RPUF also presented an exponential decrease in the compressive strength after the addition of DMMP, as the neat foam presented around 122 kPa and decreased down to around 62 kPa as seen in Figure 5.12.

Scanning electron microscopy (SEM) is an important tool for visualizing a material's morphology. Hence, the cellular structure of a PU can be analyzed in terms of pore size and the effect that an additive has on the morphology. SEM analysis for the CF-based RPUF was performed (Figure 5.13). It can be observed that the set of foams containing increasing quantities of DMMP presents a relatively regular and uniform cell size as the concentration of DMMP increases. On top of that, there was an increase in the cell size. The latest observation correlates with the decreases in density and compressive strength since, usually, the increase in cell size leads to a decrease in cell number. Because of that, there is less support from the PU matrix when it is submitted to mechanical compression, leading to a deterioration of properties as observed in the case of the CF-based RPUF. Other studies also presented the same tendency as described by Zhang *et al.* [28] which synthesized an RPUF originating from orange peels. In their work, DMMP was also employed as a non-reactive FR and led to an increase in cell size. Also, Liu *et al.* [27] made an RPUF and blended both DMMP and ammonium polyphosphate (APP). It was observed that the addition of DMMP led to a disruption of the cellular structure which caused a reduction in the mechanical properties.

It is noticeable that there is a challenge regarding maintaining, or improving, the mechanical and physical properties of an RPUF while incorporating other aspects such as flame retardancy. With this in mind, it is noted that RPUFs are materials that combine properties such as a relatively low density along with appreciable mechanical properties. Such a combination is observed partially due to its porous and rigid cellular structure. However, these characteristics also make PUs susceptible to fire, which makes it necessary to incorporate materials or chemical active groups that can improve their resistance against fire by suppressing the release of heat and smoke quickly. For that DMMP was blended into the CF-based RPUF to incorporate flame-retardant properties. The horizontal burning test was performed and assessed relative to the time taken for the sample to quench the fire, along with the weight loss of material that was combusted. For that, the CF-based RPUF was exposed to a direct flame for 10 s. The visual aspects of the CF-based RPUF are presented in

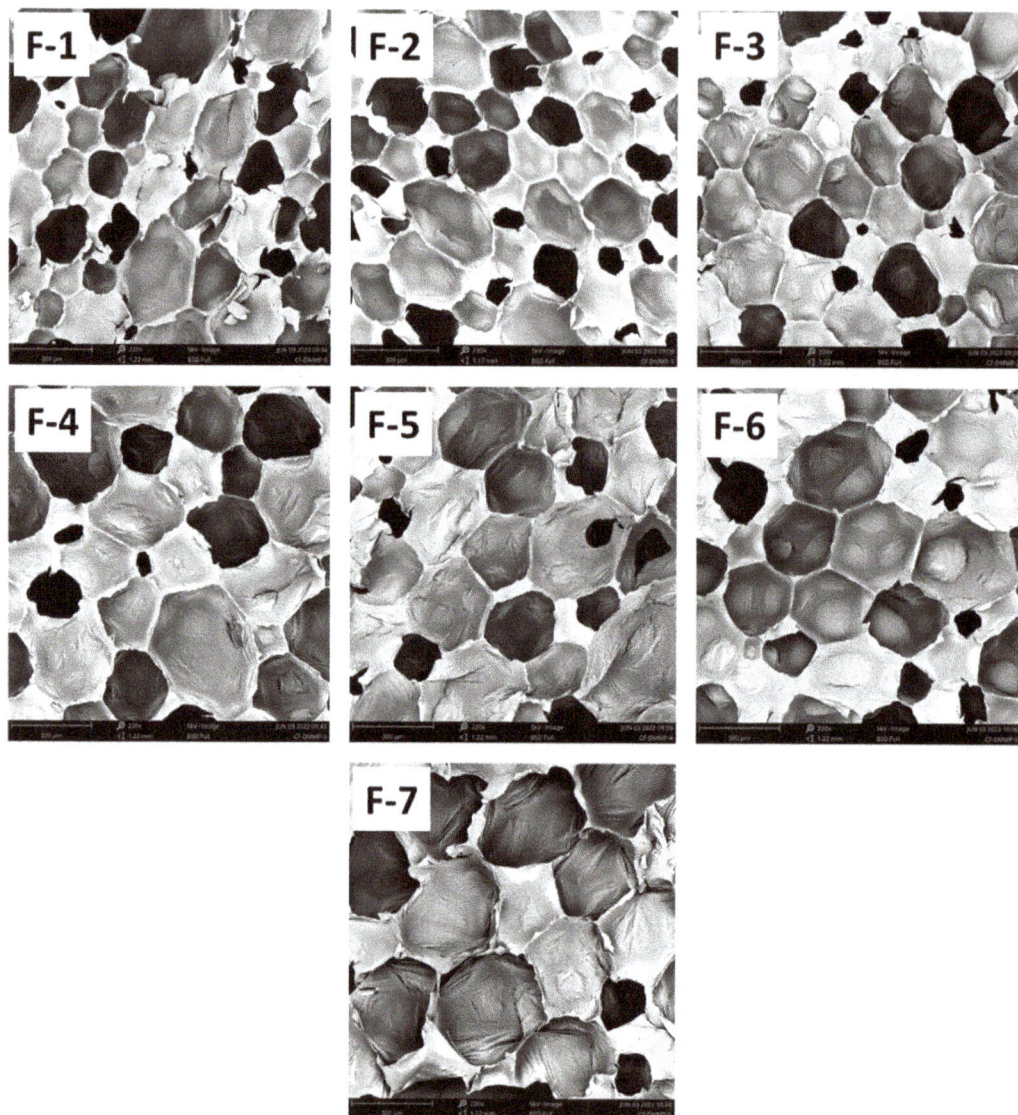

FIGURE 5.13 SEM images of polyurethane foams having varying amounts of DMMP. The scale bar is 300 μm.

Figure 5.14. As expected, the burnt area decreases with increasing concentration of DMMP, showing an improvement in flame-retardancy properties.

Figures 5.15 and 5.16 present the weight loss (%) and burning time (s), respectively. It is observed that incorporation of DMMP into the CF-based RPUF promoted a drastic decrease in weight loss as the control sample (F-1) lost around 60% of its initial weight, whereas the sample F-7 lost only about 10%. The same pattern was observed in terms of burning time as it decreased from around 87 s (F-1) to around 10 s (F-7). The addition of DMMP was effective in improving the flame retardancy of the CF-based RPUF.

Based on the presented results, it seems important to understand the FR mechanism related to DMMP. During the combustion of a PU, there is the release of heat and highly reactive radical fragments, such as H• and OH•. Once these reactive fragments make contact with the PU matrix,

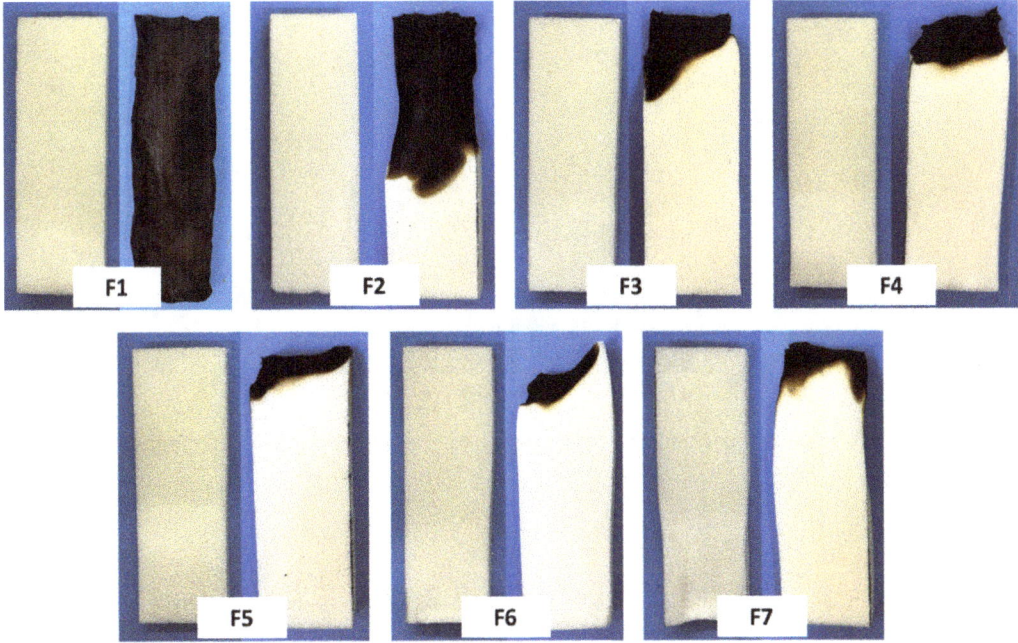

FIGURE 5.14   Digital photos of the polyurethane foams before and after the horizontal burning test.

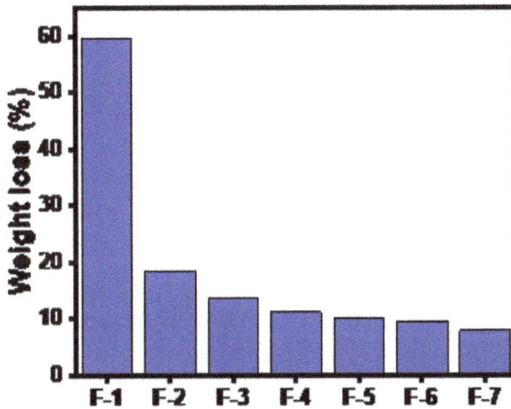

FIGURE 5.15   Weight loss of polyurethane foams having varying amounts of DMMP during the horizontal burning test.

there is the release of heat which further catalyzes the fire, along with decomposing of the polymer. The FR active material should interrupt that process by removing at least one of the fire agents which are heat, oxidant agent (oxygen and reactive radical fragments), or fuel (combustible polymeric matrix). The fire-quenching process can be divided into solid-phase or gas-phase mechanisms [29–32]. The solid-phase mechanism consists of a material that can form a solid and compact protective layer that prevents the diffusion of oxygen or reactive species, causing the flame to be extinguished. Some examples of materials that act through solid-phase mechanisms are EG, metal oxides, metal hydroxides, and minerals, among others. The gaseous-phase mechanism is performed by materials that, when exposed to fire, promote the release of gaseous fragments that can dilute the oxygen and reactive species from the neighborhood.

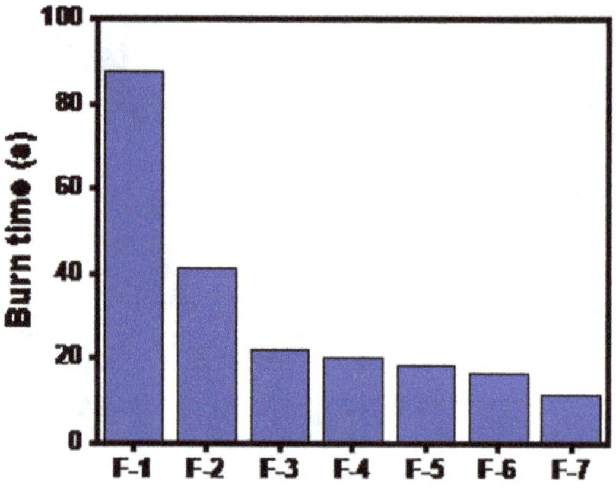

FIGURE 5.16 Burn time of polyurethane foams having varying amounts of DMMP during the horizontal burning test.

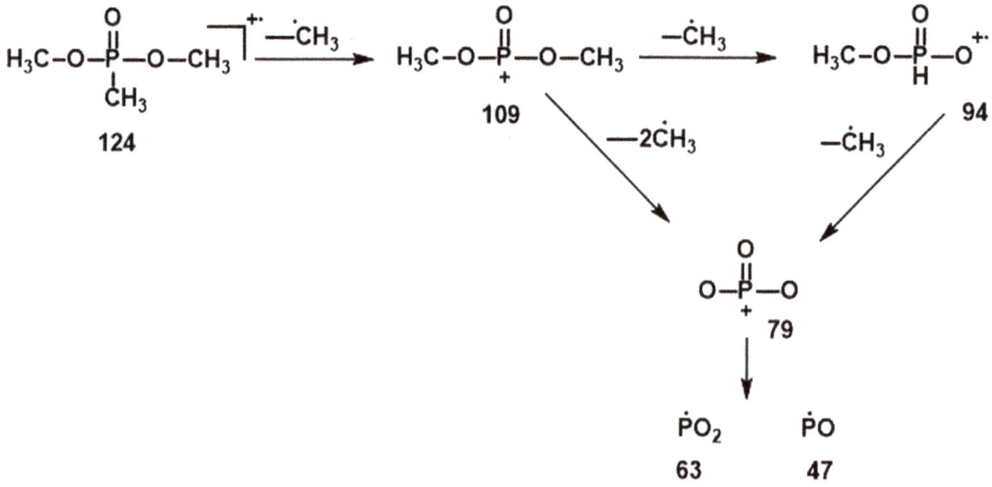

FIGURE 5.17 Thermal decomposition mechanism of DMMP. Adapted with permission [26]. Copyright (2013), John Wiley and Sons.

Some examples of materials that act mostly through gas-phase mechanisms are organophosphates, melamine, and halogen-based FR. Within this type of mechanism, some compounds can decompose into gaseous radical fragments that react with the H˙ and OH˙ species derived from the PU's combustion before they can react with the PU matrix and further catalyze the fire due to the high exothermicity of their reaction. Through this effect, the fire can be quickly quenched as the oxidative agents are removed from the system. This type of radical scavenger mechanism is known to be performed by DMMP as it releases PO_2˙ and PO˙ components which are the species responsible for capturing the reactive fragments such as H˙ and OH˙ that are formed during the PU's combustion. The reaction mechanism of the thermal decomposition of DMMP is presented in Figure 5.17. At the same time, DMMP can also generate phosphoric acid during its thermal decomposition which can promote a dehydration reaction on the PU matrix leading to the formation of a double bond. Under the heat, this double bond can promote cross-linking with the neighboring polymer chains, leading to the formation of a carbonaceous char layer. Even though this effect is known to be less prominent than the release of radical scavengers, it still aids in fire quenching [26,33,34].

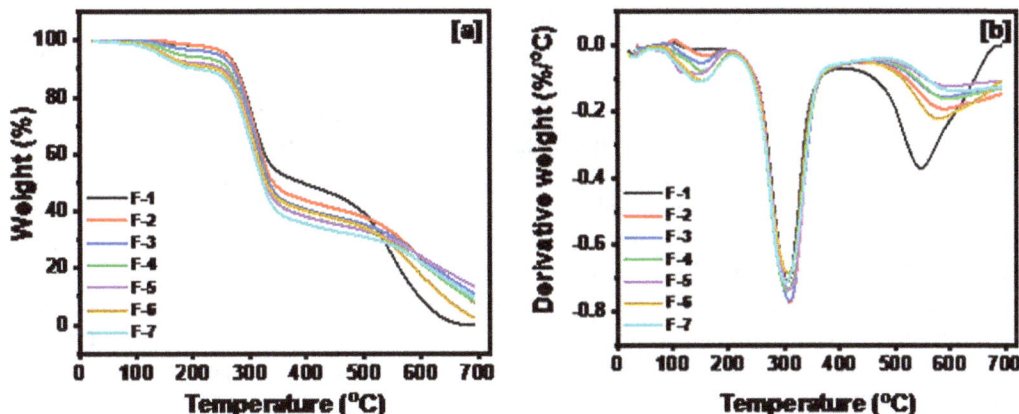

FIGURE 5.18   (a) TGA and (b) derivative TGA plots of polyurethane foams having different amounts of DMMP.

The thermal degradation behavior of RPUF provides valuable information in terms of the flame retardancy efficiency and thermal stability of the PU. Thermal gravimetric analysis (TGA) of the CF-based RPUF, with increasing concentrations of DMMP, was performed. Based on the thermal degradation studies of RPUFs it has been consistently reported in the literature that their decomposition occurs in a two- or three-step process [35]. Usually, the first degradation step consists of the disruption of the urethane linkage back into polyol and isocyanate along with the possible formation of amines, unsaturated fragments, and CO_2. In addition to that, if there is a higher number of soft segments derived from the polyol then the rate of the first degradation tends to decrease [28,35]. Following that, the second and third degradation steps involve the decomposition of the soft segments. Factors such as branching and tridimensional arrangement may influence the thermal decomposition process. Based on these concepts, Figure 5.18 displays the TGA analysis for the CF-based RPUF. Two major thermal degradations can be observed, where the first degradation occurred at around 150 °C, which is related to the partial disruption of the urethane bonds [26]. The second thermal degradation occurred at around 300 °C. It is noted that the control sample (F-1) presented a relatively milder degradation process leading to a smaller weight loss compared to the samples with increasing amounts of DMMP. Hence, the second thermal degradation could be attributed to the partial evaporation and/or pyrolysis of DMMP, since its boiling temperature is around 181 °C [36]. Because of that, gaseous species from DMMP can be released, which is correlated with the more abrupt weight loss. Yet, during the third thermal degradation step which might be attributed to the complete disruption of the RPUF, it is observed that the samples containing higher quantities of DMMP presented a higher residue percentage. This effect suggests that the gaseous species from DMMP could perhaps induce the formation of a carbonaceous char layer that would partially prevent the degradation of the RPUF. In this sense, DMMP would perform some action in the solid phase to prevent further decomposition of the CF-based RPUF. Thus, through the TGA analysis of the CF-based RPUF, it was noted that the addition of DMMP induced some improvement in the thermal stability as was expected based on the results obtained from the horizontal burning test.

6   CONCLUSION AND PERSPECTIVE

Throughout this chapter, it has been discussed that PU is an extremely versatile polymer that can cover a broad range of applications in several sectors including construction, footwear, automotive, coating protection, packaging, and biomedical, among others. Most commercial PUs are prepared using non-renewable sources, and implementing renewable sources for their preparation is, in general, a challenging process as it requires that the final product must be as competitive as the already established petrochemical-based materials. An attractive factor for research in the case of PU is its versatility in terms of its manufacturing, processing, synthetic routes, and materials that can be employed, mostly for the synthesis of polyols. Some of the discussed strategies such as epoxidation/ring-opening, hydroformylation/reduction, ozonolysis/reduction, and thiol-ene coupling, among others, are among the technologies that can be performed to introduce hydroxyl groups into unsaturated compounds. Even though few of these methods can be practically used on a large scale, there is some versatility in each process since a starting material may yield a considerably different polyol based on the synthetic procedure that is adopted.

As discussed in this chapter, there are several types of bio-based materials that can be chemically converted into polyols, such as vegetable oils derived from soybean, corn, sunflower, and castor plant, alongside terpenes such as limonene, carvone, myrcene, and others. This broad range of bio-based materials, as well as methodologies, provides a vast ground of research for the scientific community which has led to fruitful results that have made it into the market. One of the bio-based materials that has been seldomly explored within the field of PUs is animal waste such as chicken fat. Based on the discussion presented in this chapter it should be noted that the synthesized CF-based polyol served as a viable material for the making of FR RPUFs as it delivers satisfactory properties. The incorporation of this type of waste material showed novel possibilities for using animal waste by the polyurethane industry. In addition, satisfactory flame-retardancy properties can be easily introduced into CF-based RPUFs through blending of the FR. This also opens a broad range of possibilities to obtain bio-based materials that can be incorporated with virtually any type of FR material aside from P-based, which also includes C-based, N-based, intumescent, minerals, metal oxides, metal hydrides, metal hydroxides, among others. Thus, based on the results and examples presented in this chapter, it is notable that there are a plethora of possibilities in terms of novel materials that can be employed for the synthesis of polyols to make PU foams with satisfactory performance. Since the industry has been trying to shift toward the use of renewable sources, utilizing bio-waste materials, such as chicken fat, could lead to value-added products and offers interesting possibilities in that regard.

ACKNOWLEDGMENTS

The authors are grateful to the National Institute of Standards and Technology (NIST award number 70NANB20D146) and the U.S. Economic Development Administration (US-EDA award number 05-79-06038) for providing research infrastructure funding.

REFERENCES

[1] F.M. de Souza, M. Arnce, R.K. Gupta, Enhanced synergistic effect by pairing novel inherent flame-retardant polyurethane foams with nanolayers of expandable graphite for their applications in automobile industry, in: H. Song, T. Nguyen, G. Yasin, N. Singh, R. Gupta (Eds.), Nanotechnol. Automot. Ind., Elsevier, Amsterdam, Netherlands, 2022.

[2] F.M. de Souza, P.K. Kahol, R.K. Gupta, Introduction to Polyurethane Chemistry, in: R.K. Gupta, P.K. Kahol (Eds.), Polyurethane Chem. Renew. Polyols Isocyanates, American Chemical Society, Washington, D.C., 2021: p. 1.

[3] F.M. de Souza, R.K. Gupta, Waterborne Polyurethanes for Corrosion Protection BT - Sustainable Production and Applications of Waterborne Polyurethanes, in: Inamuddin, R. Boddula, A. Khan (Eds.), Springer International Publishing, Cham, 2021: pp. 1–27.

[4] Z.S. Petrović, Polyurethanes from vegetable oils, Polym. Rev. 48 (2008) 109.

[5] M.R. Islam, M.D.H. Beg, S.S. Jamari, Development of vegetable-oil-based polymers, J. Appl. Polym. Sci. 131 (2014) 9016–9028.

[6] F. M. de Souza, J. Choi, S. Bhoyate, P.K. Kahol, R.K. Gupta, Expendable Graphite as an Efficient Flame-Retardant for Novel Partial Bio-Based Rigid Polyurethane Foams, C — J. Carbon Res. 6 (2020) 27.

[7] A. Lee, Y. Deng, Green polyurethane from lignin and soybean oil through non-isocyanate reactions, Eur. Polym. J. 63 (2015) 67–73.

[8] S. Ramanujam, C. Zequine, S. Bhoyate, B. Neria, P. Kahol, R. Gupta, Novel Biobased Polyol Using Corn Oil for Highly Flame-Retardant Polyurethane Foams, C. 5 (2019) 13.

[9] A. Guo, Y. Cho, Z.S. Petrović, Structure and properties of halogenated and nonhalogenated soy-based polyols, J. Polym. Sci. Part A Polym. Chem. 38 (2000) 3900–3910.

[10] M. Zieleniewska, M. Auguścik, A. Prociak, P. Rojek, J. Ryszkowska, Polyurethane-urea substrates from rapeseed oil-based polyol for bone tissue cultures intended for application in tissue engineering, Polym. Degrad. Stab. 108 (2014) 241–249.

[11] Z.S. Petrović, A. Guo, I. Javni, I. Cvetković, D.P. Hong, Polyurethane networks from polyols obtained by hydroformylation of soybean oil, Polym. Int. 57 (2008) 275–281.

[12] C.Q. Zhang, Y. Xia, R.Q. Chen, S. Huh, P.A. Johnston, M.R. Kessler, Soy-castor oil based polyols prepared using a solvent-free and catalyst-free method and polyurethanes therefrom, Green Chem. 15 (2013) 1477.

[13] A. Zlatanić, C. Lava, W. Zhang, Z.S. Petrović, Effect of structure on properties of polyols and polyurethanes based on different vegetable oils, J. Polym. Sci. Part B Polym. Phys. 42 (2004) 809–819.

[14] Z.S. Petrović, W. Zhang, I. Javni, Structure and properties of polyurethanes prepared from triglyceride polyols by ozonolysis, Biomacromolecules. 6 (2005) 713–719.

[15] C.E. Hoyle, T.Y. Lee, T. Roper, Thiol-enes: Chemistry of the past with promise for the future, J. Polym. Sci., Part A Polym. Chem. 42 (2004) 5301–5338.

[16] R.K. Gupta, M. Ionescu, D. Radojcic, X. Wan, Z.S. Petrovic, Novel Renewable Polyols Based on Limonene for Rigid Polyurethane Foams, J. Polym. Environ. 22 (2014) 304–309.

[17] N. Arastehnejad, F. De Souza, R.K. Gupta, Highly flame-retardant and efficient bio-based polyurethane foams via addition of melamine-based intumescent flame-retardants, in: S. Kanwar, A. Kumar, T.A. Nguyen, S. Sharma, Y. Slimani (Eds.), Biopolym. Nanomater., Elsevier, 2021: pp. 497–515.

[18] A. Guo, I. Javni, Z. Petrovic, Rigid polyurethane foams based on soybean oil, J. Appl. Polym. Sci. 77 (2000) 467–473.

[19] Y. Lu, R.C. Larock, Soybean oil-based, aqueous cationic polyurethane dispersions: Synthesis and properties, Prog. Org. Coatings. 69 (2010) 31–37.

[20] S. Bhoyate, M. Ionescu, D. Radojcic, P.K. Kahol, J. Chen, S.R. Mishra, R.K. Gupta, Highly flame-retardant bio-based polyurethanes using novel reactive polyols, J. Appl. Polym. Sci. 135 (2018) 46027.

[21] M.F. Sonnenschein, Polyurethanes: Science, technology, markets, and trends, John Wiley & Sons, New Jersey, United States, 2014.

[22] J.O. Akindoyo, M.D.H. Beg, S. Ghazali, M.R. Islam, N. Jeyaratnam, A.R. Yuvaraj, Polyurethane types, synthesis and applications-a review, RSC Adv. 6 (2016) 114453–114482.

[23] M. Thirumal, D. Khastgir, N.K. Singha, B.S. Manjunath, Y.P. Naik, Effect of foam density on the properties of water blown rigid polyurethane foam, J. Appl. Polym. Sci. 108 (2008) 1810–1817.

[24] P. Mondal, D. V Khakhar, Hydraulic resistance of rigid polyurethane foams. II. Effect of variation of surfactant, water, and nucleating agent concentrations on foam structure and properties, J. Appl. Polym. Sci. 93 (2004) 2830–2837.

[25] J.C. Grumo, L.J.Y. Jabber, A.A. Lubguban, R.Y. Capangpangan, A. Alguno, Synthesis and Characterization of Bio-Based Rigid Polyurethane Foams with Varying Amount of Blowing Agent, Key Eng. Mater. 803 (2019) 346–350.

[26] F. Feng, L. Qian, The flame retardant behaviors and synergistic effect of expandable graphite and dimethyl methylphosphonate in rigid polyurethane foams, Polym. Compos. 35 (2014) 301–309.

[27] F. Liu, X. Ding, Y. Su, Properties of rigid polyurethane foams produced by the addition of phosphorus compounds, Am. J. Mater. Res. 1 (2014) 14–19.

[28] C. Zhang, S. Bhoyate, M. Ionescu, P.K. Kahol, R.K. Gupta, Highly flame retardant and bio-based rigid polyurethane foams derived from orange peel oil, Polym. Eng. Sci. 58 (2018) 2078–2087.

[29] F.M. de Souza, R.K. Gupta, P.K. Kahol, Recent Development on Flame Retardants for Polyurethanes, in: R.K. Gupta, P.K. Kahol (Eds.), Polyurethane Chem. Renew. Polyols Isocyanates, American Chemical Society, Washington, 2021: pp. 187–223.

[30] Y. jin Chung, Y. Kim, S. Kim, Flame retardant properties of polyurethane produced by the addition of phosphorous containing polyurethane oligomers (II), J. Ind. Eng. Chem. 15 (2009) 888–893.

[31] J. Troitzsch, Flame Retardants., Kunststoffe – Ger. Plast. 77 (1987) 90–91.

[32] P.M. Visakh, A.O. Semkin, I.A. Rezaev, A. V Fateev, Review on soft polyurethane flame retardant, Constr. Build. Mater. 227 (2019) 116673.

[33] Z. Xu, L. Duan, Y. Hou, F. Chu, S. Jiang, W. Hu, L. Song, The influence of carbon-encapsulated transition metal oxide microparticles on reducing toxic gases release and smoke suppression of rigid polyurethane foam composites, Compos. Part A Appl. Sci. Manuf. 131 (2020) 105815.

[34] S. Bhoyate, M. Ionescu, P.K. Kahol, R.K. Gupta, Castor-oil derived nonhalogenated reactive flame-retardant-based polyurethane foams with significant reduced heat release rate, J. Appl. Polym. Sci. 136 (2019) 1–7.

[35] D.K. Chattopadhyay, D.C. Webster, Thermal stability and flame retardancy of polyurethanes, Prog. Polym. Sci. 34 (2009) 1068–1133.

[36] A. Lorenzetti, M. Modesti, S. Besco, D. Hrelja, S. Donadi, Influence of phosphorus valency on thermal behaviour of flame retarded polyurethane foams, Polym. Degrad. Stab. 96 (2011) 1455–1461.

6 Self-Healing Polymers

Giulio Malucelli

Department of Applied Science and Technology and Local INSTM Unit, Politecnico di Torino, Viale Teresa Michel 5, 15121, Alessandria, Italy

1 INTRODUCTION

Despite their specific features that justify wide use in several application fields, most either structural or functional polymeric materials, as time goes by, show an irreversible trend toward aging and deterioration, as they are usually subjected to mechanical and environmental stresses (such as changes of chemical structure and morphology, erosion, corrosion, creep, crack formation, and wear, among a few to mention). These phenomena may be very dangerous, as the degraded polymers usually cannot be employed for the same use they were designed for. Conversely, nature shows several good examples regarding the capability of some living materials to heal moderate damage, hence recovering the envisaged functionality: in particular, it is well known that bone cracks can be repaired as the bone materials regenerate; further, a skin cut can be mended by blood flowing into the lesion and by coagulation of clots, which seal the wound, thus triggering the repair mechanism [1–3]. In this context, several efforts have been carried out in the last 15–20 years, to provide man-made synthetic polymer systems with self-healing capabilities: for this purpose, many different strategies have been designed and successfully exploited, hence making real the self-healing concept [4].

Self-healing has also profitably been applied to metals and ceramics; however, at present polymeric materials represent the main target [5]. Although the self-healing concept dates back to the 1950s, one of its first achievements only occurred in 2001, when White and co-workers set the first (and most famous) self-repairing system, based on the use of microcapsules embedding a repairing agent and homogeneously distributed into a composite (i.e., exploiting so-called "autonomic healing" [6]). Because of the crack propagation, the microencapsulated healing agent can diffuse within the polymer matrix and polymerize after having interacted with the catalyst, also dispersed in the composite system, hence adhering to the crack faces. An overall scheme of the autonomic repairing concept [7] is depicted in Figure 6.1.

Later, hollow glass fibers (Figure 6.2) started to be exploited as suitable reservoirs for containing a repairing agent [8], which, because of damage that induces breakage of the fibers, is able to diffuse within the damaged area, hence initiating the self-healing process, similarly to some biological self-repairing systems [9].

From the first pioneering studies, remarkable progress in self-healing in polymer systems, as also witnessed by the continuously increasing number of publications in peer-reviewed journals (Figure 6.3), has been made. This chapter aims at classifying the different self-healing polymer systems, explaining the mechanisms behind the design, preparation, and characterization of highly efficient self-healing materials, and providing the reader with some interesting applications. Finally, some perspectives as far as possible on further advances for the near future will be briefly presented.

DOI: 10.1201/9781003278269-6

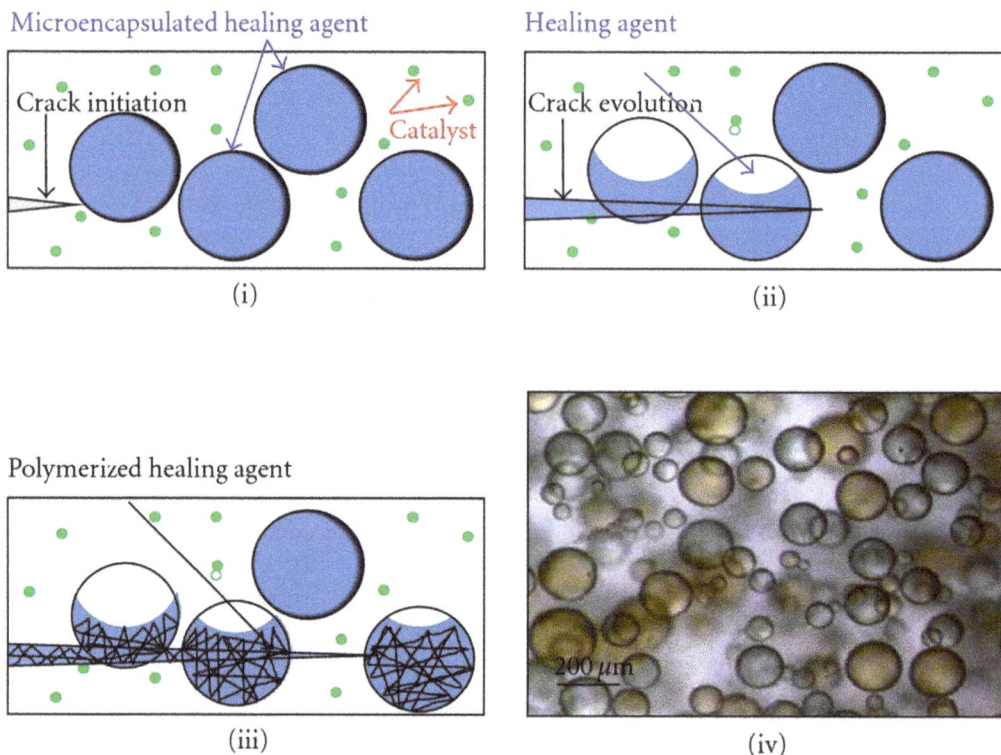

Microencapsulated healing agent

Crack initiation

Catalyst

(i)

Healing agent

Crack evolution

(ii)

Polymerized healing agent

(iii)

200 μm

(iv)

FIGURE 6.1 The self-healing concept through microencapsulation. A microencapsulated repairing agent is embedded in a structural composite matrix containing a catalyst able to polymerize the repairing agent. (i) Cracks form in the matrix wherever damage takes place; (ii) the crack breaks the microcapsules, which, in turn, release the healing agent into the crack plane through capillary action; (iii) the healing agent interacts with the catalyst, initiating polymerization that bonds together the crack faces. Reprinted from Reference [7]. Copyright 2012, Hindawi Publishing Corporation. Distributed under a Creative Commons Attribution License 4.0 (CC-BY).

2   CLASSIFICATION OF SELF-HEALING POLYMER SYSTEMS

From a very general perspective, self-healing may be autonomic (when it takes place without any external intervention/triggering) or nonautonomic (when human intervention or external triggering is mandatory for activating the repairing process). In addition, it is worth noting that each different type of polymeric system possesses its own peculiar self-healing mechanisms; therefore, various approaches can be successfully exploited for the design and preparation of self-healing materials, namely: (i) release of repairing agents; (ii) use of reversible cross-links; or (iii) exploitation of "miscellaneous" methods, including migration of nanoparticles, conductivity, and electrohydrodynamics, among a few.

2.1   RELEASE OF REPAIRING AGENTS

This autonomic mechanism exploits the use of hollow fibers, microcapsules, or channels, which act as reservoirs of liquid active agents (i.e., monomers and hardeners) necessary to carry out the healing process; these reservoirs are inserted into the polymer systems during their manufacturing, and break during a crack propagation; because of the presence of capillary forces, the liquid active

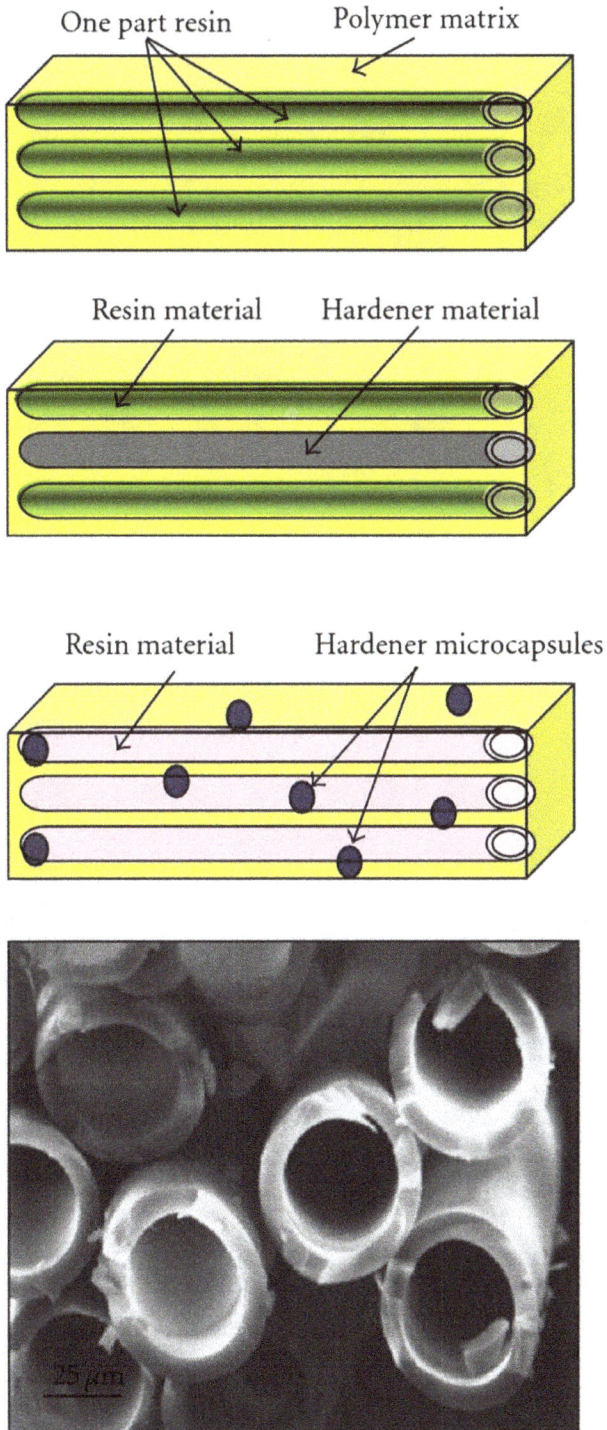

FIGURE 6.2 The self-healing concept through healing hollow fibers. The exploitation of single-part, two-part resin, and hardener, or resin with a catalyst/hardener. Reprinted from Reference [7]. Copyright 2012, Hindawi Publishing Corporation. Distributed under a Creative Commons Attribution License 4.0 (CC-BY).

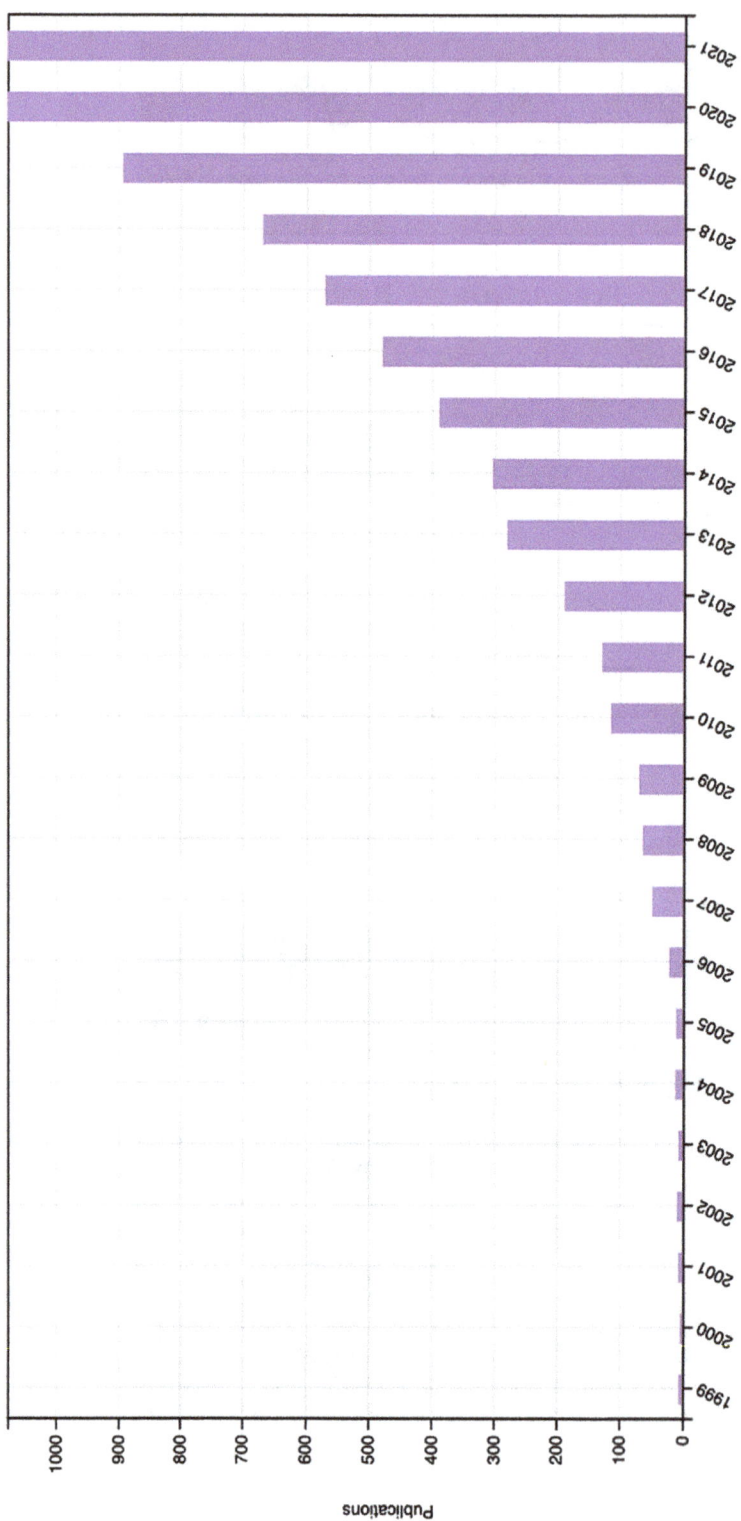

FIGURE 6.3 Number of publications (from 1999 to 2021) in peer-reviewed journals, dealing with self-healing polymer systems. From Web of Science™ database, accessed on 27 December 2021.

agents are released into the crack faces, and solidify when becoming into contact with a suitable catalyst, previously dispersed into the polymer matrix, and finally repair the cracks. It is noteworthy that the main driving force of the mending process refers to the crack propagation: indeed, without this latter factor, it is not possible at all to trigger the repairing process, as the healing liquid agents cannot be released in the damaged site. Conversely, suitable stresses should be freed from the cracks, this phenomenon, undoubtedly, may limit the effectiveness of the healing process.

2.1.1   Microencapsulation

Microencapsulation is a very well-known approach suitable for obtaining microcapsules, i.e., spherical (or even irregular) structures consisting of a core and a shell, extending from about one micron to several hundred microns in size and usually embedding solid microparticles or liquid droplets [10].

Specifically referring to the design of self-healing polymer systems, the first pioneering works dealt with the utilization of microcapsules containing self-repairing agents and dispersed into a polyester [11]; however, these first attempts failed to obtain efficient self-healing polymer systems, and it took some time (until 2001) to gather a practical demonstration of the self-repairing effectiveness of microcapsules in an epoxy network [4]. To this aim, dicyclopentadiene (DCPD) was chosen as the microencapsulated liquid repairing agent, and bis(tricyclohexylphosphine) benzylidine ruthenium (IV) dichloride (i.e., the Grubbs' catalyst) was dispersed into the epoxy system; this self-healing process is based on the occurrence of a ring-opening metathesis polymerization (ROMP) [12] involving DCPD and the catalyst, which gives rise to the formation of cross-linked polycyclopendiene between the propagating cracks. In a further research effort [13], the catalyst (i.e., di-n-butyltin dilaurate) was encapsulated into polyurethane microcapsules, while phase-separated droplets made of polydiethoxysiloxane and hydroxyl-terminated polydimethylsiloxane were embedded in a vinyl ester system. Because of the crack propagation, the catalyst is released from the microcapsules and triggers the polycondensation self-healing reaction.

The main key factors affecting the self-repairing processes based on microcapsules are listed in Table 6.1.

2.1.2   Utilization of Hollow Fibers

One of the major limitations concerning the use of microencapsulated healing systems relies on the limited amount of the employed repairing agents, as well as on the impossibility to control the completeness and/or repeatability of the repairing process when multiple (i.e., sequential) damages occur in the same polymer matrix. For these reasons, some progress has been made for the design of more effective self-healing polymer systems, thanks to the setup of new types of reservoirs of healing agents that are able to release larger amounts of the latter, hence allowing multiple repairing processes. The first attempts, though generally unsuccessful because of either the high viscosity of the employed healing agents or the slow curing process, were proposed by different research groups [14]. However, further enhancements were then obtained, optimizing the fabrication of hollow borosilicate glass fibers (diameter: 30–100 μm; hollowness: up to 55%; an example is shown in Figure 6.4), hence achieving acceptable self-healing capabilities [15]. More specifically, after healing it was possible to recover up to 97% of the initial flexural strength of composite boards when hollow fibers embedding the repairing agent were employed. The pros and cons of this self-healing strategy are collected in Table 6.2.

2.1.3   Utilization of Microvascular Systems

Several animals and plants exploit vascular systems to effectively heal damaged areas: this finding has suggested replicating these biological facilities for designing self-repairing microvascular structures [16]. A microvascular system distributes the repairing agents using a centralized network,

TABLE 6.1

Key Factors for the Development of Microencapsulated Self-Repairing Polymer Systems

Parameter	Key Factors Involved
Monomers	• Their viscosity should be low, as they have to flow within the crack surfaces • Their volatility should be limited, in order to provide enough time for the pol ymerization reactions
Catalysts	• They should be soluble in the monomers • They should not agglomerate with the polymer matrices
Microcapsules	• They should not react with the encapsulated repairing agents • They should exhibit an acceptable self-existence • They should be easily dispersed within the polymer matrix • Their shell wall should be weak enough to break when cracks propagate; therefore, strong interactions with the surrounding polymer matrix are needed, in order to transfer the stresses to the microcapsules and allow them to be broken • They should easily get in contact with the catalyst in the damaged areas
Polymerization reactions	• They should be very fast for a quick healing process • Possibly, they should occur at ambient temperature • They should not promote any shrinkage or stress relaxation after healing, in order to obtain stable repaired systems
Healing	• It should be as fast as possible • It should allow multiple repairing processes • It should be cost-effective

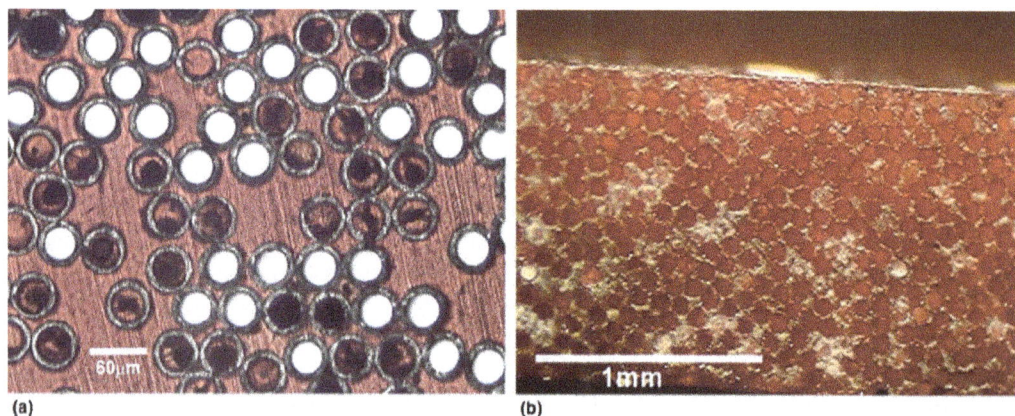

FIGURE 6.4 Optical micrographs of hollow glass fiber-reinforced composites: (a) hollow glass fibers (60 μm external diameter and 50% hollowness) and (b) the same fibers within an epoxy matrix. Adapted with permission from Reference [15]. Copyright, 2005 Elsevier.

fabricated on purpose: to this aim, a 3D array is built using selected organic inks, then an epoxy resin is infiltrated in the pores located between the printed lines of the 3D array and is subsequently cured. After curing, the organic ink is removed, giving rise to the formation of the microvascular system that can be exploited as a reservoir of the healing agents [17].

TABLE 6.2

Advantages and Drawbacks Concerning the Development of Hollow Fiber Self-Repairing Polymer Composites

Pros	Cons
Availability of high volumes of repairing agent	Multistep fabrication processes of the composites are necessary
Possibility of using various types of resins	Hollow fiber infiltration requires healing agents with low viscosity
Feasibility of visual inspection of damaged sites	The repairing agents are released only when fibers brake
Possibility of mixing and tailoring hollow fibers with conventional reinforcements	The coefficient of thermal expansion of hollow glass fibers differs from that of carbon fibers, hence leading to a certain incompatibility when self-healing glass hollow fibers are employed in carbon fiber-reinforced composites

2.2 USE OF REVERSIBLE CROSS-LINKS

Cross-linking is a permanent process, which allows enhancing the mechanical behavior and chemical resistance of different polymer systems. Apart from these advantages, the formation of cross-links (hence a polymer network) usually represents an obstacle to reprocessing the polymers, which, at the same time, may become quite brittle, significantly losing their toughness. Introducing reversible cross-links is not only an interesting strategy to facilitate reprocessing and make the related polymer systems recyclable, but it can also be exploited for the design and fabrication of nonautomatic self-healing systems. The following subsections summarize the current state-of-the-art related to reversible cross-linked self-repairing polymers.

2.2.1 Diels–Alder and Retro-Diels–Alder Reactions

Diels–Alder (DA) reactions are 4 + 2 cycloadditions usually exploited for designing thermally reversible polymer networks, such as polymers bearing maleimide pendant groups at low temperature or maleimide-containing furanic polymers. Conversely, retro-Diels–Alder reactions, taking place at high temperatures, allow breaking of the formed bonds, hence reversing the cross-linking processes. The first pioneering studies about the possibility of exploiting these types of reactions for self-healing polymer systems were proposed by Chen and co-workers [18], who demonstrated the achievement of 80% strength recovery in maleimide-containing furanic polymers (Figure 6.5).

Pursuing the research, Liu and co-workers [34] employed epoxy precursors for synthesizing trimaleimide and trifuran compounds able to undergo Diels–Alder and retro-Diels–Alder reactions, hence confirming the suitability of the designed systems for self-healing purposes. While the retro-Diels–Alder reactions were found to occur by heating the specimens at 170°C for 0.5 h, the full-scale mending was achieved in mild conditions (24 h at 50°C). The first example of light-induced crack repairing dates to 2004, when Chung and co-workers exploited [2 + 2] photochemical cycloaddition of cinnamoyl groups to attain self-mending features [20]. In particular, 1,1,1-tris-(cinnamoyloxymethyl)ethane, an UV-curable cinnamate monomer, was synthesized starting from 1,1,1-tris(hydroxymethyl)ethane and cinnamoyl chloride. The self-repairing capabilities were demonstrated by irradiating the samples with UV above 280 nm, as shown in Figure 6.6.

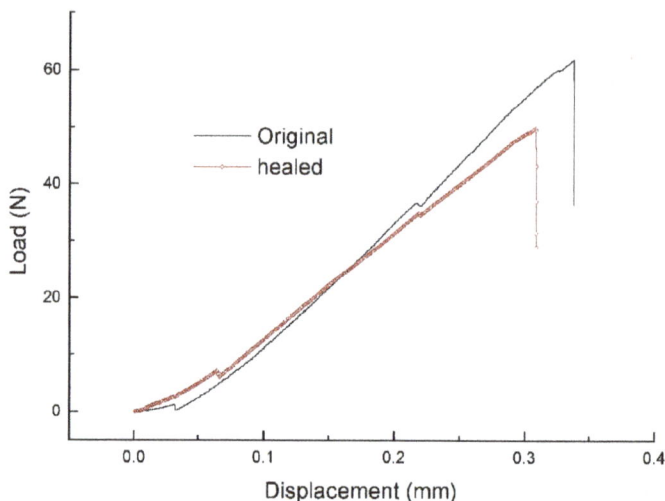

FIGURE 6.5 A typical load vs. displacement diagram of fracture toughness testing of compact tension test specimens of a maleimide-containing furanic polymer. The original and healed fracture toughness (80% strength recovery) were determined by the propagation of the starting crack along the middle plane of the specimen at the critical load. Adapted with permission from Reference [18]. Copyright 2003, American Chemical Society.

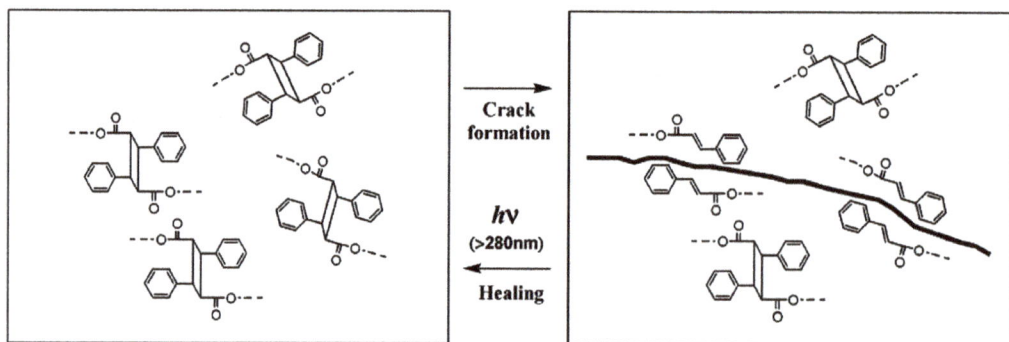

FIGURE 6.6 Scheme of the crack healing process. Self-healing is performed using the re-cycloaddition of cinnamoyl groups. Adapted with permission from Reference [20]. Copyright 2004, American Chemical Society.

2.2.2 Ionomers

Ionomeric polymers or ionomers are materials with a hydrocarbon backbone and bearing pendant acid groups; the latter can be fully or partially neutralized to salts [21]. Although it may vary over a broad range, the ion content of ionomers does not exceed 15 mol%. Their synthesis can be carried out either by copolymerizing slightly functionalized with unsaturated olefinic monomers or by post-functionalizing saturated pre-formed polymers. Generally, the types of interactions taking place in ionomers rely on electrostatic forces that involve metal cations (transitional metals or elements of Groups 1A and 2A of the periodic table) and such anions as sulfonates or carboxylates. These interactions represent a sort of physical cross-linking (i.e., reversible), which can be successfully exploited for designing self-healing polymer systems [22]: in particular, a special class of thermoplastic poly(ethylene-co-methacrylic acid) ionomers has been identified as an efficient full-scale autonomic and rapid self-repairing material as a consequence of damage

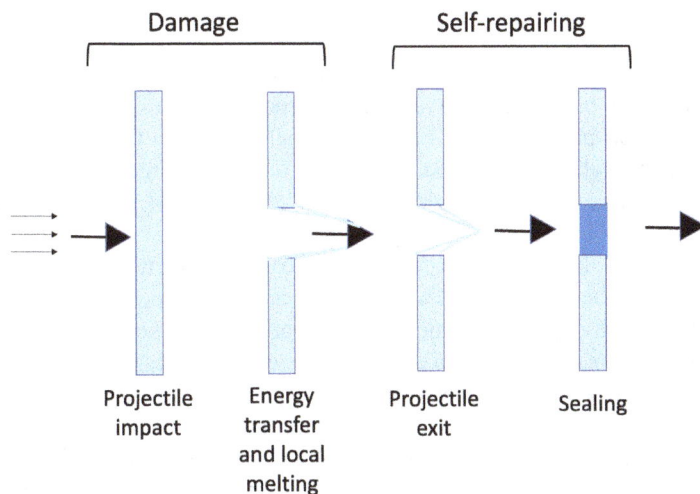

FIGURE 6.7 Scheme of self-healing process because of damage provided by the ballistic puncture. The projectile impacts the self-repairing material and transfers its energy to this latter. The friction and the high deformation, the ionomeric polymer undergoes, allow melting of the puncture area; as a consequence, the projectile passes through the polymer. Then, self-healing occurs because of the high elasticity of the ionomer in the molten state, which favors the bonding of the molten surfaces, hence sealing the damaged area in a very short time.

provided by ballistic puncture [23]. A schematic representation is depicted in Figure 6.7: the projectile impacts the self-repairing material and transfers its energy to the latter. The friction and the high deformation that the ionomeric polymer undergoes allow melting of the puncture area; consequently, the projectile passes through the polymer. Then, self-healing occurs because of the high elasticity of the ionomer in the molten state, which favors the bonding of the molten surfaces, hence sealing the damaged area in a very short time. Therefore, self-repair can be carried out only if the ionomer melts within the damaged area and the melt elasticity of the resulting material is enough to seal the hole.

2.2.3   Supramolecular Interactions

Quite recently, it was possible to design assemblies made of low-molecular-weight monomers or oligomers, which exploit reversible noncovalent interactions (such as hydrogen bonds, metal–ligand, and π–π interactions, among a few), hence giving rise to the formation of materials showing polymer-like mechanical and rheological features [24]. These materials are called "supramolecular polymers" [25], which can be successfully utilized as highly efficient self-healing materials.

One of the first examples of using supramolecular hydrogen bond interactions for the design of self-healing rubbers dates to 2008: in particular, the authors first condensed the acid groups of fatty diacids and triacids derived from renewables with diethylenetriamine, then reacted the condensed acid groups with urea [26]. The so-prepared elastomers showed interesting self-repairing features, as it was possible to heal two cut ends of these materials by simply putting them in contact with a minimum pressure at ambient temperature, without the necessity to use an external heat source. The literature is also rich in good examples dealing with the exploitation of metal–ligand interactions for the design of supramolecular self-repairing polymers. More specifically, the use of terpyridine-based metal–ligand complexes as efficient self-healing systems was successfully demonstrated [27].

2.3   Utilization of "Miscellaneous" Methods

There exist some new approaches for effectively designing self-healing polymer systems. The first strategy, taking advantage of the principle of electrohydrodynamics, mimics the blood-clotting procedure using aggregation phenomena of colloidal particles close to the damaged area, upon the application of an external electric field [28]. Indeed, the resulting increase in current density within the defected area promotes the local agglomeration of the colloidal particles, hence healing the damage.

Other research efforts have been made on electrically conductive self-repairing polymeric materials: as nicely described by Williams and co-workers [29], it is possible to design and produce self-healing systems made of transition metals and organometallic polymers (i.e., N-heterocyclic carbene derivatives). In the case of a damage event, the electrical resistance close to the defected area increases, hence leading to heat generation that is capable of healing the material and restoring the pristine electrical conductivity. In addition, it is worth noting that conductivities as high as 1 S/cm should be reached to practically exploit the designed systems for self-repairing purposes.

Although the approach is still quite far from being efficient and fully demonstrated, the migration of nanoparticles (such as, for example, carbon nanotubes) located in a polymer fluid could be employed to allow their segregation in damaged areas, where attraction forces between the nanoparticles and the crack surfaces take place because of crack formation [30]. In addition, molecular dynamics simulations are usually utilized for predicting the self-healing capabilities of the designed nanoparticle-based systems.

3   APPLICATIONS OF SELF-HEALING POLYMER SYSTEMS

The following subsections summarize the most common application sectors of self-repairing polymers.

3.1   Applications in 3D Printing

One of the current applications of self-healing polymer systems refers to 3D printing, i.e., an additive manufacturing technique that allows depositing successive layers of a material on a substrate, hence giving rise to the formation of a three-dimensional object, even with complex geometry and shape [31]. In particular, among the different materials able to undergo 3D printing processes, selected hydrogels (usually in the form of hydrogel inks) exhibiting self-healing features have been proposed for biomedical applications. These types of hydrogels are especially valuable for biomedical purposes, as they are capable of imitating or replicating the structures of cellular environments and producing architectures suitable for replacing, increasing, or modeling organic tissues. Cytocompatibility is a mandatory requirement. The scientific literature highlights the possibility of using different systems for the design of 3D printable inks, namely based on poly(ethylene glycol), DNA-polypeptide complexes, alginates, and interpenetrating polymer networks, among a few.

Conversely, specifically referring to self-healing materials, there are only a few works on this topic. One concerns the setup of a hydrogel-based 3D printing method that allows printing of a shear-thinning hydrogel (*ink*) directly into a self-healing counterpart (*support*); ink and support hydrogels consist of hyaluronic acid host–guest complexes modified with β-cyclodextrin and adamantine, and stabilized by covalent cross-linking of methacrylates (via UV-curing), added to the system on purpose [32]. The presence of supramolecular interactions in both hydrogel systems allows their use in 3D printing processes, as the applied shear stresses can disrupt the noncovalent and reversible links, hence favoring the layer-by-layer deposition; once the stresses are removed, these links are easily reformed, and the hydrogel structure recovered.

3.2 APPLICATIONS IN CORROSION-RESISTANT COATINGS

Corrosion of metals and metal alloys represents a severe issue, and its direct cost may impact the GDP of each country by up to 5%. Substantial efforts have to be carried out to substantially limit the negative effects of this phenomenon; in this context, the easiest way is to prevent corrosion by using tailored protective coatings (polymeric, inorganic, or metallic), which are capable of protecting the underlying metal substrates from corrosion [33]. Though these coatings are very efficient, they require a good overall knowledge about their correct selection and application on the chosen metal substrate; in addition, the potential defects of these coatings have to be taken into consideration, not-withstanding that their effectiveness could be remarkably reduced or even compromised as a result of external damage, with the appearance of microcracks that can expose the inside metal surface to the oxygen action, hence leading to corrosion phenomena. To this aim, the design of self-repairing autonomic polymer coatings has allowed to remarkably ameliorate the corrosion resistance of metals and metal alloys, as these special coatings are resistant to environmental stresses (both mechanical and chemical) and self-repair when damage occurs [34]. These coatings are usually filled with micro- or nanocapsules containing film-former compounds, able to heal possible damage when the coatings are subjected to scratches.

Nowadays, siloxane-based coatings are widely being utilized as self-healing anticorrosion systems, thanks to their air, water, and thermal stability that allows their use in the case of thermally curable coatings. Two main strategies are currently employed, namely: (i) the dispersion of siloxanes as phase–phase separated drops, while the catalyst is embedded in the microcapsules, or (ii) the use of microcapsules containing siloxanes or the catalyst. This latter approach is usually exploited when the coating matrix can react with the repairing agent [35]. Despite the elastomeric character of cross-linked siloxanes, which provides them with limited mechanical features, this issue does not represent a problem in the formulation of self-healing coatings, as the mechanical behavior of the latter is not of primary importance; in this context, both the exceptional passivating capability and chemical inertness of cross-linked siloxanes are more important.

3.3 APPLICATIONS IN AEROSPACE

The materials designed for aerospace applications have to work in very critical conditions that very often may lead to severe damage; these may not always be easily detected and promptly "manually" repaired, hence weakening the mechanical strength and decreasing the overall safety of the aircraft structures [36]. In this context, the design and utilization of self-healing systems become of crucial importance. The first self-repairing systems that were developed specifically for aircraft operations were based on the use of microencapsulated self-healing agents (such as dicyclopentadiene) dispersed in the polymer matrix (usually an epoxy resin), together with a catalyst (i.e., the Grubb's catalyst) [4]. It is worth highlighting that as aircraft operate between −50 and + 60°C, self-repair should be possible in this temperature range: therefore, the choice of appropriate self-healing systems for aerospace should be taken based on the thermal stability of the self-mending agents (monomers, microcapsules, catalysts), on their chemical stability, as well as on their repairing efficiency at low temperatures [37].

3.4 APPLICATIONS IN ELECTRONICS

Electronic applications, such as robotics, communication devices, artificial intelligence, and personal health monitoring, among a few, take a remarkable advantage from the use of self-repairing systems [38]; in fact, this way it is possible to extend the lifetime of devices, making them more robust and reliable.

Two types of self-repairing materials for electronics are available: one type exploits dynamic reversible bonds, while the other uses the incorporation of microencapsulated healing agents. The main drawback of the latter is in regard to the draining of the repair system after a single repair process, hence making the healing not repeatable should successive damages occur. Conversely, as the utilization of dynamic reversible bonds, such as those taking place in organometallic polymer systems (i.e., as an example, between transition metals and N-heterocyclic carbenes [38]), makes the self-healing process easier and repeatable, it represents the most preferred way for providing self-repairing features to electronic devices.

The scientific literature is rich with different examples of self-healing materials for electronics, which exploit silver nanowires covered by polymer electrolyte multilayers, supramolecular polymers with Ni particles, supramolecular elastomers with Ga-In microchannels, polyurethane systems containing silver paste capsules, and poly(urea-formaldehyde) microcapsules with tetrathiafulvalene-tetracyanoquinodimethane charge-transfer salts, to mention just a few.

3.5 Applications in Tissue Engineering

Developing organ and tissue replacements for restoring, preserving, or even increasing functionality because of possible damage is one of the main goals of tissue engineering. In this high-tech application field, the need for designing and producing biomaterials able to regenerate or replace injured tissues is of primary importance, therefore, self-repairing biomaterials started to be explored and engineered [40]. In particular, supramolecular hydrogels, which usually exploit hydrogen bonding and can self-mend spontaneously or under a physiological trigger, are currently the most promising systems for tissue engineering applications; indeed, they are responsive to different environmental stimuli (i.e., enzymes, pH, and temperature), and they prolong the lifetime of biomaterials, making them the best candidates for withstanding repeated mechanical stress or injection processes [41].

3.6 Applications in Drug Delivery

Very often, drug delivery is carried out by using injectable hydrogel-based systems, which represent a reliable way of controlling the release of different types of therapeutics [42]. Unfortunately, these injectable hydrogels may show some detrimental effects: when their gelation is slow, part of the embedded drug could be released and diffuse from the target location, meanwhile, when gelation is very fast, premature undesired solidification may take place. These issues can be solved by employing self-healing drug-delivery hydrogels that, when employed as carriers, can be injected and self-repaired *in vivo* to accurately distribute drugs or cells at high concentrations in the desired location, preventing unwanted delivery failure or burst release [43]. Among the different possible solutions, collagen–gold hybrid hydrogels, hydrogels based on glycol, chitosan, and difunctional poly(ethylene) glycol, and cross-linked hydrogels from alginate and Ca^{2+} ions have been successfully proposed.

4   RECENT ADVANCES IN SELF-HEALING POLYMER SYSTEMS

In the following, the most recent progress concerning the design and production of self-repairing polymer systems is summarized.

Significant efforts have been made by Gao and co-workers [44], who designed self-repairing polymer systems based on microcapsules containing bisphenol A epoxy acrylate and trimethylolhexane triacrylate as a healing mixture. These microcapsules, exhibiting a dielectric constant higher than that of the polymer matrix (i.e., an epoxy resin), were found to be able to fully recover the dielectric properties of the material when subjected to electrical tree degradation, because of the application of strong electrical stresses. More specifically, the electroluminescent effects *in situ* produced during electrical treeing of the epoxy system were exploited for triggering the cross-linking reactions of the microencapsulated repairing mixture.

Cao et al. [45] recently developed self-repairing epoxy foams containing bi-layered calcium–alginate capsules embedding the healing agents (i.e., epoxy resin and hardeners, encapsulated within the inner and the outer layer, respectively) and obtained using a multi-stage encapsulating process. As assessed by mechanical tests (high-speed soft impact and three-point bending tests) the self-healing effects provided by the bi-layered capsules were satisfactory, even at room temperature, and the presence of the capsules did not provide significant changes in the stiffness of the undamaged materials. Zhang and co-workers [46] recently developed castor oil-derived polyurethane multifunctional materials containing hindered urea bonds. The obtained systems, apart from being processable and with shape-memory features, exhibited very good self-healing properties: upon scratching, it was possible to reduce the crack widths by about 90–100% after a heat treatment performed at 60–100°C for 10 min.

An efficient self-repairing polymer nanocomposite was designed by Guadagno et al., by incorporating carbon nanotube-based carboxyl methylcellulose in highly amorphous vinyl alcohol containing murexide salt [47]. Mild conditions (i.e., 30°C, 80% relative humidity, and 48 h) were enough for the autonomous repair of damaged samples; in addition, as assessed by dynamic-mechanical tests performed on healed specimens, the repairing efficiency showed a temperature-dependence behavior: the storage modulus E' was fully recovered beyond 50°C. Chen and co-workers [48] exploited a novel strategy to produce polymer blends by covalently cross-linking a thermosetting vitrimer (also defined as a covalent adaptable network polymer) bearing dynamic covalent networks with a thermoplastic polyurethane: the resulting material, thanks to bond exchange reactions occurring during the blending process, showed good self-repairing features, as well as reprocessability and flexibility.

Yang and co-workers [49] recently synthesized hydrogels made of poly(acrylic acid) and polypyrrole cross-linked with polyfunctional trypan blue; the resulting materials, suitable for wearable sensing applications, exhibited outstanding self-healing capabilities, as they were able to heal more than 60% of the damaged areas within just 10 s. Song et al. [50] proposed self-healing polymer blends made of polyurethane and poly(dopamine methacrylamide) (as a dispersed phase); the presence of the latter remarkably affected the healing efficiency that achieved about 90% for the blends containing at least 20 wt.% of the dispersed phase.

5 CONCLUSIONS AND FUTURE PERSPECTIVES

Undoubtedly, the fast development recently undergone by the self-repairing polymer systems has determined a strong increase of their potential applications. Several factors contribute to the overall self-healing performances: the proper methods employed for the preparation of the self-healing polymer systems, the type of healing mechanism involved, the type and "reactivity" of the selected healing agents, as well as the healing triggering methods (if needed). Apart from the advantages clearly described in this chapter, the use of self-repairing polymers still has some drawbacks and limitations, for which possible solutions can be foreseen in the coming years.

First, many raw materials employed for synthesizing self-healing polymers are very expensive, and the syntheses are relatively complicated and, in some cases, suitable only at a lab/research scale. Apart from some supramolecular polymer systems that can heal at low temperatures and in very short times, for many self-repairing polymers more severe healing conditions are required (as an example, high temperatures and long reaction times are usually mandatory). In addition, it is worth noting that self-repairing processes generally occur with quite low healing rates: for practical and industrial applications, shorter reaction times, together with high healing efficiencies, should be accomplished.

Another current critical issue is in reference to the worsening of the self-healing performances with multiple repairing having to be carried out on the same material; also, as a consequence of repeated healing processes, the physico-mechanical properties of the repaired materials are prone to deteriorate, which is a critical issue that needs further investigation. It is expected that the self-healing polymers of the future will be suitable for a wider range of applications with respect to the current uses, and that medical devices and equipment, military equipment, transportation industry, electronics and civil engineering will surely take advantage of the further developments in this scientific topic. Therefore, it is expected that these challenging materials will be protagonists in science and technology also in the coming years.

REFERENCES

1. S.C. Burgess, Reliability and safety strategies in living organisms: potential for biomimicking. J. Proc. Inst. Mech. Eng. 216 (2002) 1–13.
2. M. Sarikaya, I.A. Aksay, Biomimetics: design and processing of materials. AIP Series in polymers and complex materials. AIP, Woodbury, New York (1995).
3. P. Fratzl, Biomimetic materials research: What can we really learn from nature's structural materials? J. R. Soc. Interface 4 (2007) 637–642.
4. M.Q. Zhang, M.Z. Rong, T. Yin, Self-healing polymers and polymer composites. In: Ghosh, S. K. (ed.) Self-healing materials: fundamentals, design strategies, and applications. pp. 29–72. Wiley WCH, Weinheim (2009).
5. H.M. Anderson, M.W. Keller, J.S. Moore, H.R. Sottos, S.R. White, Self-healing polymers and composites. In: van der Zwaag, S. (ed.) Self-healing materials – an alternative approach to 20 centuries of materials science. pp. 19–44. Springer, Dordrecht, The Netherlands (2007).
6. S.R. White, N.R. Sottos, P.H. Geubelle, J.S. Moore, M.R. Kessler, S.R. Sriram, E.N. Brown, S. Viswanathan, Autonomic healing of polymer composites. Nature, 409 (2001) 794–797.
7. B. Aïssa, D. Therriault, E. Haddad, W. Jamroz, Self-healing materials systems: Overview of major approaches and recent developed technologies. Adv. Mater. Sci. Eng. (2012) Article ID 854203.
8. R.S. Trask, G.J. Williams, I.P. Bond, Bioinspired self-healing of advanced composite structures using hollow glass fibres. J. R. Soc. Interface 4 (2007) 363–371.
9. J.W.C. Pang, I.P. Bond, "Bleeding composites"- damage detection and self-repair using a biomimetic approach. Compos. – A: Appl. Sci. Manuf. 36 (2005) 183–188.
10. S. Jyothi Sri, A. Seethadevi, K. Suria Prabha, P. Muthuprasanna, P. Pavitra, Microencapsulation: a review. Int. J. Pharma Bio Sci. 3 (2012) 509–531.
11. D. Jung, A. Hegeman, N.R. Sottos, P.H. Geubelle, S.R. White, Self-healing composites using embedded microspheres. The American Society for Mechanical Engineers Materials Division 80 (1997) 265–275.
12. C.W. Bielawskia, R.H. Grubbs, Living ring-opening metathesis polymerization. Prog. Polym. Sci. 32 (2007) 1–29.
13. S.H. Cho, H.M. Andersson, S.R. White, N.R. Sottos, P.V. Braun, Polydimethylsiloxane-based self-healing materials. Adv. Mater. 18 (2006) 997–1000.
14. M. Motuku, U.K. Vaidya, G.M. Janowski, Parametric studies on self-repairing approaches for resin infused composites subjected to low velocity impact. Smart Materials and Structures 8 (1999) 623–638.
15. J.W.C. Pang, I.P. Bond, A hollow fibre reinforced polymer composite encompassing self-healing and enhanced damage visibility. Compos. Sci. Technol. 65 (2005) 1791–1799.
16. K.S. Toohey, N.R. Sottos, J.A. Lewis, J.S. Moore, S.R. White, Self-healing materials with micro-vascular networks. Nature Materials 6 (2007) 581–585.
17. S. Kim, S. Lorente, A. Bejan, Vascularized materials: Tree-shaped flow architectures matched canopy to canopy. J. Appl. Phys. 100 (2006) 063525(1–8).
18. X. Chen, F. Wudl, A.K. Mal, H. Shen, S.R. Nutt, New thermally remendable highly cross-linked polymeric materials. Macromolecules 36 (2003) 1802–1807.
19. Y.-L. Liu, C.-Y. Hsieh, Crosslinked epoxy materials exhibiting thermal remendablility and removability from multifunctional maleimide and furan compounds. J. Polym. Sci. A Polym. Chem. 44 (2006) 905–913.

20. C.-M. Chung, Y.-S. Roh, S.-Y. Cho, J.-G. Kim, Crack healing in polymeric materials via photochemical [2+2] cycloaddition. Chem. Mater. 16 (2004) 3982–3984.

21. A. Eisenberg, J.S. Kim, Introduction to Ionomers. John Wiley & Sons, New York (1998).

22. C.X. Sun, M.A.J. van der Mee, J.G.P. Goossens, M. van Duin, Thermoreversible cross-linking of maleated ethylene/propylene copolymers using hydrogen-bonding and ionic interactions. Macromolecules 39 (2006) 3441–3449.

23. S.J. Kalista, T.C. Ward, Z. Oyetunji, Self-healing of poly(ethylene-co-methacrylic acid) copolymers following projectile puncture. Mech. Adv. Mater. Struct. 14 (2007) 391–397.

24. J.M. Lehn, Supramolecular polymer chemistry—scope and perspectives. Polym. Int. 51 (2002) 825–839.

25. L. Brunsveld, B.J.B. Folmer, E.W. Meijer, R.P. Sijbesma, Supramolecular polymers. Chem. Rev. 101 (2001) 4071–4098.

26. P. Cordier, F. Tournilhac, C. Soulie'-Ziakovic, L. Leibler, Self-healing and thermoreversible rubber from supramolecular assembly. Nature 451 (2008) 977–980.

27. S. Bode, R.K. Bose, S. Matthes, M. Ehrhardt, A. Seifert, F.H. Schacher, R.M. Paulus, S. Stumpf, B. Sandmann, J. Vitz, A. Winter, S. Hoeppener, S.J. Garcia, S. Spange, S. van der Zwaag, M.D. Hager, U.S. Schubert, Self-healing metallopolymers based on cadmium bis(terpyridine) complex containing polymer networks. Polym. Chem. 4 (2013) 4966–4973.

28. M. Trau, S. Sankaran, D.A. Saville, I.A. Aksay, Electric-field-induced pattern formation in colloidal dispersions. Nature 374 (1995) 437–439.

29. K.A. Williams, A.J. Boydston, C.W. Bielawski, Towards electrically conductive, self-healing materials. J. R. Soc. Interface 4 (2007) 359–362.

30. S. Gupta, Q. Zhang, T. Emrick, A.C. Balazs, T. Russell, Entropy-driven segregation of nanoparticles to cracks in multilayered composite polymer structures. Nat. Mater. 5 (2006) 229–233.

31. N. Shahrubudin, T.C. Lee, R. Ramlan, An overview on 3D printing technology: Technological, materials, and applications. Procedia Manuf. 35 (2019) 1286–1296.

32. C.B. Highley, C.B. Rodell, J.A. Burdick, Direct 3D printing of shear-thinning hydrogels into self-healing hydrogels. Adv. Mater. 27 (2015) 5075–5079.

33. Z.W. Wicks, F.N. Jones, S.P. Pappas, D.A. Wicks, Organic Coatings: Science and Technology. 3rd Edition, Wiley, Hoboken (2006).

34. S.H. Cho, S.R. White, P.V. Braun, Self-healing polymer coatings. Adv. Mater. 21 (2009) 645–649.

35. B. Zhang, P. Zhang, H. Zhang, C. Yan, Z. Zheng, B. Wu, Y. Yu, A transparent, highly stretchable, autonomous self-healing poly(dimethyl siloxane) elastomer. Macromol. Rapid Commun. 38 (2017) 1–9.

36. G. Zhou, The use of experimentally-determined impact force as a damage measure in impact damage resistance and tolerance of composite structures. Compos. Struct. 42 (1998) 375–382.

37. T.C. Mauldin, M.R. Kessler, Self-healing polymers and composites. Int. Mater. Rev. 6608 (2013) 317–346.

38. K.A. Williams, A.J. Boydston, C.W. Bielawski, Towards electrically conductive, self-healing materials. J. R. Soc. Interface. 4 (2007) 359–362.

39. Y. Li, S. Chen, M. Wu, J. Sun, Polyelectrolyte multilayers impart healability to highly electrically conductive films. Adv. Mater. 24 (2012) 4578–4582.

40. F. Gelain, D. Silva, A. Caprini, F. Taraballi, A. Natalello, O. Villa, K.T. Nam, R.N. Zuckermann, S.M. Doglia, A. Vescovi, BMHP1-derived self-assembling peptides: hierarchically assembled structures with self-healing propensity and potential for tissue engineering applications. ACS Nano 5 (2011) 1845–1859.

41. L. Saunders, P.X. Ma, Self-healing supramolecular hydrogels for tissue engineering applications. Macromol. Biosci. 19 (2019) 1–11.

42. J.H. Lee, Injectable hydrogels delivering therapeutic agents for disease treatment and tissue engineering. Biomater. Res. 22 (2018) 1–14.

43. T.R. Hoare, D.S. Kohane, Hydrogels in drug delivery: Progress and challenges. Polymer 49 (2008) 1993–2007.

44. L. Gao, Y. Yang, J. Xie, S. Zhang, J. Hu, R. Zeng, J. He, Q. Li, Q. Wang, Autonomous self-healing of electrical degradation in dielectric polymers using *in situ* electroluminescence. Matter 2 (2020) 451–463.

45. S. Cao, W. Zhu, T. Liu, Bio-inspired self-healing polymer foams with bilayered capsule systems. Compos. Sci. Technol. 195 (2020) 108189.

46. J. Zhang, C. Zhang, F. Song, Q. Shang, Y. Hu, P. Jia, C. Liu, L. Hu, G. Zhu, J. Huang, Y. Zhou, Castor-oil-based, robust, self-healing, shape memory, and reprocessable polymers enabled by dynamic hindered urea bonds and hydrogen bonds. Chem. Eng. J. 429 (2022) 131848.

47. L. Guadagno, L. Vertuccio, G. Barra, C. Naddeo, A. Sorrentino, M. Lavorgna, M. Raimondo, E, Calabrese, Eco-friendly polymer nanocomposites designed for self-healing applications. Polymer 223 (2021) 123718.

48. Z. Chen, Y.-C. Sun, J. Wang, H.J. Qi, T. Wang, H.E. Naguib, Flexible, reconfigurable, and self-healing TPU/vitrimer polymer blend with copolymerization triggered by bond exchange reaction. ACS Appl. Mater. Interfaces 12 (2020) 8740–8750.

49. C. Yang, J. Yin, Z. Chen, H. Du, M. Tian, M. Zhang, J. Zheng, L Ding, P. Zhang, X. Zhang, K. Deng, Highly conductive, stretchable, adhesive, and self-healing polymer hydrogels for strain and pressure sensor. Macromol. Mater. Eng. 305 (2020) 2000479.

50. S. Song, H. Yang, Y. Cui, Y. Tang, Y. Chen, B. Yang, J. Yuan, J. Huang, Mussel-inspired, self-healing polymer blends. Polymer 198 (2020) 122528.

7 Emerging Applications of Photocurable Polymers

Zehra Yildiz and Emine D. Kocak

Department of Textile Engineering, Faculty of Technology,
Marmara University, RTE Campus, Aydinevler District, Uyanik Street,
no:6, T1/319, 34840, Istanbul, Turkey

1 DEFINITION OF PHOTOCURING

The conversion of the liquid to the solid state in a coating formulation is depicted in terms of "curing." During this conversion, chemical bonds are formed via cross-linking, in other words, a polymerization mechanism. Hardening occurs after this curing stage. Photopolymerization is an effective way to covalently cross-link the polymer chains, which can be further used for various purposes, i.e., biomedical, packaging, and automotive-related materials. Photocuring means that light energy is used to instantly harden/cure the resin instead of heat. Typically, photocurable coating formulations consist of 100% reactive monomers/oligomers without any volatile solvents compared to conventional coating systems. The photocuring mechanism depends on the photo-sensitive chemical components (photoinitiators) which absorb light in certain wavelengths [1].

2 ADVANTAGES AND LIMITATIONS OF PHOTOCURING TECHNOLOGY

Photocuring technology presents many advantages in different categories: ecological, economical, and performance. Among them, the ecological advantages are the most important reasons which have led researchers and industry workers to investigate and design photocurable formulations. In a photocuring system, the coating formulations, inks, adhesives, etc. are instantly hardened in a few seconds/minutes, which leads to a high production rate. The curing process is performed at elevated temperatures, generally at room temperatures, resulting in an energy-saving property and providing the ability to apply to heat-sensitive materials. The photocurable formulation requires none/less solvent and lacks any environmental pollutants, thus it represents an environmentally friendly coating process with a volatile organic content (VOC) reduction. Considering the performance advantages, photocurable formulations show good adhesion on various substrates with controlled elasticity and high curing rates without any loss of coating thickness/volume, and they present high stability during storage. Furthermore, photocured coatings give high scratch and chemical resistance properties with enhanced optical clarity and superior toughness [2–5].

Besides the described advantages, photocuring technology also presents some limitations, such as higher material costs compared to conventional coating technologies, difficulties in the curing stage in the presence of UV stabilizers, oxygen inhibition during the curing stage, difficulties during application on complex shapes, sensitivity to moisture, poor weatherability due to the continuous absorption of UV light, and lower curing rates on colored/pigmented and thick surfaces [6,7]. In the literature, and on an industry scale, research is mainly related to improving the adhesion of photocurable resins on metal and plastic surfaces, reducing the odor of the formulations, improving cost-effective photoinitiators, designing reactive oligomers with desired functionalities, and investigating the suitability of the formulations in direct food contact [7].

DOI: 10.1201/9781003278269-7

TABLE 7.1
Components of a Photocurable Formulation with Their Functions

Component	% in the Formulation	Function
Photoinitiators	1–8	Initiation
Reactive diluents	15–60	Adjustment of the viscosity and control the cross-linking density
Oligomers	25–90	Main component, film formation
Additives	1–50	Pigments, stabilizers, surfactants, fillers, etc.

3 COMPONENTS OF PHOTOCURING TECHNOLOGY

A typical photocurable formulation consists of three main components: a reactive diluent, a reactive oligomer, and a photoinitiator. Reactive diluents are used to lower the viscosity of the photocurable formulation and to improve the overall structural properties of the coating by adjusting the cross-linking density. The reactive oligomers are macromolecules that can be polymerized by the photolysis of photoinitiators via exposure to UV light. Photoinitiators are responsible for the initiation stage of the curing system upon exposure to UV light in various wavelengths. In a typical photocurable system, a liquid oligomer and reactive diluent with a small percent of photoinitiator inclusion are all hardened (cross-linked) together in a few seconds by exposure to UV light. Table 7.1 summarizes the photocurable formulation components with their main functions [7].

3.1 PHOTOINITIATORS

A photoinitiator is an important component of a photocurable formulation as it directly adjusts the curing rate. They are thermally stable compounds having the capability of absorbing incident light in various wavelengths with high absorption coefficients. After the absorption of light, they split up into fragments to form free radicals to initiate the photopolymerization mechanism, generally via H atom abstraction. Figure 7.1 illustrates the most commonly used photoinitiator types. The absorbency properties of the photoinitiator should be to the radiation characteristics of the UV source. A proper photoinitiator selection is dependent on the following factors:

- Coating thickness
- Absence/presence of dyes, fillers, or pigments
- Curing speed of the system
- Desired surface characteristics (gloss, hardness, etc.)
- Desired thermal properties (glass transition, decomposition, etc.)
- Being cost-effective
- Presenting a nontoxic quality, with low migration and without any odor properties
- Pot stability of the photocurable formulation [1,5,7].

3.2 REACTIVE DILUENTS

In photocurable coating formulations, reactive diluents are employed to control the cross-linking density, lower the viscosity (from 10,000 to 100 cps), and enhance the mechanical/chemical/thermal properties. Reactive diluents should be selected based on the following properties: the number of functional groups, volatility, toxicity, photoresponse, odor, cost, desired properties (hardness, ductility, brittleness, etc.) of the end product, and solvation rates. The number of functional groups of a

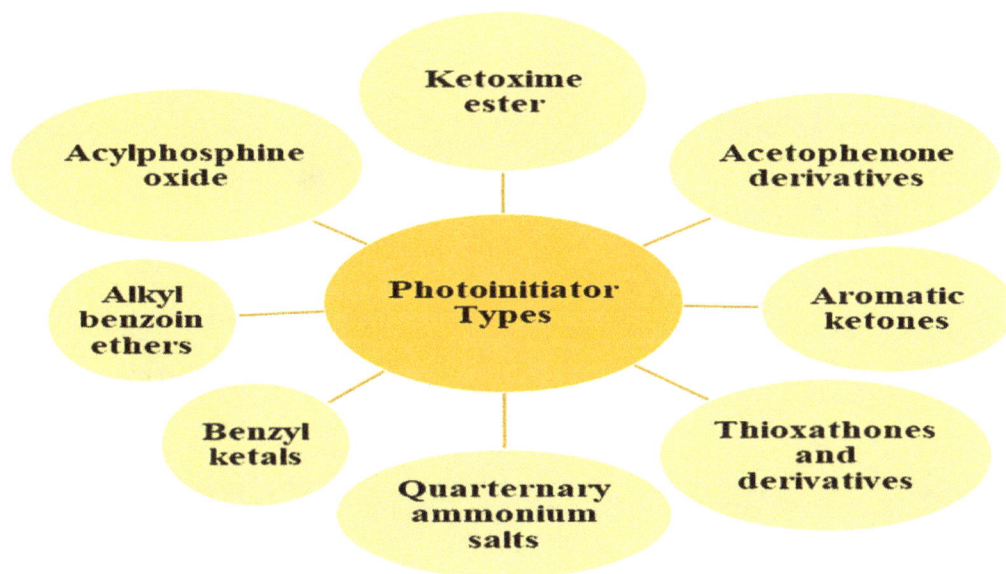

FIGURE 7.1   Most commonly used photoinitiator types.

reactive diluent is directly proportional to the curing rate of the photocurable formulation. The most commonly used reactive diluent types are given in Figure 7.2 [1,5,7].

3.3   OLIGOMERS

Researchers are mainly focused on the design of reactive oligomers with desired functionalities by the reaction of some chemical compounds with the proper functional groups. Epoxy acrylate (EA), polyester acrylate, acrylated melamines, acrylated vegetable oils, polyurethane acrylate (PUA), acrylated polyethers, and silicone acrylate are the most preferred reactive oligomers in photocurable coatings [1,5,7].

3.4   ADDITIVES

In a photocurable coating formulation, additives are used to enhance the overall coating performance of the formulation by giving additional functional properties. For instance, plasticizers are employed to adjust the rheological property and thus the viscosity of the formulations. The most used additives are as follows: flame retardants, thermal/UV stabilizers, antioxidants, color pigments, antifogging/antistatic agents, tackifiers, etc. [8,9].

4   CURING MECHANISM OF PHOTOCURABLE OLIGOMERS

Photocurable reactive oligomers can be cured via a cationic or free radical mechanism. In free-radical photopolymerization, aromatic ketones are employed to form the free radicals which will further initiate the photopolymerization reaction via the step-growth addition route. Meanwhile, in the cationic photopolymerization reaction, a proton acid is formed via the photolysis of diaryliodonium salts or triarysulfonium to initiate the photopolymerization of vinyl ethers or epoxies. Currently, free radical polymerization is favored due to the existence of a wide range of acrylate monomers and their higher reactivity levels [5,7].

Vinyl based diluents

- Methyl styrene
- Vinyl toluene
- Vinyl acetates
- N-vinyl pyrrolidone

Acrylics

- Monoacrylates (n-butyl acrylate, 2-ethyl hexyl acrylate, iso decyl acrylate, iso bornyl acrylate, 2-hydroxy ethyl acrylate, 2-hydroxy propyl acrylate)
- Diacrylates (1,4-butanedioldiacrylate, 1,6-hexanedioldiacrylate, neopentylglycoldiacrylate, diethyleneglycoldiacrylate)
- Triacrylates (pentaerythritoltriacrylate, trimethylolpropanetriacrylate)
- Tetracrylates (pentaerythritoltetracrylate)
- Pentacrylates (dipentaerythritol (monohydroxy) pentaacrylate)

Allylic monomers

- Triallyl cyanurate
- Trimethylol propane triallyl ether

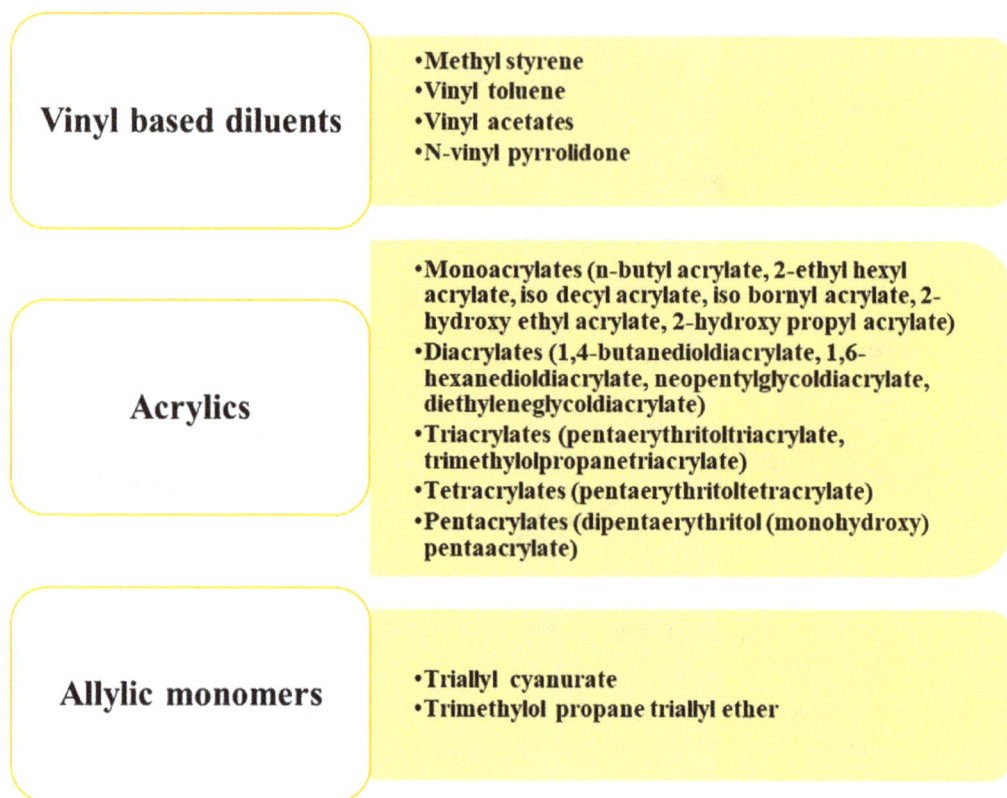

FIGURE 7.2   Commonly used reactive diluent types.

5   APPLICATIONS OF PHOTOCURABLE POLYMERS

Photocuring technology is favored in both academic and industrial fields due to its previously mentioned superior properties. It has widespread application areas, such as aerospace, automotive, packaging, biomedical, dentistry, microelectronics, etc. Photocurable formulations are used as printing inks, varnishes, adhesives, lacquers, etc. In this study, the application areas of photocuring technology are divided into two main categories in terms of the raw material of the substrate and the type of intended industry.

5.1   APPLICATIONS IN TERMS OF THE RAW MATERIAL OF THE SURFACE

5.1.1   On Wooden Surfaces

Wood coatings include the coatings of furniture and floorings, and commonly represent two basic roles: decorative and protective. The protective role aims to protect the wood surfaces from the possibility of swelling fibers/fibrils of wood and surface deformation that may be caused by the liquid environment during water/solvent-based coating processes. Photocurable coatings support the protective role by allowing the usage of a less or no solvent/water-based formulation instead of conventional water/solvent-based coatings. Considering the decorative role, photocurable coatings must possess a clear and transparent look without causing any changes to the aesthetic and natural appearance of the wood surface by remaining invisible on the inherent wood color and texture.

In a previous study, the effects of photocurable polyurethane acrylate (PUA) coating amount on the morphology, permeability, chemical changes, and surface energy of oak surfaces have been researched. Photocurable formulations have been preferred to water-based conventional wood coatings to obtain a fast curing system by reducing the drying time from >35 min to 22 min. The results have proved that with the increase of coating amount, the oak surface becomes gradually smoother and the filling depth of the coating layer in the oak surface is increased. In another study, photocurable PUA-based coating formulations were optimized in terms of the UV irradiation time, talcum powder, and $CaCO_3$ amounts and were applied on Eucalyptus and Poplar wood surfaces. According to the optimal results, the highest impact strength of 40 kg.cm with grade-1 adhesion and 3H hardness results with acceptable gloss value have all been obtained when the contents of $CaCO_3$ and talcum powder were set as 1% and 2%, by applying 1 min UV light exposure. To reduce the usage of petrochemical-based chemicals in UV-curable formulations, a bio-based reactive diluent has been synthesized by the reaction of castor oil (CO), diethanolamine, and allyl chloroformate. Then, the obtained bio-based reactive diluent was included in various proportions (5–25%) in photocurable coating formulations containing urethane acrylate oligomer and a photoinitiator, then applied on wood panels and photocured in 20 s. The optical, mechanical, and stain resistance properties have all been improved with an increasing amount of bio-based reactive diluent content in the formulation [10]. In a previous study, sweet chestnut wood panels were coated with a photocurable coating formulation containing bisphenol A dihydroxyethyldiacrylate (BHEDA) as an oligomer, and fluorinated reactive diluents and photoinitiator. Fluorinated compounds have been employed to enhance hydrophobicity, weathering resistance, and chemical stability by lowering the water permeability, refractive index, and friction coefficient. Accordingly, due to the presence of fluorinated compounds in the formulation, sweet chestnut wood panels have shown high oleophobicity and hydrophobicity, with enhanced chemical and scratch resistance properties [11]. Cellulose nanocrystals (CNCs) have been also included in epoxy acrylate-based photocurable coating formulations as reinforcing agents on wood coatings. Improved modulus of elasticity, abrasion resistance, hardness, and tensile strength have all been achieved with the inclusion of CNCs into the formulation [12]. Antimicrobial photocurable epoxy acrylate-based coatings, having citric acid (CA) as an antimicrobial additive, have been applied on wood panels. The results have illustrated that excellent solvent- and soil-resistant coatings with good adhesion and hardness properties and remarkable antimicrobial activity have all been obtained with CA inclusion into the formulation [13].

5.1.2   On Metal Surfaces

In a conventional water/solvent-based metal coating technique, the liquid that contacts the metal surface may cause corrosion on metal surfaces. Due to the elimination of the corrosion potential, photocurable coating formulations have been improved for metal surfaces.

In a previous study, vegetable oils (grapeseed oil and rosehip seed oil) have been epoxidized and included in photocurable formulations, then they were cationically photocured on metal surfaces to be used as anticorrosive coatings. Accordingly, the formulation with rosehip seed oil and the highest double-bound content presented the best hardness, solvent, and corrosion resistance properties due to having the highest cross-linking density [14]. Superhydrophobic UV-curable formulations have been prepared by using PUA oligomer and nano-SiO_2 grown in situ on the surfaces of carboxylated carbon nanotubes (CNTs) reinforcing agents. The obtained composite particles have been designated as CNTs@SiO_2. Then, the obtained formulations have been applied on aluminum plates and cured with UV light. Accordingly, the CNTs@SiO_2 reinforced and interfacially adhered PUA-based photocurable coatings have shown excellent adhesion and superhydrophobic properties even after 1500 cycles of abrasion, and improved sand and water droplet resistance [15].

5.1.3   On Glass Surfaces

UV-curable adhesives/coatings are preferred on glass surfaces due to their showing excellent optical clarity and mechanical properties (hardness, impact resistance, shear strength, tensile strength, etc.) on the end product that are responsible for the aesthetic properties with higher optical transmission, specifically in optical lenses, laminated safety glass, and decorative crystal glass, etc. Lamination of glass materials is an important issue considering the probability of the sudden and total structural collapse of glass layer(s). Thus, lamination is needed to maintain the required integrity of glass layer(s) by giving a load-bearing capacity via retaining the broken sharp elements in case of a collapse [16,17].

In a previous study, photocurable epoxy/vinyl ester resin blends were used as matrices in glass fiber-reinforced composites, allowing a quicker and cleaner manufacturing process compared to the conventional thermal-based resin systems without any post-curing stages. The epoxy/vinyl blends have supplied an excellent interlaminar shear strength and better mechanical properties and have been proposed to be used as bulletproof vests [18,19]. In another study, photocurable epoxy acrylate/ cycloaliphatic epoxy blends have been used as matrices in glass fiber-reinforced composites to obtain enhanced interlaminar properties by following the degree of polymerization during the curing stage [20]. Hydroxyl and acrylate functionalized polyaniline oligomers have been synthesized by a photografting method and then screen-printed onto the glass surface to obtain conductive lines for use in electronic devices. The photografting surface modification technique has been employed due to it presenting a rapid, wasteless, and corrosive acid-free manufacturing route [21].

5.2   APPLICATIONS BASED ON THE INDUSTRY

5.2.1   Applications in Dentistry

In dentistry, photocurable formulations are mainly used as adhesive materials, aesthetic restorative materials, implants, prosthetic materials, dental crowns, and conservative materials (inlay fillings, etc.). In the literature, UV-curable dental-related research has been mainly addressed to solving the shrinkage problem during polymerization. Shrinkage of a polymer can be defined as the reduction of intermolecular distances between the monomers due to the formation of a covalently bonded cross-linked thermoset structure upon UV exposure. The UV-curable dental material is adhered to the tooth wall during photopolymerization and creates shrinkage stress which results in internal cracks and further defects. The type and composition of the acrylate-based resins, reactive oligomer functionality and overall double bond concentration, type and amount of the photoinitiator, and UV exposure duration and dose are the key parameters that affect the shrinkage of a UV-curable dental material. All these parameters have been researched to investigate the polymerization shrinkage during UV curing in dental applications. Accordingly, when the functionality and double bond concentration of the reactive oligomer are increased, the shrinkage also is increased. Due to the lower reactivity that decreases the double bond conversion over time, the shrinkage behavior is decreased in methacrylate-based oligomers compared to acrylate-based formulations. A higher level of conversion causes greater shrinkage and contraction stress, whereas a low level of conversion leads to poor biocompatibility, inadequate physical/mechanical properties, and a reduction in product life and quality. Thus a precise double bond conversion level should be determined via optimization [22–24].

In an earlier study, a dental crown with low shrinkage (2.58%) and enhanced flexural and compressive strength was manufactured via direct ink writing 3D printing using an acrylic-based photocurable formulation containing multi-scale inorganic particles (silanized silica) [25]. In another study, photocurable formulations were prepared using zirconia ceramics and graphene oxide (GO) powders in an acrylic-based resin for use as 3D-printed dental implants. The GO powders were employed to intensify the light scattering for the prevention of curing of the ceramic paste, resulting in an adequate curing depth and width. The photoinitiator concentration has been also optimized in photocurable methacrylate-based formulations for dental applications based on the curing depth, mechanical properties (modulus of elasticity, hardness), and double bond conversion. Accordingly, the optimum concentration of the camphorquinone (CQ) has been found to be 0.5%, with the best mechanical properties and maximum curing depth [26].

5.2.2 Applications in the Electronics Industry

UV-curing technology can be employed in the electronic industry to be used in printed circuits, multi-layered circuits, smartphones, flexible electronic devices, and tablet and television screens. During the conventional circuit manufacturing process based on the use of solvent-based inks, microstructures of the circuits having pores and grooves can be filled by the ink or solvent which further causes some problems such as short-circuiting, distortion in the circuit geometry, and inconsistent circuit performance, etc. To overcome these problems, the use of solvent-free photocurable formulations in electronics is needed. In a 3D printing system, conductive ink with silver nanoparticles and a dielectric tripropyleneglycol diacrylate ink have been cured together by UV curing with the ability of design flexibility for multilayered circuits [27,28]. In an earlier study, polyurethane acrylate-based photocurable formulations containing silver-coated copper flakes have been produced to be used as highly flexible and conductive composite electronics. These formulations have also been screen printed, and showed excellent durability up to 1000 rolling cycles, and can adhere to PET film and office paper [29].

5.2.3 Applications in the Automotive Industry

Photocurable coatings in the automotive industry are mainly used as clear coatings to provide decorative and protective properties to the substrate. High damage, mar, and scratch resistance with a high glossy first-class look are the key factors to determining the quality of automotive clear coatings. Higher scratch and abrasion resistance properties of a coating are associated with higher cross-linking densities. Silane-based cross-linkers and inorganic fillers can be incorporated into photocurable formulations to obtain an interpenetrating polymer network by improving the cross-linking densities. Besides refinish and clear coatings, UV-curing technology is also used to improve the weather stability of vehicles and to supply extremely rapidly drying paints for automotive applications, thereby providing remarkable time-saving advantages to the industry [30].

In a previous study, polyurethane clear coats for automotive applications were prepared by the reaction between the hydroxyl-terminated acrylic resin and isocyanates. Tri- and tetra-functional polycaprolactone oligomers have been introduced to the coatings in various amounts to enhance the mar and scratch resistance of clear coatings. Clear coating formulations also have been prepared using four types of acrylic-melamine formaldehyde-based resins with different hydroxyl contents. The effect of cross-linking density of the coatings on water vapor transmission rate in terms of the hydroxyl content has been researched. A better coating hardness property with the least hydrophilicity and water vapor transmission rate has been achieved by lowering the hydroxyl content of the formulations [31]. Acrylate and urethane acrylate-based oligomers also have been used for automotive interior plastic parts with the required hardness and scratch resistance properties [32].

5.2.4 Applications in the Textile Industry

In a previous study, cotton, silk, and synthetic fabrics were coated with chitosan in acetic acid solution in the presence of a photoinitiator, and then cured by a UV lamp. The antibacterial activity against *Escherichia coli* (*E. coli*) has been investigated. It has been found that 12 h impregnation time of fabrics in the coating formulations showed better antibacterial activity without any loss of fabric handling quality [33]. In another study, UV-curing technology has been employed in cotton, cotton/polyester, and polyester fabrics for pigment coloration using UV-curable binders (aliphatic urethane acrylate, epoxy acrylate, and polyester acrylate). It has been found that a high colorfastness could be achieved by using photocuring technology with an acceptable fabric handling quality [34]. UV curing has also been suggested as an alternative to thermal curing in pigment prints of textile fabrics. Cotton fabrics have been coated by both thermal- and UV-curable coating formulations, and then compared in terms of colorfastness to rubbing and washing. It has been illustrated that UV curing can be acceptable as an effective method of durable pigment printing of textile fabrics [35].

The conventional fixation of the capsules is performed by a thermal process which is caused by the evaporation or swelling of the material inside the capsule, resulting in a loss of active core material or breaking of the shell. To overcome these problems, photocuring technology has been suggested as an alternative to thermal fixation for the encapsulation technology. To prepare pigmented photocurable inkjet inks for textile coloration, nanoscale organic pigments have been encapsulated into a UV-curable resin (1,6-hexanediol dimethacrylate and polyester tetraacrylate) using a mini-emulsion technique. The shelf-life stabilities of the encapsulated pigments and washing/ light-fastness of the printed UV-cured cotton/viscose/polyester fabrics have been researched. All the printed fabrics showed good fastness properties and a soft handling quality. In a previous study, the washing fastness results of cotton fabrics with an encapsulated aroma finish have been increased from 25 cycles to 50 cycles using UV-curable resin (unsaturated polyurethane and tripropylene glycol diacrylate) in the fixation of capsules instead of thermal energy [36].

Photocurable formulations also have been used as an adhesive in the textile industry in several studies. For instance, as an alternative to the resorcinol formaldehyde latex (RFL)-based coating industry, formaldehyde-free dual curable (UV/thermal) polyurethane methacrylate-based oligomers (TDI:HEMA adduct) (Figure 7.3) have been synthesized and then included in coating formulations in the presence of various reactive diluents [trimethylolpropane trimethacrylate (TMPTMA) and tricyclodecane dimethanol diacrylate (TCDDA)]. The formulations were then applied on polyester cord fabrics to adhere to styrene-butadiene rubber (SBR) for use in vehicle tires. The results have illustrated that the highest adhesion strength of 103 N/cm is achieved when the NCO:OH ratio in the oligomer was set at 4 and by using the TCDDA as a reactive diluent [37].

In order to be used between cord fabric/rubber surfaces as an adhesive, UV-curable epoxyacrylates (EAs) (Figure 7.4) have been synthesized and included in adhesive formulations in the presence of vinyl phosphonic acid (VPA). The formulations were then applied on polyester/polyamide cord fabrics before adhesion on SBR surfaces. The highest peel strength value of 50.8 N/cm was recorded in a 10% VPA-included formulation with the best flame resistance property [38].

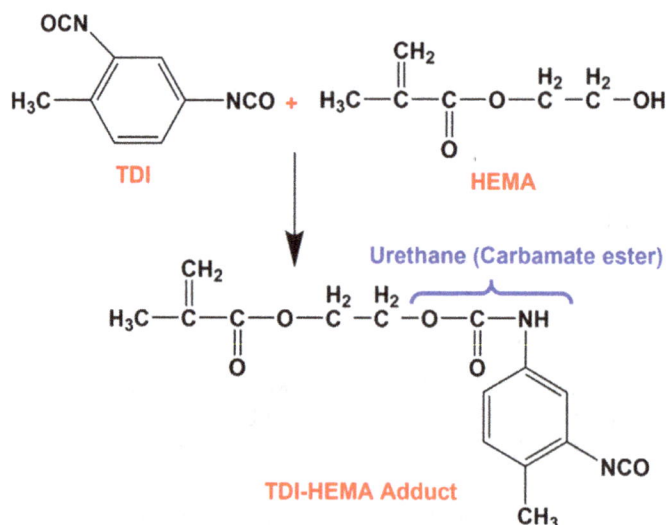

FIGURE 7.3   A synthesis scheme of the TDI:HEMA adduct.

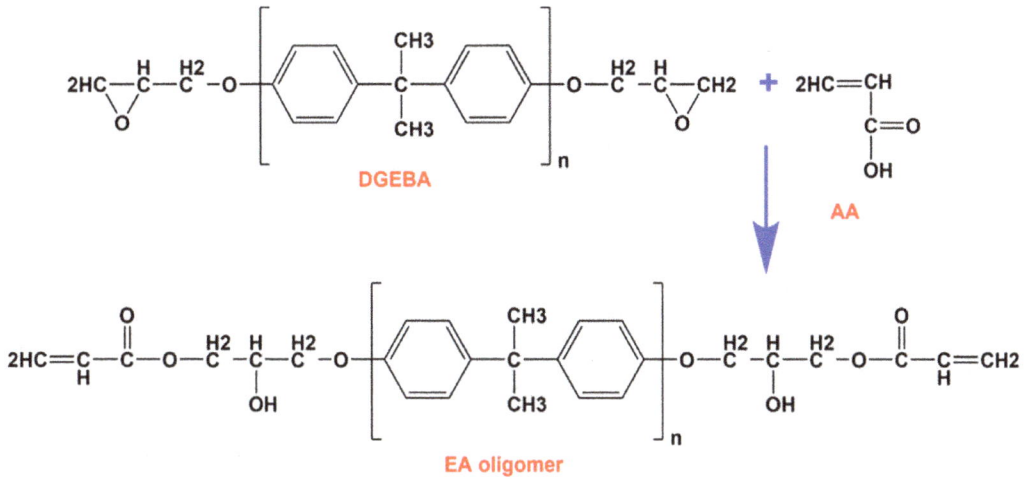

FIGURE 7.4 A synthesis scheme of the EA oligomer.

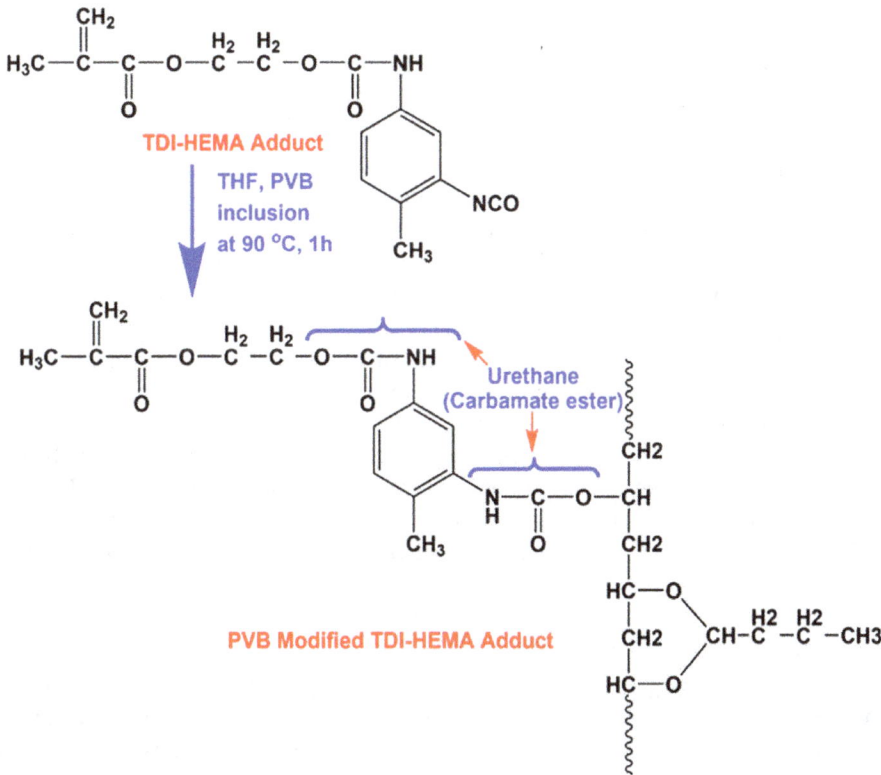

FIGURE 7.5 Synthesis scheme of the PVB-modified TDI-HEMA adduct.

Formaldehyde-free photocurable adhesive formulations have been also designed by synthesis of an oligomer [polyvinyl butyral (PVB) modified TDI-HEMA adduct] (Figure 7.5) and applied between polyester cord fabrics and SBR for the tire industry. The adhesion strength between the cord and rubber surfaces has been researched. The results showed that the highest peel strength value of 94.7 N/cm was recorded in a sample with 5% PVB in the coating formulation [39].

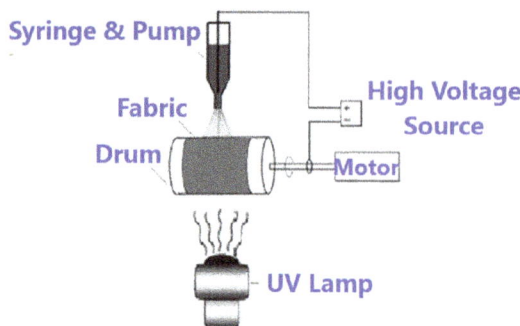

FIGURE 7.6   Scheme of the electrospinning equipped with a UV lamp.

UV-curable coating formulations have been prepared for cotton fabrics to give a super-hydrophobic property to the fabrics. Polyurethane/silica hybrid solutions have been applied on cotton fabrics by the sol–gel method via an electrospinning technique. The electrospun formulation on cotton fabrics was cured layer by layer using a UV-lamp that was positioned in front of the electrospinning cabinet (Figure 7.6). According to the results, a contact angle value of 154.5° was recorded in cotton fabric surfaces with a formulation with a 50% inorganic part inclusion [40].

Photocurable coatings also have been used to enhance the washing fastness of cotton fabrics with a water-repellent finish. Photocurable oligomer has been synthesized by the reaction between the cellulose acetate butyrate (CAB) and TDI-HEMA adduct (Figure 7.7). The obtained oligomer was included in photocurable coating formulations to protect the water-repellent finish on cotton fabric against washings. The results showed that after application of the photocurable coating, the water-repellent behavior could be protected for up to 10 washings without any loss. In addition, the UV-cured coating also increased the abrasion resistance of cotton fabrics [41]. In another study, electrospun fibrous mats were collected on cotton fabrics by the same oligomer in Figure 7.5 via an electrospinning technique with the combination of layer-by-layer UV-light exposure. According to the results, the water vapor and air permeability of the fibrous mats were increased with an increasing amount of TDI-HEMA molar amount in the oligomer due to the polarity of carbamate ester groups, which makes the fibrous layer more permeable against air and water vapor [42]. The oligomer in Figure 7.5 also has been employed as a coating layer on cotton fabrics to obtain coated cotton fabrics that are durable in outdoor environments. Accordingly, the TDI-HEMA-modified CAB oligomer (Figure 7.7) could be used in photocurable coating formulations on cotton fabrics to increase the abrasion resistance with enhanced mechanical properties and required air/water vapor permeability [43].

5.2.5   Applications in Printing Technology

Today, additive manufacturing, in other words rapid prototyping, is rapidly growing by implementing various techniques/processes to allow the production of complex shapes based on the designed patterns. Photocurable polymers are excellent candidates to be used in additive manufacturing due to their rapid curing mechanism, dimensional accuracy, high precision, and ability to provide a wide range of properties. Considering all the described advantages, the use of photocurable polymers in printing technologies (2D, 3D, 4D) has emerged. Photocurable 3D printing technology is divided into various categories based on the pattern formation principle and control system. These are liquid crystal display (LCD), digital light processing (DLP), stereolithography appearance (SLA), multi-jet printing (MJP), two-photon 3D printing (TPP), continuous liquid interface production (CLIP), and holographic 3D printing [44].

In an earlier study, hydrogel–polymer hybrid 3D structures were produced by acrylamide-polyethylene glycol diacrylate (PEGDA) hydrogel and methacrylate-based photocurable polymer

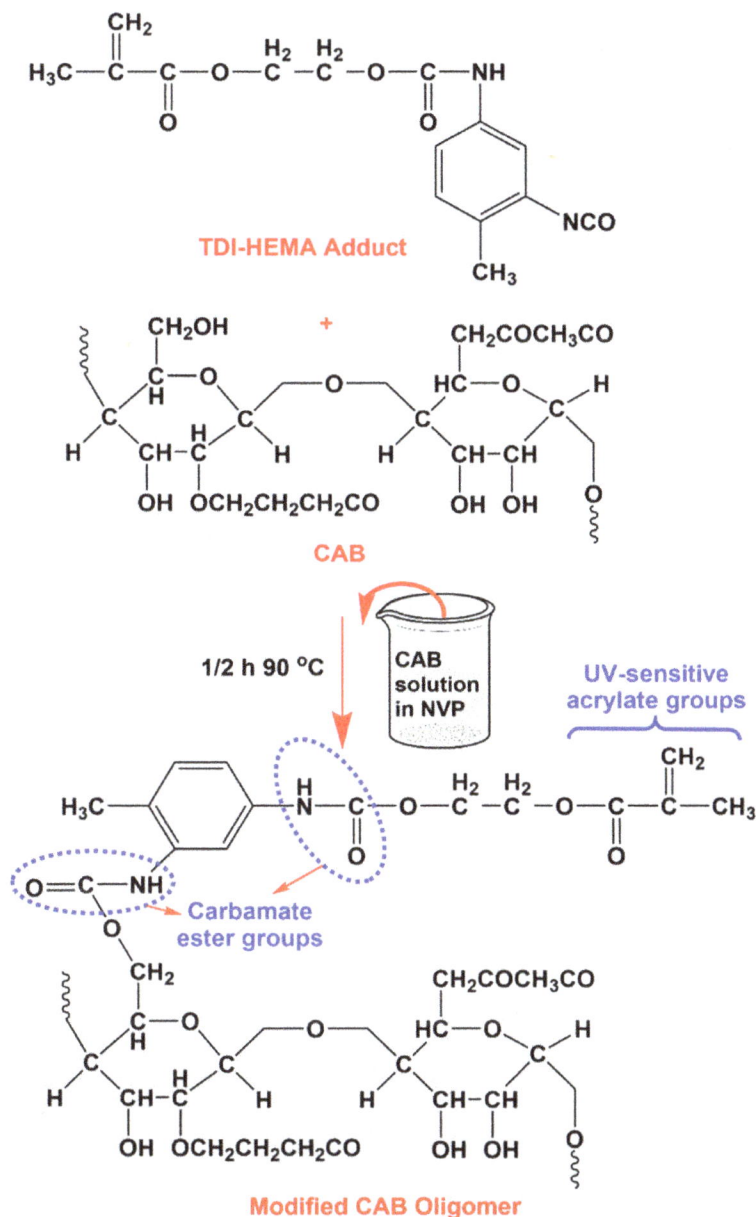

FIGURE 7.7 A synthesis scheme of the TDI-HEMA-modified CAB oligomer.

for use in highly stretchable strain sensors [45]. In another study, water-based photocurable conductive inkjet inks containing silver nanocolloids have been developed and applied on polyimide (PI) films for use in flexible photonics and electronics [46]. Photocurable polymers (off-stoichiometry thiol-ene, which is the combination of thiols and allyls) have been also used in the fabrication of inkjet-printed digital microfluidic chips by enabling the advantages of low-cost and ability to manufacture in a non-cleanroom environment [47].

In another study, a post-print UV-curing process by acrylic-based resins has been applied to 3D materials that are printed by a TPP technique, to increase the cross-linking density and enhance the mechanical properties (modulus and ultimate strength) of the materials [48]. The use of UV-curable polymeric systems in nanoimprint lithography can be accepted as an alternative to the hot embossing systems, as photocuring does not require the use of heat, pressure, or a post-annealing step. In a previous study, imprinted nano- and micrometer-scale patterns have been obtained by methacrylate-based UV-curable formulations on a spin-coater. The feature sizes of the patterns were recorded as 30 nm to several microns. The use of a photocurable formulation instead of a hot embossing system shortened the process by giving a low residual layer thickness [49].

5.2.6 Applications in the Packaging Industry

Packaging materials are mainly produced for food quality and safety issues. An active packaging system involves the incorporation of active compounds such as preservatives, antimicrobial agents, antioxidants, nutrients, etc. into the conventional packaging material. In the packaging industry, edible films, also known as active packaging materials, have great importance as they can improve food safety and quality by acting as a barrier against gas, moisture, aroma, bacteria, etc. Edible films are commonly synthesized by nontoxic, biodegradable lipids, proteins, and polysaccharides. They can be used for two main purposes: to protect food and to reduce packaging waste. The use of photocuring in the packaging industry has attracted the attention of many researchers as the conventional heat- and electron-beam-based manufacturing techniques involve the use of harmful metallic particles and cross-linking agents (formaldehyde, glutaraldehyde, etc.), which affect human health and the environment negatively. To employ a photocurable coating formulation for food packaging, several key factors have to be considered: proper processing is required, use of low migration inks/resin is needed, and migration confirmation has to be done by appropriate analysis.

In a previous study, an ecofriendly and biodegradable chitosan/polyvinyl alcohol film containing sulfosuccinic acid as a cross-linking agent has been manufactured for use in the packaging industry. The results showed that the biodegradability of the films after 220 days was recorded as about 40–65%. To evaluate the packaging performance, an apple was coated by the photocurable chitosan-based film, then, after 70 days, the uncoated and coated apples were evaluated visually. The study revealed that the coated apple was not decomposed, whereas the uncoated apple had decomposed and started to rot [50].

6 CONCLUSIONS AND FUTURE PROSPECTS

The use of UV technology offers many advantages to various industries, such as textiles, electronics, packaging, printing, dentistry, automotive, etc. In the mentioned fields, UV-curable formulations are used as inks, adhesives, reinforcing layers, lacquers, varnishes, etc. Due to the excellent performance of the UV-cured coatings with high productivity levels and low environmental impact, UV technology has gained increased attention from both industry and the academic world. The main drawback of photocuring is that the existing manufacturing lines in conventional solvent-based systems have to be converted to the method of UV-curable applications, which requires high investments. If this obstacle can be solved, future advances will be recorded in this emerging technology.

REFERENCES

[1] V. Shukla, M. Bajpai, D. Singh, M. Singh, and R. Shukla, Review of basic chemistry of UV-curing technology, *Pigment & Resin Technology,* 2004.

[2] J. Moon, Y. Shul, H. Han, S. Hong, Y. Choi, and H. Kim, A study on UV-curable adhesives for optical pick-up: I. Photo-initiator effects, *International Journal of Adhesion and Adhesives,* vol. 25, no. 4, pp. 301–312, 2005.

[3] J.-S. Hwang, M.-H. Kim, D.-S. Seo, J.-W. Won, and D.-K. Moon, Effects of soft segment mixtures with different molecular weight on the properties and reliability of UV curable adhesives for electrodes protection of plasma display panel (PDP), *Microelectronics Reliability,* vol. 49, no. 5, pp. 517–522, 2009.

[4] A. Giessmann, *Coating substrates and textiles: a practical guide to coating and laminating technologies*. Springer Science & Business Media, 2012.

[5] W. Schnabel, *Polymers and Light: Fundamentals and Technical Applications*. John Wiley & Sons, 2007.

[6] M. Bajpai, V. Shukla, and A. Kumar, Film performance and UV curing of epoxy acrylate resins, *Progress in Organic Coatings,* vol. 44, no. 4, pp. 271–278, 2002.

[7] R. Schwalm, *UV Coatings: Basics, Recent Developments and New Applications*. Elsevier, 2006.

[8] P. Cognard, *Handbook of Adhesives and Sealants: Basic Concepts and High Tech Bonding*. Elsevier, 2005.

[9] L. F. M. da Silva, A. Öchsner, and R. D. Adams, *Handbook of Adhesion Technology*. Springer, 2011.

[10] D. S. Tathe and R. Jagtap, Biobased reactive diluent for UV-curable urethane acrylate oligomers for wood coating, *Journal of Coatings Technology and Research,* vol. 12, no. 1, pp. 187–196, 2015.

[11] R. Bongiovanni *et al.*, High performance UV-cured coatings for wood protection, *Progress in Organic Coatings,* vol. 45, no. 4, pp. 359–363, 2002.

[12] A. Kaboorani, N. Auclair, B. Riedl, and V. Landry, Mechanical properties of UV-cured cellulose nanocrystal (CNC) nanocomposite coating for wood furniture, *Progress in Organic Coatings*, vol. 104, pp. 91–96, 2017.

[13] Dixit, K. Wazarkar, and A. S. Sabnis, Antimicrobial UV curable wood coatings based on citric acid, *Pigment & Resin Technology,* 2021.

[14] Noè, L. Iannucci, S. Malburet, A. Graillot, M. Sangermano, and S. Grassini, New UV-curable anticorrosion coatings from vegetable oils, *Macromolecular Materials and Engineering,* vol. 306, no. 6, p. 2100029, 2021.

[15] X. Zhang *et al.*, Fabrication of mechanically stable UV-curing superhydrophobic coating by interfacial strengthening strategy, *Journal of Alloys and Compounds,* vol. 886, p. 161156, 2021.

[16] B. Goss, Bonding glass and other substrates with UV curing adhesives, *International Journal of Adhesion and Adhesives,* vol. 22, no. 5, pp. 405–408, 2002.

[17] X. Centelles, J. R. Castro, and L. F. Cabeza, Experimental results of mechanical, adhesive, and laminated connections for laminated glass elements–A review, *Engineering Structures,* vol. 180, pp. 192–204, 2019.

[18] A. Di Pietro and P. Compston, Resin hardness and interlaminar shear strength of a glass-fibre/vinylester composite cured with high intensity ultraviolet (UV) light, *Journal of Materials Science,* vol. 44, no. 15, pp. 4188–4190, 2009.

[19] J. Ramli, A. Jeefferie, and M. Mahat, Effects of UV curing exposure time to the mechanical and physical properties of the epoxy and vinyl ester fiber glass laminates composites, *ARPN J. Eng. Appl. Sci,* vol. 6, pp. 104–109, 2011.

[20] Abulizi, Y. Duan, D. Li, and B. Lu, A new method for glass-fiber reinforced composites manufacturing: Automated fiber placement with in-situ UV curing, in *2011 IEEE International Symposium on Assembly and Manufacturing (ISAM)*, 2011: IEEE, pp. 1–4.

[21] A. B. Çiğil, E. A. Kandırmaz, H. Birtane, and M. V. Kahraman, Thermal, optical and electrical properties of UV-curing screen-printed glass substrates, *Polymer Bulletin,* vol. 76, no. 9, pp. 4355–4368, 2019.

[22] J. Świderska, Z. Czech, W. Świderski, and A. Kowalczyk, Reducing of on polymerization shrinkage by application of UV curable dental restorative composites, *Polish Journal of Chemical Technology,* vol. 16, no. 3, 2014.

[23] T. Y. Kwon, R. Bagheri, Y. K. Kim, K. H. Kim, and M. F. Burrow, Cure mechanisms in materials for use in esthetic dentistry, *Journal of Investigative and Clinical Dentistry,* vol. 3, no. 1, pp. 3–16, 2012.

[24] H. J. Staehle and C. Sekundo, The origins of acrylates and adhesive technologies in dentistry, *The Journal of Adhesive Dentistry,* vol. 23, no. 5, pp. 397–406, 2021.

[25] M. Zhao, Y. Geng, S. Fan, X. Yao, M. Zhu, and Y. Zhang, 3D-printed strong hybrid materials with low shrinkage for dental restoration, *Composites Science and Technology,* vol. 213, p. 108902, 2021.

[26] K. Wang, B. Li, K. Ni, and Z. Wang, Optimal photoinitiator concentration for light-cured dental resins, *Polymer Testing,* vol. 94, p. 107039, 2021.

[27] E. Saleh, F. Zhang, Y. He, J. Vaithilingam, J. L. Fernandez, R. Wildman, I. Ashcroft, R. Hague, P. Dickens, C. Tuck 3D inkjet printing of electronics using UV conversion, *Advanced Materials Technologies,* vol. 2, no. 10, p. 1700134, 2017.

[28] L. Goh, H. Zhang, T. H. Chong, and W. Y. Yeong, 3D printing of multilayered and multimaterial electronics: A review, *Advanced Electronic Materials,* vol. 7, no. 10, p. 2100445, 2021.

[29] Li, S. Chen, Y. Wei, K. Liu, Y. Lin, and L. Liu, Facile, low-cost, UV-curing approach to prepare highly conductive composites for flexible electronics applications, *Journal of Electronic Materials,* vol. 45, no. 7, pp. 3603–3611, 2016.

[30] K. Maag, W. Lenhard, and H. Löffles, New UV curing systems for automotive applications, *Progress in Organic Coatings,* vol. 40, no. 1–4, pp. 93–97, 2000.

[31] M. Nasirzadeh, H. Yahyaei, S. Lashgari, and M. Mohseni, Attributing the crosslinking density to water vapor transmission rate of an acrylic-melamine automotive clearcoat, *Journal of Coatings Technology and Research,* vol. 18, no. 1, pp. 239–246, 2021.

[32] S. H. Kim *et al.,* Relationship between crosslinking and surface mechanical properties of UV curable coatings for automotive interior plastic parts, *Polymer (Korea),* vol. 44, no. 1, pp. 38–48, 2020.

[33] F. Ferrero and M. Periolatto, Antimicrobial finish of textiles by chitosan UV-curing, *Journal of Nanoscience and Nanotechnology,* vol. 12, no. 6, pp. 4803–4810, 2012.

[34] S. Li, H. Boyter, and N. Stewart, Ultraviolet (UV) curing processes for textile coloration, *AATCC Review,* vol. 4, no. 8, 2004.

[35] B. Neral, S. Šostar-Turk, and B. Vončina, Properties of UV-cured pigment prints on textile fabric, *Dyes and Pigments,* vol. 68, no. 2–3, pp. 143–150, 2006.

[36] S. Li, H. Boyter Jr, and L. Qian, UV curing for encapsulated aroma finish on cotton, *Journal of the Textile Institute,* vol. 96, no. 6, pp. 407–411, 2005.

[37] Z. Yildiz, H. A. Onen, A. Gungor, Y. Wang, and K. Jacob, Effects of NCO/OH ratio and reactive diluent type on the adhesion strength of polyurethane methacrylates for cord/rubber composites, *Polymer-Plastics Technology and Engineering,* vol. 57, no. 10, pp. 935–944, 2018.

[38] Z. Yildiz, A. Onen, and A. Gungor, Preparation of flame retardant epoxyacrylate-based adhesive formulations for textile applications, *Journal of adhesion science and Technology,* vol. 30, no. 16, pp. 1765–1778, 2016.

[39] Z. Yildiz, H. Aysen Onen, A. Gungor, Y. Wang, and K. Jacob, Synthesis and application of dual-curable PVB based adhesive formulations for cord/rubber applications, *Journal of adhesion science and Technology,* vol. 31, no. 17, pp. 1900–1911, 2017.

[40] M. Cakir, I. Kartal, and Z. Yildiz, The preparation of UV-cured superhydrophobic cotton fabric surfaces by electrospinning method, *Textile Research Journal,* vol. 84, no. 14, pp. 1528–1538, 2014.

[41] Zehra Yildiz, H. Aysen Onen, Ozan G. Dehmen, Atilla Gungor, and E. D. Kocak, Improvement The Washing Fastness of Fabrics with Water Repellent Finish Via UV-Curable Coatings, presented at the 17th National and 3rd International The Recent Progress Symposium on Textile Technology and Chemistry, Bursa, Turkey, 2019.

[42] O. G. Dehmen, H. A. Onen, Z. Yildiz, and A. Gungor, Chemical, mechanical, and thermal properties of UV-curable cellulose acetate butyrate-based oligomers and their electrospun fibrous mats, *Journal of Coatings Technology and Research,* vol. 17, no. 4, pp. 1043–1052, 2020.

[43] O. G. Dehmen, H. A. Onen, Z. Yildiz, A. Gungor, and Y. Boztoprak, Synthesis and characterization of UV-curable cellulose acetate butyrate-based oligomers and their cotton fabric coatings, *Journal of Coatings Technology and Research,* vol. 18, no. 4, pp. 1075–1085, 2021.

[44] Quan, T. Zhang, H. Xu, S. Luo, J. Nie, and X. Zhu, Photo-curing 3D printing technique and its challenges, *Bioactive Materials,* vol. 5, no. 1, pp. 110–115, 2020.

[45] Q. Ge *et al.,* 3D printing of highly stretchable hydrogel with diverse UV curable polymers, *Science Advances,* vol. 7, no. 2, p. 4261, 2021.

[46] D. Zhai, T. Zhang, J. Guo, X. Fang, and J. Wei, Water-based ultraviolet curable conductive inkjet ink containing silver nano-colloids for flexible electronics, *Colloids and Surfaces A: Physicochemical and Engineering Aspects,* vol. 424, pp. 1–9, 2013.

[47] G. Sathyanarayanan, M. Haapala, C. Dixon, A. R. Wheeler, and T. M. Sikanen, A digital-to-channel microfluidic interface via inkjet printing of silver and UV curing of thiol–enes, *Advanced Materials Technologies,* vol. 5, no. 10, p. 2000451, 2020.

[48] S. Oakdale, J. Ye, W. L. Smith, and J. Biener, Post-print UV curing method for improving the mechanical properties of prototypes derived from two-photon lithography, *Optics Express,* vol. 24, no. 24, pp. 27077–27086, 2016.

[49] M. Vogler *et al.*, Development of a novel, low-viscosity UV-curable polymer system for UV-nanoimprint lithography, *Microelectronic Engineering,* vol. 84, no. 5–8, pp. 984–988, 2007.

[50] Y.-H. Yun, C.-M. Lee, Y.-S. Kim, and S.-D. Yoon, Preparation of chitosan/polyvinyl alcohol blended films containing sulfosuccinic acid as the crosslinking agent using UV curing process, *Food Research International,* vol. 100, pp. 377–386, 2017.

8 Hyperbranched Polymers and Their Emerging Applications

Saeed Bastani,[1,2] Morteza Ganjaee Sari,[3] and
Samane Jafarifard[4]

[1] Surface Coatings and Corrosion Dept., Institute for Color Science and Technology, Vafamanesh St., Tehran, 1654-16765, Iran
[2] Printing Ink Science and Technology Dept., Institute for Color Science and Technology, Vafamanesh St., Tehran, 1654-16765, Iran
[3] Nano Technology Dept., Institute for Color Science and Technology, Vafamanesh St., Tehran, 1654-16765, Iran
[4] Polymer Engineering and Color Technology Dept., Amirkabir University of Technology, Hafez St., Tehran, 15875-4413, Iran

1 INTRODUCTION

Dendritic polymers are considered to be one of the most attractive polymer categories. Schematic structures of the most popular dendritic polymers are given in Figure 8.1. As can be seen, these polymers commonly consist of three distinctive parts: core, branches, and end groups. However, there exist various types of dendritic polymers with different structures according to their branch controllability and molecular symmetry. Given that, the polydispersity index (PDI) of the dendritic polymers may range between 1 (M_w/M_n = 1.01–1.0001) for monodisperse polymers known as dendrimers and higher values (M_w/M_n >1.1) for polydisperse structures usually known as hyperbranched polymers.

1.1 HYPERBRANCHED POLYMERS

Hyperbranched polymers belong to the dendritic polymer categories. They possess a highly branched structure and an asymmetrical pseudo-spherical shape. The synthesis of HBPs, in comparison with other types of dendritic polymers, is simpler and usually doable in a one-step process. This feature can lower the final price of the product at large-scale production, enabling economical HBPs to be used in industrial applications. Perstorp Group in Sweden, DSM in the Netherlands, and BASF in Germany are the main manufacturers of HBPs across the world. Although the lack of symmetry and higher PDI in HBPs limit their utilization in high-tech applications, they have a multitude of uses where other specifications such as physical and functional properties of the dendritic polymers are needed. Table 8.1 provides a comparison of some properties of HBPs, dendrimers, and their equivalent linear polymers.

2 SYNTHESIS OF HBPS

A prolonged multi-step synthesis method of dendrimers which contains a complex purification process after each step will result in high-cost products that limit the application of these polymers. To find a solution to this challenge, in 1952 the idea of HBPs synthesis via a one-step or one-pot process emerged according to Florie's theory. Forty years later, in 1990, HBPs were successfully synthesized through a one-step polycondensation polymerization method. This was the first

DOI: 10.1201/9781003278269-8

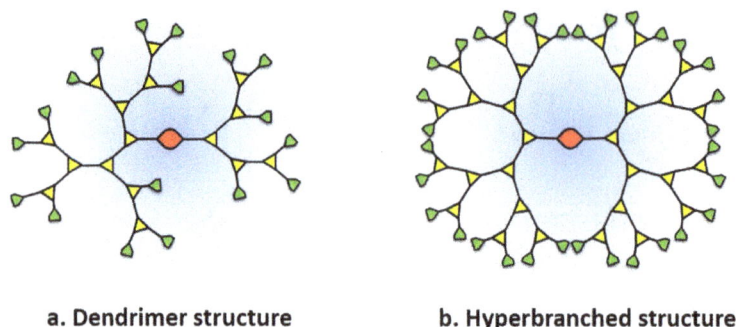

a. Dendrimer structure   b. Hyperbranched structure

FIGURE 8.1   Schematic molecular structure of dendritic polymers.

TABLE 8.1
Comparison of Some Properties of HBPs, Dendrimers, and Their Equivalent Linear Polymers

Property	Linear Polymers	HBPs	Dendrimers
Shape	Random chain	Asymmetrical sphere	Symmetrical sphere
Melt viscosity	High	Low	Low
Intrinsic viscosity	High	Fairly high	Low
Crystallinity	High	Amorphous	Amorphous
Reactivity	Low	High	High
Structure controllability	Low	Medium	High
Compressibility	High	Low	Low

successful and reliable synthesis of HBPs [1]. Self-condensing vinyl polymerization (SCVP) and free radical polymerization (FRP) methods were also utilized to synthesize HBPs. In these methods, monomers with an AB_x molecular formula, where x is equal to or more than 2, are used in the one-step synthesis of HBPs as long as A only reacts with B and creates an A–B chemical bond. These reactions cause the polymer to grow and create a 3D and branched structure. Although the main backbone of the synthesis has remained unchanged, the methods have been progressively developed as industrialization advances, and higher efficiencies and delayed gelation are now possible [2–4]. The degree of branching (DB) is one of the most important specifications in the synthesis of HBPs. This parameter can be calculated according to Equation (1). Here D is the number of dendritic units, T is the number of terminal units, and L is the number of linear units.

$$DB = \frac{D+T}{D+T+L} \tag{1}$$

DB spans between 0 and 1 and can be estimated quantitatively according to the peak area in NMR spectra. $DB = 0$ means that the polymer is linear and when DB is unity, it means the polymer is a dendrimer. The DB for HBPs is between 0.4 and 0.7.

Steric hindrance of the growing chains and the reactivity of functional groups are the most important factors affecting the DB of HBPs [5].

3 TYPES OF HBPS

As shown in Figure 8.2, HBPs may be categorized according to different aspects of the HBP properties such as physical and chemical molecular structures and polarity. These categorizations are discussed below.

3.1 BASED ON STERIC MOLECULAR STRUCTURE

The steric molecular structure of HBPs depends on the core type and repeated branches uniting. The steric molecular structure of HBPs is divided into two categories: compact and segmented HBPs. A schematic of these HBP structures can be seen in Figure 8.3. The compact HBPs have few repeated units between every two branching points, which leads to a compact structure of branches and terminal groups in the outer layer of the polymer structure. Meanwhile, in segmented HBPs, repeated units are longer, causing a greater distance between the branching points. Therefore, the branches are farther from each other and the functional groups may be located on the outer or inner layer of these HBPs [6]. From another aspect, HBPs can be divided into linear and branched HBPs. There are some linear parts in linear HBPs, while branched HBPs consist of various branched parts that grow over or alongside each other creating an asymmetrical structure. An HBP nanoparticle is another possible type of steric structure in which several HBP molecules attach through a chemical bond and create a nanoparticle [6].

3.2 BASED ON THE POLARITY OF TERMINAL GROUPS

The polarity of terminal groups is an important property that affects the interaction of HBPs with their surroundings. The polarity of HBPs can be altered by changing their terminal groups. HBPs fall into two sections based on their terminal groups' polarity as follows.

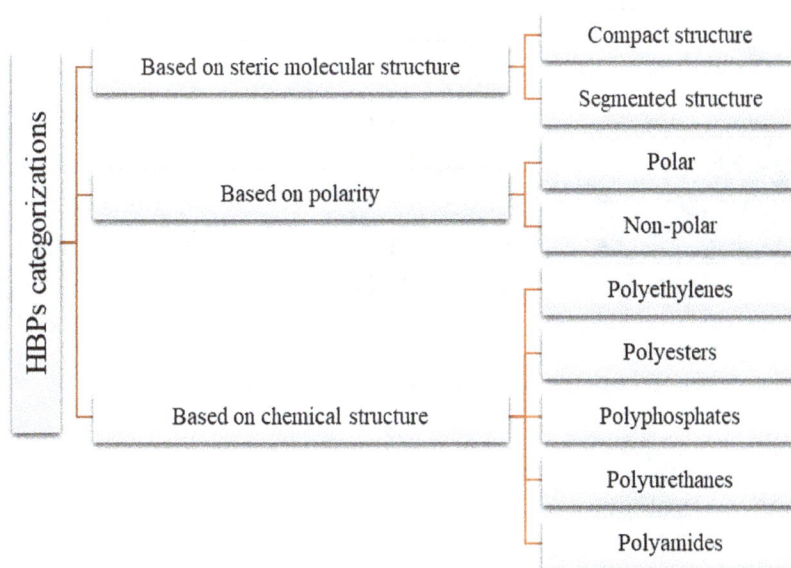

FIGURE 8.2 HBP categories.

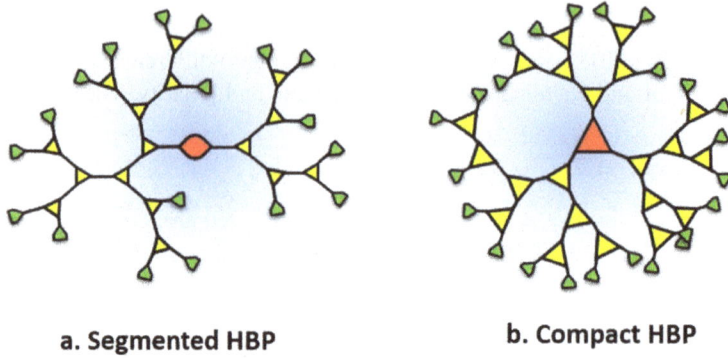

a. Segmented HBP   **b. Compact HBP**

FIGURE 8.3   Schematic of segmented and compact HBP structures.

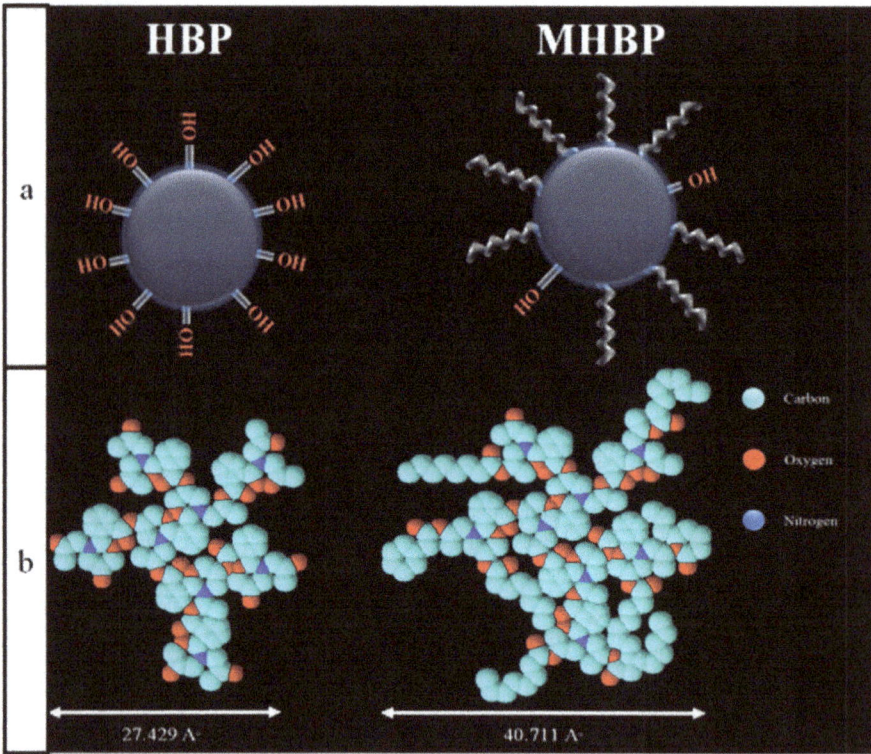

FIGURE 8.4   HBP with hydroxyl terminal groups and fatty acid modified HBP. Adapted with permission from Reference [7]. Copyright 2016, Elsevier.

3.2.1   Non-Polar HBPs

Non-polar HBPs are HBPs with a low polarity which is a result of their non-polar functionalities in the terminal groups. The hydrocarbon chains are one of the most common groups used to modify HBPs to lower the polarity [7]. The molecular structure of HBP modified with saturated fatty acid with seven carbon lengths before (HBP) and after (MHBP) modification is shown in Figure 8.4. Substitution of polar groups by short hydrocarbon chains leads to a reduction of intermolecular interactions and T_g also. The reduction in T_g causes the physical state of HBPs in room temperature to change from solid to a viscous liquid. In the literature, a 50–60°C reduction in T_g has been reported which has been shown to have a direct correlation with the length of the carbon chain [7–9].

3.2.2 Polar HBPs

Accordingly, the existence of polar functional groups as the terminal groups of HBPs will result in polar HBPs. For instance, hydroxyl functionalities are the most common terminal groups seen in polar HBPs. Furthermore, the hydroxyl functionalities are capable of the formation of hydrogen bonds that strengthen intermolecular interactions. HBPs with hydroxyl functionalities can be used as rheological modifiers and mechanical reinforcing agents [10–12].

3.3 BASED ON THE CHEMICAL STRUCTURE OF HBPs

Since various monomers can be used for the synthesis of HBPs, different types of HBPs based on the chemical structure of the monomers are attainable. Therefore, categorization based on the chemical structure seems reasonable. Some are explained next.

3.3.1 Polyethylene HBPs

Polyethylene HBPs have a hydrophobic structure and are soluble in solvents like hexane, THF, and chloroform. These HBPs are synthesized as low-density polyethylene (LDP). The result of a pressure reduction in the synthesis of LDP is a hyperbranched polyethylene structure. The degree of branching and molecular weight of the polyethylene HBPs can be controlled by adjusting the pressure, temperature, and metal catalyst [13–15].

3.3.2 Polyester HBPs

Polyester HBPs are synthesized from monomers containing ester or carboxylic acid functionalities with the ester bonds being between the structure of the HBPs. Hydroxyl and carboxyl functionalities are the most frequent terminal groups of polyester HBPs. Therefore, they are usually soluble in solvents like acetone, THF, and dimethylformamide. Polyester HBPs show high thermal resistance. Considering the type of ester monomers, two types of polyester HBPs are imagined: (a) aromatic polyester HBPs consisting of aromatic monomers; and (b) aliphatic polyester HBPs consisting of an aliphatic ester structure. In some cases, aromatic polyester HBPs are modified by hydrocarbon chains resulting in HBPs having both flexible aliphatic parts along with rigid aromatic segments.

3.3.3 Polyphosphate HBPs

Polyphosphate HBPs are biocompatible and biodegradable polymers. These HBPs may degrade by hydrolysis or enzymatic cleavage of phosphate bonds. The terminal groups of the polyphosphate HBPs are usually hydroxyl functionalities. These terminal groups provide higher biocompatibility with the polyphosphate HBPs. In addition to bioapplications, polyphosphate HBPs can be used as fire-retardant agents in related applications [16,17].

3.3.4 Polyurethane HBPs

Polyurethanes are composed of the reactions between isocyanate and alcohol or amines. Polyurethane HBPs consist of monomers with an alkyl hydroxyl group and two inactivated isocyanate functionalities. The product of such a reaction is the formation of the –NH–C(=O)–O– group that bonds the repetitive units [18].

3.3.5 Polyamide HBPs

The amide groups are molecular bridges of the repetitive units in polyamide HBPs. In bioscience, polymers in which the constituent monomers are attached through amide bonds are also called peptides. The peptides are smaller chains of amino acids. Therefore, polyamide HBPs have peptide structures and can be called peptide HBPs. Peptide HBPs have unique thermal, mechanical, and chemical properties. Usually, repetitive units in these HBPs have rigid structures. Also, relatively strong hydrogen bonds between the molecular structures cause poor solubility in organic solvents at room temperature.

Peptide HBPs, like the other types of HBPs, can be synthesized with either aromatic or aliphatic structures according to the chemical structure of the monomers. In addition to the amide groups, other functional groups such as amine and ether could be created during the HBP synthesis process. These new structures have more diversified features compared to the polyamide HBPs. Polyamidoamine and polyamide ether HBPs are examples of these HBP structures.

4 PROPERTIES OF HBPS

As mentioned previously, the HBP structure consists of a core, branches, and terminal groups. Reactivity, flexibility, radius of gyration, and other physical-chemical properties of HBPs are strongly dependent on these constituents. The main differences between HBPs and linear polymer structures are the absence of chain entanglements, globular steric structure, and a multitude of terminal groups of HBPs.

As the generation number of HBPs increases, the molecular radius linearly increases; however, the number of terminal groups grows exponentially. Hence, lower generations of HBPs have more flexible and mobile structures, while higher generations of HBPs behave like rigid spheres with limited mobility and lower flexibility [15]. The latter is mostly related to a phenomenon called a "tethered" surface in which the numerous terminal groups on the surface somehow entangle with each other due to a lack of space and make a hard surface and the whole structure becomes like a hard sphere. Also, it must be mentioned that HBPs with a lot of terminal groups show different properties. The most common terminal groups of HBPs are hydroxyl, carboxyl, amino, thiol, and halide functionalities [19]. These features result in a myriad of unique characteristics not found in other polymer structures that make HBPs an appropriate candidate for a vast range of applications.

4.1 THERMAL PROPERTIES

HBPs are amorphous polymers mainly due to the branching structures that prevent the orientation of the chains and hence crystallization. Therefore, glass transition temperature, T_g, is one of the most important properties of these polymers. T_g regulates the application areas of HBPs in different industries such as powder coating, rheological modifier, etc. For example, if T_g is lower than ambient temperature, long-range rotational and longitudinal chain movements, along with short-range movements of side chains and terminal groups, are possible at ambient temperature.

The number and type of terminal groups are the most important parameters affecting T_g and hence the thermal properties of HBPs. A literature review shows that terminal groups with different polarities alter the T_g of HBPs across a wide range of temperatures.

The molecular weight of HBPs is another influencing parameter changing the T_g of HBPs. On the other hand, molecular weight is directly related to the generation number of HBPs. Therefore, higher generations mean higher T_g. Another probable cause that increases T_g is the restrictions in chain motions as a result of a large number of branches.

4.2 MECHANICAL PROPERTIES

Despite linear polymers, HBPs possess no chain entanglements. The lack of chain entanglements causes these polymers to have poor mechanical properties. The behavior of HBPs under mechanical stress is more likely to resemble ductile metals. HBPs are almost intolerant to tensile forces because of their globular structure and inability to show chain orientation, resulting in very poor elongation properties. Intermolecular forces like hydrogen bonds and van der Waals interactions are the only forces acting among HBP molecules. However, despite their poor intrinsic mechanical properties, thanks to their numerous terminal groups that can be functional, it is possible for them to form extensive physical and chemical bonds with other polymeric chains acting as reinforcing agents in polymer blends [20].

4.3 RHEOLOGICAL PROPERTIES

There are many researchers studying the properties of dilute and semidilute solutions of HBPs. HBPs have low viscosity, smaller hydrodynamic volume, and a low radius of gyration to hydrodynamic radius ratio compared to their linear counterparts in the solution state. Unlike linear polymers, HBPs have many terminal groups that regulate molecular interactions. However, in HBP solutions, the lack of molecular entanglements causes a Newtonian rheological behavior and lowers the solution viscosity in comparison with the linear counterparts.

The viscosity of the HBP solution depends on the generation and type of terminal groups. The lower the polarity of the terminal groups, the weaker the molecular interactions and the lower the solution viscosity. Also, higher generations have higher solution viscosity. The melt viscosity of HBPs, nonetheless, is different from their correlated linear polymers. As the molecular weight increases in the linear polymers, the molecular entanglements become more probable, and hence the melt viscosity rapidly increases. The nonexistence of molecular entanglements in HBPs results in a lower melt viscosity. Furthermore, the viscosity grows much more slowly as the molecular weight increases. In addition to the molecular weight, the polarity of the terminal groups has also a significant effect on the melt viscosity. A reduction in polarity of the terminal groups weakens the intermolecular interactions, leading to easing of the mobility of the molecules and hence a reduction of the melt viscosity. The melt viscosity of HBPs is well explained by the WLF model.

5 APPLICATIONS

Considering the unique properties of HBPs, diversified special applications for these polymeric structures are imaginable. A large number of terminal groups, the spherical steric structure, the inexistence of molecular entanglements, and the amorphous structure are the most characteristic properties of HBPs. Therefore, HBPs can have a wide range of applications in various areas including photoelectricity, biotechnology, composites, coatings, adhesives, and different types of modifiers. Nevertheless, most HBP applications are not greater than the laboratory scale. Some companies like Boltorn™ (HPE, Perstorp Co., Sweden), Lupasol® (hyperbranched polyethylenimine), and BASF Co. (Germany) produce these polymers at an industrial scale.

5.1 HBPs IN OPTICAL AND ELECTRONIC APPLICATIONS

HBPs possessing electron transfer properties are used in light-emitting diodes (LED). Polyfluorines (PFs), for example, are used in blue LEDs. The existence of triazole, truxene, oxadiazole, or carbazole building units improves the electron transfer abilities. Also, diodes consisting of HBPs show very desirable luminous intensities [21–23]. Nowadays, poly(3,4-ethylenedioxythiophene) polystyrene sulfonate (PEDOT/PSS) is the most common material used in LEDs. However, some studies have shown diodes made of HBPs that have more radiation power in comparison with the diodes containing PEDOT/PSS [24]. Also, HBPs have been employed as non-linear optoelectronic (NLO) materials. Donor-π-acceptor chromophores in NLO materials play an important role in the electro-optic area. Dipole–dipole interactions of the chromophores should be eliminated to obtain higher efficiency for NLO materials. The related studies show that placing the chromophores at the main or side branches or the periphery of HBPs can reduce the dipole—dipole interactions of the chromophores [25].

5.2 HBPs for Nanocrystals

Nanocrystals (NCs) can be found in a wide range of materials including insulators, semiconductors, and metal crystals. The physical and chemical properties of NCs mostly depend on the crystal size. However, the accumulation of NCs reduces the performance efficiency. HBPs may be of good use to solve this problem. There are several methods to create HBPs on the surface of NCs. Figure 8.5 shows these methods schematically. 3-D and voluminous structures of HBPs prevent the accumulation of NCs. The stabilizing property of HBPs may be increased by altering the terminal groups, leading to enhanced stability of NCs in various media.

Also, quantum dots are considered to be semi-conductive NCs that have a wide range of applications in biosensors and optical devices. These NCs possess higher thermal resistance and hence lifetime compared to the other organic compounds. Many studies have investigated quantum dot stabilization by HBPs. Polyglycerol and polyethyleneimine HBPs are examples used for the stabilization of quantum dots [6,25]. Also, HBPs can be used to stabilize metal NCs. HBPs with amine and sulfur functionalities are more applicable for this purpose. These functional groups can create strong interactions with metal NCs [26,27].

5.3 HBPs for Supramolecular Self-Assembly

Molecular self-assembly can be carried out using surfactants, linear block copolymers, and dendritic polymers, especially HBPs. The strong interactions, e.g., hydrogen bonding, electrostatic attraction, etc., ease the self-assembly process of supramolecules consisting of HBPs. Supramolecules may have 0-D, 1-D, 2-D, or 3-D structures. 0-D supramolecules are micelles formed of amphiphilic HBPs. Their size is less than 10 nm. An amphiphilic HBP has a hydrophilic core of which hydrophobic branches are pointed out or vice versa.

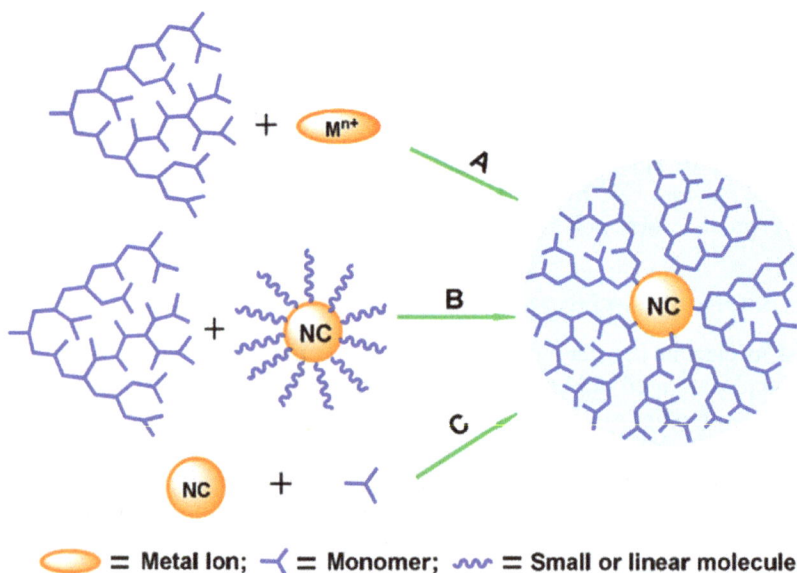

FIGURE 8.5   Three methods for the synthesis of NCs. Adapted with permission from Reference [6]. Copyright 2015, Royal Society of Chemistry.

1-D supramolecules are nanofibers or nanotubes made of HBPs. Nanofibers are synthesized by a polyester HBP consisting of alkyl chain branchings and carboxylic acid terminal groups. The nanofiber supramolecule is formed at the interface of air and water [28]. Also, studies have shown that polyether HBPs with hydroxyl terminal groups can be used in making nanotube structures. Vesicles and membranes are examples of 2-D HBPs self-assembly. Micelles, fibers, tubes, or membranes created at the microscale are products of the 3-D self-assembly of HBPs. A common practice in the self-assembly of HBPs is the use of amphiphilic HBPs. The latter structures have a polarity gradient from the core to the surface. Molecular weight, number of hydrophilic functions, an inner structure of HBPs, and concentration of HBPs in the medium are effective factors in the self-assembly process of HBPs [29,30].

Biotechnology is one of the most hi-tech application areas for these structures. For example, nanovesicle self-assembly HBPs can be used as a replacement for liposomes that have higher stability in comparison with liposomes [31]. Drug encapsulation is another application of self-assembled HBPs. This encapsulation is used to control the drug-delivery rate. Drugs can be placed inside the internal free space of a single-molecule micelle. HBP micelles are more stable compared to linear block copolymer micelles [19]. HBP self-assembly also can be used in encapsulating other chemical compounds such as dyestuffs. The encapsulation of dyestuffs improves the dying process of the fibers [32].

5.4 HBPs in Biological Applications

HBPs have a wide range of applications in bioscience. The existence of a variety of biocompatible and biodegradable HBPs makes them suitable in the bioscience area. The phosphates, peptides, and saccharides are the building units of lipids, proteins, and polysaccharides, respectively, and HBPs consisting of phosphates, peptides, and saccharides can be biocompatible and biodegradable. Nowadays, polyglycerol, polyethylene oxide, polyester, polyphosphate, and polypeptide HBPs are being used in bioapplications. Drug delivery, bioscaffolds, tissue engineering, and diagnosis methods like bio-imaging are examples of HBP usage in biotechnology [33].

5.5 HBPs in Polymers and Composites

HBPs have a multitude of applications in the polymer and composite industry. Polyurethane HBPs, for instance, are applicable in the production of shape-memory polymers. Short chains of HBPs have mobility suitable for self-healing applications. HBPs containing azide or alkyne terminal groups can be used for self-healing applications. Polyurethane HBPs embed enhanced mechanical properties in elastomers. The presence of polytriazol HBPs in epoxy adhesive can increase the adhesion strength by up to seven times.

Polyolefin is a type of polymer showing desired physical and mechanical properties. However, some properties such as hydrophobicity, nonpolar structure, and high crystallinity limit their applications. HBPs can be used to modify the properties of polyolefins. Amphiphilic HBPs containing lots of polar terminal groups and a 3-D steric structure can improve the polarity of these polymers and reduce crystallinity. These changes can improve, for example, the dye-ability of the polyolefin filaments, which extends their applications in the textile industry [34–36].

5.6 HBPs in Nanocomposites

Nowadays, nanocomposites are being used to improve the mechanical or thermal properties or even to produce fire-retardant or electrically conductive materials. The stability of nanoparticles inside the medium and strong interaction of the nanoparticles with the surrounding molecules is the greatest challenge in the nanocomposite production process. Surface modification of nanoparticles

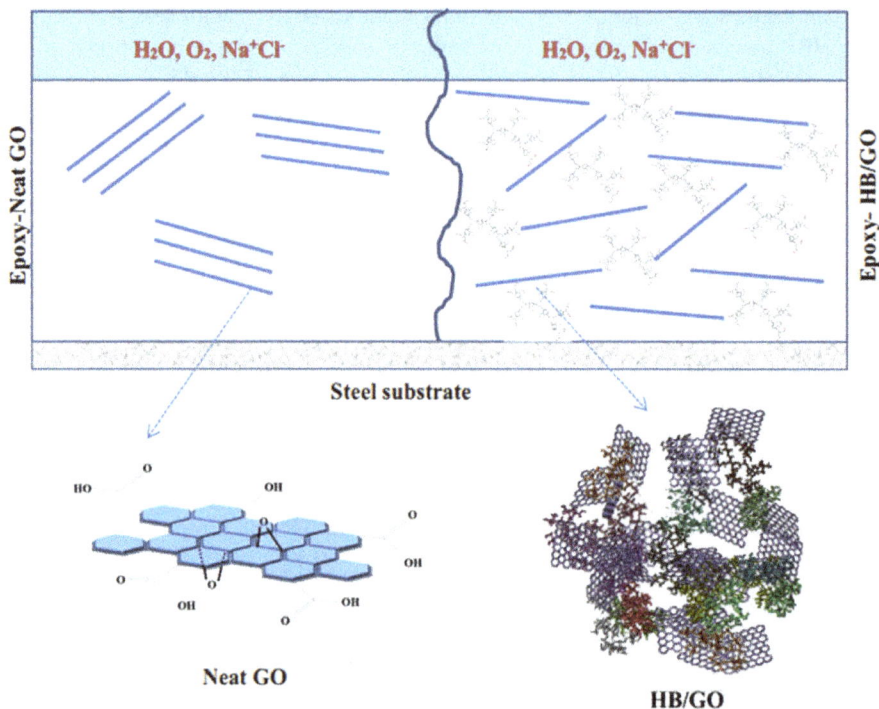

FIGURE 8.6   Corrosion protection mechanism of HB/GO in epoxy coating matrix. Adapted with permission from Reference [37]. Copyright 2019, Elsevier.

is the most common solution to improving the stability of nanoparticles inside the nanocomposites. The use of HBPs for surface modification of nanoparticles is a method that has been investigated by many researchers. For example, modification of graphene oxide (GO) with HBPs can improve dispersion via a voluminous steric structure among of carbon sheets. As can be seen in Figure 8.6, this improvement in GO dispersion can improve corrosion protection of an epoxy coating filled with HBP-modified GO.

3-D structure, lack of chain entanglements, and lots of terminal groups are the desirable features that improve the stability of nanoparticle dispersions. Some terminal groups of HBPs can react with the polymeric medium molecules. In this case, the reacted terminal groups can act as chemical bridges between a nanoparticle and the dispersing medium. Therefore, the interaction between the nanoparticle and the medium improves and hence the properties of the resulting nanocomposite are enhanced. The surface modification of nanoparticles can be carried out via in situ polymerization of HBPs in the presence of the nanoparticles. The formation of HBPs on the nanoparticle surface requires a core existing on the surface of the nanoparticle or should be created on it. HBPs have been used for the modification of silica, polyhedral oligomeric silsesquioxane (POSS), clay, graphene, and CNT nanoparticles [6,37–40].

5.7   HBPs in 3D Printing

3-D printing is a newly emerging manufacturing method that has many applications in various industries. One of the most widely used methods in 3-D printing is the UV curing procedure. In this method, UV irradiation is used for curing each layer. HBPs with a globular steric structure can improve the mobility of the reactive specimens and reduce the internal stresses of the cured system. On the other hand, the reactive functionalities on the HBP surface increase the average functionality, and hence the kinetics of the reactions increases, resulting in improving the mechanical properties of the printed objects [41,42].

5.8   HBPs in Coatings

The globular steric structure that contains no crystalline part in HBPs makes these polymers transparent. Therefore, HBPs are useful in the coating industries.

5.8.1   UV-Curable Coatings

UV-curable coatings can be cross-linked in lower temperatures and a shorter time in comparison with thermal setting systems. These coatings can be used for thermally sensitive substrates as they cause no deformation in the substrate. Monomers and oligomers with acrylate functional groups are the most common materials used in UV-curable coatings. The acrylate functional groups can participate in the chain photoreactions and become a part of the cross-linked network.

HBPs with acrylate terminal groups are recommended to be used in UV-curable coating formulations. HBPs can either act as the main resin of the UV-curable formulation or as a modifier to improve the final properties of the coating. UV-curable coatings that contain HBPs show enhanced curing and mechanical properties. HBPs can also be used in special fire-retardant, hydrophobic, water-based, and powder UV-curable formulations [41].

5.8.2   Anti-Corrosion Coatings

Anti-corrosion coatings are applied on metal surfaces to prevent corrosion and provide scratch protection. It is well known that cross-linking density affects the performance and durability of the anti-corrosion coatings. The presence of HBPs in anti-corrosion coatings can improve the protection function. The numerous functional groups on HBPs can increase the cross-linking density of the cured coating. HBPs with hydroxyl terminal groups are often used in these functional coatings. Researchers have illustrated protective coatings containing HBPs that have enhanced mechanical and anti-corrosion properties with higher stability and durability in comparison with linear polymers [43].

5.8.3   Printing Inks

Printing inks are coatings applied at a low thickness on various substrates. Printing inks must show appropriate adhesion with the substrate and have a good appearance and durability. Also, these formulations must have controlled rheological properties per the printing process. HBPs with globular structures and a large number of terminal groups can adjust the properties of printing inks. For example, HBPs with nonpolar terminal groups can tune the rheological properties of UV-curable inkjet inks [7]. As can be seen in Figure 8.7, HBPs with a nonpolar group can act like ball bearings, facilitate system mobility, and reduce trapped tension in the network after curing. HBPs with hydroxyl functional groups can also improve the colorfastness of inks [44].

5.9   HBPs as Modifiers

Modifiers are used in small quantities in formulations to improve various properties such as thermal, physical, and mechanical characteristics. Because of the unique mentioned properties, HBPs can also be used as modifiers with multiple influences on the formulations.

5.9.1   Toughening or Reinforcing Agents

Toughening or reinforcing agents are modifiers that improve the mechanical properties of a formulation after curing. The strength of the polymeric network after curing depends on the cross-linking density. A high cross-linking density will result in greater stiffness, hardness, and scratch resistance.

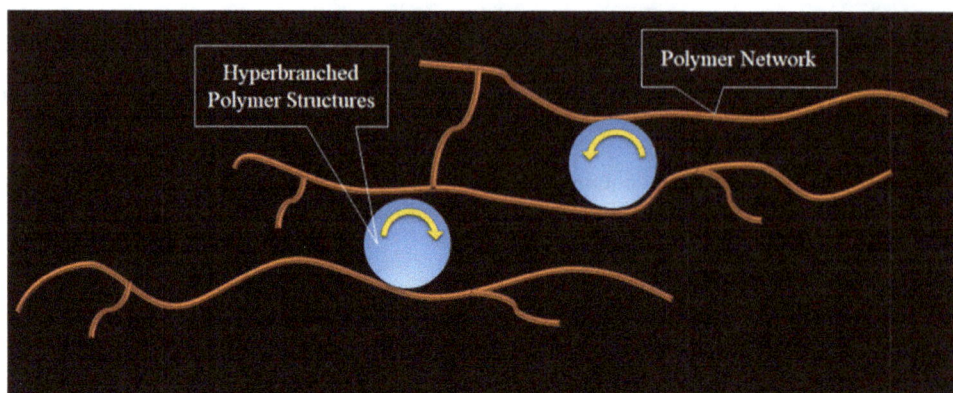

Nevertheless, high cross-linking density may increase the brittleness of the polymeric network. Therefore, the strength and flexibility of the polymeric network must be compromised. Generally, two important features of HBPs make them suitable as reinforcing agents. First is the low viscosity and lack of chain entanglements. Second is a large number of functional groups. HBPs can also act as a lubricant during the curing process and cause higher toughness after the curing process. Controlling the type and number of terminal groups of HBPs is necessary to adjust the cross-linking density and hence the flexibility of the polymeric network. For instance, reactive functionalities can form a chemical bond between the hyperbranched polymer and the polymeric network. These connections facilitate tension transfer through the polymeric network and as a result, the toughness of the final product is improved.

The shrinkage of the thermosetting resins is usually frequent because of the formation of chemical bonds. This can cause internal stresses in the polymer network. Before reaching the gel point, the chain mobility is high enough to dissipate stresses. However, at the gel point, chain mobility is restricted and the internal stresses will remain in the system. This may result in the creation of microcracks in the network. Microcracks are the main reason for network fragility that reduces the performance of the material in microelectronic materials. Researchers have shown that HBPs can reduce network shrinkage after the curing process [7].

Diglycidyl ether of bisphenol A (DGEBA), which usually is cured by a multifunctional amine, is one of the most important resins among the thermoset resins. The rigid aromatic ring in their backbone chain leads to a brittle structure that limits their applications. It has been shown that a hyperbranched polymer with epoxide terminal groups can be used as a modifier to improve the toughness of the cured DGEBA network [45].

5.9.2   Rheology Modifiers

The globular structure and lack of chain entanglements of HBPs make them a good agent as a rheology modifier in polymeric blends. Also, the terminal groups play an important role in the compatibility of the polymer with the surrounding molecules that may provide the desired rheological properties. HBPs can reduce the melt viscosity of the polymers. Therefore, HBPs can improve polymer processability [46]. Notwithstanding, by altering the interactions of HBPs with the surrounding polymer, HBPs can be used as thickening agents that increase viscosity. Thickening agents are often used to modify the rheological properties of oils, coatings, cosmetics, and drugs. They usually consist of many polar functionalities that increase the molecular interactions, leading to an increase in viscosity. The large number of terminal groups of HBPs makes them suitable to act as thickening agents also. According to the available studies, HBP thickening agents often have a special structure. For example, star-like polyurethane HBPs having only six branches can be used as a thickening agent. Having a higher or lesser number of branches reduces the thickening agent performance [47].

6 CONCLUSION AND FUTURE PROSPECTS

This chapter described hyperbranched polymers as a category of dendritic polymers. HBPs have attracted attention because of their unique properties such as lower degree of entanglement, a significant chain-end effect, and low viscosity of solution and the molten state compared to their linear counterparts. HBPs are prepared by a one-pot synthesis technique that is relatively simple and consists of rapid polymerization reactions. HBPs can be categorized according to steric molecular structure (segmented and condense structures), polarity (polar and non-polar), and chemical structure (such as polyphenylenes, polyethers, polyesters, etc.). HBPs are used in different areas of research, such as optics and electronics, nanocrystals, biological science, polymers and nanocomposites, and modifiers. Properties of HBPs such as the globular structure and the large number of functional end groups make these polymers appropriate for use as rheological modifiers. Although large advances have been achieved in the past few decades in various areas, most HBP applications are not greater than the laboratory scale.

It is expected that HBPs will become one of the most applicable advanced materials in the future. This expectation is expected to be fulfilled with reduced production cost of HBPs with novel synthesis procedures and the infrastructure to use HBPs in related area being provided.

REFERENCES

[1] C. J. Hawker, R. Lee, and J. M. J. Fr, One-step synthesis of hyperbranched dendritic polyesters, *J. Am. Chem. Soc.,* vol. 4588, no. log c, pp. 4583–4588, 1991.

[2] D. Wilms, S.-E. Stiriba, and H. Frey, Hyperbranched polyglycerols: from the controlled synthesis of biocompatible polyether polyols to multipurpose applications, *Acc. Chem. Res.*, vol. 43, no. 1, pp. 129–141, 2010.

[3] J. Kolomanska *et al.*, Design, synthesis and thermal behaviour of a series of well-defined clickable and triggerable sulfonate polymers, *RSC Adv.*, vol. 5, no. 82, pp. 66554–66562, 2015.

[4] I. A. Barker, J. El Harfi, K. Adlington, S. M. Howdle, and D. J. Irvine, Catalytic chain transfer mediated autopolymerization of divinylbenzene: toward facile synthesis of high alkene functional group density hyperbranched materials, *Macromolecules*, vol. 45, no. 23, pp. 9258–9266, 2012.

[5] K. Inoue, Functional dendrimers, hyperbranched and star polymers, Prog. Poly. Sci., vol. 25, 2000.

[6] F. Paquin, J. Rivnay, A. Salleo, N. Stingelin, and C. Silva, Hyperbranched polymers: Advances from synthesis to applications, *Chem. Soc. Rev.*, vol. 44, pp. 4091–4130, 2015.

[7] S. Jafarifard, S. Bastani, A. Soleimani Gorgani, and M. Ganjaee Sari, The chemo-rheological behavior of an acrylic based UV-curable inkjet ink: Effect of surface chemistry for hyperbranched polymers, *Prog. Org. Coatings*, vol. 90, pp. 399–406, 2016.

[8] P. D. Hamilton, D. Z. Jacobs, B. Rapp, and N. Ravi, Surface hydrophobic modification of fifth-generation hydroxyl-terminated poly(amidoamine) dendrimers and its effect on biocompatibility and rheology, *Materials (Basel).*, pp. 883–902, 2009.

[9] D. Schmaljohann and L. Ha, Modification with alkyl chains and the influence on thermal and mechanical properties of aromatic hyperbranched polyesters, *Macromol. Chem. Phys.*, vol. 57, pp. 49–57, 2000.

[10] E. Dzunuzovic, S. Tasic, B. Bozic, D. Babic, and B. Dunjic, UV-curable hyperbranched urethane acrylate oligomers containing soybean fatty acids, *Prog. Org. Coatings*, vol. 52, pp. 136–143, 2005.

[11] Z. Jiao, Q. Yang, X. Wang, and C. Wang, UV-curable hyperbranched urethane acrylate oligomers modified with different fatty acids, *Polym. Bull.*, vol. 74, no. 12, pp. 5049–5063, 2017.

[12] K. Han, W. Li, C. Wu, and M. Yu, Study on hyperbranched polyesters as rheological modifier for Spandex spinning solution, *Polym. Int.*, vol. 55, no. 8, pp. 898–903, 2006.

[13] J. Wang, M. Kontopoulou, Z. Ye, R. Subramanian, and S. Zhu, Chain-topology-controlled hyperbranched polyethylene as effective polymer processing aid (PPA) for extrusion of a metallocene linear-low-density polyethylene (mLLDPE), *J. Rheol. (N. Y. N. Y).*, vol. 52, no. 1, pp. 243–260, 2008.

[14] Z. Guan, P. M. Cotts, E. F. McCord, and S. J. McLain, Chain walking: A new strategy to control polymer topology, *Science*, vol. 283, no. 5410, pp. 2059–2062, 1999.

[15] Z. Dong and Z. Ye, Hyperbranched polyethylenes by chain walking polymerization: Synthesis, properties, functionalization, and applications, *Polym. Chem.*, vol. 3, no. 2, pp. 286–301, 2012.

[16] H. Wang, Q. Wang, Z. Huang, and W. Shi, Synthesis and thermal degradation behaviors of hyperbranched polyphosphate, *Polym. Degrad. Stab.*, vol. 92, no. 10, pp. 1788–1794, 2007.

[17] J. Liu, W. Huang, Y. Zhou, and D. Yan, Synthesis of hyperbranched polyphosphates by self-condensing ring-opening polymerization of HEEP without catalyst, *Macromolecules*, vol. 42, no. 13, pp. 4394–4399, 2009.

[18] M. Jikei and M. Kakimoto, Hyperbranched polymers: a promising new class of materials, *Prog. Polym. Sci.*, vol. 26, pp. 1233–1285, 2001.

[19] D. Wang, T. Zhao, X. Zhu, D. Yan, and W. Wang, Bioapplications of hyperbranched polymers, *Chem. Soc. Rev.*, vol. 44, no. 12, pp. 4023–4071, 2015.

[20] D. J. Massa, K. A. Shriner, S. R. Turner, and B. I. Voit, Novel blends of hyperbranched polyesters and linear polymers, *Macromolecules*, vol. 28, no. 9, pp. 3214–3220, 1995.

[21] T. Guo et al., Highly efficient, red-emitting hyperbranched polymers utilizing a phenyl-isoquinoline iridium complex as the core, *Macromol. Chem. Phys.*, vol. 213, no. 8, pp. 820–828, 2012.

[22] Y. Wu, X. Hao, J. Wu, J. Jin, and X. Ba, Pure blue-light-emitting materials: hyperbranched ladder-type poly (p-phenylene) s containing truxene units, *Macromolecules*, vol. 43, no. 2, pp. 731–738, 2010.

[23] L.-R. Tsai and Y. Chen, Novel hyperbranched polyfluorenes containing electron-transporting aromatic triazole as branch unit, *Macromolecules*, vol. 40, no. 9, pp. 2984–2992, 2007.

[24] T. Lee, Y. Kwon, J. Park, L. Pu, T. Hayakawa, and M. Kakimoto, Novel hyperbranched phthalocyanine as a hole injection nanolayer in organic light-emitting diodes, *Macromol. Rapid Commun.*, vol. 28, no. 16, pp. 1657–1662, 2007.

[25] W. B. Wu et al., Second-order nonlinear optical hyperbranched polymer containing isolation chromophore moieties derived from both 'h'-type and star-type chromophores, *Chinese J. Polym. Sci. (English Ed.*, vol. 31, no. 10, pp. 1415–1423, 2013.

[26] H. H. Nguyen et al., Mesomorphic ionic hyperbranched polymers: effect of structural parameters on liquid-crystalline properties and on the formation of gold nanohybrids, *Nanoscale*, vol. 6, no. 7, pp. 3599–3610, 2014.

[27] Z. Yuan, N. Cai, Y. Du, Y. He, and E. S. Yeung, Sensitive and selective detection of copper ions with highly stable polyethyleneimine-protected silver nanoclusters, *Anal. Chem.*, vol. 86, no. 1, pp. 419–426, 2014.

[28] M. Ornatska, K. N. Bergman, M. Goodman, S. Peleshanko, V. V Shevchenko, and V. V Tsukruk, Role of functionalized terminal groups in formation of nanofibrillar morphology of hyperbranched polyesters, *Polymer (Guildf).*, vol. 47, no. 24, pp. 8137–8146, 2006.

[29] B. Guo, Z. Shi, Y. Yao, Y. Zhou, and D. Yan, Facile preparation of novel peptosomes through complex self-assembly of hyperbranched polyester and polypeptide, *Langmuir*, vol. 25, no. 12, pp. 6622–6626, 2009.

[30] Y. Zhou, W. Huang, J. Liu, X. Zhu, and D. Yan, Self-assembly of hyperbranched polymers and its biomedical applications, *Adv. Mater.*, vol. 22, no. 41, pp. 4567–4590, 2010.

[31] Y. Zhou and D. Yan, Real-time membrane fusion of giant polymer vesicles, *J. Am. Chem. Soc.*, vol. 127, no. 30, pp. 10468–10469, 2005.

[32] Z. Weng, Y. Zheng, A. Tang, and C. Gao, Synthesis, dye encapsulation, and highly efficient colouring application of amphiphilic hyperbranched polymers, *Aust. J. Chem.*, vol. 67, no. 1, pp. 103–111, 2013.

[33] H. Jin, W. Huang, X. Zhu, Y. Zhou, and D. Yan, Biocompatible or biodegradable hyperbranched polymers: from self-assembly to cytomimetic applications, *Chem. Soc. Rev*, vol. 41, no. 7, pp. 5986–5997, 2012.

[34] M. G. Sari, N. Stribeck, S. Moradian, A. Zeinolebadi, S. Bastani, and S. Botta, Correlation of nanostructural parameters and macromechanical behaviour of hyperbranched-modified polypropylene using time-resolved small-angle X-ray scattering measurements, *Polym. Int.*, vol. 62, no. 7, pp. 1101–1111, 2013.

[35] M. G. Sari, S. Moradian, S. Bastani, and N. Stribeck, Modification of poly(propylene) by grafted polyester-amide-based dendritic nanostructures with the aim of improving its dyeability, *J. Appl. Polym. Sci.*, vol. 124, no. 3, pp. 2449–2462, 2012.

[36] M. G. Sari, N. Stribeck, S. Moradian, S. Bastani, and E. Bakhshandeh, Dynamic mechanical behavior and nanostructure morphology of hyperbranched-modified polypropylene blends, *Polymer International*, vol. 63, p. 195, 2014.

[37] M. G. Sari and B. Ramezanzadeh, Epoxy composite coating corrosion protection properties reinforcement through the addition of hydroxyl-terminated hyperbranched polyamide non-covalently assembled graphene oxide platforms, *Constr. Build. Mater.*, vol. 234, p. 117421, 2020.

[38] S. Ghiyasi *et al.*, Hyperbranched poly(ethyleneimine) physically attached to silica nanoparticles to facilitate curing of epoxy nanocomposite coatings, *Prog. Org. Coatings*, vol. 120, pp. 100–109, 2018.

[39] L. Ghazi Moradi, M. Ganjaee Sari, and B. Ramezanzadeh, Polyester-amide hyperbranched polymer as an interfacial modifier for graphene oxide nanosheets: Mechanistic approach in an epoxy nanocomposite coating, *Prog. Org. Coatings*, vol. 142, p. 105573, 2020.

[40] M. Lotfi, H. Yari, M. G. Sari, and A. Azizi, Fabrication of a highly hard yet tough epoxy nanocomposite coating by incorporating graphene oxide nanosheets dually modified with amino silane coupling agent and hyperbranched polyester-amide, *Prog. Org. Coatings*, vol. 162, no. July 2021, p. 106570, 2022.

[41] F. Mirshahi, S. Bastani, and M. Ganjaee, Studying the effect of hyperbranched polymer modification on the kinetics of curing reactions and physical/mechanical properties of UV-curable coatings, *Prog. Org. Coatings*, vol. 90, pp. 187–199, 2016.

[42] Z. Wang, J. Zhang, J. Liu, S. Hao, H. Song, and J. Zhang, 3D printable, highly stretchable, superior stable ionogels based on poly(ionic liquid) with hyperbranched polymers as macro-cross-linkers for high-performance strain sensors," *ACS Appl. Mater. Interfaces*, vol. 13, no. 4, pp. 5614–5624, 2021.

[43] A. K. Mishra, R. Narayan, and K. Raju, Structure–property correlation study of hyperbranched polyurethane–urea (HBPU) coatings, *Prog. Org. coatings*, vol. 74, no. 3, pp. 491–501, 2012.

[44] Z. Żołek-Tryznowska and J. Izdebska, Flexographic printing ink modified with hyperbranched polymers: Boltorn™ P500 and Boltorn™ P1000, *Dye. Pigment.*, vol. 96, no. 2, pp. 602–608, 2013.

[45] L. Luo, Y. Meng, T. Qiu, and X. Li, An epoxy-ended hyperbranched polymer as a new modifier for toughening and reinforcing in epoxy resin, *J. Appl. Polym. Sci.*, vol. 130, no. 2, pp. 1064–1073, 2013.

[46] Y. Hong, J. J. Cooper-White, M. E. Mackay, C. J. Hawker, E. Malmström, and N. Rehnberg, A novel processing aid for polymer extrusion: Rheology and processing of polyethylene and hyperbranched polymer blends, *J. Rheol. (N. Y. N. Y.).*, vol. 43, no. 3, pp. 781–793, 1999.

[47] D. Zhang and D. Jia, Toughness and strength improvement of diglycidyl ether of bisphenol-A by low viscosity liquid hyperbranched epoxy resin, *J. Appl. Polym. Sci.*, vol. 101, no. 4, pp. 2504–2511, 2006.

9 Advanced Polymers for Defense Applications

Nektarios K. Nasikas and Dionysios E. Mouzakis

Department of Military Studies, Division of Mathematics and
Engineering Sciences, Hellenic Army Academy, Vari, 16673, Attica, Greece

1 INTRODUCTION

Polymers are among the most widely used materials in our everyday lives with their applications seeming to be almost endless [1]. Ever since synthetic chemists were able to create structures that had a repeatedly positioned structural unit, called a monomer, polymers and polymer chemistry were about to take off [2]. Even the word polymer describes just that. Polymer is derived from the Greek word "πολυμερής" composed of the words "πολύ" meaning "many" and "μέρος" meaning "part." Therefore, a polymer is something that has many parts. Polymers are typically macromolecules that are essentially comprised of a variety of chemical units which in turn create a chain-like structure. This repetition causes the polymer to exhibit a very high molecular weight, which is simply a multiple of the monomers' molecular weight. The links between the monomers are essentially covalent bonds that hold these macromolecules together.

It is important to say that one should not confuse a monomer with a structural unit that just repeats itself. A monomer is the small starting molecule which a polymer can be constructed with. The structural unit that repeats itself will eventually produce the polymeric chain. How long this polymeric chain will be is of course governed by the number of structural units that will repeat themselves, and that number is widely termed the degree of polymerization $(\overline{X}_n)$. where X denotes the unit that repeats itself and n denotes the number of repetitions.

Of course, the process to form polymeric chains dictates that the monomers must have at least one chemically reactive site for them to form bonds with other monomers and through this process create a polymer. Here the monomer could or could not be the basic structural unit that repeats itself. Monomers are connected through chemical reactions and thus this chemical process to synthesize polymers is called a polymerization process. We attempt to clarify the above with an example shown in Figure 9.1. Figure 9.1 shows the polymerization process of the simplest polymer, namely polyethylene (PE). Polyethylene is derived from the polymerization process of ethylene which is, of course, a monomer to PE polymer.

The polymerization procedure shown in Figure 9.1, clearly, is not the only possible polymerization process. Figure 9.2 shows other polymerization processes that can create a variety of polymers, namely, alternating copolymers, block copolymers, graft copolymers, and random copolymers.

The above-shown copolymers constitute a classification of polymers depending on the type of monomer participating. This classification is analyzed below.

1.1 POLYMER CLASSIFICATION

Copolymers: Polymers that are derived from more than one type of monomer. The way these monomers are connected can create subdivisions or subgroups for these copolymers:

- Alternating copolymers: Copolymers that have two different types of monomers that are positioned in an exact alternating sequence within the polymeric chain (Figure 9.2a).

DOI: 10.1201/9781003278269-9

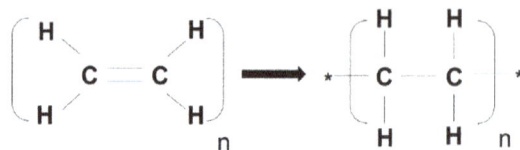

FIGURE 9.1   The polymerization process of ethylene.

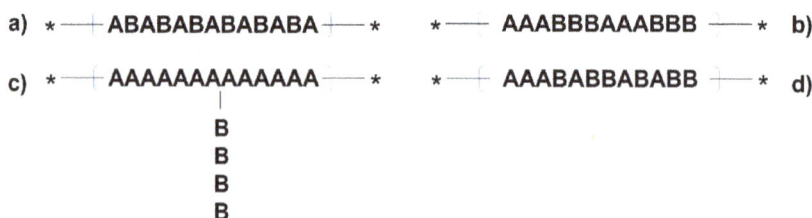

a) * ┤ABABABABABABA├ *   * ── AAABBBAAABBB ── * b)

c) * ┤AAAAAAAAAAAAA ── *   * ┤ AAABABBABABABB ├ * d)
              │
              B
              B
              B
              B

FIGURE 9.2   Graphical representations of (a) alternating copolymers, (b) block copolymers, (c) graft copolymers, and (d) random copolymers.

- Block copolymers: Copolymers that have two or more linear sequences of different blocks of monomers creating a long polymeric chain (Figure 9.2b).
- Graft copolymers: Copolymers that have an attached polymer chain onto another polymer chain. Both polymer chains are usually block-type polymers of a specific chemical kind (Figure 9.2c).
- Random copolymers: A randomly created polymeric chain that consists of a varying number and type of polymeric units that are attached (Figure 9.2d).
- Homopolymers: Polymers that simply derive from the same type of monomer.

Polymers can also occur naturally and are found on numerous occasions in Nature. They can occur in flora and fauna, and have a unique role to play on each occasion. They are essential to sustain life itself since they can be found in starch, cellulose, silk, proteins, collagen, nucleic acids, and, of course, natural rubber.

The Industrial Revolution, however, has given rise to chemical methods and synthetic processes that have allowed for the synthesis of an extremely large variety of synthetic polymers for various applications and needs. Synthetic polymers are, as the name implies, artificially synthesized polymers that derive from chemical synthetic methods. The industrial applications are very important and have offered some extremely important solutions and corresponding products in many fields ranging from packaging and specialty textiles, all the way to pharmaceutical products and medical devices. The most notable products in the above context are polyethylene, polyvinylchloride most commonly known as PVC, and polyurethane.

In the interphase between naturally occurring polymers and synthetic polymers, one could identify an interim category which could be described as semisynthetic polymers. These can be described as basically naturally occurring polymers that have undergone some kind of chemical treatment or chemical alteration of their structure.

1.2   STRUCTURAL CHARACTERISTICS OF POLYMERS

The structural characteristics of polymers, as expected, vary a lot and can attain various patterns. Below, we describe the most notable structural characteristics that characterize them.

Linear polymers: As the name implies, these polymers have repeatable structural units that are held together and as a result create a long linear chain. Of course, use of the word linear may give the impression that this line is completely straight. On the contrary, this linear arrangement is anything but straight in most cases and mostly resembles a spaghetti-like chain.

Branched polymers: Again, branched polymers are simply linear polymers (in most cases) where the linear chain has attracted branches of other linear chains attaching to the main linear chain. This type of polymer type is very useful when it comes to functionalized polymers when certain branches can play important roles giving special characteristics and functions to meet specific needs.

Cross-linked polymers: The structural characteristics of this type of polymer are represented by a three-dimensional network of polymers where multiple linear polymer chains are blended, thus creating a complex structure. This interconnected network is comprised of the connectivity of specific functional groups existing within the polymeric chains and is formed during the polymerization process.

1.3 MOLECULAR CHARACTERISTICS OF POLYMERS

Polymers in general can also be distinguished based on the molecular forces governing their synthesis and/or polymerization process. These molecular forces play a key role in polymer chemistry as the corresponding properties of polymers are tied to the molecular forces that hold the polymeric chains connected.

In general, we can distinguish the following broad polymer families based on the molecular forces governing their polymeric chain interconnection.

Elastomers: These polymers are generally amorphous concerning their structural characteristics exhibiting a high degree of elasticity (as the name also implies). This is due to the weak interaction forces between the polymeric chains existing in this type of polymer. An elastomer, in general, should exhibit elastic deformation under stress or strain, meaning that it holds its original shape when an external force is applied. It should be noted here that a lot of naturally occurring polymers are in fact elastomers.

Fibrous polymers: This type of polymer is characterized by strong intermolecular forces applied between the polymeric chains leading to high tensile strength, low elasticity, and resistance to deformation. Of course, their ability to create fibers is of paramount importance for defense applications and as we will discuss in depth in the following pages of this chapter, some of the most important polymers used in defense and armor applications are in the form of fibers.

Thermoplastic polymers: Thermoplastics are probably one of the most, if not the most, extensively used categories of polymers when grouped by their molecular forces. In thermoplastics, the intermolecular forces are somewhat intermediate in strength and thus can undergo a lot of processing with respect to their intended applications. When thermoplastics are heated the existing linear and branched type chains become increasingly soft but can regain their rigidness when the temperature is lowered. This process allows their mechanical treatment with heating and melting. The most important aspect of thermoplastic materials processing is that, in the above-mentioned procedure, thermoplastics can be reshaped by reapplying heat and pressure. This is because the thermal treatment does not cause any alterations to the chemical composition of the polymer, nor does it create any kind of new bonding between the polymeric chains. The most notable thermoplastic polymers are polystyrene, polyvinyl chloride (PVC), and Teflon.

Thermoset polymers: The most notable difference between thermoplastic and thermoset polymers is that the latter cannot be melted and reshaped when heat is applied. This is due to the extensive cross-linking reactions that are initiated upon heating. Concomitantly with applying heat to the polymer, irreversible chemical reactions take place leading to irreversible changes in the polymer's chemical composition. Hence the heating process of this type of polymer must be the last stage in their processing as any kind of alterations caused to the polymer by heat are permanent. The most notable example of a thermoset polymer is Bakelite.

One of the most distinctive characteristics of polymers and bearing in mind all of the above is the high molecular weight they exhibit. Oligomers, which are molecules that consist of a few units that repeat themselves, exhibit molecular weights that are in the vicinity of a few hundred to a few thousand of g per mole. Higher molecular weight polymers, which are also very important for several industrial as well as defense applications, exhibit molecular weights that are several orders of magnitude higher than the above, usually between 10^3 and 10^8 g per mole. The usual polymerization mechanisms out of which the above-mentioned polymers can be synthesized vary between step-growth polymers and chain polymers.

The step-growth polymers are different from their corresponding monomers and can be synthesized by utilizing multifunctional molecules and having them react with each other. In this way, through a subsequent condensation process, macromolecules can be produced by eliminating other small molecules such as polyesters. Chain polymers, on the other hand, can be produced by chain reactions between doubly bonded monomers by creating a chain-like structure on the opening bond.

Finally, tacticity is another characteristic that is frequently used to classify polymers. Hence, based on the tacticity, polymers can be grouped into atactic, syndiotactic, and isotactic polymers. Atactic polymers are those whose pendant groups are arranged in a random (atactic) way, syndiotactic polymers are those whose pendant groups are attached interchangeably, and finally, isotactic polymers are those whose pendant groups are arranged on one side of the polymer's main chain.

2 ADVANCED POLYMERS USED IN ARMOR

Ever since man started engaging in conflicts throughout history the concept of having the ability to use some kind of protective equipment has been of paramount importance. Conflicts were not just limited to war as we would imagine it, but could also have the form of protection against wild animals, especially when people would hunt for food and share hunting grounds with other animals. Therefore, the idea was fairly simple. We had to use some kind of protection for our vulnerable bodies, whether the threat was a knife, a spear, an arrow, a bullet, or a pair of teeth.

The first armor-type materials were natural, and mostly leather taken from pray animals. Leather was thick and could be stitched on top of clothing or headcovers. As long as the threats could not penetrate the thick leather of the protective gear it continued to be used. However, with the advancements of the available arms, leather protective gear could no longer fulfill its goal, which was to stop incoming threats from penetrating and thus inflicting wounds and/or fatal injuries. When craftsmanship around metallic materials created hard metallic weapons, such as spearheads, swords, knives, and many other metallic components used as weapons, it became clear that protective gear should also change to adequately face any threat coming from a metallic object.

The material of choice aiming to stop a threat that came from a metallic object was, naturally, another metal, in a "fight fire with fire" approach. Therefore, we entered an era where most of the armor used for protection was made of some kind of metal [3]. Figure 9.3 shows two characteristic metallic helmets that are now exhibited in the museum of Ancient Olympia in Greece.

In Figure 9.3, the Assyrian bronze helmet bears the following inscription: "ΔΙΙ ΑΘΕΝΑΙΟΙ ΜΕΔΟΝ ΛΑΒΟΝΤΕΣ" meaning "The Athenians dedicated to Zeus this booty that they took from the Medes," while the bronze helmet of Miltiades who was the Athenian General leading the Athenians to the victorious battle of Marathon (490 BC) reads: "ΜΙΛΤΙΑΔΕΣ ΑΝΕ[Θ]ΕΚΕΝ [Τ]ΟΙ

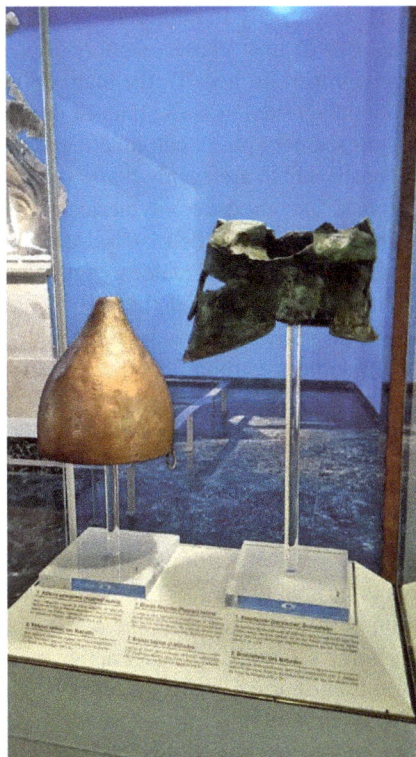

FIGURE 9.3 Bronze Assyrian (Persian) helmet taken from the Persian wars, beginning in the 5th century BCE (left) and the bronze helmet of Miltiades (right). Photo taken from the personal archive of NKN.

ΔΙ" meaning "Miltiades offered to Zeus," and was dedicated by Miltiades after his victory over the Persians in the battle of Marathon. As can be easily understood, both of these helmets could provide adequate protection against incoming threats for the heads of those who wore them (after all, they are still here after all these centuries) but it's safe to assume that they were very heavy and after long use could create severe fatigue to the fighter using them. The technological advancements regarding armor materials were not very fast, and metallic helmets are used even today for head protection by many armies around the world. Modern warfare requires that the fighter has maximum protection against threats but also has maximum movement availability while minimizing fatigue.

2.1 BODY ARMOR

Although we used helmets as an example above to emphasize the importance of protecting vital parts of the human body against threats, body armor is also extremely important and constitutes possibly the most significant type of armor for today's fighters. If we link the above to the metallic body armor, imagine how difficult it would be to have to move on the battlefield wearing an extremely heavy metallic body armor. It's not by chance that those who wore metallic body armor mostly in the Middle Ages moved about by riding horses. No wonder that the cavalry is today's armored army, equipped with tanks.

In the ever-evolving quest for lighter materials with upgraded protective capabilities, polymer or polymeric materials have played a key role. With the advancements of the available fabrication techniques at an industrial scale, a significant number of composite materials have been created for various defense applications. In this regard, polymeric fiber composites [4] have attracted a lot of attention and have been used extensively in numerous applications. Composite materials are generally composed of the at least two distinct phases, namely the matrix phase and the reinforcement phase. For polymer composite materials, usually, the matrix phase consists of a thermoplastic polymer, such as polycarbonate or polyamide, or some kind of duroplastic resin, such as an epoxy or polyester. Additionally, the reinforcement phase can take the form of fibers, carbon, aramid, or hybrid, woven fabrics, and glass beads which are evenly dispersed in the matrix phase.

The role of the reinforcement phase in composite polymeric material is to be able to sustain large external loads without failing. The properties of the material used in the reinforcement phase are extremely important for the above function. High tensile strength and Young's modulus, low elongation at the breakpoint, and low density are some of the key characteristics these materials used for the reinforcement phase should possess. Some of the most widely used fibers as key elements for the reinforcement phase are glass fibers, carbon fibers, polymeric fibers such as p-aramids, and ultra-high-molecular-weight polyethylene (UHMWPE). These fibers are in general characterized by an inverted correlation between their tensile strength and Young's modulus with respect to their diameter. that is, as their diameter decreases, their tensile strength and Young's modulus increase. Usually, the most common fibers used as reinforcement phases in composite materials are glass, carbon, and aramid fibers [5]. Some of the most important properties of the above-mentioned fibers are summarized in Table 9.1.

Body armor, in general, needs to be able to intercept and completely stop an incoming projectile that has been fired from a weapon or a fragment that is the result of the detonation of a bomb or a land mine. Ammunition used in today's assault rifles can achieve very high velocities after being fired from a weapon. Typical ammunition used in several NATO countries today is the 7.62 mm × 51 mm. A similar bullet is the 7.62 mm × 39 mm for the AK-47 rifle that is widely used by eastern countries. The 7.62 mm × 51 mm bullet can achieve a velocity of 850 m/s, leading to an extremely high kinetic energy upon impact, while the 7.62 mm × 39 mm can achieve a velocity of 641.3 m/s.

A bulletproof vest usually consists of several layers of different materials that are brought together to form a protective shield that will be impermeable to an incoming projectile (bullet or fragment). The matrix phase of the composite used as an insert in a bulletproof vest is usually some kind of resin or epoxy which is pressed and thermally treated to form a durable panel. The reinforcement phase could be any of the fibers presented in Table 9.1. More recent advancements in bulletproof vest technologies have made use of an outer ceramic panel that is described as the "striking face." The reason for using a ceramic as a striking face is to reduce as much as possible the velocity and kinetic energy of the incoming projectile.

TABLE 9.1

Mechanical and Physical Properties of Reinforcement Phase Fibers Used in Composite Materials for Armor Protection [6]

Reinforcement Phase Fiber	Density (g/cm³)	Tensile Strength (GPa)	Young's Modulus (GPa)	Elongation at Break (%)
E glass fiber	2.63	3.5	68.5	4.0
S glass fiber	2.48	4.4	90.0	5.7
Carbon fiber (Celton)	1.80	4.0	230.0	1.8
p-Aramid (Kevlar 149)	1.47	3.5	179.0	1.6
m-Aramid (Nomex)	1.40	0.7	17.0	22.0
UHMWPE (Dyneema SK76)	0.97	3.6	116.0	3.8
Zylon AS	1.54	5.8	180.0	3.5
Zylon HM	1.56	5.8	270.0	2.5
Boron fiber	2.64	3.5–4.2	420.0–450.0	3.7
Silicon carbide	2.80	4.0	420.0	0.6

Another very important component of a bulletproof vest is the final panel, called the "backing face," which is the surface of the insert that is against the body. This part of the protective insert usually gets the least attention but, on the contrary, its function is very important for preserving the safety of those who wear it. Conventionally, we think that every bulletproof vest that has an insert that can completely stop a bullet will prevent any harm to the bearer of the vest. Unfortunately, that is only half the truth. The incoming projectile, as it travels with an extremely high velocity through the atmosphere will carry with it a shock wave in the form of a cone from the projectile. This is most important when it comes to supersonic bullets (as is the case with most bullets used by today's firearms). This shock wave. as it hits the bulletproof vest along with the projectile. can cause severe deformation of the protective materials and ultimately cause blunt trauma.

This type of trauma is especially dangerous precisely because there is no readily observed wound or penetration mark on the body. This can lead to the mistaken assumption that no harm has been done by the impact. However, the mechanism of blunt trauma propagates the shock wave of the incoming bullet into the human body, which can lead to internal bleeding and subsequent death, even though the incoming projectile *did not* penetrate the bulletproof vest. The backing face is usually some kind of a foamy polymer, such as neoprene (polychloroprene foam) or Nitrile rubber. It is important to bear in mind that no penetration of the protective insert of a bulletproof vest does not necessarily mean that the wearer is out of harm's way.

2.2 Vehicle Armor

For decades, if not centuries, the material that has been most extensively used as protection against incoming threats and projectiles has been metallic. Large cast iron and steel plates were used to cover the panels of chariots in the Middle Ages and continued to protect tanks and armored vehicles in WWI and II.

The use of large metallic panels though comes with a price, which is energetic. Large metallic and steel panels are quite heavy, which means that the armored vehicle needs to be equipped with a large engine to move, which in turn requires a large amount of energy which is directly linked to its fuel consumption. An increased fuel consumption affects directly the fighting capability of an armored vehicle or tank. in the field, and especially in combat situations, refueling can be challenging. Fuel stations are almost nonexistent and fuel trucks are easy targets and are targeted frequently to deprive the fighting vehicles of getting refueled, thus becoming immobilized and correspondingly out of service.

Polymer solutions in armored vehicles have been revisited over the last few decades, primarily due to their reduced weight as compared with more traditional armor options such as steel [7]. The use of either large plates or smaller tiles in the form of composite materials, making use of polymeric fiber-reinforced epoxy or resin matrices, has provided a very adequate solution for enhancing the ballistic performance of armored vehicles. These plates have been primarily used on top of the large metallic surfaces of tanks and armored fighting vehicles (AFVs) or armored personnel carriers (APCs). Originally, very large plates were used to serve as a striking face which would reduce the kinetic energy of the incoming projectile, thus minimizing its ability to penetrate the steel plate. However, the use of large panels had significant drawbacks, with the main drawback being the following.

Upon impact of the incoming projectile, the kinetic energy would dissipate in the form of cracks and deformation of the polymeric striking face material. The propagation of cracks, however, was almost uncontrollable, and after the material had sustained as few as three hits it would start to significantly degrade, resulting in completely losing its properties and thus stopping it from performing as a protective layer. To overcome this, instead of using large "monolithic" panels, the use of separate tiles made an appearance. This had the profound benefit that when an incoming projectile hit the striking face, it would hit just one tile and the cracking propagation would be limited to just that one. This would leave the other tiles completely intact and thus perfectly capable of providing the highest level of protection against any other incoming projectiles.

2.3   STRUCTURAL ARMOR

2.3.1   Polymer Concrete

When it comes to armor we have been used to referring mostly to armor that is used to protect either personnel or vehicles. However, when it comes to protection, especially in an armed conflict, sensitive or important structures such as buildings, bridges, or storage facilities must be well protected against any threat. The kind of threats these kinds of structures usually face, in most cases, originate from bombing, missile hits, and less commonly from small arms and assault rifles. Nevertheless, advanced polymers have also found their way into providing enhanced protection for large structures and infrastructures.

Although the first polymer concrete synthesis dates to the 1950s and 1960s, recent advancements in novel structural materials have given rise to so-called polymer concrete. Polymer concrete occurs when we completely replace the cement hydrate binders that exist in conventional concrete with polymer binders, thus creating a new composite material, namely polymer concrete. Consequently, the binder phase of the polymer concrete consists only of polymers and is deprived of the presence of ordinary cement hydrates, as is the case for ordinary concrete. The polymerization of polymer cement usually takes place at ambient temperatures and the resulting material is hardened to the desired point. A comparison between the properties of ordinary concrete and polymer concrete is summarized in Table 9.2.

As is evident from the comparison shown in Table 9.2, polymer concrete has significant advantages when compared to ordinary cement concrete and, as expected, it has a few disadvantages. Most of the above-mentioned properties and especially those that relate to the use of polymer concrete for armor applications are proving to be very useful when it comes to protecting buildings and other facilities from threats.

The above should be taken together with the fact that using polymer cement instead of regular cement can reduce the weight of the structures and their fabrication cost. This is also very important if considered in a circular economy framework, where the recycling of plastics is a critical issue. However, the severe effect of temperature on the mechanical properties of polymers and thus polymer concrete has significantly reduced the extensive use of polymer concrete versus ordinary concrete for most building facilities by the defense sector [8].

TABLE 9.2
Comparison of Some Basic Properties of Polymer Concrete and Ordinary Concrete

Properties	Concrete Type	
	Polymer Concrete	**Ordinary Concrete**
Strength	☑	☒
Adhesion	☑	☒
Water-tightness	☑	☒
Chemical resistance	☑	☒
Abrasion resistance	☑	☒
Thermal resistance	☒	☑
Fire resistance	☒	☑

Note: ☑: Superior properties; ☒: inferior properties.

2.3.2 Fiber-Reinforced Polymer (FRP) Composites

One very useful application of polymers to enhance the structural armor of buildings and other sensitive facilities is the use of FRP composites. Usually, this procedure entails the retrofitting of FRP composites to the cement concrete structure of a building to enhance its ability to withstand large explosions or other hits. The way these FRP composites work does not differ significantly from the way any armor aims to defend whatever is behind it. That is, to reduce the kinetic energy of the incoming threat, whether that is a bullet, a fragment, or even a large explosion. In the above-mentioned procedure of applying FRP composites to enhance the durability and ballistic performance of ordinary concrete, the most common materials used are glass fiber-reinforced polymer (GFRP) and carbon fiber-reinforced polymer (CFRP) composites. In some cases, we also find aramid fiber-reinforced polymer (AFRP) composites, mainly due to their ability to resist impact and penetration from high-kinetic-energy projectiles. AFRPs are very important in instances where buildings need to be protected from bomb explosions that aim at weakening the building's structural integrity.

The application of FRPs can also occur with the concomitant application of spray-on polyurethanes to enhance their structural integrity upon impact and provide better flame resistance and temperature stability. The application can occur with various techniques, with the most common being the application of tile-shaped FRPs. To protect columns existing in buildings, which are very important for the structural stability and integrity of the whole building, the application of FRPs in the form of wrapping has been utilized extensively. The above-mentioned techniques and methods to enhance the ballistic performance of buildings and other structures are not only important for purely defense applications.

High-kinetic-energy projectiles, which can be tree logs, cars, or even other houses, are commonly found to be "flying around" in extreme tornadoes or flooded areas. The kinetic energy of this kind of debris can cause significant damage to other buildings and civic structures. This was extremely evident in Hurricane "Katrina" that struck a very large area of the southeast United States of America and the Bahamas [9]. Figure 9.4 shows the path followed by Hurricane Katrina, a category 5 major

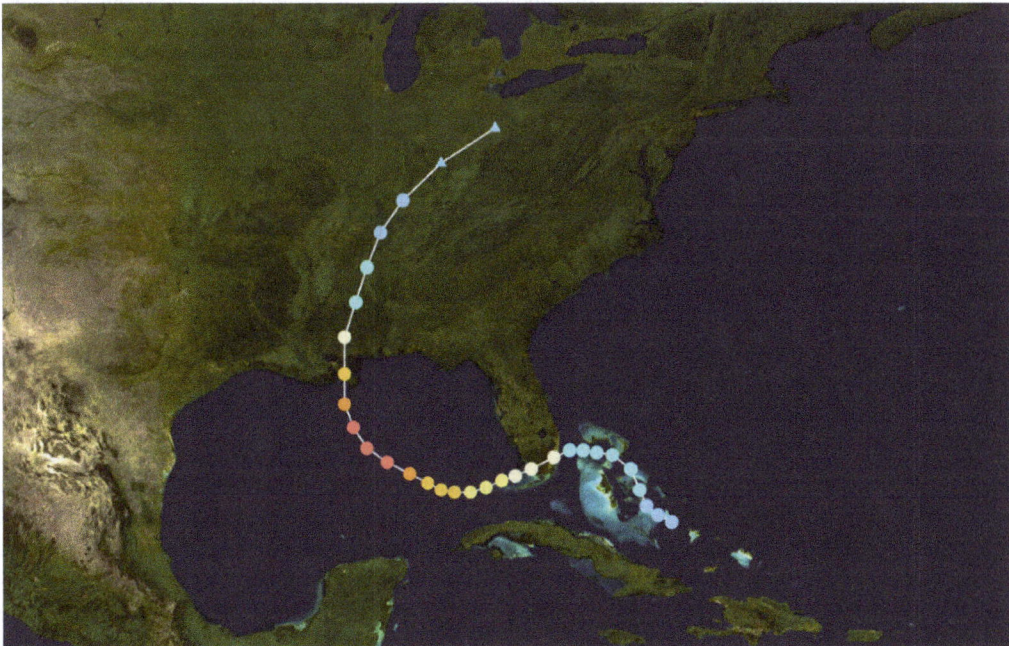

FIGURE 9.4 The path of Hurricane "Katrina." The circles show the location of the storm at 6-hour intervals. Adapted with permission from Reference [9]. Copyright (2015), Publisher Supportstorm.

hurricane, that lasted from August 23, 2005, until August 31, 2005 [9]. Hurricane Katrina had a major impact on the ground, and resulted in the loss of 1836 lives and $125 billion (2005 prices) in damage, recording it as the costliest tropical cyclone ever [9]. Therefore, it is always useful to remember that reinforcing buildings to withstand high-kinetic-energy impacts is always important, whether in an armed conflict or at the mercy of extreme weather phenomena.

3 ADVANCED POLYMERS USED IN PERSONAL PROTECTIVE EQUIPMENT (PPE)

As discussed earlier in this chapter, the need to protect vital parts of the body is something that dates back many centuries. Figure 9.3 shows two very distinct helmet types that were used in the Battle of Marathon, indicating the importance of head protection. Personal protective equipment (PPE) can be found in many types and functionalities from head protection using suitable helmets, to eye protection with the use of proper eyewear and many other parts of the body such as the neck, torso, and groin area.

3.1 HELMETS

The use of a helmet has profound importance in safety and armor materials as it protects probably the most important part of the human body. The head, anatomically, is where the brain is situated, and also the eyes, ears, mouth, and nose. All of these are important for providing a human with the necessary senses. Of course, the brain is the "king" of all organs as in the form of an "orchestra conductor" coordinating the function of our systems and preserving life through the intentional and unintentional continuous function of every organ of the body.

One of the main concerns of the helmet design and manufacturing industry was and still is to maintain a delicate balance. The ballistic performance protects the wearer and reduced weight minimizes fatigue from usage [10]. The latter is also a key factor on the battlefield. A heavy helmet will cause not only fatigue but also loss of concentration and a reduced optical field, and in some cases will be discarded by the user because it becomes unbearable. This exposes the helmet wearer to elevated danger but also introduces an appalling effect toward helmet usage.

In helmet design today, the approach is that of a composite material using fibers such as UHMWPE and aramid. These fiber-reinforced composites have been proven to be very successful in terms of ballistic protection combined with reduced weight when compared with traditional metallic and steel helmets. The most notable fibers used today in ballistic applications and more specifically in helmets are aramid-type fibers such as Kevlar or Twaron, which are considered some of the most widely used fibers in composite materials used in helmets [11,12]. Recently, novel advancements in the available fibers have given way to Dyneema (UHMWPE), Spectra (liquid polymer), and Endumax fibers (also (UHMWPE), which have been shown to exhibit superior ballistic protection as compared to Kevlar or Twaron. Most helmets used by the military today, or related applications, are characterized by a fibrous structure, regarding the above-mentioned fibers that can take various shapes. The most notable are woven, stitched, and unidirectional (UD) [13].

The way these fibers are woven or stitched together plays a very important role in the ballistic behavior of the composite material that is used to makes the helmet. The usual 90^0 weave between fibers and the unilateral stacking of fibers in a composite laminate structure essentially play the role of a trapping surface for the incoming projectile, whether that is a bullet or fragment. The way these fibers are positioned with respect to each other and the multiple combinations of their directions are of critical importance also, as the dissemination of the kinetic energy of the incoming projectile occurs in multiple directions other than the direction of the incoming projectile, thus minimizing its penetration capability. Of course, the head along with other parts of the body needs to be protected from blunt trauma as well. No penetration of the helmet does not mean necessarily that the bearer is completely out of harm's way. Deformation of the helmet must be small because the cranial cavity can sustain fractures and fatal injuries, even though no penetration may have occurred.

3.2   EYEWEAR AND FACE MASKS

As stated earlier, head protection is of vital importance on the battlefield and is not limited just to the cranial cavity [14]. All or most of the sensory functions, such as our eyes and the nasal and oral cavities, are in the head, and thus more modern protection systems tend to integrate the traditional helmet with eyewear and a face mask. These "full-face" helmets aim to provide several levels of protection and not just ballistic protection. In today's battlefields, and with traditional warfare having evolved to essentially hybrid warfare, the fighter needs to protect themselves from various dangers. Eye protection is essential for the fighter to maintain a clear view of the field ahead and accurate knowledge of their surroundings. Also, the variety of terrains that troops need to operate in calls for adequate eye protection. This was extremely important in military operations in desert environments where large amounts of dust were present in the atmosphere which could impair or hinder the fighter's vision.

in addition to the environmental factors affecting the fighter's vision, it is important to note that eye protection is also essential in occasions where chemical agents are used in the form of gases. The best known type of gas that is aimed at impairing the fighter's vision is tear gas, which irritates the eyes and causes extreme pain to the exposed person. Finally, it is also important to protect the eyes from extreme UV radiation when exposed to extreme sunlight. Polymers in the form of filters have played a significant role in the above circumstances and have provided viable solutions that can guarantee that the fighter will always have the highest level of eye protection [15].

It is also important to acknowledge the importance of a full-face helmet within the framework of protecting the nasal and oral cavities and hence the whole respiratory system of the helmet wearer. If we consider that the hybrid warfare that we mentioned above can also have a biological warfare aspect and bearing in mind the tremendous effects of the COVID-19 pandemic on our societies, modern helmet designs are increasingly focusing on providing a full-face helmet to be able to protect the wearer not just from incoming projectiles but also various other threats, such as chemical and biological warfare. Combining all the above functionalities in a full-face helmet of course adds weight as compared to an ordinary helmet that aims to protect mainly the head, leaving all the other parts such as the face exposed [16]. However, in turn, we must always bear in mind the basic equilibrium. This equilibrium is summarized in the sense of maintaining optimum or maximum protection for the head while keeping the helmet's weight low or at a bearable level [17]. In the above context, polymers have a unique role to play as they can take multiple shapes, attain enhanced mechanical properties, and provide adequate protection in many ways and against many different threats.

3.3   BREASTPLATES

Recently, lightweight ultra-high-molecular-weight polyethylene (UHMWPE) bulletproof vest chest plates have made their appearance in the defense markets. UHMWPE is the highest quality polyethylene available for use in harsh working environments and a variety of applications. Its main processing method is derived from powder metal technology and it is also extruded. Composite plates also are made by autoclaving [18]. A molecular weight of 3–6 million is achieved, ensuring that ultra-high-molecular-weight polyethylene has sufficient strength to sustain wear and impact resistance that is impossible to obtain with other lower polymer products. The product is tough and lightweight and can float on water. Mainly, it is used in bulletproof vests, body armor, bulletproof backpacks, and plate carriers, etc.

Unlike traditional solutions, advanced polymer ballistic plates are designed to absorb the kinetic kick delivered from a bullet by engulfing the projectile—simultaneously eliminating ricochet. The most significant advantage in the absence of ricochet is that it protects both the operator and their comrades, as well as the environment of the team, from secondary damage and injuries. Already available in the market are Level III UHMWPE Small Arms Protective Insert (SAPI) plates [19]. Made from advanced UHMWPE composite, these plates offer protection from the most common rifles with minimum carry weight. The shape mimics the chest and the back curvatures, providing optimal comfort.

As an example, novel body armor plates stand alone at 1.5 kg weight and will stop the penetration of the following projectile types and velocities:

- 7.62×51 mm NATO ball ammunition, 847 m/s, at 10 meters
- 7.62×39 mm mild steel core, 790 m/s, at 10 meters
- 5.56×45 mm M139 ball, 830 m/s, at 10 meters.

4   CONCLUSIONS

In this chapter, versatility and universality of using advanced polymers to provide adequate protection in defense have been described. The world has witnessed increasing armed conflicts while also trying to tackle the rise of extremism that is expressed by terrorist actions. The above call for novel solutions that can provide maximum security for the fighter, combined with new functionalities and, of course, reduced weight. The advancements of the available weapons and the many different types of ammunition have in turn created the need for the development of high-performing polymers and composite materials that are starting to almost completely replace traditional armor materials such as steel. The cost reduction and the optimization of polymer production, the ease in machining and treatment, and the large variety of chemical compositions which are manifested in various properties have made polymers extremely useful for defense applications.

The future will call for even better-performing polymers that will have the ability to withstand even greater impacts from incoming projectiles or fragments. They will possess even lower weight and augmented mechanical properties. If we consider the advancements in additive manufacturing, termed 3D printing, where the polymers have the most significant role to play as the most commonly used materials, a real revolution in the use of polymers in defense is about to take place [20].

REFERENCES

1. Fakirov S, Fundamentals of Polymer Science for Engineers. Wiley – VCH Verlag GmbH, Weinheim, Germany (2017).
2. Koltzenburg S, Maskos M, Nuyken O, Polymer Chemistry. Springer – Verlag GmbH, Berlin, Germany (2017).
3. Williams S, The Knight and the Blast furnace, A history of the metallurgy of armour in the Middle Ages and early modern period. Ed. Brill, History of Warfare (2003).
4. Lim J, Zheng J, Masters K, Chen W, Mechanical behavior of A265 single fibers. Journal of Materials Science. 45 (2009) 652–661.
5. Abtew MA, Boussu F, Bruniaux P, Loghin C, Cristian I. Ballistic impact mechanisms – A review on textiles and reinforced composites impact responses. Compos. Struct. 223 (2019) 110966.
6. Czech K, Oliwa R, Krajewski D, Bulanda K, Oleksy M, Budzik G, Mazurkow A, Hybrid polymer composites used in the arms industry: A review. Materials. 14 (2021) 3047.
7. Ash RA, Vehicle armor. In Lightweight Ballistic Composites; Elsevier: Amsterdam, The Netherlands, pp. 285–309 (2016).
8. Fowler D W, State of the art in concrete polymer materials in the U.S., Proc 12th Int Congr Polymers in Concrete, Vol. 1, Chuncheon (Korea), Kangwon National University, 29–36 (2007).
9. https://en.wikipedia.org/wiki/Hurricane_Katrina
10. Tan L, Bin, Tse KM, Lee HP, Tan VBC, Lim SP, Performance of an advanced combat helmet with diferent interior cushioning systems in ballistic impact: Experiments and fnite element simulations. Int. J. Impact Eng. 50 (2012) 99–112.

11. Dixit D, Pal R, Kapoor G, Stabenau M, Lightweight Composite Materials Processing. Elsevier (2016).
12. Folgar F. Thermoplastic Matrix Combat Helmet with Carbon-epoxy Skin for Ballistic Performance. Elsevier (2016).
13. Chen X, Hearle JWS, Structural Hierarchy in Textile Materials: An Overview. Woodhead Publishing Limited (2010).
14. Tobin L, Iremonger M, Modern Body Armour and Helmets: An Introduction. Argros Press, Canberra, Australia. (2006).
15. Breeze J, Gibbons AJ, Opie NJ, Monaghan A, Maxillofacial injuries in military personnel treated at the Royal Centre for Defence Medicine June 2001 to December 2007. British Journal of Oral and Maxillofacial Surgery 48 (2010) 613–616.
16. Breeze J, McVeigh K, Lee JJ, Monaghan AM, Gibbons AJ, Management of maxillofacial wounds sustained by British service personnel in Afghanistan. International Journal of Oral and Maxillofacial Surgery 40 (2011) 483–486.
17. Rosenblatt M, Analysis of fragmentation protection for two commercial helmets. In: Personal Armour Systems Symposium 2010, Fairmont Le Chateau Frontenac, Quebec City, Canada. (2010).
18. Roth, M, "Effects of Autoclaving on the Ballistic Performance of Ultra High Molecular Weight Polyethylene Composites." PhD diss., Worcester Polytechnic Institute (2021).
19. Umair, M, Nawab Y, and Zeeshan Ul H, "Personal and structural protection." In Composite Solutions for Ballistics, pp. 109–136. Woodhead Publishing (2021).
20. Wang Y, Li L, Hofmann D, Andrade JE, Daraio C, Structured fabrics with tunable mechanical properties. Nature 596 (2021) 238–243.

10 Polymers for Additive Manufacturing

Rossella Arrigo and Giulio Malucelli

Department of Applied Science and Technology, Politecnico di Torino, and Local INSTM Unit, Viale Teresa Michel 5, 15121, Alessandria, Italy

1 INTRODUCTION

AM is currently experiencing great success, as a novel technology suitable for producing parts and structures with high complexity, in a fast and relatively inexpensive way, with high freedom of design, waste minimization, and mass customization. AM is based on a layer-by-layer approach, through which consecutive layers of printed material are deposited on top of each other, resulting in the buildup of the final 3D part or component [1–3]. Initially, additive manufacturing was broadly employed by designers and architects to manufacture either functional or aesthetic prototypes, thanks to its fast, strongly customizable, and profitable prototyping capability. The current application fields of AM range from construction, to biomedical, to the transportation industry. In the past, each AM process was set on a specific apparatus that had to work with specific material with particular characteristics tuned to the type of apparatus itself; in other words, each seller of the apparatus was also the supplier of the material to be processed by the AM unit, hence significantly limiting the suitable materials to a restricted number. Conversely, nowadays, quite a wide choice of materials (polymers, ceramics, metal alloys, and composites) are available for AM processes so that any end-user has many possibilities for selecting the most suitable and best performing material according to the envisaged final application. This happened also as far as polymer-based AM processes are concerned: in the last 10 years, significant progress has been made and, at present, a rich selection of different materials is available for such AM techniques as fused deposition modeling (FDM), selective laser sintering (SLS), stereolithography (SLA), and 3D inkjet printing (IJP). This chapter is aimed at summarizing the current state-of-the-art regarding the materials suitable for these AM processes, highlighting advantages and drawbacks as a benchmark for the research and development of the forthcoming years. The chapter is organized according to different sections, each dealing with a particular AM process involving specific polymer-based materials.

2 POLYMER SYSTEMS FOR FUSED DEPOSITION MODELING

Fused deposition modeling is part of the material extrusion AM technologies, and was first developed by Scott Grump (Stratays) at the end of the 1980s [4]. FDM apparatuses generally include a heated extruder, in which the polymer filament, fed into by a system of rollers and wheels, melts and passes through the printing nozzle. The extrudate is then deposited layer by layer on the printing platform, where the polymer cools down and resolidifies. Once the deposition of a layer is completed, the platform moves downward to print subsequent layers to generate printed parts with specific geometries as required by the design (Figure 10.1a).

The most commonly used FDM machines consist of one extrusion head; dual extruder apparatuses are generally exploited to print at the same time the build thermoplastic and the support material, and are frequently required to provide support for hollow or porous objects or overhang sections [5]. FDM has several advantages, including the non-requirement of post-processing treatments and the cost-effectiveness of both machines and materials. However, this process also presents some

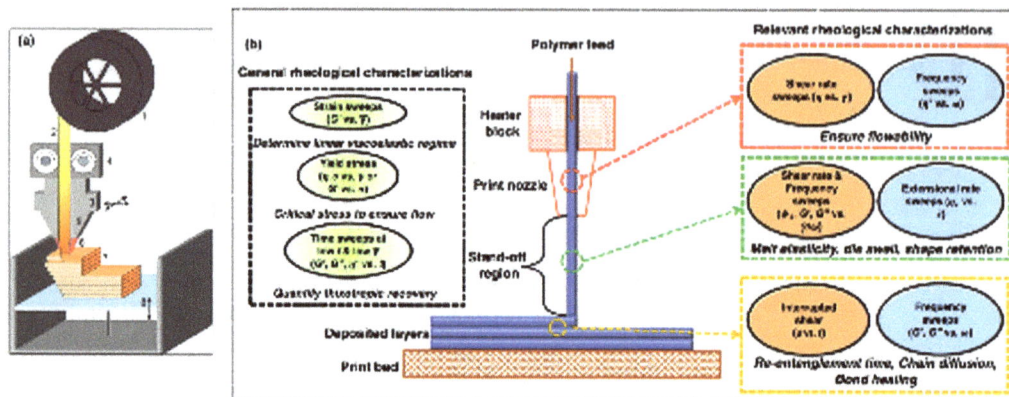

FIGURE 10.1   (a) Schematic of a typical FDM process and (b) overview of the rheological behavior of the polymer melt during the different processing stages. Reprinted with permission from Reference [7]. Copyright (2021) American Chemical Society.

limitations, mostly related to the low resolution (limited by the nozzle dimensions) and the high anisotropy of the printed objects.

From a general point of view, any thermoplastic-based formulation that can be melted at an adequate temperature (considering the polymer degradation and the constructive limits of the used FDM apparatus) is a suitable material for FDM processing. There are several requirements, associated with the mechanical, rheological, and thermal characteristics of the candidate thermoplastic, which need to be fulfilled to classify a polymer as "FDM-printable." First, the filament has to be spooled during its production and despooled during processing; therefore, the material should present a proper balance between stiffness and flexibility, which is achieved when the filament presents a minimum strain at a yield of about 5% [6].

Conversely, the extrudability of the filament through the printing nozzle, as well as the shape stability of the extrudate in the stand-off region and during the layer deposition, are governed by the material rheological behavior. In particular, as shown in Figure 10.1b, low viscosity values are required during the extrusion step, to ensure polymer melt flowability and to reduce the extrusion pressure. Furthermore, low viscosities allow for minimizing the issues related to filament buckling phenomena, which could compromise successful processing. Then, the filament must show a rapid viscosity increase at the nozzle exit, where the shear rate drops exponentially, to avoid material oozing and to facilitate the retention of the extrudate shape in the stand-off region. Therefore, polymer melts showing a marked shear-thinning and, possibly, a yield stress behavior (involving a sudden increase of the melt viscosity in the zero-shear region) are required. Finally, the relaxation dynamics of the polymer macromolecules dictate the welding between the subsequently deposited layers in the post-deposition stage. To guarantee a good level of interlayer adhesion, the polymer chains lying at the interface between adjacent deposited layers must be able to inter-diffuse across the interface, which requires adequate macromolecular mobility [7]. This is a critical stage of FDM processing, since usually printed parts tend to have poorer mechanical properties as compared to samples produced through traditional processing technologies such as injection molding, because of the weak interlayer adhesion along the axis perpendicular to the print direction.

As far as the requirements related to the material thermal properties are concerned, filaments with amorphous microstructures or with a low degree of crystallinity and slow crystallization kinetics are preferred. This is to minimize the issues associated with the volumetric contraction experienced by semicrystalline polymers during crystallization, which causes the entrapment of high residual

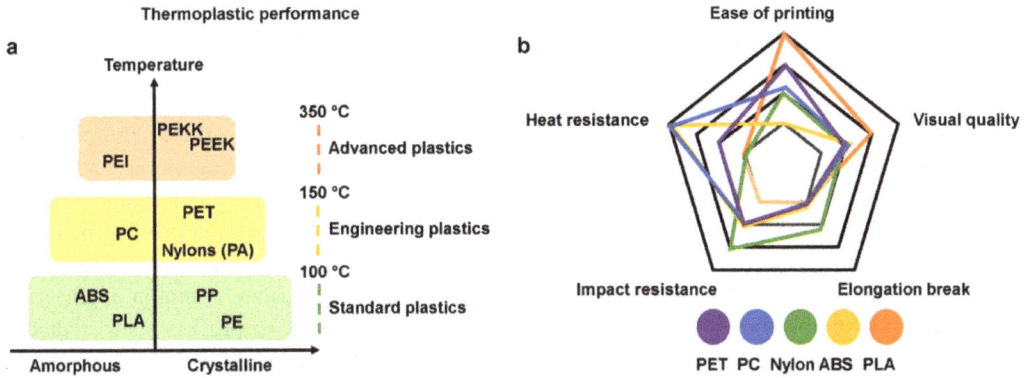

FIGURE 10.2 (a) Thermoplastics as feedstock materials for FDM and (b) Rader plot graph showing polymer properties. Reprinted with permission from Reference [12]. Copyright (2021) Elsevier.

TABLE 10.1
Average Price of Thermoplastic Filaments for FDM [13]

Filament	ABS	PLA	PA	PC	PEI	PEEK
Price ($/kg)	17–25	15–25	30–70	30–70	140–220	400–700

stresses within the printed part, resulting in warpage and shrinkage, or even delamination and cracking [8].

All these features are challenging for the choice of a thermoplastic suitable for FDM; nevertheless, pure polymers are usually modified with additives and/or fillers selected specifically, which allows a proper modulation of the properties of the base material, making them suitable for FDM. For example, a common methodology exploited for increasing the yield strain of polypropylene (PP)-based filaments involves the addition of small amounts of amorphous polymers such as poly(vinyl chloride) or amorphous polyolefins [9]. Similarly, it has been demonstrated that shrinkage and warpage phenomena typically encountered during the 3D printing of semicrystalline thermoplastics can be minimized by incorporating talc or expanded perlite particles [10] or via slight modifications of the processing setup. Zein et al. [11] employed FDM to produce scaffolds with different honeycomb-like patterns using a polycaprolactone (PCL)-based filament with variable circular cross-sections, achieving undeformed printed parts endowed with acceptable mechanical properties.

Nowadays, a great variety of thermoplastics is available in the form of filament as feedstock for FDM processes. Figure 10.2a shows the most commonly employed materials, classified according to their performance and processing temperature; Figure 10.2b presents the FDM processability, printing quality, and mechanical properties of the considered polymers.

Among them, polylactic acid (PLA) and acrylonitrile butadiene styrene (ABS) are the most used at a commercial level, likely due to their relatively low cost (Table 10.1) [13]. PLA is extensively considered for its ease of print, and exploited to produce printed devices with applications in the biomedical field; its utilization in the last few years has shown a steadily increasing trend, due to its lower environmental impact as compared to traditional fossil fuel-derived thermoplastics. ABS has been widely used to build functional models for real-scale characterization. despite its amorphous microstructure, the high shrinkage factor of ABS causes warping issues during the solidification stage, making it difficult for optimization of the processing conditions to obtain undeformed objects. Several studies have demonstrated that the mechanical properties of ABS printed parts, and especially the tensile strength, are highly affected by the air gap and the raster orientation of the layers [14].

Filaments based on polycarbonate (PC) and polyamide 6 (PA6) are typically exploited when high mechanical strength and impact resistance are required; PC exhibits a wide thermal resistance range (from –150 to +140°C), which enables applications in automotive and aerospace industries. However, both materials are sensitive to moisture and should be kept in a dry environment before processing. A further limitation of PC in FDM applications is the required high printing bed temperature, to ensure sufficient bed adhesion and minimize distortions in the printed part [12]. Also, polyethylene terephthalate (PET) and glycol-modified PET (i.e., PETG) are common engineering thermoplastics suitable for FDM. In addition, FDM can be employed to process such advanced thermoplastics as polyether-ether-ketone (PEEK), which is usually employed in high-performance applications in extreme conditions, requiring high service temperatures and excellent mechanical properties. Nevertheless, the processability of PEEK is difficult, due to its high melting temperature (about 340°C), which is above the capabilities of a standard FDM apparatus, thereby further increasing the overall cost of the process [15].

As briefly discussed earlier, in addition to the build material, in some cases, depending on the geometry of the printed part, the introduction of a support material, easily removable at the end of the processing, is required. Thermoplastics typically used in FDM as support materials are poly(vinyl alcohol), which is water-soluble, highly flexible, and has lower cost as compared to other polymers with similar properties, and high-impact polystyrene that is usually exploited in combination with ABS, owing to their similar characteristics and the easy selective dissolution using limonene as solvent [16].

Despite the original thermoplastic feedstocks being typically unfilled polymers, in recent years, composite and nanocomposite filaments have been developed, thus widening the capability of FDM, which is currently limited by the poor selection of thermoplastics for high-performance applications [5]. Depending on the final purpose of the printed part, different micro- and, even more, nano-fillers have been exploited, including carbon-based material, biomaterials, ceramic, and metal particles [12]. As an example, carbon-based fillers are typically used as reinforcement to produce lightweight and conductive parts, ceramic particles are embedded in filaments for the 3D printing of biomedical and electronic products, while metal/thermoplastics filaments are actively employed in energy storage devices, semiconductors, and circuits. The utilization of composite and nanocomposite filaments usually brings about an enhancement of the mechanical performance of the printed parts. Weng et al. [17] compared the tensile properties of FDM-printed ABS/organo-modified montmorillonite nanocomposites with those of injection-molded samples, showing that with the introduction of 5 wt.% of nanoclays the gap between the two processing technologies was significantly reduced. Xu et al. [18] developed PCL/hydroxyapatite filaments that were employed for producing artificial bones through FDM; the printed parts showed compressive strength and modulus very close to those of natural bone, with good biocompatibility and biodegradability. ABS-based feedstock filaments for FDM containing iron or copper microparticles were produced by Nikzad et al. [19]; the printed samples exhibited improved storage modulus and thermal conductivity values, as compared to the unfilled matrix, as well as a reduced thermal expansion coefficient. Nevertheless, Dorigato et al. [20] observed a decrease in ductility and electrical conductivity in FDM-printed ABS/carbon nanotube nanocomposites as compared to compression-molded samples, due to the preferential orientation of the embedded nanofillers along the printing direction and the strong anisotropy of the 3D-printed parts. In addition, the embedded filler significantly impacts the rheological behavior and, hence, the processability of thermoplastics. The introduction of low amounts of micro-sized fillers and, even more, nanofillers, usually increases the viscosity of the polymer matrix, thereby affecting the extrudability of the filament. Furthermore, the embedded particles often influence the relaxation processes of the matrix macromolecules, restricting their mobility; this feature may adversely influence the inter-layer adhesion after the deposition step. Finally, nanoparticles are prone to form aggregates at high loadings, further compromising the material extrudability, as the formed agglomerates may lead to the obstruction of the printing nozzle causing the FDM process to fail.

Recently, to reduce the costs associated with the feedstock materials and to adhere to the new economic paradigm of the circular economy, filaments based on recycled thermoplastics are attracting increasing interest. In particular, new formulations of filaments originating from homogeneous plastic wastes derived from 3D-printed parts in PLA and ABS, PET bottles, low-density polyethylene bags, and PP woven bags have been studied; also, some recycled filaments are already commercially available [21]. Cafiero et al. [22] reported on the formulation of filaments for FDM originating from the mechanical recycling of the plastic fraction of waste from electrical and electronic equipment (WEEE). Preliminary physico-chemical characterizations performed on the collected WEEE wastes disclosed a high heterogeneity of the starting material, involving the presence of 11 different polymers or blends. Selected samples were subjected to a cleaning procedure, aiming at removing foreign materials, reduced into small flakes, and extruded in filaments suitable for FDM. The quality control of the printed objects proved that FDM-printed specimens had no remarkable deviations from the model design as compared to the same objects produced using commercial filaments.

3   POLYMER SYSTEMS FOR SELECTIVE LASER SINTERING

Selective laser sintering is a powder-based AM technology that allows the free generation of complex 3D objects and components through the layer-by-layer solidification of the starting powder material. The solidification is achieved by selectively melting or sintering specific areas in each layer, using the thermal energy provided by a focused layer radiation system. In a typical SLS process, a single layer of powder is first deposited into the build chamber; then, the laser beam sinters or melts selective regions on the powder bed to form a solid layer according to the original part design. Once the first layer is completed, the building platform is moved downward, and the process continues with the deposition of a new layer of powder that is sintered and simultaneously bonded with the previous layer. These steps are then repeated until the final part is achieved [5]. Unlike other AM processes, SLS does not require support material, as the subsequently sintered layers are surrounded and supported by the residual powder. Additionally, it permits efficient utilization of the material because the non-sintered powder can be recycled several times without significant variations of the polymer properties.

During SLS, different complex dynamic and transient phenomena are experienced by the polymer, due to the very fast heating and cooling processes: the consolidation of the powder occurs by heating the polymer particles above their glass transition (T_g) or melting (T_m) temperature, depending on the amorphous or semicrystalline nature of the material, respectively. After the initial polymer softening, two adjacent particles form a neck at the contact interface, which progressively grows until the intermediate interface disappears. The initial particle pair is then gradually transformed into a single particle through coalescence processes. The success of this step is strongly dependent on the rheological characteristics of the polymer and in particular its zero-shear viscosity [23]. Low zero-shear viscosity values are essential in promoting adequate coalescence and in ensuring high coalescence rates, allowing the production of sintered parts with low porosity, high density, and strength. SLS, unlike traditional injection molding, cannot provide additional compacting of the material during the part production, making even more crucial the requirement for low viscosity. Therefore, usually, semicrystalline thermoplastics are preferred over amorphous polymers for SLS applications, as generally the viscosity of amorphous polymers is still high above the Tg, compromising proper coalescence and, hence, the final properties of the sintered part.

SLS processes of amorphous thermoplastics are difficult also due to their non-favorable thermal properties. Indeed, as amorphous polymers do not present a specific melting point but a wide range of temperatures in which the glass transition occurs, the temperature of the powder bed should be maintained below T_g to avoid the powder particles sticking together. Nevertheless, if the temperature is too low, the high viscosity of the material impedes a proper coalescence. Therefore, a very high-energy laser is required to rapidly heat the polymer. At the same time, this can induce polymer degradation and smoke production during the processing, thereby affecting the density and mechanical properties of the produced part [24]. For semicrystalline polymers, when the laser selectively melts the powder, a complete coalescence between adjacent particles and a proper adhesion with the previously sintered layers are needed. This implies that the occurrence of crystallization events must be repressed until appropriate interconnections between particles and between adjacent layers are established. Thus, the process temperature must be precisely set between the onset temperature for the melting and the onset temperature for the crystallization. This region is the so-called "sintering window" for SLS processing (Figure 10.3) [24]. If the processing temperature is too close to the onset temperature for crystallization and the polymer shows slow crystallization kinetics, premature crystallization phenomena may occur, resulting in curling and warpage in the printed sample. For this reason, polymers exhibiting a wide sintering window are preferred.

Further, a polymer suitable for SLS should present adequate optical properties. In particular, it should be able to effectively absorb the energy from the laser source at a specific wavelength. Typically, commercial SLS apparatuses use CO_2 lasers with a wavelength of about 10.6 μm. As common thermoplastics contain a large amount of C-H bonds, they exhibit an FT-IR absorption peak at 943 cm^{-1} (corresponding to a wavelength of 10.6 μm), which ensures the absorption of a relevant portion of the incident radiation. The higher the absorption peak intensity, the stronger the light absorption ability and the more easily the polymer is melted. In the case of poor absorption characteristics, the photothermal effect can be enhanced by increasing the laser energy or through the introduction of light absorbents, such as carbon-based fillers [25].

At present, the most commonly used semicrystalline thermoplastics for SLS processes are polyamides (PAs), and, in particular polyamide 12 (PA12), polyamide 11 (PA11), and polyamide 6 (PA6). These all possess a remarkably low melt viscosity value that allows the achievement of sintered parts with low porosity and mechanical properties practically comparable to those obtained through traditional manufacturing processes. Among PAs, PA12 is the most widely employed, having a market share of more than 90%, due to its excellent sintering processability and the high mechanical strength of the sintered parts. PA11 is often considered a cheap alternative to produce more ductile parts; however, its processing is more difficult as compared to PA12, due to the narrow sintering window. Also, PA6 finds several applications and has a wide market demand; however, it

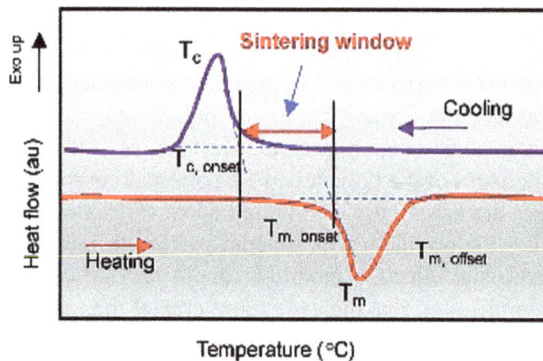

FIGURE 10.3 Typical thermogram of a semicrystalline polymer obtained through DSC analysis with the indication of the sintering windows for SLS. Adapted with permission from Reference [24] under Creative Commons Attribution License 4.0 (CC BY).

has a high melting point and high viscosity, resulting in difficult processability [26]. In addition, PA-based composite powder containing a great variety of fillers (such as glass beads, clay, organoclays, nano-silica, carbon-based particles, talc) has been recently developed. The sintered parts obtained using composite powders exhibit in general enhanced stiffness, strength, and thermal stability, but also reduced ductility and flexibility.

Besides PAs, many other semicrystalline polymers are available for SLS processes, including commodity thermoplastics [PP and high-density polyethylene (HDPE)], which in the near future may reach the popularity of PA12 due to their lower cost. However, the processing of these materials, and especially of HDPE, remains problematic and the production of dense and mechanically stable sintered parts is extremely difficult. To overcome this drawback, Salmoria et al. proposed the formulation of PA12/HDPE blend powders that, through proper modulation of the weight ratio of the two polymers, can be used to produce printed samples with a controlled variation of porosity and functionally graded structures [27].

Some studies report SLS processing of amorphous polymers, like polystyrene, poly(methyl metacrylate), and polycarbonate [26]. The produced sintered parts show high porosity and poor mechanical performance, hence requiring mandatory post-processing treatments for obtaining final products with adequate quality.

In recent years, thermoplastic elastomers have been gaining increasing interest, as they permit a widening of the capability of the SLS technique, especially concerning the possibility to produce elastomeric functional parts for applications such as sports equipment or patient-specific orthopedic devices [28].

Finally, powders based on biocompatible and resorbable aliphatic polyesters [such as PLA, PCL, and poly(hydroxybutyrate-co-hydroxyvalerate)] have been exploited for SLS processes. Several studies have documented good processability for either unfilled polymers or composite powders containing different types of fillers selected specifically, highlighting the potential of SLS in the fabrication of structures and scaffolds with applications in the biological and biomedical fields [25].

4 POLYMER SYSTEMS FOR STEREOLITHOGRAPHY

Stereolithography (also known as *vat photopolymerization*) represents the pioneering AM process, invented and developed in 1986 by Charles Hull [29]. In general, this process exploits a photopolymerization reaction for achieving very fast and spatially controlled solidification of a liquid resin (also known as photoresist, or photosensitive resin). To this aim, either a digital light projector equipped with a computer-driven building platform or a computer-controlled laser beam can be successfully employed for creating an illuminated pattern on the surface of a resin located in a vat (Figure 10.4).

Consequently, the resin in the pattern solidifies to a definite depth, making it adherent to a support platform. In a stereolithographic bottom-up system (Figure 10.4, left), after the creation of the first solidified resin layer, the support platform moves downward, hence allowing the built layer to be re-coated with the liquid resin. Again, a second layer is solidified according to the irradiated pattern: as the platform step height is slightly smaller than the curing depth, it is possible to ensure good adhesion between the two layers, exploiting the polymerization of unreacted reactive groups located on the surface of the solidified first layer with those of the irradiated resin in the second layer. This procedure is then repeated until the buildup of the 3D part or component. Finally, the excess resin is drained and washed off and the manufactured object (called "green structure") needs further treatment (i.e., post-curing) carried out in UV ovens, to complete the conversion of the reactive groups still present within and on the surface of the object, hence enhancing the overall mechanical features of this latter.

In a stereolithographic top-down approach (Figure 10.4, right), the radiation is projected on a transparent plate that forms the bottom of the resin vat and does not adhere to the resin itself too much, while the build platform is immersed into the resin vat from the top. The only drawback of

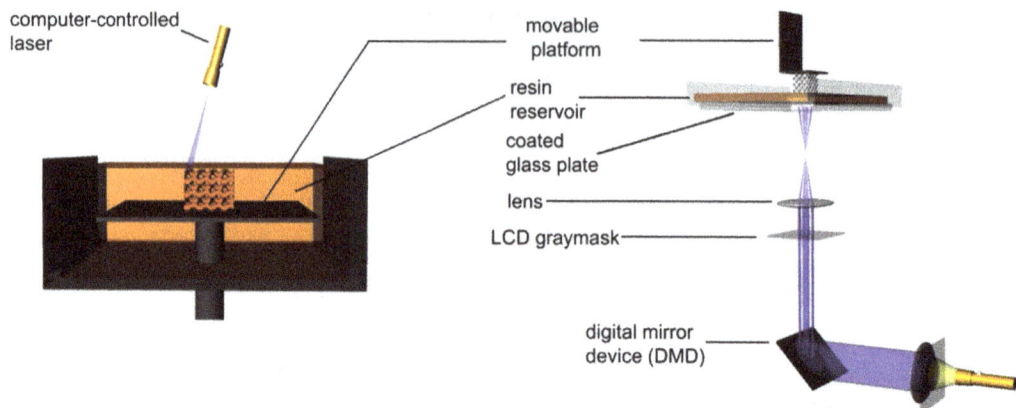

FIGURE 10.4 Scheme of the stereolithography process. Left: a bottom-up system with a scanning laser. Right: a top-down setup with digital light projection. Reprinted with permission from Reference [30]. Copyright (2010) Elsevier.

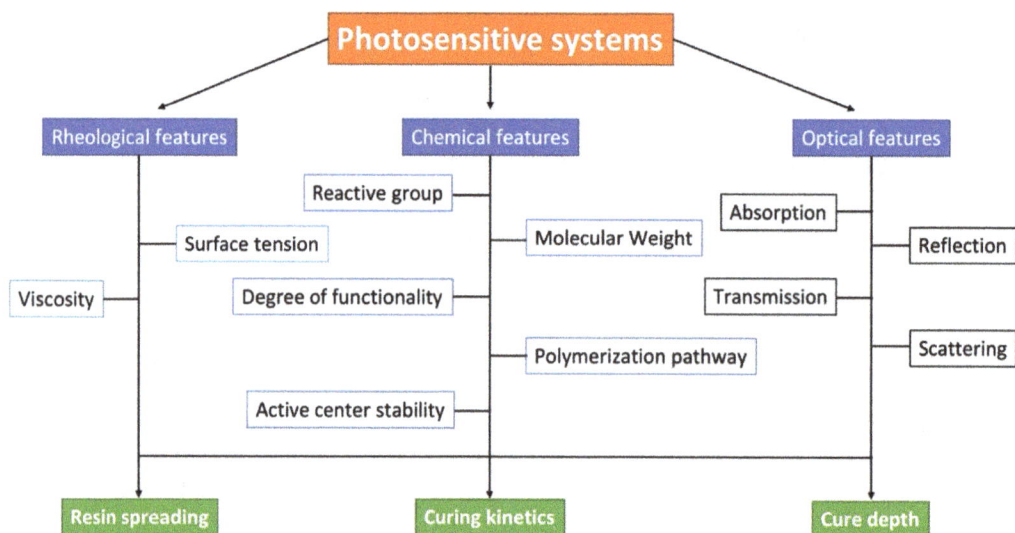

FIGURE 10.5 Critical requirements for the formulations of photosensitive systems (monomers/oligomers, photoinitiators, additives/fillers) employed for stereolithography.

this configuration refers to the need to apply some mechanical stresses to detach the solidified layers from the bottom of the resin vat; among the advantages, the solidified layer does not need to be re-coated with the liquid resin, and the irradiated layer is not exposed to air, hence limiting the oxygen inhibition on the photopolymerization reaction. In addition, the amount of resin required in the vat is much less than that needed for the bottom-up system.

The key properties that have to be fulfilled by the photosensitive resins suitable for SLA purposes are schematized in Figure 10.5. All these features can be precisely adjusted by changing the photo-sensitive recipe, i.e., the type and quantity of monomers, oligomers, photoinitiators, additives, and fillers. Monomers are low-molecular-weight molecules bearing one or multiple reactive groups that can take part in the curing process upon irradiation. As the photopolymerization process in an SLA equipment usually occurs according to a radical or cationic mechanism, monomers having different

TABLE 10.2

Physicomechanical Properties of Two Commercially Available Photosensitive Resins for Stereolithography

	Protogen™ (Epoxy Resin)	Waterclear™ (Acrylic Resin)
Density (g/cm³)	1.16	1.13
Glass transition temperature	57–59	44
Tensile ultimate strength (MPa)	42.2–43.8	56
Tensile modulus (MPa)	2180–2310	2880
Elongation at break (%)	8–16	7.5
Flexural strength (MPa)	66.7–70.5	84
Flexural modulus (MPa)	1990–2130	2490
Izod impact (notched) (J/m)	0.2–0.22	25

functional groups are selected. The monomers generally employed for radical photopolymerization contain vinyl functional groups and are mostly acrylates [31]. Some efficient vinyl monomers for SLA include bisphenol A acrylates, urethane acrylates, and poly(ethylene glycol) acrylates. The main general disadvantage of using these acrylates is their oxygen sensitivity which inhibits the free radical propagation reactions, hence considerably lowering the conversion of the reactive double bonds. The monomers employed for cationic photopolymerization bear epoxy or vinyl ether functionalities; among them, cycloaliphatic epoxy monomers that usually combine suitable viscosity values (i.e., appropriated for the resin spreading during the process) with the highest curing kinetics among the epoxides, are employed. Epoxy systems exhibit good mechanical features, low shrinkage upon photopolymerization, and limited warpage during post-curing steps [32]. Bisphenol A diglycidyl ethers and trimethylolpropane triglycidyl ether are among the most utilized epoxy monomers in SLA. Quite recently, hybrid epoxy/acrylate systems based on dual-cure (i.e., undergoing both radical and cationic photopolymerization) processes have been proposed [33]; in this way, it is possible to combine in a one-pot formulation the advantages of the two photopolymerization mechanisms, i.e., the high reactivity of acrylates with the low shrinkage, high mechanical behavior, and limited warpage of epoxy resins.

Oligomers are important components of the recipes for SLA; they show higher molecular weights and viscosity values, as well as higher degrees of functionality with respect to monomers. All these features are key issues for obtaining high-performing parts with good mechanical strength, toughness, and surface finish. Table 10.2 lists some typical physico-mechanical values for standard commercially available photosensitive resins for SLA processes [34].

The key limitation of stereolithography has frequently been attributed to the limited choice of commercially available photosensitive resins. The first resin formulations developed for SLA included low-molecular-weight polyacrylates or epoxides, which gave rise to the formation of rigid and brittle parts. From that point forward, several photosensitive systems, also including elastomeric formulations, were developed [35].

Stereolithography has been employed for creating polymer–ceramic composite objects: to this aim, different ceramic particles (like hydroxyapatite and alumina) can be exploited for obtaining a homogeneous suspension in the photocurable system [36]. These filled systems require quite a high control of the rheological behavior of the resulting suspensions, as the viscosity of the photosensitive system can remarkably increase with the incorporation of the ceramic filler; usually, the maximum ceramic filler loadings are around 50 wt.% [37]. Also, the filler size must be carefully controlled, as it should not exceed the thickness of each deposited and photopolymerized layer. The polymer–ceramic composite objects are usually much stiffer and with higher mechanical strength with respect to the all-polymer counterparts.

TABLE 10.3
Main Effects of Different Particle Reinforcements into Photosensitive Resins for Stereolithography

System	Main Outcomes	Ref.
Epoxy resin + FeO	- Good mechanical properties for the printed parts with a layer thickness below 80 µm - Irregular mechanical properties beyond 100 µm	[38]
Acrylic ester resin + multiwalled carbon nanotubes	- The incorporation of the nanofiller increases the radar absorptivity - Due to printability issues, the maximum filler loading is 1.6 wt.%	[39]
Poly(ethylenglycol)diacrylate + SiO_2 nanoparticles	- Increase of tensile strength in the presence of 1 wt.% silica. - Monotonic increase of tensile modulus with increasing the filler loading (up to 5 wt.%) - Agglomeration of silica nanoparticles at loadings beyond 5 wt.%	[40]
Acrylic resin + SiO_2, organic montmorillonite, and attapulgite	- Contrary to silica, both attapulgite and organic montmorillonite hinder the curing process - The curability of the printed parts is remarkably affected by the filler geometry - The viscosity of the resin is critical for accurate printability	[41]

Table 10.3 highlights the main effects of different particle reinforcements into photosensitive resins for SLA.

At present, the impossibility to use more than one resin at a time in stereolithography is challenging; some attempts for patterning multiple photosensitive resins within a single layer or at least in a multilayered structure have been carried out, although the precise control of sequential photopolymerization and rinsing steps is very difficult and would be remarkably implemented by the development of an automated system, able to take the uncured resin away and swap the resin reservoirs [42].

5 POLYMER SYSTEMS FOR 3D INKJET PRINTING

3D inkjet printing (also known as *material jetting*) represents a printing process similar to traditional 2D inkjet printing, as it exploits liquid resin systems (namely, photosensitive inks, waxes, thermoplastics), which have suitable inkjet print-heads that carry out layer-by-layer deposit onto a build platform, utilizing a drop-on-demand or continuous method; after the deposition step, each deposited layer is solidified using photopolymerization (photosensitive ink) or cooling (wax).

Specifically referring to photosensitive inks, 3D Systems and Stratasys, have designed, manufactured, and commercialized their own 3D inkjet printing systems, known as MultiJet printing and PolyJet, respectively. Although the working philosophies of the two methods are very similar, PolyJet utilizes gel-like support materials soluble in water, and hence is removable by selective chemical dissolution or by applying high-pressure water jets; conversely, a paraffin wax support material, that is heat-removable, is employed by MultiJet printing. Figure 10.6 shows a general scheme of 3D inkjet printing.

Layer height, intensity of UV radiation, polymer spray rate, and printing orientation represent the most important factors that may affect the quality of the printed parts [43,44]. Compared to other techniques, 3D inkjet printing is very fast, due to the possibility of using multiple nozzles for simultaneously spraying the photosensitive inks. In addition, it is possible to print parts with multimaterials, with a very high resolution, and without the need for post-curing treatments [45,46]; the only disadvantage is related to the high cost of printing with a single photosensitive ink.

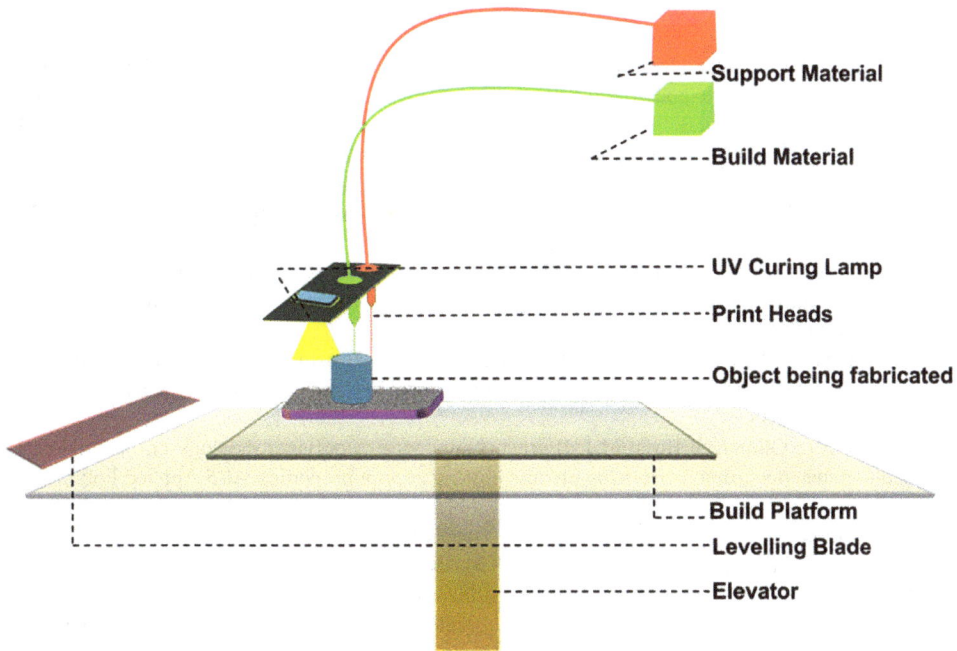

FIGURE 10.6   Scheme of 3D inkjet printing. Reprinted from Reference [47] under CC BY 4.0 license.

6   CONCLUSIONS AND FUTURE PERSPECTIVES

Undoubtedly, AM technologies that use either thermoplastic polymers or photosensitive resins have continuously grown very rapidly over the last decade. The achieved advances in polymer-based AM technologies have been exceptional; in fact, better-performing polymer systems and larger, increasingly efficient, apparatuses are frequently being introduced into the market.

Looking at the whole market for polymeric materials for AM processes, it is worth noting that most of the photosensitive resins, which represent its largest proportion, have proprietary formulations, and are supplied by a few companies only; this is a clear limitation, that is currently unsolved.

Though polymer powders, which are in the second place in the ranking of the most employed polymeric materials for AM, are at present mainly polyamide-sourced, current research is exploring the potential and suitability of other powders, based on commodities, specialty polymers, and also thermoplastic elastomeric materials.

Finally, the interest in designing novel viscous inks for additive manufacturing processes is growing constantly, thanks to the usefulness and potential of these materials as energy storage devices, flexible electronics, and biomedical uses. Despite all these efforts, the potential of new polymeric materials for additive manufacturing processes has not been fully disclosed and further research work needs to be performed, at the academic or industrial level.

Finally, to reach a deep understanding and optimize the control of each AM process, it is very important to fully understand those properties of the employed polymer systems, which are pressure-, temperature-, and time-dependent: this could be achieved using *in situ* [48] and *offline* measurements. In this way, it would be possible to exploit the gathered data for a deeper knowledge of the AM process fundamentals that, in turn, could be used for the design and development of more appropriate polymer systems for the AM processes.

REFERENCES

[1] T.D. Ngo, A. Kashani, G. Imbalzano, K.T.Q. Nguyen, D. Hui, Additive manufacturing (3D printing): A review of materials, methods, applications and challenges. Comp. B Eng. 143 (2018) 172–196.

[2] I. Gibson, D.W. Rosen, B. Stucker, Additive Manufacturing Technologies. Rapid Prototyping to Direct Digital Manufacturing. Springer Science Business Media, New York, United States (2010).

[3] A. Gebhardt Understanding Additive Manufacturing. Carl Hanser Verlag, Munich, Germany (2011).

[4] S.S. Crump Apparatus and method for creating three-dimensional objects. US patent US5121329A (1992).

[5] H. Wu, W.P. Faky, S. Kim, H. Kim, N. Zhao, L. Pilato, A. Kaki, S. Bateman, J.H. Koo, Recent developments in polymer/polymer nanocomposites for additive manufacturing. Progr. Mater. Sci. 111 (2020) 100638

[6] M. Spoerk, C. Holzer, J. Gonzalez-Gutierrez, Material extrusion-based additive manufacturing of polypropylene: A review on how to improve dimensional inaccuracy and warpage. J. Appl. Polym. Sci. 137 (2020) 48545.

[7] A. Das, E.L. Gilmer, S. Biria, M.J. Bortner, Importance of polymer rheology on material extrusion additive manufacturing: correlating process physics to print properties. ACS Applied Polym. Mater. 3 (2021) 1218–1249.

[8] M. Bertolino, D. Battegazzore, R. Arrigo, A. Frache, Designing 3D printable polypropylene: Material and process optimization through rheology. Add. Manufact. 40 (2021) 101944.

[9] R. Sharma, R. Singh, R. Penna, F. Fraternali, Investigations for mechanical properties of Hap, PVC and PP based 3D porous structures obtained through biocompatible FDM filaments. Comp. B Eng. 132 (2018) 237–243.

[10] M. Spoerk, J. Sapkota, G. Weingrill, T. Fischinger, F. Arbeiter, C. Holzer, Shrinkage and warpage optimization of expanded-perlite-filled polypropylene composites in extrusion-based additive manufacturing. Macromol. Mater. Eng. 302 (2017) 1700143.

[11] I. Zein, D.W. Hutmacher, K.C. Tan, S.H. Teoh, Fused deposition modeling of novel scaffold architectures for tissue engineering applications. Biomaterials 23 (2002) 1169–1185.

[12] S. Park, K. Fu, Polymer-based filament feedstocks for additive manufacturing. Compos. Sci. Technol. 213 (2021) 108876.

[13] A. Dey, I.N.R. Eagle, N. Yodo, A review on filament materials for fused filament fabrication. J. Manuf. Mater. Process. 5 (2021) 69.

[14] S.H. Ahn, M. Montero, D. Odell, S. Roundy, P.K. Wright, Anisotropic material properties of fused deposition modeling ABS. Rapid Prototyp. J. 8 (2002) 248–257.

[15] S. Pollack, C. Venkatesh, M. Neff, A.V. Healy, G. Hu, E.A. Fuenmayor, J.G. Lyons, I. Major, D.M. Devine, Polymer-based additive manufacturing: Historical developments, process types and material considerations. In: Polymer-Based Additive Manufacturing. Biomedical Applications, Devine DM Ed., Springer Nature, Cham, Switzerland, pages 1–22 (2019).

[16] C.M. Gonzales-Henriquez, M.A. Sarabia-Vallejos, J. Rodriguez-Hernandez, Polymers for additive manufacturing and 4D-printing: materials, methodologies, and biomedical applications. Progr. Polym. Sci. 94 (2019) 57–116.

[17] J. Weng, J. Wang, T. Senthil, L. Wu, Mechanical and thermal properties of ABS/montmorillonite nanocomposites for fused deposition modeling 3D printing. Mater. Des. 102 (2016) 276–283.

[18] N. Xu, X. Ye, D. Wei, J. Zhong, Y. Chen, G. Xu, D. He, 3D artificial bones for bone repair prepared by computed tomography-guided fused deposition modelling for bone repair. ACS Appl. Mater. Interfaces 6 (2014) 14952–14963.

[19] M. Nikzad, S.H. Masood, I. Sbarski, Thermo-mechanical properties of a highly filled polymeric composites for fused deposition modeling. Mater. Des. 32 (2011) 3448–3456.

[20] A. Dorigato, V. Moretti, S. Dul, S.H. Unterberger, A. Pegoretti, Electrically conductive nanocomposites for fused deposition modelling. Synth. Met. 226 (2017) 7–14.

[21] N.E. Zander, Recycled polymer feedstocks for material extrusion additive manufacturing. In: Polymer-Based Additive Manufacturing: Recent Developments, Seppala JE, Kotula EP, Snyder CR Eds., ACS symposium series 1315, ACS Publications, Washington, United States, pages 37–52 (2019).

[22] L. Cafiero, D. De Angelis, M. Di Dio, P. Di Lorenzo, M. Pietrantonio, S. Pucciarmati, R. Terzi, L. Tuccinardi, R. Tuffi, A. Ubertini, Characterization of WEEE plastics and their potential valorisation through the production of 3D printing filaments. J. Environ. Chem. Eng. 9 (2021) 105532.

[23] L.J. Tan, W. Zhu, K. Zhou, Recent progress on polymer materials for additive manufacturing. Adv. Funct. Mater. 30 (2020) 2003062.

[24] A. Kafle, E. Luis, R. Silwal, H.M. Pan, P.L. Shrestha, A.K. Bastola, 3D/4D Printing of polymers: Fused deposition modelling (FDM), selective laser sintering (SLS), and stereolithography (SLA). Polymers 13 (2021) 3101.

[25] P. Liu, V. Kunc, Effect of 3D printing conditions on the micro and macrostructure and properties of high-performance thermoplastic composites. In: Structure and Properties of Additive Manufactured Polymer Components, Friedrich K, Walter R Eds., Woodhead Publishing Series in Composites Science and Engineering, Elsevier, Cambridge, United States, pages 65–82 (2020).

[26] S.C. Ligon, R. Liska, J. Stampfl, M. Gurr, R. Mülhaupt, Polymers for 3D printing and customized additive manufacturing. Chem. Rev. 117 (2017) 10212–10290.

[27] G.V. Salmoria, J.L. Leite, C.H. Ahrens, A. Lago, A.T.N. Pires, Rapid manufacturing of PA/HDPE blend specimens by selective laser sintering: Microstructural characterization. Polym. Test. 26 (2007) 361–368.

[28] Z. He, C. Ren, A. Zhang, J. Bao, Preparation and properties of styrene ethylene butylene styrene/polypropylene thermoplastic elastomer powder for selective laser sintering 3D printing. J Appl Polym Sci. 138 (2021) e50908.

[29] C. Hull, Apparatus for production of three-dimensional objects by stereolithography. US Patent 4,575,330 (1986).

[30] F.P.W. Melchels, J. Feijena, D.W. Grijpma, A review on stereolithography and its applications in biomedical engineering. Biomaterials 31(2010) 6121–6130.

[31] B.E. Kelly, I. Bhattacharya, H. Heidari, M. Shusteff, C.M. Spadaccini, H.K. Taylor, Volumetric additive manufacturing via tomographic reconstruction. Science 363 (2019) 1075.

[32] C. Esposito Corcione, R. Striani, F. Montagna, D. Cannoletta, Organically modified montmorillonite polymer nanocomposites for stereolithography building process. Polym. Adv. Technol. 26 (2015) 92–98.

[33] R. Yu, X. Yang, Y. Zhang, X. Zhao, X. Wu, T. Zhao, Y. Zhao, W. Huang, Three-dimensional printing of shape memory composites with epoxy-acrylate hybrid photopolymer. ACS Appl. Mater. Interfaces 9 (2017) 1820–1829.

[34] I. Jasiuk, D.W. Abueidda, C. Kozuch, S. Pang, F.Y. Su, J. Mckittrick, An overview on additive manufacturing of polymers. JOM 70 (2018) 275–283.

[35] A. Bens, H. Seitz, G. Bermes, M. Emons, A. Pansky, B. Roitzheim, E. Tobiasch, C. Tille, Non-toxic flexible photopolymers for medical stereolithography technology. Rapid Prototyping J. 13 (2007) 38–47.

[36] J.W. Lee, G. Ahn, D.S. Kim, D.W. Cho, Development of nano- and microscale composite 3D scaffolds using PPF/DEF-HA and micro-stereolithography. Microelectron. Eng. 86 (2009) 1465–1467.

[37] C. Hinczewski, S. Corbel, T. Chartier, Ceramic suspensions suitable for stereolithography. J. Eur. Ceram. Soc. 18 (1998) 583–590.

[38] E.B. Joyee, L. Lu, Y. Pan, Analysis of mechanical behavior of 3D printed heterogeneous particle-polymer composites. Compos. Part B Eng. 173 (2019) 106840.

[39] Y. Zhang, H. Li, X. Yang, T. Zhang, K. Zhu, W. Si, Z. Liu, H. Sun, Additive manufacturing of carbon nanotube-photopolymer composite radar absorbing materials. Polym. Compos. 39 (2018) E671–E676.

[40] J.R.C. Dizon, Q. Chen, A.D. Valino, R.C. Advincula, Thermo-mechanical and swelling properties of three-dimensional-printed poly (ethylene glycol) diacrylate/silica nanocomposites. MRS Commun. 9 (2019) 209–217.

[41] Z. Weng, Y. Zhou, W. Lin, T. Senthil, L. Wu, Structure-property relationship of nano enhanced stereolithography resin for desktop SLA 3D printer. Compos. Part A Appl. Sci. Manuf. 88 (2016) 234–242.

[42] K. Arcaute, B.K. Mann, R.B. Wicker, Stereolithography of three-dimensional bioactive poly(ethylene glycol) constructs with encapsulated cells. Ann. Biomed. Eng. 34 (2006) 1429–1441.

[43] A. Cazòn, P. Morer, L. Matey, PolyJet technology for product prototyping: Tensile strength and surface roughness properties. Proc. Inst. Mech. Eng. Part B J. Eng. Manuf. 228 (2014) 1664–1675.

[44] H. Miyanaji, N. Momenzadeh, L. Yang, Effect of printing speed on quality of printed parts in Binder Jetting Process. Addit. Manuf. 20 (2018) 1–10.

[45] A.T. Gaynor, N.A. Meisel, C.B. Williams, J.K. Guest, Multiple-material topology optimization of compliant mechanisms created via PolyJet three-dimensional printing. J. Manuf. Sci. Eng. 136 (2014) 061015.

[46] W.S. Tan, S.R. Suwarno, J. An, C.K. Chua, A.G. Fane, T.H. Chong, Comparison of solid, liquid and powder forms of 3D printing techniques in membrane spacer fabrication. J. Membr. Sci. 537 (2017) 283–296.

[47] A.S.K. Kiran, J.B. Veluru, S. Merum, A.V. Radhamani, M. Doble, T.S.S. Kumar, S. Ramakrishna, Additive manufacturing technologies: an overview of challenges and perspective of using electrospraying. Nanocomposites 4 (2018) 190–214.

[48] B.H. Jared, M.A. Aguilo, L.L. Beghini, B.L. Boyce, B.W. Clark, A. Cook, B.J. Kaehr, J. Robbins, Additive manufacturing: toward holistic design. Scr. Mater. 135 (2017) 141.

11 Emerging Applications of Polymers for Automobile Industries

Çiğdem Gül[1] and Emine Dilara Kocak[2]

[1] Marmara University, Institute of Pure and Applied Sciences, Goztepe Campus, The Buildings of Institutes, Floor: 2, 34722, Kadıköy/İstanbul, Turkey
[2] Marmara University Faculty of Technology, Recep Tayyip Erdogan Kulliye Aydınevler Mah. Idealtepe Yolu no:15, 34854, Maltepe/İstanbul, Turkey

1 INTRODUCTION

The automotive industry uses the latest developments in science and technology. Improving the durability and reliability of automobile parts is the most important of the fundamental problems of material science. The growth of this industry requires the creation of new designs, increasing material quality and safety. The main subjects that the modern automobile industry focuses on are:

- the creation of smart cars without driver intervention; and
- the development of cars with alternative energy sources [1].

Today, the production of automobile parts from polymeric materials is constantly increasing and this increasing trend is expected to continue in the future. The main factors in the selection of polymeric materials in automobiles are appearance, functionality, economy, and low fuel consumption. Although reducing the mass of parts is one of the main reasons for choosing polymeric materials, their increasing use in the future will pave the way for new applications such as comfort, safety, and component integration in automobiles [2]. The ultimate goal of automobile designers in their work to increase speed and power is to create the lightest cars possible. The purpose of producing a light car is to increase performance and efficiency, while maintaining safety and comfort [3]. A change in vehicle weight directly affects energy consumption. For example, if the vehicle weight is reduced by 10%, fuel economy increases by about 7% [4].

The performance of cars is constantly improving because the engine efficiency is increasing, the body is more aerodynamic, the transmission is improving, and the rolling resistance of the tires is reduced. When designing a car, reducing its mass allows for maintaining the basic characteristics of the car, which consumes less fuel and emits less harmful substances into the atmosphere. Reducing the weight of the car reduces the load on the suspension parts, which in turn extends their life [1]. The materials used in vehicle manufacturing are classified as ferrous metals, non-ferrous metals, and polymers. In this chapter, the polymers used in the automobile industry and their reasons for use are reviewed.

DOI: 10.1201/9781003278269-11

2 MATERIALS IN THE AUTOMOBILE INDUSTRY: AN OVERVIEW

The automobile industry provides employment opportunities to millions of people and makes significant contributions to the economies of countries. Automobile production is sometimes seen as an indicator of economic growth. In the United States, there was a 143% increase in the number of cars registered per 100 people between 1950 and 1996. This situation showed similar growth trends in Asia and China [5].

A large number of materials are used to manufacture cars. The main materials used in the production of automobiles and their parts are steel, aluminum, glass, plastics and polymer composites, rubber, magnesium, copper, and carbon fiber. The materials to be used in the vehicle body are expected to be resistant to heat, chemical and mechanical impacts, easy to manufacture, and durable. In addition, the price of the material is an important factor in vehicle production [6]. The introduction of lightweight materials in the automotive industry stems from our need for higher efficiency. With the introduction of thermoplastics into the automotive industry in 1950, this need began to be met. In the following years, various researches were made on advanced, high-performance plastics or polymers and the materials used in automobile production began to change. Initially plastics were used for their good mechanical properties, excellent appearance, self-coloring, etc. Today, plastics are used in automobiles for durability, toughness, design flexibility, corrosion resistance, and weight reduction with minimal cost [7].

3 POLYMERIC MATERIALS USED IN THE AUTOMOBILE INDUSTRY

Recent advances in automobile designs show that material selection is critical to product performance. The use of environmentally friendly, green materials is encouraged in order to reduce the carbon footprint, with very strict legal regulations. Previously, most auto parts were made of metals or different metal alloys, but today many auto parts are made from polymers or plastics. Today, 59% of the average vehicle weight consists of iron and steel; this rate was 60% in 2010, 65% in 2000, and 70% in 1990 [5].

Polymer is a Greek word, formed from the combination of "poly" (many) and "mer" (repeat units). Many polymers are composed of carbon and hydrogen bonds, and may contain oxygen, nitrogen, sulfur, and fluorine. Polymers are composed of their smallest components, called monomers. Polymer chains can be of different lengths and differ in the number of cross-links between their molecular structures. In general, polymers can be classified into three categories: thermoplastics, thermosets, and elastomers. Thermoplastics are polymers that are solid at room conditions and connected by van der Waals bonds, which are not very strong when their inner chain structures are examined. When thermoplastics are heated, their viscosity decreases, their atomic chains break apart, and they become fluid. When they cool, the broken chains solidify again and the material becomes solid. Because thermoset plastics are cured or cross-linked compounds, they are durable and heat-resistant. Materials in this group cannot be reworked or shaped. Three-dimensional cross-links are formed between polymer chains. Three-dimensional meshes do not flow, even when heated and pressurized, and the polymer cannot be restored by cooling and heating.

Elastomers are usually thermoset plastics. During curing, cross-links occur between long polymeric chains. Their most important features are that they are flexible and elastic. When stretched at room temperature, they stretch to at least twice their length and immediately return to their original size when the tension is removed. Generally, thermoplastics are preferred over thermosets in the manufacture of auto parts because thermoplastics are simpler and faster to form. The material to be used in the parts related to the automobile engine should be chosen among those with high thermal resistance. In the production of external and structural parts, materials with high mechanical strength should be considered. Thermal resistance and mechanical properties in thermoplastics are limited compared to thermosets. Therefore, it can be applied after some modification to produce high-performance polymers [8]. Thermoplastic polymers are frequently used in automobile bumpers, seating, dashboard, and fuel systems, including panels, under-bonnet components, interior trim, lighting, electrical components, distribution systems, upholstery, liquid reservoirs. Thermoplastic polymers are preferred in the automobile industry due to their low density (between 0.9 and 2.1 g/cm^3) and low cost. Thermoplastic polymers are non-corrosive and have low thermal and electrical conductivity [9]. Polymers are used in the automobile industry due to their low cost, light weight, corrosion resistance, and good mechanical properties. Polymers used in the automobile industry can be classified as commercial or engineering polymers. The proportions of plastics and polymers used in cars with different properties produced between 2016 and 2020 are as follows: 44% polypropylene (PP), 4% polybutylene terephthalate (PBT) and polyethylene terephthalate (PET), 5% acrylonitrile butadiene styrene (ABS), 7% polyethylene (PE), 8% polyamide (PA), 9% polyurethane (PUR), and 23% others [10].

4 TYPES OF POLYMERS USED IN THE AUTOMOBILE INDUSTRY

Since engineering polymers and plastics replaced metals in cars, more than 50% of car interior components are made from plastics. The use of polymeric materials continues to increase to produce lighter, economical, easily produced, recyclable auto parts.

4.1 POLYPROPYLENE

Polypropylene (PP) is a thermoplastic "addition polymer" made from the combination of propylene monomers by using a Zeigler-Natta catalyst. PP polymers can be made with an atactic, isotactic, or syndiotactic chain configuration. PP is widely used in automobile parts due to its superior mechanical properties, lightness, easy processing, recyclability, and economic applicability [11]. The use of PP in the automotive industry is as follows: exterior applications 34%, interior applications 34%, under the hood 24%, and electrical applications 12% [12]. The main applications of PP are: bumpers, air ducts, battery boxes, all fender linings, interior trim, dashboard, and door trims [13]. The properties of polypropylene can be improved using natural fibers and synthetic fibers.

4.2 POLYURETHANE

Polyurethane (PUR) is a monomer-free polymer and is mostly formed by urethane bonds. PURs are synthesized by a polycondensation reaction between diols or polyols and isocyanates in the presence of catalysts and additives. PU foams are one of the most researched materials in the automobile industry. Due to their low density, PU foams reduce the overall weight and gas consumption of the car. Reducing vehicle weight improves performance and reduces greenhouse gas emissions. PU foams can be flexible or rigid. Thanks to these features, vehicles with mechanical shock protection can be produced. PU foams also protect the vehicle against the risk of fire due to its flame-retardant feature [14]. Demand for lightweight and high-performance materials is increasing daily in emerging industries such as the automotive sector. They are also used for insulation and acoustic damping. PP foams are used as sandwich structures or materials in some ultra-luxury sports cars. This is because PU foams increase impact resistance and provide superior thermal insulation [15].

They are one of the important components used in almost all automobiles worldwide. Polyurethane foams are mostly found in car seats, armrests, and headrests, and their cushioning helps reduce the fatigue and stress associated with driving. Their durability and lightness, combined with their strength, also provide insulation against the heat and noise of the engine in the car bodies [15,16].

4.3 Polycarbonates

The production of lighter automobile bodies with plastic materials replacing metal and glass has gained importance. The use of polycarbonate (PC), an engineering plastic, in automobile windows offers greater transparency, and impact and heat resistance [17].

Polycarbonate (PC) is high heat engineered thermoplastic polymer. It is prepared by two different processes:

1. The Schotten-Baumann reaction consists of an amine-catalyzed interfacial polycondensation of phosgene and an aromatic dihydroxy compound.
2. The base-catalyzed transesterification of a bisphenol and diphenyl carbonate [18].

PCs melt at about 265°C, and these polymers have great impact strength. PCs are resistant to water and most organic compounds, however, they gradually hydrolyze under alkaline conditions. They are transparent, colored plastic. The mechanical properties (hardness and toughness) of PCs are good. Polycarbonates also show good abrasion, high impact strength, creep resistance, UV resistance, and excellent optical and electrical properties. It is known that reducing fuel consumption is related to vehicle weight. That is why car designers trying to reduce vehicle weight are increasingly turning to polycarbonate as an alternative to metal and glass. Polycarbonate, a much lighter yet strong polymer, plays a key role in improving fuel economy and reducing CO_2 emissions. Lighter cars brake more easily and have less collision impact, thus providing drivers with a safer driving experience. As a result, polycarbonate has become a growing force in the automotive industry. A typical vehicle contains 10 kilograms of polycarbonate [19].

4.4 Polyamide

Polyamide (PA), which was the first commercial thermoplastic engineering polymer, is commonly known as Nylon. PA66 was produced by Wallace Carothers in 1928 from the condensation polymerization of hexamethylene diamine (HMDA) and adipic acid. The structure of polyamides can be aliphatic, semiaromatic, or aromatic polyamide (aramids). Nylons produced from aromatic diamines and aromatic dicarboxylic acids are called aromatic nylons (aramids). Poly(p-phenylenerephthalamide) (PPTA), produced by Dupont with the tradename Kevlar, is also used in automotive transmission parts and tires. Semiaromatic polyamides are produced of reaction of poly(hexamethylene terephthalate) (HMDA) and terephthalic acid [20]. In the automobile industry, different types of PAs are widely used in the production of auto parts. Manufacturing of parts under the hood using glass fiber-reinforced polyamide (PA) is the main application. In addition, nylon cord fabric is used in the production of automobile tires. PA is used in the manufacture of the throttle body and cylinder heads, cooling systems, air intake manifolds, and car engine parts. PA reduces production cost by 30% and part weight by 50% compared to conventional materials (e.g. metals). PA provides a better surface appearance and strength, and thinner designs. PA is used in the engine and cylinder head cover and door and tailgate handles, rearview mirror, and hubcaps, power trims, and headlight bezels [21,22].

4.5 Polyvinyl Chloride

Polyvinyl chloride (PVC) is a high-gloss, excellently flexible, and thermally stable polymer. PVC is one of the most widely used and least expensive thermoplastics worldwide. It is not resistant to organic solvents but shows good resistance to alkalis, salts, and concentrated polar solvents. Since PVC has chlorine in its structure, it also illustrates flame-retardant properties [22]. Soft PVC especially has made a significant contribution to lightening cars and increasing their performance and efficiency. PVC polymers are also used in instrument panels, electrical cables, pipes, door panels, etc.. It reduces vehicle weight while increasing the comfort and performance of modern vehicles.

Depending on the amount of plasticizers used, PVC products can be rigid or flexible. The vinyl component in PVC improves the tensile strength and resistance to chemicals and solvents [22]. For instance, layers of flexible vinyl are used for stonechip protection, as sound-dampening material, as sealants, as well as in protective coatings to cover widely exposed areas such as underbodies, wheel arches, and rocker panels. In today's automobiles, the PVC compound is used in instrument panels and sills, sun visors, synthetic leather seat covers, headliners, gaskets, mudguards, noise- and vibration-dampening components, floor coverings, exterior side moldings, and protective strips. Because PVC components weigh less than conventional materials, vehicles consume less energy and provide a lower carbon footprint. PVC is used in automotives because of the following properties: controlled oxygen and water vapor transmission, cost-efficient, soft and scratch-resistant coatings for panels, cold temperature resistance, UV stability, durability, and lightweight [23].

4.6 POLYSTYRENE

Polystyrene (PS) is an aromatic hydrocarbon formed by the polymerization of styrene monomer, discovered by Eduard Simon in 1839. It is produced in the presence of styrene, ethylene, and benzene. PS is clear and colorless and its optical properties and high stiffness are very good. Commercial PS is an atactic structure and is an amorphous, glassy polymer. It is generally hard and relatively cost-effective. PS is not much affected by acids, alkalis, or oxidizing or reducing agents. Rubber or butadiene copolymer can be added to the polymer to produce high-impact polystyrene (HIPS), increasing the toughness and impact resistance of the PVC [24]. It has superior chemical and electrical resistance properties. PS is used in the manufacture of car fixing equipment, equipment housing, buttons, and the automobile display base [15].

4.7 POLYETHYLENE

Polyethylene (PE) is the largest commercial polymer obtained by repetition of ethylene (CH_2) units. PE has various flexibility levels depending on the manufacturing process and is the toughest of the high-density materials. It is the polymer of choice when products need to be particularly moisture-resistant and cost-effective. Polyethylenes can be of different densities. Low-density polyethylene (LDPE) has a density value in the range of 0.91–0.925 g/cm³, linear low-density polyethylene has a range of 0.918–0.94 g/cm³, and high-density polyethylene (HDPE) has a range of 0.935–0.96 g/cm³ [25]. PE has properties such as toughness, near zero moisture absorption and low moisture permeability, excellent resistance to chemicals, superior electrical insulation, non-toxicity, and low cost [22]. It is not used much in large automotive applications because most melt-processed PE parts remain below the required modulus and temperature resistance. HDPE is used in fuel tanks and fuel systems in automobiles. LDPE is used in automotive cable covering. Ultra-high-molecular-weight polyethylene (UHMWPE) is designed to reduce automobile interior and exterior noise. PE is also used in the automobile dashboard, upholstery, and other reservoirs [26].

4.8 ACRYLONITRILE BUTADIENE STYRENE

Acrylonitrile butadiene styrene (ABS) is a commercial copolymer obtained by copolymerization of acrylonitrile, butadiene, and styrene monomers [27]. The properties of ABS can be adjusted by changing the ratio of monomer units. It is derived from acrylonitrile, propylene, and ammonia and provides thermal stability and resistance to chemicals. Acrylonitrile monomer offers the ABS very good chemical resistance, resistance to aging, hardness, and gloss. The butadiene monomer gives ABS toughness, ductility at low temperatures, flexibility, and good melt strength. The styrene monomer provides ABS with good workability, gloss, and hardness [28]. Acrylonitrile butadiene styrene is used for the manufacture of different automotive body parts, helmets, dashboards, wheel covers, exterior components, interior parts, door trim, door handles, loudspeaker grilles, consoles, and navigation system housings [22].

4.9 Polyoxymethylene or Polyacetals

Polyoxymethylene (POM) is a kind of plastic that is derived from formaldehyde and is used extensively in the automobile industry. POM is an engineering thermoplastic used in the manufacture of precision parts that require high rigidity, low friction, and dimensional stability. It is naturally opaque white due to its highly crystalline nature, but may be of any other color [29]. POM is highly resistant to chemicals and has a high crystal content. It has excellent short-term mechanical properties. The density of POM ranges between 1.38–1.44 g/cm^3, its tensile strength is 5,000–9,000 psi, and its modulus is 447 psi [24]. POM is resistant to most chemicals and organic solvents at room temperature [28]. The high crystallinity of POM provides good dimensional stability and high modulus. POM is also used in the manufacture of bearings, seating systems, gears, fuel tanks, sunroof systems, window guides, lighters, door lock systems, speaker grilles, electrical systems, automotive interior and exterior trims, mirrors, and wiper systems [22,30].

4.10 Polymethyl Methacrylate

Polymethyl methacrylate (PMMA) is a polymer produced from the polymerization of methyl methacrylate and has a density of 1.19 g/cm^3. PMMA has excellent optical properties which are very similar to glass transparency with 92% light transmittance. It has high hardness, rigidity, and strength; good electrical properties; and resistance to weak acids and alkaline solution [31]. Due to the very good dimensional stability of the products produced from PMMA, it shows low moisture absorption capacity. PMMA is resistant to lots of chemicals but dissolves in organic solvents. It has attracted the attention of automakers because it offers numerous tinting options, from clear to deep, and is lighter than conventional glass. In addition to its lightness, PMMA also has properties such as transparency, resistance to bad weather conditions, good acoustic properties, and allowing for new designs. It is used in exterior, rear, and indicator light covers, interior light covers, exterior panels, trim, bumpers, fenders and other molded parts, light guides, and fascia [22]. PMMA is widely used in the production of durable and highly visible plates [32].

4.11 Polybutylene Terephthalate

PBT is a thermoplastic (semi)crystalline engineering polymer that is synthesized by polycondensation of terephthalic acid or dimethyl terephthalate with 1,4-butanediol using special catalysts [24]. The main properties of PBT are excellent mechanical and thermal properties, dimensional stability, resistance to chemicals and solvents, and flame retardancy [33]. PBT is used in exterior automobile components, fog lamp reflectors and housings, sun-roof front parts, central locking system housings, door handles, bumpers, mudguards (rear and front), radiator grilles, mirror housings, and carburetor components [22,25].

4.12 Polyethylene Terephthalate

Polyethylene terephthalate (PET) is a condensation polymer obtained by the esterification of ethylene glycol, terephthalic acid, or dimethyl terephthalate. It is a widely used thermoplastic polyester and is called "polyester." PET can be either amorphous (transparent) or semicrystalline structure (opaque and white). Semicrystalline polyester has good strength, ductility, hardness, and rigidity. Amorphous polyester has better ductility, but less stiffness and hardness. Since PET is hydrophobic, it absorbs almost no water [33]. It has excellent electrical insulation, dimensional stability, and high strength. Polyester has a higher heat distortion temperature (HDT) as compared to PBT, and PET's fracture resistance is very good. PET is extensively used in the engine cover, wiper arm and housings, headlight, headliners, boot liners, door panels, parcel shelves, and connector housings [25,34]. Carpets used in automobiles are expected to have high colorfastness, resistance to abrasion, aesthetic appearance, and insulating noise [34].

5   POLYMER BLENDS USED IN AUTOMOBILES

Today's car manufacturers place great emphasis on aesthetics, efficiency, and safety when designing cars. The properties sought in materials in the automobile industry are lightness, rigidity, ductility, thermal stability, flame retardancy, and impact resistance. To achieve these properties, the importance given to studies on blending different polymers continues to increase [18]. Automobile parts are classified as interior parts, exterior parts, under the hood, chassis, electrical systems, and safety [10]. In bodywork applications, polycarbonates/thermoplastic polyester/impact modifiers, modified polybutylene terephthalate, polypropylene/ethylene propylene diene monomer (EPDM), and modified polyamides are preferred because of their excellent impact resistance, dimensional stability, hardness, and easy workability. There are many kinds of polymer systems used in exterior applications where the functional demands are almost identical. Basic requirements such as toughness, hardness, heat resistance, fuel efficiency, and weather resistance are extremely important in material selection. The only plastic body part that can be left unpainted or painted in mass-produced vehicles is the bumper. Polycarbonates/polybutylene terephthalate/modifiers and polybutylene terephthalate/modifier polymer blends are preferred in the production of automobile bumpers. On the other hand, polypropylene/ethylene propylene diene monomer and polyamide/modifier mixtures are used in bumper covers and backed bumpers. Other applications have been developed such as wings or front fenders molded in polycarbonates/acrylonitrile butadiene styrene blends and painted off-line. Dashboards, glove box flaps, consoles, and steering gear shaft covers in the passenger compartment are made from thermoplastic compounds. Thermoplastic blends such as ABS, polycarbonates/ABS, and poly(2, 6-dimethyl phenylene oxide)/PS are used in automobile interiors. The use of thermoplastics in the manufacture of automobile interior parts has decreased, as the blends are showing expanding application in body parts [22].

Thermoplastic polyolefins (TPOs) are widely used in the manufacture of auto parts due to their machinability and improved mechanical properties. TPOs are most commonly used in the automotive industry. Recent emerging developments have shown that TPOs are also used in under-hood applications in the automobile industry. TPOs are used for exterior body parts such as bumpers, rocker panels, body gaskets, doors, and windows [36]. PPO provides for enhanced heat distortion temperature (HDT). It is a measure of the resistance of a polymer to change under a given load at elevated temperature. PS material is extremely economical and is an ideal material for automobile parts that are required to be easily processed and shaped. PS and PPO blends are used in automotive panels and pump components. In the automotive industry, heating system components, air filter housings, radiators, protection carters, lighting system components, fans, and electrofan supports are produced from glass-reinforced PP. Also known as acrylic styrene-acrylonitrile or acrylonitrile styrene acrylate (ASA), it is a thermoplastic widely used in the automobile industry, developed as an alternative to ABS. PC/ASA blends are used in the manufacture of automobile exterior components due to their improved chemical, mechanical, and thermal properties [22].

6   AUTOMOBILE PARTS

6.1   EXTERIOR PARTS

Exterior parts of automobiles consist of bumpers, wheels, mirrors and housings, lenses, and sunroofs. Exterior structures in automobiles are expected to have good strength, high impact resistance, and a smooth optical surface coating. In order to ensure fuel economy and environmental sustainability, importance is given to the production of automobiles from light and recyclable materials. Desired properties when manufacturing exterior components are lightness, part consolidation, minimum cost, easy design, impact resistance, and surface aesthetics [37].

6.1.1 Automobile Bumpers

Bumpers are important because they help protect critical auto parts (e.g. headlights, taillights, etc.) and costly parts (e.g. hoods, front and rear fenders, exhaust and cooling systems, etc.). The front and rear bumpers of the car protect the car body like a shield, helping to prevent or minimize the damage caused by the impact. The bumper system consists of three parts: the outer plastic cover that controls the airflow; energy-absorbing material that reduces shock effects; and a reinforcing bar/beam made of steel, aluminum, or fiber-reinforced composite that absorbs impact energy and protects the car body [37].

Different polymer blends are used in the production of tampons. Polypropylenes, polyurethanes, and polycarbonates are used in front bumpers due to their good strength, low density, rigidity, and adhesion to the bumper body. Polypropylenes, polyurethanes, and low-density polyethylenes are used in mechanical energy absorbers. Although TPOs have also gained importance recently, PP, PUR, and LDPE continue to be used for mechanical energy absorbers. These materials can be used as reinforcing elements to improve shock absorption ability, strength, and stiffness [22].

6.1.2 Automobile Headlamps/Rear Lights and Their Housings

Headlamps are lights that are installed at the front of the vehicle to illuminate the driving path in the dark and to see objects better. Until the 1980s, both the headlight and taillight were made of glass, due to its transparency and inexpensiveness. However, the biggest disadvantages of glass were the difficulty in design and shaping. In the following years, efforts were made to lighten the body, lens, and reflectors to improve the fuel economy and the low resistance of glass against scratches and impacts. Therefore, PMMA and PC have started to be used in headlights due to their transparent structure [37].

One of the developments in the field of automobile lighting is the optical lenses of headlights made of PC. Headlights made of PC are much lighter and more resistant to breakage than conventional ones, providing freedom in headlight design. The surface of the headlights is covered with a protective layer to increase resistance to scratches [2]. For other drivers to easily perceive the presence and movement of the car, rear lights and park lights are placed behind the vehicle. Headlights/rear light housings are very important in addressing the rigors of the weather and dark [22]. Headlight lamps are also made of PS, PC polymer [35]. Reflectors for rear lights are usually made of ABS or PP as they are economical [37].

Headlight systems generate a large amount of heat, so headlight covers must be heat resistant. Rear lighting housings do not need to be made of high heat-resistant polymers or polymer blends. ABS, ASA, and PP polymers or their blends can be used in rear lamp housings. Due to their superior properties, polyphenylene ether (PPE)/high-impact polystyrene (HIPS), and PA/ABS blends are used in headlight housings. Acrylonitrile butadiene styrene/PC blends and PMMA/PC blends are used for automotive lenses [22].

6.1.3 Wheel Covers

The functions of wheel covers are to give the wheel a pleasing aesthetic appearance and also to prevent the wheel nuts from corroding and falling off. Plastics are resistant to salt, chemicals, heat, or cold. They can be painted easily and have good corrosion resistance [22]. Plastics are rarely used in car wheels, as they cannot dissipate brake heat due to their low thermal conductivity. PC/ABS blends are used in wheel covers due to their high impact resistance and heat resistance and minimum cost. The use of ASA polymer in Mercedes-Benz Vito is due to its greater resistance to ultraviolet (UV) radiation [37].

6.1.4   Body-in-White

Body-in-white (BiW) is the name given to the car body's sheet excluding moving parts (e.g. hood, front and rear fenders, etc.), trims (e.g. glass, seats, etc.), or chassis subassemblies. Usually, BiW accounts for around 27% of a vehicle curb weight. BiW must have bending, torsion, static and dynamic high rigidity, and high tensile strength. In addition, according to the Federal Motor Vehicle Safety Standard No. 208, it should ensure the safety of both the vehicle body and the passengers against all kinds of impacts that may come from all over the vehicle and even the vehicle overturning. BiW must be able to protect passengers from noise, vibration, and/or harshness. In addition, it should be easily welded, shaped, painted, and designed [37].

6.1.5   Chassis

The chassis is the most important main structural component of the car. By using polymers in chassis production, it is expected that the weight of the automobile will be reduced by 30%. Polymer chassis are not only lightweight, but also contribute to safety, design, and performance improvement. As an alternative to aluminum, POM, PA matrix reinforced with glass fiber and carbon fiber composites is used in the body. PA polymer is used in the production of the chassis due to its resistance to very high temperatures [35].

6.2   INTERIOR COMPONENTS

6.2.1   Instrument Panel (IP)

The instrument panel, dashboard, or fascia is a control panel set within the central console of an automobile. Instrument panels are divided into three parts as rigid, covered, and foamed panels according to the production techniques and features. Stiff dashboards are seen in lower segment automobiles. Foamed instrument panels have a soft layer between the support and the covering. They have good aesthetics. Dashboards should be eco-friendly and safe. They should be lightweight for fuel-efficiency and ultraviolet (UV)-resistant. The instrument panel materials must be durable and resistant to wear. Stiff dashboards are manufactured from PP or ABS/PC blends. While laminated PVC was used in the covered dashboard in the past, today TPOs are widely used. Ultra-high-molecular-weight polyethylene (UHMWPE) and PUR are frequently used in automotive dashboards. Natural fibers such as kenaf, hemp, flax, jute, and sisal with a thermoplastic and thermoset matrix can be used to produce sustainable automobile panels [38].

6.2.2   Door Panels

Today's door panels have evolved from a simple two-piece latch system and rolling mechanism to a more complex system. Door panels consist of multiple components such as electronic windows, automatic locking system, and speakers. They consist of a foamed structure covered with textile or plastic. The most used polymers are PVC and PUR. In addition, ABS/PP blends are used in door panels due to their high mechanical strength and chemical resistance [37].

6.2.3   Seats and Related Components

Automobile seats consist of various parts, such as the armrest, backrest, headrest, seat base, and seat track position sensor. Seat belts and airbags are safety requirements. PC/ABS mixtures are used on the seat backs. Seat covers must be resistant to temperature, abrasion, UV radiation, and moisture [3]. In the 1960s, PVC was replaced by woven nylon and polyester fabric. The car seat fabric is 90% polyester, and the face and lining fabric usually have a PUR foam layer [40]. Thermoplastic polymers such as PE, PP, PET, and PA are used in seats and belts [11,34].

6.2.4   Under the Hood Components

Components such as the electrical and mechanical parts, powertrain, transmission, driveshaft, and fuel system of the engine are covered by the hood. Because under-hood components often operate at high temperatures, it becomes difficult to use plastic in under-hood applications. [35]. When choosing materials for under-hood applications, some important factors are considered, such as good machinability, good thermal aging, low specific gravity, UV resistance, dimensional stability, resistance to chemicals, and high modulus at high temperature. Plastics that are used under the hood must be resistant to corrosion, wear and tear, vibration, damping, and high temperatures. The job of the powertrain is to transmit the engine's power to the wheels. Automatic transmission techniques in cars must be able to apply high torque to maintain high speeds. Engineering plastics and PUR are used to manufacture powertrains. Radiators are made of PA and PP, batteries are made of PET and PP, air ducts are made of PA, and air filters are made of plastic such as PP and PUR foams [35]. PET/PC and PPE/HIPS blends are used for ignition compartments. PET/PC, PA/PPE, PPS/ PEI, and other polymer blends can be used in underbody applications [41].

7   POLYMER COMPOSITES USED IN AUTOMOBILES

The properties of the polymer can be changed by incorporating fibers, inorganic fillers, organic fillers, etc. Composites are the best choice for producing stronger, lighter, and more cost-effective automobile components. In fiber-reinforced polymer composites, the matrix consists of polymeric materials and the reinforcement elements consist of fibers. Reinforced polymer composites have been used to produce low-density, stiffness, and strength materials in automotive applications. Some examples of fiber-reinforced polymer composites in automotive applications can be given. PP with 45% glass fiber-filler blend is used in sunroof modules, flax-reinforced PES composite is used in sound insulation, *Abaca* fiber-reinforced polypropylene is used in spare tire well covers, and flax/sisal fiber-reinforced PU is used in door panels. PE and polyvinylchloride-based wood composites are used in the windows and door frames of automobiles [35].

8   CONCLUSION

Polymers and polymer blends are used in various parts of automobiles including the interior, exterior, and under-hood parts. The main goal of automotive manufacturers should be to develop advanced materials using developing technologies to meet consumer demand. In this chapter, detailed information is given about different polymers and polymer mixtures used in automobiles. In addition, it is emphasized that the use of polymers in various components such as the interior, exterior, under the hood, and chassis of the car will reduce vehicle weight and provide fuel efficiency. In the automobile industry of the future, lightweight materials produced from polymer blends will be of greater importance.

REFERENCES

[1] Hovorun T. P., Berladir K. V., Pererva V. I., Rudenko S. G., Martynov A. I. Modern materials for automotive industry. Journal of Engineering Sciences, Vol. 4(2) (2017) 8–18.
[2] Štrumberger, N., Gospočić, A., Bartulić, Č. Polymeric materials in automobiles. *Promet-Traffic&Transportation*, *17*(3) (2005) 149–160.
[3] https://blogs.autodesk.com/advanced-manufacturing/2017/02/08/materials-used-lightweight-cars/ Accessed: 15.05.2022
[4] Lyu, M. Y., Choi, T. G. Research trends in polymer materials for use in lightweight vehicles. *International Journal of Precision Engineering and Manufacturing*, *16*(1) (2015). 213–220.
[5] Srivastava, V., & Srivastava, R. Advances in automotive polymer applications and recycling. *International Journal of Innovative Research in Science, Engineering and Technology*, 2(3) (2013) 744–746.

[6] Todor, M. P., Kiss, I. Systematic approach on materials selection in the automotive industry for making vehicles lighter, safer and more fuel-efficient. *Applied Engineering Letters*, *1*(4) (2016) 91–97.

[7] Szeteiova, K. (2010). *Automotive Materials Plastics in Automotive Markets Today*. Institute of Production Technologies, Machine Technologies and Materials, Faculty of Material Science and Technology in Trnava, Slovak University of Technology, Bratislava.

[8] Lyu, M. Y., Choi, T. G. Research trends in polymer materials for use in lightweight vehicles. *International Journal of Precision Engineering and Manufacturing*, *16*(1) (2015) 213–220.

[9] Mallick, P. K. (2021). Thermoplastics and thermoplastic–matrix composites for lightweight automotive structures. In *Materials, Design and Manufacturing for Lightweight Vehicles* (pp. 1–36). Woodhead Publishing.

[10] Ladhari, A., Kucukpinar, E., Stoll, H., Sängerlaub, S. Comparison of properties with relevance for the automotive sector in mechanically recycled and virgin polypropylene. *Recycling*, *6*(4) (2021) 76.

[11] Agarwal, J., Sahoo, S., Mohanty, S., Nayak, S. K. Progress of novel techniques for lightweight automobile applications through innovative eco-friendly composite materials: A review. *Journal of Thermoplastic Composite Materials*, *33*(7) (2020) 978–1013.

[12] Jansz J. Polypropylene in automotive applications. *Polypropylene*, *2* (1999) 643–651.

[13] Emilsson, E.; Dahllöf, L.; Ljunggren, M. (2019) *Plastics in Passenger Cars: A Comparison Over Types and Time*; IVL Swedish Environmental Research Institute: Stockholm.

[14] de Souza, F. M., Choi, J., Ingsel, T., Gupta, R. K. (2022) High-performance polyurethanes foams for automobile industry. In *Nanotechnology in the Automotive Industry* (pp. 105–129). Elsevier.

[15] Das, A., Mahanwar, P. A brief discussion on advances in polyurethane applications. *Advanced Industrial and Engineering Polymer Research*, *3*(3) (2020) 93–101.

[16] www.polyurethanes.org/en/where-is-it/automotive/ Accessed: 15.05.2022

[17] Hotaka, T., Kondo, F., Niimi, R., Togashi, F., Morita, Y. Industrialization of automotive glazing by polycarbonate and hard-coating. *Polymer Journal*, *51*(12) (2019) 1249–1263.

[18] Sweileh, B. A., Al-Hiari, Y. M., Kailani, M. H., Mohammad, H. A. Synthesis and characterization of polycarbonates by melt phase interchange reactions of alkylene and arylene diacetates with alkylene and arylene diphenyl dicarbonates. *Molecules*, *15*(5) (2010) 3661–3682.

[19] https://polymerdatabase.com/polymer%20classes/Polycarbonate%20type.html Accessed: 15.05.2022

[20] Peters, E. N. (2017) Engineering thermoplastics—materials, properties, trends. In *Applied Plastics Engineering Handbook* (pp. 3–26). William Andrew Publishing.

[21] https://polymerdatabase.com/polymer%20classes/Polyamide%20type.html Accessed: 15.05.2022

[22] Begum, S. A., Rane, A. V., Kanny, K. (2020) Applications of compatibilized polymer blends in auto-mobile industry. In *Compatibilization of Polymer Blends* (pp. 563–593), Elsevier.

[23] https://vinyl.org.au/automotive Accessed: 15.05.2022

[24] Greene, J.P., *Commodity Plastics, Automotive Plastics and Composites*, William Andrew Publishing, 2021, 83–105.

[25] Patil, A., Patel, A., Purohit, R. An overview of polymeric materials for automotive applications. *Materials Today: Proceedings*, *4*(2) (2017) 3807–3815.

[26] Sadiku, R., Ibrahim, D., Agboola, O., Owonubi, S. J., Fasiku, V. O., Kupolati, Jamiru, T., Eze, A.A, Adekomaya, O.S., Varaprasad, K., Agwuncha, S.C., Reddy, A.B., Manjula, B., Oboirien, B., Nkuna, C., Dludlu, M., Adeyeye, A., Osholona, T. S., Phiri, G., Drowoju, O., Olubambi, P.A., Biotidara, F., Ramakokovhu, M., Shongwe, B., Ojijo, V. (2017) Automotive components composed of polyolefins. In *Polyolefin Fibres* (pp. 449–496). Woodhead Publishing.

[27] Shi, Y., Yan, C., Zhou, Y., Wu, J., Wang, Y., Yu, S., Chen, Y. (2021) Polymer materials for additive manufacturing—powder materials Editor(s): Shi, Y., Yan, C., Zhou, Y., Wu, J., Wang, Y., Yu, S., Chen, Y., In *3D Printing Technology Series, Materials for Additive Manufacturing*, Academic Press, 9–189.

[28] Pious, C.V.,Thomas, S. (2016) 2-Polymeric materials—Structure, properties, and applications, Editor(s): Joanna Izdebska, Sabu Thomas, *Printing on Polymers*, William Andrew Publishing, 21–39.

[29] Greene, J.P. (2021) 8 Engineering plastics, Editor(s): Joseph P. Greene, In *Plastics Design Library, Automotive Plastics and Composites*, William Andrew Publishing, 107–125.

[30] www.americanchemistry.com/industry-groups/formaldehyde Accessed: 15.05.2022

[31] Pouzada, A.S (2021) Selection of thermoplastics, Editor(s): Antonio Sergio Pouzada, In *Plastics Design Library, Design and Manufacturing of Plastics Products*, William Andrew Publishing, 87–140.

[32] www.pmma-online.eu/applications/automotive-lighting/#:~:text=PMMA%20is%20used%20to%20create,it%20high%20gloss%20or%20matt Accessed: 15.05.2022

[33] McKeen, L.W. (2014) 5 – Polyesters, Editor(s): Laurence W. McKeen, *The Effect of Long Term Thermal Exposure on Plastics and Elastomers*, William Andrew Publishing, 85–115

[34] Saricam, C., Okur, N. (2018) Polyester usage for automotive applications. *Polyester-Production, Characterization and Innovative Applications, 1st ed.;* Camlibel, NO, Ed, 69–85.

[35] Girijappa, Y. G., Ayyappan, V., Puttegowda, M., Rangappa, S. M., Parameswaranpillai, J., Siengchin, S. (2020) Plastics in automotive applications. In S. Hashmi. *Reference Module in Materials Science and Materials Engineering.* UK: Elsevier 1–11.

[36] www.marketsandmarkets.com/Market-Reports/thermoplastic-polyolefin-market-256331319.html#:~:text=TPOs%20are%20used%20for%20exterior,largest%20market%20share%20in%202020 Accessed: 20.05.2022

[37] Pradeep, S. A., Iyer, R. K., Kazan, H., Pilla, S. (2017) Automotive applications of plastics: past, present, and future. In *Applied Plastics Engineering Handbook* (pp. 651–673). William Andrew Publishing.

[38] Shinde, N. G., Patel, D. M. (2020) A short review on automobile dashboard materials. In *IOP Conference Series: Materials Science and Engineering* (Vol. 810, No. 1, p. 012033). IOP Publishing. (2020).

[39] Fung W. *Coated and Laminated Textiles.* 1st ed (2002). Boca Raton: CRC Press LLC and Woodhead Publishing.

[40] Maxwell J. (1994) *Plastics in the Automotive* Industry. Cambridge: Woodhead Publishing.

[41] Utracki, L. A., & Wilkie, C. A. (Eds.) (2002) *Polymer Blends Handbook* (Vol. 1, p. 2). Dordrecht: Kluwer Academic Publishers.

12 Polymeric Nanocomposites for Toxic Waste Removal

Pragati Chauhan,[1] Mansi Sharma,[1] Rekha Sharma,[1] and Dinesh Kumar[2]

[1] Department of Chemistry, Banasthali Vidyapith, Rajasthan 304022, India
[2] School of Chemical Sciences, Central University of Gujarat, Gandhinagar-382030, India

1 INTRODUCTION

The environment has supplied sufficient resources for sustaining and developing species around the world. Freshwater, existing even before the emergence of life, is among the most vital of elements. Accessibility to safe drinking freshwater is among the most fundamental humanitarian aims, and it continues to be a large global concern in the twenty-first century. Water covers roughly 71% of the Earth's surface. A World Health Organization report and other publications have shown that the quantity of water on the planet is 97% seawater, and the remaining 3% is primarily clean water. Of this, 70% is clean water in the form of sea ice, glacial ice, and ice caps, with the remainder mostly being moisture in the soil or that in aquifers. Therefore, freshwater accessible for usage accounts for about 1% of the entire groundwater on the planet, which is quite small [1].

The quality of freshwater is degraded for a number of reasons, including urbanization, population development, commercial output, and climate variability. Due to rapid industrialization and urbanization, the fast discharge of various pollutants has harmed groundwater quality, resulting in water contamination. Various contaminants can pollute water, including potentially hazardous elements, dyes, phenolic compounds, pesticides, herbicides, medications, personal care items, etc. These contaminants have the potential to be bioaccumulative, persistent, carcinogenic, mutagenic, and harmful to aquatic species, flora, and fauna. Scientists have a significant issue to address since water pollution poses a growing hazard to people's well-being and the environment. As a result, removing impurities from freshwater has become essential. Again, for water treatment, a range of methodologies are used, including precipitation, coagulation, flocculation, incineration, ion exchange, reverse osmosis, membrane filtration, electrochemistry, photoelectrochemistry, advanced oxidation processes, adsorption, and biological methods [2]. It is critical to eliminate harmful chemicals from sewage to a healthy level while also accomplishing it quickly, reliably, and cost-effectively. Nanotechnology has the potential to play a significant contribution in this respect. Nanotechnology, defined as the science and art of controlling things at the nanoscale (1–100 nm), can develop innovative nanoparticles for hazardous remediation materials for use in polluted surface water, groundwater, and wastewater. Nanomaterials have gained a lot of interest due to the manufactured nanoscale materials.

Regarding treating wastewater, nanomaterials outperform nanoparticles in terms of adsorption capacity, selectivity, and stability. Modern processes, including solution blending, melt blending, layer-by-layer deposition, in situ polymerization, electro-polymerization, and surface-initiated polymerization, have created a wide range of conductivity nanomaterials [3]. Polymer nanocomposites are being utilized to remove pollutants from water using various methods. This chapter gives an overview of current breakthroughs in the use of polymeric nanocomposites in treating wastewater.

DOI: 10.1201/9781003278269-12

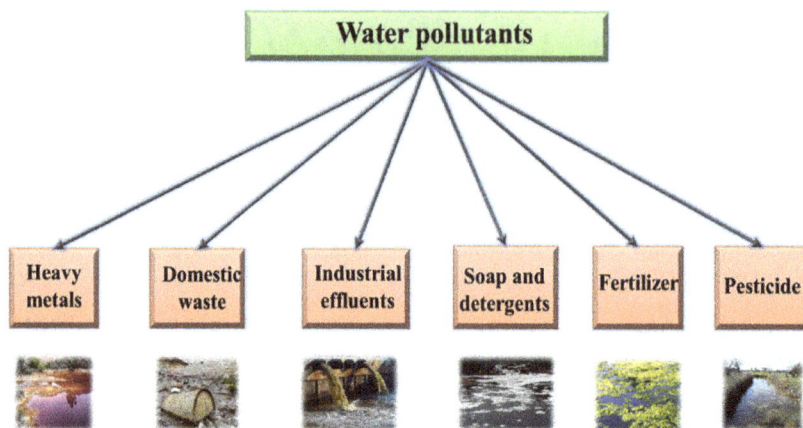

FIGURE 12.1   Sources of water pollutants.

We also discuss techniques used to synthesize polymer nanocomposites, their properties, and their application.

2   WATER POLLUTANTS

Figure 12.1 depicts the sources of water contaminants. There are two forms of pollution of water:

1. The most prevalent form of resource is static resources. Sewer lines, industries, coal, electricity factories, and oil wells are just a few examples. About 300–400 million tons of toxic substances, chemicals, sludge, and other contaminants are dumped into the Earth's freshwater every year, creating a severe risk to health [4].
2. Nonpoint contaminants are scattered over a vast area, and their contamination cannot be traced back to a single source.

3   POLYMER NANOCOMPOSITES

The primary goal is to create polymer nanomaterials with huge interlayer interactions between the nanomaterials and the polymer. That's only possible if the nanoparticles are evenly distributed across the polymeric matrices. For this, ultrasonic pressures and physical ripping are routinely applied. On the other hand, the availability of appropriate chemical groups between the nanomaterials and the polymer matrix is critical. Consequently, polymer biocomposite processing is important to achieving excellence. Melt mixing techniques and in situ polymerization are often used to generate materials in research and industry. Other methods, including template, microemulsion, sol–gel, etc., are also utilized, depending on the requirements and suitability [4,5]. Figure 12.2 depicts polymer nanostructured manufacturing procedures. Consequently, a short rundown of a few key polymer nanocomposite production techniques is provided here.

3.1   SOLUTION METHODOLOGY

Physical and ultrasonication combing are utilized to make a homogeneous mixture of a polymeric mixture or a suitable nanoparticle. The nanocomposites are homogenized and swollen in an appropriate fluid solution. In contrast, the polymer solutions are prepared simultaneously in a flammable solution with nanomaterials used in a fluid solution. The unwanted solvent or liquid is removed

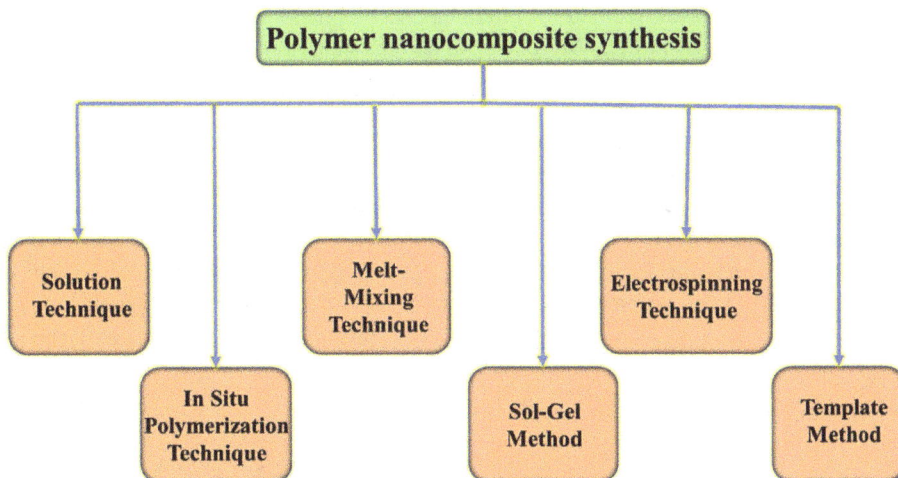

FIGURE 12.2 Synthesis techniques of polymer nanocomposites.

from the homogeneous solution by precipitation or evaporation of the nanostructured material polymer matrices. The intermarriages between nanomaterials and polymerization are determined by the kinetics of contacts and mixing between the constituents in the final mixture [6]. Interactions between nanoparticle strands must be greater than contacts between the polymer-solvent and nanoparticle to enable homogeneous dispersal of nanoparticles in the polymer matrices. Polymers can infiltrate through nanoparticle formations, even while incorporating link components into the nanomaterial interface, culminating in the desired composite [7]. This method can yield simultaneously uncapped nanomaterials and interaction between both ingredients, depending on the level of dispersal. The composite must be thermally stable, having advantageous enthalpic and dynamic characteristics. Therefore, because the structural movement of nanoparticles is improved as the interfacial distance increases, the enthalpic component is lowered whenever contacts with rigid nanoparticles restrain movement of the polymer. However, this impact is modest. The Gibbs' energy gradually becomes advantageous, although the entropic loss is much higher due to the strong contacts between the nanocomposites and the matrices. Generally, this technique has several disadvantages, such as high cost, ignitability, and health and environmental issues owing to minimal contact rates and toxic organic solvents. On the other hand, the solutions technique is simple because ready-to-use industrial polymers may be used right away, and it's much better if freshwater can be used as the solvent [8].

3.2 IN SITU POLYMERIZATION TECHNIQUE

In this technique, the polymeric is created in the context of distributed nanocomposites. The nanocomposites scattering and expanding in the polymer's lower viscous monomer produces efficient nanocomposites. The polymers facilitate the debonding of the nanoparticles throughout the polymer's procedure and the monomer molecules' adsorption on the surfaces of scattered nanocomposites.

The method for creating aqueous polyurethane nanoparticles containing carbon quantum dots is standardized [9]. Consequently, the connections between nanomaterials and polymers can be quite powerful in many cases. The nanomaterials are evenly scattered in the polymer matrices due to homogeneous dispersion. This method may result in exfoliated nanocomposites due to the lower viscosity of the homogeneous distribution of nanoparticles, prepolymer or monomer, and a good interaction.

Consequently, it is the ideal method for producing several nanomaterials, notably plastics, since it needs no or very minimal liquids. Moreover, functional compounds in nanoparticles may influence the polymerization reaction and also the bridging process for thermoplastics nanomaterials. This is particularly tempting when employing the spontaneous emulsification technique because the carrier is freshwater. An appropriate monomer is cross-linked to use a bifunctional activator such as 2,20-azobis (2-methyl propionamide) dihydrochloride bifunctional clay. Consequently, the liquid bifunctional clay can operate as an activator (azo-compound) and a nanoparticle [10]. The needed polymeric nanoparticles are obtained by blending the dispersion after it has created a monomer using the standard emulsion polymerization technique. Whenever an emulsified solution can be employed effectively, this is extremely useful. In situ polymerization is generally preferred, particularly in laboratories, due to the outstanding efficiency of the produced polymeric composites.

3.3 MELT-MIXING TECHNIQUE

This approach combines nanoparticles in a polymeric matrix with typical melting facilities to produce the needed polymeric nanocomposites. Consequently, the thermal method avoids the need for a liquid and the influence of polymerization or cross-linking processes, making it a popular choice among industries. The commercial polymeric production technique may be used immediately, with no extra procedures or equipment. Microcrystalline materials can be converted into amorphous polymeric materials by blending devices (such as an injector, bartender plastic order, metal injection, kneader, etc.). After this, integrating nanocomposites under robotic sheer accomplishes homogeneous distribution on the nanocomposites. Mild interlayer nanomaterials are created in the bulk of cases due to the excessive melt temperature of the prepared polymer. A number of organizations value the manufacture of polymeric composites employing this method because the technique is more environmentally friendly [11,12].

3.4 SOL–GEL METHOD

The sol–gel process is also utilized to produce nanocomposites. This approach produces nanomaterials in a polymer matrix. A greater temperature of aqueous gels or solution containing the nanocomposites' building components or precursors and the polymeric solution is used. During the procedure, the polymer aids in the crystal growth of the nanocomposite. As a consequence of this procedure, the nanocomposites are confined inside the polymer matrices. The bulk of aqueous nanomaterials is made using this process. Even though this method has the potential to promote nanocomposite distribution in a single movement without the need for additional energies, it has several limitations [13]. For example, in a polymer nanoparticle, the synthesis of clay particles needs high heating that could melt the polymer matrices. A harsh handling setup may result in the nanocomposites clumping together. Consequently, compared to processes described earlier, this method is much less common.

3.5 ELECTROSPINNING TECHNIQUE

Polymers nanoparticle nanofibers are made by electrospinning. This is a standard method to produce polymer threads. A mechanized mechanical needle with a small hair needle, a higher voltage supply, and a collection linked to electric earthing is used. Physical shears accompanied by ultrasonic treatment are utilized, with the same solution approach to combine the liquid solution in an appropriate polarized liquid or polarized polymers melted with the required nanoparticle distribution. A power sourse is required to be applied on the tip of the needle to create nanofibers [14]. The supplied power must be high enough to resist interfacial tension and extend the flow, leading to the development of nano-dimensional structures. The polymeric fibers must stiffen as the strands travel toward the collecting surface. The polymeric solution's polarization influences the nanowires' form, uniformity, and size, the melt's viscosity and content, the syringe's hydrostatic fluid, electric field strength, fluid velocity, and tip-to-collector length tips diameter, as well as other rotating factors. The shape and size of the collections are also determined by the collectible geometries [15].

3,6 TEMPLATE METHOD

Polymers are often used as templates in the templating method to synthesize nanoparticles using their polymer solution, culminating in nanocomposites being formed in situ. As per the hypothesis, this method can increase nanoparticle distribution in the polymer matrices in a single movement. While templates synthesizing components are important for the construction of inorganic biomaterials, the process doesn't really perform very well for polymeric composites [16]. As a consequence, this approach is rarely used. Organizations have begun to adopt a typical polymeric master-batch technique to enhance specific nanomaterials recently.

4 PROPERTIES

The development of suitable polymeric composites improves some critical aspects of virgin polymerics, even while possibly introducing new features at the end-stage. The level of increase is influenced by the nanomaterial's size, shape, aspect ratio, state of dispersion, and interfacial interactions with the polymer matrices in these cases. In Figure 12.3, some of the fundamental characteristics of polymeric composites are outlined.

4.1 PHYSICAL

The incorporation of nanomaterials has a minor influence on the intrinsic properties of the material, such as density, crystallinity, solubility, etc. However, these properties are very similar in many conditions to polymer nanocomposites. Because the increased density of polymeric composites is

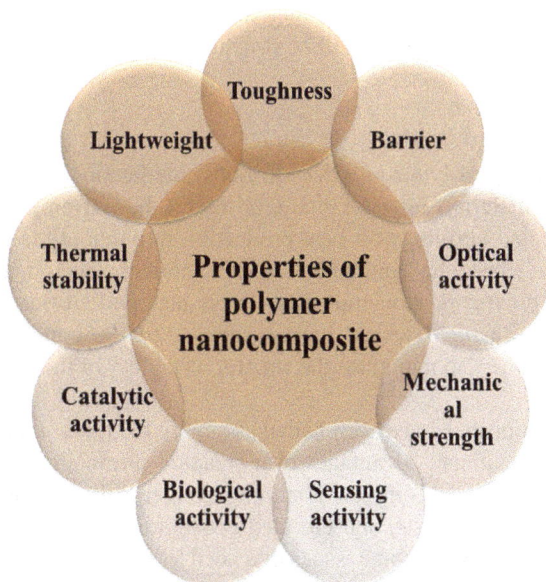

FIGURE 12.3 Properties of polymer nanocomposites.

minimal, the compact qualities of polymers are almost always retained. Among the major advances in polymer nanocomposites is the incorporation of nanomaterials. The light weight of such substances involves mastering, and makes carrying easier while lowering the cost per unit of volume. The impact of nanostructures on the crystallization of polymeric composites is variable, but in most cases, it is negligible. The dissolution rate of polymeric nanoparticles is generally difficult, resulting from reactions with nanomaterials; most of these are hydrophobic [17]. Nanocomposites are not completely dissolved in either solution.

4.2 Mechanical

Physical properties, notably the rigidity of a polymeric matrix, improve considerably whenever suitable nanomaterials are introduced, at quite low dosage levels. Furthermore, increasing nanocomposite loading enhances the mechanical energy levels up to a certain dose level, whereby it may deteriorate due to nanocomposites clumping in the polymer matrices. The more nanomaterials engage with polymer matrices, the greater the gain in structural rigidity. In both situations, highly porous nanocomposites possess greater value increases than interpenetrating nanomaterials, but the overall intensity has always been better than for standard macrocomposites and pure materials. The mechanical properties of polymeric composites are increased because the load is efficiently transmitted from the physically weak polymer matrices to the strong and stiff nanoparticles due to the higher surface modification of nanocomposites [18,19].

4.3 Thermal

Among the most noticeable limits of polymers is their thermal durability with rising temperatures, except for specially created super partially purified compounds. Applying appropriate nanoparticles to a pure polymer can dramatically increase their high thermal stability. The inclusion of nanostructures enhances thermal properties. Since nanocomposites function as heat absorbers, they serve as a mass transfer barrier for volatile matter generated during the dissolution procedure and vary the overall dissolution route. The glass transition temperature increases when rigid nanomaterials limit the movement of the polymer network. The creation of polymeric nanocomposites also may increase the crystallization process enthalpy and crystalline melting temperature. In such settings, nanostructures act as foaming agents, encouraging the crystalline structure of the crystallized polymeric matrix [20].

4.4 Flame Retardancy

Nanoparticles are mixed into polymer matrices to boost the heating value and delay decomposition. Fireball composite materials impede, retard, or inhibit the warming, degradation, ignition of toxic gases, breakdown, and burning with heat production throughout the process of combustion. Reduced graphene oxide, silica, clay, graphene, carbon nanofibers, and other materials improve the fireball performance of polymer nanocomposites. These nanoparticles are essentially crushed or turned into non-flammable charcoal, which slows the process of heat transfer during the heating and burning procedure. The flame also maintains the structural stability of the nanocomposites, stopping the fire from spreading.

Consequently, nanocomposites have been hailed as among the most significant advancements in heat resistance, offering significant advantages over standard formulations [21]. The regulation of poisonous flame-resistant chemical compounds starting to emerge through burning, infecting people and the environment, additional costs, and elevated concentrations of load capacity to attain the intended amount of this property are some issues that need to be taken care of. The processing difficulties and, among many other things, the degradation of many other useful characteristics affiliated with the following a systematic system are among them.

4.5 CHEMICAL AND BARRIER RESISTANCE

If suitable nanocomposites are created, pure polymer barrier characteristics and chemical stability increase considerably. Nanostructures with large aspect ratios have a huge surface area that prevents certain penetration chemicals from diffusing. The diffusion coefficient of a piercing substance is influenced by the level of dispersal of the nanomaterials. The serpentine or zigzag path concept may be used to explain the excellent improvement in barrier characteristics provided by two-dimensional nanoparticles since their high surface area greatly limits permeability. The deformation ratio is computed by multiplying the real path with the shortest route they could use if there was no impediment [22]. Molecular fumes or ions generated in distinct molecular conditions have greater difficulties communicating as the dispersion becomes more complicated.

4.6 BIOLOGICAL

Such characteristics are equally important in polymeric composites as they are in nanomaterials' biological activities. Biocompatibility and biodegradability are among the most important biological properties of polymer nanomaterials. The cellular survival rate in the area of the evaluated polymer composites is used to establish the initial in situ biomaterials of polymer composites. Greater bioavailability may be obtained by altering the bioactivity and shape of polymeric composites. Polyamide, polyurethane, and polyester nanocomposites are the most investigated biodegradable polymers [23,24]. When tested in vivo, some metals and metal oxide nanomaterials have proven dangerous to various types of cells. However, those made to use in a green process have shown promise as a means of ensuring cell compatibility. Moreover, the toxic effects of nanocomposites such as carbon nanotubes, graphene, and other materials are dramatically decreased after incorporation and derivatization into a polymeric matrix. Cell adhesion, proliferation, and growth in polymeric composites have also been shown to be advantageous for biopolymers. Biological material is defined as manufactured or organic fibers that come into contact with bodily or cellular samples and are designed for therapeutic, prosthetic, diagnostic, or storage uses while damaging living organisms. Polyglycolic acid, polyurethane, polylactic acid, chitosan-based polysaccharides, polycaprolactone, polyphosphazenes, polyorthoester, polyfumarate, poly(glycerol sebacate), polyarylates, polypyrrole, poly(amido amine), poly(ether ester amide), and other matrix composites are used to stabilize numerous cell types. Antimicrobial nanocomposites are particularly beneficial in material science and biological research [25].

5 POLYMER NANOCOMPOSITES: POTENTIAL WATER TREATMENT MATERIALS

Different methods used in the treatment of industrial wastewater are described below and shown in Figure 12.4.

5.1 ADSORBENTS

Due to their cheap cost and good performance, adsorbent technologies are often used in water purification. Although activated charcoal is the most commonly used synthetic adsorption material in water purification, other activated carbons are used because of their high cost. Virtually every type of contamination may be effectively extracted from sewage with the correct adsorbent materials. Many different kinds of adsorbent materials are used to remove pollutants from sewage. The type and quantity of contaminants in the sewage influence the type of adsorption that should be employed to filter it. The adsorption used in the procedure should have good specificity for the pollutant since the amount of cleaned water is frequently large [26].

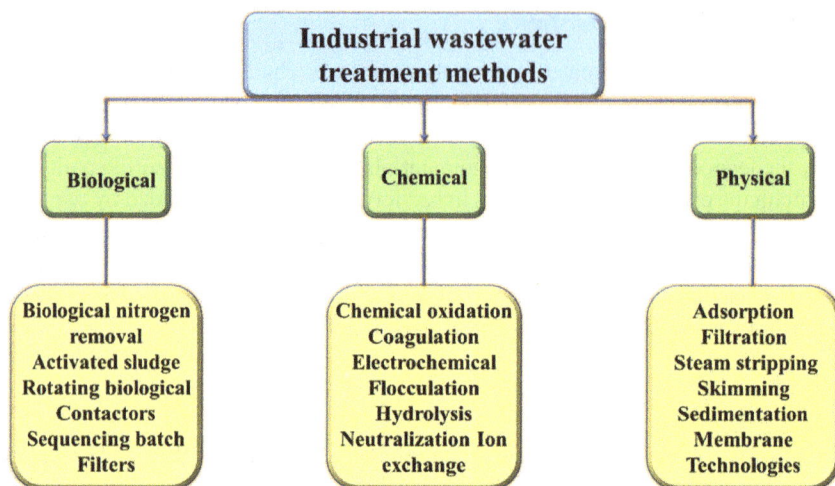

FIGURE 12.4 Methods of treatment of industrial wastewater.

It must also be safe, transformable, affordable, easy to recover, and freely accessible. Among the engineered adsorbent materials, nanocomposites are the most efficient. Nanocomposites are effective adsorbents for the catalysts, removal, and sensors due to their large effective surface area and high sensitivity. Nanoparticles with a high surface-area-to-mass ratio can significantly improve the adsorbed properties of adsorbent materials [27]. Magnetic nanocomposites are particularly intriguing because they are simple to identify and remove from treated wastewater. When nanoparticles are employed in packed beds or even other water movement devices, their small particle size creates significant force decreases and challenges with separation and recycling. Mixed nanomaterials were formed by inseminating microscopic nanoparticles onto larger fine materials to get around these restrictions. Because of the nanoparticles' contact, the resultant biocompatible nanoparticle retains the essential properties of nanoparticles while improving the processability, stability, and other advantages [28]. The surface area of nanostructures, the essence of the polymer nanostructure, intensity of toxins, pH of the liquid solution, dosages of adsorbent materials, temperature, contact time, and other factors all impact the rate of adsorption. Equilibrium adsorption systems are important when it comes to sorption on adsorbent materials. At a fixed pH and temperature, an absorption spectrum is a helpful graph representing the processes of a product's absorption or movement from aqueous porous materials or freshwater systems to a solid matrix. Adsorption isotherms can be used to determine the surface properties, adsorption process, and degree of attraction among adsorbents [29].

5.2 MEMBRANES

Composite films are presently the most widely utilized membranes used in water purification. Still, they are hampered by various difficulties, along with a tradeoff among selectivity, permeability, and limited clogging tolerance. Figure 12.5 depicts the various membrane processes. Nanocomposite membranes created by combining synthetic polymers with nanomaterials promise to offer a potential solution to these issues. Nanomaterials are used to develop customized membranes to satisfy all of the problems of desalination and sewage treatment, particularly fouling and biofouling, while also increasing the membrane's lifespan by boosting the physical strength and resistance to cleansing regimes [30]. Upgraded nanocomposite membranes could be constructed to meet particular water purification purposes by modifying the design and physicochemical features. The four types of

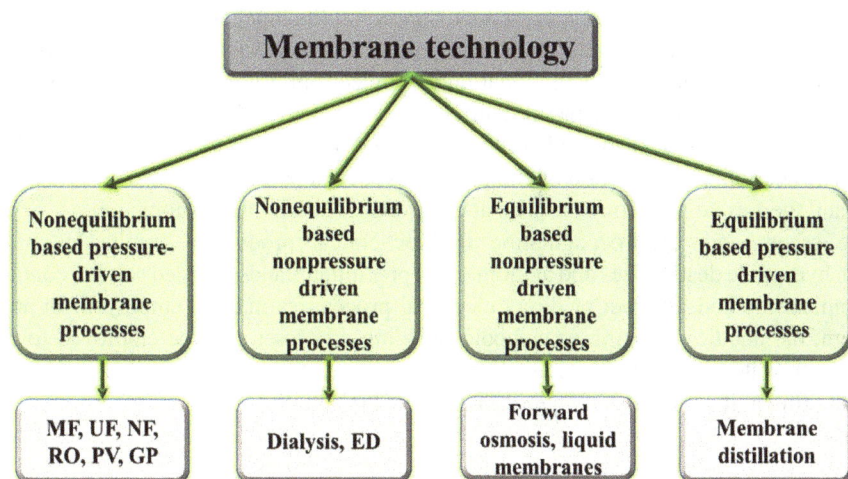

FIGURE 12.5 Schematic representation of membrane processes.

membranes are (1) thin-film nanocomposite, (2) conventional nanocomposite, (3) surface-located nanocomposite, and (4) thin-film composite on a nanocomposite substrate. Due to their ionic interaction, nanocomposites have significantly better membranes with specific distinct nanoparticle traits and the possibility for novel capabilities and characteristics. This makes a significant contribution to the creation of high-performance antifouling biocomposites. Polymeric membranes can filtrate the entire range, using ultrafiltration, microfiltration, reverse osmosis, and nanofiltration [31,32]. However, the practical use of polymeric membranes for process purification remains in the initial phases, and more studies are needed to determine polymeric membranes' cost and superior quality products and services.

5.3 COAGULANTS

Coagulants and flocculation procedures eliminate colloid contaminants such as clay, silt, bacteria, and organic materials from contaminated water. Massive groups of aggregate can be extracted from groundwater during the clarification and cleaning treatments. Al (III) compounds are commonly utilized as coagulant compounds in water purification. Still, they also have a number of disadvantages, such as their inability to control the kind of coagulation process and competing with those other procedures. One strategy for improving the coagulation rate is to use a more positive coagulation factor. An aluminum-silicate polymer nanocomposite is a better coagulant. Clay-based products have been extensively used for wastewater remediation and contaminant elimination. Nanomaterials consisting of an anchored component and a polymer are used as flocculants to effectively and swiftly decrease the overall suspended materials or visibility in sewage [33]. These nanoparticles mix the effects of the flocculent and coagulant by reducing the colloidal matter charges and connecting them and attaching them to a larger component, increasing the sediment. Preparation is achieved in a simple treatment process and is highly efficient and rapid.

5.4 ELECTRODE MODIFICATION FOR ELECTROCHEMICAL THERAPY

An electrochemical treatment process is an interesting option for eliminating inorganic and organic pollutants in sewage. The two main areas of innovation are chemical and physical procedures. A primary interaction at the electrodes or an intermediate reaction with just an electrochemical reaction-generated species are examples of the biochemical technique. At the same time, electroflotation, electrodialysis, and electrocoagulation are examples of the conventional process. Electroplating is a physical method that involves electromigrating ions over alternating cation and anion exchange barriers to separate and concentrate them [34]. The vapors produced at an anode (O_2 and H_2) propel the dispersed pollutants to the level at which they are readily gathered and removed. The neutralization reaction of charges on suspended particles due to their proximity to the electrode and the formation of tiny droplets at electrodes induces this phenomenon.

On the other hand, electrocoagulation in biochemical production results in coagulation. Aluminum hydroxide destabilizes and agglomerates precipitates and adsorbed soluble contaminants. In the chemical method, indirect or direct electrical procedures may be employed. In an electrolytic system, the anodic oxidizing or cathodic reducing processes may be employed to eliminate pollutants from water efficiently. Chemical oxidizing is used to remove chemical material from an agricultural effluent that employs oxygen contained in the solution [35].

6 APPLICATIONS

One of the major reasons for such a desire to grow polymer materials would be that the technique could be used for commercial purposes. Figure 12.6 illustrates some of the most common applications for polymer nanocomposites.

6.1 AUTOMOBILES

Polymers nanomaterials are used in a variety of automobile parts, this includes mirror housings, timing belt covers, engine covers, door handles, and other components. In 1991, clay/nylon-6 nanomaterials were employed as starter motor coverings for Toyota cars, marking the very first commercial use of polymeric composites. Clay/polyolefin nanomaterials were employed as a step assist element in the GMC Safari and Chevrolet Astro vans in 2001. After that, these nanomaterials were to be employed in Chevrolet Sedan doors. Nobel composites produced clay/polypropylene nanomaterials for mechanical seat backsides. At the same time, clay/nylon-12 nanomaterials were produced for vehicle gasoline pipes and gas components of the system. Soon after, nylon-6 nanomaterials for

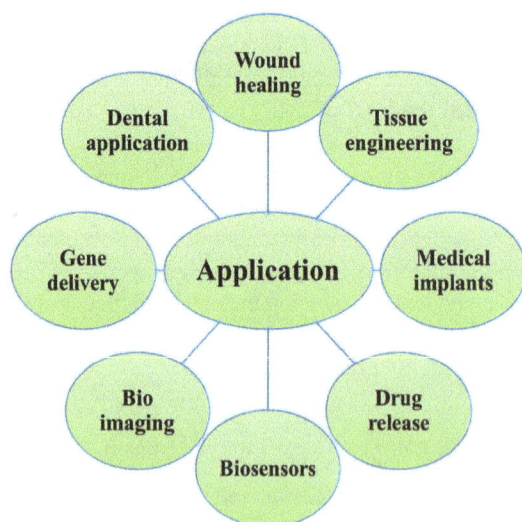

FIGURE 12.6   Applications of polymer nanocomposites.

motor covering for motors were available. Recently, clay/polypropylene nanomaterials have been produced by Noble polymers for Honda Acura mechanical back seats. In contrast, clay/nylon-12 nanocomposites were produced for vehicle gasoline lines and fuel elements of the system [36].

6.2   Containers

The inclusion of relatively small quantities of nanomaterials has considerably increased the evaluation of the progress of polymeric composites. Nanocomposites were used to improve the barrier rigidity. One concept is multilayered clay/polymer nanomaterials used as wall-lining elements in enclosures. One study dealt with developing commercialized clay/nylon-6 composite compounds for beverage packaging. Mitsubishi's gas chemical and nanocore branch has created Nylon-MXD6 nanomaterials as a multilayered poly(ethylene terephthalate) container. Packaged food uses, including load-bearing ones, have aroused great interest in polymeric composites with strong resistance qualities, especially those founded on nanoclay. Some examples include processed meats, cheese, cereals, confectionery, and boil-in-the-bag packaging foods. Nanoparticle compositions are also expected to greatly prolong the storage life of several foods. The higher insulating properties of nanocomposites have also been exploited in energy conservation. An illustration of this is the decreased liquid conduction in polymers, including polyamides, after nanoclay implantation. As a response, such metals are gaining popularity in automotive gas tanks and gasoline components [37]. The price reductions associated with reduced energy-transmitted power contribute to the attraction of this use.

6.3   Paint and Coatings

Polymer nanomaterials have found usage in the paint and coating industries as combination thin films. Such coverings, which combine the simplicity and fluidity of polymer manufacturing with the toughness of inorganic compounds, have been effectively applied to substrates. Such hybrid coverings are very clear, have a suitable surface appearance, thermal stability, good adhesion, chemical or corrosion resistance, and increased scratch resistance properties of polymeric surfaces [38].

6.4   Miscellaneous

Suction cleaners' centrifugal pumps, rotors, mower hoods, power tool housings, and covers for wearable digital gadgets, including mobile telephones and fax machines, have been investigated. Nanocomposites might also be used in a range of manufacturing areas, including bumpers, gas tanks, exterior and interior panels, aviation for fire-resistant panels, construction structure sections, and also have an important role in preserving high-performance components, among many others, due to their considerably elevated fuel economy and reduced weight. Antibacterial, insect resistance, self-cleaning, crease resistance, and other qualities of nanocomposite fabrics have recently been in high demand. One example is the carbon nanofiber epoxy matrix seen in sports equipment. Some nanoparticles are combined with polymeric membranes to eliminate chemical contaminants and microorganisms from water, as water purification procedures that have been used in the past have been ineffective in providing sufficient safe drinking water [39].

Nevertheless, nanoscale materials that are adaptable and extremely effective provide affordable water purification choices. In advanced waterways, nanoparticles assist in creating efficient wastewater purification techniques. In this respect, nanocomposites have captivated interest. Matrix composites may be designed for certain thermal, physical, and other characteristics, allowing nanotechnology to be used more effectively in various applications. In respect of their features, composite materials outperform typical nanoscopic composites, and they can be manufactured using low-cost and simple processes. Nanocomposites have proven to be a great alternative for portable water purification processes. Composite materials with nanostructured constituents are generally suitable with conventional processing techniques and can be simply integrated into regular components. Among the most important advantages of nanomaterials over conventional water systems is their ability to combine various characteristics. As a result, multipurpose technologies like nanostructured membranes can retain nanoparticles and eliminate contaminants.

Moreover, polymeric materials boost productivity due to the inherent distinctive features of NPs, including a fast reaction rate [37,38]. Bifunctional polymers having NPs embedded or deposited on their interface have hazardous risks, as NPs can be released into the atmosphere, in which they can concentrate after a time. Moreover, nanostructured water treatments are rarely easily scalable, and in many cases, conventional treatments available are currently more cost-effective. In contrast, polymeric nanocomposites offer a great deal of promise for water resource improvements, particularly for distributed treatment processes [40].

7   POLYMER NANOCOMPOSITE RECYCLING AND RECOVERY

Sustainability is an important part of the waste managerial hierarchy. In reality, it includes everything, from item recycling to thermally recovery. Figure 12.7 depicts several mechanical recycling procedures. The term recycling corresponds to the reusing of a material like a polymer. Industrial effluent recovery and recycled waste recovery are two options for recovering water purification nanocomposites. The latter comprises collecting and reusing waste polymers created during the manufacturing process.

This could be due to nanoparticle discharge into the atmosphere or the matrices' polymeric stage. Although there has been significant progress in applying polymeric composite materials for water purification, there are gaps in knowledge on the hazards posed by nanoparticles and polymer matrices. Inflammation, oxidative stress, acute toxicity, tissue barrier crossing, DNA damage, and disruption are the major biologic factors that could influence the potentially adverse effects of nanoparticles. The kind of particulate matter released impacts the likelihood of unfavorable repercussions. At high concentrations, nano-ZnO has been shown to be dangerous. Al_2O_3 nanoparticles of size 20 nm induced irritation in the lungs of a rat. AgNPs are toxic to living organisms at high levels, and continuous contact with silver has been related to argyria in humans [40]. In keratinocytes, single-walled

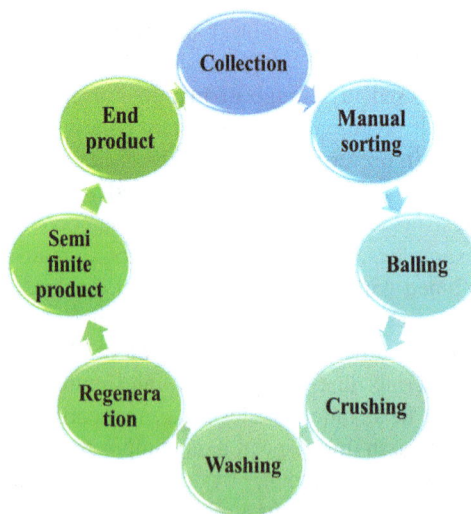

FIGURE 12.7   Steps of mechanical recycling.

carbon nanotubes (SWCNTs) lowered cell survival, causing structural and morphological alterations in sebocytes. Normal platelet clumping was influenced by SWCNTs and multiwalled carbon nanotubes (MWCNTs).

8   CONCLUSIONS AND FUTURE PROSPECTS

This chapter has thoroughly demonstrated the concepts of polymeric nanocomposites. Polymeric composites have a significant influence on the size, shape, and aspect ratio of nanomaterials and their interfacial interactions and composition. Moreover, uncoated nanoparticles' toxicity, stability, and safety are key scientific problems. The synthesis of polymeric composites is not only a viable alternative in this situation but also a technique that takes advantage of their combinatorial effects. In the present scenario, new water purification techniques are necessary to ensure good water quality, speed up industrial effluent manufacturing techniques, and eliminate biological and chemical impurities. In this sense, nanotechnology is among the greatest options for sophisticated sewage treatment techniques. Membranes, coagulants, adsorbents, and modified electrodes formed of polymeric composites have a great deal of potential for electrolytic wastewater purification. Toxins of different sorts can be virtually eliminated from wastewater by polymeric nanomaterials cost-effectively. Recyclability, toxicology, and recovery all require greater attention. This study has looked at basic nanocomposite development for water purification, focusing on the challenges faced.

REFERENCES

1. N. Karak, Fundamentals of nanomaterials and polymer nanocomposites. In Nanomaterials and Polymer Nanocomposites (2019) 1–45. Elsevier, Amsterdam.
2. J. T. Choi, D. H. Kim, K. S. Ryu, H. I. Lee, H. M. Jeong, C. M. Shin,... & B. K. Kim, Functionalized graphene sheet/polyurethane nanocomposites: effect of particle size on physical properties. Macromol. Res. 19 (2011) 809–814.
3. P. C. Ma, N. A. Siddiqui, G. Marom, & J. K. Kim, Dispersion and functionalization of carbon nanotubes for polymer-based nanocomposites: A review. Compos. – A: Appl. Sci. Manuf. 41 (2010) 1345–1367.
4. S. Palit, & C. M. Hussain, Frontiers of application of nanocomposites and the wide vision of membrane science: a critical overview and a vision for the future. In Nanocomposites for Pollution Control (2018) 441–476. CRC Press, Boca Raton, Florida.
5. S. Palit, & C. M. Hussain, Carbon-based polymer nanocomposite and environmental perspective. In Emerging Carbon-Based Nanocomposites for Environmental Applications (2020) 121–145. John Wiley, Ney Jersey.
6. Z. A. Karim, G. P. Sean, & A. F. Ismail, Nanocomposite Membranes for Heavy Metal Removal from Wastewater. In Nanocomposites for Pollution Control (2018) 361–402. CRC Press, Boca Raton, Florida.
7. M. A. Shannon, P. W. Bohn, M. Elimelech, J. G. Georgiadis, B. J. Marinas, & A. M. Mayes, Science and technology for water purification in the coming decades. Nanoscience and Technology: A Collection of Reviews from Nature Journals 452 (2010) 337–346.
8. B. I. Kharisov, O. V. Kharissova, & U. Ortiz-Mendez, (Eds.). CRC Concise Encyclopedia of Nanotechnology (2016). CRC Press, Boca Raton, Florida.
9. S. Palit, & C. M. Hussain, Green sustainability, nanotechnology and advanced materials: A critical overview and a vision for the future. In Green and Sustainable Advanced Materials: Applications, 1 (2018). John Wiley, Ney Jersey.
10. T. Rogers-Hayden, & N. Pidgeon, Nanotechnologies and the Royal Society and Royal Academy of Engineering's inquiry. Public Underst Sci 16 (2007) 345–364.
11. R. R. Hoque, & S. Balachandran, Handbook of Environmental Materials Management. (2019). C. M. Hussain (Ed.). Cham, Switzerland: Springer.
12. Q. H. Tran, & A. T. Le, Silver nanoparticles: synthesis, properties, toxicology, applications and perspectives. Adv. Nat. Sci.: Nanosci. Nanotechnol 4 (2013) 033001.

13. P. Das, S. Barua, S. Sarkar, N. Karak, P. Bhattacharyya,, N. Raza,... & S. S. Bhattacharya, Plant extract-mediated green silver nanoparticles: Efficacy as soil conditioner and plant growth promoter. J. Hazard. Mater. 346 (2018) 62–72.

14. S. Barua, P. Chattopadhyay, M. M. Phukan, B. K. Konwar, J. Islam, & N. Karak, Biocompatible hyperbranched epoxy/silver–reduced graphene oxide–curcumin nanocomposite as an advanced anti-microbial material. RSC Adv. 4 (2014) 47797–47805.

15. P. A. Clausen, V. Kofoed-Sørensen, A. W. Nørgaard, N. M. Sahlgren, & K. A. Jensen, Thermogravimetry and mass spectrometry of extractable organics from manufactured nanomaterials for identification of potential coating components. Mater. 12 (2019) 3657.

16. B. Das, P. Chattopadhyay, D. Mishra, T. K. Maiti, S. Maji, R. Narayan, & N. Karak, Nanocomposites of bio-based hyperbranched polyurethane/funtionalized MWCNT as non-immunogenic, osteoconductive, biodegradable and biocompatible scaffolds in bone tissue engineering. J. Mater. Chem. A B, 1 (2013) 4115–4126.

17. R. Duarah, & N. Karak, Facile and ultrafast green approach to synthesize biobased luminescent reduced carbon nanodot: an efficient photocatalyst. ACS Sustain. Chem. Eng. 5 (2017) 9454–9466.

18. V. K. Gupta, P. J. M. Carrott, M. M. L. Ribeiro Carrott, & Suhas, Low-cost adsorbents: growing approach to wastewater treatment—a review. Crit Rev Environ Sci Technol 39 (2009) 783–842.

19. N. Karak, Vegetable Oil-Based Polymers: Properties, Processing and Applications. (2012) Elsevier, Amsterdam.

20. J. Parvole, I. Chaduc, K. Ako, O. Spalla, A. Thill, S. Ravaine,... & E. Bourgeat-Lami, Efficient synthesis of snowman-and dumbbell-like silica/polymer anisotropic heterodimers through emulsion poly-merization using a surface-anchored cationic initiator. Macromolecules, 45 (2012) 7009–7018.

21. R. Konwarh, N. Karak, & M. Misra, Electrospun cellulose acetate nanofibers: the present status and gamut of biotechnological applications. Biotechnol. Adv. 31 (2013) 421–437.

22. J. Parvole, I. Chaduc, K. Ako, O. Spalla, A. Thill, S. Ravaine,... & E. Bourgeat-Lami, Efficient synthesis of snowman- and dumbbell-like silica/polymer anisotropic heterodimers through emulsion poly-merization using a surface-anchored cationic initiator. Macromolecules, 45 (2012) 7009–7018.

23. S. Thakur, & N. Karak, Green reduction of graphene oxide by aqueous phytoextracts. C. 50 (2012) 5331–5339.

24. X. Qu, P. J. Alvarez, & Q. Li, Applications of nanotechnology in water and wastewater treatment. Water Res. 47 (2013) 3931–3946.

25. C. Su, Environmental implications and applications of engineered nanoscale magnetite and its hybrid nanocomposites: A review of recent literature. J. Hazard. Mater. 322 (2017) 48–84.

26. R. F. Gomes, A. C. N. de Azevedo, A. G. Pereira, E. C. Muniz, A. R. Fajardo, & F. H. Rodrigues, Fast dye removal from water by starch-based nanocomposites. J. Colloid Interface Sci. 454 (2015) 200–209.

27. M. S. Zoromba, M. I. Ismail, M. Bassyouni, M. H. Abdel-Aziz, N. Salah, A. Alshahrie, & A. Memic, Fabrication and characterization of poly (aniline-co-o-anthranilic acid)/magnetite nanocomposites and their application in wastewater treatment. Colloids Surf. 520 (2017) 121–130.

28. N. El Badawi, A. R. Ramadan, A. M. Esawi, & M. El-Morsi, Novel carbon nanotube–cellulose acetate nanocomposite membranes for water filtration applications. Desalination, 344 (2014) 79–85.

29. B. Mu, J. Tang, L. Zhang, & A. Wang, Preparation, characterization and application on dye adsorption of a well-defined two-dimensional superparamagnetic clay/polyaniline/Fe_3O_4 nanocomposite. Appl. Clay Sci. 132 (2016) 7–16.

30. M. A. Olatunji, M. U. Khandaker, Y. M. Amin, & H. N. M. Ekramul Mahmud, Development and Characterization of Polypyrrole-Based Nanocomposite Adsorbent and Its Applications in Removal of Radioactive Materials. ICIBEL (2015) 30–35. Springer, Singapore.

31. J. Fawaz, & V. Mittal, Synthesis of polymer nanocomposites: review of various techniques. In Synthesis Techniques for Polymer Nanocomposites (2015) 1–30. Wiley.

32. J. H. Jhaveri, & Z. V. P. Murthy, A comprehensive review on anti-fouling nanocomposite membranes for pressure driven membrane separation processes. Desalination, 379 (2016) 137–154.

33. M. M. Khin, A. S. Nair, V. J. Babu, R. Murugan, & S. Ramakrishna, A review on nanomaterials for environmental remediation. Energy Environ. Sci. 5 (2012) 8075–8109.

34. K. Y. Foo, & B. H. Hameed, Insights into the modeling of adsorption isotherm systems. Chem. Eng. J. 156 (2010) 2–10.

35. G. Crini, Kinetic and equilibrium studies on the removal of cationic dyes from aqueous solution by adsorption onto a cyclodextrin polymer. Dyes Pigm. 77 (2008) 415–426.
36. B. Y. Gao, H. H. Hahn, & E. Hoffmann, Evaluation of aluminum-silicate polymer composite as a coagulant for water treatment. Water Res. 36 (2002) 3573–3581.
37. T. I. Farhana, M. Y. A. Mollah, M. A. B. H. Susan, & M. M. Islam, Catalytic degradation of an organic dye through electroreduction of dioxygen in aqueous solution. Electrochim. Acta 139 (2014) 244–249.
38. A. E. Chávez-Guajardo, J. C. Medina-Llamas, L. Maqueira, C. A. Andrade, K. G. Alves, & C. P. de Melo, Efficient removal of Cr (VI) and Cu (II) ions from aqueous media by use of polypyrrole/maghemite and polyaniline/maghemite magnetic nanocomposites. Chem. Eng. J. 281 (2015) 826–836.
39. Z. M. Badruddoza, Z. B. Z. Shawon, W. J. D. Tay, K. Hidajat, & M. S. Uddin, Fe_3O_4/cyclodextrin polymer nanocomposites for selective heavy metals removal from industrial wastewater. Carbohydr. Polym. 91 (2013) 322–332.
40. S. M. Lam, J. C. Sin, A. Z. Abdullah, & A. R. Mohamed, Degradation of wastewaters containing organic dyes photocatalysed by zinc oxide: a review. Desalination Water Treat. 41 (2012) 131–169.

13 Polymeric Adsorbents for Toxic Waste Removal

Md. Sadiqul Islam Sheikh,[1,2] Md. Mahinur Islam,[1]
Md. Saddam Hossain,[2] and Md. Mominul Islam[1]

[1] Department of Chemistry, University of Dhaka, Dhaka-1000, Bangladesh
[2] Department of Chemistry, Khulna University of Engineering & Technology, Khulna-9203, Bangladesh

1 INTRODUCTION

With the rapid urbanization and industrial revolution, environmental pollution, particularly water pollution, has become one of the utmost serious environmental issues that endangers living beings all over the world, especially in industrialized and developing countries. Synthetic organic dyestuffs have been widely used as coloring agents in different industries. Dyes, pharmaceuticals, industrial additives including heavy metals, and agrochemicals, e.g., pesticides and herbicides, have been contaminating our aquatic system. Adsorption, coagulation-flocculation, filtration, ion-exchange, reverse osmosis, oxidation, chemical precipitation, electrochemical techniques, etc. have been practiced for the treatment of wastewater [1,2].

Adsorption has been considered one of the most efficient and facile techniques, where the pollutants are attached to the surface of an adsorbent. It has several specific advantages over other approaches, these include: (1) low operational cost, (2) ease of operation of materials used, (3) high efficiency, (4) environment friendliness, (5) wide adaptability, (6) generation of lower secondary pollutants, and (7) the possibility of recycling the adsorbent. However, the applicability of the adsorption process mainly depends on the nature and capacity of the adsorbent, residual waste management, and capital costs [3,4].

Polymer-based adsorbents of different forms, such as particles, membranes, fibers, films, gels, and nanocomposites, have been considered as promising for the treatment of wastewater. Natural and synthetic polymers including chitosan, poly(vinyl alcohol) (PVA), poly(ethylene oxide), starch, polycyclodextrin, cellulose acetate, polyacrylonitrile, and polyaniline (PAni) are examples [5–10]. Carbohydrate adsorbents, e.g., starch, wheat flour, rice flour, graham flour, etc. have been used to encapsulate toxic dyes [6,7,10]. The cost-effectiveness of starch-based adsorbents has been realized in removing dyes from aqueous solutions [10]. The recovery of an organic dye, methylene blue (MB), adsorbed on wheat flour has been shown to be possible through the conversion of adsorbent into alcohol [7].

Conducting polymers, particularly PAni and polypyrrole (PPy), have attracted increased interest as adsorbents due to the possibilities of tuning morphology and functional groups, low cost, excellent stability, and unique doping/dedoping opportunity. Polymers have also presented the opportunity to enhance properties like chemical stability, hydrophobic–hydrophilic balance, and functionalities in composites with carbonaceous materials, metal oxides, and natural polymers [8]. A PAni-modified graphite electrode (PAni/GE) has been employed for potential-driven sorption of heavy metals from aqueous solutions as a simple, cost-effective, and environmentally benign technique [9].

This chapter focuses on the removal of toxic pollutants from wastewater. The fundamentals of different methods followed in synthesizing polymer-based nanocomposites (PNCs) are described.

DOI: 10.1201/9781003278269-13

The removal scenarios for several toxic contaminants using PNCs are summarized and extensively discussed by highlighting their adsorption capacity with information taken from recent literature. Attempts at the recovery and reuse of adsorbents are also described.

2 FABRICATION OF POLYMER-BASED ADSORBENTS

Several methods, including sol–gel, self-assembly, construction or dispersion of nano-building blocks, hierarchical architecture, interpenetrating networking, and so on have been followed to synthesize polymer-based materials. Synthetic routes are commonly typed as direct compounding and *in situ* synthesis [11]. PNCs can be synthesized *via* sol–gel, self-assembly process, assembling or dispersion of nano-building blocks, hierarchical structuring, and interpenetrating network formation [12]. Some of these methods are described below.

2.1 DIRECT COMPOUNDING METHOD

This is a commonly used method to prepare PNCs because of its ease of usage, low cost, and appropriateness for large-scale manufacturing. In this method, nanofillers and polymer materials are initially prepared separately, and then combined using a solution, emulsion, fusion, or mechanical forces [11]. The key challenge to this method is to provide uniform distribution, dispersion, and the mixing of fillers in the polymer matrix since nanoparticles tend to form larger aggregates. Various surface treatments are performed in the synthetic step and other factors such as temperature, time, shear force, and reactor layout are adjusted to obtain excellent dispersion of nanoparticles in PNCs. Suitable dispersants are sometimes used to increase particle dispersion and adhesion between the nanoparticles and the matrix. Single screw extruders (SSEs), twin-screw extruders, extension flow mixtures (EFMs), and master-batch methods are used to enhance the dispersion of micro- and nano-scale fillers in different polymer matrices [11]. Gardner *et al.* synthesized a high-density polyethylene/ultrafine cellulose composite by the master-batch method combined with SSE and EFM processes [12]. The master-batch method through SSE and EFM processes shows better dispersion of cellulose over the polymer matrix as well as enhancing the mechanical properties.

2.2 IN SITU SYNTHESIS

This method is very popular for developing PNCs with various transition metals, sulfide and halide, wherein nanoparticles of metal counterparts are incorporated within a polymer framework or matrices (Figure 13.1). *In situ* synthesis is of different types, as follows [11]:

1. Metal ions that act as the precursor of nanoparticles are preloaded for their uniform distribution within the polymer matrix (Figure 13.1). Poly(methyl methacrylate) (PMMA), PMMA–SiO_2, and PMMA–TiO_2 nanocomposites have been prepared by following this method [13]. Free radical suspension polymerization has been followed to synthesize PMMA using a water medium where *in situ* gel transformation occurs. PNCs with the expected particle distribution and an amorphous nature have been achieved [13].
2. This approach is gaining popularity because it allows the creation of nanocomposites with customized physical characteristics. In this case, nanoparticles are disseminated in the monomers or precursors before polymerization in the presence of a suitable catalyst. Starting materials can be both the monomers of the polymeric hosts and the target nanofillers. Uniform dispersion of nanoparticles in precursors in a liquid medium prevents agglomeration in the polymer matrix that enhances the interfacial interactions between the two phases.
3. Nanoparticles and polymers could be generated concurrently by combining nanoparticle precursors and polymeric monomers with an initiator in a suitable solvent. Wang *et al.*

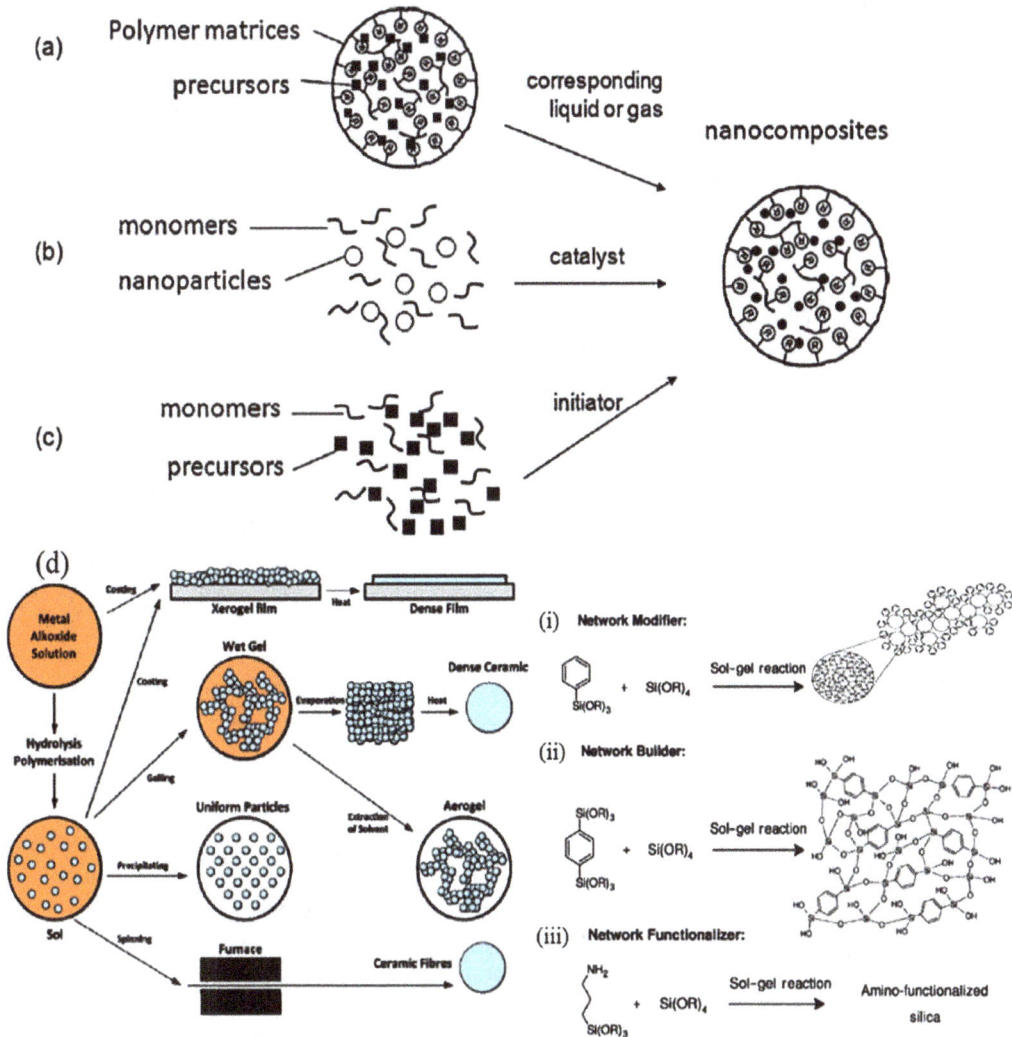

FIGURE 13.1 A typical example of *in situ* synthesis of PNCs: (a) preloading of metal ions within a polymer matrix, (b) addition of monomers of polymeric hosts and the target nanofillers, and (c) simultaneous formation of nanoparticles and polymers in the presence of initiator (Adapted with permission from Reference [11]. Copyright (2011) Elsevier). (d) Organic formation of functionalized products by sol–gel process. (Adapted with permission from Reference [12]. Copyright (2014) MDPI).

have followed this strategy in preparing modified nano-Al_2O_3-filled composites by mixing all components at one time, such as ethylene propylene diene monomer, nano-Al_2O_3, and Si69 at room temperature [14]. The dispersion of nano-Al_2O_3 particles in the composite can be enhanced by following this *in situ* modification technique with fewer agglomerates (Figure 13.1).

2.3 SOL–GEL PROCESS

This is a low-cost method for producing translucent and homogeneous solid materials with outstanding purity. This process occurs in water and organic solvents, using metal halides dissolved in organic solvents and metal alkoxides such as $Si(OR)_4$, $SiR'(OR)_3$, $Ti(OR)_4$, and others as precursors.

These are subjected to a series of hydrolysis and condensation reactions, culminating in the forma-tion of sols *via* nucleophilic substitution, as shown in Figure 13.1 [12]. Sol is a colloidal solution in which individual particles periodically touch one another and eventually combine into an integrated network (wet gel). This structure forms a gel after multiple drying treatments. Incorporating organic groups (R) into the structure of materials created is a potential strategy for modifying their proper-ties. For the generation of silica-based composites, the sol–gel technique has been widely studied. By polymerizing their precursors or by grafting appropriate functional groups onto parent materials these compounds can be synthesized by the sol–gel technique [12].

2.4 SELF-ASSEMBLY PROCESS

This is the most promising, practically low-cost, and high-throughput approach for the fabrication of nanocomposites. In this process, individual, pre-existing components are organized themselves. Self-assembly must be driven by molecular interactions to govern the formation of a nanostructure. One-dimensional non-covalent interactions like hydrogen bonding and π-stacking, which are weak in water, drive assembly. As a result, hydrophobic forces drive assembly in the polar phase, water. A balance of forces drives the self-assembly of such one-dimensional structures, rather than crystal-lization. The liquid crystal templating method is used to produce mesoporous materials, as shown in Figure 13.2a, b [12]. When the concentration of surfactant becomes larger than that of their critical micelle concentration (CMC), the self-assembly of precursor molecules builds walls in the gap between micellar rods in the lyotropic liquid crystal phase through the route, as represented in Figure 13.2a. On the other hand, below the CMC, the mesostructure is generated as a cooperative self-assembly of the precursor and surfactant species through the pathway shown in Figure 13.2b [12].

2.5 DISPERSION OF NANO-BUILDING BLOCKS

In this method, two opposite parts of hybrid organic–inorganic materials forming the composites are pre-formed separately and then allowed to assemble *via* van der Waals, hydrogen bond, and electric or magnetic dipole interactions. For the creation of organic–inorganic hybrid polymers, a set of chemicals known as nano-building blocks (NBB) is utilized (Figure 13.2c). The dispersal of well-defined NBB, which are fully calibrated prefabricated items, retains their integrity in the final composition. This is an appropriate method for obtaining a more accurate characterization of the organic–inorganic component. Organically pre- or post-functionalized metallic oxides, metals, chalcogenides, clusters, layered compounds, core–shells such as oxides, layered double hydroxides, clays, lamellar phosphates, and chalcogenides enable intercalating organic components are all examples of NBB [15]. These NBBs are capped with ligands or linked by organic spacers such as telechelic compounds, polymers, or functional dendrimers. The diversity of nano-building pieces including nature, structure, functionality, and linkages enables the creation of a dizzying array of diverse structures and organic–inorganic interfaces. Furthermore, the stepwise synthesis of these materials generally provides a greater degree of control over their semi-local structure [12,15].

2.6 HIERARCHICAL STRUCTURES FORMATION

A hierarchy is a self-assembly feature in which fundamental building blocks are linked together to form more sophisticated structures (Figure 13.2d). Self-assembly as opposed to template-directed assembly is a bottom-up technique for the synthesis of nanostructured materials. The assembling of components into a specific design that performs a certain function is a characteristic of an integrated chemical, biological, or other system. Through molecular recognition mechanisms, these hierarch-ical structures with programmable functions can enable the development of novel vectorial chem-istry [12,15].

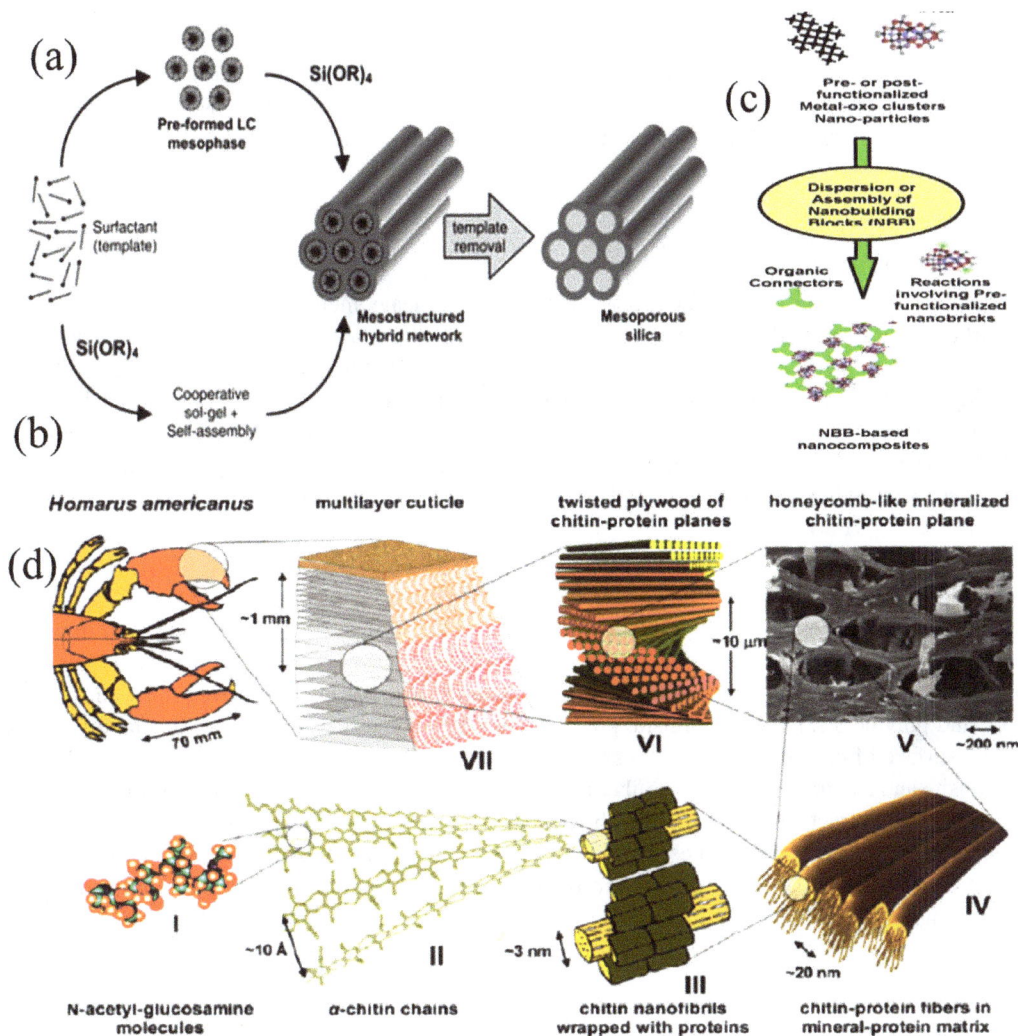

FIGURE 13.2 (a, b) Routes of self-assembly process. (Adapted with permission from Reference [12]. Copyright (2014) MDPI). (c) Steps associated with the dispersion of nano-building blocks. (Adapted with permission from Reference [15]. Copyright (2005) RSC). (d) Cuticle of lobster representing hierarchical structures. (Adapted with permission from Reference [12]. Copyright (2014) MDPI).

2.7 INTERPENETRATING NETWORKS FORMATION

In this method, two counterparts of the PNCs formed by the combination of organic and inorganic substances are interpenetrated with each other at the molecular level. The interpenetrating networks (IPNs) are generated by (1) forming a secondary network within the main network, (2) forming two networks simultaneously, and (3) connecting the components of IPNs through covalent bonds, resulting in a dual organic–inorganic hybrid polymer [12]. Macroscopically, they appear uniform and have weak interactions between the moieties. However, polymerization in a sol–gel process in the presence of premade polymer is used in the first technique to create IPNs. For the development of different types of networks, bifunctional precursors or a premade inorganic or organic polymer are functionalized by the appropriate functional groups [16].

3   REMEDIATION OF TOXIC WASTES

Efficient removal of toxic wastes such as heavy metals, organic dyes, and other organic compounds from wastewater has been carried out using PNCs. Typical examples of the adsorption capacity of pollutants by different PNC adsorbents are systematically described below.

3.1   ENCAPSULATION OF HEAVY METALS

3.1.1   Copolymer Composites

Copolymer composites exhibit better performance toward heavy metal removal than single polymer adsorbents because of their abundant surface groups, better stability, and mechanical feasibility [17]. Composite films of poly(3,4-ethylenedioxythiophene)/polystyrene sulfonate (PEDOT/PSS) and lignin (LG) are highly stable and efficient hybrid materials that have been employed for the potential-driven remediation of metals [18]. By applying a negative potential, Pb^{2+} could be adsorbed on PEDOT/PSS films with a maximum adsorption capacity (q_{max}) of 245.5 mg/g and subsequently desorbed by applying a positive potential. The q_{max} is almost doubled to 452.8 mg/g after introducing LG into PEDOT/PSS.

3.1.2   Polymer/Clay Composites

SiO_2 has been widely used as an adsorbent for the removal of heavy metal ions dissolved in water, being not just a porous material, but also having a large surface area and good mechanical properties. It is known that SiO_2 is not suitable for the chelation of metal ions. However, the composite of SiO_2 and polymers exhibits synergistic properties where polymers have abundant chelating groups on the surface of composites [19]. Wu *et al.* employed thiol-functionalized mesoporous PVA/SiO_2 composite and electrospun PVA nanofiber membranes for the removal of Cu^{2+} ions from an aqueous solution. The characteristics of the membranes used and features of adsorption are represented in Figure 13.3 [20]. The membranes of thiol-functionalized mesoporous PVA/SiO_2 nanofiber composite exhibit better q_{max} toward Cu^{2+} ion than pristine PVA nanofiber membrane having q_{max} of 489.12 mg/g at 303 K.

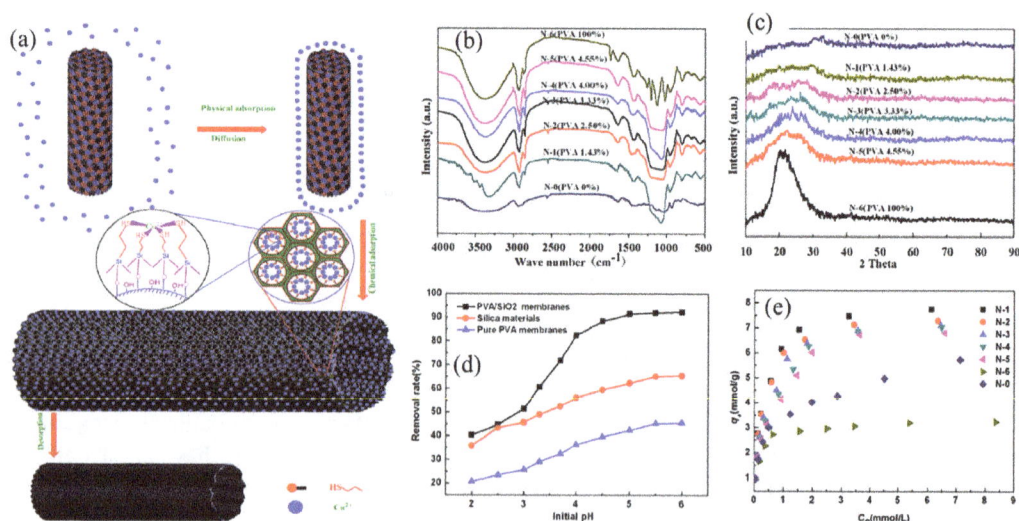

FIGURE 13.3   (a) The mechanism of adsorption of Cu^{2+} ions, (b) FTIR spectra, (c) XRD pattern, (d) effects of pH, and (e) adsorption isotherm on/off PVA/SiO_2 composite nanofibers at different PVA contents. (Adapted with permission from Reference [20]. Copyright (2010) Elsevier).

3.1.3 Polymer/Carbon Material Composites

Carbon-based materials, e.g., carbon nanotubes (CNTs), activated carbon, graphene, graphene oxide (GO), and multiwalled CNTs (MWCNTs) are regarded as promising for forming PNCs due to their compatibility with the polymeric structure, resulting in a high dispersion into the polymers ensuring strong interaction and adhesion. Functionalized carbon nanofillers-based composite provides different functional groups for the adsorption of various types of pollutants [3]. PNCs of GO preclude the accumulation of GO nanofillers, which is helpful to introduce more functional groups and active sites to encore metal ions on nanocomposites. PAni/GO nanocomposites that have been synthesized by chemical oxidative polymerization have exhibited an efficient adsorption property toward Cr^{6+} ions. The q_{max} of PAni/GO has been reported to be 1149.4 mg/g. The different functional groups containing oxygen and nitrogen on the surface of PAni/GO nanocomposite have been considered to act as chelating groups [3]. In another study, electropolymerized PAni/GE has been employed for potential-driven adsorption of Fe^{2+} from an aqueous solution. In this case, the q_{max} has been reported to be 0.212 mg/g [9].

3.1.4 Polymers/Metal or Metal Oxide Nanocomposites

Recently, PNCs of metal or metal oxide nanoparticles have been demonstrated as promising materials for the adsorption of heavy metals. Typical examples of adsorption capacities of polymers and PNCs toward heavy metals are given in Table 13.1. The removal of metal ions utilizing Mn_2O_3-doped PAni nanocomposites has been studied and the q_{max} of PAni/Mn_2O_3 nanocomposites toward Pb^{2+}, Ni^{2+}, and Cd^{2+} ions have been found to be 437, 494 and 480 mg/g, respectively [21].

3.1.5 Fiber-Based Materials

Fiber-based materials show promise to be used for the treatment of wastewater since they can form PNCs with high porosity, mechanical stability, and large surface area. MoS_2-enabled fiber mats have been fabricated from the co-polymerization of polyacrylonitrile and polystyrene mats for the removal of mercury from water [22]. The q_{max} of the synthesized nanocomposites is 6258.7 mg/g, which is almost 2.5 times greater than the theoretical value. The anchoring of mercury occurs by adsorption followed by a redox reaction between surface-confined MoS_2 and mercury [22]. It follows pseudo-second-order kinetics.

TABLE 13.1

Typical Examples of Adsorption Capacities of Polymers and PNCs toward Heavy Metals

Materials	Pollutants	Adsorption Capacity (mg g^{-1})	Removal Efficiency (%)	Ref.
PAni/rGO	Hg^{2+}	1000	94	23
Chitosan/GO	Au^{2+}	1077	99	24
Poly(N-vinylcarbazole)/GO	Pb^{2+}	983	83	25
Chitosan/clinoptilolite	Cu^{2+}	719	–	26
Poly(aniline-co-5-sulfo-2-anisidine)	Ag^+	2034	99	27
PAni/GO	Cr^{6+}	1149	98	28
PAni/Mn_2O_3	Pb^{2+}, Ni^{2+}, Cd^{2+}	437, 494, 480	95, 94, 91	29
Guar gum-graft-poly(acrylamide)/SiO$_2$	Cd^{2+}	2000	99	30
Poly(p-phenylenediamine)/Fe$_3$O$_4$	Cr^{6+}	2750	84	31
PMSAE–TTDD	Ni^{2+}	1519	17	32
Poly(1,5-diaminonaphthalene)	Ag^{2+}	1975	84	33

Note: PMSAE–TTDD: Poly(methyl salicylate acrylic ester) and 4,7,10-trioxa-1,13-tridecanediamine.

3.2   Encapsulation of Organic Dyes

3.2.1   Natural Polymers-Based Nanocomposites

Innumerable nanocomposites based on natural polymers and their derivatives show promise to be utilized for the treatment of wastewater because of their low cost, renewable nature, sustainable sources, considerable effectiveness, biodegradability, and nontoxicity [3,33]. Several research groups have employed starch [6], dextrin [1], chitosan [42], rice husks [5], sawdust [2], jute fibers [34], etc. as bioadsorbents for the removal of dyes from an aqueous solution. In addition, the composites of natural polymers with conducting polymers or metal oxides or carbon materials, especially PAni/starch composite and wheat flour originated starch, have been extensively studied for the removal of dyes through adsorption [6,7,10].

Yeamin *et al.* investigated the remediation of different anionic and cationic dyes with PAni and PAni/starch composites. The resulting composite materials exhibited better dye adsorption capacities compared to their individual components [10]. The adsorption performance of modified jute fibers has been investigated toward encapsulation of an organic dye from an aqueous solution. The q_{max} of the modified fibers was found to be 23.08 mg/g with 78% removal efficiency [34]. PAni/lignocellulose nanocomposite has been utilized for the removal of congo red (CR) from an aqueous solution. In this case, the q_{max} was 1672.5 mg/g [35].

3.2.2   Polymer/Magnetic Nanocomposites

PNCs possessing magnetic behavior have been recognized as important adsorbents in the decontamination of water due to their ease of operation, high efficiency, and cost-effectiveness. Conducting polymers like PAni, PPy, polythiophene, etc., with iron-based nanoparticles have been employed for the adsorption of dyes from wastewater. Increased surface area to volume ratio and the complex interactions between dyes and metal oxide of PNCs have been considered to result in enhanced adsorption capacity [3]. The characteristics studies of Fe_3O_4/terpolymer of aniline/*m*-aminobenzoic acid/m-phenylenediamine (Fe_3O_4/PAmABAmPD-TCAS) nanocomposites and their isotherms of adsorption of MB and malachite green (MG) dyes are represented in Figure 13.4. The q_{max} of the nanocomposites toward MB and MG have been revealed to be 31.64 and 29.07 mg/g, respectively [36]. The adsorption follows a pseudo-second-order kinetic model and occurs *via* the electrostatic attractions between the aromatic site of MB and MG with sulfonate, amine, and phenoxide groups of adsorbents.

3.2.3   Polymeric Membranes

Membrane technology has boosted wastewater treatment in the last couple of decades. This technology significantly reduces the size of equipment, consumption of energy, and capital cost [37]. Catechol-polyethyleneimine (Cc-PEI) nanocomposite-deposited membranes have been employed for water purification as a low-cost material [38]. A schematic diagram of the preparation and characterization of Cc-PEI-deposited membrane and its adsorption characteristics of MB dye is represented in Figure 13.5. It retains superhydrophobicity and underwater superoleophobicity, which are helpful for anchoring dye through an electrostatic interaction. This membrane exhibits superior stability in a strongly alkaline solution with a high efficiency of adsorption, even after 30 cycles.

3.3   Removal of Pharmaceuticals

Medicinal compounds present in wastewater pose a great risk to human health and natural ecosystems. These include hormonal interference of aquatic animals, endocrine disruption, genotoxicity, and immune toxicity. Glutaraldehyde-cross-linked chitosan nanoparticles (Chi-NPs) have been employed for the removal of diclofenac, an active pharmaceutical ingredient, from water.

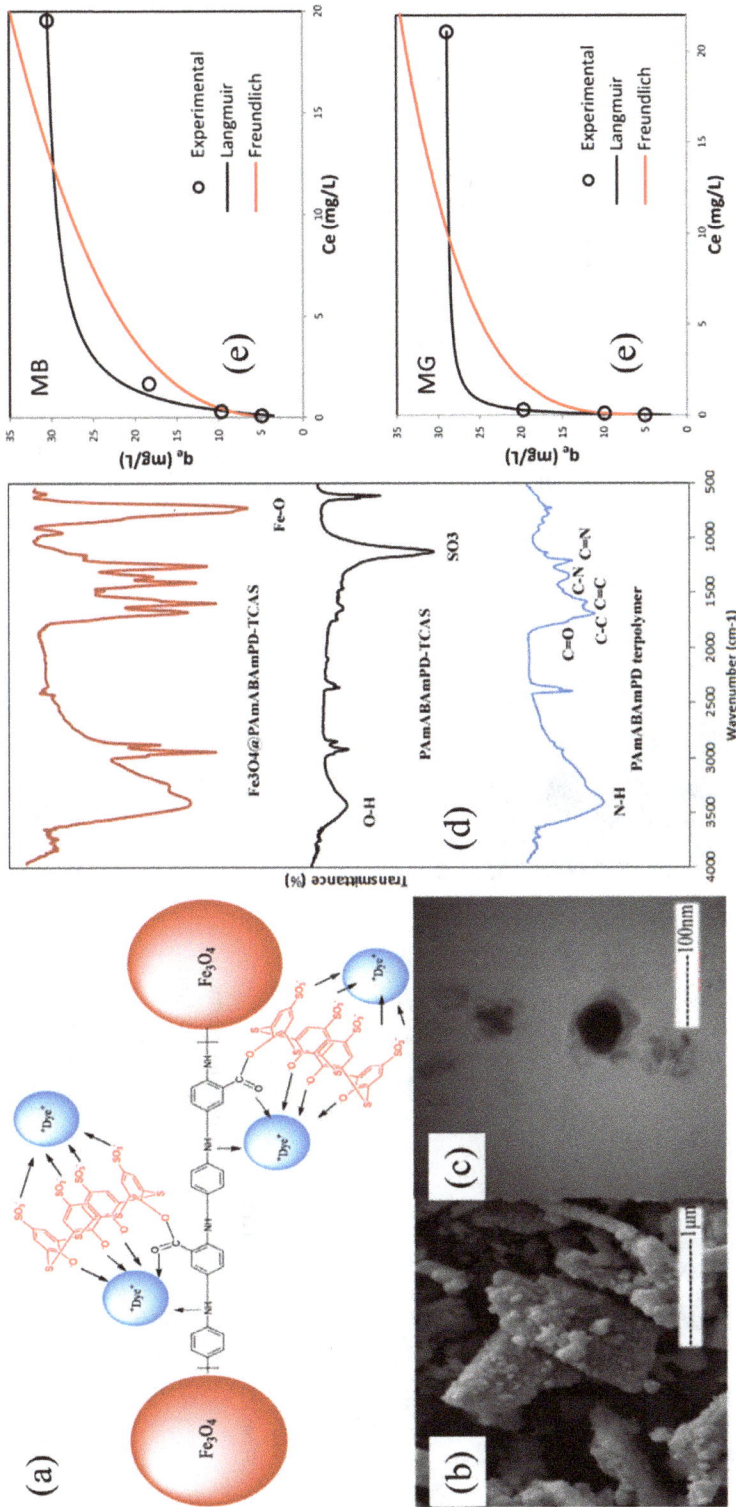

FIGURE 13.4 (a) The adsorption mechanism, (b) SEM image, (c) TEM image, (d) FTIR spectra, and (e) adsorption isotherms of Fe_3O_4/PAmABAmPD-TCAS nanoadsorbent for MB and MG dyes. (Adapted with permission from Reference [36]. Copyright (2015) American Chemical Society).

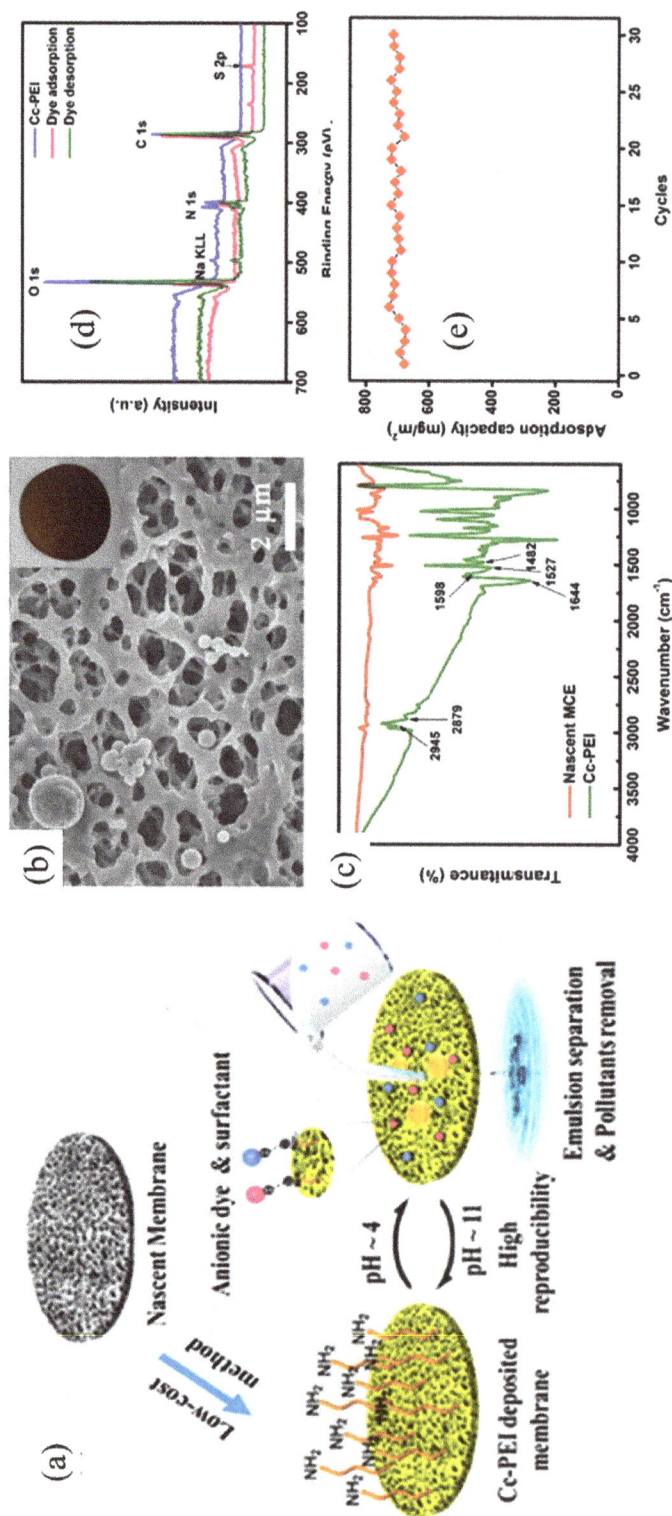

FIGURE 13.5 (a) Schematic diagram of preparation, (b) FESEM image, (c) FTIR spectra, (d) XPS spectra, and (e) variation in adsorption of MB of the Cc-PEI deposited membrane. (Adapted with permission from Reference [38]. Copyright (2017) American Chemical Society).

A casting and phase inversion process has been employed to fabricate membranes by processing Chi-NPs into porous polyethersulfone microfiltration membranes [39]. Modeling of adsorption isotherms resulted in a q_{max} of 358.3 mg/g. Selective removal of tetracycline has been successfully carried out with cryogel composite made by embedding tetracycline-imprinted poly(hydroxyethyl methacrylate-N-methacryloyl-L-glutamic acid methyl ester) particles into poly(hydroxyethyl methacrylate). The q_{max} of the imprinted composite cryogel has been determined as 680 mg/g at 25°C (pH 5.0) [40]. Zhou *et al.* synthesized PAni-coated Fe_3O_4 to eliminate bisphenol A (BPA), α-naphthol, and β-naphthol, which are endocrine-disrupting compounds, from water samples. The q_{max} values of core–shell materials for BPA, α-naphthol, and β-naphthol have been reported to be 23.09, 28.73, and 9.13 mg/g, respectively. The adsorption process follows a pseudo-second-order kinetic model and the free energy changes associated with the adsorption of BPA, α-naphthol, and β-naphthol are negative, demonstrating that the adsorption of these compounds is spontaneous [41]. Typical examples of the adsorption capacities of some polymers and PNCs toward different dyes, and medicinal and organic pollutants are given in Table 13.2.

TABLE 13.2

Adsorption Capacities of Typical Polymers and Their Composites toward Different Organic Pollutants

Materials	Target Pollutants	Adsorption Capacity (mg g⁻¹)	Removal Efficiency (%)	Ref.
Dyes				
Chitosan/γ–Fe_2O_3/SiO_2 composite	Methyl orange	3429	–	42
PMSAE–TTDD	MB	1601	42	32
Poly(amidoamine-co-acrylic acid)	Direct red 31, Direct red 80 and Acid blue 25	3400, 3448, 3500	88, 90, 86	43
Poly(N-vinyl caprolactam-co-maleic acid)	Rhodamine 6G, MB	2012, 1441	95	44
PAni/LC composite	CR	1672.5	99	35
Pharmaceuticals				
Chitosan nanoparticles	Diclofenac	502	90	39
Polyacrylic resin/Fe_3O_4	Ibuprofen	206	97	46
PAni/magnetic GO	Ciprofloxacin	97.70	97	47
Antibiotics				
PHEMA-MAGA cryogel	Tetracycline	680		40
Microporous triazine polymer	Sulfamethoxazole	483	95	48
Poly(1-trimethylsilyl-1-propyne) and poly(4-methyl-2-pentyne)	Amoxicillin	213	80	49
Endocrine-disrupting compounds				
PAni core@shell/Fe_3O_4	BPA	23	95	41
β–Cyclodextrin	17 β–estradiol	126	90	50
Polyethersulfone-attapulgite	BPA	23	95	51
Other pollutants				
XAD–4 (Polystyrene)	Carbon tetrachloride	2600		52
XAD–7 (Poly(acrylic ester))	Diethyl phthalate	480		52
PAni-based nanotubes	Perflourooctane sulfonate	1651		53
	Perfluorooctanoate	1100		
PAni-derived porous carbon	Triclosan, Oxybenzone	1318, 906		54
	p-Chloro-m-xylenol	840		

3.4 ENCAPSULATION OF OTHER ORGANIC POLLUTANTS

The polymeric adsorbents also exhibit a high affinity toward the adsorption of toxic organic species [52]. Multifunctional interactions, namely hydrogen bonding, hydrophobic interaction, electrostatic attraction, and complex formation are the driving force for the encapsulation of organic pollutants, although adsorption of metal ions occurs mainly *via* electrostatic interaction. Polystyrene resins have been widely employed for the exclusion of hydrophobic organic pollutants because of their nonpolar and hydrophobic characteristics. The removal of benzene, toluene, chlorobenzene, *p*-xylene, carbon tetrachloride, trichloroethylene, and chloroform from aqueous solutions has been achieved on polystyrene-based resin with considerable efficiency. Effective remediation of triclosan, oxybenzone, and *p*-chloro-*m*-xylenol from water by highly porous carbon synthesized from PAni has been carried out. Apart from hydrogen bonding, especially at high pH, the hydrophobic interactions are perhaps the plausible mechanism for the significant adsorption of certain organic pollutants on PAni [54].

4 FACTORS AFFECTING THE REMOVAL PROCESS

Factors including pH, dose, contact time, temperature, etc., are known to affect the adsorption of pollutants. It is generally recommended to optimize these parameters before assessing the removal efficiency of a particular adsorbent.

4.1 EFFECT OF SOLUTION pH

The solution pH is an important variable since changing pH may alter the charges of both the adsorbent and adsorbate [8,42]. An efficient adsorbent should be able to adsorb toxic waste from water in a wide pH range. Interactions between the adsorbate and adsorbent vary with the change in pH because it leads to changes in the surface charge as well as the ionization level of the adsorbent [8]. The polymeric adsorbents usually contain pH-sensitive functional sites that undergo dissociation when the solution pH becomes higher than the pKa of a particular site. Generally, the PNC adsorbents have functional groups such as $-NH_2$, $-COOH$, $-OH$, etc., in their structure and at lower pH, most of which are protonated and become positively charged, whereas, at higher pH, they are deprotonated and hence become negatively charged. Similarly, pH may also alter the charges of particularly organic pollutants to be adsorbed. Therefore, the solution pH not only shows a synergistic role but also has an imperative factor affecting the capacity of an adsorbent in water treatment [8,42].

The remediation of Cr^{6+} with PPy-PAni nanofibers has been revealed to be pH-dependent because it controls the surface charge of the adsorbent and affects the speciation of the metal ion species [60]. The removal efficiency decreases with an increase in pH from 2 to 12. The decrease in removal efficiencies at alkaline pH up to 8 is due to the competition between CrO_4^{2-} and OH^- ions for the anion exchange sites of the adsorbent. The removal of Cr^{6+} species at higher pH is probably associated with the effect of the reduction of Cr^{6+} to Cr^{3+} by the electron-rich polymer matrix.

4.2 EFFECT OF TEMPERATURE

Temperature, a critical variable, affects the adsorption performance of PNCs especially. In general, the diffusion of intraparticles increases with an increase in temperature, and hence the small micropores present in an adsorbent excavate and widen as well as additional active sites on the adsorbent being created [8, 55]. However, the physical change of the adsorbent may also occur at high temperatures and, consequently, result in a drastic decrease in its adsorption performance. This is because of the damaging active sites of the adsorbent and thus weakening of the attractive forces [8,55]. On the other hand, desorption of the species adsorbed from the adsorbent surface increases with an increase in temperature. Moreover, the viscosity of the medium decreases with increasing temperature, which then improves the diffusion rate of adsorbate species through the outer boundary layer and in the inner openings of the adsorbents [55].

4.3 ADSORBENT DOSE

This is an essential controlling parameter of the adsorption process. Generally, increasing the amount of adsorbent leads to an increase in the adsorption sites, and then a limited increase in the removal of pollutants occurs [8,42,56]. The adsorption efficiency may also reduce under the same condition, because adsorbent particles may agglomerate, resulting in reducing the active sites, and hence enhancing the number of adsorbates in contact with the fixed weight of the adsorbent [42,56].

4.4 CONTACT TIME

It is common that the removal of adsorbate by PNCs initially occurs more rapidly and then the rate of adsorption slows down until equilibrium of adsorption is attained [5,8]. This is due to the presence of large numbers of available vacant active sites on the adsorbent surface at an earlier stage of the adsorption [5]. With time the active sites become occupied. As a result, the saturation of active sites occurs and hence no further adsorption takes place. Maximum adsorption is attained at the equilibrium time. The rate of binding and removal of pollutants also depend on the contact time [5].

5 RECOVERY OF ADSORBENTS

Removal of toxic species *via* adsorption generates solid residues that pose a problem for secondary waste management. The regeneration of the used adsorbent is of immense importance in wastewater treatment. Designing effluent treatment plants that can be operated to clean industrial effluents without generating secondary wastes remains a very challenging task [3]. Different physical, chemical, and biological techniques have been investigated to regain the adsorption capacity of adsorbents. These include chemical, thermal, oxidative, steam, pressure swing, ozone, vacuum, microwave, ultrasound, and bio- and electrochemical regenerations. Thermal regeneration is currently used in many industrial plants for the regeneration of activated carbon and clay adsorbents [57]. In this method, the adsorbent is heated up to a certain temperature to release the bonds between the adsorbate and adsorbent. The recovery of MB adsorbed on activated carbon modified with iron oxide nanoparticles has been studied [58].

The recovery can be accomplished by the photocatalytic oxidation method, wherein reactive free radicals are generated to degrade diversified organic pollutants. The regeneration of iron- and copper-modified clay from indigo blue has been studied by using a photo-Fenton process [57]. In this case, photosensitizers remove the organic pollutants in layers and further decompose them. Finally, the plausible oxidation products including oxalic, formic, acetic acid, sulfate, nitrate ions, etc. are formed. The recovery of dyes with iron- and copper-modified clays is 90% up to the repeated use for four cycles [57].

Microbial regeneration of an adsorbent is ascribed to renewing the adsorbent utilizing biodegradation of the remaining organics by microbial activities. It can be performed by mixing microorganisms, such as bacteria, with adsorbents. Biological degradation can be attained either by mixing bacteria in offline systems or it can be done in the course of biological treatments. Microbe nutrients and dissolved oxygen are combined with pollutant-loaded adsorbents loaded with waste materials in a batch system for offline bio-regeneration [59]. In another study, conversion of starch-based adsorbents to alcohol has been practiced in treating secondary residue generated during the removal of MB [7]. The maximum recovery of MB is more than 99% within less than 10 h by converting starch adsorbent into useful chemicals in the presence of yeast at 37°C [7]. The switching of the electric potential applied for electrosorption of heavy metals has been proved to be a useful tool for the recovery of PNCs [9,18].

6   CONCLUDING REMARKS

Polymeric adsorbents of both types, natural and synthetic, with different forms, e.g., particles, membranes, films, nanocomposites, and hydrogels are a great choice in an application for eliminating contaminants from water owing to their environmental and mechanical stability, ease of synthesis, large surface area, porous structure, tunable functional sites, and low cost. The organic dyes, pharmaceuticals, personal care products, stimulants, industrial additives, and agrochemicals, e.g. pesticides and herbicides, possess complex molecular structures with aromatic or aliphatic structures. Moreover, the functionality of polymeric adsorbents can be tuned by choosing suitable backbones or functional substituents of aliphatic and aromatic structures. These would enable the creation of multiple opportunities for interactions such as ionic, hydrophobic, hydrogen bonding, π–π stacking, etc. between the adsorbents and toxic pollutants for adsorption. Factors affecting adsorption capacity, particularly solution pH and temperature, would be the key tools to use for the recovery of synthetic PNC adsorbents. Recovery of synthetic polymeric adsorbents with greater efficiency is important for repeated use of adsorbents as well as making the secondary waste management facile. The natural polymers are biodegradable and can be converted into alcohol, a useful chemical, through enzymatic reaction [7]. Moreover, reversing electrode potential for the recovery of PNCs adsorbents would be a simple, environmentally friendly, and sustainable methodology for the treatment of industrial effluents [9,18].

REFERENCES

1. S. Alipoori, H. Rouhi, E. Linn, H. Stumpfl, H. Mokarizadeh, M.R. Esfahani, A. Koh, S. T. Weinman, E.K. Wujcik, Polymer-based devices and remediation strategies for emerging contaminants in water, ACS Appl. Polym. Mater. 3 (2021) 549–577.
2. P. Arunachalam, M. Jawaid, M.M. Khan, Polymer-based nanocomposites for energy and environmental applications, 2018, Netherlands, Elsevier.
3. Momina, K. Ahmad, Study of different polymer nano-composites and their pollutant removal efficiency: Review, Polymer 217 (2021) 123453.
4. A. Samadi, M. Xie, J. Li, H. Shon, C. Zheng, S. Zhao, Polyaniline-based adsorbents for aqueous pollutants removal: A review, Chem. Eng. J. 418 (2021) 129425.
5. E.N. Zare, A. Motahari, M. Sillanpää, Nanoadsorbents based on conducting polymer nanocomposites with main focus on polyaniline and its derivatives for removal of heavy metal ions/dyes: a review, Environ. Res. 162 (2018) 173–195.
6. H.M. Munjur, M.N. Hasan, M.R. Awual, M.M. Islam, M.A. Shenashen, J. Iqbal, Biodegradable natural carbohydrate polymeric sustainable adsorbents for efficient toxic dye removal from wastewater, J. Mol. Liq. 319 (2020) 114356.
7. S. Saha, M.Y.A. Mollah, M.A.B.H. Susan, M.M. Islam, Treatment of wastewater containing organic dyes: Recovery of dye adsorbed on starch-based materials through conversion of adsorbent into alcohol, Dhaka Univ. J. Sci. 63 (2015) 119–124.
8. G. Zhao, X. Huang, Z. Tang, Q. Huang, F. Niu, X. Wang, Polymer-based nanocomposites for heavy metal ions removal from aqueous solution: A review, Polym. Chem. 9 (2018) 3562–3582.
9. S. Sultana, M.S. Hossain, M.A.B.H. Susan, M.M. Islam, Electrosorption of heavy metal from aqueous solution on polyaniline modified graphite electrode, Bangladesh J. Sci. Res. 31–33 (2020) 1–6.
10. M.B. Yeamin, M.M. Islam, A.N. Chowdhury, M.R. Awual, Efficient encapsulation of toxic dyes from wastewater using several biodegradable natural polymers and their composites, J. Clean. Prod. 291 (2021) 125920.
11. X. Zhao, L. Lv, B. Pan, W. Zhang, S. Zhang, Q. Zhang, Polymer-supported nanocomposites for environmental application: A review, Chem. Eng. J. 170 (2011) 381–394.

12. B. Samiey, C.H. Cheng, J. Wu, Organic-inorganic hybrid polymers as adsorbents for removal of heavy metal ions from solutions: A review, Materials 7 (2014) 673–726.
13. S. Ahmad, S. Ahmad, S.A. Agnihotry, Synthesis and characterization of in situ prepared poly(methyl methacrylate) nanocomposites, Bull. Mater. Sci. 30 (2007) 31–35.
14. Z.H. Wang, Y.L. Lu, J. Liu, Z.M. Dang, L.Q. Zhang, W. Wang, Preparation of nanoalumina/EPDM composites with good performance in thermal conductivity and mechanical properties, Polym. Adv. Technol. 22 (2011) 2302–2310.
15. C. Sanchez, B. Julián, P. Belleville, M. Popall, Applications of hybrid organic-inorganic nanocomposites, J. Mater. Chem. 15 (2005) 3559–3592.
16. Q. Deng, R.B. Moore, K.A. Mauritz, Novel nafion/ORMOSIL hybrids via in situ sol-gel reactions. 1. Probe of ORMOSIL phase nanostructures by infrared spectroscopy, Chem. Mater. 7 (1995) 2259–2268.
17. M.T. Wu, Y.L. Tsai, C.W. Chiu, C.C. Cheng, Synthesis, characterization, and highly acid-resistant properties of crosslinking β-chitosan with polyamines for heavy metal ion adsorption, RSC adv. 6 (2016) 104754–104762.
18. F. Checkol, A. Elfwing, G. Greczynski, S. Mehretie, O. Inganäs, S. Admassie, Highly stable and efficient lignin-PEDOT/PSS composites for removal of toxic metals, Adv. Sustain. Syst. 2 (2018) 1700114.
19. Y. Yu, Z. Hu, Z. Chen, J. Yang, H. Gao, Z. Chen, Organically-modified magnesium silicate nanocomposites for high-performance heavy metal removal, RSC Adv. 6 (2016) 97523–97531.
20. S. Wu, F. Li, H. Wang, L. Fu, B. Zhang, G. Li, Effects of poly(vinyl alcohol) (PVA) content on preparation of novel thiol-functionalized mesoporous PVA/SiO$_2$ composite nanofiber membranes and their application for adsorption of heavy metal ions from aqueous solution, Polymer 51 (2010) 6203–6211.
21. K. Rajakumar, S.D. Kirupha, S. Sivanesan, R.L. Sai, Effective removal of heavy metal ions using Mn, Nanosci. Nanotechnol. 13 (2013) 1–10.
22. C.L. Fausey, I. Zucker, D.E. Lee, E. Shaulsky, J.B. Zimmerman, M. Elimelech, Tunable molybdenum disulfide-enabled fiber mats for high-efficiency removal of mercury from water, ACS Appl. Mater. Interfaces 12 (2020) 18446–18456.
23. R. Li, L. Liu, F. Yang, Preparation of polyaniline/reduced graphene oxide nanocomposite and its application in adsorption of aqueous Hg (II), Chem. Eng. J. 229 (2013) 460–468.
24. L. Liu, C. Li, C. Bao, Q. Jia, P. Xiao, X. Liu, Q. Zhang, Preparation and characterization of chitosan/graphene oxide composites for the adsorption of Au (III) and Pd (II), Talanta 93 (2012) 350–357.
25. Y.L.F. Musico, C.M. Santos, M.L.P. Dalida, D.F. Rodrigues, Improved removal of lead (II) from water using a polymer-based graphene oxide nanocomposite, J. Mater. Chem. A 1 (2013) 3789–3796.
26. M.V. Dinu, E.S. Dragan, Evaluation of Cu^{2+}, Co^{2+} and Ni^{2+} ions removal from aqueous solution using a novel chitosan/clinoptilolite composite: kinetics and isotherms, Chem. Eng. J. 160 (2010) 157–163.
27. X.G. Li, H. Feng, M.R. Huang, Redox sorption and recovery of silver ions as silver nanocrystals on poly(aniline-co-5-sulfo-2-anisidine) nanosorbents, Chem.-A Eur. J. 16 (2010) 10113–10123.
28. S. Zhang, M. Zeng, W. Xu, J. Li, J. Li, J. Xu, X. Wang, Polyaniline nanorods dotted on graphene oxide nanosheets as a novel super adsorbent for Cr(VI), Dalton Trans. 42 (2013) 7854–7858.
29. K. Rajakumar, S.D. Kirupha, S. Sivanesan, R.L. Sai, Effective removal of heavy metal ions using Mn$_2$O$_3$ doped polyaniline nanocomposite, J. Nanosci. Nanotechnol. 14 (2014) 2937–2946.
30. V. Singh, S. Pandey, S.K. Singh, R. Sanghi, Removal of cadmium from aqueous solutions by adsorption using poly(acrylamide) modified guar gum–silica nanocomposites, Sep. Purif. Technol. 67 (2009) 251–261.
31. S. Yang, D. Liu, F. Liao, T. Guo, Z. Wu, T. Zhang, Synthesis, characterization, morphology control of poly(p-phenylenediamine)-Fe$_3$O$_4$ magnetic micro-composite and their application for the removal of Cr$_2$O$_7^{2-}$ from water, Synth. Met. 162 (2012) 2329–2336.
32. X. Zhang, Z. Li, S. Lin, P. Théato, Fibrous materials based on polymeric salicyl active esters as efficient adsorbents for selective removal of anionic dye, ACS Appl. Mater. Interfaces 12 (2020) 21100–21113.
33. X.G. Li, M.R. Huang, Y.B. Jiang, J. Yu, Z. He, Synthesis of poly(1,5-diaminonaphthalene) microparticles with abundant amino and imino groups as strong adsorbers for heavy metal ions, Microchim. Acta 186 (2019) 1–14.

34. R. Rahman, M.S.I. Sheikh, M.S. Miran, M.M. Alamgir, M.A.B.H. Susan, M.M. Islam, Functionalization of jute fibers by reactive oxygen species for encapsulation of an organic dye from aqueous solution, Bangladesh J. Sci. Res. 33 (2020) 66–72.

35. S. Debnath, N. Ballav, A. Maity, K. Pillay, Development of a polyaniline-lignocellulose composite for optimal adsorption of congo red, Int. J. Biol. Macromol. 75 (2015) 199–209.

36. M.M. Lakouraj, R.S. Norouzian, S. Balo, Preparation and cationic dye adsorption of novel Fe_3O_4 supermagnetic/thiacalix[4]arene tetrasulfonate self-doped/polyaniline nanocomposite: Kinetics, isotherms, and thermodynamic study, J. Chem. Eng. Data 60 (2015) 2262–2272.

37. E. Obotey Ezugbe, S. Rathilal, Membrane technologies in wastewater treatment: a review, Membranes 10 (2020) 89.

38. N. Liu, Q. Zhang, R. Qu, W. Zhang, H. Li, Y. Wei, L. Feng, Nanocomposite deposited membrane for oil-in-water emulsion separation with in situ removal of anionic dyes and surfactants, Langmuir 33 (2017) 7380–7388.

39. B.R. Riegger, R. Kowalski, L. Hilfert, G.E. Tovar, M. Bach, Chitosan nanoparticles via high-pressure homogenization-assisted miniemulsion crosslinking for mixed-matrix membrane adsorbers, Carbohydr. Polym. 201 (2018) 172–181.

40. E. Yeşilova, B. Osman, A. Kara, E.T. Özer, Molecularly imprinted particle embedded composite cryogel for selective tetracycline adsorption, Sep. Purif. Technol. 200 (2018) 155–163.

41. Q. Zhou, Y. Wang, J. Xiao, H. Fan, Adsorption and removal of bisphenol A, α-naphthol and β-naphthol from aqueous solution by Fe_3O_4@polyaniline core–shell nanomaterials, Synth. Met. 212 (2016) 113–122.

42. H.Y. Zhu, R. Jiang, Y.Q. Fu, J.H. Jiang, L. Xiao, G.M. Zeng, Preparation, characterization and dye adsorption properties of γ-Fe_2O_3/SiO_2/chitosan composite, Appl. Surf. Sci. 258 (2011) 1337–1344.

43. N.M. Mahmoodi, F. Najafi, A. Neshat, Poly(amidoamine-co-acrylic acid) copolymer: synthesis, characterization and dye removal ability, Ind. Crops Prod. 42 (2013) 119–125.

44. Popescu, D.M. Suflet, Poly(N-vinyl caprolactam-co-maleic acid) microparticles for cationic dye removal, Polym. Bull. 73 (2016) 1283–1301.

45. S.A. Agnihotri, N.N. Mallikarjuna, T.M. Aminabhavi, Recent advances on chitosan-based micro-and nanoparticles in drug delivery, J. Control. Release 100 (2004) 5–28.

46. G. Zhang, S. Li, C. Shuang, Y. Mu, A. Li, L. Tan, The effect of incorporating inorganic materials into quaternized polyacrylic polymer on its mechanical strength and adsorption behaviour for ibuprofen removal, Sci. Rep. 10 (2020) 1–11.

47. M.K. Mohammadi Nodeh, S. Soltani, S. Shahabuddin, H. Rashidi Nodeh, H. Sereshti, Equilibrium, kinetic and thermodynamic study of magnetic polyaniline/graphene oxide based nanocomposites for ciprofloxacin removal from water, J. Inorg. Organomet. Polym. Mater. 28 (2018) 1226–1234.

48. D. Juela, M. Vera, C. Cruzat, X. Alvarez, E. Vanegas, Adsorption properties of sugarcane bagasse and corn cob for the sulfamethoxazole removal in a fixed-bed column, Sustain. Environ. Res. 31 (2021) 1–14.

49. M.N. Alnajrani, O.A. Alsager, Removal of antibiotics from water by polymer of intrinsic microporosity: Isotherms, kinetics, thermodynamics, and adsorption mechanism, Sci. Rep. 10 (2020) 1–14.

50. P. Tang, Q. Sun, Z. Suo, L. Zhao, H. Yang, X. Xiong, H. Pu, N. Gan, H. Li, Rapid and efficient removal of estrogenic pollutants from water by using beta-and gamma-cyclodextrin polymers, Chem. Eng. J. 344 (2018) 514–523.

51. J. Yu, H. Shen, B. Liu, Adsorption properties of polyethersulfone-modified attapulgite hybrid microspheres for bisphenol A and sulfamethoxazole, Int. J. Environ. Res. Public Health 17 (2020) 473.

52. E.J. Simpson, R.K. Abukhadra, W.J. Koros, R.S. Schechter, Sorption equilibrium isotherms for volatile organics in aqueous solution. Comparison of head-space gas chromatography and on-line UV stirred cell results, Ind. Eng. Chem. Res. 32 (1993) 2269–2276.

53. C. Xu, H. Chen, F. Jiang, Adsorption of perflourooctane sulfonate (PFOS) and perfluorooctanoate (PFOA) on polyaniline nanotubes, Colloids Surf. A: Physicochem. Eng. Asp. 479 (2015) 60–67.

54. D.K. Yoo, H.J. An, N.A. Khan, G.T. Hwang, S.H. Jhung, Record-high adsorption capacities of polyaniline-derived porous carbons for the removal of personal care products from water, Chem. Eng. J. 352 (2018) 71–78.

55. A.A. Adeyemo, I.O. Adeoye, O.S. Bello, Adsorption of dyes using different types of clay: a review, Appl. Water Sci. 7 (2017) 543–568.

56. S. Netpradit, P. Thiravetyan, S. Towprayoon, Adsorption of three azo reactive dyes by metal hydroxide sludge: Effect of temperature, pH, and electrolytes, J. Colloid Interface Sci. 270 (2004) 255–261.

57. M. Shahadat, S. Isamil, Regeneration performance of clay-based adsorbents for the removal of industrial dyes: A review, RSC Adv. 8 (2018) 24571–24587.

58. B. Zargar, H. Parhan M. Rezazade, Fast removal and recovery of methylene blue by activated carbon modified with magnetic iron oxide nanoparticles, J. Chin. Chem. Soc. 58 (2011) 694–699.

59. T. C. Drage, K.M. Smith, C. Pevida, A. Arenillas, C.E. Snape, Development of adsorbent technologies for post-combustion CO_2 capture, Energy Procedia 1 (2009) 881–884.

60. M. Bhaumik, A. Maity, V.V. Srinivasu, M.S. Onyango, Removal of hexavalent chromium from aqueous solution using polypyrrole-polyaniline nanofibers, Chem. Eng. J. 181 (2012) 323–333.

14 Smart Polymers for Food and Water Quality Control and Safety

Saúl Vallejos,[1] Miriam Trigo-López,[1] Ana Arnaiz,[1,2] Álvaro Miguel,[1,3] and José M. García[1]

[1] Departamento de Química, Facultad de Ciencias, Universidad de Burgos, Plaza de Misael Bañuelos s/n, 09001 Burgos, Spain
[2] Universidad Politécnica de Madrid, Calle Ramiro de Maeztu 7, 28040 Madrid, Spain
[3] Universidad Autónoma de Madrid, Calle Einstein 3, 28049 Madrid, Spain

1 INTRODUCTION

The World Health Organization reported that 10% of the population falls ill after eating contaminated food every year, causing over 420,000 deaths and 550 million cases of diarrheal disease, as well as 200 other diseases. Spoiled food can occur at any point in the manufacturing chain, and is undesirable for consumption as it changes the food's texture, appearance, odor, and flavor. It is estimated that around 30% of food is wasted [1] between production and consumption due to poor food management and contamination [2], with a concomitant socio-economic impact.

Food safety refers to the procedures regarding the handling, preparation, and storage of different foods to prevent foodborne illnesses. In this sense, food quality implies numerous factors, including nutritional value, stability, consumer standards, and food health. Although factors vary according to different countries [3], the expiration date is taken as a reference to estimate foods' and beverages' end-life. However, this reference does not account for the actual food state and whether it is safe or not to eat, since no information about the handling, manufacture, and transportation conditions are available. For this reason, raising real-time food quality control systems is required to ensure consumer health and avoid unnecessary food waste.

Several physical and chemical quality indicators can be measured, both qualitatively and quantitatively, to ensure food and beverage safety, such as different gases, humidity, temperature, pH, microorganisms, pesticides, biotoxins, persistent organic pollutants, etc. Early detection of these real food state indicators would reduce the number of foodborne diseases [4]. However, traditional measuring techniques are expensive, time-consuming, are not available to consumers, and, above all, they do not allow real-time and *in situ* determination. As a result, researchers have directed their interest to developing and applying different food quality indicator measuring techniques to ensure food and water quality, including smart polymers.

Polymer science and technology is currently oriented to the development of polymers with special functionalities to be used in many fields such as aeronautics, packaging, electronics, construction, medical, etc. Polymers are tunable materials with a wide range of physical and chemical properties derived from their structure. As defined by García et al. [5], a smart polymer generates a response to a stimulus through a specific mechanism. For example, they can be sensitive to their microenvironment modifying their properties and produce a measurable response to specific targets. Smart polymers in the food industry are a user-friendly tool that allows obtaining real-time information

DOI: 10.1201/9781003278269-14

about food state and safety, that can either be part of the packaging or be used to measure food quality parameters in the food.

This chapter shows the potential of smart polymers according to their application in food quality and safety control. The chapter comprises two main sections, the first of which describes the concept of smart polymers, while the second describes the uses of smart polymers for quality control and food safety based on quality indicators.

2 SMART POLYMERS

As mentioned in the Introduction, a smart polymer responds to a stimulus through a specific mechanism. Consequently, they can be classified according to the response, stimulus, or mechanism by which the response is produced, as depicted in Figure 14.1. Among the different classifications, we can classify them as (i) stimuli-responsive polymers and (ii) sensory polymers. The former responds to the stimulus with an action, such as a variation in the volume or shape, in the interaction with solvents, generation/break of chemical bonds, etc. The latter responds with an alert, that is, useful analytical information. However, both types of smart polymers can respond to any type of stimulus, through any mechanism, and even show visual macroscopic signals (see examples in Figure 14.1).

2.1 STIMULI-RESPONSIVE POLYMERS

Stimuli-responsive polymers show a drastic change in a physical property caused by an answer to a small environmental factor variation, such as temperature, pH, light, the presence of target molecules (chemical or biochemical), electromagnetic field, etc.; and they are related to solvent–polymer interactions, physical state, hydrophilic/lipophilic rates, or solubility, among others. As a result, they cause different answers in the polymer, such as formation/breakage of secondary interactions, simple reactions (oxidation, reduction, acid–base, hydrolysis, etc.), changes in the solubility or size, or conformational variations in the structure, also leading to absorption or releasing processes. Stimuli-responsive polymers are unique materials since they can be tailored by selecting the polymer backbone and the functional groups to control the different processes. For this reason, they have been widely investigated for controlled drug-delivery systems also [6].

The number of stimuli-responsive polymers is vast, and only a few examples will be mentioned here. On the one hand, physical stimuli modify the polymer's chain dynamics, such as the polymer/solvent system. In this sense, the most widely studied are temperature-responsive polymers, such as lower critical solution temperature (LCST) polymers. Poly(N-isopropylacrylamide) (PNIPAM) is the most representative example, which varies from transparent to opaque or vice versa at about 32°C due to a phase transition resulting from intra- or interchain weakening/strengthening of hydrogen bonds. Another common physical stimulus is pH. The pH-responsive polymers have ionizable functional groups that protonate/deprotonate upon environmental pH variations, such as acrylic acid and N,N-dimethylaminoethyl methacrylate (DMAEMA). Moreover, light-responsive polymers show isomerization or bond-breaking upon light irradiation, with common examples including azobenzene monomer-containing polymers. On the other hand, chemical stimuli vary the molecular interactions between the polymer and the solvent or between polymer chains. The most studied are biological stimuli-responsive polymers that modify the functioning of molecules, i.e., the interaction with the biomolecules results in polymeric network cross-linking and/or ionization [6].

2.2 SENSORY POLYMERS

Sensors are based on two integrated subunits, one of them acts as the receptor subunit, able to interact with an analyte or a target molecule, and the other is the indicator subunit, which delivers a readable signal. If the interaction is reversible, they are called chemical sensors or chemosensors,

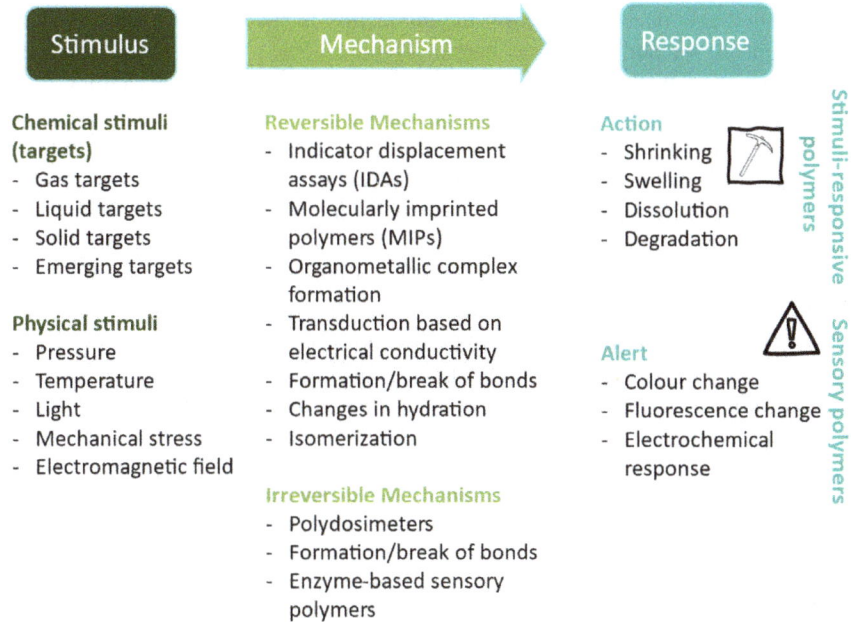

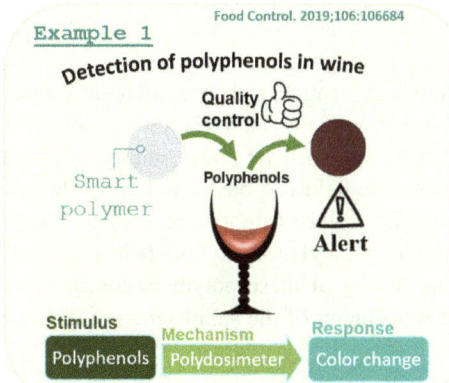

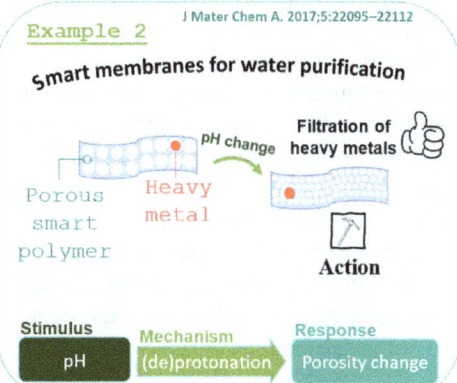

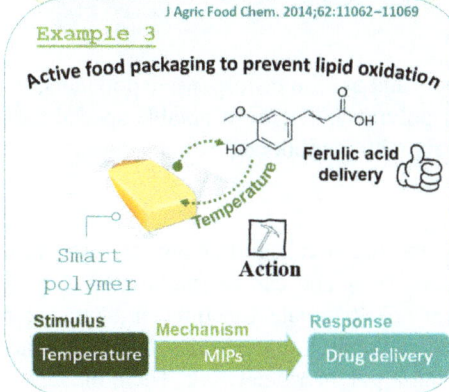

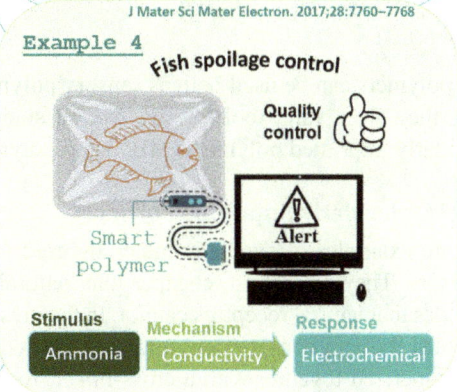

FIGURE 14.1 Classification of smart polymers according to the stimulus, the responsive mechanism, or the produced response.

while they are called chemical dosimeters if the interaction is not reversible. However, we use the term chemical sensors for both situations herein. The detection phenomena rely on the indicator subunit or transducer, that transforms the interaction into a measurable property (color, fluorescence, conductivity, etc.).

Most of the research related to chemical sensors is based on discrete molecules with low molecular weight, low chemical and thermal resistances, and which are difficult to recover. The immobilization of these chemical sensors in polymeric matrices allows for obtaining solid sensory materials that do not migrate, with good mechanical properties, and that are tunable in shape and properties. However, more importantly, polymer sensors are low-cost, simplified sensing counterparts of complex analytical techniques adapted for non-specialized users. In this sense, polymeric chemosensors can be designed to have functional monomers in their structure to detect discrete molecules (anions, cations, or neutral molecules), or they can even be further functionalized by covalently bonding larger biological macromolecules on the polymeric substrate, such as enzymes or antibodies.

A relevant group of polymer sensors is optical sensors, namely, optodes. These devices allow the identification of substances of interest by modifications of the transducer's optical properties, once the reaction between the receptor and the analyte takes place. Among the most widely used optical detection techniques are colorimetry, absorbance, fluorescence, luminescence, chemiluminescence, energy transfer, or reflectance by measuring light intensity in different spectrum regions (from UV to IR). In some of these techniques, determination can be made by the naked eye or with the help of simple and inexpensive equipment, which is very useful for non-specialized personnel, especially those of colorimetry or luminescence.

Another recognition and transduction mechanism worth mentioning here is related to conductive polymers due to their sensitivity. They show an electric response, a variation of their electrical properties, in the presence of a stimulus, which can be analyzed by amperometry, voltammetry, conductometry, etc. However, they are also used as electric field-responsive polymers. In addition, these polymers can also show a visual response through a color or fluorescence change. Common conductive polymers include polyaniline (PAni), poly(3,4-ethylenedioxythiophene) (PEDOT), or polypyrrole (PPy) [7]. Also, the electron-conducting ability of these polymers contributes either by enhancing or quenching the luminescence as the transductor of the smart sensing devices in the presence of the desired molecules.

2.3 SPECIAL CASES

Some polymers can be used both as sensory polymers and as stimuli-responsive polymers, in other words, they are a valid tool for any type of smart polymers. The most notable special cases are molecularly imprinted polymers (MIPs) and polymers with immobilized biomolecules.

2.3.1 Molecularly Imprinted Polymers

MIPs are examples of synthetic materials used for molecular recognition and are used to mimic biosensors. They are robust, cheaper than natural materials, and can be synthesized to recognize molecules that natural receptors cannot. In the presence of a template, functional monomers forming a complex through non-covalent interactions (hydrogen bonds, van der Waals, or π–π interactions) are polymerized together with a cross-linker, forming a polymeric network. Then, the template is removed, leaving recognition sites within the polymer (Figure 14.2). The template should be similar to (or the same as) the target molecule in chemical functionality, shape, and size to rebind with the target. MIPs serve as adsorption sites to coat the electrodes' surfaces and increase their selectivity in techniques such as potentiometry or voltammetry [8].

The most popular functional monomers used to prepare MIPs are methacrylic acid (MMA), 3-aminopropyl-triethoxysilane (APTES), and phenyltrimethoxysilane (PTES). These functional

FIGURE 14.2 Schematic representation of the preparation of MIPs.

monomers are always co-polymerized with cross-linkers, such as ethylene glycol dimethacrylate (EGDMA) and tetraethyl orthosilicate (TEOS). Other interesting monomers for electropolymerization without a cross-linker are o-phenylenediamine, pyrrole, dopamine, *o*-aminophenol, and *p*-aminothiophenol.

2.3.2 Polymers with Immobilized Biomolecules

The immobilization of biomolecules in polymeric materials has been used for decades. It is a resource used in different fields of science, although its benefits are known especially in medicine and food/beverage control. There are many types of polymers for the immobilization of biomolecules, and there are several ways to carry out this immobilization. García et al. [5] classified these polymers as type 1, type 2, pseudo-type 2, and type 3 smart polymers, based on their complexity level.

Typical immobilized biomolecules include antibodies (for the detection of antigens), proteases (enzymes for the proteolysis of substrates), DNA, RNA, etc. and are intended for the recognition of or to produce a response in the presence of large biomolecules. The most remarkable advantage of this type of smart polymer is that they allow easy handling. Without the support offered by polymers, these biomolecules are generally more unstable, and their management involves more complex experimental procedures.

2.4 Application of Smart Polymers

These smart polymers have several applications, such as drug-delivery systems, tissue engineering, cell culture supports, antifouling coatings, sensors, and actuator systems among others, and they can be prepared in the shape of films, fibers, coatings, micelles, etc. according to their final applications. Regarding their use in food quality control and safety, the most convenient shape they are applied as is films or coatings, to be integrated as part of the food packaging. Traditional food packaging is designed to be as inert as possible, to protect, communicate, and be convenient. The packaging is a shield against external environmental conditions such as heat, light, moisture, gas emission, pathogens, etc. The information printed on the surface communicates valuable information about the food inside, and they are also meant to save time for consumers and to facilitate different sizes and shapes [10]. On the contrary, active or smart food packaging are far more than an informative barrier against pernicious external factors. They are designed to provide additional protection to the consumer by revealing the foodstuff's state before being directly exposed to it, after the interaction with a specific compound or change in the package environment. For example, triggered by an external stimulus, active packaging can change its conditions to extend the shelf life or to improve the quality and safety of the food inside by the controlled release of encapsulated components such as salt, CO_2, or by controlling the moisture, temperature, or gaseous concentration [11]. On the other hand, sensory polymers can selectively interact with a target molecule and transform that information into an optical or electrochemical response. When used in food packaging, sensory polymers become a simple, rapid, and cheap means to visually obtain information about the food state (Figure 14.3).

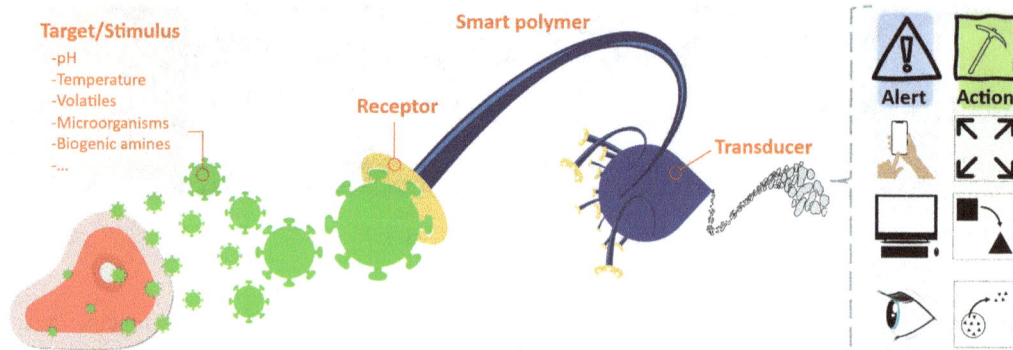

FIGURE 14.3   Detection of food quality indicators by a smart polymer and transduction.

Food quality and safety indicators

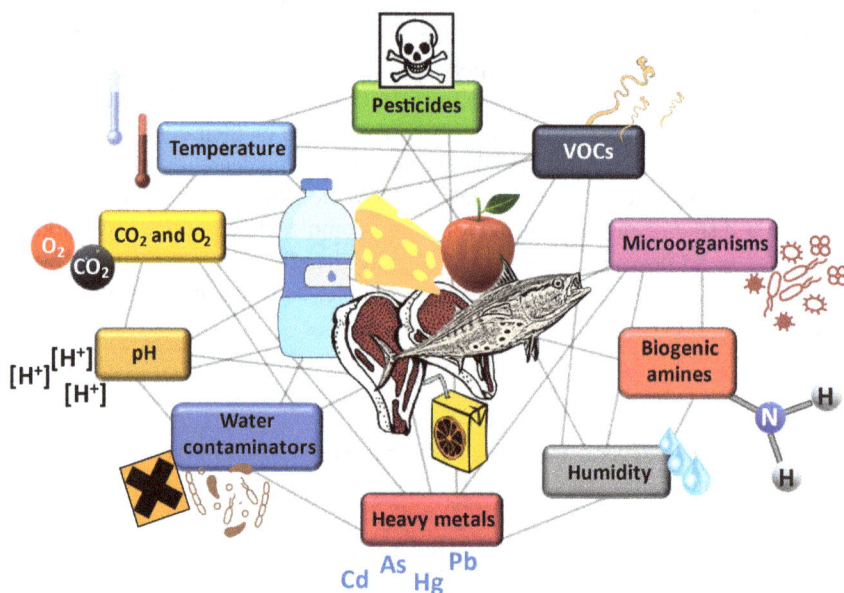

FIGURE 14.4   Main food and water quality and safety indicators, closely related and dependent.

In general, targets for food quality control and safety can either be a physical (temperature, light, mechanical stress, etc.) or chemical stimulus (pH, chemical species, redox oxidation, etc.), or even a biomolecule (enzyme, pathogen). Although we do not mention all of them here, the main stimuli that are useful food quality indicators are described in the next section.

3   FOOD QUALITY AND SAFETY TARGETS FOR SMART POLYMERS

As previously mentioned, contaminated food causes a great number of deaths and diseases every year, and specific indicators are needed to evaluate the state of the food product. This section discusses and explains the main indicators of the quality and safety of food and beverages (Figure 14.4). Although we made up a classification based on the type of indicator, it is important to highlight that most of the time all of them are closely related, since when a food product deteriorates, multiple markers alert simultaneously about the state of that product.

The main characteristics and functioning of the developed smart polymers for the detection or elimination of these analytes are also included with some examples. In addition, commercialized smart polymers for food quality and safety are also included.

3.1   GAS INDICATORS

The main food quality gas targets are carbon dioxide (CO_2), oxygen (O_2), volatile organic compounds (VOCs), and biogenic amines (BAs). These types of indicators are especially important in modified atmosphere packaged (MAP) food products, with CO_2 being the most used gas in the food industry.

3.1.1   Carbon Dioxide and Oxygen

High CO_2 concentrations are used in MAP food as it inhibits bacterial and fungal growth and decreases the environmental pH. Its antimicrobial activity mechanism is based on avoiding decarboxylation due to the anaerobic atmosphere. Amino acid decarboxylation produces the release of BAs, which are another significant quality and safety food indicator described later. In addition, CO_2 accumulation disturbs the membrane permeability of some microorganisms [4], and fresh food bacteria increase CO_2 in the environment owing to breathing processes. Therefore, CO_2 concentration reduction in fresh food indicates the drop-off of modified atmosphere packaging, and, on the other hand, the increase in the CO_2 levels causes food freshness to be reduced [12].

Another key gas for food quality is O_2, which is crucial for the biological activity of several microorganisms. The increase of O_2 in MAP causes oxidative reactions in food, such as fat oxidation, browning reactions, and pigment oxidation [4], prompting flavor deterioration, odor changes, aerobic bacteria growth, and decreasing the nutritional food value. On the other hand, a decrease in O_2 can be associated with aerobic microorganisms' activity; therefore, alterations in the O_2 concentration in fresh food indicate a food quality loss, accelerating food spoilage [13].

3.1.2   Volatile Organic Compounds

VOCs include a wide range of natural and synthetic chemicals such as alcohols, hydrocarbons, carbonyls, sulfur, and nitrogen-containing compounds. The presence of VOCs in food arises from different sources: (i) they are released by microbial activity in perishable foods such as milk, meat, seafood, fruits, and vegetables; (ii) they are used in product processing; and (iii) they are derived from different steps in the food industry, fermentation, cooking, cleaning, disinfection, and others.

Ethanol, lactic acid, and acetic acid are the main secondary metabolites produced in fermentation. An increase in ethanol levels and simultaneous low O_2 and high CO_2 concentrations indicate anaerobic bacteria growth in fresh food. In addition, both acetate and acetic acid levels rise with the storage time in fresh fish and meat under MAP [12].

The major volatile basic nitrogen compounds produced by microbial activity in fresh fish are mainly trimethylamine, ammonia, and dimethylamine, and their presence is directly related to the microbial deterioration of food [14]. On the other hand, volatile sulfur compounds are naturally generated in food and have a high impact on food aroma in many products such as bread, wine, vegetables, coffee, and chicken, and contribute to food flavor in chocolate, cheese, tropical fruits, and cereals [15]. They are an important indicator of food fresh quality also. In addition, volatile sulfur compounds are responsible for food off-flavor.

3.1.3   Biogenic Amines

BAs are basic nitrogen compounds with a low molecular weight that are metabolized by the normal physiological function of microorganisms, plants, and animals. In food and beverages, enzymatic activity on proteins and amino acids of raw materials and amination of aldehydes and ketones are responsible for the formation of BAs, although the main source of these compounds is microbiological activity and amino acid decarboxylation [16]. Small amounts of BAs have a role in the normal physiology of living organisms, while a high intake of BAs can produce toxic or adverse effects on human health, such as allergies, and gastrointestinal and/or vascular disorders [17], and they are also able to cause more serious diseases such as cancer and even death. BAs are found in many foods and beverages, including meat, fish, vegetables, fruits, and dairy products.

Due to the high impact that gases have on food quality and safety, many different approaches have been carried out to develop intelligent polymers to detect them. Most of these smart polymers have been developed to be included in food packaging systems whose response can be easily observed through optical changes, although they also exist with an electrochemical response. Increased CO_2 levels in packaged foods can be detected directly or indirectly by smart polymers in combination with pH-responsive dyes. Natural polymer-based coatings containing different mixtures of pH-sensitive dyes (bromothymol blue, methyl red, bromocresol green, and/or phenol red) respond to changes in CO_2 concentration since their presence lowers the pH [18]. Although they are not strictly smart polymers, since the polymer is not the one with the sensing ability but the dye, we have included them here because they are very commonly used as food quality indicators. A relevant approach is the use of an intelligent polymer that modifies its transparency with an increasing concentration of CO_2 [19]. When CO_2 is dissolved in water, it forms carbonic acid, lowering the pH, which modifies the opacity of the chitosan-based polymer. In general, common strategies to measure CO_2 to control food quality are based on a pH change, and further discussion can be found in this sense later in this chapter.

In addition, most smart polymers developed for O_2 sensing have been based on redox-type reactions using LDPE polymers and redox dyes such as methylene blue [20]. However, intelligent polymers also detect O_2 in commodity plastic films by treating PE/PP polymer films with a solvent crazing process to form nanometric pores in a well-defined network and filling the pores with a sensor that changes its color in the presence of O_2 [21].

The main smart polymers developed to detect VOCs are based on colorimetric detection methods using pH-dye indicators that react with the chemical compounds released by the microbial activity in food spoilage. One of the most used strategies consists of the entrapment or dispersion of pH-sensitive molecules in a film in which a color change occurs due to the halochromic nature of the indicators that open or close their sulfoxide ring in the presence of OH^- or H^+ [22]. These types of intelligent polymers are highly sensitive to the presence of basic nitrogen compounds such as trimethylamine, dimethylamine, and ammonia in seafood products. In addition, intelligent polymers can be developed to monitor fresh milk following a similar approach.

Due to their basic nature, most BA sensors developed are also based on pH indicators dispersed in polymeric films. Moreover, smart polymers with a color response also have been used for the detection of BAs such as histamine, putrescine, cadaverine, and tyramine, among others. For example, acrylic and aramid smart polymers (films or coated textiles) containing sensory motifs have been developed, which undergo a color change when interacting with BAs in the gas phase in beef or tuna. The sensing mechanism is based on the formation of Meisenheimer complexes in the acrylic polymers or through a nucleophilic substitution in the aramid-sensing motifs [23,24]. Another interesting alternative is the use of PANi films as chemical sensors to monitor fish spoilage through color variation from green to blue due to a pH increase as a response to BA release [25].

Commercial food freshness indicators related to fish or meat spoilage can also be found. For example, Senso Q™ (DSM), based on anthocyanins dispersed in a polymer matrix, was used to determine BAs in fish and poultry. Unfortunately, this one was not commercially successful, and the same authors developed a new colorimetric sensor, Raflatac. This time the sensor based on a silver nanolayer (brown and opaque) turns transparent when silver sulfide is formed after reacting with hydrogen sulfide (decomposition of cysteine) [26].

On the other hand, it is important to highlight that enzyme-based sensory polymers have been developed to detect BAs in combination with an electrochemical response using conducting polymers, providing a simple, rapid, and very sensitive solution to quantify them [27]. One typical useful response when using conducting polymers (PAni, PEDOT, or PPy) is electrical resistance when the smart polymer interacts with the target analyte. Gaseous species like NH_3, H_2S, CO, CO_2, alcohols, or environmental humidity are usually detected using resistive techniques [28]. In addition, the use of MIPs in combination with electrochemical sensors is also applied to detect different BAs. This polymer-electrode smart disposition allows performing electrochemical measurements and is the most widely used voltammetry method [29].

Despite the large number of smart polymers developed to detect different gaseous indicators, further research is still needed because there are still molecules that we cannot distinguish by these intelligent polymers because they are very similar to each other. For example, differentiating the levels of methanol from ethanol in beverages such as wine samples using a smart polymer remains a challenge.

3.2 HUMIDITY, TEMPERATURE, AND pH

Food is drastically affected by sudden temperature, humidity, and pH changes. Generally, alterations in one of these parameters produce modifications in another since they are closely related, especially in packaged products. Therefore, the control and monitoring of these parameters during the food production and distribution chain is essential to guarantee food quality and safety.

Food temperature control from the production process to the consumer's arrival is a critical parameter affecting food spoilage since temperature determines the storage time. Variations in the temperature during the food distribution chain trigger microbiological activity, decreasing the shelf life, and boosting food waste with an important economic impact. Time–temperature control during food distribution is important since breaks in the cold chain cause an increase in bacterial growth, loss of nutrients, protein denaturation or carbohydrate fermentation, among others, leading to food safety issues, and sensory attributes, especially important for dairy products, meat and fish products, and vegetables [4,13].

Also, an optimal humidity degree is needed for preserving the food quality and shelf life of products. For instance, fresh-cut fruit and vegetables require high relative humidity to maintain food properties and avoid weight loss, although these high humidity levels trigger rapid microorganism and fungi growth. In addition, the type of materials used for food packaging has a key role in its permeability and sorption characteristics for controlling humidity [4,30].

Regarding the pH, most food shows a pH between 3.5 and 7, and this directly affects the food pigments (responsible for fruit, vegetable, and meat color) and the food texture (associated with the pH of fish and meat muscles). Modifications in food environmental pH are directly related to the metabolic activity of microorganisms, indicating microbial proliferation and food spoilage. These metabolites can be liquids or gases, such as lactic acid, acetic acid, volatiles, or CO_2, which decrease the environmental pH once dissolved. Therefore, identifying and knowing the pH of the food and maintaining it at its normal values is critical to producing and consuming safe and healthy products [4,13]. The detection of changes in pH using smart polymers has been carried out mainly by pH-dye polymers, as mentioned in the gas indicator section, due to both gas and pH indicators being closely related.

Thermo-responsive polyurethanes (TSPU) can control vapor permeation through the phase transition temperatures. They can be used as packaging materials since they exhibit close–open behavior of the free volume holes as a function of the temperature. Below the phase transition temperature, the polymer chains are frozen and in a glassy state, and the free volume holes are very small. When the temperature increases, chains can move, and the free volume holes increase their size, changing the breathability and water vapor transmission, and thus, the humidity [31]. In addition to changes in volume, TSPU nanofibers can be deposited over a PET nonwoven nanofiber mat. The material is opaque when refrigerated and becomes irreversibly transparent at room temperature. This change in the optical properties reveals a hidden warning alert behind the TPU [32].

Also, dual-responsive materials can be prepared, e.g., temperature- and pH-responsive. A typical thermo-responsive material is poly(N-isopropylacrylamide) (PNIPAM), which has both hydrophilic and hydrophobic side groups. When the temperature increases, the hydrophobic interactions dominate over the hydrophilic, collapsing the polymeric structure. This fact is used to encapsulate substances, to be released as the temperature rises in the food package. One of these substances can be pimaricin, which has antifungal properties. If a hydrophilic monomer such as acrylic acid is included in the formulation of PNIPAM, the material is responsive both to the temperature and the pH, with a higher degree of collapse at low pH [31].

3.3 Microorganisms

Microorganisms are sometimes used to process foods, but they are probably the most important food quality and safety indicator. Although microorganism detection methods are highly sensitive and reliable, they are usually time-consuming, require qualified personnel, and in some situations, show problems in the identification of specific microorganisms [33]. Microorganisms present in food can be part of the food's flora (associated with microbial quality and food spoilage), or pathogenic microorganisms of external origin that show up after poor handling or contamination of the food product [4].

Pathogenic microorganisms are the main cause of illness and death from food poisoning, so their detection is essential for public health. Pathogenic microorganisms include bacteria (*Escherichia coli*, *Listeria monocytogenes*, or *Salmonella* spp.), viruses (Hepatitis A viruses, Noroviruses), parasites (*Cyclospora cayetanensis*, *Toxoplasma gondii*), yeasts (*Candida alimentaria*, *Pichia fermentans*), and molds (*Aspergillus flavus*, *Penicillium expansum*) and they can be found in meat, seafood products, dairy products, and water. Moreover, these microorganisms can produce serious illnesses due to their rapid growth and colonization of new niches inside the human body, and they can produce hazardous mycotoxins for human health [34].

Polymeric biosensors that respond to pathogenic microorganisms are very specific but still challenging for researchers. There are some examples of the use of smart polymers based on the covalent binding of highly specific ligands to polymeric matrices. These ligands are molecules (antibodies, peptide ligands, enzymes, aptamers, etc.) that specifically recognize other molecules present in microorganisms such as proteins or polysaccharides or even their genetic material. One of the first commercialized smart polymers to detect pathogenic microorganisms was Toxin Guard™ (Toxin Alert, Canada), which is prepared by immobilizing different antibodies onto a flexible polyethylene packaging. This sensor is an example of a unique composite material capable of detecting and identifying multiple biological materials in a single package, such as *Salmonella* sp., *Campylobacter* sp., *E. coli*, and *Listeria* sp., using a distinctive icon to visually identify the biological material through conjugated dyes, photoactive compounds, etc. [35]. More recently, a cyclo-olefin polymer with a covalently attached RNA-cleaving fluorogenic DNAzyme probe can detect the presence of *E. coli* colonies in meat and juice by emitting a fluorescent signal [36]. Other commercial strategies with less success were also based on colorimetric indicators (FreshTag®), or biosensors in the barcode to detect pathogens (Sentinel System™). In addition, some MIPs have also been described for the specific detection of these microorganisms [29].

Responsive materials also find application in food packaging by encapsulating active components with antimicrobial activity, vitamins, or antioxidants. Normally, these active compounds are released through variations in the swelling and water uptake of the films. However, they are not extensively used due to a lack of consistency with the extrapolated results from the laboratory to the industry, based on the use of food simulants in the former, while the latter contains more salts, lower water activity, nutrients, proteins, and fats, that interact with the released encapsulated components [37].

3.4 MISCELLANEA

3.4.1 Chemicals Derived from Pesticides

According to the European Commission, a pesticide is "something that prevents, destroys or controls a harmful organism (pest) or disease, or protects plants or plant products during their production, storage, and distribution." This term includes herbicides, insecticides, fungicides, acaricides, molluscicides, nematicides, growth regulators, and repellents, among others [38].

For years, pesticides have increased crop yields to feed the world's population. However, their improper, abusive, and uncontrolled use cause the emergence of very harmful residues for humans and other living beings, mainly in drinking water. Human exposure to these pesticide residues can cause headaches, neurotoxicity, respiratory distress, chronic diseases, cancer, and/or death [39]. Because of this, the European Commission in the 396/2005 regulation, established the maximum limits of pesticide residues that can be found in food, and it also presents an annual report evaluating the levels of these residues in the European marketed food, finding several non-accepted pesticides of illegal use in high values. Pesticide residue levels above the established limits have also been identified, especially in spinach [40]. Identification and detection of pesticide residues are complex due to the high amount of chemical pesticide derivatives and the time-consuming techniques.

Recently, MIPs have been finding applications for the determination of pesticides. MIPs allow both the pre-concentration and purification of the sample and the extraction of the pesticide from the media in a selective manner, which is very important when interferents are present that affect the determination. The use of MIPs in water quality control serves for *in situ* monitoring of carbofuran, or organophosphorus pesticides such as diazinon in combination with other electrochemical techniques such as cyclic voltammetry or impedance measurements [41]. Other interesting analytical procedures include the coating of paper with MIPs and the use of colorimetry or fluorescence to visually observe the presence of pesticides in fruits such as tomatoes and apples [42].

Also, biosensors based on enzyme immobilization in a polymeric matrix in screen-printed electrodes are raised as real-time analytical systems for pesticides, especially dichlorvos and methylparaoxon in aqueous matrices [9].

3.4.2 Nitrates and Nitrites in Water

Industrial manufacturer and agriculture runoff primarily cause nitrate and nitrite water contamination. When consumed in drinking water, both nitrates and nitrites are rapidly absorbed by the digestive system, distributed along the human body, and finally excreted in the urine. However, higher doses of these compounds can cause methemoglobinemia (hemoglobin oxidation by nitrite inhibiting the normal oxygen transport across the organism), producing cyanosis and asphyxia. Furthermore, nitrates and nitrites can cause cancer and affect the thyroid system [43].

MIPs-modified electrodes are also being used to determine many compounds in water and food, including nitrates. MIPs (ion-imprinted polymers particularly) have been used as a coating to develop a low-cost sensor with selective recognition combined with other techniques such as electrochemical impedance spectroscopy to measure nitrate in water [44].

3.4.3 Heavy Metals

Heavy metals (Hg, Cd, Pb, Cr, As, etc.), mainly derived from atmospheric deposition, industrialization, and intensive agricultural practices, contaminate soil and water and are absorbed by crops, thus passing into the food chain. All these pollutants seriously affect health by causing metabolic disorders and affecting the physiological function of organs. Various polymeric materials have been described for the detection of heavy metals in aqueous media, mainly with an optical response [45]. Acrylic-based polymers can detect by a colorimetric change and eliminate Hg (II) in seafood and water [46]. In addition, as adsorbents, both pH- and temperature-responsive polymers have proven to be effective in removing heavy metals in water as a function of the pH of the media, both for domestic use and for wastewater treatment. A variation in the pH or temperature results in variations in the filtration ability through an alteration of their pore sizes and flux through shrinking and swelling, and even the possibility to separate oil from wastewater. Also, these polymeric materials can change their working mechanism triggered by the nature and concentration of an electrolyte/salt ion in the surroundings through control of the pore conformation [47].

3.4.4   Other Food Quality Parameters

There are other substances present in food and not necessarily harmful to humans that guarantee food authenticity or provide beneficial qualities. Polyphenols are an important example and are present in different foods such as fruits, vegetables, and beverages derived from them and that have antioxidant and antimicrobial activities. Intelligent polymers capable of detecting this type of compound have also been developed by a colorimetric response using acrylic-based smart polymers [48]. Furthermore, halal verification can be performed by using a naked-eye ethanol PANi biosensor. Alcohol oxidase is immobilized onto PANi, and when alcohol reacts with it, a color variation from green to blue is produced due to the release of acetaldehyde and hydrogen peroxidase, which oxidizes PANi [26].

4   CONCLUSIONS AND FUTURE PERSPECTIVES

Food quality analytical tools are currently expensive, require specialized personnel, are time-consuming, and are not readily available for *in situ* and real-time evaluation. On the other hand, smart polymers can be a key tool to reduce diseases and food waste, since they allow the determination of food quality indicators in a fast, easy, simple, and cheap way. Additionally, they can prevent food adulteration, and improve the distribution chain knowledge during storage and transportation. Although these systems could open a revolutionary approach to food packaging and measuring techniques, some major factors still need to be considered, as observed in the limited number of commercialized systems. In this sense, research directed this way will provide industries, consumers, and regulatory authorities acceptance, environmental effectiveness, and economic benefits, by decreasing foodborne illnesses and food waste, and contributing to a more sustainable world.

ACKNOWLEDGMENTS

We gratefully acknowledge the financial support provided by FEDER (Fondo Europeo de Desarrollo Regional), the Spanish AEI (State Research Agency, PID2020-113264RB-I00/AEI/10.13039/ 501100011033 and PID2019-108583RJ-I00/AEI/10.13039/501100011033), and "La Caixa" Foundation (under agreement LCF/PR/PR18/51130007). We also acknowledge the financial support provided by the Spanish Ministerio de Universidades (Plan de Recuperación, Transformación y Resiliciencia, European Union-NextGenerationEU, Universidad Politécnica de Madrid (RD 289/ 2021) and Universidad Autónoma de Madrid (CA1/RSUE/2021-00409)).

REFERENCES

[1] O.A. Odeyemi, N.A. Sani, A.O. Obadina, C.K.S. Saba, F.A. Bamidele, M. Abughoush, A. Asghar, F.F.D. Dongmo, D. Macer, A. Aberoumand, Food safety knowledge, attitudes and practices among consumers in developing countries: An international survey, Food Res. Int. 116 (2019) 1386–1390.

[2] J. Gustavsson, C. Cederberg, U. Sonesson, R. V. Otterdijk, A. Meybeck, Global food losses and food waste – Extent, causes and prevention, Food Agric. Organ. United Nations. Rome; 2011.

[3] A.M. Giusti, E. Bignetti, C. Cannella, Exploring New Frontiers in Total Food Quality Definition and Assessment: From Chemical to Neurochemical Properties, Food Bioprocess Technol. 1 (2008) 130–142.

[4] H. Yousefi, H.-M. Su, S.M. Imani, K. Alkhaldi, C.D. M. Filipe, T.F. Didar, Intelligent Food Packaging: A Review of Smart Sensing Technologies for Monitoring Food Quality, ACS Sensors 4 (2019) 808–821.

[5] J.M. García, F.C. García, J.A. Reglero Ruiz, S. Vallejos, M. Trigo-López, Smart Polymers. Principles and Applications, De Gruyter; 2022.

[6] M. Wei, Y. Gao, X. Li, M.J. Serpe, Stimuli-responsive polymers and their applications, Polym. Chem. 8 (2017) 127–143.

[7] Q. Lin, Y. Li, M. Yang, Polyaniline nanofiber humidity sensor prepared by electrospinning, Sensors Actuators B Chem. 161 (2012) 967–972.

[8] M. Khadem, F. Faridbod, P. Norouzi, A. Rahimi Foroushani, M.R. Ganjali, S.J. Shahtaheri, R. Yarahmadi, Modification of Carbon Paste Electrode Based on Molecularly Imprinted Polymer for Electrochemical Determination of Diazinon in Biological and Environmental Samples, Electroanalysis 29 (2017) 708–715.

[9] I. Yaroshenko, D. Kirsanov, M. Marjanovic, P.A. Lieberzeit, O. Korostynska, A. Mason, I. Frau, A. Legin, Real-Time Water Quality Monitoring with Chemical Sensors, Sensors 20 (2020) 3432.

[10] K.B. Biji, C.N. Ravishankar, C.O. Mohan, T.K. Srinivasa Gopal, Smart packaging systems for food applications: a review, J. Food Sci. Technol. 52 (2015) 6125–6135.

[11] R. Ahvenainen, Active and intelligent packaging, In: Ahvenainen R, editor. Nov. Food Packag. Tech. Cambridge: Woodhead Publishing; 2003. pp. 5–21.

[12] M. Smolander, Freshness Indicators for Food Packaging, In: Kerry J, Butler P, editors. Smart Packag. Technol. Fast Mov. Consum. Goods John Wiley & Sons; 2008. p. 111–127.

[13] M. Weston, S. Geng, R. Chandrawati, Food Sensors: Challenges and Opportunities, Adv. Mater. Technol. 6 (2021) 2001242.

[14] M. Loughran, D. Diamond, Monitoring of volatile bases in fish sample headspace using an acidochromic dye, Food Chem. 69 (2000) 97–103.

[15] R.J. Mcgorrin, The Significance of Volatile Sulfur Compounds in Food Flavors An Overview, Volatile Sulfur Compd. Food American Chemical Society; 2011. pp. 3–31.

[16] Y. Özogul, F. Özogul, Chapter 1 Biogenic Amines Formation, Toxicity, Regulations in Food, Food Chem. Funct. Anal. (2019) 1–17.

[17] D. Doeun, M. Davaatseren, M.S. Chung, Biogenic amines in foods, Food Sci. Biotechnol. 26 (2017) 1463–1474.

[18] C. Rukchon, A. Nopwinyuwong, S. Trevanich, T. Jinkarn, P. Suppakul, Development of a food spoilage indicator for monitoring freshness of skinless chicken breast, Talanta 130 (2014) 547–554.

[19] J. Jung, P. Puligundla, S. Ko, Proof-of-concept study of chitosan-based carbon dioxide indicator for food packaging applications, Food Chem. 135 (2012) 2170–2174.

[20] A. Mills, K. Lawrie, J. Bardin, A. Apedaile, G.A. Skinner, C. O'Rourke, An O2 smart plastic film for packaging, Analyst 137 (2011) 106–112.

[21] R.N. Gillanders, O. V. Arzhakova, A. Hempel, A. Dolgova, J.P. Kerry, L.M. Yarysheva, N.F. Bakeev, A.L. Volynskii, D.B. Papkovsky, Phosphorescent Oxygen Sensors Based on Nanostructured Polyolefin Substrates, Anal. Chem. 82 (2010) 466–468.

[22] B. Liu, P.A. Gurr, G.G. Qiao, Irreversible Spoilage Sensors for Protein-Based Food, ACS Sensors 5 (2020) 2903–2908.

[23] J.L. Pablos, S. Vallejos, A. Muñoz, M.J. Rojo, F. Serna, F.C. García, J.M. García, Solid polymer substrates and coated fibers containing 2,4,6-trinitrobenzene motifs as smart labels for the visual detection of biogenic amine vapors, Chem. - Eur. J. 21 (2015) 8733–8736.

[24] L. González-Ceballos, B. Melero, M. Trigo-López, S. Vallejos, A. Muñoz, F.C. García, M.A. Fernandez-Muiño, M.T. Sancho, J.M. García, Functional aromatic polyamides for the preparation of coated fibres as smart labels for the visual detection of biogenic amine vapours and fish spoilage, Sensors Actuators B Chem. 304 (2020) 127249.

[25] B. Kuswandi, Jayus, A. Restyana, A. Abdullah, L.Y. Heng, M. Ahmad, A novel colorimetric food package label for fish spoilage based on polyaniline film, Food Control 25 (2012) 184–189.

[26] E. Poyatos-Racionero, J.V. Ros-Lis, J.-L. Vivancos, R. Martínez-Máñez, Recent advances on intelligent packaging as tools to reduce food waste, J. Clean. Prod. 172 (2018) 3398–3409.

[27] K. Mitsubayashi, Y. Kubotera, K. Yano, Y. Hashimoto, T. Kon, S. Nakakura, Y. Nishi, H. Endo, Trimethylamine biosensor with flavin-containing monooxygenase type 3 (FMO3) for fish-freshness analysis, Sensors Actuators, B Chem. 103 (2004) 463–467.

[28] J. Reglero Ruiz, A. Sanjuán, S. Vallejos, F. García, J. García, Smart polymers in micro and nano sensory devices, Chemosensors 6 (2018) 12.

[29] D. Elfadil, A. Lamaoui, F. Della Pelle, A. Amine, D. Compagnone, Molecularly imprinted polymers combined with electrochemical sensors for food contaminants analysis, Molecules 26 (2021) 4607.

[30] V. Siracusa, Food packaging permeability behaviour: A report, Int. J. Polym. Sci. 2012 (2012) 11.

[31] S. Purkayastha, A.K. Biswal, S. Saha, Responsive systems in food packaging, J. Packag. Technol. Res. 1 (2017) 53–64.

[32] S. Choi, Y. Eom, S. Kim, D. Jeong, J. Han, J.M. Koo, S.Y. Hwang, J. Park, D.X. Oh, A self-healing nanofiber-based self-responsive time–temperature indicator for securing a cold-supply chain, Adv. Mater. 32 (2020) 1907064.

[33] J.W.F. Law, N.S.A. Mutalib, K.G. Chan, L.H. Lee, Rapid methods for the detection of foodborne bacterial pathogens: Principles, applications, advantages and limitations, Front. Microbiol. 5 (2014) 770.

[34] T. Bintsis, Foodborne pathogens, AIMS Microbiol. 3 (2017) 529–563.

[35] W.T. Bodenhamer, Method and apparatus for selective biological material detection (US6051388A), United States; 1999.

[36] H. Yousefi, M.M. Ali, H.M. Su, C.D.M. Filipe, T.F. Didar, Sentinel wraps: Real-time monitoring of food contamination by printing DNAzyme probes on food packaging, ACS Nano 12 (2018) 3287–3294.

[37] B. Malhotra, A. Keshwani, H. Kharkwal, Antimicrobial food packaging: potential and pitfalls, Front. Microbiol. 6 (2015) 611.

[38] European Commission. Food Safety. Pesticides [Internet]: https://food.ec.europa.eu/plants/pestic ides_en.

[39] V.L. Zikankuba, G. Mwanyika, J.E. Ntwenya, A. James, Pesticide regulations and their malpractice implications on food and environment safety, Cogent Food Agric. 5 (2019) 1601544.

[40] L. Carrasco Cabrera, P. Medina Pastor, The 2019 European Union report on pesticide residues in food, EFSA J. 19 (2021) 6491.

[41] A.R. Cardoso, A.P.M. Tavares, M.G.F. Sales, In-situ generated molecularly imprinted material for chloramphenicol electrochemical sensing in waters down to the nanomolar level, Sensors Actuators B Chem. 256 (2018) 420–428.

[42] M. Vodova, L. Nejdl, K. Pavelicova, K. Zemankova, T. Rrypar, D. Skopalova Sterbova, J. Bezdekova, N. Nuchtavorn, M. Macka, V. Adam, M. Vaculovicova, Detection of pesticides in food products using paper-based devices by UV-induced fluorescence spectroscopy combined with molecularly imprinted polymers, Food Chem. 380 (2022) 132141.

[43] World Health Organization, Nitrate and nitrite in Drinking-water. Background document for development of WHO Guidelines for Drinking-water Quality, Guidel. Drink. Qual. 2003. p. WHO/SDE/WSH/ 04.03/56.

[44] M.E.E. Alahi, S.C. Mukhopadhyay, L. Burkitt, Imprinted polymer coated impedimetric nitrate sensor for real- time water quality monitoring, Sensors Actuators B Chem. 259 (2018) 753–761.

[45] A.M. Sanjuán, J.A. Reglero Ruiz, F.C. García, J.M. García, Recent developments in sensing devices based on polymeric systems, React. Funct. Polym. 133 (2018) 103–125.

[46] S. Vallejos, J.A. Reglero, F.C. García, J.M. García, Direct visual detection and quantification of mercury in fresh fish meat using facilely prepared polymeric sensory labels, J. Mater. Chem. A 5 (2017) 13710–13716.

[47] K. Dutta, S. De, Smart responsive materials for water purification: an overview, J. Mater. Chem. A 5 (2017) 22095–22112.

[48] S. Vallejos, D. Moreno, S. Ibeas, A. Muñoz, F.C. García, J.M. García, Polymeric chemosensor for the colorimetric determination of the total polyphenol index (TPI) in wines, Food Control 106 (2019) 106684.

15 Polymers and Their Nanocomposites for Corrosion Protection

Vishwa Suthar,[1,2] Felipe M. de Souza,[2]
Magdalene A. Asare,[1,2] and Ram K. Gupta[1,2]

[1] Department of Chemistry, Pittsburg State University, Pittsburg, KS 66762, USA
[2] National Institute for Materials Advancement, Pittsburg State University, Pittsburg, KS 66762, USA

1 INTRODUCTION

Polymeric nanocomposites (PNCs) are a broad group of high-performance materials that function by incorporating the advantageous properties of their components. PNCs can be employed in several fields for the fabrication of high-end products such as water purification membranes, high abrasion resistance loaded rubbers, dielectric materials for energy storage applications, and fibrous supplemented thermoset composites for high mechanical and thermal properties, with high corrosion resistance [1]. These properties can be implemented into the polymeric composite through the incorporation of certain nanomaterials based on silicon, carbon, transition metal oxide, etc. Through this process many properties such as hydrophobicity, chemical and thermal stability, and conductivity, among others, can be introduced into the polymeric matrix. These properties can be useful in terms of applications for anti-corrosion coatings. Several synthetic techniques such as electrodeposition, powder coating, dry coating, vapor corrosion inhibitors (VCI), electrospinning, plasma sputtering, chemical vapor deposition (CVD), physical vapor deposition (PVD), and others can be used to synthesize PNCs that can widen their applications [2].

For making composites, nanofillers are usually added. Nanofillers should be introduced into the polymeric matrix in such a way that they can add the desired property without jeopardizing the performance of the polymers. One of the main aims lies in obtaining a structure that can properly hinder the permeation or diffusion of reactive species such as certain anions or oxygen for effective corrosion protection. For that, it is necessary to obtain a structure that can be properly packed over the substrate to create a barrier effect or labyrinth effect. The packed structure of polymer along with the proper distribution of nanoparticles can play an important role in protecting surfaces against corrosion [3]. By successfully preventing this, several issues such as contamination of products, chemical leakage, breaking of pipelines, failure of bridges, early equipment replacement, and casting pits can be avoided. These issues are likely to occur in any country, and can lead to an exorbitant amount of funds to repair the damage caused by corrosion [4]. Hence, the complications caused by corrosion make it an urgent issue to be solved, which reinforces the need for research in this area to further optimize the current technology [5,6]. It is recommended that the materials utilized for anti-corrosion are derived from renewable sources along with environmentally friendly aspects, cost efficiency, and competitive performance. Based on these requirements this chapter provides some

DOI: 10.1201/9781003278269-15

recent examples from the current literature that can aid the reader to design novel ideas to tackling such issues.

2 METHODS FOR STUDYING CORROSION

To prevent corrosion, it is very important to understand its cause and its evaluation. The characterization and analysis of synthesized materials provide valuable information to elucidate their properties and the mechanisms of corrosion protection. Several characterization techniques such as salt spray test (SST), X-ray diffraction (XRD), open circuit potential (OCP), electrochemical impedance spectroscopy (EIS), weight loss test, and surface morphology analysis such as scanning electron microscopy (SEM) and transmission electron microscopy (TEM) can be performed to understand the corrosion process.

2.1 SALT SPRAY TEST

This test consists of exposing the protective coating against a corrosive brine of usually 3 wt.% NaCl for an extended time and analyzing how the sample responds to the aggressive environment. For example, Chen *et al.* [7] synthesized a polymeric nanocomposite based on polyaniline/partially phosphorylated poly(vinyl alcohol) (PANI/P-PVA) that was mixed with epoxy film for a corrosion protective coating. Through the empirical result provided from the salt spray test, it was observed that the PANI/P-PVA composite presented the most efficient protection against high concentrations of NaCl solutions when compared to neat epoxy film and emeraldine PANI (PANI ES). It was found that the formation of a passive and compact layer prevented the diffusion of oxygen and permeation of ions to protect the coated surface. This layer was composed of Fe_2O_3 formed through a redox process between PANI with the steel underneath, along with the formation of $FeHPO_4$ and $Fe_3(PO_4)_2$ due to the presence of the partially phosphorylated groups. The empirical aspect of this process is depicted in Figure 15.1.

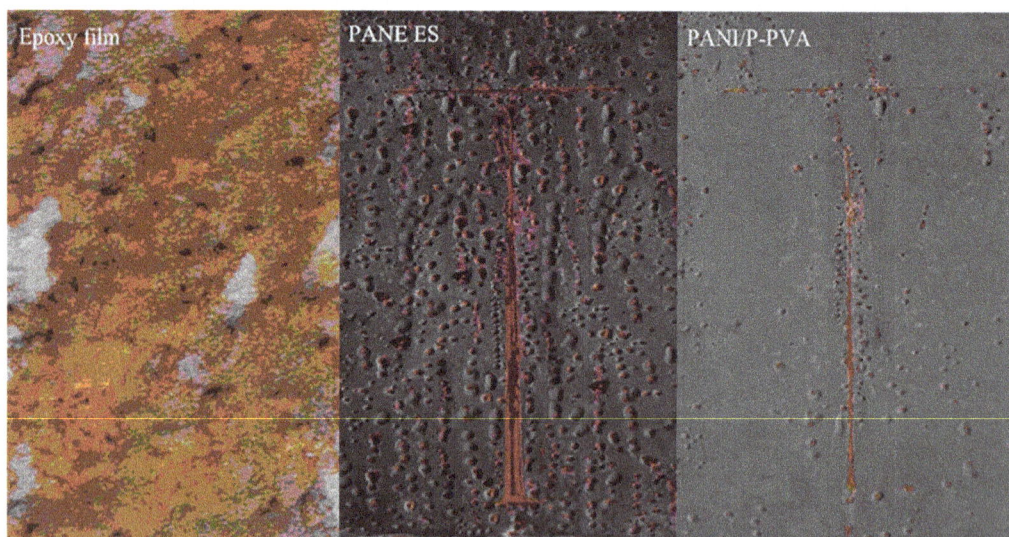

FIGURE 15.1   Images of the salt spray test performed on the steel surface coated with pure epoxy, 2.5 wt.% PANI-ES and 2.5 wt.% PANI-PVA after being exposed for 30 days to a 3 wt.% NaCl solution. Adapted with permission [7]. Copyright (2011), American Chemical Society.

2.2 X-RAY DIFFRACTION

The XRD technique is a widely employed characterization method to determine the crystallographic structure of a material. XRD allows the precise identification of crystalline materials. The measurement is based on the intensities of a scattered X-ray that is irradiated upon the sample at different angles. XRD analysis is employed in most publications as a characterization process. For example, Liu *et al.* [8] utilized acrylic acid-allylpolyethoxy carboxylate copolymer (AL 15) as an inhibitor of corrosion in mild steel surfaces. The mechanism of action was based on the barrier effect of the polymer. Anti-scale properties were also observed when the material was exposed to a seawater simulation environment. AL 15 could promote the encapsulation of Ca^{2+} due to the presence of carboxyl groups, leading to the spontaneous formation of AL 15–Ca complexes. Through that, the Ca^{2+} acted as ties by simultaneously linking CO_3^{2-} or SO_4^{2-} in the seawater media. The presence of these ions leads to the formation of $CaCO_3$, which is incorporated into the AL 15 polymeric matrix. The presence of PEG segments improved the hydrophilicity and prevented the $CaCO_3$ or $CaSO_4$ from agglomerating. The XRD analysis was performed to elucidate this mechanism. Figure 15.2 shows the XRD spectrum for the $CaCO_3$. The diffraction peak intensity of $CaCO_3$ was stronger when the blank sample was analyzed, suggesting that under the absence of AL 15 there was a larger amount of $CaCO_3$ attributed to the stronger peak at 29.24°, which is characteristic for 104 calcite's face. On the other hand, under the presence of 20 mg/L of AL 15, the XRD characterization revealed the presence of valerite crystal. It was observed that AL 15 induced a modification in the crystal's structure which became more soluble in water. Thus, this helped the coat to be preserved and avoid the formation of a thick crust of $CaCO_3$. This example shows a proper description of the use of XRD to not only elucidate the crystal structures over the coating but also to describe part of the anti-corrosion mechanism.

2.3 X-RAY PHOTOELECTRON SPECTROMETRY

X-ray photoelectron spectrometry (XPS) is an important technique used to analyze a material's surface in such a way that it can identify the elements along with their oxidation state. This characterization provides a spectrum that consists of intensity usually in counts (s) in the function of binding energy (eV). Along with that, it can also be used to determine the elements quantitively. This characterization is extremely handy for several situations including the elucidation of anti-corrosion

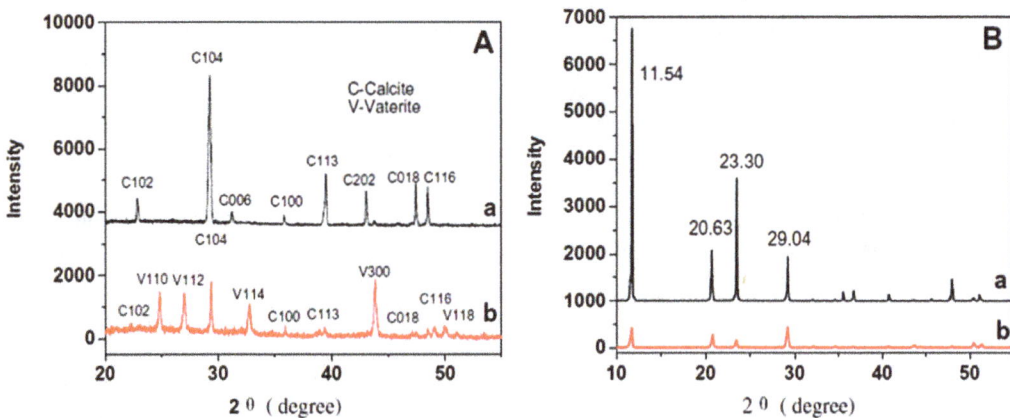

FIGURE 15.2 (A) $CaCO_3$ crystals with XRD patterns: (a) uncoated steel and (b) coated with 20 mg/L of AL15. (B) $CaSO_4$ crystals patterns: (a) uncoated steel and (b) coated with 20 mg/L of AL 15. Adapted with permission [8]. Copyright (2017), Elsevier.

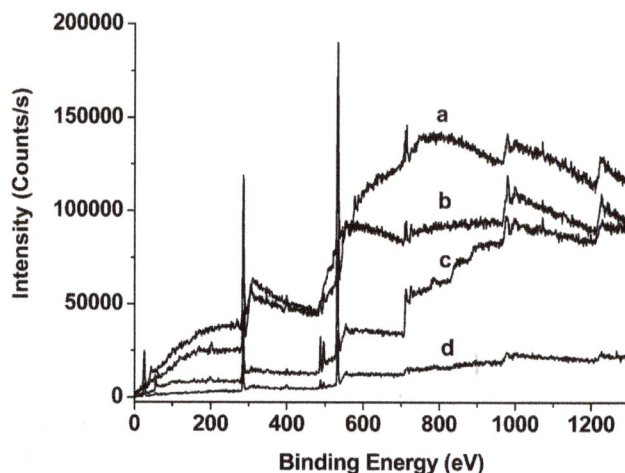

FIGURE 15.3   XPS spectra for the steel substrate of (a) uncoated, (b) epoxy coated, (c) PANI ES 2.5 wt.%, and (d) PANI/P-PVA 2.5 wt.%. Adapted with permission [7]. Copyright (2011), American Chemical Society.

mechanisms. Anti-corrosion properties of a composite are based on spherical nanoparticles of PANI with partially phosphorylated poly(vinyl alcohol), PANI/P-PVA were analyzed for the elemental characterization and elucidation of the anti-corrosion mechanism using the XPS technique [7]. Figure 15.3 displays the XPS spectra for (a) pristine steel, (b) epoxy coated, (c) PANI, and (d) PANI/P-PVA-based composite. As can be seen, the steel coated with PANI/P-PVA presented the lowest peaks of Na 1s and Cl 1s, along with the expected highest peaks for N 1s and P 2p. Therefore, it was concluded that there was less permeation of NaCl in the structure compared with the other samples, thus reinforcing the better anti-corrosion properties.

2.4   ELECTROCHEMICAL IMPEDANCE SPECTROSCOPY

Electrochemical impedance spectroscopy (EIS) consists of the measurement of the overall resistance, which aids in understanding of electrochemical processes along with mass and species transports, which helps in the interpretation of the mechanisms involved in the electrochemical process at the material's surface. Through this, the redox process can be understood along with the possibility to propose the equivalent electric circuit of the system, along with other aspects. EIS studies have been performed to understand the electrochemical properties of PANI/P-PVA coatings in terms of anti-corrosion properties [7]. Figure 15.4 displays the variation of impedance as a function of applied frequency for samples after being exposed for 30 days to an NaCl solution. It was observed that the 2.5 wt.% PANI/P-PVA coating was able to maintain a considerably high impedance of around 3.4×10^7 Ωcm^2 even after 30 days of exposure to 3.0 wt.% NaCl. Such results could be attributed to the compact structure of the composite along with the redox properties of PANI which reduced the reactivity of the corrosive species and therefore prevented the oxidation of the steel underneath.

2.5   OPEN CIRCUIT POTENTIAL

Open circuit potential (OCP) is a passive measurement of a system's potential at zero current. It provides some information on the resting potential that can appear in an electrochemical cell between the working and reference electrodes. OCP can reveal the potential exhibited between the metal surface and surroundings, cooperated by the reference electrode. For example, Odewunmi *et al.* [9] compared the corrosion protection properties of a monomer called

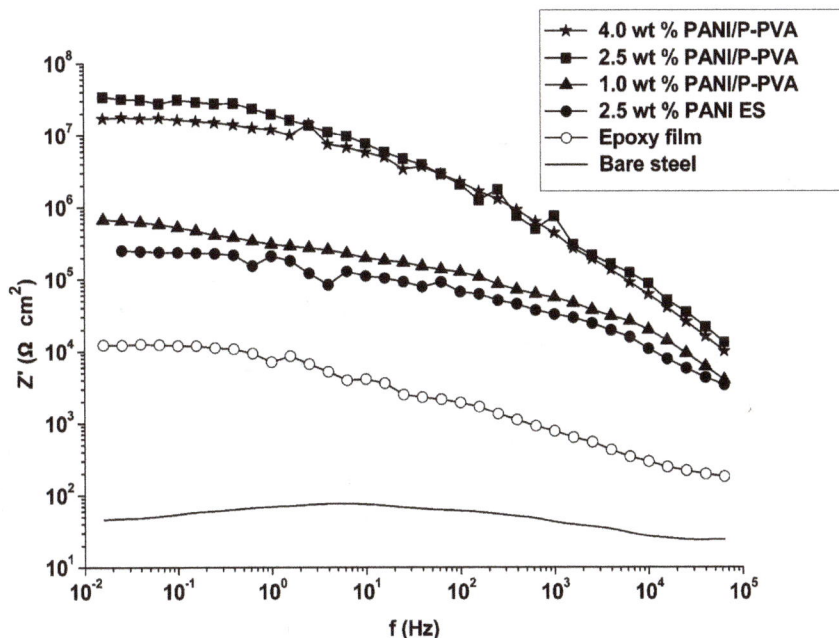

FIGURE 15.4 EIS plot for the uncoated and coated steel samples in 3.0 wt.% NaCl after 30 days' exposure to a salt spray test. Adapted with permission [7]. Copyright (2011), American Chemical Society.

N^1,N^1-diallyl-N^6,N^6,N^6-tripropylhexane-1,6-diaminium chloride (NDTHDC) against its respective polymer (poly-NDTHDC) over carbon steel. The variation of OCP was measured then the system was exposed to a 15 wt.% HCl solution. The variation in the OCP can be seen in Figure 15.5. Through that, it could be seen that the OCP began at −476 mV. The variations in potential through time could be attributed to the corrosion process that formed an oxide layer. Also, the observed shift toward anodic potentials is related to oxidizing effects. In this sense, it was observed that poly-NDTHDC presented a stronger anodic behavior as compared to NHTHDC.

2.6 Weight Loss

The rate of corrosion is an important factor to test the long-term efficiency of a polymeric nanocoating against corrosion. This process can be analyzed through the weight loss of the coating as a predic-tion of the coating's usage term. Taking the previous literature example studied by Odewunmi *et al.* [9], it can be seen that the uncoated steel presented a corrosion rate of 5.80 mm/year, whereas 100 mg/NDTHDC and 100 mg/poly-NDTHDC presented corrosion rates of 3.25 and 0.83 mm/year, respectively. Through these results, 100 mg/NDTHDC and 100 mg/poly-NDTHDC presented corrosion inhibition coefficients of 43.9% and 85.5%, respectively. The difference in performance from the monomer compared to the respective polymer could be attributed to the proper packing of the polymer over the substrate as it was more likely to cover active sites for corrosion. Through this effect the corrosion rate could be calculated based on the weight loss (CR_{WL}) (mm/y), which can be described by equation (1):

$$CR_{WL} = \frac{87600 \times \Delta M_m}{S_m \times T_m \times \rho_m} \tag{1}$$

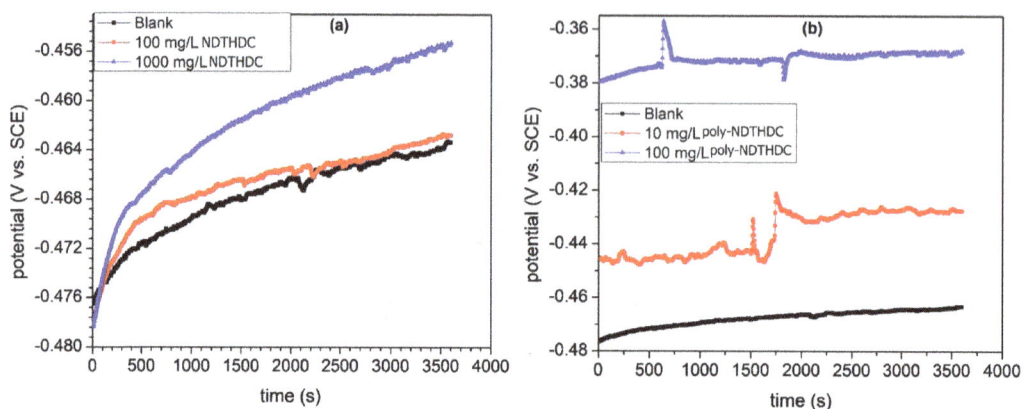

FIGURE 15.5 OPC variation for a steel substrate exposed to 15 wt.% HCl at room temperature in the absence and presence of either (a) NDTHDC or (b) poly-NDTHDC. Adapted from Reference [9]. Copyright (2020), American Chemical Society. This is an open-access article published under a Creative Commons Attribution (CC-BY) License.

where ΔM_m is the weight loss observed after the immersion test (g), S_m is the sample's actively exposed area (cm^2), T_m is the immersion time (h), and ρ_m is the density of the specimen (g/cm^3).

2.7 SCANNING ELECTRON MICROSCOPY

SEM is an important technique as it provides empirical data through an image of the material's surface. This allows scientists to study variations in the morphology after functionalizing a material, making a composite, exposing the material to certain environments, and so on. Based on the uses of SEM within the field of PNC for anti-corrosion, Ubaid *et al.* [10] synthesized a self-healing epoxy-based coating that contained TiO$_2$ nanotubes along with dodecylamine (DDA) that functioned as the self-healing and anti-corrosion agent. The SEM analysis was performed to track the self-healing process over time. After the coatings were artificially scratched it was observed that the plain epoxy coating did not present an apparent self-healing process (Figure 15.6a). On the other hand, after incorporation of DDA and TiO$_2$ into the epoxy coating it was observed that the scratch's width decreased considerably, showing a nearly fully healed coating, as can be observed in Figure 15.6b.

2.8 TRANSMISSION ELECTRON MICROSCOPY

TEM forms an image based on a beam of electrons that is transmitted through the sample. For that, the sample should be a thin film of around 100 nm thickness or a suspension in a grid. Then, the image is magnified and focused on a substrate that can form the image such as a fluorescent screen. Hence, similarly to SEM, TEM can provide useful information regarding the structure as it also reaches ranges in the nanoscale. An example of the use of TEM can be seen in the work of Balakrishnan *et al.* [11], who performed a thermal treatment on the sunflower to form a protective coating over the Fe substrate. The polymeric film formed over the mild steel was analyzed through TEM. Based on that, the TEM analysis revealed that non-spherical and partially agglomerated structures were formed which were on average around 200–250 nm in size (Figure 15.7a). Also, nanofibrils were formed which presented a width and diameter in the range of 150 nm and 35 nm, respectively (Figure 15.7b). It was believed that the agglomerated structure was formed due to the π–π stacking of the nanofibrils. Such a packed and agglomerated structure could be one of the

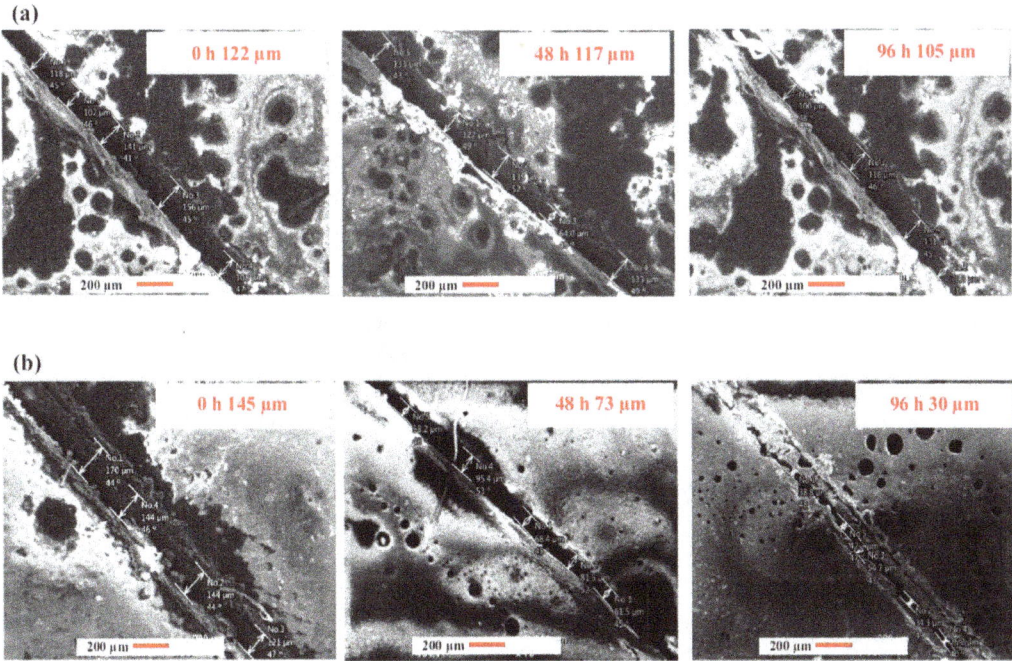

FIGURE 15.6 SEM micrographs displaying the scratched coatings with (a) plain epoxy and (b) functionalized with TiO$_2$ and DDA. Adapted with permission [10]. Copyright (2019), Elsevier.

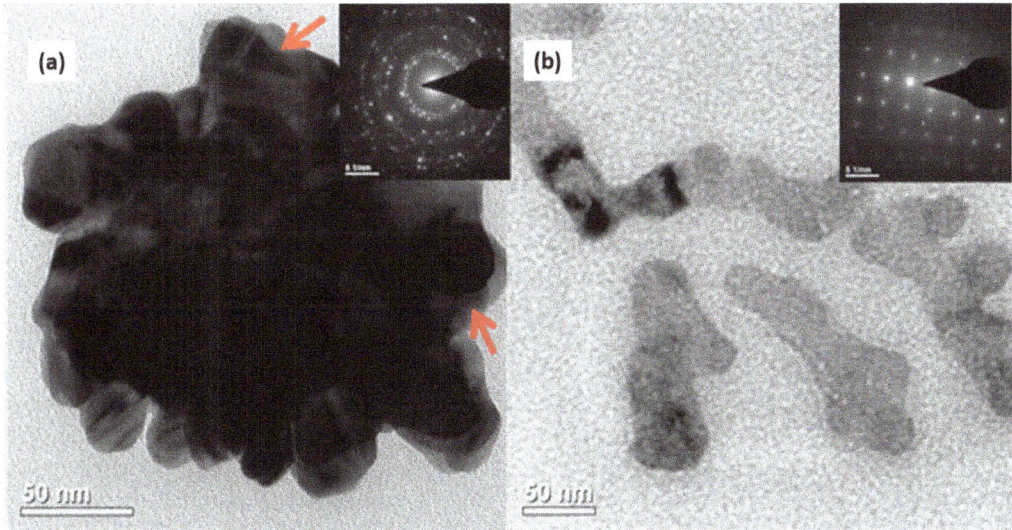

FIGURE 15.7 TEM images for the polymerized sunflower oil-based coating and SAED insets for each of the structures. Adapted with permission [11]. Copyright (2015), American Chemical Society.

factors that contributed to the satisfactory performance of the polymerized sunflower film against corrosion. Also, the insets show the selected area electron diffraction (SAED) which demonstrated a polycrystalline and single-crystal nature for the non-spherical and nanofibrils structures, respectively (Figure 15.7a, b).

3   CORROSION PROTECTIVE COATINGS

The necessity for novel technologies for anti-corrosion coatings is an important field of research since the damage caused due to corrosion can be a considerable percentage of any given country's gross national product. It is therefore necessary to develop protective coatings that can tackle this issue effectively. Several types of coatings, such as epoxy, acrylic, siloxane, alkyd, and polyurethane, among others, can be employed for corrosion protection. Further, compositing these polymers with nanoparticles can lead to further enhancement of their protective properties. In this sense, anti-corrosion coatings can be roughly classified as either conductive or non-conductive. The conducting coatings can be used on metallic substrates as they usually function by forming a metal oxide passive layer of the respective metal underneath the conducting coating due to the redox process. This effect can lead to adhesion of the coating along with increasing the packing density, which prevents or slows down the permeation of reactive species. On the other hand, a non-conductive coating functions by physically preventing the corrosive species from coming into contact with the substrate by forming a physical barrier or through a labyrinth effect due to an inherent tortuosity in its morphology [12,13]. Examples of both cases are provided in the following section.

3.1   CONDUCTING POLYMER-BASED CORROSION PROTECTIVE COATING

3.1.1   Polyaniline

Polyaniline (PANI) belongs to a semi-flexible polymer group. It has three recognizable oxidation states, which are emeraldine, pernigraniline, and leucoemeraldine. Emeraldine is a partially oxidized state, whereas pernigraniline and leucoemeraldine represent the states in which PANI is completely oxidized or reduced, respectively. Aside from these characteristic redox properties, PANI can be synthesized through either electrochemical or chemical routes along with different procedures, making it a versatile conducting polymer, while aniline is a relatively low-cost monomer. PANI also presents relatively high chemical stability, which makes it attractive for anti-corrosion applications [14,15].

Chen *et al.* [16] developed a composite based on PANI doped with perfluorooctanoic acid (PFOA/PANI) through oxidation polymerization using the ultrasonication method. The nanoparticles were highly dispersible in ethanol and presented high hydrophobicity, which allowed them to be evenly distributed in an epoxy resin (EP) leading to an effective anti-corrosion composite coating PFOA/PANI/EP for steel. The corrosion protection mechanism of this composite could be explained through three effects: (1) a barrier, (2) passivation, and (3) corrosion inhibition. The barrier effect was observed due to the high hydrophobicity of the coating along with the good mixability between the PFOA/PANI nanoparticles with the EP, which provided an even distribution that prevented the formation of pores in the coating. Hence, the corrosive species could be hindered from reaching the metallic substrate underneath. Second, a thin layer of passive Fe_2O_3 was formed due to the redox properties of the PFOA/PANI which reinforced the protection against corrosive species [17–19]. Third, PFOA presents a hydrophilic carboxyl group as well as a hydrophobic alkyl long chain. The latter can be adsorbed on the metallic surface forming a hydrophobic film. On the other hand, the PFOA molecule can be de-doped from PFOA/PANI, which leads to the formation of a hydrophobic anti-corrosion layer on the metal's surface which acts as another anti-corrosion mechanism.

3.1.2   Polypyrrole

Polypyrrole (PPy) is a widely researched conducting polymer which has several high-end applications that include sensors and electronic devices [20,21]. It is also applicable to the development of anti-corrosion coatings on metallic surfaces [22]. The monomer, pyrrole, can be easily polymerized either through a chemical or electrochemical route, which is also an attractive aspect of this material. In terms of corrosion protection mechanisms, PPy can act in two ways. One is based on forming a physical barrier, by preventing the contact of the substrate with the corrosive species. The other occurs through anodic protection, based on the conducting properties of PPy which cause the metallic substrate to oxidize leading to the formation of a more chemically stable thin layer of metal oxide. These factors are influenced by the conductivity of PPy, which induces a higher degree of oxidation of the metallic layer underneath. Yet, as the PPy coating is chemically reduced its conductivity decreases along with the oxidation degree of the metallic substrate underneath. However, the PPy's conductivity and metal oxide layer can be refreshed if an external oxidant in the environment reoxidizes the PPy.

3.1.3 Poly(vinylcarbazole)

Poly(vinylcarbazole) (PVK) is an interesting conducting polymer able to form films that present satisfactory mechanical and thermal stability along with enabling proper dispersibility of nanoparticles such as ZrO_2, SiO_2, TiO_2, Al_2O_3, CdS, etc. Elangovan *et al.* [23] proposed the fabrication of a PVK-Al_2O_3 nanocomposite coating for protection against corrosion. It was observed that when 2 wt.% of Al_2O_3 was incorporated into PVK the best performance against corrosion was achieved. Alongside that, a further improvement in mechanical and thermal properties was observed. The lower amount of Al_2O_3 nanoparticles added led to a smaller percentage of porosity in the coating's morphology when compared to different wt.% of Al_2O_3. In that sense, the 2 wt.% of Al_2O_3 was more likely to hinder the permeation of corrosive species. Hence, compositing PVK with Al_2O_3 in such amounts led to the optimum morphology which was a highly influential factor for the proper efficacy of the composite coating.

3.2 NANOCOMPOSITES OF CONDUCTING POLYMERS FOR CORROSION PROTECTIVE COATING

3.2.1 Carbon Nanoparticles

Carbon nanomaterials are highly versatile due to their broad range of properties which generally include high thermal and electronic conductivity, surface area, porosity, tuneability in terms of functionalization, and variation in morphology among many other aspects. Some examples of carbon-based nanomaterials include graphene, carbon nanotubes, carbon black, acetylene black, expandable graphite, etc. These aspects, along with its versatile range of geometry that can go from 0D to 3D, grant them applications in several areas including energy storage devices, sensors, and flame-retardants, among many others [24–26]. Aside from that, these materials can also be employed to fabricate protective coating composites as their high surface area allows further functionalization with nanoparticles. Also, the layered morphology, such as for 2D carbon-based nanomaterials, can induce a satisfactory barrier effect. Finally, the inherent conductivity can promote the passivation layer, leading to more stable coatings.

One example of the incorporation of carbon-based materials to improve the performance of coatings was studied by Ghasemi-Kahrizsangi *et al.* [27] who incorporated 2.5 wt.% carbon black nanoparticles into an epoxy coating that were evenly dispersed through the use of sodium dodecyl sulfate (SDS) as surfactant. The composite coating was efficient against UV radiation followed by immersion in a corrosive media of 3.5 wt.% NaCl. Through that it was observed that carbon black nanoparticles were able to absorb UV radiation and decrease the degradation of the epoxy coating, preventing the formation of cracks that would allow permeation of the corrosive species. On top of that, the proper dispersion of carbon black also aided in the prevention of microcracks as the carbon-based nanoparticles could properly fit into the epoxy matrix's micropores.

In another study, Chen *et al.* [28] fabricated a composite coating by using a carbon powder functionalized with polydopamine (PDA), named C-PDA. After that, PPy was coated over a steel surface through electrodeposition in the presence of the C-PDA. Hence, a PPy/C-PDA nanocomposite coating was obtained which could withstand immersion in a 0.1 M H_2SO_4 solution for up to 720 h while maintaining its properties. The optimized performance could be attributed to the barrier effect that arose from the π–π interactions between PPy and PDA which improved the structure packing. There was also a redox process performed by PPy, as described in equations 2 and 3. Through that, it is notable that PPy undergoes a redox process as it is reduced by a counter anion A⁻ and re-oxidized by O_2. This system maintains the conductibility of PPy, which aids in the anodic protection along with catalyzing the formation of a passive film in the interface between the composite and metallic substrate.

$$PPy^+A^- + e^- \rightarrow PPy^0 + A^- \tag{2}$$

$$PPy^0 + O_2 + 4H^+ \rightarrow PPy^+ + 2H_2O \tag{3}$$

The composite acquired superior properties such as high conductivity, adhesion strength, and chemical stability. However, the process lost its efficiency over time as, eventually, the corrosive agents were able to permeate through the composite coating.

3.2.2 Graphene

Graphene is a 2D single-layer structure of sp^2 carbons forming a hexagonal arrangement that has had a great impact on the scientific community due to its inherent properties, such as high conductivity, flexibility, chemical stability, transparency, and hydrophobicity, among others. Because of such properties, graphene has been widely researched in several fields including energy storage, flexible materials, magnetic shielding devices, etc. Within this area, graphene can also be employed as a material for protection against corrosion due to its high chemical stability, packed structure, conductivity, and hydrophobicity. Even though the latter is important to enhance the barrier effect in aqueous corrosive media, it also makes it challenging to obtain stable aqueous-based coatings. Such a factor is important as it prevents the use of volatile organic compounds (VOCs), making the coating more eco-friendly and less toxic. Hence, to address this issue as well as to preserve the anti-corrosion performance, Qiu et al. [22] fabricated a composite coating based on graphene and PPy dispersed in a waterborne resin. The procedure consisted of obtaining PPy nanocolloids that acted as a surfactant able to disperse the graphene nanosheets when exposed to ultrasonic vibration. Such an effect was observed due to the π–π interactions that were formed between the graphene nanosheets and PPy. Alongside that, the polar groups present in PPy improved the dispersion of graphene in an aqueous solution that reached up to 5 mg/mL. Thereby, a synergistic effect was observed from the barrier effect and chemical stability from graphene along with the redox properties of PPy which led to the formation of a Fe_2O_3 passive layer. The schematics for the anti-corrosion protection mechanism are depicted in Figure 15.8.

3.2.3 MXene Nanosheets

Another class of 2D nanomaterials, MXenes, presents the formula $M_{n+1}X_nT_x$ where M is a transition metal, X is either C or N, and T is the terminal group which can be –F, –O, or –OH. Most of its applications are devoted to batteries, electrocatalysts, electromagnetic shielding, and purification. However, MXenes have been rarely explored in terms of anti-corrosion due to their tendency to oxidize. Despite that, Zhao et al. [29] fabricated a composite based on $Ti_3C_2T_x$ noncovalently functionalized with an ionic liquid (IL) to form an IL@MXene that was incorporated into a waterborne epoxy resin (WEP). Through this there was a combination of a barrier effect along with the formation of a passive metallic layer due to the action of MXene along with the self-healing properties derived from the IL, leading to a satisfactory synergistic effect. The schematics of the IL@MXene-WEP coating and its efficiency against aqueous corrosive media with salt are provided in Figure 15.9.

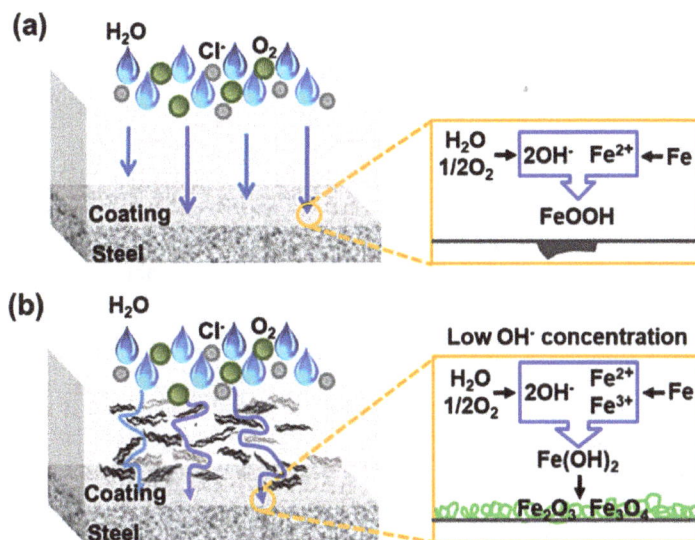

FIGURE 15.8 Scheme for the anti-corrosion protection mechanism in steel coated with (a) pristine epoxy resin and (b) PPy-G composite coating. Adapted with permission [22]. Copyright (2017) American Chemical Society.

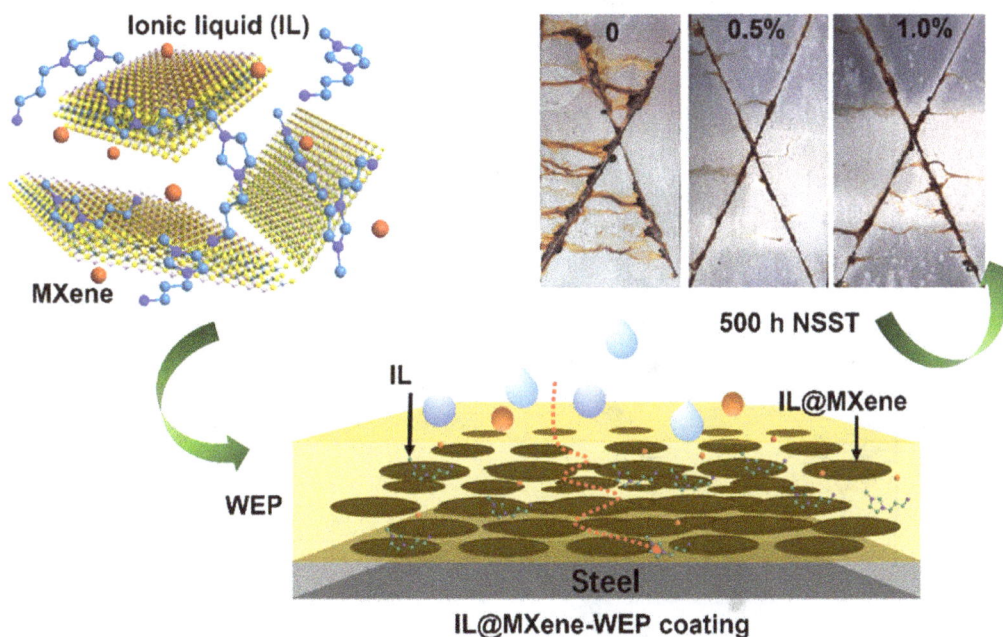

FIGURE 15.9 Schematics of the synthesis of the IL@MXene-WEP coating along with photocopies of neutral spray salt test (NSST) after 500 h for the neat WEP and coatings with 0.5% and 1% of IL@MXene. Adapted with permission [29]. Copyright (2021), American Chemical Society.

3.3   NON-CONDUCTING POLYMERIC-BASED NANOCOMPOSITE COATINGS FOR CORROSION PROTECTION

There is a relatively high number of non-conducting polymers that can be incorporated in coatings for protection against corrosion as they would function mostly through the barrier effect. However, some of these commercialized coatings may require large amounts of VOCs, which can make the process costly and hazardous to the environment. The scientific community and industry are pushing toward an eco-friendlier approach with a search to obtain coatings derived from bio-renewable sources that can be competitive against the currently established ones available in the market.

3.3.1   Sunflower Oil-Based Coating

For the use of bio-based materials for corrosion protection, Balakrishnan et al.[11] used sunflower oil to form a uniform film over Fe by performing a straightforward thermal treatment. This process yielded a networked structure that presented a strong adhesion to the substrate. The polymeric film of sunflower formed over the steel substrate demonstrated satisfactory performance against corrosion as it led to a resistance higher than 10^9 ohm $\times$ cm^2 along with low capacitive values of around $< 10^{-10}$ F/cm^2. The formation of the passive layer could be eluded due to the OCP testing at 1 M HCl solution which provided a corrosion potential of +165 mV, whereas the uncoated Fe presented –445 mV, meaning that the presence of sunflower oil made it less reactive. It was observed that the sunflower coating was able to prevent the permeation of corrosive agents such as O_2, Cl_2, and H_2O due to the formation of a compact film along with the formation of a passive layer of metal oxide underneath it, which further enhanced the film's adhesion and corrosion protection. A scheme and description of the phenomenon of the physical barrier and passivation are presented in Figure 15.10.

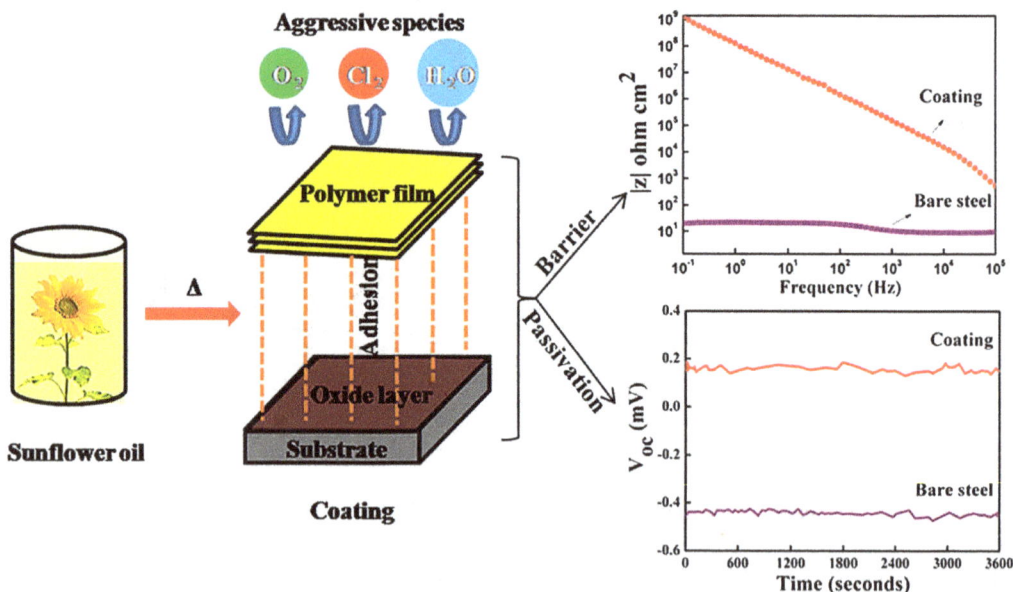

FIGURE 15.10   Schematic of the thermal treatment of sunflower oil followed by the Bode (upper) and OCP (downer) plots representing the barrier and passivation effects. Adapted with permission [11]. Copyright (2015), American Chemical Society.

FIGURE 15.11 Empirical aspects of the corrosion after performing the NSST for 720 h for the AEJO-based coatings with concentrations of ZnO of 0, 1, 3, 5, 7, and 9 wt.%. Adapted with permission [30]. Copyright (2020), American Chemical Society.

3.3.2 Jatropha Oil-Based Coating

Aung *et al.* [30] synthesized a binder originating from a Jatropha oil-based epoxy acrylate (AEJO). After that, ZnO nanoparticles in different concentrations were incorporated to add anti-corrosion protection properties. After the incorporation of 5 wt.% of ZnO, there was an increase in OCP from –0.508 to 0.159 V comparing the neat AEJO against AEJO with 5 wt.% ZnO (AEJO-5 wt.% ZnO). Also, pull-off adhesion increased around 45.7 to 133 psi when comparing the AEJO and AEJO-5 wt.% ZnO. Lastly, NSST was performed to analyze the empirical effect of corrosion in the coatings. Through that, it could be observed that AEJO-5 wt.% ZnO presented the least degraded surface, which could be due to the proper dispersion and barrier effect when compared with the lower concentration which would not hinder the corrosive agents. On the other hand, larger concentrations prevented the cross-linking of Jatropha oil, which would lead to a less compact structure and therefore facilitated the permeation of corrosive agents. The NSST test for the AEJO-based coatings is illustrated in Figure 15.11.

3.3.3 Lignin-Based Coating

Lignin is a macromolecule that can be found in most plants. It is a widely available bio-renewable resource as it is one of the byproducts of the papermaking and pulp industry. Hence, due to its abundance and low cost, there has been a push to find some applications for this material. Some of these may include its use in adhesives, surfactants, and fillers for polymeric materials, for instance.

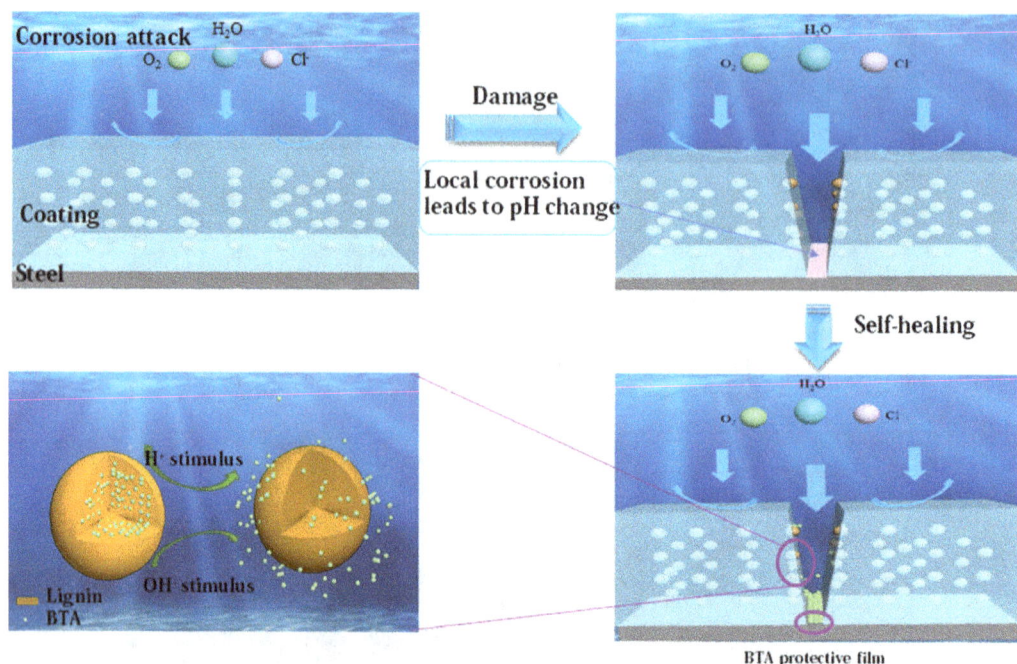

FIGURE 15.12   Scheme for the self-healing mechanism of LMS@BTA-WEP. Adapted with permission [31]. Copyright (2020), American Chemical Society.

One of the recent attempts to find uses for lignin was based on using it as an encapsulation agent of corrosion inhibitors to fabricate a self-healing coating with anti-corrosion properties. In addition, WEP is another eco-friendly coating matrix that can be employed. However, the evaporation of water from the coating can lead to the formation of microcracks and micropores, which would facilitate the permeation of corrosive species when using the plain coating. Tan *et al.* [31] fabricated lignin microspheres (LMS) that were loaded with benzotriazole (BTA). Aside from corrosion inhibition, BTA also provides self-healing properties and activation based on the pH of the environment. Then the LMS@BTA was mixed with WEP to fabricate the nanocomposite coating LMS@BTA-WEP for steel protection. Based on that, when the metal surface was oxidized it caused the release of H^+ which triggered the release of BTA from the LMS. Then, LMS could adsorb on the metal surface and promote its reduction, forming a protective film as well as partially reestablishing the reduced metal. This self-healing mechanism is shown in Figure 15.12.

4   CONCLUSION AND OUTLOOK

Corrosion is a recurrent concern that causes a considerable impact on most countries' economies, creating a strong demand for quick and effective solutions. It has been noted that the scientific community has made considerable progress in the prevention of corrosion in metallic surfaces. In that sense, this new generation of coatings for protection against corrosion can be produced with neither large quantities of VOCs nor Cr-based materials which are inherently toxic and hazardous to the environment. Based on the examples provided in this chapter it has been observed that conducting polymers such as PANI, PPy, and PVK, for instance, can greatly enhance anti-corrosion properties. Alongside that, carbon-based materials such as graphene, carbon nanotubes, and even MXenes are gaining attention in this area. Combining these materials with nanoparticles of metal oxides or corrosion inhibitor agents can yield polymeric nanocomposites with high performance against corrosion. Thereby, a two-fold protection mechanism can be introduced which is based on the barrier effect and formation of a passive layer. In addition, another important aspect is that bio-based materials such as vegetable oils and lignin can also be used as promising materials for the fabrication of protective coatings. Such materials can add environmental credentials, low cost, and competitive properties, which makes them highly attractive for further research. Despite this progress, there remains the need to address some issues such as fabricating coats with a smooth application over a metallic surface, proper adhesion, even distribution of corrosion inhibition agents, and utilizing eco-friendlier materials that can match the performance of commercial products. Based on these factors, this chapter provides novel ideas for readers to tackle the current challenges regarding corrosion on metallic surfaces with polymeric nanocomposites.

REFERENCES

[1] S.K. Kumar, B.C. Benicewicz, R.A. Vaia, K.I. Winey, 50th Anniversary perspective: Are polymer nanocomposites practical for applications?, Macromolecules. 50 (2017) 714–731.

[2] H. Fischer, Polymer nanocomposites: From fundamental research to specific applications, Mater. Sci. Eng. C. 23 (2003) 763–772.

[3] S. Jafarzadeh, E. Thormann, T. Rönnevall, A. Adhikari, P.E. Sundell, J. Pan, P.M. Claesson, Toward homogeneous nanostructured polyaniline/resin blends, ACS Appl. Mater. Interfaces. 3 (2011) 1681–1691.

[4] A. Behera, P. Mallick, S.S. Mohapatra, Nanocoatings for anticorrosion, Corros. Prot. Nanoscale. (2020) 227–243.

[5] E.M. Fayyad, K.K. Sadasivuni, D. Ponnamma, M.A.A. Al-Maadeed, Oleic acid-grafted chitosan/graphene oxide composite coating for corrosion protection of carbon steel, Carbohydr. Polym. 151 (2016) 871–878.

[6] A.M. Vaysburd, P.H. Emmons, How to make today's repairs durable for tomorrow – corrosion protection in concrete repair, Constr. Build. Mater. 14 (2000) 189–197.

[7] F. Chen, P. Liu, Conducting polyaniline nanoparticles and their dispersion for waterborne corrosion protection coatings, ACS Appl. Mater. Interfaces. 3 (2011) 2694–2702.

[8] G. Liu, M. Xue, H. Yang, Polyether copolymer as an environmentally friendly scale and corrosion inhibitor in seawater, Desalination. 419 (2017) 133–140.

[9] N.A. Odewunmi, M.M. Solomon, S.A. Umoren, S.A. Ali, Comparative studies of the corrosion inhibition efficacy of a dicationic monomer and its polymer against API X60 steel corrosion in simulated acidizing fluid under static and hydrodynamic conditions, ACS Omega. 5 (2020) 27057–27071.

[10] F. Ubaid, A.B. Radwan, N. Naeem, R.A. Shakoor, Z. Ahmad, M.F. Montemor, R. Kahraman, A.M. Abdullah, A. Soliman, Multifunctional self-healing polymeric nanocomposite coatings for corrosion inhibition of steel, Surf. Coatings Technol. 372 (2019) 121–133.

[11] T. Balakrishnan, S. Sathiyanarayanan, S. Mayavan, Advanced anticorrosion coating materials derived from sunflower oil with bifunctional properties, ACS Appl. Mater. Interfaces. 7 (2015) 19781–19788.

[12] T. Ohtsuka, Corrosion protection of steels by conducting polymer coating, Int. J. Corros. 2012 (2012) 915090.

[13] P.A. Sørensen, S. Kiil, K. Dam-Johansen, C.E. Weinell, Anticorrosive coatings: a review, J. Coatings Technol. Res. 6 (2009) 135–176.

[14] D.E. Tallman, G. Spinks, A. Dominis, G.G. Wallace, Electroactive conducting polymers for corrosion control: Part 1. General introduction and a review of non-ferrous metals, J. Solid State Electrochem. 6 (2002) 73–84.

[15] G.M. Spinks, A.J. Dominis, G.G. Wallace, D.E. Tallman, Electroactive conducting polymers for corrosion control: Part 2. Ferrous metals, J. Solid State Electrochem. 6 (2002) 85–100.

[16] H. Chen, H. Fan, N. Su, R. Hong, X. Lu, Highly hydrophobic polyaniline nanoparticles for anti-corrosion epoxy coatings, Chem. Eng. J. 420 (2021) 130540.

[17] P.J. Kinlen, Y. Ding, D.C. Silverman, Corrosion protection of mild steel using sulfonic and phosphonic acid-doped polyanilines, Corrosion. 58 (2002) 490–497.

[18] I. Šeděnková, J. Prokeš, M. Trchová, J. Stejskal, Conformational transition in polyaniline films – Spectroscopic and conductivity studies of ageing, Polym. Degrad. Stab. 93 (2008) 428–435.

[19] S.R. Moraes, D. Huerta-Vilca, A.J. Motheo, Corrosion protection of stainless steel by polyaniline electrosynthesized from phosphate buffer solutions, Prog. Org. Coatings. 48 (2003) 28–33.

[20] Y.Y. Wang, H.F. Zhang, D.H. Wang, N. Sheng, G.G. Zhang, L. Yin, J.Q. Sha, Development of a uricase-free colorimetric biosensor for uric acid based on PPy-coated polyoxometalate-encapsulated fourfold helical metal-organic frameworks, ACS Biomater. Sci. Eng. 6 (2020) 1438–1448.

[21] S. Zhang, Y. Jiang, H. Bai, J. Yang, Cable-like β-AgVO3@PPy nanowires as novel anode materials for lithium-ion batteries, J. Phys. Chem. C. 124 (2020) 19467–19475.

[22] S. Qiu, W. Li, W. Zheng, H. Zhao, L. Wang, Synergistic effect of polypyrrole-intercalated graphene for enhanced corrosion protection of aqueous coating in 3.5% NaCl solution, ACS Appl. Mater. Interfaces. 9 (2017) 34294–34304.

[23] N. Elangovan, A. Srinivasan, S. Pugalmani, N. Rajendiran, N. Rajendran, Development of poly(vinylcarbazole)/alumina nanocomposite coatings for corrosion protection of 316L stainless steel in 3.5% NaCl medium, J. Appl. Polym. Sci. 134 (2017) 44937.

[24] E. Mitchell, J. Candler, F. De Souza, R.K. Gupta, B.K. Gupta, L.F. Dong, High performance supercapacitor based on multilayer of polyaniline and graphene oxide, Synth. Met. 199 (2015) 214–218.

[25] F.M. de Souza, R.K. Gupta, P.K. Kahol, Recent development on flame retardants for polyurethanes, in: R.K. Gupta, P.K. Kahol (Eds.), Polyurethane Chem. Renew. Polyols Isocyanates, American Chemical Society, Washington, 2021: pp. 187–223.

[26] W. Deng, Q. Fang, H. Huang, X. Zhou, J. Ma, Z. Liu, Oriented arrangement: The origin of versatility for porous graphene materials, Small. 13 (2017) 1701231.

[27] A. Ghasemi-Kahrizsangi, H. Shariatpanahi, J. Neshati, E. Akbarinezhad, Degradation of modified carbon black/epoxy nanocomposite coatings under ultraviolet exposure, Appl. Surf. Sci. 353 (2015) 530–539.

[28] Z. Chen, G. Zhang, W. Yang, B. Xu, Y. Chen, X. Yin, Y. Liu, Superior conducting polypyrrole anti-corrosion coating containing functionalized carbon powders for 304 stainless steel bipolar plates in proton exchange membrane fuel cells, Chem. Eng. J. 393 (2020) 124675.

[29] H. Zhao, J. Ding, M. Zhou, H. Yu, Air-stable titanium carbide MXene nanosheets for corrosion protection, ACS Appl. Nano Mater. 4 (2021) 3075–3086.

[30] M.M. Aung, W.J. Li, H.N. Lim, Improvement of anticorrosion coating properties in bio-based polymer epoxy acrylate incorporated with nano zinc oxide particles, Ind. Eng. Chem. Res. 59 (2020) 1753–1763.

[31] Z. Tan, S. Wang, Z. Hu, W. Chen, Z. Qu, C. Xu, Q. Zhang, K. Wu, J. Shi, M. Lu, PH-responsive self-healing anticorrosion coating based on a lignin microsphere encapsulating inhibitor, Ind. Eng. Chem. Res. 59 (2020) 2657–2666.

16 Polymers for Smart Coatings

Alireza Fatahi,[1] Abbas Mohammadi,[1] and Mohamad Reza Sarfjoo[2]

[1] Department of Chemistry, University of Isfahan, Isfahan 81746-73441, I.R. Iran
[2] Department of Chemistry, Isfahan University of Technology, Isfahan 415683111, I.R. Iran

1 INTRODUCTION

Today, due to a decrease in volatile organic contents (VOC), reducing the use of expensive oil-based solvents, and improving energy efficiency both from an economic and environmental standpoint, the demand for polymer coatings has shifted toward smart coatings. Smart coating technology is also driven by a combination of demands for improved performance, longer product life, and a reduction in maintenance costs [1].

These coatings can be responsive to stimuli and have expanded to many applications due to their characteristics in various industries such as textile, military, medical, building, transport, electronics, and aviation industries for different applications such as self-cleaning, antifouling, antimicrobial, antifogging, antifingerprint, anti-icing, and fire-retardant coatings [2]. A smart coating detects changes in its environment, interacts with them, and responds to them while keeping its compositional integrity. Respondent polymers can change their surface and color properties in response to a signal, in some cases releasing encapsulated agents for various applications. Moreover, they can expand, shrink, bend, and even degrade in response to different factors including light, pH, biological factors, pressure, temperature, polarity, etc. Smart coatings may be categorized based on the type of functional components (such as nanomaterials, enzymes, pigments, and different resins), manufacturing methods, stimulus, function, and application [3].

The purpose of this chapter is to provide a classification of smart coatings based on their application and the type of response to stimuli. The properties, additives, or agents required to create each of these coatings in specific applications are then described. In addition, the methods of preparing these smart coatings and examples from studies have been elaborated. In addition, some recent advances in smart coatings for different applications are also reported.

2 EASY-TO-CLEAN COATINGS

2.1 SELF-CLEANING COATINGS

Self-cleaning coatings offer many advantages, such as lower maintenance costs, increased durability, no snow or ice adhesion, as well as pollution protection. Self-cleaning coatings can be widely used in windows, solar panels, cement, and paint. Depending on the surface contact angles with water droplets, self-cleaning coatings are categorized as hydrophilic or hydrophobic. With exposure to water, both hydrophilic and hydrophobic self-cleaning coatings can clean themselves. A range of applications can be achieved with hydrophobic coatings, including self-cleaning fabrics, stain-resistant fabrics, and anticorrosion coatings. The most important characteristic of a self-cleaning coating is the water contact angle (WCA) of more than 90°, while for a superhydrophobic coating,

DOI: 10.1201/9781003278269-16

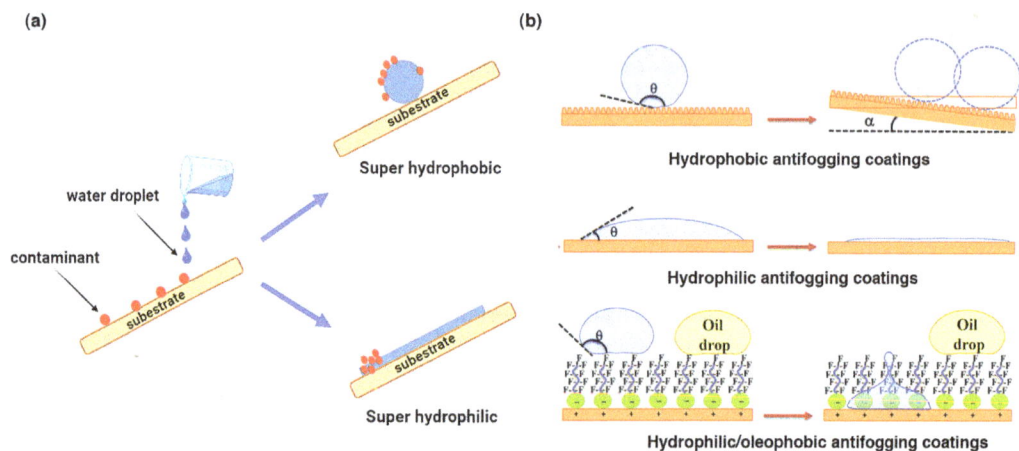

FIGURE 16.1 (a) Self-cleaning mechanism of hydrophobic and hydrophilic coatings. (b) classification of antifogging coatings based on wetting properties of the surface and their antifogging mechanism. Adapted with permission from Reference [11]. Copyright (2019), Elsevier.

the water contact angle is more than 150°. As shown in Figure 16.1a, in hydrophobic self-cleaning coatings, the high contact angle value of the water droplet with the surface allows the water quickly to flow away and eliminate dirt particles from the surface. Regarding the hydrophilic coatings with a water contact angle of less than 90°, water droplets spread on the surface and clean the surfaces by collecting the soil (Figure 16.1a). It is also called "active cleaning" because the water spreads on the surface and picks up the soil. If the water contact angle is below 10°, it has a superhydrophilic property [4].

Self-cleaning surfaces can be elucidated using two distinct approaches. The first approach involves the application of a photocatalytic coating to the substrate surface, in which the effect of the sun's ultraviolet rays catalytically breaks down organic dirt. In recent studies on coatings and photocatalysts, titania has been found to have many applications, some of which are discussed below. A topic of interest in transparent materials such as glass is the possibility of self-cleaning, which can be accomplished with titanium (TiO_2) to eliminate bacteria and organic matter. In addition, titanium has photocatalytic activity and can decompose bacteria and organic contaminants from the surface under ultraviolet (UV) illumination [5].

Functionalization of surface structures with fluorinated alkyl and fluorinated alkyl silanes is another method to develop self-cleaning materials. Another example of making a superhydrophobic surface is the combination of TiO_2 and polytetrafluoroethylene (PTFE) that is applied by radio-magnetron frequency (RF-MS) sputtering deposition. In addition, this procedure suppresses the occurrence of surface wetting changes to a hydrophilic state and achieves sufficient photocatalytic activity for self-cleaning [6].

In another study, a hydrophobic coating was prepared based on tetraethoxysilane (TEOS) and different molar ratios of methyltrimethoxysilane (MTMS) as a co-precursor. It was observed that the contact angle gradually increased from 30 ° (in a molar ratio of 0) to 135 ° (in a molar ratio of 1.57) [7].

2.2 ANTIFINGERPRINT COATINGS

Antifingerprint coatings are another type of easy-to-clean coating that can reduce dust, oil, water, or fingerprints and improve the overall optical performance. These coatings have a variety of applications in optical lenses as well as screens on smartphones, televisions, and other devices.

It has been found that both parameters of surface energy and surface roughness have a key role in the development of antifingerprint coatings with water and oil repelling. Low surface energy and roughness reduce intermolecular attraction and adhesion, and thereby decrease the impact of fingerprints. Fluorinated components are commonly used to make antifingerprint coatings due to their low surface energy. However, coatings containing only fluorinated components have problems such as toxicity, poor adhesion, and low durability [8]. Synthesized fluoroalkyl silane copolymer contains SiO_2 nanoparticles that can be used as silane compounds to improve adhesion, abrasion resistance, and hardness. Increasing the content of SiO_2 nanoparticles increase the surface roughness and contact angle. In addition, nanoparticles usually can improve the adhesion and hardness of coatings [9].

In response to fluorinated components' toxicity, non-fluorinated compounds have been developed to coat glass surfaces with self-assembling monolayers of organosilanes with different functionalities. The result showed that methyl-terminated organosiloxane can provide amphiphilic and antifingerprint properties [10].

2.3 ANTIFOGGING COATINGS

Fogging occurs when water droplets accumulate on surfaces with a temperature below the dew point of the environment. Surface fogging has adverse effects on many applications such as eyeglasses, windshields, and agricultural films, and consequently reduces the efficiency of solar cells, medical instruments, and the visual appearance of food industry packaging. Two methods have been reported for preparing antifogging coatings. The first method involves the use of stimulus-responsive coatings, such as electrothermal coatings, which increase the surface temperature by changing the voltage. In the second method, the surface structure and morphology are changed by chemical modification and surface roughness, respectively. According to the wetting properties of the surface, they can be classified into hydrophilic, hydrophobic, or hydrophilic/oleophobic surfaces, as demonstrated in Figure 16.1b. Hydrophilic surfaces (with a contact angle of less than 40–50°) spread water droplets on the surface and allow light to pass through without scattering, even under powerful fogging conditions. While hydrophobic surfaces need to be tilted to prevent condensation, they are less commonly used for antifogging surfaces than for hydrophilic surfaces. Hydrophilic/oleophobic coatings are made from fluorosurfactant polymers. In these coatings, fluorinated chains are placed outwards and the hydrophilic part is placed inwards and repels water and oil droplets. These coatings are suitable for cases that require self-cleaning and antifogging [11].

2.4 ANTI-ICING COATINGS

Since ice particles accumulate on the surfaces of roads, aircraft, power lines, offshore oil platforms, wind turbines, and ships in cold regions, traditional methods such as thermal and mechanical methods have been used to remove them, which were not suitable due to energy consumption and environmental pollution. In this regard, smart anti-icing coatings were introduced to overcome this existing problem [12]. At first, it was believed that superhydrophobic coatings were suitable for preparing anti-icing coatings, but these coatings could not be used in high humidity conditions and their surface structure is destroyed over time and during icing and de-icing cycles. Smart anti-icing coatings reduce the formation of ice and ice adhesion on surfaces by responding to environmental stimuli. This coating can be classified as thermo-responsive, electrosensitive, magneto-sensitive, electromechanical, self-healing, phase-change, and self-lubricating coatings based on the type of stimuli and response. Thermo-responsive and phase-change types cause anti-icing action by converting stimuli to heat; in electromechanical smart anti-icing coating, the mechanical pulses reduce ice adhesion to the surface. In the self-lubricating coating, a slippery liquid is infused into porous surfaces to obtain an anti-icing application. Oil (perfluorinated, silicon oil) and an aqueous lubricating layer can be used as infused liquids to reduce ice nucleation sites. Although these coatings provide good anti-icing performance, evaporation or exit of infused liquids reduces their performance and the need for re-infused liquids. Generally, low surface energy coatings reduce the ice adhesion to surfaces by reducing the molecular interactions of water and surfaces. Silicon, fluoride, and fluorosilicone-based polymer coatings are described in the literature for anti-icing applications due to their low surface energy [13].

3 SELF-HEALING COATING

Self-healing coatings possess the capability to repair the damage, either by themselves or with some external stimulation (such as pH change, metal ions, heat, photo, and/or electrochemical stimulation). A self-healing material has a long service life and lowers the cost of repairs. It can be used for various coatings, and even for tissue engineering. There are two main categories of these materials based on the manufacturing method: extrinsic (capsule-based, vascular) and intrinsic [14].

3.1 EXTRINSIC SELF-HEALING COATINGS

Many self-healing coatings are produced by the addition of a healing agent to existing materials. This is defined as "extrinsic" healing. This type of coating can sense cracks, and repair is initiated when microcapsules or vascular networks contained in the coating rupture and the healing agent are released [14].

3.1.1 Capsule-Based Self-Healing Coatings

Capsules consist of two parts, the core (10–90%) that can contain a solid, liquid, or gas and a layer. The capsules have a diameter of 3–800 µm. Capsules can be prepared in different ways such as coacervation, extrusion, sol–gel, interfacial polymerization, and in situ polymerization. Creating self-healing coatings can be achieved by dispersing the healing micro/nanocapsules within a uniform distribution and trapping them within the coating matrix. Microcapsules can also be sensitive to environmental stimuli such as temperature, pressure, and pH, and can rupture as a result of these stimuli. As shown in Figure 16.2a, healing agents are usually accumulated in a microcapsule and when the coating is broken or torn, the healing agents release into the matrix and a catalyst accelerates the polymerization of the healing agent at the damaged surface. When polymerization

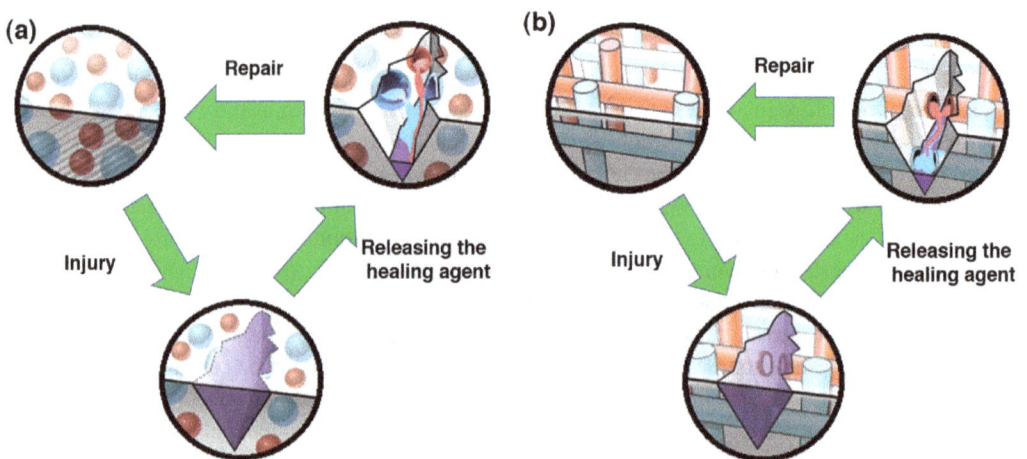

FIGURE 16.2 (a) The capsule-based healing mechanism and (b) the vascular self-healing mechanism.

occurs, it causes a spontaneous reaction leading to the repair of the crack, and the crack faces are closed at the end of the process [15].

Besides healing agents, other agents (such as corrosion inhibitors) can also be loaded into capsules for various applications. Self-healing coatings with anticorrosion activity are suitable to apply on surfaces of zinc, galvanized steel, and aluminum alloys. Urea–formaldehyde microcapsules containing $NaNO_3$ as a corrosion inhibitor and linseed oil as a healing agent have been prepared. These prepared microcapsules were loaded into halloysite nanotubes and dispersed in the polymer matrix. The results showed that these coatings are responsive to pH changes and release healing and anticorrosion agents [16].

3.1.2 Vascular and Hollow Fibers

The use of thermosetting polymers and polymer composites has seen a significant increase over the past several decades. The development of a self-healing vascular network increases the durability and reliability of thermosets. This technology offers other advantages as well as the capability of refilling the healing agent indefinitely. It can also be used to repair various types of injuries. The points that are mentioned above can overcome the current limitations of capsule-based self-healing systems [17]. Initially, the system was intended to be used in other areas, such as the repair of concrete cracks and the restoration of mechanical properties, but it was used later for composite polymeric materials. Healing agents are first encapsulated in capillaries (such as nanofibers and hollow channels). The capillaries in a network-like structure are interconnected, which means several local therapeutic events will receive healing agents [18].

When the mentioned networks are damaged, they can be connected in different ways, for example, they can be connected in a one-dimensional, two-dimensional, or even three-dimensional manner. As shown in Figure 16.2b, if the region has been damaged and the healing agent has been delivered, the arteries can be filled with a healthy and connected network area or an external source of a healing agent in preparation for the next possible network damage. Multiple local healing events can be triggered by this refilling action [18].

3.2 INTRINSIC SELF-HEALING COATING

In the intrinsic self-healing process, the polymer itself heals molecular and macroscopic cracks. Intrinsic self-healing systems can also be classified based on their underlying type of healing mechanism. The two main categories are physical interaction and chemical bonding. Noncovalent chemistry includes melting of thermoplastic phases, supramolecular interactions, and metal–ion binding. The typical dynamic bonds responsible for intrinsic self-healing include the Diels–Alder bond, disulfide exchange, and ester exchange [14].

3.2.1 Self-Healing by Dispersed Thermoplastic Polymers

Self-healing thermoset materials can be prepared by the dispersion of thermoplastic polymers in their matrix. Self-healing occurs when thermoplastic materials are melted, then dispersed in the cracks. The amount of melted material increases, subsequently filling the cracks, and eventually, it can be seen that these matrix materials are mechanically intertwined and locked together [14]. Poly(ε-caprolactone) (PCL) can be dispersed in the epoxy matrix as thermoplastic polymers to prepare a self-healing and shape-memory coating. To repair cracks, melting of thermoplastic phases and flow of PCL fibers are used [19].

3.2.2 Supramolecular Self-Healing Materials

Supramolecular polymers possess dynamic properties resulting from non-covalent bonds, which has drawn considerable attention to them. They can also be used to highlight the highly dynamic nature of hydrogen bonds and how they respond to external stimuli within supramolecular polymers, along

(a)　　　　　　　　　　　　　　　　　　　**(b)**

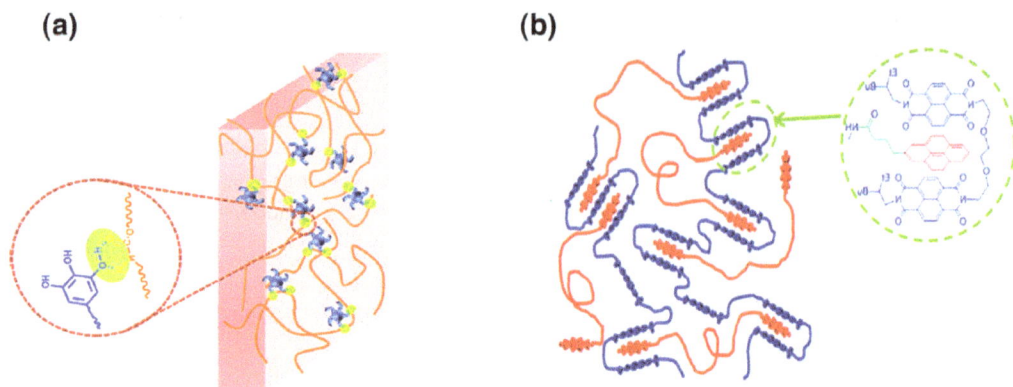

FIGURE 16.3　Self-healing polyurethane coating is based on (a) hydrogen bonds (Adapted with permission from Reference [20]. Copyright (2021), Elsevier) and (b) π–π stacking.

with their ability to be adjusted and guided. The host–guest interactions and π–π interactions are other non-covalent bonds found in these materials. Self-healing polymers such as polyurethane (PU) offer super-molecular polymer structure and hydrogen bonds. Based on hydrogen bonds between tannic acid (TA) and linear polyurethane domains, cationic self-healing polyurethanes have been developed (Figure 16.3a) [20].

The use of supramolecular interactions (π–π stacking and hydrogen bonding) was used to produce an elastomeric and tough polyurethane between chain folding polyamide and telechelic polyurethane with the end group of pyrenyl (Figure 16.3b) [21].

3.2.3   Host–Guest Chemistry

A method to design self-healing materials has been considered using host–guest chemistry in recent years. Cyclodextrins contain specific guest groups such as azobenzene, adamantane, and ferrocene, which have high selectivity and participate in host–guest chemical reactions through reversible non-covalent interactions. Host–guest chemistry has been used to synthesize intrinsic self-healing epoxy materials. The beta-cyclodextrin/graphene complex is attached to the unsaturated epoxy resin by free-radical copolymerization. Throughout the dynamic host–guest interaction, the complex mentioned can reestablish broken bonds caused by damage through a photothermal agent and macro-cross-linking agent [22].

3.2.4   Disulfide Bonds

Disulfide bonds have been used for self-healing studies recently because disulfide exchange reactions can occur at low temperatures. Disulfide bonds are weaker than carbon bonds, which results in opening disulfide bonds with less energy. The disulfide opening reaction is performed under heat, mechanical, light (UV), and redox conditions. As shown in Figure 16.4a, a disulfide bond can be exchanged with another disulfide bond on the cut surface to form a new bond "in the mobile phase." As a consequence of disulfide breaking, thiyl radicals present can be expected to exchange rapidly with other disulfide bonds [23].

To prepare the multifunctioning smart coating, disulfide bonds have been shown in research to be used for creating hydrophobic and self-healing coatings. To achieve this task, polyurethane with disulfide bonds and acrylic end groups was prepared. Modified Al_2O_3 nanoparticles with hydrophobic groups were also used to prepare a hydrophobic coating. These coatings can heal cracks in the presence of heat and UV light, and the results also showed that increasing temperature and UV light irradiation increases the speed of repair [24].

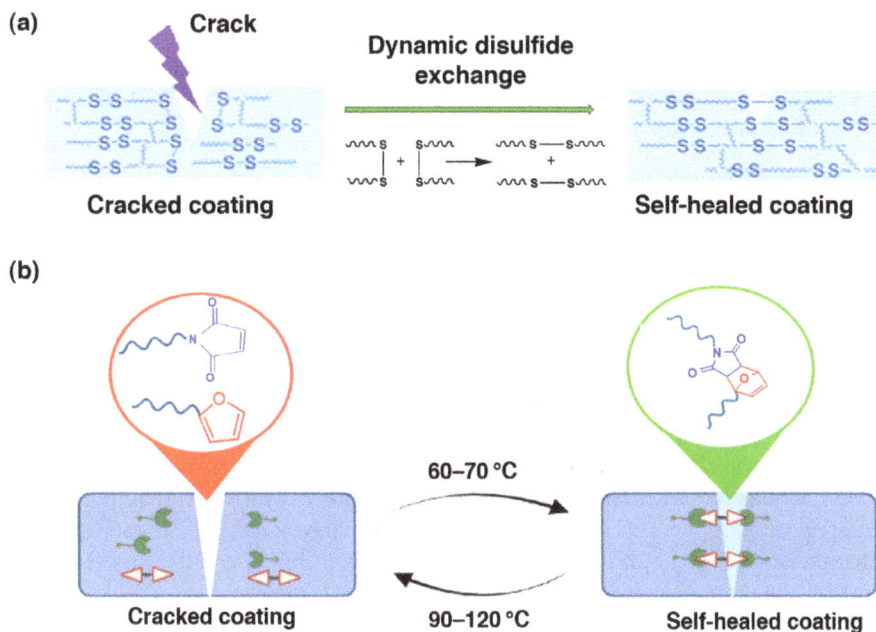

FIGURE 16.4 Self-healing mechanism based on (a) disulfide exchange reactions (Adapted with permission from Reference [28]. Copyright (2016), John Wiley and Sons). (b) Diels-Alder reaction.

3.2.5 Diels–Alder Reactions

Diels–Alder reactions have received attention in recent studies because of the necessity of no byproducts or catalysts during the process of preparing intrinsic self-healing polymers. In the Diels–Alder (DA) reaction, a six-membered ring is formed by the (4+2) cycloaddition between an electron-rich diene (furan) with an electron-deficient dienophile (maleimide). The DA cycloaddition can be accomplished at or slightly above room temperature, while the DA reaction is performed at elevated temperatures (60–70 °C). The retro Diels–Alder reaction and the opening of the six-membered rings take place in the temperature range of 90–120 °C. Due to the properties and thermal reversibility of Diels–Alder reactions, they can be used in self-healing applications in the presence of furan and maleimide groups in polymer chains [25]. The healing mechanism based on Diels–Alder bonds is shown in Figure 16.4b, first, with the opening of the Diels–Alder bonds (retro Diels–Alder reaction) at a temperature of 90–120 °C, small chains have formed that move and fill the crack, then Diels–Alder bonds are formed at 60–70 °C and restore the mechanical properties.

However, by using the Diels–Alder reaction, it is possible to produce biocompatible bio-based hydrogels with enhanced performance such as stimulus-responsivity, self-healing property, and the in situ forming ability for a variety of applications. Diels–Alder reactions were used to prepare natural hydrogels which had a 78% recovery rate by using furan-modified cellulose and polyethylene glycol with maleimide end groups [26]. Fortunato et al. prepared a self-healing coating based on the Diels–Alder reaction from methacrylate with furan pendant groups and aliphatic bismaleimides. Due to the presence of DA bonds, these coatings showed thermal reversibility and the ability to repair mechanical damage. In addition, these coatings had good transparency and adhesion, which makes them suitable for optical applications [27].

4   SMART CHROMIC COATINGS

Stimuli-chromism is a new phenomenon that occurs when smart materials change color under external stimuli, such as light, temperature, pH, mechanical stress, polarity, electrical potential, and metal ion coordination, which is the cause of reversible color change. It is more important to use light-responsive materials than other materials, as light is fast and clean [29].

4.1 Photochromic Smart Coatings

Materials with photochromic properties are noted for being able to change their chemical structure when exposed to external light, as well as for their reversible color change. A polymer can contain these materials, which can be used for a variety of applications, such as photochromic eyeglasses, windows, and dyes. Photochromic dyes can be used in various applications such as the plastic and cosmetics industries, and printing ink for cloth or paper. Materials with photochromic capabilities include metal oxides (titanium dioxide, zinc oxide, tungsten oxide, silicon dioxide, and stannic oxide) and organic photochromic compounds (azobenzene, spiropyran, spiroxazine, diarylethene, and fulgide) [30].

Spiropyran has been used as an attractive photochromic material in recent studies. Under the stimulus of light, it can change its color and structure between hydrophobic and hydrophilic forms. Styrene-butyl acrylate copolymers containing spiropyran have been fabricated, taking advantage of this property of spiropyran to produce a coating with photoswitchable wettability. The results showed that the contact angle of the coating under UV light (365 nm) differs reversibly from the range of 60–93° (hydrophobic and discolored form) to 55–86° (hydrophilic and colored form) [31].

4.2 Thermochromic Smart Coatings

As another type of stimuli-chromism, thermochromic coatings change color reversibly (transmission/reflection of infrared irradiation) at a certain temperature. Thermochromic materials can be classified into liquid crystals, leuco dyes (consisting of a dye, color developer, and an organic solvent), and pigments. Recent studies have shown that buildings containing thermochromic coatings are very useful for storing energy. These coatings in buildings respond to the reversible color change of the sun, reflecting light in summer and cooling the inside of the building, while absorbing more heat in winter and making the inside of the building warmer [32].

In thermochromic smart coatings, vanadium dioxide (VO_2) is commonly used as a component because of its metal-insulator phase transition (MIT) at 68°C and reversible nature. In addition, its optical properties are also different at high temperatures and below 68°C. At temperatures above 68°C, it reflects NIR light and at below 68°C, it transmits NIR light. However, in thermochromic windows, the use of vanadium is limited due to the reflection of visible light. A two-layer coating of VO_2 and mesoporous SiO_2 has been prepared, in which the SiO_2 layer reduces surface reflection [33].

4.3 Electrochromic Smart Coatings

Electrochromic coatings change color reversibly when their electric fields are changed. A wide range of electronic devices can be made from electrochromic materials, including optical devices, smart windows (for buildings and automobiles), monitors, and sensors. Numerous studies have been performed on the preparation of electrochromic coatings for creating windows that prevent energy loss. Different organic (such as viologens and conductive polymer) and inorganic (transition metal oxides such as WO_3, V_2O_5, TiO_2, Nb_2O_5) materials are used to prepare these coatings, which change their optical properties by changing between the oxidation and reduction states. An inorganic compound called cathode electrochromic is colorless when oxidized and colored when reduced [34]. WO_3 is one of the most widely used cathodic electrochromic materials in studies. A WO_3 coating can be prepared by different methods such as sputtering, sol–gel, spraying, and electron beam evaporation. A WO_3 coating showed limited applications only in the visible range, degradation, and corrosion during application. For this purpose WO_3 coatings are usually used on TiO_2 and SnO_2 nanostructures [35].

5 ANTICORROSION SMART COATINGS

Due to the increasing costs of corrosion in industries (pipes and tubes) and potential risks, anticorrosion coatings have received a lot of attention recently. Smart anticorrosion coatings are more effective than the traditional methods because detection of nano-cracks in the traditional methods is difficult and their corrosion resistance is limited. Meanwhile, in smart anticorrosion coatings, in addition to improving corrosion resistance, they can also release other agents such as healing agents. Anticorrosion coatings prevent the cathodic and anodic reactions of corrosion and reduce the corrosion rate [36]. Corrosion locations are associated with pH and electrochemical changes, therefore it is important to use pH- and electrochemical-responsive materials to identify and prevent corrosion. Corrosion causes acidic pH in the anode region and alkaline pH in the cathode region. Methods of corrosion inhibition include the use of microcapsules, conductive polymers, chromate-rich surface treatments, pigments based on chromates, and surface-modified nanoparticles. As mentioned in the self-healing coatings section, microcapsules have many advantages, including the ability to encapsulate corrosion indicators (such as fluorescent compounds and color dyes), and corrosion inhibitors (such as cerium, benzotriazole, and 8-hydroxyquinoline), and dyes in solid or liquid phases. In addition, they can encapsulate healing agents and multi-action smart coatings that are synthesized. pH changes cause the release of corrosion inhibitors and healing agents from the capsules (Figure 16.5a) [37].

A smart coating has been made up of microcapsules containing melamine–formaldehyde and cerium nitrate as cathodic inhibitors that were synthesized and incorporated into epoxy resins. In an alkali pH, these microcapsules are broken and release cerium ions. The results showed that in addition to preventing corrosion, they also repair damaged areas [38]. To identify the sites of corrosion in the early stages, fluorescent compounds and color dyes can be used as corrosion indicators, which interact with corrosion products and cause discoloration or fluorescent light. Hydroxyquinoleine (as a corrosion indicator) and cerium acetate (as a corrosion inhibitor) were added to melamine–formaldehyde microcapsules to identify corrosion sites. 8-Hydroxyquinoleine forms chelates with Fe^{2+} or Fe^{3+} ions, which produce fluorescent light at corrosion sites [39].

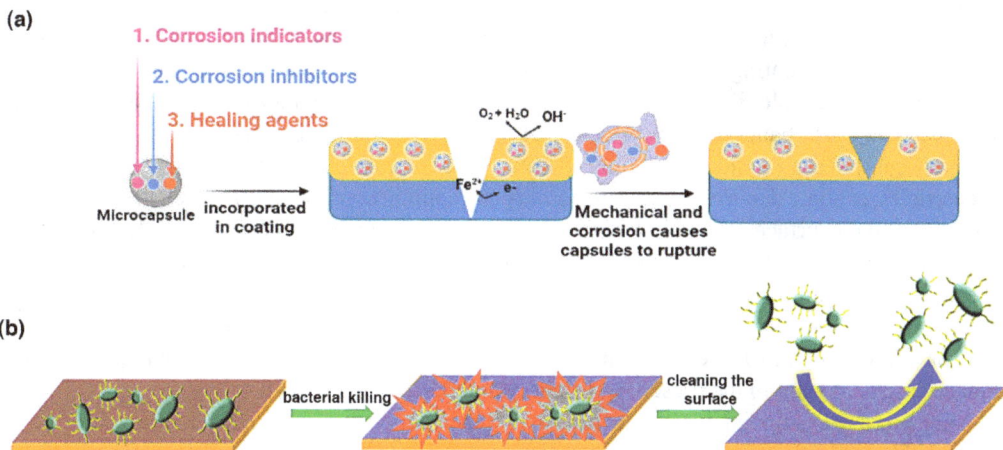

FIGURE 16.5 (a) Methods of corrosion inhibition based on microcapsules with self-healing ability. (b) Antibacterial mechanism of smart coatings. Reproduced with permission from Reference [14].

Using superhydrophobic surfaces with self-healing ability is another way to create smart anticorrosion coatings. A two-layer coating of shape-memory epoxy-containing benzotriazole (to inhibit corrosion) and hydrophobic polydimethylsiloxane was used to achieve this. The healing process of this smart coating was performed according to the shape memory of the epoxy layer during thermal treatment or exposed to sunlight. In addition, benzotriazole acts as an anticorrosion agent, and also, the results showed that the presence of benzotriazole improves healing efficiency [40].

6 BIOACTIVE SMART COATINGS

6.1 ANTIFOULING COATINGS

Fouling can be categorized into two types: inorganic fouling and biofouling. Inorganic fouling is due to the absorption and accumulation of non-living particles such as dust, ice, corrosion products, crystals, and suspended inorganic particles on the surface. Inorganic fouling has many negative effects in many applications, such as fingerprint and power generation industries. It is the absorption, agglomeration, and growth of microbes on wet surfaces that cause biofouling, which has the greatest impact on the maritime and shipping industries due to an increase in submarine fuel consumption and the blocking of submarine pipelines. Smart antifouling coatings have been introduced to reduce the effects of fouling. Smart antifouling coatings can be prepared from super-hydrophobic coatings, self-polishing coatings, and fouling-release coatings based on coatings containing fluorine and silicones [41]. A smart antifouling coating based on silicone prevents the absorption of microbes due to their low surface energy and low surface roughness. Antifouling silicone-based polyurea coatings were synthesized with fluorinated nanodiamond. Fluorinated nanodiamonds increase the mechanical properties and reduce surface energy. In addition, hydrogen bonds between urea chains caused the self-healing properties of these coatings at room temperature [42].

6.2 ANTIMICROBIAL COATINGS

Health risks arise from the growth of bacteria in applications such as the medical field, hospital equipment, implants, food packaging, and drinking water pipes. Initially, traditional antibacterial coatings were used to prevent the adsorption of bacteria on the surfaces of coatings. The use of these coatings was limited due to the release of antibacterial agents even in the absence of bacteria and the accumulation of dead bacteria on the surface of these coatings. Due to the problems of traditional coatings, smart antimicrobial coatings were introduced with the ability to respond to stimuli and gradually release antibacterial agents and clean the surface, as demonstrated in Figure 16.5b. Smart antimicrobial coatings are classified into two categories based on the type of stimuli: coatings that respond to non-biological stimuli such as light, heat, magnetism, redox changes, and biological stimulus due to pH changes in the presence of bacteria. Smart antibacterial coatings in response to stimuli can release biocides including antimicrobial peptides, hydrogen peroxide, metal oxide nanoparticles, cationic dendrimers, and quaternary ammonium compounds or switch the antibacterial properties of their surface [43]. Additives used to prepare antibacterial coatings include antibacterial nanomaterials (such as TiO_2, ZnO, MgO, CuO, Al_2O_3, Au, and CNTs), metal cations (such as silver, copper, lead, cadmium, zinc, and tin), photothermal agents (such as polyaniline, polydopamine(, and antibiotics. New nanomaterials, called quantum dots, which are light-responsive and antibacterial, have also been introduced to smart antibacterial coatings. Hydrophobic carbon quantum dots produce singlet oxygen in the presence of light, which kills bacteria [44].

Using phase-transitioned lysozyme film as a "sacrificial" layer, a smart antibacterial hybrid coating with gold nanoparticles and near-infrared laser-induced destruction of bacteria could be prepared. Gold nanoparticles kill bacteria by absorbing near-infrared laser irradiation and converting it to heat. Then a phase-transitioned lysozyme film was destroyed by storage in a vitamin C solution, the dead bacteria on the surface were removed, and the surface regenerated [45].

7 ANTIREFLECTIVE SMART COATINGS

The term antireflective coating is used to describe an application of a thin film to an optical device in order to reduce the glare or images that are produced. Such coatings are used in many applications including glasses, lasers, windscreens, and lenses. Antireflective materials include silicon, TiO_2, gallium, carbon, organic, and polymer-based materials. Antireflective coatings can be classified into two categories. In the first category, they are divided into single-layer, double-layer, and multi-layer based on the layer composition [46]. TiO_2/SiO_2 multilayer films have been used as antireflective coatings in solar panel applications. TiO_2 films have some properties such as semiconducting and a high index of refraction, while SiO_2 films have shown a low refractive index and good transparency [47]. In another category, antireflective coatings are classified based on surface properties to (1) porous antireflective coatings, it was shown that water-treated aluminum oxide nanoparticles can be used as porous nanoparticles in the preparation of antireflective coatings; (2) antireflective coatings based on biomimetic photonic nanostructures such as butterfly eyes; and (3) textured surface antireflective coatings. These coatings reduce light reflection by trapping light and internal reflections. Recent studies have reported honeycomb structures with the ability to trap light. (4) Antireflection grating creates antireflective properties over a large range of wavelengths (even up to the terahertz) based on the formation of a continuous gradient of refractive index [48,49]. Recently, antireflective coatings have been prepared from polydimethylsiloxane by etching Cu on the surface. Through the enhancement of the lateral surface area and aspect ratio of micro/nanostructures, etching Cu on a surface improved optical performance in the visible light region [50].

8 FIRE-RETARDANT SMART COATINGS

Fire-retardant smart coatings protect the organic surface of a material from fire without changing the chemical structure of the internal layers. Coatings of this type should have properties such as minimizing toxic gases, adhering to surfaces well, and resisting abrasion. Additives used to prevent flames in coatings include halogen, phosphorus, nitrogen, minerals, oxides, intumescents, and nanofillers. Based on their fire-retardant mechanisms, these coatings can be categorized into two categories: non-intumescent and intumescent coatings. Non-intumescent coatings contain fire retardant additives and their fire-retardant mechanism is based on slowing down the spread of fire and smoke. Coatings that form an intumescent layer create a protective charred layer that prevents heat from transferring to the inner layers [51].

To prepare non-intumescent smart fire-retardant coatings, core-double-shells were synthesized from graphite, polyaniline, and silver and then aluminum phosphate as a fire-retardant additive was loaded into these microcapsules. Surfaces requiring electrical conductivity and polypropylene can be coated with this coating [52].

The first multifunctional smart coating for cotton fabric (intumescent fire retardant, self-healing, and self-cleaning) was developed to achieve this, and a three-layer coating of poly(ethylenimine), ammonium polyphosphate, and fluorinated-decyl polyhedral oligomeric silsesquioxane (F-POSS) was prepared. A layer of char formed on these coatings with indirect flame exposure, which made them fire-resistant. These coatings can also be washed several times without losing their flame-retardant properties due to the hydrophobic properties of F-POSS [53].

9 CONCLUSION

Recently, highly remarkable attention has been attracted by smart coatings due to the increased use of polymer coatings in a variety of applications (buildings, medical devices, packaging, etc.) and the difficulties associated with traditional coatings. Smart coatings, due to their advantages in various applications such as energy storage, reduction of repair costs, and environmental pollution, increase the service time of materials, response to stimuli, and properties such as anticorrosion, antimicrobial, antifogging, anti-icing, antifingerprint, and fire-retardancy.

The investigation of various types of smart coatings has shown that one strategy can be useful for various applications. The strategy of preparing hydrophobic surfaces is also used in self-cleaning and anticorrosion coatings. Reducing surface energy and roughness are other strategies to prevent the adsorption of dust, ice, fingerprints, and bacteria on surfaces. Self-healing coatings have found wide applications due to the increased lifetime of coatings, and their strategies are often used to provide multi-functional smart coatings. In addition, studies have shown that nanoparticles are important components for the preparation of smart coatings such as chromic, anticorrosion, antifouling, antimicrobial, and antireflective coatings. However, in this regard, significant progress can be seen in the development of smart coatings that have high durability, low cost, and multiple functions.

REFERENCES

[1] Baghdachi J. Smart coatings. ACS Symposium Series, 1002 (2009) 3–24.

[2] Ulaeto SB, Pancrecious JK, Rajan TPD, Pai BC. Smart coatings. Noble Metal-Metal Oxide Hybrid Nanoparticles: Fundamentals and Applications (2019) 341–372.

[3] Nagappan S, Moorthy MS, Rao KM, Ha CS. Stimuli-responsive smart polymeric coatings: an overview. Industrial Applications for Intelligent Polymers and Coatings (2016) 27–49.

[4] Ganesh VA, Raut HK, Nair AS, Ramakrishna S. A review on self-cleaning coatings. Journal of Materials Chemistry. 21 (2011) 16304–16322.

[5] Zhang XT, Sato O, Taguchi M, Einaga Y, Murakami T, Fujishima A. Self-cleaning particle coating with antireflection properties. Chemistry of Materials, 17 (2005) 696–700.

[6] Kamegawa T, Shimizu Y, Yamashita H. Superhydrophobic surfaces with photocatalytic self-cleaning properties by nanocomposite coating of TiO_2 and polytetrafluoroethylene. Advanced Materials. 24 (2012) 3697–3700.

[7] Ganbavle V v., Bangi UKH, Latthe SS, Mahadik SA, Rao AV. Self-cleaning silica coatings on glass by single step sol–gel route. Surface and Coatings Technology, 205 (2011) 5338–44.

[8] Belhadjamor M, el Mansori M, Belghith S, Mezlini S. Anti-fingerprint properties of engineering surfaces: a review. 34 (2016) 85–120.

[9] Luo ZK, Chen PQ, Wang F, Pang Y, Xu YH, Hong YR, et al. Preparation of amphiphobic coating by combining fluoroalkyl silane with nano-SiO_2. physica status solidi (a). 212 (2015) 259–264.

[10] Siriviriyanun A, Imae T. Anti-fingerprint properties of non-fluorinated organosiloxane self-assembled monolayer-coated glass surfaces. Chemical Engineering Journal. 246 (2014) 254–259.

[11] Durán IR, Laroche G. Water drop-surface interactions as the basis for the design of anti-fogging surfaces: Theory, practice, and applications trends. Advances in Colloid and Interface Science. 263 (2019) 68–94.

[12] Zhang S, Huang J, Cheng Y, Yang H, Chen Z, Lai Y. Bioinspired surfaces with superwettability for anti-icing and ice-phobic application: Concept, mechanism, and design. Small. 13 (2017) 1701867.

[13] Shamshiri M, Jafari R, Momen G. Potential use of smart coatings for icephobic applications: A review. Surface and Coatings Technology. 424 (2021) 127656.

[14] Blaiszik BJ, Kramer SLB, Olugebefola SC, Moore JS, Sottos NR, White SR. Self-healing polymers and composites. 40 (2010) 179–211.

[15] Samadzadeh M, Boura SH, Peikari M, Kasiriha SM, Ashrafi A. A review on self-healing coatings based on micro/nanocapsules. Progress in Organic Coatings. 68 (2010) 159–164.

[16] Habib S, Khan A, Nawaz M, Sliem MH, Shakoor RA, Kahraman R, et al. Self-healing performance of multifunctional polymeric smart coatings. Polymers. 11 (2019) 1519.

[17] Shields Y, de Belie N, Jefferson A, van Tittelboom K. A review of vascular networks for self-healing applications. Smart Materials and Structures. 30 (2021).

[18] Lee MW, An S, Yoon SS, Yarin AL. Advances in self-healing materials based on vascular networks with mechanical self-repair characteristics. Advances in Colloid and Interface Science. 252 (2018) 21–37.

[19] Luo X, Mather PT. Shape memory assisted self-healing coating. ACS Macro Letters. 2 (2013) 152–156.

[20] Peng H, Du X, Cheng X, Wang H. Room-temperature self-healable and stretchable waterborne polyurethane film fabricated via multiple hydrogen bonds Progress in Organic Coatings, 151 (2021) 106081.

[21] Burattini S, Greenland BW, Merino DH, Weng W, Seppala J, Colquhoun HM, et al. A healable supramolecular polymer blend based on aromatic π-π Stacking and hydrogen-bonding interactions. J Am Chem Soc. 132 (2010) 12051–12058.

[22] Hu Z, Zhang D, Lu F, Yuan W, Xu X, Zhang Q, et al. Multistimuli-responsive intrinsic self-healing epoxy resin constructed by host–guest interactions. Macromolecules. 51 (2018) 5294–5303.

[23] Xu Y, Chen D. A novel self-healing polyurethane based on disulfide bonds. Macromolecular Chemistry and Physics . 217 (2016) 1191–1196.

[24] Zhao D, Du Z, Liu S, Wu Y, Guan T, Sun Q, et al. UV light curable self-healing superamphiphobic coatings by photopromoted disulfide exchange reaction. ACS Applied Polymer Materials. 1 (2019) 2951–2960.

[25] Fang Y, Li J, Du X, Du Z, Cheng X, Wang H. Thermal- and mechanical-responsive polyurethane elastomers with self-healing, mechanical-reinforced, and thermal-stable capabilities. Polymer (Guildf). 158 (2018) 166–175.

[26] Shao C, Wang M, Chang H, Xu F, Yang J. A self-healing cellulose nanocrystal-poly(ethylene glycol) nanocomposite hydrogel via Diels-Alder click reaction. ACS Sustainable Chemistry and Engineering. 5 (2017) 6167–6174.

[27] Fortunato G, Tatsi E, Rigatelli B, Turri S, Griffini G, Fortunato G, et al. Highly transparent and colorless self-healing polyacrylate coatings based on Diels-Alder chemistry. Macromolecular Materials and Engineering. 305.2 (2020) 1900652

[28] Yang W, Tao X, Zhao T, Weng L. Antifouling and antibacterial hydrogel coatings with self-healing properties based on a dynamic disulfide exchange reaction. Polymer Chemistry. 6.39 (2015) 7027–7035.

[29] Abdollahi A, Roghani-Mamaqani H, Razavi B. Stimuli-chromism of photoswitches in smart polymers: Recent advances and applications as chemosensors. Progress in Polymer Science. 98 (2019) 101149.

[30] Zhang Y, Luo L, Li K, Morsümbül S, Akçakoca Kumbasar EP. Photochromic textile materials. IOP Conference Series: Materials Science and Engineering. 459 (2018).

[31] Nezhadghaffar-Borhani E, Abdollahi A, Roghani-Mamaqani H, Salami-Kalajahi M. Photoswitchable surface wettability of ultrahydrophobic nanofibrous coatings composed of spiropyran-acrylic copolymers. Journal of Colloid and Interface Science. 593 (2021) 67–78.

[32] Zheng S, Xu Y, Shen Q, Yang H. Preparation of thermochromic coatings and their energy saving analysis. Solar Energy. 112 (2015) 263–271.

[33] Zhang J, Wang J, Yang C, Jia H, Cui X, Zhao S, et al. Mesoporous SiO2/VO2 double-layer thermochromic coating with improved visible transmittance for smart window. Solar Energy Materials and Solar Cells. 162 (2017) 134–141.

[34] Shchegolkov AV, Jang SH, Shchegolkov AV, Rodionov YV, Sukhova AO, Lipkin MS. A brief overview of electrochromic materials and related devices: A nanostructured materials perspective. Nanomaterials. 11 (2021) 2367.

[35] Duy Tam N, Pin Yeo L, Amanda Ong J, Mandler D, Duy Nguyen T, Jiamin Ong A, et al. Electrochromic smart glass coating on functional nano-frameworks for effective building energy conservation. Materials Today Energy. 18 (2020) 100496

[36] Cui G, Bi Z, Wang S, Liu J, Xing X, Li Z, et al. A comprehensive review on smart anti-corrosive coatings. Progress in Organic Coatings. 148 (2020) 105821.

[37] Li W, Hintze P, Calle LM, Buhrow J, Curran J, Muehlberg AJ, et al. Smart Coating For Corrosion Indication And Prevention: Recent Progress. CORROSION. (2009). https://onepetro.org/NACEC ORR/proceedings-abstract/CORR09/All-CORR09/NACE-09499/126624

[38] Ghahremani P, Sarabi AA, Roshan S. Cerium containing pH-responsive microcapsule for smart coating application: Characterization and corrosion study. Surface and Coatings Technology. 427 (2021) 127820.

[39] Eivaz Mohammadloo H, Mirabedini SM, Pezeshk-Fallah H. Microencapsulation of quinoline and cerium based inhibitors for smart coating application: Anti-corrosion, morphology and adhesion study. Progress in Organic Coatings. 137 (2019) 105339.

[40] Qian H, Xu D, Du C, Zhang D, Li X, Huang L, et al. Dual-action smart coatings with a self-healing superhydrophobic surface and anti-corrosion properties. Journal of Materials Chemistry A. 5.5 (2017): 2355–2364.

[41] He Z, Lan X, Hu Q, Li H, Li L, Mao J. Antifouling strategies based on super-phobic polymer materials. Progress in Organic Coatings. 157 (2021) 106285.

[42] Xie Q, Liu C, Lin X, Ma C, Zhang G. Nanodiamond reinforced poly(dimethylsiloxane)-based polyurea with self-healing ability for fouling release coating. ACS Applied Polymer Materials. 2.8 (2020) 3181–3188.

[43] Tallet L, Gribova V, Ploux L, Vrana NE, Lavalle P. New smart antimicrobial hydrogels, nanomaterials, and coatings: Earlier action, more specific, better dosing? Advanced Healthcare Materials. 10.1 (2021) 2001199.

[44] Kováčová M, Kleinová A, Vajdʹák J, Humpolíček P, Kubát P, Bodík M, et al. Photodynamic-active smart biocompatible material for an antibacterial surface coating. Journal of Photochemistry and Photobiology B: Biology. 211 (2020) 112012.

[45] Qu Y, Wei T, Zhao J, Jiang S, Yang P, Yu Q, et al. Regenerable smart antibacterial surfaces: full removal of killed bacteria via a sequential degradable layer. Journal of Materials Chemistry B. 6.23 (2018) 3946–3955.

[46] Raut HK, Ganesh VA, Nair AS, Ramakrishna S. Anti-reflective coatings: A critical, in-depth review. Energy & Environmental Science. 4 (2011) 3779–3804.

[47] Zambrano DF, Villarroel R, Espinoza-González R, Carvajal N, Rosenkranz A, Montaño-Figueroa AG, et al. Mechanical and microstructural properties of broadband anti-reflective TiO_2/SiO_2 coatings for photovoltaic applications fabricated by magnetron sputtering. Solar Energy Materials and Solar Cells. 220 (2021) 110841.

[48] Raut, Hemant Kumar, et al. Anti-reflective coatings: A critical, in-depth review. Energy & Environmental Science. 4.10 (2011) 3779–3804.

[49] Reuna J, Aho A, Isoaho R, Raappana M, Aho T, Anttola E, et al. Use of nanostructured alumina thin films in multilayer anti-reflective coatings. Nanotechnology. 32.21 (2021) 215602

[50] Lee H, Yi A, Choi JG, Ko DH, Jung Kim H. Texturing of polydimethylsiloxane surface for anti-reflective films with super-hydrophobicity in solar cell application. Applied Surface Science. 584 (2022) 152625.

[51] Mariappan, Thirumal. Fire retardant coatings. New Technologies in Protective Coatings, [Carlos Giudice and Guadalupe Canosa, IntechOpen]. (2017) 101–122

[52] Zhang B, Jiang Y. Highly electrically conductive and smart fire-resistant coating. Journal of Materials Science: Materials in Electronics. 29 (2018) 16378–16387.

[53] Chen S, Li X, Li Y, Sun J. Intumescent flame-retardant and self-healing superhydrophobic coatings on cotton fabric. ACS Nano. 9 (2015) 4070–4076.

17 Polymers for Smart Sensors

Matineh Ghomi,[1] Mahtab Yadolahi,[1] Farnaz Dabbagh Moghaddam,[2] and Ehsan Nazarzadeh Zare[1]

[1] School of Chemistry, Damghan University, Damghan, 36716-41167, Iran
[2] Institute for Photonics and Nanotechnologies, National Research Council, Via Fosso del Cavaliere, 100, 00133, Rome, Italy

1 INTRODUCTION

"Smart" materials are materials that can respond to external or internal stimuli. Polymers are one of the main groups of "smart" materials. They are inexpensive materials and more easily personalized than metals or ceramics. Smart polymers can be categorized into temperature-, light-, electric-, pH-, and multi-responsive groups. These materials have been used in different areas such as water treatment, tissue engineering, drug/gene delivery, and sensors. Smart polymers can self-assemble or undergo phase/morphology changes through physicochemical changes in response to small external/internal changes in their surroundings [1].

A smart sensor is a system that receives a response from the physical surroundings and uses built-in computing resources to make predefined roles upon finding definite input and then method data before passing it on. They allow for additional accurate and automated group of environmental data with less erroneous noise between the perfectly recorded information. Smart sensors are employed for checking and controlling mechanisms in an extensive diversity of surroundings containing smart grids, examination, and numerous scientific applications.

2 RESPONSIVE POLYMERS

2.1 pH-RESPONSIVE POLYMERS

Over the last two decades, the synthesis of pH-responsive polymers with their wide applications has attracted much commercial attention. pH-responsive polymers are a class of stimulus-responsive materials that respond well to the pH of a solution by having special properties such as solubility, surface activity, structural changes in the chain, and configuration. Typically, pH-responsive polymers have ionizable acidic/basic groups, and their ionization depends on the pH of the solution. Thus, pH-responsive polymers are polyelectrolytes that protonate or deprotonate with weakly acidic/basic groups in response to changes in ambient pH. These polymers have functional groups such as pyridine, carboxylic, sulfonic, phosphate, tertiary amines, etc. In general, the ionization of acidic or basic groups leads to changes in pH and polymer structure. In addition, ionization of pH-responsive polymers can regulate hydrophilicity/hydrophobicity, self-assembly, and phase separation capabilities. The unique properties of pH-responsive polymers make them interesting candidates for different applications such as sensors, drug/gene delivery, biotechnological, chromatography, surfaces, and membranes [2].

pH changes in pH-responsive polymers can deprotonate functional groups in the polymer backbone and some cases can cause the polymer chain to flocculate or collapse. They can also cause self-assembly formation such as micelle, gel, or swelling, etc. Star-like and branched copolymers can

DOI: 10.1201/9781003278269-17

also be classified as pH-responsive polymers that exhibit surface-active behaviors as the pH changes. In addition, hydrogel-like structures also exhibit swelling or de-swelling behavior by changing the pH. Thus, polymer-modified surfaces create the conditions for ionic surfaces to change the pH (Figure 17.1) [3]. pH-responsive polymers can also be prepared between the pH range. Generally,

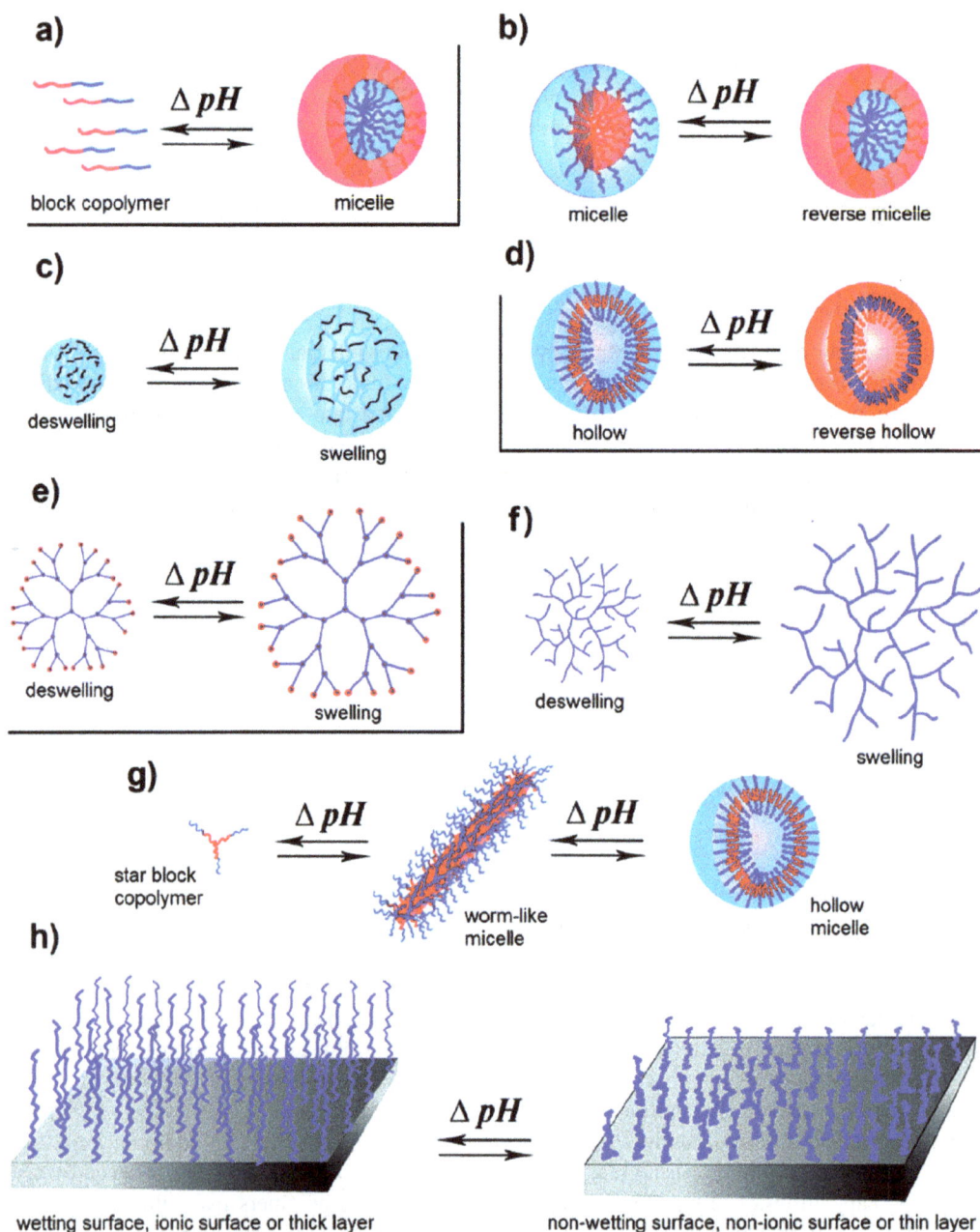

FIGURE 17.1   pH-responsive polymers with various structures: (A) unimer-micelle, (B) micelle-reverse micelle, (C) nanogels or microgels, (D) hollow-reverse hollow, (E) dendrimer, (F) hyper-branched, (G) micelle morphology changes (from wormlike to hollow), and (H) polymer brushes. Adapted with permission from Reference [3]. Copyright (2017) Royal Chemical Society.

pH-responsive polymers are categorized into cationic and anionic polymers. Cationic polymers synthesized with basic monomers operate under acidic circumstances and anionic polymers synthesized with acidic monomers operate under basic conditions. These two kinds of polymers are used alone or in combination for different applications. However, natural polymers have been used to respond to pH more than synthetic polymers due to their degradability, abundance in nature, and biocompatibility. The most important polypeptides for pH-responsive synthetic polymers are poly(aspartic acid) (PASA), poly(histidine) (PHIS), and poly(glutamic acid) (PLGA), which, like natural pH-responsive polymers, are biodegradable [4]. In this section, the types of pH-responsive polymers, synthesis approaches, structures, and their applications are reviewed.

2.1.1   pH-Responsive Acidic Polymers

These are polymers with weak acidic/basic chains that accept or release H^+ at low or high pH, respectively. Due to the different pKa values of pH-responsive polymers at different pH, their polyelectrolytic nature also changes, which leads to changes in hydrophilicity/hydrophobicity in the aqueous phase. These changes also lead to the deposition or dissolution of polymer chains. Polymers with weak acidic/basic chains are introduced as pH-responsive polymers that accept or release protons at low or high pH, respectively. Due to the different pKa values of pH-responsive polymers at different pH, their polyelectrolytic nature also changes, which leads to changes in hydrophilicity/hydrophobicity in the aqueous phase. These changes also lead to the deposition or dissolution of polymer chains. Some functional groups, e.g., $COOH$, SO_3H, $PO(OH)_2$, and $B(OH)_2$, affect the response of polymers, as discussed below. The carboxylic groups in polyacids such as poly(acrylic acid) (PAAc) and poly(methacrylic acid) (PMAAc) are converted to carboxylate ions (polyelectrolytes) at basic pH values, while at acidic pH the carboxylic groups are converted to an uncharged macromolecule. These polyacids can be synthesized by polymerization of acrylic acid and methacrylic acid monomers *via* various polymerization methods. Sometimes, protective groups are also employed during polymerization and followed through deprotection chemistry to achieve polymers with carboxylic groups [5]. The most important and widely used pH-responsive polyacids containing sulfonic acid group are poly(4-styrene sulfonic acid) (PSSA) and poly(2-acrylamide-2-methylpropane sulfonic acid) (PAMPS) [6]. pH-responsive polyacids containing the sulfonic acid group are commonly used in the preparation of pH-responsive hydrogels. pH-responsive hydrogels with sulfonic acid groups swell easily at pH above the pKa of the acidic groups. This indicates that they are more hydrophilic in their anionic type. pH-responsive hydrogels containing phosphorus methacrylate usually swell easily at low pH values. Synthetic and natural polymers containing phosphonic acids are widely used to synthesize pH-responsive polymers. pH-responsive polymers containing boronic acid groups are widely used in glucose sensors and self-healing gel. In recent years, pH-responsive polymers containing phenylboronic acid [$Ph-B(OH)_2$] groups have been widely reported [7]. These polymers contain boronic acid groups and are commonly applied in self-healing gel synthesis and some sensors. Phenylboronic acid moieties are common among them. There are various polymerization methodologies for boronic acid-based monomers [8].

2.1.2   pH-Responsive Basic Polymers

Polymers with base groups (polybases or polycations), which are ionized or deionized from pH 7 to 11, are used as pH-responsive base polymers. pH-responsive basic polymers with chains containing amino groups are introduced as basic polyelectrolytes that accept protons at low pH. Today, methacrylate, methacrylamide, and vinyl polymers based on tertiary amine groups have also been considered [9]. Basic polymers based on tertiary amine methacrylate such as poly[(2-diethylamino)ethyl methacrylate] (PDEA), poly[(2-diisopropylamine)ethyl methacrylate] (PDPA), and poly[2-(dimethylamino)ethyl methacrylate] (PDMA) are the most preferred pH-responsive basic polymers. PDMA is one of the most pH-responsive basic polymers, which is also a thermo-responsive polymer and widely available in nature. Basic polymers based on polyvinyl pyridine are also pH-responsive basic polymers which are poly(2-vinyl pyridine) (P2VP) and poly(4- vinyl pyri-dine) (P4VP). These perform phase transition due to the protonation of pyridine groups at pH values greater than 5 [10]. Other pH-responsive polymers have functional groups such as pyrrolidine, piperazine imidazole, and morpholino [29]. Poly[(2-N-morpholino) ethyl methacrylate] (PMEMA) is an important polymer that has morpholino groups and responds to the pH, temperature, and ionic strength of the environment. The synthesis of several PMEMAs based on polymers and the study of their performance in solutions have been reported [11]. Dendrimers such as poly(propylene imine) (PPI), poly(amidoamine) (PAMAM), and poly(ethyleneimine) (PEI) can also be classified as pH-responsive polymers. They can be modified with different materials after synthesis and can be attached to different polymers [12].

2.1.3 pH-Responsive Natural Polymers

Over the past decade, scientists have achieved many results on synthetic biodegradable polymers for different biomedical applications. In addition, the use of natural biocompatible polymers has attracted much attention due to their good biodegradability and easy post-modification by con-ventional chemicals. Natural pH-responsive polymers have also been reported in self-healing gel systems [13]. The most important natural polymers used are chitosan, hyaluronic acid, alginic acid, gelatin, and dextran. These polymers can provide better conditions for drug delivery. Today, the bonding of pH-responsive polymers into polysaccharides can have a wide range of applications in these polymers. In many studies today, hydrogel-based pH-responsive polymers have been synthesized using cross-linking. To date, chitosan is the most widely used species in this model of polymers [14]. Several acidic, basic, and natural pH-responsive polymers are shown in Table 17.1.

2.2 THERMO-RESPONSIVE POLYMERS

A class of smart polymers with thermoresponsive performance are thermo-responsive polymers. This property makes them useful polymers for many applications, which has led to widespread scientific attention. As is well known, the temperature of different tissues of the human body is always set in the temperature range of 35–37°C. Therefore, thermo-responsive polymers are of particular importance for biological and biomedical applications Thermo-responsive polymers can exist naturally or be synthetic. Two important classes of smart thermo-responsive polymers are poly(2-oxazoline)s (POXs) and poly(N-isopropyl acrylamide)s (PNIPAAms). In general, cellulosic polymers and other polymers that have an upper critical solution temperature (UCST) or lower crit-ical solution temperature (LCST) are also classified as thermo-responsive polymers.

In general, smart thermo-responsive polymers demonstrate a temperature-dependent solubility range in aqueous solutions. This temperature range is shown in temperature phase diagrams rela-tive to the polymer fraction volume. This solubility range creates hydrophobic interactions between polymer chains that allow polymers to self-assemble or accumulate in aqueous solutions. In the phase diagram (Figure 17.2), the minimum and maximum temperatures are indicated as being the LCST and UCST, respectively [15].

Typically, UCST and LCST temperatures emerge when phase separation occurs in a smart thermo-responsive polymer solution below or above a certain temperature range [17]. Thermo-responsive polymers are described by a critical solution temperature. The critical solution tempera-ture is a limited temperature range in which hydrophilic or hydrophobic interactions occur between the polymer chains and the aqueous solution. These interactions can lead to swelling or collapse of polymer chains. A solvent–polymer mixture that is converted from a low-temperature single-phase system to a high-temperature two-phase system is described by the LCST temperature. On the other hand, a solvent–polymer mixture consisting of a high-temperature single-phase system and a low-temperature two-phase system is described by the LCST temperature. The phase transfer tempera-ture of thermo-responsive polymers can also depend on factors such as concentration and polymer

TABLE 17.1
Chemical Structures of pH-Responsive Acidic, Basic, and Natural Polymers

Polymer Name	Type of pH-Responsive	Structure	Functional Groups	References
Poly[2-(dimethylamino)ethyl methacrylate] (PDMA)	Basic polymers		Tertiary amine	[9]
Poly [(2-N-morpholino) ethyl methacrylate] (PMEMA)			Morpholino	[11]
Poly(4-vinylpyridine) (P4VP)			Pyridine	[10]
Poly (amidoamine) (PAMAM)			Amidoamine	[12]

(continued)

TABLE 17.1 (Continued)

Chemical Structures of pH-Responsive Acidic, Basic, and Natural Polymers

Polymer Name	Type of pH-Responsive	Structure	Functional Groups	References
Poly (methacrylic acid) (PMAAc)	Acidic polymers		–COOH	[5]
Poly (ethylene glycol) acrylate (PEGAP)			Phosphonic acid	[7]
Poly(2-acrylamido-2-methylpropane sulfonic acid) (PAMPS)			Sulfonic acid	[6]
Poly(3-acrylamido phenylboronic acid (PAAPBA)			Boronic acid	[8]
Chitosan	Natural polymers		–NH$_2$, –OH	[14]
Hyaluronic acid			–COOH, –OH	[5]

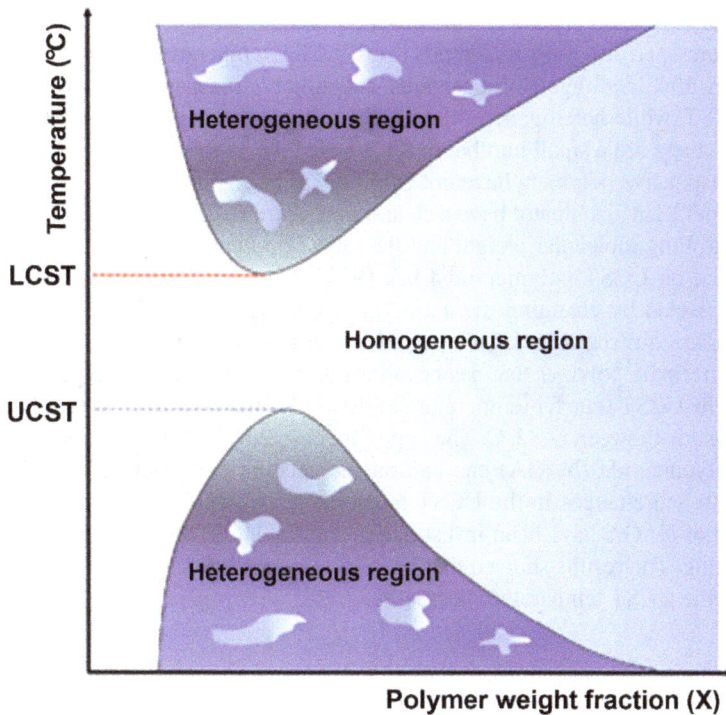

FIGURE 17.2 Phase diagrams of LCST- and UCST-type polymer aqueous solutions (temperature versus polymer weight fraction). Adapted with permission from Reference [16]. Copyright (2017) Royal Chemical Society.

molecular weight [18]. In recent years, LCST and UCST smart polymers have been extensively studied by scientists for the preparation and development of smart sensors.

2.2.1 LCST-Thermoresponsive Polymers

They show fast variations in reversible-level transitions due to variations in the temperature. The solubility of these polymers in aqueous solutions depends on the increase or decrease in the temperature changes. The changes in solubility lead to the deposition of polymer chains at temperatures above LCST and their hydration at temperatures under LCST. Therefore, the LCST temperature is also known as the cloud point temperature. The cloud point temperature is once a polymer solution shows a phase change in the macroscopic state from clear to cloudy [19]. The most popular smart thermo-responsive polymers are PNIPAAms [20]. Smart PNIPAAms exhibit variable aquatic solubility in cooling and heating procedures. The change in solubility is due to the LCST temperature of the polymers in the range of 30–35°C and the polymer chains undergo a reversible phase switch from a hydrophilic coil to a hydrophobic sphere. One of the advantages of PNIPAAms is that their LCST temperature is close to the physiological temperature of the human body, which can be very beneficial. One of the factors that allows the LCST of smart polymers to be adjusted to the desired temperature is the replacement of the isopropyl group with hydroxyl, amide, and carboxyl groups such as 2-hydroxyisopropylacrylamide (HIPAAm) [21], 2-amino isopropyl acrylamide (AIPAAm) [54,55], and 2-carboxyl isopropyl acrylamide (CIPAAm) [22], respectively. An interesting method to achieve the desired LCST temperature of the smart polymers is copolymerization with these functional monomers. These agents can be used to create smart polymers that respond to pH, ions, or salt concentrations in addition to temperature.

2.2.2 UCST-Type Thermo-Responsive Polymers

UCST temperatures perform very differently from LCST temperatures. In this way, they exhibit phase separation after cooling. Homogeneous and highly transparent smart polymer solutions have a lower LCST, while heterogeneous and highly cloudy smart polymer solutions have a lower UCST. However, there are a small number of UCST-type thermo-responsive polymers. In addition, UCST thermo-responsive polymers have not exhibited as good phase transitions as LCST thermo-responsive polymers and also do not have a clear temperature range [23]. UCST temperature can be adjusted by controlling molecular weight and the ratio of polymers. Poly(allylurea-co-allylamine) (PU-Am) is based on UCST polymers and has UCST temperatures from 8 to 65°C. The UCST temperature is selected by changing the amine group. Polybetaines (PBs) are also a well-known group of smart thermo-responsive UCST polymers. Poly(pentafluorophenyl acrylate) (PFPA) is a highly polar zwitterionic polymer that has been introduced as smart UCST-type thermo-responsive polymers [24]. The UCST transfer temperature of these polymers is related to the chemical structure of the polymer and is between 6–82°C. The copolymer systems of N-acetylacrylamide (NAcAAm) and N-acryloylglycinamide (NAGA) can exhibit the specific performance of UCST behavior in water [25]. Controlled changes in the UCST temperature by copolymerization of NAcAAm with varying amounts of NAGA have been investigated, and the UCST temperature was studied during cooling and heating. The results showed that by increasing the amount of NAGA in the copolymer, the sharpness of the UCST temperature decreases.

2.3 PHOTO-RESPONSIVE POLYMERS

The properties of photo-responsive polymers change when exposed to a light stimulus.

The structure and properties of these polymers alter after exposure to visible, ultraviolet, and near-infrared light. In general, various molecular properties of photo-responsive polymers such as conjugation, conformation, optical chirality, amphiphilicity, and polarity can be adjusted with light [1]. The light radiation into photo-responsive polymers also leads to macroscopic changes such as conductivity, solubility, optical properties, adherence, wettability, and configuration. Photo-responsive polymers have inherent advantages over thermal, electrical, and pH-reactive polymers, such as response strength, remote application, and temporal/local response resolution. In general, the performance of these polymers after light irradiation is generally defined by measuring parameters including the rate of change in the polymer behavior, the rate of occurrence of this variation, and the process's reversibility. Generally, photo-responsive polymers, depending on their application to light irradiation, should show rapid and sharp changes with appropriate modulation [1]. To obtain photo-responsive polymers, chromophore-containing molecules must be included in the polymer chain, which, related to the kind of chromophore applied (reversible or irreversible), the polymer can have a reversible or irreversible behavior. Reversible chromophores or molecular switches cause a reversible isomerization reaction due to light irradiation at a special wavelength, therefore photochromic isomerization between two isomeric forms swaps the properties of these polymers by light irradiation at two diverse wavelengths. The properties of reversible photo-responsive polymers can alternate between two photostationary states and are used as artificial muscles, stimuli, information storage, and switches [1]. The nature of photo-responsive polymers depends on the choice of photo-responsive moieties. One of the important features of photo-responsive moieties is the reversible or irreversible processes under light radiation. Several photo-responsive moieties with reversible and irreversible transformation under different light irradiation are reported in Table 17.2. Photo-responsive moieties cause specific photochemical reactions such as isomerization, dimerization, and self-destruction. Coumarin derivatives can be subjected to reversible light-induced intermolecular dimerization to form stable isomers [26]. Light-sensitive structures such as azobenzene and spiropiran derivatives can also produce a reversible isomerization reaction to convert *cis* configuration to *trans* configuration and vice versa when exposed to light radiation [27]. Dimerization is

TABLE 17.2

Examples of Reversible/Irreversible Light-Response Groups

Number	Name of Reversible Photo-Responsive Moieties	Reaction
1a	Azobenzene	Trans-form / cis-form
2a	Coumarin	Single / Dimer
3a	Diarylethene	Open-form / Closed-form
4a	Spiropyran	Closed-form / Open-form
1b	2-Napthoquinone -3-methide	
2b	Coumarin-4-ylmethyl	
3b	o-Nitrobenzyl	
4b	Pyrenylmethyl	

used to regulate the LCST temperature of polymers and the stability of intermolecular or intramolecular cross-linking in polymersomes.

Chromophore materials with an irreversible nature are generally employed in targeted drug-delivery systems as photodegradable materials. The main advantage of these materials is the photoconversion rate of about 100%, meanwhile no equilibrium between two situations is involved. This causes the operative release of drugs or a strong reduction of the molecular weight in applications requiring degradation [1]. Examples of irreversible chromophores such as 2-naphthoquinone-3-methides, coumarin-4-ylmethyls, O-nitrobenzyl, and phenylmethyl groups are shown in Table 17.2 [28]. 2-Naphthoquinone-3-metides under irreversible light radical reaction produce radical intermediates that can react with vinyl compounds containing an electron donor group (e.g. O or N) through the Diels–Alder cycloaddition reaction, leading to ring formation. An example of this reaction is given in Table 17.2b [29].

O-Nitrobenzyl derivatives under an irreversible light-induced intramolecular oxidation reaction lead to the synthesis of nitrosocarbonyl by-products, and coumarin-4-yl-methyls under irreversible light-induced nucleophilic attack with alcohol derivatives produce coumarin by-products (Table 17.2b and 17.3b). The polymer chains containing these compounds perform very well in light-induced reactions and make them very strong in photolithography.

In general, chromophores separate from the polymer chain when exposed to light radiation and depending on their location in the polymer chain, perform light-induced reactions such as click reaction, depolymerization, formation of active molecules, and charge generation in the polymer [1]. Polymer chains containing light-sensitive functional groups can lead to the synthesis of responsive polymers in which light causes changes in the physical properties and morphology of the compound [30]. Photo-responsive moieties have also been widely used in the synthesis of organic compounds. For example, o-nitrobenzyl derivatives are used as protectors for carboxylic acid groups in organic compounds. Light irradiation also leads to the deprotection of compounds containing o-nitrobenzyl after completion of the reaction [31]. Azobenzene derivatives in poly(ethylene glycol)-poly(dimethylamino-azobenzene) block copolymer (PEG-b-PMDA-Azo) lead to the synthesis of a photo-responsive polymer that causes reversible isomerization under light radiation. Also, the polymerization of o-nitrobenzyl derivatives as a cross-linker with poly(N-acryloyl morpholine) (PAcM) leads to the synthesis of a photocycle polymer [32].

One of the most important parameters that affect the performance of a photo-responsive polymer or copolymer is the position of the photo-responsive moiety in the polymer or copolymer. Figure 17.3 shows the effect of the photo-responsive moiety position on the performance of block copolymer micelles in solution under light irradiation. Block copolymer micelles can have four different reversible or irreversible processes inside groups or major chains, such as cross-linking, shifting hydrophilic/hydrophobic equilibrium, main chain degradation, and breaking block junction under light irradiation [33].

2.4   ENZYME-RESPONSIVE POLYMERS

These are a novel class of responsive polymers that can respond to biological molecules to control the function of natural compounds. Enzyme-responsive polymers are particularly effective in the environment; therefore, their performance is only for biological processes. Enzyme-sensitive oligopeptides (dipeptides or tripeptides) are a group of enzyme-responsive polymers that alter the physicochemical properties of oligopeptides when they are affected by enzymes. One of the advantages of these polymers is that they usually operate under mild conditions such as a temperature of 37° C, pH 5–8, and water solution. Therefore, they exhibit high selectivity in biological processes. To date, many enzyme-sensitive oligopeptides have been prepared. For instance, biodegradable hydrogels consisting of oligopeptide-dextran and PEG were prepared based on enzyme-sensitive oligopeptides and their degradation was investigated in the presence of papain and dextranase enzymes. In another report, enzyme-sensitive copolymer hydrogels containing acrylamide and PEG polymers were prepared and their selective enzymatic hydrolysis was investigated in a liquid sample. In addition, an enzyme-sensitive biopolymer containing poly(allylamine) (PAAm) and acetyl-protected dialanine was prepared and their selective enzymatic degradation was also investigated [34].

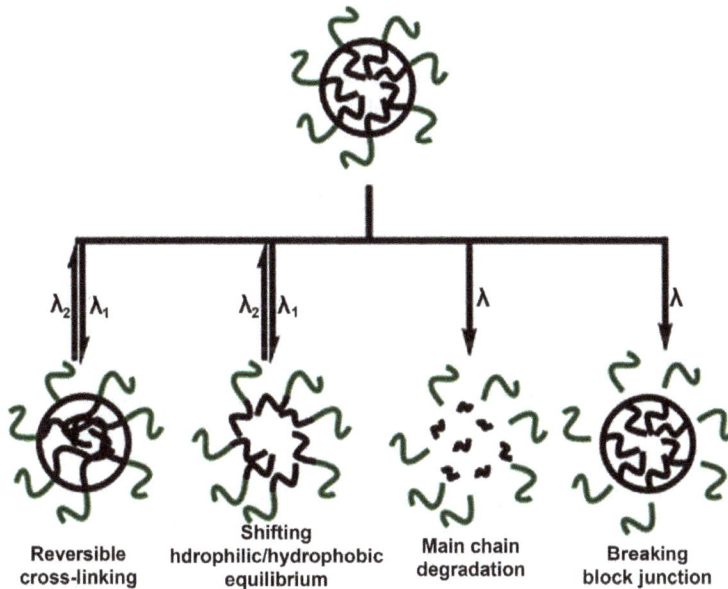

FIGURE 17.3 Effect of the photo-responsive moieties on the behavior of micelles in solution under irradiation Adapted with permission from Reference [33]. Copyright (2012) American Chemical Society.

In general, the processes of living organisms are controlled by the catalytic function of enzymes. The catalytic function of enzymes is usually regulated by the availability of cofactors or by controlling enzyme expression levels. The reactions performed by enzyme catalysts are usually in equilibrium, therefore enzymes are very useful for the development of functional biomaterials. Hence, enzymes have a variety of biological applications due to their inherent presence in biological systems.

2.4.1 Electric-Responsive Polymers

Electric-responsive polymers (ERPs) display form or size variations in response to different electrical stimuli. They convert electrical energy into mechanical energy, for example in soft robots or synthetic muscles. ERPs are categorized as inherent conductive, ionic, and voltage-responsive.

2.4.1.1 Inherent Conducting Polymers

These are conjugated polymers that behave as conducting materials. Conducting polymers include polyaniline, polypyrrole, and polythiophene. The mechanism of conduction in these polymers is due to the polarons and bipolarons as mobile charge carriers. Polarons are π-conjugated systems that have non-degenerate ground states. On the other hand, bipolarons are types with two charges in the same unit cell [36]. Doping is a process for the improvement of the electrical conductivity of conductive polymers. This process is affected by several physicochemical and mechanical properties of the conductive polymers. Chemical-, electrochemical- and photo-doping are common methods of doping process in conductive polymers [37]. Generally, electrical conductivity in conductive polymers depends on the type of initiator and dopant, medium acidity, molar ratio of initiator to monomer, monomer concentration, and temperature [38]. It has been reported that the electrical conductivity of prepared polypyrrole by iron trichloride or ammonium persulfate improved with the reaction temperature [39]. The electrically conductive polymers are prepared by two common methods: chemical and electrochemical [40]. The synthesis of polymers, in small scale and large scale, can be used for electrochemical and chemical methods, respectively. In the chemical method, a defined concentration of monomers is initiated by the addition of initiators, e.g., $FeCl_3$, $(NH_4)_2S_2O_8$, $K_2S_2O_8$,

H_2O_2 etc. in an aqueous solution. The electrochemical approaches consist of electrode coating and co-deposition methods. In the first process, all electrodes, such as reference and working one, are employed in one cell enclosing the electrolyte and the solution of monomer. In the second process, a host polymer (insulator) is dissolved in an electrolyte solution including the monomer [41].

2.4.1.2 IONIC POLYMERS

These are among the current-responsive materials established by Oguro and Asaka [42]. They used a Nafion membrane covered with Au or Pt. The ionic polymers bend to the anode by employing an electric field. The action mechanism of these polymers depends on the variance of swelling degree at both edges produced through the electrophoretic carrying of protons with H_2O molecules. The ionic polymers are flexible and appropriate for economizing that is accomplished dynamically at low voltages (0.5–3 V), which can be used in medical areas e.g., catheters and guide wires. Nevertheless, most ionic polymers work in a solution of electrolyte or a swollen state [43].

2.4.1.3 VOLTAGE-RESPONSIVE

These are dielectric gels and elastomers that have high proficiency and toughness that are predictable owing to a small electric current [44]. The first reported dielectric gels were based on PVA in dimethyl sulfoxide. It was declared that poly(vinyl alcohol) dielectric gels show rapid contraction (8%) within 0.1 s under an electric field (250 V/mm). Moreover, these gels demonstrate amoeba-like pseudopodia deformations and are employed in an electro-active synthetic pupil. Pelrine et al. reported dielectric elastomers applying rubbers (acrylic or silicone) with Ag or C grease as a compliant electrode. They have great distortion of the dielectric elastomers (> 100%) under an electric field. Dielectric elastomers have advantages such as simple structure and diversity of elastomeric materials. In addition, they possess disadvantages, e.g., high driving voltages (>1,000 V), low elastic and stretchable electrodes, and restraints applied before working the dielectric elastomer actuators [44].

2.4.2 Multi-Responsive Polymers

These are materials that respond to a combination of signals, as revealed by researchers and engineers in numerous areas of medicine, biology, chemistry, and physics. They have included micelles, gels, vesicles, and films.

Micelles: These polymers are prepared from amphiphilic copolymers through self-assembly. Micelles are employed as carriers in gene and drug systems [45].

Gels: These are cross-linked hydrophilic polymers on the micro- or nanoscale that respond to external stimuli by altering their physicochemical properties, e.g., penetrability, adjustable size, hydrophilicity, hydrophobicity, and morphology. They have good potential in various areas such as biomedical (wound healing and wound dressing), water purification, coatings, and catalysis. Gels are prepared by different methods such as mini/microemulsion, graft copolymerization, ionic/chemical cross-linking, distillation–precipitation, and dispersion polymerization [46].

Vesicles: These are usually called polymersomes, and are bags on a microscopic scale that surround a volume with a molecular membrane prepared from amphiphilic block copolymers. Recently, functional vesicles have gained great attention as their membrane penetrability can be activated via different stimuli. Multi-responsive vesicles possess definite potential for extensive applications in drug and gene delivery, reactors, and material templates on the micro/nanoscale [47].

Films: The common and simple technique for the preparation of polymer films is solvent casting. In this method, two or more polymers are solved in similar or miscible solvents separately and mixed to form thin-layer films. They have great potential in different areas, such as tissue

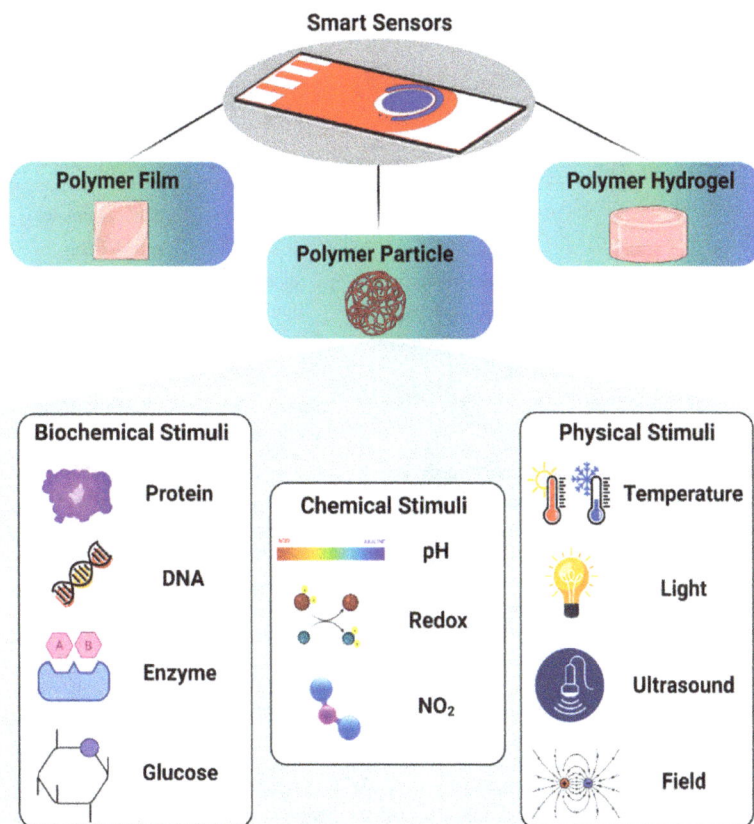

FIGURE 17.4 Types of polymer-responsive materials in smart sensors.

engineering, sensors, and coatings. Smart films have the capability of a controlled response to single or dual external signals [48]. Figure 17.4 shows the types of stimulation-responsive polymers that are employed for the fabrication of smart sensors.

2.5 SMART SENSORS

Currently, smart sensors made of responsive polymers with inherent properties such as fast response time, reusability, and high sensitivity have attracted increased attention. Thermo-responsive polymers with various LCST temperatures such as copolymers containing N-isopropyl acrylamide (NIPAM), *N*-tert-butylacrylamide (NTBAM), and N-isopropylmethacrylamide (NIPMAM), are used as the basis for nanoparticle thermometers based on copolymers. Poly(DBD-AE-*co*-DMAPAM-*co*-NTBAM) containing NTBAM, *N*, *N*-dimethylaminopropylacrylamide (DMAPAM), and 4-N-(2-acryloyloxyethyl)-N-methylamino-7-N, N-dimethylaminosulfonyl- 2,1,3-benzoxadiazole (DBD-AE), are also used as fluorescent thermometers based on this copolymer. The strong intrinsic fluorescence emission intensity of this copolymer is exhibited only above its LCST temperature of 35°C, at which the thermal phase transition goes from the soluble state to the nanoparticles aggregating by DBD hydrophobicity causing an important increase in the fluorescence yield of the thermometer [49]. A new smart sensor (Pyr-PSDM-AuNP) based on responsive polymers also has been designed and prepared using poly(sulfadimethoxine) (PSDM), pyrene, and Au NPs, which can be used as a pH sensor.

TABLE 17.3
Different Sensors and Their Applications

Smart Sensor Types	Polymer Type	Applications	Fabrication Techniques	References
Pressure sensor	PANI	Smart textile	Immersion and printing	[51]
Pressure sensor	Polystyrene (PS), polymethyl methacrylate (PMMA), polydimethylsiloxane (PDMS), poly(dimethylsiloxane)-graft-polyacrylates (PDMSg-PAA)	Detection of human motion	Breath figure method	[52]
Piezoresistive sensors	Poly(vinylidene fluoride)	Sensor applications	Spray printing	[53]
Electro-responsive sensors	N-isopropylacrylamide, N,N'-methylene-bis-acrylamide, methacrylamide hydrochloride, 2,2,2-trifluoroethyl methacrylate, 2-hydroxyethyl methacrylate	Insulin detection	Solid-phase approach	[54]
Chemiresistive sensor	A copolymer of methacrylic acid and dimethacrylate	Aceton detection	Precipitation polymerization method	[55]

Smart sensors are widely used in wearable electronic systems. They can detect, collect, analyze, and transmit physiological information such as body temperature, heart rate, location information, etc. So far, different traditional sensors have been made according to human needs. Textile sensors have been made that can be integrated with conductive compounds such as metals or carbon. Today, polymers or copolymers are used to provide flexible smart textile sensors due to their different properties to environmental conditions such as changes in pressure, temperature, humidity, etc. These sensors exhibit a sensory nervous system in smart clothing by detecting signals and transmitting data to the processor [50]. Table 17.3 shows the applications of some smart sensors.

3   CONCLUSION

Responsive materials, specifically polymers, are widely applied to expand smart sensors. The functionality of smart sensors arises from their capability to respond to environmental stimuli. The diverse natures of these responses depend on the type of responsive polymers. To this end, types of responsive polymers, and their properties for employment in smart sensors have been summarized in this chapter.

REFERENCES

[1] J. Cui, A. Del Campo, Photo-responsive polymers: properties, synthesis and applications, Smart polymers and their applications, Elsevier 2014, pp. 93–133.
[2] S. Dai, P. Ravi, K.C. Tam, pH-Responsive polymers: synthesis, properties and applications, Soft Matter, 4 (2008) 435–449.
[3] G. Kocak, C. Tuncer, V. Bütün, pH-Responsive polymers, Polym. Chem., 8 (2017) 144–176.
[4] C. Alvarez-Lorenzo, B. Blanco-Fernandez, A.M. Puga, A. Concheiro, Crosslinked ionic polysaccharides for stimuli-sensitive drug delivery, Adv. Drug Delivery Rev., 65 (2013) 1148–1171.

[5] A.E. Felber, M.-H. Dufresne, J.-C. Leroux, pH-sensitive vesicles, polymeric micelles, and nanospheres prepared with polycarboxylates, Adv. Drug Delivery Rev., 64 (2012) 979–992.

[6] L. Gabaston, S. Furlong, R. Jackson, S. Armes, Direct synthesis of novel acidic and zwitterionic block copolymers via TEMPO-mediated living free-radical polymerization, Polymer, 40 (1999) 4505–4514.

[7] A. Popa, C.-M. Davidescu, P. Negrea, G. Ilia, A. Katsaros, K.D. Demadis, Synthesis and characterization of phosphonate ester/phosphonic acid grafted styrene–divinylbenzene copolymer microbeads and their utility in adsorption of divalent metal ions in aqueous solutions, Ind. Eng. Chem. Res., 47 (2008) 2010–2017.

[8] G. Vancoillie, R. Hoogenboom, Synthesis and polymerization of boronic acid containing monomers, Polym. Chem., 7 (2016) 5484–5495.

[9] J. Hu, G. Zhang, Z. Ge, S. Liu, Stimuli-responsive tertiary amine methacrylate-based block copolymers: Synthesis, supramolecular self-assembly and functional applications, Prog. Polym. Sci., 39 (2014) 1096–1143.

[10] J. Seidel, V. Pinkrah, J. Mitchell, B. Chowdhry, M. Snowden, Isothermal titration calorimetric studies of the acid–base properties of poly (N-isopropylacrylamide-co-4-vinylpyridine) cationic polyelectrolyte colloidal microgels, Thermochim. Acta, 414 (2004) 47–52.

[11] V. Bütün, F.F. Taktak, C. Tuncer, Tertiary amine methacrylate-based ABC triblock copolymers: Synthesis, characterization, and self-assembly in both aqueous and nonaqueous media, Macromol. Chem. Phys., 212 (2011) 1115–1128.

[12] C. Kojima, Design of stimuli-responsive dendrimers, Expert Opin. Drug Delivery, 7 (2010) 307–319.

[13] L. Yuan, Z. Li, X. Li, S. Qiu, J. Lei, D. Li, C. Mu, L. Ge, Functionalization of an injectable self-healing pH-responsive hydrogel by incorporating a curcumin/polymerized β-cyclodextrin inclusion complex for selective toxicity to osteosarcoma, ACS Appl. Polym. Mater., 4 (2022) 1243–1254.

[14] R. Heras-Mozos, R. Gavara, P. Hernández-Muñoz, Chitosan films as pH-responsive sustained release systems of naturally occurring antifungal volatile compounds, Carbohydr. Polym. (2022) 119137.

[15] N. Morimoto, M. Yamamoto, Design of an LCST–UCST-Like Thermoresponsive zwitterionic copolymer, Langmuir, 37 (2021) 3261–3269.

[16] Y.-J. Kim, Y.T. Matsunaga, Thermo-responsive polymers and their application as smart biomaterials, J. Mater. Chem. B, 5 (2017) 4307–4321.

[17] J. Wei, H. Yu, H. Liu, C. Du, Z. Zhou, Q. Huang, X. Yao, Facile synthesis of thermo-responsive nanogels less than 50 nm in diameter via soap-and heat-free precipitation polymerization, J. Mater. Sci., 53 (2018) 12056–12064.

[18] D. de Morais Zanata, M.I. Felisberti, Thermo-and pH-responsive POEGMA-b-PDMAEMA-b-POEGMA triblock copolymers, Eur. Polym. J., 167 (2022) 111069.

[19] M.K. Yoo, Y.K. Sung, S.C. Chong, M.L. Young, Effect of polymer complex formation on the cloud-point of poly (N-isopropyl acrylamide)(PNIPAAm) in the poly (NIPAAm-co-acrylic acid): polyelectrolyte complex between poly (acrylic acid) and poly (allylamine), Polymer, 38 (1997) 2759–2765.

[20] T. Maeda, Y. Akasaki, K. Yamamoto, T. Aoyagi, Stimuli-responsive coacervate induced in binary functionalized poly (N-isopropylacrylamide) aqueous system and novel method for preparing semi-IPN microgel using the coacervate, Langmuir, 25 (2009) 9510–9517.

[21] X. Tang, A. Sun, C. Chu, C. Wang, Z. Liu, J. Guo, G. Xu, Highly sensitive multiresponsive photonic hydrogels based on a crosslinked Acrylamide-N-isopropylacrylamide (AM-NIPAM) co-polymer containing Fe_3O_4@ C crystalline colloidal arrays, Sens. Actuators, B, 236 (2016) 399–407.

[22] K.S. Soppimath, T.M. Aminabhavi, A.M. Dave, S.G. Kumbar, W. Rudzinski, Stimulus-responsive "smart" hydrogels as novel drug delivery systems, Drug Dev. Ind. Pharm., 28 (2002) 957–974.

[23] S.P. Rwei, Y.Y. Chuang, T.F. Way, W.Y. Chiang, Thermosensitive copolymer synthesized by controlled living radical polymerization: Phase behavior of diblock copolymers of poly (N-isopropyl acrylamide) families, J. Appl. Polym. Sci., 133 (2016) 43224.

[24] P.A. Woodfield, Y. Zhu, Y. Pei, P.J. Roth, Hydrophobically modified sulfobetaine copolymers with tunable aqueous UCST through postpolymerization modification of poly (pentafluorophenyl acrylate), Macromol., 47 (2014) 750–762.

[25] J. Seuring, F.M. Bayer, K. Huber, S. Agarwal, Upper critical solution temperature of poly (N-acryloyl glycinamide) in water: a concealed property, Macromol., 45 (2012) 374–384.

[26] C. Cardenas-Daw, A. Kroeger, W. Schaertl, P. Froimowicz, K. Landfester, Reversible photocycloadditions, a powerful tool for tailoring (nano) materials, Macromol. Chem. Phys., 213 (2012) 144–156.

[27] D. Xia, G. Yu, J. Li, F. Huang, Photo-responsive self-assembly based on a water-soluble pillar [6] arene and an azobenzene-containing amphiphile in water, Chem. Commun., 50 (2014) 3606–3608.

[28] G. Pasparakis, T. Manouras, P. Argitis, M. Vamvakaki, Photodegradable polymers for biotechnological applications, Macromol. Rapid Commun., 33 (2012) 183–198.

[29] S. Arumugam, V.V. Popik, Light-induced hetero-Diels– Alder cycloaddition: a facile and selective photoclick reaction, J. Am. Chem. Soc., 133 (2011) 5573–5579.

[30] L. Yang, H. Tang, H. Sun, Progress in photo-responsive polypeptide derived nano-assemblies, Micromachines, 9 (2018) 296.

[31] X. Liu, J. He, Y. Niu, Y. Li, D. Hu, X. Xia, Y. Lu, W. Xu, Photo-responsive amphiphilic poly (α-hydroxy acids) with pendent o-nitrobenzyl ester constructed via copper-catalyzed azide-alkyne cycloaddition reaction, Polym. Adv. Technol., 26 (2015) 449–456.

[32] R. Dong, B. Zhu, Y. Zhou, D. Yan, X. Zhu, "Breathing" vesicles with lellyfish-like on–off switchable fluorescence behavior, Angew. Chem., Int. Ed., 51 (2012) 11633–11637.

[33] Y. Zhao, Light-responsive block copolymer micelles, Macromol., 45 (2012) 3647–3657.

[34] J. Hu, G. Zhang, S. Liu, Enzyme-responsive polymeric assemblies, nanoparticles and hydrogels, Chem. Soc. Rev., 41 (2012) 5933–5949.

[35] M. Zelzer, R. Ulijn, Enzyme-responsive polymers: properties, synthesis and applications, Smart polymers and their applications, Elsevier 2014, pp. 166–203.

[36] E.N. Zare, T. Agarwal, A. Zarepour, F. Pinelli, A. Zarrabi, F. Rossi, M. Ashrafizadeh, A. Maleki, M.-A. Shahbazi, T.K. Maiti, Electroconductive multi-functional polypyrrole composites for biomedical applications, Appl. Mater. Today, 24 (2021) 101117.

[37] Z. Capáková, K.A. Radaszkiewicz, U. Acharya, T.H. Truong, J. Pacherník, P. Bober, V. Kašpárková, J. Stejskal, J. Pfleger, M. Lehocký, The biocompatibility of polyaniline and polypyrrole 2: Doping with organic phosphonates, Mater. Sci. Eng., 113 (2020) 110986.

[38] M. Aghelinejad, Y. Zhang, S.N. Leung, Processing parameters to enhance the electrical conductivity and thermoelectric power factor of polypyrrole/multi-walled carbon nanotubes nanocomposites, Synth. Met., 247 (2019) 59–66.

[39] A. Yussuf, M. Al-Saleh, S. Al-Enezi, G. Abraham, Synthesis and characterization of conductive polypyrrole: the influence of the oxidants and monomer on the electrical, thermal, and morphological properties, Int. J. Polym. Sci., 2018 (2018) 4191747.

[40] G. Jin, K. Li, The electrically conductive scaffold as the skeleton of stem cell niche in regenerative medicine, Mater. Sci. Eng., 45 (2014) 671–681.

[41] E.N. Zare, P. Makvandi, B. Ashtari, F. Rossi, A. Motahari, G. Perale, Progress in conductive polyaniline-based nanocomposites for biomedical applications: a review, J. Med. Chem., 63 (2019) 1–22.

[42] K. Asaka, K. Oguro, Y. Nishimura, M. Mizuhata, H. Takenaka, Bending of polyelectrolyte membrane–platinum composites by electric stimuli I. Response characteristics to various waveforms, Polym. J., 27 (1995) 436–440.

[43] T. Fukushima, K. Asaka, A. Kosaka, T. Aida, Fully plastic actuator through layer-by-layer casting with ionic-liquid-based bucky gel, Angew. Chem., Int. Ed., 44 (2005) 2410–2413.

[44] T. Hirai, H. Nemoto, M. Hirai, S. Hayashi, Electrostriction of highly swollen polymer gel: possible application for gel actuator, J. Appl. Polym. Sci., 53 (1994) 79–84.

[45] G. Jiang, T. Jiang, H. Chen, L. Li, Y. Liu, H. Zhou, Y. Feng, J. Zhou, Preparation of multi-responsive micelles for controlled release of insulin, Colloid Polym. Sci., 293 (2015) 209–215.

[46] C. Echeverria, S.N. Fernandes, M.H. Godinho, J.P. Borges, P.I. Soares, Functional stimuli-responsive gels: Hydrogels and microgels, Gels, 4 (2018) 54.

[47] Z.Q. Cao, G.J. Wang, Multi-stimuli-responsive polymer materials: Particles, films, and bulk gels, Chem. Rec., 16 (2016) 1398–1435.

[48] X. Fu, L. Hosta-Rigau, R. Chandrawati, J. Cui, Multi-stimuli-responsive polymer particles, films, and hydrogels for drug delivery, Chem., 4 (2018) 2084–2107.

[49] S. Uchiyama, N. Kawai, A.P. de Silva, K. Iwai, Fluorescent polymeric AND logic gate with temperature and pH as inputs, J. Am. Chem. Soc., 126 (2004) 3032–3033.

[50] A.M. Al-Dhahebi, J. Ling, S.G. Krishnan, M. Yousefzadeh, N.K. Elumalai, M.S.M. Saheed, S. Ramakrishna, R. Jose, Electrospinning research and products: The road and the way forward, Appl. Phys. Rev., 9 (2022) 011319.

[51] K. Liu, Z. Zhou, X. Yan, X. Meng, H. Tang, K. Qu, Y. Gao, Y. Li, J. Yu, L. Li, Polyaniline nanofiber wrapped fabric for high performance flexible pressure sensors, Polymers, 11 (2019) 1120.

[52] S. Zhang, J. Xu, Y. Sun, Construction of porous polymer films on rGO coated cotton fabric for self-powered pressure sensors in human motion monitoring, Cellulose, 28 (2021) 4439–4453.

[53] A. Ferreira, S. Lanceros-Mendez, Piezoresistive response of spray-printed carbon nanotube/poly (vinylidene fluoride) composites, Composites, Part B, 96 (2016) 242–247.

[54] A.G. Cruz, I. Haq, T. Cowen, S. Di Masi, S. Trivedi, K. Alanazi, E. Piletska, A. Mujahid, S.A. Piletsky, Design and fabrication of a smart sensor using in silico epitope mapping and electro-responsive imprinted polymer nanoparticles for determination of insulin levels in human plasma, Biosens. Bioelectron., 169 (2020) 112536.

[55] A. Jahangiri-Manesh, M. Mousazadeh, M. Nikkhah, Fabrication of chemiresistive nanosensor using molecularly imprinted polymers for acetone detection in gaseous state, Iran. Polym. J. (2022) 1–9.

18 Polymers for the Textile Industry

Emine D. Kocak and Zehra Yildiz

Department of Textile Engineering, Faculty of Technology,
Marmara University, RTE Campus, Aydinevler District, Uyanik Street,
no:6, T1/307, 34840, Istanbul, Turkey

1 POLYMERS: GENERAL KNOWLEDGE

Polymers are macromolecules that consist of small molecules (monomers) that are linked together by covalent bonding. The term "polymer" is derived from two Greek words, "poly" meaning many (numerous) and "mer" meaning units [1]. A polymer consisting of one type of monomer is called a homopolymer, and a polymer consisting of two or more monomer types is called a copolymer. The polymerization degree (DP), a key characteristic of polymers, indicates the number of repeat units (monomer) in a polymer chain and is calculated by the ratio of the molecular weight of the polymer to the molecular weight of the repeat unit [2].

There are two types of polymers: natural and synthetic. Natural polymers, occurring in nature and extracted from vegetable, animal, and mineral sources, have been around for a few billion years, such as cellulose and derivatives. Meanwhile, synthetic polymers, known as "man-made polymers," involve petrochemical compounds such as polyester, polypropylene, etc. [2, 3].

2 CLASSIFICATION OF POLYMERS

Numerous polymers exist having various properties and their classification can be done according to their origin, physical properties, and/or thermal behavior. The classification of polymers is summarized in Figure 18.1 [4]. This chapter mainly focuses on the usage of polymeric materials to produce synthetic fibers, also known as man-made fibers, and polymers that add functional properties to textile materials.

3 POLYMERS IN THE TEXTILE INDUSTRY

3.1 TERMINOLOGY AND DEFINITIONS

A textile fiber, which is the basic element of fabrics and other textiles, has a length of at least 100 times its diameter. Fibers have different chemical structures, cross-sections, surface contours, crimps, color, as well as length and diameter. They can be classified as either natural or manufactured. They can also be classified as protein, cellulosic, mineral, or synthetic based on their chemical structure. The types of fibers with their monomer are illustrated in Table 18.1.

Textile fibers have relatively small diameters, generally within the range of 0.0004–0.002 in or 11–50 μm. Their length ranges from approximately 7/8 in (2.2 cm) to almost miles. Fibers are classified according to their length as filament and staple fiber. Filaments are long, continuous fibers

DOI: 10.1201/9781003278269-18

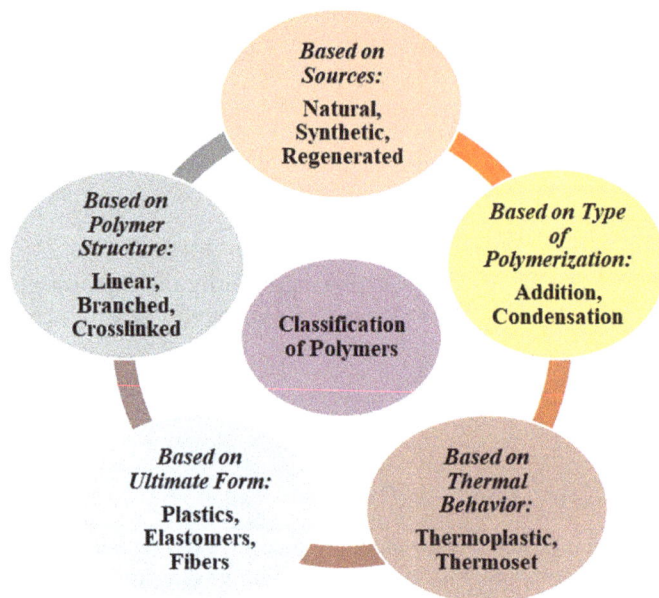

FIGURE 18.1 Scheme of the classification of polymers.

TABLE 18.1
Types and Monomers of Fibers

Type of Fiber	Monomer
Protein fibers	Amino acids
Cellulosic fibers	Carbohydrates
Mineral (inorganic) fibers	Silica for asbestos fibers
Synthetic fibers	Petrochemical-based organic macromolecules

having infinite or excessive length, such as silk or a synthetic fiber, whereas staple fibers are natural fibers excluding silk, or cut filaments with a length of 7/8–8 in (2.2–28.5 cm) [5].

3.2   FIBER-FORMING POLYMERS

Glass fibers have been known since ancient times, however, the use of lamp wicks comprised of glass spun or drawn into fine thread was patented in 1822. The first resins in the vinyl class, the first of the synthetic polymers, and vinylidene chloride were discovered by Regnault in 1838. This was followed by the discovery of nitrocellulose in 1846, cellulose acetate in 1865, acrylonitrile in 1893, polyamide in 1899, protein fibers in 1857, and vinyl acetate and alginate yarns in 1892. Most recently, the fiber-forming polyesters were discovered in 1928, and polyethylene yarns were discovered in 1936. The reproduction of products in the form of filaments or fibers took some time after most of these discoveries were made [6].

During the fiber-forming process, polymers such as polyethylene (PE), polypropylene (PP), poly-tetrafluoroethylene (PET), polyacrylonitrile (PAN), etc. in molten or dissolved states, are pumped through a spinneret with many holes in various shapes. Then the obtained polymeric filaments are solidified in cold/hot air or a coagulation bath. Table 18.2 describes fiber-spinning methods for the production of filaments, staple fibers, and nonwovens [7].

TABLE 18.2

Types of Spinning Methods Based on Material and Product

Spinning System	Raw Material	End-Product
Melt spinning	Molten polymer	Short/filament fibers, textured fiber/yarns
Dry/wet (solution) spinning	Dissolved polymer	Short/filament fibers
Drawing of preforms	Molten organic/inorganic polymer	Filament
Spun-bonding and melt-blowing	Molten polymer	Nonwoven
Film splitting process	Molten polymer	Tape filament
Centrifugal/electro-spinning	Molten/dissolved polymer	Nonwoven
Luminary aided spinning	Dissolved polymer	Nonwoven
Gelation spinning	Polymer gel in solvent	Filament

3.2.1 Polyethylene

Polyethylene or polythene (PE) is the most widely used thermoplastic polymer belonging to the polyolefin class, also called alkene, with the chemical formula of C_nH_{2n}. PE is one of the vinylic type polymers that is manufactured through the additional polymerization of ethylene through a free radical mechanism.

Low-density polyethylene (LDPE) has a less tightly packed polymer chain due to the presence of both short- and long-chain branching between the polymer chains, resulting in less crystallinity, lower strength, and higher ductility. It has a density in the range of about 0.910–0.940 g/cm³. On the other hand, linear low-density polyethylene (LLDPE) has a linear polymer chain with considerable short-chain branching in the density range of 0.915–0.925 g/cm³. Compared to LDPE, LLDPE shows higher tenacity and less ductility. High-density polyethylene (HDPE) is tough and rigid with a high density of greater than 0.941 g/cm³. It has much less branching, which makes it have a high crystal structure and higher tensile strength. Ultra-high-molecular-weight polyethylene (UHMWPE) has a highly dense and very rigid structure with a density of about 0.935 g/cm³. It does not have a packed structure like HDPE due to its high rigidity. UHMWPE has exceptional toughness and high wear and chemical resistance, giving it potential for use in a wide variety of products such as bottle/can handling machine parts, bearings, gears, artificial joints, gaskets, bulletproof vests, etc. [8]. Colored PE fiber can be fabricated using the dope dyeing technique, which involves the inclusion of pigments into the molten polymer before extrusion via the nozzle. PE is resistant to all kinds of acids, alkalis, bleaching agents, organic solvents, chlorinated hydrocarbons, and aromatic solvents, except xylene [9]. PE has the disadvantage of a low melting point, which limits its use in many high-performance applications. It has a brittleness point lower than −114°C, which allows it to maintain its flexibility at very low temperatures. When it is exposed to flame, it burns slowly and then melts before flame propagation. PE fibers are used in many products such as squeeze bottles, plastic bags/films, ropes, toys, filtration substrates, bottles, jugs, buckets, containers, protective clothing, curtains, bulletproof vests, etc.

3.2.2 Polypropylene

Polypropylene (PP) is composed of a propylene monomer and is known as a lightweight polymer compared to the other man-made polymers, and is one of the thermoplastic polyolefin polymers. It is the second-most widely produced plastic after PE [10]. In contrast to PE, the methyl group (CH$_3$) of the pendant group of PP is positioned at one of the hydrogen atoms of the PE structure, giving a slightly harder and better heat-resistant property. PP polymer is manufactured in three different structural stereostatic arrangements: isotactic, syndiotactic, and atactic polypropylene. In isotactic positioning, the methyl pendant groups are located side by side on the polymer backbone, whereas in syndiotactic positioning the methyl pendant groups are positioned in an alternating order. Methyl groups in the atactic structure are found on both sides of the polymer backbone randomly. Syndiotactic and isotactic PP are in the crystalline structure; whereas, atactic PP is characterized by an amorphous structure because of the irregular and random positioning of the polymer chain. Isotactic PP has widespread commercial production methods.

Polypropylene polymer is produced by an additional polymerization reaction from propylene gas, with the addition of catalysts at a pressure of 10 atmospheres and below 80°C. PP polymer has a melting temperature of 163–171°C, hence, PP fibers can be obtained by the melt spinning process [9]. The dope dyeing technique, in which pigments are added to the polymer solution before fibers are extruded, is suitable for PP fiber. PP fiber can be dyed with dispersed dyestuffs under certain conditions with the addition of a solubilizing agent. PP fiber, which has superior durability to alkalis, bleaches, acids, and many solvents, can be deteriorated in nitric acid at high temperature. These fibers dissolve in perchloroethylene, xylene, and 1,1,2,2 tetrachloroethane above 100°C.

PP polymer is used for the production of many products such as parts of TVs, radios, package films, ropes, storage tanks, toys, seat coverings, twine, filters, etc. It has the potential to be used in a wide variety of applications, including the automotive applications, home and wearable textiles, and medical science [11].

3.2.3 Polytetrafluoroethylene

Polytetrafluoroethylene (PTFE), which is a vinyl group, is a thermoplastic polymer that is produced from tetrafluoroethylene (TFE) monomer [12]. Differently from polyethylene, which is the major vinyl group polymer, four H atoms are replaced by F atoms. PTFE exhibits a hydrophobic character with superior chemical, abrasion, and thermal resistance and low friction characteristics. PTFE is generally synthesized by addition polymerization of tetrafluoroethylene and suspension polymerization.

PTFE has a very oriented structure with a high melting temperature of around 330°C. It is insoluble in many solvents, thus, fiber spinning cannot be carried out by dry/wet spinning methods. PTFE fibers are spun by a special method known as the dispersion spinning mechanism for the soluble and insoluble polymer. This method involves the extrusion of the dispersed emulsion of PTFE via a spinneret and sintering at 385°C to manufacture the filaments. PTFE fiber shows excellent thermal characteristics compared to the other man-made fibers. PTFE is used for many high-tech and industrial applications, including coating non-stick cookware, filtration products, laundry pads, water-resistant composites, conveyor belts, electrical insulators/tapes, etc. [13].

3.2.4 Polyvinyl Chloride

Polyvinyl chloride (PVC), that has a Cl atom in the vinylic group $(-CH_2-CHCl-)n$, is a widely used, rigid, non-toxic, inert, thermoplastic polymer produced by the polymerization of vinyl chloride. A total of 80% of PVC is produced industrially by the suspension polymerization method. The PVC fibers can be spun by a dry/melt spinning process. The melting point of PVC is about 120–130°C. Carbon disulfide or acetone are being used as solvents in the dry spinning process. PVC can be dyed with the dope dyeing method.

PVC polymer can be used in windows, braiding, pipes, curtains, filtering surfaces, artificial limbs, energy storage fabrics, accessories for different machineries, etc.

3.2.5 Polyamide

Polyamide (PA) is one of the man-made polymers in which amide bonds (–CO–NH–) are located between the repeating monomers. Polyamides can be developed into various textile fibers. The word "nylon" is used to explain the polyamide fiber derivatives [9]. There are three types of polyamide fibers based on the chemical structure, as explained below.

3.2.5.1 ALIPHATIC POLYAMIDE

Aliphatic polyamides are the most important class of engineering thermoplastics. This type of polyamide has no aromatic or benzene ring in its structure and is synthesized on a larger scale compared to other polyamides. Aliphatic polyamides have widespread uses in many applications, such as cable/wire sheaths, wearable textiles, cooling fans, brakes, power steering reservoirs, speedometers, etc. [14].

3.2.5.2 SEMI-AROMATIC POLYAMIDE

Semi-aromatic polyamide, with a semi-crystalline character, shows a thermoplastic property, and is synthesized through a condensation reaction between terephthalic acid and hexamethylene diamine. The aromatic group amount is about 55% of the total repeating units in the polymer chain [14].

3.2.5.3 AROMATIC POLYAMIDE

The aromatic polyamides, in other words "aramids," are produced by replacing the aliphatic groups with aromatic ones. They have very high melting points, excellent mechanical properties, and heat and flame resistance owing to their highly oriented, fully aromatic structure. Kevlar [poly(p-phenylene terephthalamide)] and Nomex [poly(m-phenylene isophthalamide)] are the most commonly used aramid fibers [14]. Both these fibers are dry spun because they are not suitable for melt-spinning due to their high melting temperatures. Aramid fiber has a wide variety of uses, including bulletproof vests, military and aerospace materials, ballistic composites, protective clothes, bicycle tires, electrical/thermal insulators, fiber cordage, etc. [15].

Nylon fibers are given via a numbering system. These numbers correspond to the number of carbon atoms in the monomers. If a single number is used like nylon 6, nylon 4, nylon 11, that means the monomer/s also consist of the same number of carbon atoms. If two numbers are given with nylon, the first number indicates the amount of C atoms of diamine, whereas the second number shows the amount of C atoms of diacid.

Nylon 6 polymer is synthesized via a ring-opening reaction of caprolactam, which is a white crystalline powder produced using phenol/cyclohexanol. Nylon 6 fiber can be manufactured in different forms including monofilaments, multifilaments, staples, or tow. Nylon 6 fiber shows good durability to weak acids and alkalis, while it is not resistant to mineral acids. It is soluble in formic acid, cresol, and phenol. This fiber shows a melting temperature of 213–220°C and a glass transition temperature of 29–42°C.

Nylon 6.6 fiber is manufactured in an autoclave by condensation polymerization from two monomers, one is di-amine and the other is di-carboxylic acid. Each monomer has six carbon atoms in its chemical structure. Its durability to acids and alkalis is higher than nylon 6 due to having longer polymer chains, which increases the cross-linking density. The fiber dissolves with some degree of degradation in hydrochloric acid, nitric acid, and sulfuric acid. Fiber is not soluble in organic solvents except cresol and phenol. Nylon 6.6 has a melting point range of 249–260°C, whereas its glass transition point is in the range of 29–42°C [9].

The monomer of nylon 11, namely w-amino undecanoic acid, is manufactured from petrochemical resources (castor oil). It is a melt-spun fiber, with a melting temperature of 188°C. Nylon 6.10 is fabricated by the reaction between sebacic acid and hexamethylene diamine through the step-growth reaction. Fibers are obtained at a melting point of 215ºC via melt spinning. Nylon fibers can be easily dyed due to their affinity to dyestuffs such as acids, metal complexes, chromium, and pigments.

3.2.6 Polyethylene Terephthalate

PET, a polymer in which the monomer molecules are linked together by ester linkages, is also called polyester. PET is produced by step-growth polymerization between the ethylene glycol and terephthalic acid or dimethyl terephthalate through releasing water or methyl alcohol [16]. PET polymer has a softening temperature of 225–230°C and a melting temperature of 260°C. Polyester has a hydrophobic character, therefore dyeing should be conducted with hydrophobic dyestuffs in the presence of heat. Polyester fibers are dyed with dispersed dye at high temperatures, 130–140°C, under pressure with acidic media. Tensile strength and elongation percentage vary in the ranges of 4,900–8,750 kg/cm² and 8–50% based on the chemical structure. Polyester fibers are generally resistant to organic acids, but concentrated sulfuric acid dissolves these fibers with some degradation. Fibers are also resistant to weak alkalis and weaken to strong alkalis and degrade at elevated temperatures. However, the fiber exhibits excellent durability to the oxidants and many organic solvents. The fibers also have superior durability to all microorganisms, insects, mold, and moths.

PET, one of the most widely used polyesters today, is used in packaging, engineering, electronics, ropes, home furnishing, industrial fabrics, and biomedical industries. It is widely employed as staple/filament form to produce wearable textile materials. It is also used by blending with natural fibers, commonly cellulose, to be used in a wide range of applications [17].

3.2.7   Polyacrylonitrile

Polyacrylonitrile (PAN), one of the vinyl group polymer having the CN group as R group in the chemical formula of $(-CH_2-CHR-)_n$, is also known as polyvinyl cyanide. PAN is synthesized by radical polymerization from acrylonitrile monomer via an addition mechanism. The fiber, having at least 85% by weight of acrylonitrile, is called acrylic fiber. PAN polymer can be spun into fibers through wet/dry spinning techniques. However, it is difficult to obtain the stable phase of the molten PAN polymer, hence melt spinning of PAN is not practical commercially. The most common solvent for a dry spinning process is dimethylformamide, whereas for wet spinning nitric acid, dimethylacetamide, zinc chloride, or dimethyl sulfoxide can all be used as solvent [18]. Colored acrylic fibers can be produced with the dope dyeing technique. When the tension is removed, PAN fibers can be turned to their initial length with a 90–95% elastic recovery value. Acrylic fibers show strong durability to acids, reducing and oxidizing agents, alkalis, and microorganisms. Their moisture regain is approximately 1–3%. Acrylic fibers are widely used in the production of products such as core-spun yarns, knitted fabrics, carpets, outdoor and furnishing fabrics, flocking fibers for paper printing and tufting, etc.

Modacrylic fiber, a modified acrylic fiber, is a copolymer containing 35–85% acrylonitrile units. A comonomer is used together with acrylonitrile monomer to synthesize modacrylic fibers. The dry or wet spinning technique can be used to produce modacrylic fibers. It has a melting temperature of 200–210°C and a 4% moisture recovery. Modacrylic fibers can be used in sleepwear, carpets, blankets, fur fabrics, wigs, rugs, etc. [18].

3.2.8   Spandex

Spandex is a polymer of urethane (–NH–COO–) and composed of polyurethane (at least 85%), an elastic type polymer that stretches like natural rubber. It is an elastomeric fiber and has an extension at break of more than 200%. When the tension is removed, spandex fibers tend to return to their initial shapes rapidly due to the elastic recovery. The spandex fiber contains two different segments, namely soft segments, which are oriented randomly in an amorphous character that is responsible for the stretchability property, and hard segments, which are stiffly bonded together by strong hydrogen forces. The solvent of DMF is employed to prepare the polymer solution for wet/dry spinning systems. The fibers are durable to alkalis, ozone, peroxide, bleaching agents, chlorine, and UV rays, whereas acidic media may cause color changes (yellowing) of the fibers. Spandex fiber can be dyed with dispersed dye, acid dye, and basic dye. This fiber has a low water absorption capacity, with a moisture regain of 1.3%. Spandex fiber is used in accessories, sportswear, apparel, hosiery, surgical hose, etc. [19].

3.2.9   Polyvinyl Alcohol

Polyvinyl alcohol (PVA) is a man-made vinylic-type polymer in which a hydrogen atom on the vinyl group is replaced by a hydroxyl group (–OH), and is vinyl group polymer. Due to the presence of hydroxyl groups on its structure, it is a water-soluble polymer. The solubility of PVA in hot water is greater than in cold water and it dissolves completely at higher temperatures (90°C). It is an odorless, tasteless, translucent granular powder. Since vinyl alcohol is unstable, PVA is not synthesized from its vinyl alcohol monomer, instead, it is indirectly produced by the hydrolysis of polyvinyl acetate.

PVA fiber is commonly manufactured using a wet spinning system. A 15% PVA solution in water is pumped via a nozzle into the coagulation bath containing sodium sulfate solution to solidify the PVA filaments. The dry spinning technique also can be employed to obtain PVA fibers using a concentrated polymer solution (30–50% PVA). Both dry and wet spinning techniques require a high temperature of about 240°C during the drawing of the fibers to improve the fibers' water durability by enhancing the fiber compactness. PVA fiber is commonly durable against acid and alkali environments but in concentrated acids or in acids at higher temperatures fibers can shrink. The dyeing behavior of PVA fiber, which has hydroxyl groups in its polymer structure, is very similar to the cellulosic fiber dyeing procedure. In the dyeing process of PVA fibers, sulfur, vat, reactive, direct, basic, and acid dyes can be all employed. PVA fibers are not affected by insects or microorganisms. PVA polymer is commonly used as a sizing compound to cover the warp yarns in the weaving stage. The application areas of PVA fiber are as follows: curtains, carpets, upholstery, umbrella, table cloths, sheets, etc. [12, 20].

3.3   POLYMERIC ADDITIVES FOR FIBER-SPINNING

In this chapter, polymeric additives used only in the fiber spinning line for the functionalization of textile materials are investigated.

Additives in fiber spinning are used to ease the processing, to give additional properties, to enhance the already-existing functionalities, or to improve the life span of the end product. The additives should be thermally stable, miscible, and compatible. Additives should also be used in as low a quantity and variety as possible, as they may cause blockages and fluctuations in processing. The main purposes of spinning additives are as follows: to ease the processing (processing additives), to enhance the existing properties (enhancing additives), and to add functional properties (functional additives). The most commonly used additives in fiber spinning are listed in Table 18.3.

Additives in powder forms should have present a uniform particle size distribution and present superior miscibility with a remarkable wettability property with the polymeric component. The particle and aggregation size of additives should be below 5% of the fiber diameter to provide processability, spinnability, and to avoid any loss in the mechanical performances. Additives also carry the probability of causing undesired side effects such as polymer degradation, uncontrolled side reactions, and gas release during spinning. At the end of the spinning, foreign substances that remain in the fiber may cause defects via the formation of microcracks, as well as unwanted fibrillation, that negatively affect the tensile properties, at the interphase of polymeric component and the additive particle. Additives in biodegradable polymers must also be degradable and environmentally friendly, which limits their availability. For instance, commonly used inorganic fillers such as TiO_2 may lead to problems when released into the environment in substantial quantities. With recycling, residual additives that have lost their original function are incorporated in the recycled mixed materials and impair targeted properties. The success of PET bottle flakes in fiber production is mainly due to the absence of colorants, which would turn fibers from recycled PET gray.

TABLE 18.3
Commonly Used Additives for Fiber Spinning

Function	Type
Antioxidant	Phenols, amines, phosphites
Lubricant	Stearates, wax
Processing aid	Fluoropolymers
Surfactant	PEG, stearates
Inhibitor/stabilizer for hydrolysis	Carbodiimide
Compounds supporting nucleation	Talcum, phosphate salts, boron nitrite
Plasticizer	Tributylcitrate, acetyltributylcitrate
Chain extender	Difunctional acids, anhydrides, epoxies
Flame retardant	Phosphorus and halogen derivatives
Antibacterial	Plant extract, TiO_2, Ag^+, ZnO, Zn^{2+}, Cu^{2+}, phenolic compounds
Oil/water repellency	Fluorine- and silica-based chemicals
Antistatic	Carbon black, carbon nanotubes, graphene, ZnO
Colorant	Pigments, dyes, carbon black
Thermal protection	Zirconia
Matting agent	TiO_2, ZnO, mica
UV-stabilizer	TiO_2, ZnO, carbon black

3.3.1 Processing Additives

Oxidative decomposition during melt spinning can be eliminated by using antioxidants that control the oxidative stability of the free radical polymerization reactions until consumed. For instance, antioxidants are being employed to enable the melt-blown stage of polypropylene spunbond wastes [21]. Dispersive additives are preferred to be used as plasticizer/lubricant for the polymeric phase by adjusting the viscosity, and by dissipating the heat and pressure that are required for the spinning equally [22]. Polymer processing additives such as polymers with fluorine groups can thinly coat the die wall, delaying the onset of flow instabilities, which reduces adhesion between the polymer and wall, promoting slippage [23,24]. Nucleating agents such as boron nitride, talcum powder, etc. are used to enhance the heat of crystallization, resulting in the formation of homogeneous and smaller crystals for enhanced mechanical performance of the spun filaments [25]. Stabilizers can be used to diminish the hydrolytic degradation of polyesters by reacting commonly with carboxylic groups or water molecules in order to control the hydrolysis reaction kinetics [26]. Consequently, lubricants are used to ease the spinning process and their melting points are greater than their thermal degradation levels [27].

3.3.2 Enhancing Additives

Flame-retardant additive materials can function in the both gas and condensed phases. In the gas phase, they quench the radical groups in the fire by reacting with the highly active H· and OH· radicals. Meanwhile, if they are being used in the condensed form, they promote char formation as a protective layer that can stop the spread of flames [28]. Plasticizers are mixed with polymeric substrates to increase their elasticity, flexibility, along with processability [29]. The usage of fillers may also restrict the chain mobility of polymers that further cause higher stiffness, rigidity, and creep stability [30,31]. Enhancing additives also should be compatible with other polymeric additives such as light stabilizers, fillers, and surfactants, etc. [32]. Stabilizers can extend the life cycle of polymers by suppressing degradation from UV radiation and other high-energy sources. UV absorbers act as a harmful radiation spreader since less harmful heat, hindered amines/phenols, and light stabilizers inactivate the existing free radicals [33].

3.3.3 Functional Additives

TiO$_2$ or other metal oxides are added as a matting agent and gloss reduction of man-made fibers. In melt spinning of intrinsically antistatic fibers, fillers like carbon black, carbon nanotubes, graphene, or metal powders are mainly used as electrically conductive additives to enable stable fiber melt-spinning [34–36]. Metal and metal oxide nanoparticles are commonly used as antimicrobial agents for man-made fibers. Fillers significantly affect the mechanical properties of man-made fibers [31]. Conductive fibers or a translucent fiber with star-shaped pigmented core, which prevents white swimsuits from losing their opaqueness in wet conditions, are other examples where functional additives are used [37]. In the dope dyeing method, inorganic and organic pigments can be added to the matrix polymer during extrusion in the form of colorant masterbatches to produce colored fibers. These masterbatches usually include additional additives like UV-stabilizers.

3.4 COMMONLY USED ADDITIVES FOR FIBER SPINNING

3.4.1 UV Protection

UV radiation is classified into three main groups based on the wavelength; UV-A, UV-B, and UV-C. Ozone, water vapor, and CO$_2$ in the stratosphere absorb and block the majority of high-energy UV-C and about 90% of UV-B radiation. In other words, 94% of radiation reaching earth is UV-A, which has a longer wavelength and penetrates deeper into the skin [38]. Unfortunately, anthropogenic activity, ozone depletion, and climate change cause more UV-B to penetrate the troposphere, the innermost layer of Earth's atmosphere.

The light transmitted through the fabric pores causes chemical bonds in the natural and synthetic textile fiber to break, similar to the effect of direct UV-ray (UVR) exposure. Additionally, UVR reaches the skin and generates free radicals, causing photo-degradation. Organic and inorganic UV absorbers named UV stabilizers have been frequently used in textile finishing. Incorporation of UV stabilizers in textile finishing can provide absorption of the incoming intense radiation, thereby improving the UV protection factor (UPF) value. The UV absorbers, compounds that absorb ultraviolet light in the UV region (200–400 nm), are transparent to visible light and thus act as optical filters. Organic UV absorbers are conjugated aromatic compounds that have functional groups such as hydroxyl, carboxyl, and amino groups as a chromophore and are responsible for sunscreen properties. O-hydroxy benzophenones, o-hydroxyphenyl benzotriazoles, o-hydroxyphenyl hydrazine, and salicylic acid derivatives are representative examples of this kind of UV absorbers [39].

There have been earlier reports on the direct application of the substituted benzotriazole and benzophenone in order to be used as an UV absorbers. These chemical species have been employed during or after the dyeing of polyester fabrics with synthetic dyes [40,41]. In a previous study, the substituted benzophenone, 2,4-dihydroxy benzophenone was found to be more effective than benzophenone in the UV finishing of cotton fabrics with reactive dyed [42]. The UV absorbers having reactive groups like chloride, –OH, and sulfonic acid, which enable the formation of covalent as well as hydrogen bondings with textile fibers tightly, result in improved long-term UV protecting performance. It was reported that benzophenone-containing benzotriazole groups enhance the light resistance of dyes and the UPF [43]. In another study, the reactive UV absorber, 2-(2-hydroxy-4-acryloyl)-2H-benzotriazole (BTHA), was synthesized and grafted onto polyester fabric to improve the UV resistance of PET [44]. The stabilizing effect of the combination of these UV absorbers with NTS (nickel p-toluenesulfonate), as a single oxygen quencher, is found to be more effective against photofading of dyes [45]. While the developed reactive UV absorbers provided excellent results on cellulosic fibers, some of the synthesized UV absorbers exhibited good UV protection on bamboo-viscose tricot. Bi-functional cotton fabric has been manufactured by a double coating of 2-hydroxy-4-acryloyloxy-benzophenone (HAB) on a cotton fabric followed by methacryl oxymethyl trimethylsilane (MSi) [46].

3.4.2 Flame Retardancy

Textiles are primarily composed of organic polymers, thus, they are flammable and potentially dangerous if they do not inherently possess a flame-resistant property. In this respect, researchers have focused on designing and synthesizing additives in order to impart a flame-resistant property that can inhibit or delay the flame and/or reduce the rate of flame propagation or delay ignition, or reduce the combustion rate. Combustion is a gas-phase chemical process requiring oxygen/air from the atmosphere. Textiles decompose before the combustion process takes place; some of the decomposition products obtained turn into combustible volatile species, which, together with oxygen, fuel the flame. Fibers and fabrics, if not inherently flame-retardant, have to be treated with additives that may include halogen, phosphorus, nitrogen, sulfur, boron, metals, etc. to become flame-retardant. These additives can be included into the polymer solution during spinning processes of man-made fibers, or applied on the man-made or natural textile substrates, thereby forming a protective layer on the material during the fire.

The flame-retardant (FR) additives isolate the flame from the oxygen in the air, release the flame inhibitors through the ignition temperature, reduce the generated heat below that required for the combustion process, support the char formation to prevent contact with the oxygen/air source, reduce the heat flow back to the fabric to limit further pyrolysis, and modify the pyrolysis mechanism to prevent fire propagation through the textile substrate.

FRs can also be classified according to "laundry durability" for textiles as non-durable, semi-durable, and durable FRs. A non-durable FR may have the ability to be washed immediately while resisting dry cleaning. Semi-durable FRs can withstand a few washes, while durable FRs withstand about 50 or 100 washing cycles. Studies show that new flame retardants based on phosphorus compounds have lower toxicity compared to halogen-based ones.

A propionylmethylphosphinate-based phosphorus-containing monomer has been used to impart flame-resistant properties to polyester textile materials. Another focus of current studies, for cotton and cellulose-based substrates, is the synthesis of effective non-halogenated additives for coatings and back-coated fabrics, or the use of hydroxymethylphosphonium salts or N-methylol phosphonopropionamide derivatives [47,48]. FRs should be formaldehyde-free, non-toxic, environmentally friendly, and cheap compared to currently available chemicals. They should demonstrate equivalent or superior ease of application. They should not cause any change in the shade of the dye and/or the dyeability of the textile materials. In particular, they should provide comparable properties in terms of durability, tensile, and comfort-related properties, external appearance, and aesthetics [28].

3.4.3 Water/Oil Repellency

The contact angle, "Θ," is the angle formed (tangent T) at a three-phase boundary where gas, solid, and liquid intersect. According to Thomas Young's theory, if the adhesion between a liquid and solid (textile material) is higher than the cohesion of the liquid, the contact angle is zero, so the fabric gets completely wet. If the cohesion of the liquid to solid is smaller than the cohesion of the liquid, the contact angle is higher than zero. This contact angle increases as the adhesion between the liquid and the solid decreases with respect to the cohesion of the liquid. Polydimethylsiloxane, silica, and fluorine agents chemically incorporated into fiber surfaces are water/oil-repellent agents commonly used today. In recent years, there has been a significant increase in the use of commercial fluorochemicals, especially for the partial water/oil repellency of cotton [49].

3.4.4 Antimicrobial Additives

The growth of microorganisms on textiles creates many undesired effects on both the textile itself and the user, including unpleasant odor, stains, discoloration in the fabric, reduced fabric mechanical strength, and an increased likelihood of contamination. To avoid or eliminate these undesired effects, it is aimed to minimize the growth of microbes on textile products during use and storage.

In the production of synthetic fibers, the antimicrobial active agents can be added into the polymer matrix before extrusion or blended into the fibers during their formation. This provides the best durability by allowing release of the active agent slowly during usage. Biocides such as triclosan and polyhexamethylene biguanide (PHMB) have been applied to natural and synthetic fibers by conventional exhaust and pad–dry–cure methods for antimicrobial finishing. Silicone-based quaternary agents have been applied by padding, spraying, and foam finishing. Other methods for antimicrobial finishing include the use of nanosized colloidal solutions, nanoscale shell–core particles, chemical modification of the biocide for covalent bond formation with the fiber, and cross-linking of the active agent onto the fiber using a cross-linker and polymerization grafting. Sol–gel, which has been extensively investigated for antimicrobial finishing applications, has been incorporated with biocides. Several biocides have been encapsulated in sol–gel particles, and are then coated onto textile products to provide the desired antimicrobial property [50].

4   CONCLUSIONS

In this chapter, polymers that are being used in the textile industry were divided into two categories. One is the polymers that form a textile substrate, and the other is polymers that are used as additives, specifically for usage in fiber spinning systems. Brief explanations were given about the polymers related to the physical, chemical, mechanical, and structural properties with specific examples of their application areas. The additive polymers section was limited to UV protection, flame retardancy, water/oil repellency, and microbial protection, and specific examples were given from the literature. Further market growth will be seen in the method of designing specialty polymers based on customer demands with desired functionalities.

REFERENCES

[1] V. Gowariker, N. Viswanathan, and J. Sreedhar, Individual polymer, *Polymer Science,* vol. 1, 1999.

[2] R. O. Ebewele, *Polymer science and technology*. CRC Press, 2000.

[3] G. Odian, *Principles of polymerization*. John Wiley & Sons, 2004.

[4] K. Matyjaszewski and T. P. Davis, *Handbook of radical polymerization*. John Wiley & Sons, 2003.

[5] M. M. Houck and J. A. Siegel, Chapter 15 – Textile Fibers, *Fundamentals of Forensic Science (3rd ed.). Cambridge: Academic Press*. pp. 381–404, 2015.

[6] G. Loasby, The development of the synthetic fibres, *Journal of the Textile Institute Proceedings,* vol. 42, no. 8, pp. P411–P441, 1951.

[7] M. Lewin, *Handbook of fiber chemistry*. CRC Press, 2006.

[8] J. G. Cook, *Handbook of textile fibres: man-made fibres*. Elsevier, 1984.

[9] S. Mishra, *A text book of fibre science and technology*. New Age International, 2000.

[10] M. Titow, *PVC technology*. Springer Science & Business Media, 2012.

[11] J. Karger-Kocsis, *Polypropylene: an AZ reference*. Springer Science & Business Media, 2012.

[12] S. C. Ugbolue, *Polyolefin fibres: Structure, properties and industrial applications*. Woodhead Publishing, 2017.

[13] R. J. Plunkett, The history of polytetrafluoroethylene: discovery and development, *High Performance Polymers: Their Origin and Development,* pp. 261–266, Springer, 1986.

[14] R. Meredith, *Elastomeric Fibres*. Woodhead Publishing Limited 2004.

[15] P. Reports, "Global Aramid Fiber Market Research Report 2012–2024," 2019.

[16] F. Ashrafi and M. R. B. Lavasani, Improvement of the mechanical and thermal properties of polyester nonwoven fabrics by PTFE coating, *Turkish Journal of Chemistry,* vol. 43, no. 3, pp. 760–765, 2019.

[17] P. Visakh and M. Liang, *Poly (ethylene terephthalate) based blends, composites and nanocomposites*. William Andrew, 2015.

[18] C. Andreoli and F. Freti, Reference books of textile technology: man-made fibres, *Fondazione Acimit,* vol. 6, no. 9, 2006.

[19] M. Senthilkumar, N. Anbumani, and J. Hayavadana, Elastane fabrics–A tool for stretch applications in sports, *Indian Journal of Fibre & Textile Research*, vol. 36, pp. 300–307, 2011.

[20] T. S. Gaaz, A. B. Sulong, M. N. Akhtar, A. A. H. Kadhum, A. B. Mohamad, and A. A. Al-Amiery, Properties and applications of polyvinyl alcohol, halloysite nanotubes and their nanocomposites, *Molecules*, vol. 20, no. 12, pp. 22833–22847, 2015.

[21] M. N. Subramanian, *Plastics additives and testing*. John Wiley & Sons, 2013.

[22] L. R. Rudnick, *Lubricant additives: chemistry and applications*. CRC Press, 2009.

[23] K. Migler, S. Hatzikiriakos, and K. Migler, Chapter 5 Sharkskin Instability in Extrusion, *Polymer Processing Instabilities: Control and Understanding*, Hatzikiriakos, S., Migler, K. (Eds.), Marcel Dekker, Monticello, NY, pp. 121–160, 2005.

[24] L. A. Archer, Wall slip: measurement and modeling issues, *Polymer Processing Instabilities: Control and Understanding*, pp. 73–120, 2005.

[25] G. Wypych, *Handbook of nucleating agents*. Elsevier, 2021.

[26] P. Stloukal, G. Jandikova, M. Koutny, and V. Sedlařík, Carbodiimide additive to control hydrolytic stability and biodegradability of PLA, *Polymer Testing*, vol. 54, pp. 19–28, 2016.

[27] A. Gooneie, P. Simonetti, K. A. Salmeia, S. Gaan, R. Hufenus, and M. P. Heuberger, Enhanced PET processing with organophosphorus additive: Flame retardant products with added-value for recycling, *Polymer degradation and stability*, vol. 160, pp. 218–228, 2019.

[28] K. A. Salmeia, S. Gaan, and G. Malucelli, Recent advances for flame retardancy of textiles based on phosphorus chemistry, *Polymers*, vol. 8, no. 9, p. 319, 2016.

[29] A. Marcilla and M. Beltran, Mechanisms of plasticizers action, *Handbook of Plasticizers*, pp. 119–133, 2004.

[30] Gooneie and R. Hufenus, Hybrid carbon nanoparticles in polymer matrix for efficient connected networks: Self-assembly and continuous pathways, *Macromolecules*, vol. 51, no. 10, pp. 3547–3562, 2018.

[31] M. Kotal and A. K. Bhowmick, Polymer nanocomposites from modified clays: Recent advances and challenges, *Progress in Polymer Science*, vol. 51, pp. 127–187, 2015.

[32] M. Lewin and E. D. Weil, Mechanisms and modes of action in flame retardancy of polymers, *Fire Retardant Materials*, vol. 1, pp. 31–68, 2001.

[33] M. Tolinski, *Additives for polyolefins: getting the most out of polypropylene, polyethylene and TPO*. William Andrew, 2015.

[34] M. Miao and J. H. Xin, *Engineering of high-performance textiles*. Woodhead Publishing, 2017.

[35] M. Stoppa and A. Chiolerio, Wearable electronics and smart textiles: A critical review, *Sensors*, vol. 14, no. 7, pp. 11957–11992, 2014.

[36] F. Meng, W. Lu, Q. Li, J. H. Byun, Y. Oh, and T. W. Chou, Graphene-based fibers: a review, *Advanced Materials*, vol. 27, no. 35, pp. 5113–5131, 2015.

[37] R. Hufenus, Y. Yan, M. Dauner, and T. Kikutani, Melt-spun fibers for textile applications, *Materials*, vol. 13, no. 19, p. 4298, 2020.

[38] T. N. T. Tran, J. Schulman, and D. E. Fisher, UV and pigmentation: molecular mechanisms and social controversies, *Pigment Cell & Melanoma Research*, vol. 21, no. 5, pp. 509–516, 2008.

[39] G. Wypych, *Handbook of UV degradation and stabilization*. Elsevier, 2020.

[40] E. Tsatsaroni and A. Kehayoglou, Dyeing of polyester with CI disperse yellow 42 in the presence of various UV-absorbers. Part II, *Dyes and pigments*, vol. 28, no. 2, pp. 123–130, 1995.

[41] A. Kehayoglou, E. Tsatsaroni, I. Eleftheriadis, K. Loufakis, and L. Kyriazis, Effectiveness of various UV-absorbers in dyeing of polyester with disperse dyes. Part III, *Dyes and Pigments*, vol. 34, no. 3, pp. 207–218, 1997.

[42] J. Chakraborty, Enhancing UV protection of cotton through application of novel UV absorbers, *Journal of Textile and Apparel, Technology and Management*, vol. 9, no. 1, 2014.

[43] H. Oda, Development of UV absorbers for sun protective fabrics, *Textile Research Journal*, vol. 81, no. 20, pp. 2139–2148, 2011.

[44] Li, W. Ma, and X. Ren, Synthesis and application of benzotriazole UV absorbers to improve the UV resistance of polyester fabrics, *Fibers and Polymers*, vol. 20, no. 11, pp. 2289–2296, 2019.

[45] Y. Shen, L. Zhen, D. Huang, and J. Xue, Improving anti-UV performances of cotton fabrics via graft modification using a reactive UV-absorber, *Cellulose*, vol. 21, no. 5, pp. 3745–3754, 2014.

[46] H. N. Ergindemir, A. Aker, A. Hamitbeyli, and N. Ocal, Synthesis of novel UV absorbers bisindolylmethanes and investigation of their applications on cotton-based textile materials, *Molecules,* vol. 21, no. 6, p. 718, 2016.

[47] J. Smith, B. Coston, and C. Duckett, Treated inherently flame resistant polyester fabrics, Google Patents, 2006.

[48] M. Sato, S. Endo, Y. Araki, G. Matsuoka, S. Gyobu, and H. Takeuchi, The flame-retardant polyester fiber: Improvement of hydrolysis resistance, *Journal of Applied Polymer Science,* vol. 78, no. 5, pp. 1134–1138, 2000.

[49] A. Khatton, M. N. Islam, M. Hossen, J. Sarker, H. A. Sikder, and A. S. Chowdhury, Development of water repellency on jute fabric by chemical means for diverse textile uses, *Saudi J Eng Technol,* vol. 7, no. 3, pp. 128–131, 2022.

[50] Y. Gao and R. Cranston, Recent advances in antimicrobial treatments of textiles, *Textile Research Journal,* vol. 78, no. 1, pp. 60–72, 2008.

19 Polymers for Adhesives and Sealants

S.T. Mhaske, Karan W. Chugh, Umesh R. Mahajan, and Jyotidarsan Mohanty

Department of Polymer and Surface Engineering, Institute of Chemical Technology, Nathalal Parekh Marg, Matunga (E), Mumbai-400019, India

1 INTRODUCTION

Adhesives and sealants are frequently considered in the same framework. Their characteristics are extremely dependent on how they are applied and treated since they both adhere and seal; both must be resilient to their operational environments. An adhesive is a substance that is applied to the surfaces of objects in order to permanently or temporarily connect them together through an adhesive bonding technique. By attaching adhesive to both surfaces to be bonded, the adhesive bonding process balances two equally significant forces: adhesive force and cohesive force [1].

Adhesive – a substance capable of firmly and permanently adhering at least two surfaces together.
A sealant – a substance that can attach to at least two surfaces, covering the gap between them and acting as a barrier or protective layer.

The capacity to hold and bond is why adhesives are chosen. They are typically materials with strong shear and tensile strengths. The phrase "structural adhesive" refers to any adhesive whose strength is crucial to the assembly's performance. This phrase refers to adhesives that have a high shear strength (more than 1,000 pounds per square inch or psi) and are resistant to the elements. Structural adhesives include epoxy, thermosetting acrylic, and urethane solutions. Structural adhesives are designed to survive longer than the thing on which they are applied. Non-structural adhesives are weaker and less long-lasting than structural adhesives. They're commonly utilized to link two weak substrates for temporary fastening or anchoring. Non-structural adhesives include pressure-sensitive films, wood glue, elastomers, and sealants [2–4].

Sealants are used to fill gaps, restrict relative substrate movement, and exclude or enclose another material. They are less robust but more flexible than adhesives in general. Sealants such as urethanes, silicones, and acrylic systems are commonly used. Adhesives and sealants are both essentially based on the adhesion property. The attraction of two different substances generated by intermolecular forces between them is known as adhesion. This is in contrast to cohesion, which is limited to the intermolecular attractive interactions inside a single substance. Adhesives and sealants are frequently considered in the same context because they both adhere and seal and share several characteristics. The main purpose of a sealant is not to withstand significant stress, but to seal and fill gaps between joints and to provide a barrier or protective coating [5–7].

Sealants are widely used to join and/or seal materials with varying thermal coefficients of expansion or modulus, needing sufficient flexibility and elasticity, and typically have lower strength and greater elongation than adhesives. In some cases, the substrates may even move, necessitating significant expansion and shrinkage of the sealant without losing adhesion to the substrates. The adhesive properties of a sealant are affected by water exposure, temperature, and surface cleanliness, depending on the chemistry. In some cases, a priming step may be required to improve the sealant's wetting and adhesive strength. Flexibility and adhesive strength aren't the only things that matter. A sealant must be resistant to heat, tear, ultraviolet light, moisture, and oxidation, depending on the application. Appropriate open and working time, good sag resistance, paint-over-ability, colour, self-levelling properties, and hardening time are all important application properties [7,8].

Adhesives and sealants are frequently composed of similar compounds; that is, both can be formulated with similar monomers/polymers, fillers, plasticizers, and elastomers to achieve desired properties such as cure speed, cost, flexibility, and elongation. Sealants are typically paste-like in consistency to allow for the filling of (large) gaps between substrates and have low shrinkage to prevent crack formation after cure or physical solidification. Application requirements and failure mechanisms are comparable for sealants and adhesives. However, because adhesives and sealants are designed to perform various roles and must fulfil different performance requirements, their specifications and test procedures are frequently extremely different [9–12].

2   CLASSIFICATIONS OF SEALANTS

Sealants are classified based on their ability to handle joint width movement. In general, three classes are defined based on the sealant's movement capacity:

Low-performance sealants are highly filled and have movement capabilities ranging from 0 to 5% of the joint width. These crack fillers are low cost and have a short lifespan. The foundation material can be oil, resin, bituminous, or polyvinyl acetate.

Medium-performance sealants have a joint width movement capacity of 5–12% and have a longer service life than low-performance sealants. This category includes butyl rubber, latex acrylics, and solvent-based acrylics. The shrinkage of these sealants after application ranges from 10% to 30%.

High-performance sealants are chemically curable elastomers with a movement capacity of more than 12%. This category includes polyurethane, acrylics, polyether-modified silicones, polysulphides, and silicones [7,13,14].

2.1   TYPES OF SEALANT

2.1.1   Silicone Sealants

Silicone sealant is one of the most prevalent sealants. Silicone sealants come in two varieties: neutral cure and acetoxy cure. Silicone sealants are made by polymerizing and hydrolysing siloxanes and silanes over a long period of time. Both neutral and acetoxy silicone sealants cure at room temperature and are suitable with many different materials. Acetoxy silicone sealants are less expensive and cure faster than their equivalents. Acetoxy silicones, on the other hand, are incompatible with sealing substrates that could react with acids. Neutral-cure silicone sealants cure slower and are slightly more expensive to manufacture than acetoxy silicone sealants. Silicone sealants have a 10–20- year life expectancy after application.

Due to its ease of use, durability, and vulcanizability at room temperature, silicone is the most widely used sealant in both residential and industrial settings (RTV). Silicone is commonly utilized in high-rise building sealing, bridge joints, and weatherproofing. Silicone is commonly utilized in the sealing of sanitary joints in homes due to its water resistance. Silicone is also the most widely used sealant for electrical sockets, wiring, and fire-rated joints. Silicone is also utilized in structural glazing as a load-bearing structure because of its great strength after curing [13,14] (Table 19.1).

TABLE 19.1

Types of Polymers Used as Sealants

Sr. No.	Polymers	Property	Application
01	Silicone	Room temperature and fast curing	High-rise building sealing, bridge joints, fire-rated joints
02	Epoxy	High strength, high toughness, high chemical resistance	Hybrid circuits, PCBs, and integrated circuits
03	Phenolics	High temperature endurance	Sealing of plywood, building construction, appliance industry
04	Polyurethane	Flexible, strong adhesion, abrasion and shear resistant	Home decorative applications
05	Acrylic	High environmental resistance	Doors and window frames sealing
06	Polysulphide	Lower shrinkage and UV resistant	Underwater applications

2.1.2 Epoxy Sealants

Epoxy sealants are usually sold in two packs, one with the resin and the other with the hardener. For the epoxy to conduct joint sealing, they are mixed in a predetermined ratio. Epoxy sealants are noted for their high strength, outstanding cure toughness, and resistance to weather and chemical damage. One of the few sealants that can also act as an adhesive is epoxy sealants. Epoxy sealants cure at room temperature, however they must be thermally cured in specific instances.

Epoxy is frequently utilized in industry due to its strong sealing and adhesive characteristics. Epoxy is utilized in the paint industry as a protective coating since it is so effective. Epoxy sealants are frequently employed in the automobile and aviation industries because they give good structural stability. Because of their non-conductive and quick-drying properties, epoxy sealants are widely used in the electronics industry. Epoxy is used to seal the joint gaps in hybrid circuits, PCBs, and integrated circuits. Epoxy sealants are also used to seal floor joints in high-traffic areas and around swimming pools [1,7].

2.1.3 Phenolic Sealants

Phenolic sealants are resins that have a high temperature endurance rating and provide excellent bonding. Only powder, liquid, and film phenolic sealants are available. Phenolic sealants are typically composed of the chemicals phenol and formaldehyde. Plywood adhesives and sealants, building construction, and the appliance industries all employ phenolic sealants. [6].

2.1.4 Polyurethane Sealants

A popular solution is polyurethane sealant. Polyurethane sealants are not only flexible, but also have good adhesion and are resistant to abrasion and shear. Polyurethane sealants for the home cling well to a variety of substrates and are easy to apply due to their minimal surface preparation requirements. Polyurethane sealant is a type of sealant that comes in conventional and reduced VOC versions. The higher the volatile organic component level, the more safety precautions are required during sealant application [5,7,15].

2.1.5 Acrylic Sealants

Acrylic sealants are created by catalysing an acrylic acid reaction (hence the name acrylic sealant). Acrylic sealants are extremely resistant to the elements. Acrylic sealants, on the other hand, are chemically vulnerable. Acrylic sealant can be cured by a number of methods, but thermal curing speeds up the process dramatically. Acrylic sealants have a strong holding strength and keep foreign particles out.

Doors and window frames are typically sealed with acrylic sealants. Caulking, jointing, and grouting are all frequent uses for acrylic sealants. Because they are odourless and easily paintable, acrylic sealants are the most widely utilized in sealing household joints. Because of their flexibility, acrylic sealants are not advised for use on glass panes [5,15].

2.1.6 Polysulfide Sealants

Polysulfide sealants are becoming more popular due to their flexibility and ability to retain joint suppleness even at low temperatures. They have a low shrinking rate and are UV-resistant, therefore they can be used outside. These sealants are also appropriate for use underwater.

Polysulphide sealants are usually more expensive than other types of sealants for the home. They do, however, have a 15–20-year life expectancy, which compensates for the initial investment. Another factor to consider is that polysulphide sealants have a relatively high VOC concentration. Additional safety precautions are required while using these sorts of sealant systems. Manufacturers commonly provide them in product packaging [6,7].

2.2 Curing of Sealants

Curing is the process of toughening or allowing the sealant to settle down properly after it has been applied. The type of sealant used on the joints determines the curing process and duration. Some sealants can take anywhere from a few hours to many weeks to cure. Curing techniques for various types of sealants include the following.

2.2.1 Ambient Temperature

During the curing process, the sealant is applied to the joint and allowed to cure at room temperature. The moisture in the air causes the sealant to cure. The cure period for ambient-temperature sealants is usually 30 minutes to 4 hours. The thickness of the sealant layer, as well as the ambient humidity, have a direct impact on the curing time. This procedure is commonly used to cure silicone and epoxy sealants.

2.2.2 Curing with Heat

The full toughness and hardness of thermally cured sealants is not achieved until they are heated to a certain temperature. Thermoplastic sealants and thermosetting sealants are the two forms of thermally cured sealants. Polymer sealants frequently use this method of curing.

2.2.3 Anaerobic Curing

Sealants are cured in the absence of oxygen, which is known as anaerobic curing. Anaerobic sealants are commonly utilized in metal joining when metal ions are present.

2.2.4 UV (ultraviolet) Rays/Radiation

To cure the sealant, ultraviolet light or electron beams are used instead of heat from an external source. This method of curing is beneficial since it uses less energy and takes less time to cure. This process can be used to cure acrylic sealants [1,4–7,14].

2.3 Material compatibility

It is necessary to check whether the connecting material is compatible before selecting a sealer from the available sealants. When a sealant is applied to an incompatible substance, the sealant may degrade and fail to seal the connection.

Porous surfaces: Sealants with a high viscosity or a gel-like texture function best on porous surfaces. Silicone, polymers, and epoxies are the finest sealants for porous materials.

Concrete: This is a construction material used to build buildings, walls, and other constructions. Polymer sealants are commonly used to seal concrete joints.

Metal: Silicone and polymer-based sealants are commonly used to seal metal joints. Iron, aluminium, steel, and iron compounds are chemically inert to silicone.

Ceramics: Non-metal oxides and nitrides with high melting and boiling points help compensate ceramics. Ceramics can be sealed with epoxy, silicone, or acrylic sealants.

Textiles: Silicone-based sealants are ideal for textiles.

Other polymers include polymer-derived organic, process, or synthetic materials. The most appropriate sealants are silicone and polymer [1,4–7,14].

2.4 PROPERTIES OF SEALANTS

When choosing a sealant, it is important to think about the properties that will have the biggest impact on the area of the building where it will be used. The important sealant properties to consider for your project are as follows.

2.4.1 Consistency

Pourable sealants are employed in horizontal joints because of their fluid consistency. They can also level themselves. Even when applied to vertical joints, non-sag sealants are thicker and do not run.

2.4.2 Durability

Under ideal conditions, a sealant's predicted life cycle is unlikely to match its actual lifespan, particularly if the sealant was applied inappropriately to the surface or is incompatible with the substrate to which it is applied.

Silicones have the longest service life in general (around 20 years or more). Some acrylics and butyls have a 5-year shelf life.

2.4.3 Hardness

Damage resistance is higher with a tougher sealant. Flexibility, on the other hand, reduces as hardness increases.

2.4.4 Resistance to Exposure

In the sun, temperature fluctuations, and wetness, high-performance sealants continue to operate effectively and remain flexible.

2.4.5 Capability to Move

As an example, a sealant with 10% movement capability in a 25-mm joint can stretch to 28 mm or shrink to 23 mm without failing.

2.4.6 Modulus

This is shorthand for elasticity modulus. Although this is not always the case, low-modulus sealants have a high movement capability and vice versa. Low-modulus sealants are typically employed when working with fragile substrates. Non-moving and static joints commonly utilize high-modulus sealants. Medium-modulus sealants are multipurpose materials that balance stress at the sealant's adhesion surface with sealant stiffness.

2.4.7　Adhesion

A sealant's ability to cling to the construction material is a crucial consideration. Test methods are used to assess the adherence of elastomeric sealants (for example, ASTM C794 Standard Test Methods of Adhesion-in-Peel of Elastomeric Joint Sealants). Manufacturers also supply adhesion statistics for various substrates

2.4.8　Staining

Sealant components can leak into porous substrates (like natural stone) and discolour them. Even if the sealant promises to be non-staining, you should try it first in a hidden location.

2.4.9　VOC Content

Any product that releases volatile organic compounds (VOCs) needs to be understood. The majority of sealant producers have developed low-VOC formulations. Solvent-based sealants contain more respiratory irritants and environmental pollutants than water-based sealants and should thus be avoided. The amount of VOCs in a product, however, varies substantially.

2.4.10　Ease of Application

The curing and tooling qualities (ease of achieving a smooth surface with the correct/required geometry) are crucial when considering a sealant's ease of application. It is worth mentioning that some cure rapidly, while others are created to remain uncured.

2.4.11　Cost

Being less expensive does not usually mean better, as it does with most building products. Higher priced items deliver higher results. Replacing failed sealants is nearly always more costly than selecting the right sealant in the first place. Make prudent purchases though, and concentrate your efforts on meeting the performance requirements [1,5–7].

3　TYPES OF ADHESIVES

3.1　Hot Melt Adhesives

Hot melt adhesives (HMA), often known as hot glues, are 100% solid thermoplastic resin-based compositions. They are solid at room temperature, but after being heated over their softening point, they can be softened, moulded, and dispensed. These thermoplastic adhesives often harden in seconds after dispensing and reach maximum strength once cooled to room temperature. This speeds up assembly by reducing clamp time. Hot melt adhesives are also simple to clean, have low to no toxicity, and may be applied and bonded in an automated manner. They can also fill gaps and bond a variety of porous and non-porous substrates, including pre-painted steel and polyolefin plastics. Traditional hot melts have various drawbacks, including temperature sensitivity, which means they soften and lose strength at high temperatures and become brittle at low temperatures. They are also prone to creep, which, when pressured, can lead to joint failure. Most hot melt adhesives cannot be utilized on temperature-sensitive surfaces since they are applied at temperatures ranging from 100 to 230°C (212 to 450°F). Another downside is the short open time, which refers to how long it takes the liquid adhesive to harden [16,17] (Table 19.2).

Ethyl vinyl acetate (EVA) is used for general-purpose bonding, polyolefins for difficult-to-bond plastics, styrene block copolymers (SBC) for low-temperature flexibility, high elongation, and increased heat resistance, polyamides for hostile environments, and reactive urethanes and silicones for raised temperature and/or high flexibility needs make up hot melts. Tackifying resins, waxes, plasticizers, oils, fillers, and antioxidants are commonly combined with these thermoplastics and added into basic hot melt resins to improve adhesive performance. Tackifiers (natural or synthetic)

TABLE 19.2

Types of Polymers Used as HMA Adhesives

Sr. No.	Polymer	Properties	Application
01	Ethylene-vinyl acetate	High strength, high toughness, high impact resistance	Paper, wood, plastic, rubber substrate adhesion
02	Polyamide	Flame retardancy, high thermal stability, high creep resistance	Automotive, textile, hot air filter, shoe adhesive
03	Polyurethane	Low-temperature flexibility, low glass transition temperature	Construction, woodwork, textile laminates
04	Polyolefins	Excellent barrier properties, high chemical resistance	Packaging and non-wovens industries

are often employed in adhesive formulations to alter or improve tack (adhesion), open time, surface wetting and polymer flexibility. Waxes aid in pellet blockage reduction, melt viscosity decrease, and/or adhesive tack augmentation. Antioxidants are used to keep hot melts from oxidizing and decomposing during service and processing, as well as to promote storage stability.

Hot melts are utilized in a wide range of sectors for a variety of purposes. Rubbers, plastics, metals, ceramics, glass, and wood are among the substrates they can bond. The packaging sector is one of the most important users. Carton sealing, corrugated box and paperboard carton construction, and labelling are all done with hot melts. Other commodity applications include shoemaking (toecap and sole bonding), disposable diaper and sanitary napkin bonding, and bookbinding. Hot melts are also commonly employed in the textile sector to make non-woven fabrics, and also in the furniture industry for veneer surrounds and edging. In the automotive business, hot melts are used to bond carpeting and seat covers, while in the electronics industry, hot melts are used to bond coil windings and coil ends, among other things [9,18,19]. Figure 19.1 shows the applications of polymers used in adhesives and sealants.

Hot melt adhesives made of ethylene-vinyl acetate (EVA) are one of the most popular, affordable, and commonly used commercial hot melt adhesives. They adhere to a wide range of cellulose fibres and are available in a number of formulations. Because they, like other hot melts, have a low creep resistance beneath load, they have a low production cost but a low performance. The EVA copolymer gives strength, toughness, and resilience, while the tackifier resin enhances wetting and tack, and the wax lowers viscosity and cost while improving tack. Stabilizers (antioxidants) are routinely used to enhance thermal stability or UV light resistance, while fillers are occasionally used to cut costs and change performance attributes. EVA formulations have a 1-minute open time and great adhesion to a range of substrates, including paper, woods, plastics, and rubbers [20–22]. Polyamide hot melt adhesives (HMA) are a good option for high-temperature applications. They have strong thermal stability, melting temperatures, and chemical resistance to many common organic solvents, plasticizers, and oil. In terms of creep resistance under load, flame retardancy, and heat stability, they surpass other hot melts like EVA and polyolefins. Many grades can withstand temperatures as high as 160°C. Because of the presence of strong inter-chain hydrogen bonding, polyamides have a high cohesive strength, which explain their relative strength, melting point, and high temperature and chemical resistance [16,23,24].

Polyurethane hot melt adhesives differ from standard hot melt adhesives in several ways. They are often more suitable for heat-sensitive materials than other hot melts since they are applied at a low temperature from a heated cartridge. This type of adhesive binds well to a range of substrates due to the presence of polar groups. Urethane hot melts have a wide range of softening points and relatively low glass transition temperatures. Long linear chains with flexibility, soft segments

FIGURE 19.1   Application of polymers used in adhesives and sealants.

(polyether/polyethylene) alternate with short, rigid urethane bridges created by diisocyanate reacting with polyol chain extenders to form polyurethanes. Hydrogen bonds are formed between the stiff portions. A higher soft-to-hard segment ratio enhances flexibility and reduces the glass transition, resulting in better low-temperature performance, but also reduces hardness, modulus, and resistance to abrasion. Polyurethane characteristics can be customized to specific applications using diisocyanate and polyols. Many of them are compatible, and they can be made with or without a plasticizer and resin [9,25].

3.1.1   Polyolefin Hot Melt Adhesives

Hot melt adhesives made of polyolefin are some of the most widely used. APOs adhere effectively to non-polar substrates like polyethylene and polypropylene, but they are not suggested for polar surfaces. They also have good barrier qualities, such as low moisture and water vapour permeability and great chemical resistance to polar solvents and solutions like acids, bases, esters, and alcohols, but only mild heat resistance and poor chemical resistance for non-polar solvents like alkanes, ethers, and oils. Polyolefin-based hot melts are commonly utilized in the packaging and nonwovens sectors (feminine hygiene, diapers, etc.). Paper, (olefin) plastic films, and metal foils can all be adhered to a variety of substrates with them. They have a wide range of uses in the appliance, automotive, and assembling industries due to their durability to moisture and chemicals, but also their ability to cling to difficult-to-bond plastics like typical polyolefin housings and parts [9,12,25].

3.2 STRUCTURAL ADHESIVES

Structural adhesives are utilized where great strength and long-term adhesion are required. They are capable of withstanding heavy weights. In many circumstances, a structural adhesive having good adherence to a substrate will not fail when a bonded connection is stressed towards its yield point. They are typically the primary mechanism of connection in structural applications. They are usually formed of thermosetting resins that have to be chemically cross-linked using a curing agent (hardener), a catalyst, or/and heat. Some high-strength thermosetting elastomers, such as polyurethanes, are also classified as structural adhesives. The following are details of the most prevalent polymeric resin groups used in structural adhesive composition [1,4].

3.2.1 Epoxies

Epoxies are a versatile class of structural adhesives. They stick to a wide range of surfaces and can be easily changed to obtain a variety of qualities. They have a high shear strength and are suitable for a variety of plastics, metals, and glass. When fully cured, these thermosetting adhesives offer high chemical and thermal resistance, high cohesive strength, and minimum shrinkage. Toughened epoxy adhesives are a popular choice for high-strength applications, and they are utilized extensively in the automobile, industrial, and aerospace industries [4,8,26].

3.2.2 Acrylics

Modified (meth)acrylic glue is one of the most prevalent structural adhesives. To differentiate it from other acrylic resins often used in pressure-sensitive applications, it is often referred to as reactive acrylic. These are liquids that polymerize to form strong bonds. The monomers in acrylic structural adhesives are the same in anaerobic adhesives. However, they have been carefully designed to cure even when exposed to air. At normal temperature, drying can take minutes to an hour, depending on type of acrylic resin used. Acrylics commonly bond through thin coatings of oil due to their high solvent power. Most acrylics, on the other hand, have a more powerful (and occasionally pungent) odour than many other structural adhesives (Table 19.3).

3.2.3 Acrylics That Are Anaerobic

Anaerobic acrylic adhesives have monomer systems which only cure when the resin is devoid of air. They are thin liquids made mostly of (meth)acrylic monomers that, when contained between

TABLE 19.3
Types of Polymers Used as Structural Adhesives

Sr. No.	Polymer	Properties	Application
01	Epoxy	High shear strength, high thermal and chemical resistance	Automotive, aerospace industries
02	Acrylate	High solvent power, high shear strength,	Quick setting based applications
03	Urethane	High flexibility	Bonding films, foils, and elastomers, sheet moulding composites
04	Silicones	Excellent peel and impact strength, excellent heat resistance	Polyethylene and fluorocarbon bonding

tightly fitted (metal) joints, polymerize to form durable plastic connections. Anaerobic adhesives can adhere to a variety of materials, including metals, glass, ceramics, and thermoset polymers. However, many anaerobic adhesives and sealants will only cure if exposed to metal ions, which are normally emitted by a metal surface. They're mostly utilized to improve thread or joint seals, as well as to provide such a holding force for mechanically linked components.

3.2.4 Cyanoacrylates

Superglues, commonly known as cyanoacrylate adhesives, are single-component solutions with a fast cure time and strong shear strength. Although cyanoacrylates have comparable curing qualities as anaerobic adhesives, they are far more stiff and moisture resistant. This glue is only accessible as a low-viscosity liquid which cures in seconds around room temperature and adheres to a variety of substrates. They require moisture to start polymerization and cure only within thin bond lines, like many one-part urethanes. Heat and moisture resistance, as well as peel and impact strengths, are all issues with cyanoacrylates; nevertheless, subsequent innovations have resulted in adhesives that are more durable. Cyanoacrylates are commonly employed in settings where little environmental stress is present and rapid set times are necessary [2,27].

3.2.5 Urethanes

Reactive urethane adhesives are offered in one or two portions as solids or as solvent-based solutions. Unlike most epoxies, polyurethanes can be made to be exceedingly flexible. With high peel and medium to high shear strength, they establish strong connections. They are typically a great choice for bonding films, foils, and elastomers due to their flexibility. They are frequently employed in the automotive industries due to their good adhesive qualities to sheet moulding composites (SMC). Unlike most other structural adhesives, urethane adhesives offer exceptional low-temperature characteristics. Urethanes, on the other side, are not heat-resistant and typically do not attach effectively to metals until a primer is added to the substrate before bonding [5,15,28,29].

3.2.6 Silicones

Silicone adhesives are offered as one- or two-part semi-structural adhesives or as pressure-sensitive adhesive (PSA) solvent solutions. These adhesives have (much) lower shear strength than other forms of structural adhesives, but they have great peel and impact strength, as well as high heat resistance. Solvent-based silicone adhesives cure by condensation or free-radical polymerization. These adhesives are extremely sticky and have only a medium to low peel strength, but they have outstanding high- and low-temperature performance, even after prolonged exposure to high temperatures, and they adhere to a variety of substrates. Because silicones have a low surface energy, they attach to low-surface-energy plastics like polyethylene and fluorocarbons [4,13,30]. Figure 19.2 shows the mechanism of the adhesive bonding process.

3.3 PRESSURE-SENSITIVE ADHESIVES

Tape producers frequently use pressure-sensitive adhesives, or PSAs. These adhesives require pressure to adhere to the substrate. That is why applying the correct pressure is so important for adhesion. PSAs often develop strength over time by fully soaking the substrate. The adhesive literally pours into the substrate's minuscule pores and crevices, eliminating air bubbles in the process. The tape's adhesive must not only attach to the substrate but also be viscoelastic to provide strong peel strength. A plastic, cloth, or paper backing is found on the majority of pressure-sensitive adhesives. Single-sided (adhesive on one side) or double-sided (sticky on both sides) tapes are available.

A significant disadvantage of most PSAs is their low bond strength and lack of creep resistance. As a result, structural bonding is not possible. However, certain double-faced tapes with high bond

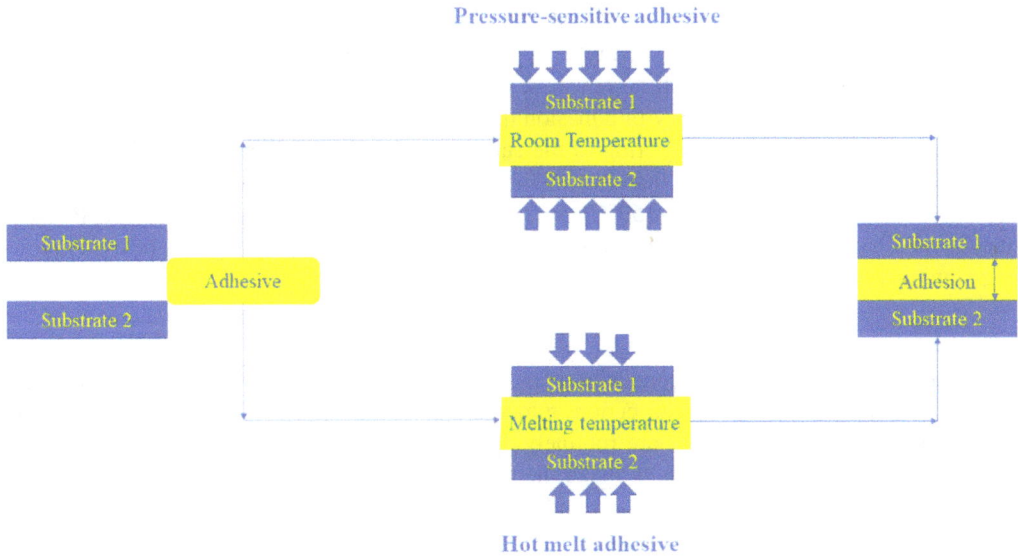

FIGURE 19.2 Mechanism of the adhesive bonding process.

TABLE 19.4
Types of Polymers Used as PSA Adhesives

Sr. No.	Polymer	Properties	Application
01	Styrene butadiene rubber	Good heat-ageing properties, high abrasion resistance, crack resistance	Tapes and floor adhesive
02	Acrylic	Heat and oxygen resistant	Low-temperature applications
03	Silicone	High thermal stability, high chemical resistance	Glass cloth, Kapton, and aluminium foils adhesion

strength for semi-structural applications have been developed in recent years. These adhesives are typically applied to a foam carrier, which helps them adhere to uneven surfaces, curves, and gaps in the joint area. Depending on the use, tape makers choose from five distinct adhesive classes. There are several variations within each adhesive type. The following is a quick description of the many sorts of PSAs, as well as some basic tips for selecting the right tape for the job [2,3,8,27] (Table 19.4).

3.3.1 Styrene-Butadiene Rubber-Based Adhesives

SBRs, or styrene-butadiene rubbers, are frequently utilized in pressure-sensitive tape manufacturing. They are synthetic elastomers made of styrene and butadiene. SBRs are found in a wide spectrum of PSAs and are developed for a number of applications, whether solvent-borne or waterborne. The amount of bonded styrene, the average molecular weight and distribution, and the presence of functional groups incorporated during the polymerization process all influence their properties. PSAs made of styrene-butadiene have greater heat-aging capabilities than those made of natural rubber, but their physical strength, durability, and low-temperature properties are often inferior to those made of natural rubber. SBRs are sometimes combined with natural rubber to attain the greatest characteristics.

3.3.2 Acrylic-Based Adhesives

Acrylic adhesives are a frequently used type of pressure-sensitive glue. Due to the lack of double bonds in the rubber, they are much more robust to heat and oxygen than rubber-based adhesives. Acrylic adhesives come in a variety of adhesion qualities, from minimal adhesion and low tack to aggressive tack and permanent adhering to the substrate. The temperature range of acrylic adhesive tapes is substantially broader than that of rubber-based PSAs. They can tolerate temperatures ranging from 250°F to 300°F (120–150°C) due to their strong heat and oxygen resistance. Because many acrylic-based PSAs have high adhesion qualities at low temperatures, they can be used in low-temperature applications.

3.3.3 Silicone-Based Adhesives

The most expensive PSAs are silicone adhesives, but they have some of the best qualities for high-performance applications. They can endure temperatures of up to 500°F, for example. Silicone-based pressure-sensitive adhesives are routinely applied to more expensive carriers such as glass cloth, Kapton, and aluminium foils to make them acceptable for higher temperature applications. Silicone-based adhesives may even withstand temperatures as low as –100°F. It is also the only PSA that can stick to low-surface-energy materials like silicone or silicone-coated items like release liners. It is chemically resistant to a considerable degree [1,2,4,31].

3.4 Waterborne and Solvent-Borne Adhesives (WSA)

WSAs can be produced with a variety of adhesive qualities and are utilized in a variety of applications, including food and pharmaceutical package sealing, labelling and lidding, weather stripping, industrial coating, laminating, and sealing. Water-based and organic solvent-based adhesives are the two basic types.

Dispersions or emulsions are common forms of waterborne adhesives. The distributed polymer (latex) particles are usually spherical, with sizes ranging from 50 to 300 nanometres. Emulsion polymerization often achieves high molecular weights due to the low concentration of developing chains within each latex particle. These adhesives offer excellent shear strength, a pleasing look, and are both cost-effective and environmentally benign. However, they have less adhesion and weather/chemical resilience than solvent-borne counterparts.

Solvent-borne adhesives contain solvents and thermoplastic or slightly cross-linked polymers such as polychloroprene, polyurethane, acrylic, silicone, and natural and synthetic rubbers (elastomers). They compete with their water-based equivalents directly. Solvent-based adhesives can be used for wet or dry bonding. These adhesives contain high levels of volatile organic compounds (VOCs), which can be dangerous and irritating and are heavily controlled around the world. Solvent-borne adhesives can be applied using a roller, brush, spray, or beads, either manually or automatically. They bond better than water-based adhesives, are more flexible (lower T_g), and can be designed to adhere to a wider range of substrates than water-based adhesives. They are less susceptible to contaminants like grease and oil on the substrate to be joined than other types of adhesives since they are manufactured with organic solvents. They are, however, typically more expensive and less cost-effective to deploy [32–35].

4 BIOBASED ADHESIVES AND SEALANTS

The rapid growth of the economy, as well as the demand for petroleum-based feedstock, increases a material's environmental impact, as well as the desire for novel and renewable feedstock. In general, the adhesive and sealant industries rely on non-renewable synthetic raw resources. Biopolymers, such as cellulose, lignin, starch, and protein-based raw materials, would be viable alternatives in the search for renewable feedstocks. Biopolymers are concerned with the material's stability and performance, which is a field that needs to be explored in terms of material property modification and optimization. In the future, the biobased raw material approach could partially or entirely replace non-renewable petroleum feedstock.

Cellulose is the most abundant biopolymer in the area, as well as a material of choice because of its promising qualities such as renewability, availability, cost effectiveness, and intriguing physiological properties. Cellulose is the most versatile biopolymer due to its crystallinity and nature of hydrogen bonding which are necessary requirements needed for adhesive applications. The first derivative of cellulose, trimethylsilylcellulose, was developed for adhesive application. It was developed by reacting cellulose with organo-chlorosilane compounds. The obtained product is highly hydrolysable [36]. Another method has been developed to derive cellulose derivatives by combining poly[dimethyl(methyl-H) siloxane) with cellulose acetate. The formed silylation product is beneficial for industrial adhesive purposes [13]. Another chemical grafting method was developed to graft the hydroxyl group on cellulose chains. The designed method involves modification of cellulose nanofibrils with aminopropyltriethoxysilane. The cellulose nanocrystals were acetylated by using acetic anhydride to impart functionality on cellulose surface. These were specifically designed for pressure-sensitive adhesives [13]. The mechanical and physicochemical properties of fibreboard panels formed with urea-formaldehyde adhesive were improved by cellulose nanocrystals treated with aminopropyltriethoxysilane silane [37]. Polypyrrole composites are made with a cellulose-based adhesive. To impart the adhesion property, an epoxy moiety was first immersed onto the cellulose surface. The electrical conductivity of this cellulose-based adhesive was found to be decreased [38]. In comparison to cellulose, starch is a less crystalline biopolymer with higher solubility. A thermoplastic elastomer was made from a polymethyl siloxane modified starch. In an aquatic environment, this biobased elastomer is biocompatible and biodegradable [14].

Although lignin has high thermal stability, biodegradability, high carbon content, and stiffness, its use in adhesives and sealants is a high priority. A lignin elastomer based on siloxane was developed with softness and a low Young's modulus [26]. A polyurethane sealant was made by incorporating lignin into a polyurethane formulation. The use of lignin improved cross-linking and amplified the hardening process [5]. The use of succinic anhydride to make a lignin-based adhesive has opened up the possibility of obtaining high-lignin-content cross-linking thermosets [27]. Due to their functionalities such as double bond and ester groups, vegetable oil-based adhesives and sealants are quite a major topic of discussion. The carboxylic acid group from rapeseed oil was combined with the siloxane molecule to create an alternative to silane [30]. A soy protein-based adhesive with a highly robust nature and outstanding water resistance has been developed for use in wood adhesive applications [3].

REFERENCES

[1] Skeist, J. Miron, Introduction to Adhesives, Handbook of Adhesives. (1990) 3–20.

[2] S. Mapari, S. Mestry, S.T. Mhaske, Developments in pressure-sensitive adhesives: a review, Springer: Berlin Heidelberg, 2021.

[3] Y. Liu, K. Li, Chemical modification of soy protein for wood adhesives, Macromolecular Rapid Communications. 23 (2002) 739–742.

[4] B. Strickland, Introduction to Adhesives UPACO Division of Worthen Industries (2013) www.turi.org/content/download/9090/161389/file/Adhesives%20and%20Sealants.Strickland.CE%20Conf.11%20April%202013.pdf.

[5] D. Feldman, M. Lacasse, R. St. J. Manley, Polyurethane-based sealants modified by blending with Kraft lignin, Journal of Applied Polymer Science. 35 (1988) 247–257.

[6] J.M. Klosowski, Sealants in Construction, 1978. Routledge, New York.

[7] J. Klosowski, A.T. Wolf, Sealants in Construction, 2016. CRC Press.

[8] N.K. Jangid, N.P. Singh Chauhan, K. Meghwal, P.B. Punjabi, Conducting polymers and their applications, Research Journal of Pharmaceutical, Biological and Chemical Sciences. 5 (2014) 383–412.

[9] G. V. Malysheva, N. V. Bodrykh, Hot-melt adhesives, Elsevier B.V., 2011.

[10] G. Phalak, D. Patil, V. Vignesh, S. Mhaske, Development of tri-functional biobased reactive diluent from ricinoleic acid for UV curable coating application, Industrial Crops and Products. 119 (2018) 9–21.

[12] William bunnelle, Hot melt adhesive, US20150351977A1-2015. https://patents.google.com/patent/US5534575A/en

[13] G. Stiubianu, C. Racles, M. Cazacu, B.C. Simionescu, Silicone-modified cellulose. Crosslinking of cellulose acetate with poly[dimethyl(methyl-H)siloxane] by Pt-catalyzed dehydrogenative coupling, Journal of Materials Science. 45 (2010) 4141–4150.

[14] L. Ceseracciu, J.A. Heredia-Guerrero, S. Dante, A. Athanassiou, I.S. Bayer, Robust and biodegradable elastomers based on corn starch and polydimethylsiloxane (PDMS), ACS Applied Materials and Interfaces. 7 (2015) 3742–3753.

[15] K.P. Ang, C.S. Lee, S.F. Cheng, C.H. Chuah, Synthesis of palm oil-based polyester polyol for polyurethane adhesive production, Journal of Applied Polymer Science. 131 (2014) 1–8.

[16] K. Chugh, G. Phalak, S. Mhaske, Fatty acid based novel precursors for polyesteramide hot melt adhesive, Journal of Adhesion Science and Technology. 34 (2020) 1871–1884.

[17] K. Chugh, G. Phalak, S. Mhaske, Preparation and characterization of a polyester-etheramide hot melt adhesive system from renewable resources, International Journal of Adhesion and Adhesives. 95 (2019) 102432.

[18] Hot melt adhesive composition EP0957147A1-1998. https://patentimages.storage.googleapis.com/97/b3/16/fa2fd85a1cd6dc/EP1976953B1.pdf

[19] Puletti et. al., Heat resistant hot melt adhesives, US4419494-1983. https://patents.google.com/patent/US4419494A/en

[20] C. Gu, M.R. Dubay, S.J. Severtson, U. States, L.E. Gwin, Hot-Melt Pressure-Sensitive Adhesives Containing High Biomass Contents, Ind. Eng. Chem. Res. 53 (27) (2014), 11000–11006.

[21] W. Li, L. Bouzidi, S.S. Narine, Current research and development status and prospect of hot-melt adhesives: A Review, Industrial & Engineering Chemistry Research. 47 (2008) 7524–7532.

[22] C. Heinzmann, C. Weder, L.M. De Espinosa, Supramolecular polymer adhesives: Advanced materials inspired by nature, Chemical Society Reviews. 45 (2016) 342–358.

[23] P. Kadam, P. Vaidya, S. Mhaske, Synthesis and characterization of polyesteramide based hot melt adhesive obtained with dimer acid, castor oil and ethylenediamine, International Journal of Adhesion and Adhesives. 50 (2014) 151–156.

[24] P.G. Kadam, S.T. Mhaske, Synthesis and properties of polyamide derived from piperazine and lower purity dimer acid as hot melt adhesive, International Journal of Adhesion and Adhesives. 31 (2011) 735–742.

[25] W.Y. Choi, C.M. Lee, H.J. Park, C.O. Akintayo, E.T. Akintayo, T. Ziegler, X. Chen, H. Zhong, L. Jia, J. Ning, R. Tang, J. Qiao, Z. Zhang, P. Kadam, P. Vaidya, S. Mhaske, W. Li, L. Bouzidi, S.S. Narine, C.R. Frihart, Z.S. Petrovi, P.E.M. Person, J.C. Logomasini, R. Measurements, A. Introduction, I.J. Zvonkina, M. Hilt, M. Company, Current research and development status and prospect of hot-melt adhesives: A review, International Journal of Adhesion and Adhesives. 47 (2008) 7524–7532.

[26] G. Stiubianu, A. Bele, C. Tugui, V. Musteata, New dielectric elastomers with improved properties for energy harvesting and actuation, Advanced Topics in Optoelectronics, Microelectronics, and Nanotechnologies VII. 9258 (2015) 925808.

[27] T. Robert, S. Friebel, Itaconic acid-a versatile building block for renewable polyesters with enhanced functionality, Green Chemistry. 18 (2016) 2922–2934.

[28] N. Gama, A. Ferreira, A. Barros-Timmons, Cure and performance of castor oil polyurethane adhesive, International Journal of Adhesion and Adhesives. 95 (2019) 102413.

[29] A. Wołosiak-Hnat, K. Zych, M. Mężyńska, M. Karłowicz, A. Dajworski, A. Bartkowiak, S. Lisiecki, The influence of type and concentration of inorganic pigments on the polyurethane adhesive properties and adhesion of laminates, International Journal of Adhesion and Adhesives. 90 (2019) 1–8.

[30] K. Szubert, Synthesis of organofunctional silane from rapeseed oil and its application as a coating material, Cellulose. 25 (2018) 6269–6278.

[31] X. Kong, G. Liu, J.M. Curtis, Characterization of canola oil based polyurethane wood adhesives, International Journal of Adhesion and Adhesives. 31 (2011) 559–564.

[32] G. Amini, S.N. Banitaba, A.A. Gharehaghaji, Imparting strength into nano fi berous yarn by adhesive bonding, International Journal of Adhesion and Adhesives. 75 (2017) 96–100.

[33] C.R. Frihart, Specific adhesion model for bonding hot-melt polyamides to vinyl, International Journal of Adhesion and Adhesives. 24 (2004) 415–422.

[34] Y. Tong, W. Huang, J.U.N. Luo, M. Ding, Synthesis and properties of aromatic polyimides derived, Polymer. (1998) 1425–1433.

[35] I.J. Zvonkina, M. Hilt, Tuning the mechanical performance and adhesion of polyurethane UV cured coatings by composition of acrylic reactive diluents, Progress in Organic Coatings. 89 (2015) 288–296.

[37] H. Khanjanzadeh, R. Behrooz, N. Bahramifar, W. Gindl-Altmutter, M. Bacher, M. Edler, T. Griesser, Surface chemical functionalization of cellulose nanocrystals by 3-aminopropyltriethoxysilane, International Journal of Biological Macromolecules. 106 (2018) 1288–1296.

[38] C.O. Baker, X. Huang, R.B. Kaner, Polyaniline nanofibers: broadening applications for conducting polymers, Chemical Society Reviews. 46 (2017) 1510–1525.

20 Polymers for Foams and Their Emerging Applications

Adrija Ghosh,[1] Jonathan Tersur Orasugh,[1,3,4,5]
Dipankar Chattopadhyay,[1,2] and Suprakash Sinha Ray[4,5]

[1] Department of Polymer Science and Technology, University of Calcutta, 92 A.P.C. Road, Kolkata – 700 009, West Bengal, India
[2] Center for Research in Nanoscience and Nanotechnology, Acharya Prafulla Chandra Roy Sikhsha Prangan, University of Calcutta, JD-2, Sector-III, Saltlake City, Kolkata-700098, West Bengal, India
[3] Department of Chemical Sciences, University of Johannesburg, Doorfontein, Johannesburg 2028, South Africa
[4] Department of Textile Technology, Kaduna Polytechnic, P. M. B. 2021, Tudun-Wada, Kaduna, Nigeria
[5] DST-CSIR National Centre for Nanostructured Materials, Council for Scientific and Industrial Research, Pretoria 0001, South Africa

1 INTRODUCTION

Polymeric foams are used for various applications owing to their plethora of unique properties such as lightweight, low thermal conductivity, heat insulation, etc. Polymer foam is fundamentally made up of a blend of gas and solid phases giving rise to a microcellular structure. This is done by mixing both gaseous and solid phases rapidly. The foams formed have air bubbles integrated into them. They can be flexible or rigid, depending upon their cell geometry. The gas used in the foam-making process is called a blowing agent. The blowing agent can be either physical or chemical. Chemical blowing agents participate in the chemical reaction, or undergo decomposition during the process and emit chemicals. Meanwhile, physical blowing agents do not react during the foaming process and do not affect the polymer matrix. Both include the use of organic and inorganic materials as blowing agents. Inorganic physical blowing agents include the use of nitrogen, air, or carbon dioxide, while freon, pentane, or hexane are used as organic physical blowing agents. Isocyanate compounds can be used as organic chemical blowing agents and carbonate or zinc powder are used as inorganic chemical blowing agents. Expandable beads are another class of material that facilitates foaming. They have cores comprising void spheres of alkanes and shells made of acrylic resin polymer. These expandable beads increase in size on heating and ameliorate the foaming effect. Expandable polythene or polypropylene are examples of expandable beads.

A polymer foam can be made up of plastic or rubber. The plastics can be either thermoplastic or thermosetting. Thermoplastic plastics can be softened on heating but hardened on cooling. Thermoplastics can be remolded and thus are recyclable. On the other hand, thermosetting plastics cannot be softened on heating and cannot be remolded. The thermosetting resins include epoxy resin, polyester, polyurethane, or phenolic resins. Examples of thermoplastic raw materials used for making polymeric foams include polythene, polystyrene, polyvinyl, etc. Rubbers used for making polymeric foams include natural rubber, silicone rubber, styrene-butadiene rubber, etc.

Polymer foams are widely used for packaging materials, construction purposes, sports equipment, insulation, making furniture, and much more [1–9]. Polymer foams have been successfully used for dye adsorption. Feng et al. [10] used trimethylammonium-grafted cellulose foams for adsorption of Eosin Y. The foam exhibited a maximum adsorption capability value of 364.22 mg/g. Despite their

DOI: 10.1201/9781003278269-20

various applications, recyclability and disposal of polymer foams remain a burning issue that needs to be addressed. With increasing demand and ongoing progress, the future holds greater possibilities for developments in this area.

2   HISTORICAL BACKGROUND

The year 1914 marked the emergence of the first cellular polymer on the market. It was a sponge rubber formed by the addition of acid to natural rubber consisting of gas-producing agent. In the late 1920s, latex foams were produced from natural rubber consisting of 40% solids. These were stabilized with soap, beaten with air, and further refined by drying in an oven. Also, in the 1920s, the oldest known rigid cellular plastic was produced. This was called ebonite. The year 1928 marks the year of the origination of the Dunlop latex foam process. This involves gelling and vulcanization of foam rubber. In the mid-1930s, latex foam rubber produced by the Talalay process was originated and eventually it was commercialized following World War II.

In 1937, Professor Otto Bayer and his colleagues discovered a new polymerization reaction that produced polyurethanes and polyureas as products [11,12]. This was the diisocyanate polymerization reaction. In due course, German scientists working in the laboratory of I.G. Farbenindustrie developed the first-ever polyurethane foam [13]. These foams were used during World War II for making aircraft, submarines, and tanks.

The discovery of phenolic foams dates back to the time when phenolic resins were produced. Baekeland discovered the formation of "spongy" and "porous" products during the production of phenol-formaldehyde resins. He found these products to be unsuitable for commercialization. These foams were put to use during World War II as an alternative to balsa wood in aircraft under the tradename Troporit P [14–17]. At the same time, the expanded Rubber Company pioneered the production of Thermozote phenolic foams in Great Britain. Later, phenolic foams were used in countries including the United States, France, and Germany.

"Foamed polystyrene" was patented by Swedish inventors Munters and Tandberg on August 20, 1931, although it was not commercialized until early 1940. The Dow Chemical Co. introduced an extrusion process for the commercial production of polystyrene foams in 1942 [18]. Commercial use of polyethylene foams for insulation purposes started in the 1950s. Urea-formaldehyde (U-F) foams were discovered by researchers at the German company I.G. Farbenindustrie. Thermalon, Ltd. in England first commercially produced U-F foams for insulation purposes. Other than these, various other foams have been produced. Observing the large number of foams that have been reported in various journals and patents, we can predict that foams can be prepared from roughly every polymer employing one or another technique.

3   TYPES OF POLYMER FOAMS

3.1   Classification Based on Foaming Structure: Open- and Closed-Cell Foams

Polymer foams can be widely differentiated based on their cell hole. As the name suggests, open-cell foams have an array of open cells that are linked to each other. Such open cells allow easy passage of air between the cells. This also espouses its capability to entrap foreign particles. Epoxy foams are one type of open-cell foam. In contrast, in closed-cell foams, every cell is separate and encircled by connected faces. This makes them leakage-proof and a superior moisture barrier; examples include polyethylene foam, ethylene-vinyl acetate foam, neoprene form, etc.

3.2   Classification Based on Hardness: Soft, Semi-Rigid, and Rigid Polymer Foams

The hardness of polymer foams can be estimated from their elastic modulus. They are generally called soft foams if their elastic modulus is less than 68.6 MPa at a temperature of 23°C and humidity of 50%. Meanwhile foams having an elastic modulus in the range of 68.6–686 MPa are classified as semi-rigid foams and those having more than 686 MPa of elastic modulus are known as rigid polymer foams.

3.3 CLASSIFICATION BASED ON DENSITY: LOW-, MEDIUM-, AND HIGH-FOAMING POLYMER FOAMS

Polymer foams having a density less than 0.1 g/m^3 are regarded as high-foaming polymer foams, while those having a density of more than 0.4 g/m^3 are considered to be low foaming. On the other hand, if the density lies between these extreme values then they are regarded as medium foaming.

4 BASIC PRINCIPLES FOR MAKING FOAMS

The basic method of foam formation involves three stages: bubble formation, bubble growth, and bubble stabilization.

4.1 BUBBLE FORMATION

This stage involves the addition of a blowing agent to a molten polymer which initiates a chain of chemical reactions giving rise to a bulk quantity of gas. This produces a polymer/gas solution. A gradual increase in the amount of gas leads to the formation of a supersaturated solution and escape of gas. Eventually, the released gas forms the bubble nucleus.

4.2 BUBBLE GROWTH

The gaseous pressure inside the bubble instigates its growth. When two bubbles are in close vicinity to each other the gas spreads, leading to their fusion. This multiplies the number of cells and ultimately causes bubble growth.

4.3 BUBBLE STABILIZATION

The incessant growth of bubble size makes its walls thinner and causes instability. Stabilizing foams is important, as rapid rupture may cause foam collapse. The bubbles are stabilized by cooling, which reduces the pressure within them.

5 METHODS TO MAKE FOAMS

5.1 MECHANICAL FOAMING

In this process, gases are mechanically stirred in a liquid polymer which leads to the entrapment of gases in it (Figure 20.1). Ultimately, the polymer solution hardens due to catalysis or heating forming foam. This is a green method that does not involve the use of any additional foaming agent.

5.2 PHYSICAL FOAMING

This method involves mixing a liquid of low-boiling point with the polymer solution. The liquid is volatilized within the polymer system by applying heat or pressure (Figure 20.2). This is an environment-friendly method that requires low-cost raw materials.

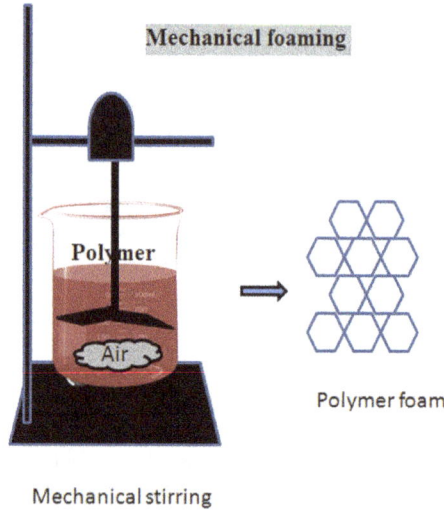

FIGURE 20.1   Schematic representation of mechanical foaming.

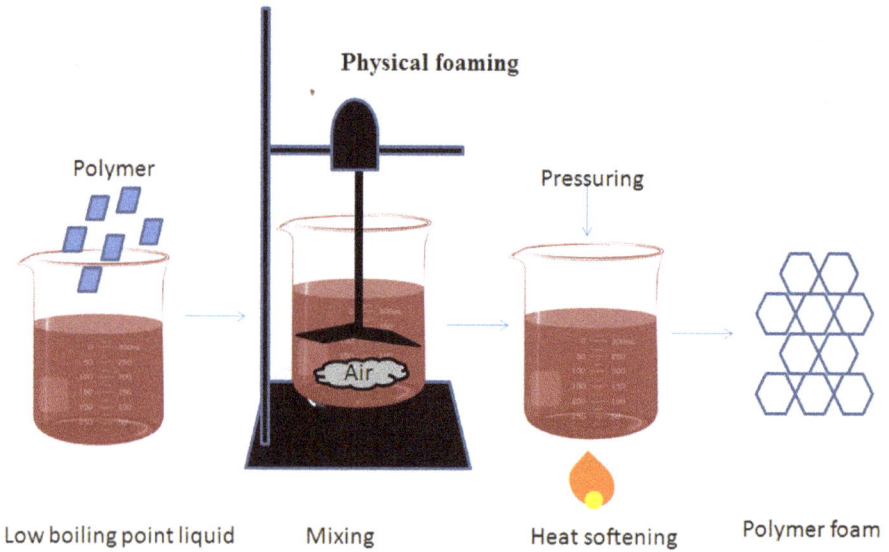

FIGURE 20.2   Schematic representation of physical foaming.

5.3   CHEMICAL FOAMING

This process involves the use of chemical blowing agents such as carbonates, azides, bicarbonates, hydrazides, etc. Thus, these agents can be either organic or inorganic in nature. The blowing agents are added to the fluid polymer, which is thermally decomposed into nitrogen or carbon dioxide within the melt. The heat required is either formed due to exothermic polymerization or is provided externally. It is a costly method to use. Figure 20.3 illustrates a chemical foaming method.

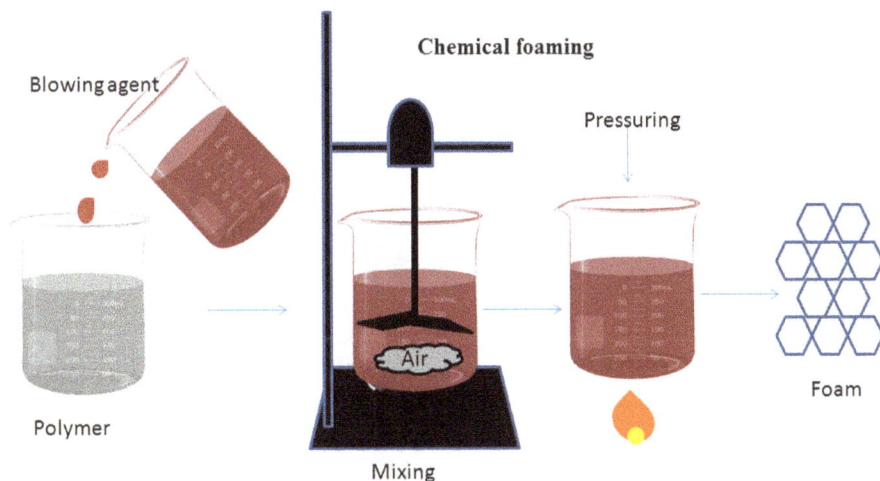

FIGURE 20.3 Schematic representation of chemical foaming.

6 TECHNIQUES FOR MAKING POLYMER FOAMS

6.1 BATCH FOAMING

Batch foaming can be done in two ways (Figure 20.4). The first method is a pressure-induced way in which the polymer is mixed with a blowing agent in an autoclave until it becomes saturated and the nucleation is initiated by reducing the pressure of the system to atmospheric pressure. Ultimately, the cell is cooled for stabilization. The second method involves a temperature-induced approach. Here the initiation steps are similar to those of the previous method, except that the process is carried out at lower temperatures. Once the polymer is saturated, it is transferred into a hot oil bath at a temperature range of 80–150°C until nucleation is commenced. Once complete, the cell is cooled by either placing it in a cool water bath or by using a solvent. This requires a very small amount of material for processing. It produces uniform celled and good-quality products. It is a relatively cheap method as compared to other techniques.

6.2 FOAM EXTRUSION MOLDING

In this method, supercritical fluids like carbon dioxide (CO_2) or nitrogen (N_2) are utilized as blowing agents. The extruder is set at a certain temperature for the molding process. The polymer is introduced at the hopper and melted under processing conditions. The supercritical fluid is injected into the polymer melt. Cells are formed once the polymer leaves the die. Eventually, cell density increases and cellular foams are formed.

This technique can involve either a physical or chemical foaming process. Pressure and temperature influence the cell morphology largely. While adequate temperature is solely responsible for the decomposition of the foaming agent, high pressure keeps the gas dissolved in the polymer melt. This technique generally gives products with uniform cells and good surface quality. Figure 20.5 presents an extrusion molding machine schematically.

6.3 FOAM INJECTION MOLDING

Various injection molding parameters need to be fixed before commencing the injection molding process. The required polymer is plasticized and the supercritical fluid is dissolved in it by introducing it to the barrel (Figure 20.6). Once the polymer solution is injected into a mold cavity there is

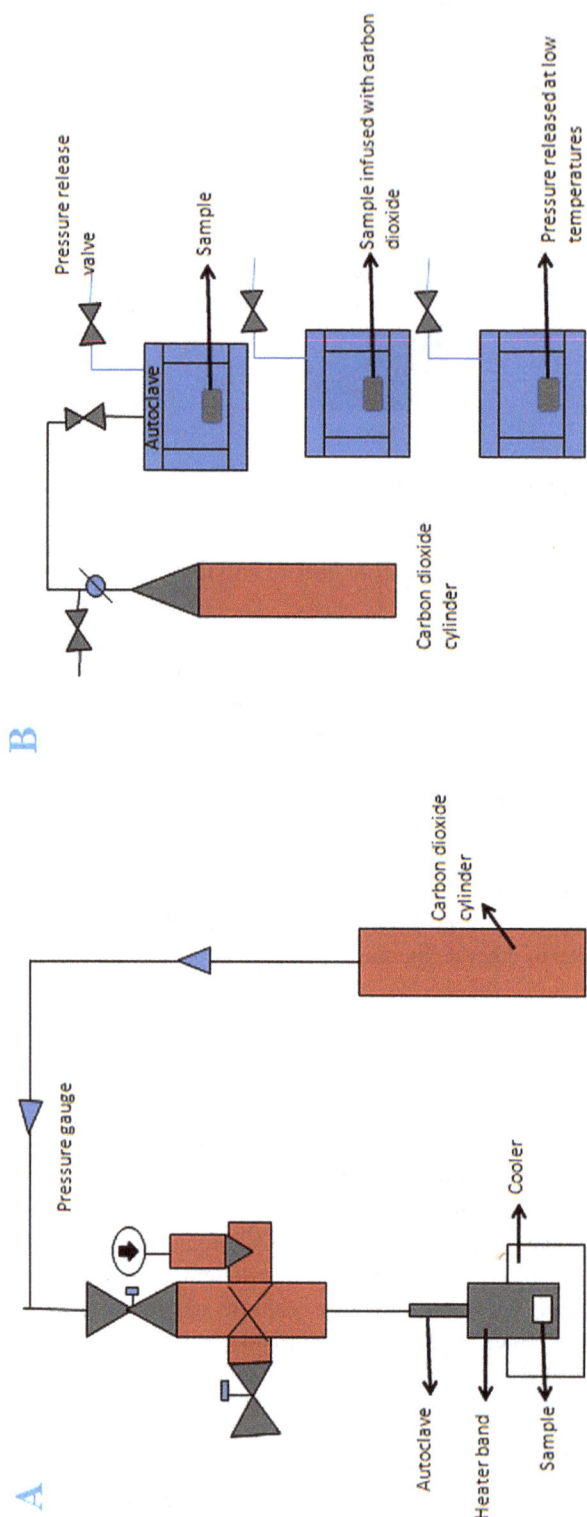

FIGURE 20.4 Schematic representation of (A) pressure-induced batch foaming and (B) temperature-induced batch foaming.

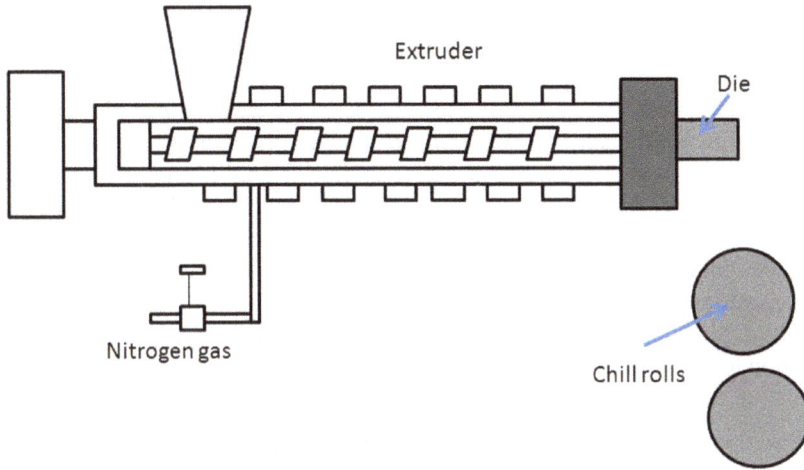

FIGURE 20.5   Schematic representation of extrusion molding.

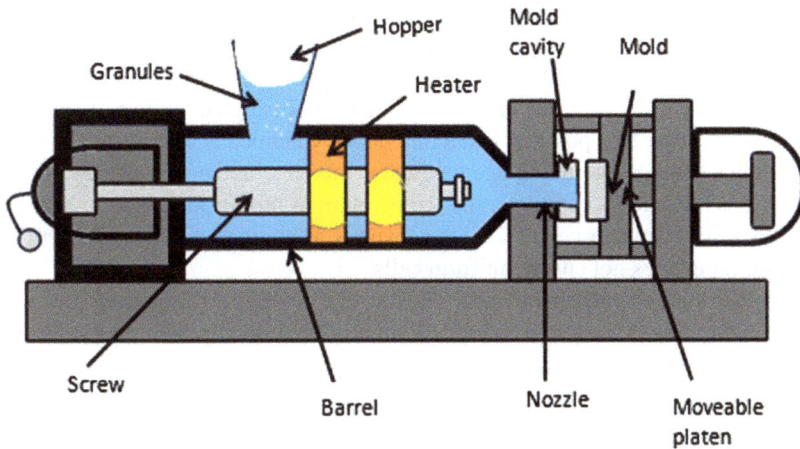

FIGURE 20.6   Schematic representation of foam injection molding.

a major drop in the pressure which induces bubble nucleation and, eventually, foaming. The morphology and quality of the cells largely depend upon the mold design and processing conditions.

This can also involve physical or chemical foaming processes. While the physical foaming process involves the use of costly equipment, the chemical foaming process is much more easy and convenient. It gives rise to cells of density in the range of 10^4–10^8 cells/cm^3, although this process is unsuitable for making uniform cells and gives products with poor surface quality. Molds used in this process increase the cost of the machinery required for this technique.

7   WIDELY USED POLYMERIC FOAMS

7.1   POLYURETHANE FOAMS (PUF)

Polyurethanes are synthesized by the addition reaction of polyisocyanates and compounds having active hydrogen in the presence of a catalyst. They contain the urethane linkage -NH-CO-O-. They can be prepared in three ways. The first is a single-step process where all the raw materials are mixed to form foam in a single shot. This is a fast and cost-effective method. In the second procedure, a quantity of polyol is reacted with the complete amount of isocyanate. The product obtained consists of free isocyanate units amounting up to 32 wt.%. Then, the rest of the raw materials along with the residual polyol are added to finally obtain the foam. In the third method, the hydroxyl reactant is reacted with an excess quantity of isocyanate to obtain a prepolymer with free isocyanate unit amounting up to 15 wt.%. PUFs can exist as either flexible or rigid foam. Generally, flexible PUFs have a low load-bearing capability in comparison to rigid foams. Depending on the nature of polyol used in the reaction, the resultant foam can be either ester or ether foam. Rigid foams were earlier utilized as a core component of aircraft or buoyant materials, while flexible foams are generally used for cushioning, packaging, and making carpets and furniture. While flexible foams can be prepared by a slabstock or molding technique, rigid foams can be prepared also by the sprayed foam technique in addition to the earlier described methods.

7.2   POLYETHYLENE FOAMS (PEF) AND POLYPROPYLENE FOAMS (PPF)

PEFs are closed-cell foams. Low-density polyethylene foam (LDPE) is one of the most important classes of PEFs which is used widely. They generally have a semi-rigid structure. This is a thermoplastic that is blown into foam. It has many attractive properties. It has high compression rates which make it capable of bearing a high load and maintaining its strength after repeated use. Owing to their lower density and high elasticity, LDPE foams are generally used as packaging materials or sports parts. They can be prepared by batch foaming, extrusion, or injection molding techniques. The morphology of the foam depends largely upon the processing conditions, which should be set according to the viscosity of the polymer and temperature. Chemically cross-linked LDPE foams are produced by using a solid cross-linking agent which is mixed with the polymer by exposure to temperature. Once cross-linking is complete, the temperature is further increased to achieve foaming. Cross-linked PEFs have smaller cells and softer textures as compared to extruded PEFs. However, chemical cross-linking does not ensure uniform cells.

PPFs are another kind of polyolefin foam. They have a closed-cell structure. The manufacturing process used is in this case is similar to that for PE foams. The blowing agents are selected according to the density of foam to be prepared.

7.3   POLYSTYRENE FOAMS (PSF)

Polystyrene is a man-made aromatic polymer that is formed by the polymerization of styrene. Polystyrene foams are generally found in two forms: expanded polystyrene foam (EPS) and extruded polystyrene foam (XPS). EPS is a white-colored foam that is known for its good insulating properties. It is a closed-cell rigid polymer foam. EPS foams are prepared from expanded polystyrene beads that are heated, aged, molded, and cooled. EPS foams generally consist of 2–4% styrene and around 96–98% air. On the other hand, XPS foams are made by the extrusion method and can be of different colors. They are much stronger and more expensive than EPS foams.

8   IMPORTANT APPLICATIONS OF POLYMERIC FOAMS

8.1   PRODUCT PACKAGING

The most common polymers used for packaging electronics or consumer products are polystyrene, polypropylene, polyethylene, etc. Owing to their light weight, durability, and chemical resistance, polymeric foams are used as packaging materials. Bubblewrap sheets are made from welding LDPE foams.

The use of biodegradable polymeric foams for food packaging has attracted the attention of researchers. Starch is often used in combination with other polymers to enhance the properties of the final product. Although biodegradable polymeric foams show good mechanical properties, high hydrophilicity in comparison to synthetic foams is a major issue that needs to be addressed. Water sensitivity is an important factor while packaging food materials with high moisture or foods that need to be stored in high humidity.

8.2 SPORTS EQUIPMENT

Polymer foams are used to make many types of sports equipment. They are used to make helmets which are worn as protective head covers in many sports. These helmets generally have two layers. The outer layer acts as a protective shell, while the inner layer which is made of polymer foams, helps in cushioning and energy absorption. Low-density foams are also used to make yoga mats or gymnastic mats as they provide impact protection. They are also used to make light-weight, abrasion-resistant sports shoes. Sometimes, antibacterial agents are also added to the foams to combat foot infections.

8.3 SANDWICH PANEL

Sandwich panels are widely used nowadays as they have high strength and good absorption capability. They consist of a porous core and two stiff metal faces. Polymer foams are highly preferred for sandwich panel applications in construction and transport. Panels for boats generally require materials of high compressive strength and robustness. High-density specialized polyvinyl chloride (PVC) foams are used for making these. On the other hand, panels used in buildings are meant for thermal insulation. Therefore, low-density polymer foams are used for this purpose. The inside walls of refrigerators are made of foams with a thermoplastic skin.

9 CURRENT RESEARCH TRENDS

9.1 RECENT ADVANCEMENTS IN FOAM-MAKING METHODS

Huang et al. [19] developed an industrial-feasible and proficient technique to prepare polypropylene/intumescent flame-retardant (PP/IFR) composites. This technique involved the use of supercritical CO_2 ($scCO_2$) as a plasticizer and a foaming agent while extruding the premix of dried PP/IFR. It was further defoamed by hot-pressing. It was observed that the presence of $scCO_2$ aided in the good dispersion of IFR in the prepared composite. Also, biaxial bubble stretching influenced the dispersion of IFR. The larger the bubble stretching force, the better is the dispersion of IFR. Visser et al. [20] reported for the first time a novel additive manufacturing technique to prepare polymer foams called direct bubble writing. This technique involves the use of a 3D printer with a core–shell nozzle. The polymer foams had a liquid shell–gas core bubble structure. The cell structure depended upon the type of gas used. Oxygen-rich gases favored the formation of open-cell foams, while oxygen-deficient gases favored the formation of closed-cell foams. The bubbles could easily retain their structures as the outer shell was composed of less viscous monomers which polymerized fast during the printing process.

Wang et al. [21] reported a method to prepare polypropylene/polytetrafluoroethylene (PP/PTFE) nanocomposite by amalgamating two techniques: *in situ* fibrillation and injection molding. This is the first approach to preparing nanocellular foam using injection molding. The foamed composites had enhanced strength and were light in weight. L. Wang et al. [22] developed a new injection molding machine that used air in place of CO_2 or N_2 as a physical blowing agent. This machine used a gas cylinder instead of a pump or injection valve to introduce air into PP resins. PP foams fabricated by air foaming were much more flexible and had finer cells compared to N_2 or CO_2 foaming.

9.2 Recent Progress in Making Foams with Superior Properties

Yang et al. [23] proposed an idea for making PP foams with more uniform cell sizes. These authors used hollow molecular-sieve (MS) particles as nucleating agents while using the scCO$_2$ foaming method of PP. The resultant foam had uniform cell-size distribution, enhanced mechanical strength, and cell density. The cell density increased 10 times and the tensile strength increased two times. Wen et al. [24] devised a method to improve the mechanical properties of microcellular ethylene-propylene-diene monomer (EPDM) foams by blending silicone rubber (SR) along with EPDM to prepare SR/EPDM foams by scCO$_2$ foaming technique. As SR has a different cross-linking temperature compared to EPDM, it influences its mechanical properties greatly. The tensile strength of SR/EPDM foam increased by 461% and the compression strength increased by 283%.

Sun et al. [25] proposed the use of ultra-high-molecular-weight polyethylene (UHMWPE) to ameliorate the foaming capability and mechanical strength of high-density polyethylene (HDPE) foams. UHMPE/HDPE composite foam was prepared by compression molding using azodicarbonamide (ADC) as a foaming agent and dicumyl peroxide (DCP) as a cross-linking agent.

9.3 Emerging Applications of Polymer Foams

Researchers often use polymer foams for tissue engineering purposes. Mishra et al. [26] developed a gelatin-polyvinylpyrrolidone (PVP) scaffold by an *in situ* foaming method which could be used as a substitute for a bone graft. Gelatin and PVP were blended using glutaraldehyde as a cross-linking agent. It was further lyophilized and characterized. The cytocompatibility and biocompatibility of the prepared scaffold were confirmed using mesenchymal stem cells and chorioallantoic membrane, respectively. Hou et al. [27] prepared poly(ε-caprolactone) (PCL) scaffold for tissue engineering by combining the scCO$_2$ foaming technique with polymer leaching. The polymer used for leaching was poly(ethylene)oxide. Highly porous yet smaller pore-sized PCL scaffolds were fabricated using this method. Xie et al. [28] prepared a polyurethane/hydroxyapatite shape-memory polymer foam (SMP) which could be used for bone regeneration. Histological analysis and computed tomography tests confirmed that the scaffold could be used for bone remodeling and vascularization.

Li et al. [28] prepared a bilayered polymer foam for solar steam generation. The procedure involved the initial production of hot-pressed melamine foam and immobilization of polypyrrole onto it to generate a bilayered polymer foam. The polypyrrole layer is responsible for absorbing light and evaporating water, while the melamine foam layer is responsible for transporting water and insulating heat. The proposed device is capable of producing freshwater under normal sunlight. He et al. [30] fabricated a foam from a cross-linked aromatic polymer by a one-step hydrothermal process. Styrene was used as the monomer, while divinylbenzene was used as the cross-linking agent. The foam was thus prepared and further modified using sodium alginate and polypyrrole (PPy-coated M-PDVB-PS). The prepared foam had a strong light absorption capability and had a conversion efficiency of around 87.6% under 1-sun irradiation.

Lu et al. [31] fabricated polyurethane pressure sensors which could detect human motion and be used as electronic skin. Initially, a PU foam was prepared by a freeze-drying technique which was coated with graphene oxide followed by chemical reduction. The foam had good sensitivity and stability. Experimental results confirmed that the foam could be used in the pressure detection range of 20 kPa to 1.94 MPa.

10 DRAWBACKS AND POSSIBLE SOLUTIONS

(1) Non-biodegradable polymeric foams are a major concern nowadays. For instance, foams made from petroleum contribute largely to environmental pollution. Petroleum-based polymer foams are largely used as packaging material as they protect devices during transportation. Discarding these

foams is a major problem as they do not degrade. EPS foams fragment into smaller pieces, forming microplastics that cause water pollution endangering thousands of species of marine animals.

A possible solution to this problem is to use vegetable oils for making polymer foams. Both vegetable oil and modified vegetable oil have been used for making foams. Of these, acrylate epoxidized soybean oil-based foams have gained significant recognition among researchers. X. Huang et al. [32] fabricated a biodegradable epoxy polymeric foam by reacting epoxidized soybean oil (ESO) with fumaropimaric acid. N,N-dimethylbenzylamine was used as a catalyst and sodium bicarbonate was used as a foaming agent in this process. The foam prepared had low apparent density and thermal conductivity. Dicks et al. [33] prepared polymeric foams from maleated castor oil glycerides using styrene and isobornyl methacrylate as diluents. The foams showed excellent compressive strength when styrene was used as a diluent, while its strength decreased with the integration of isobornyl methacrylate. The foams exhibited rapid weight loss in an aerobic soil milieu, as confirmed by SEM images.

Fabrication of biodegradable polymer foams using polymers like polylactic acid (PLA) or polycaprolactone (PCL) is another way of encouraging the use of eco-friendly products. Y. Li [34] synthesized a biodegradable composite foam based on polylactic acid and carbon nanotubes by the CO_2 foaming method. García-Casas et al. [35] prepared polycaprolactone foams and incorporated quercetin in them by the batch foaming technique. The foam was efficient enough to deliver quercetin.

Another strategy to combat this problem is to reuse polymeric foam wastes. PU foam wastes can be reused as dry material in cement or gypsum. Coppola et al. [36] investigated the effect of aggregated polymeric foam waste on cement mortar. Different properties of the mortar were studied by replacing sand in it with foam waste by 10%, 25%, and 50% of volume. The presence of aggregated foam waste reduced mortar density and thermal conductivity. Farhan et al. [37] prepared carbon foams from powdered aggregates of rigid PU foam waste (50 wt.%), novolac resin (33 wt.%), and coal-tar pitch (17 wt.%). Another variation of this foam was fabricated by replacing novolac with waste-resole. The second variety showed improved results with higher porosity and better compressive strength. The fabricated foam did not blaze.

(2) Many chemical foaming agents are used to make polymeric foams. Although most of these gases are captured by gas filters, some of these become trapped within the foams which releases them over time. These gases are harmful both to human health and the environment.

The use of deodorizing agents is a possible solution to this problem. Pozzolan is a natural deodorizing agent used for making eco-friendly foams. Lee et al. [38] prepared a hybrid deodorant that could be coated on poly(vinyl chloride) (PVC) foams to mask harmful ammonia gas. It consisted of a UV-curable resin and porous materials such as zeolite, fumed silica, etc.

(3) The majority of polymer foams are extensively flammable and release toxic gases on burning. For instance, polyurethane foams catch fire instantly and give off dense fumes and harmful gases. Activities like drilling and welding produce toxic vapors and dust.

Wu et al. [39] prepared a flame-retardant polymer foam coated with silicone resin by the dip-coating technique. The foam thus prepared had good thermal stability and mechanical strength. The minimum amount of oxygen required to completely combust the foam increased. While the limiting oxygen index (LOI) was found to be 14.6% for pure foam, it increased to 26% for the coated foam. C. Wang et al. [40] prepared 2,2-diethyl-1,3-propanediol phosphoryl melamine (DPPM) and added it as an additive to rigid PU foam (DPPM-RPUF). The synthesized foam showed an LOI value of 29.5%.

(4) Polymer foams have lower stiffness and are brittle. A possible solution to this problem is adding tougheners to polymeric foams. Song et al. [41] elevated the toughness of phenolic foams by adding tung oil, which comprises P and Si. The incorporation of the toughener in the polymer matrix increased its compression strength by 180% and bending strength by 198%. It exhibited lower thermal conductivity and enhanced flame resistance. Liu et al. [42] also proposed a method to enhance the robustness of phenolic foam. In this process, poplar fiber was acetylated using an

TABLE 20.1

Effect of Toughening Agents on the Mechanical Properties of Polymer Foams

Polymer Foam	Toughening Agent	Compressive Strength	Bending Strength	References
Phenolic foam	Tung oil-based siloxane	0.361 MPa	0.624 MPa	[41]
Lignin-based phenolic foams	Multiwalled carbon nanotubes	0.326 MPa	0.483 MPa	[44]
Phenol–urea–formaldehyde foam	Nano ZnO	2.63 MPa	0.28 MPa	[43]
Polymethacrylimide	MMT	2.05,	1.17,	[45]
	Kaolin	1.89, and	1.46, and	
	Talc	1.94 MPa	1.43 MPa	
Phenolic foam	Acetoacetic ester-terminated polyether	0.203 MPa	0.305 MPa	[46]
Phenolic foam	Bio-oil	0.28 MPa	0.36 MPa	[47]
Melamine-formaldehyde foam	Ethylene glycol (EG) and carbon fiber (CF)	355.3 kPa	0.44 MPa	[48]

acetylation regent and the acetylated fiber was added as a toughening agent to the phenolic foam. The compressive strength of the modified phenolic foam was amplified by 28.5% as compared to pure phenolic foam. In addition, the foam showed better thermal and flame resistance. Z. Wang et al. [43] synthesized a urea formaldehyde and phenol formaldehyde (UF/PF) composite foam and enhanced its mechanical properties by adding ZnO nanoparticles using *in situ* polymerization. The strength of the foam increased significantly with the addition of 0.9 wt.% of nano-ZnO. The compression strength increased by 200% as compared to pure UF/PF foams. The addition of ZnO not only enhanced the mechanical properties but also increased the flame resistance of the foam (Table 20.1).

11   CONCLUSION

Polymer foams have been used for many years for various purposes such as insulation, packaging, cushioning, etc. due to their unique properties. Fabrication of these foams can be done by processes like batch foaming, mold injection, and extrusion. Recently, new methods like microwave-assisted processes and freeze-drying also have been used to make polymeric foams. However, over-use of oil-based polymeric foams has increased concern among environmentalists as they are non-biodegradable and break down into microplastics leading to water pollution. This issue has inculcated researchers to prepare biodegradable polymeric foams. Also, the mechanical properties and flame resistance properties of most polymeric foams are poor enough to restrict their use. Ongoing research works in this field will help to divulge new domains and thus expand its areas of applications.

REFERENCES

[1] S. Mali, Biodegradable foams in the development of food packaging, in: T. Gutiérrez (eds) *Polymers for Food Applications*. Springer, Cham, 2018 https://doi.org/10.1007/978-3-319-94625-2_12

[2] S. Vorawongsagul, P. Pratumpong, C. Pechyen, Preparation and foaming behavior of poly (lactic acid)/poly (butylene succinate)/cellulose fiber composite for hot cups packaging application, Food Packaging and Shelf Life. 27(2021) 100608 https://doi.org/10.1016/j.fpsl.2020.100608.

[3] Z. Ma, X. Liu, X. Xu, L. Liu, B. Yu, C. Maluk,G. Huang, H. Wang & P. Song, Bioinspired, Highly adhesive, nanostructured polymeric coatings for superhydrophobic fire-extinguishing thermal insulation foam, ACS Nano, 15 (7) (2021) 11667–11680 10.1021/acsnano.1c02254

[4] B. Furet, P. Poullain, S. Garnier, 3D printing for construction based on a complex wall of polymer-foam and concrete, Additive Manufacturing. 28 (2019) 58–64 https://doi.org/10.1016/j.addma.2019.04.002

[5] M. Tomin, Á. Kmetty, Polymer foams as advanced energy absorbing materials for sports applications— A review, Journal of Applied Polymer Science 139(9) (2022) 51714. https://doi.org/10.1002/app.51714

[6] O. Duncan, T. Shepherd, C. Moroney, L. Foster, P. D. Venkatraman, K. Winwood, T. Allen, and A. Alderson, Review of auxetic materials for sports applications: Expanding options in comfort and protection, Applied Sciences 8(6) (2018) 941. https://doi.org/10.3390/app8060941

[7] S. Khojasteh-Khosro, A. Shalbafan, and H. Thoemen, Development of ultra-light foam-core fibreboard for furniture application, European Journal of Wood and Wood Products 79(6) (2021) 1435–1449. https://doi.org/10.1007/s00107-021-01723-0

[8] S. Signetti, M. Nicotra, M. Colonna, N.M. Pugno, Modeling and simulation of the impact behavior of soft polymeric-foam-based back protectors for winter sports, Journal of Science and Medicine in Sport 22 (2019) S65—S70. https://doi.org/10.1016/j.jsams.2018.10.007

[9] S.F.A. Zaidi, E. U. Haq, K. Nur, N. Ejaz, M. Anis-ur-Rehman, M. Zubair, and M. Naveed, Synthesis & characterization of natural soil based inorganic polymer foam for thermal insulations, Construction and Building Materials 157 (2017) 994–1000. https://doi.org/10.1016/j.conbuildmat.2017.09.112

[10] C. Feng, P. Ren, M. Huo, Z. Dai, D. Liang, Y. Jin, and F. Ren, Facile synthesis of trimethylammonium grafted cellulose foams with high capacity for selective adsorption of anionic dyes from water, Carbohydrate Polymers 241 (2020) 116369. https://doi.org/10.1016/j.carbpol.2020.116369

[11] O. Bayer .The Odyssey of an Invention Charles Goodyear Medal Address—1975. Rubber Chemistry and Technology, 48 (3) (1975) 73–81.

[12] R.B. Seymour, H.F. Mark, L. Pauling, C.H. Fisher, G.A. Stahl, L.H. Sperling, C.S. Marvel, C.E. Carraher Jr. Otto Bayer Father of Polyurethanes, in: R.B. Seymour (eds) Pioneers in Polymer Science. Chemists and Chemistry, 10 Springer, Dordrecht, 1987 https://doi.org/10.1007/978-94-009-2407-9_22

[13] K.C. Frisch, History of science and technology of polymeric foams, Journal of Macromolecular Science: Part A – Chemistry: Pure and Applied Chemistry. 15:6 (1981) 1089–1112, 10.1080/00222338108066455

[14] A.J. Papa, and W.R. Proops, Phenolic Foams. In: Frisch K. C. and Saunders J. H. (eds) Plastic Foams, Part II. Dekker, New York, 1973

[15] J.W. Yao, Z.M. Hou, Y.S. Yao., Application of phenolic foam plate in the exterior wall thermal insulation, Applied Mechanics and Materials, 174–177 (2012) 1363–1366. doi:10.4028/www.scientific.net/amm.174-177.1363

[16] A. Cooper, A.K. Unsworth and A. Hill, Cellular Rubber and Plastics, P. B. 93,484, Technical Information and Documents Unit, London, 1947, 62–63.

[17] L. Pilato, Introduction, in: L. Pilato (eds). Phenolic Resins: A Century of Progress. Springer, Berlin, Heidelberg, 2010 https://doi.org/10.1007/978-3-642-04714-5_1

[18] K.W. Suh, A.N. Paquet, Rigid polystyrene foams and alternating blowing agents, in: Scheirs, J., Priddy, D. (eds) Modern Styrenic Polymers: Polystyrenes and Styrenic Copolymers, 6 (2003) 203–231, https://doi.org/10.1002/0470867213.ch10

[19] P. Huang, F. Wu, Y. Pang, M. Wu, X. Lan, H. Luo, B. Shen, and W. Zheng, Enhanced dispersion, flame retardancy and mechanical properties of polypropylene/intumescent flame retardant composites via supercritical CO2 foaming followed by defoaming, Composites Science and Technology 171 (2019) 282–290. https://doi.org/10.1016/j.compscitech.2018.12.029

[20] C.W. Visser, D. N. Amato, J. Mueller, and J. A. Lewis, Architected polymer foams via direct bubble writing, Advanced materials 31(46) (2019) 1904668. https://doi.org/10.1002/adma.201904668

[21] G. Wang, G. Zhao, L. Zhang, Y. Mu, and C. B. Park, Lightweight and tough nanocellular PP/PTFE nanocomposite foams with defect-free surfaces obtained using in situ nanofibrillation and nanocellular injection molding, Chemical Engineering Journal 350 (2018) 1–11. https://doi.org/10.1016/j.cej.2018.05.161

[22] L. Wang, Y. Hikima, M. Ohshima, A. Yusa, S. Yamamoto, and H. Goto, Unusual fabrication of lightweight injection-molded polypropylene foams by using air as the novel foaming agent, Industrial & Engineering Chemistry Research 57(10) (2018) 3800–3804. 10.1021/acs.iecr.7b05331

[23] C. Yang, M. Wang, Z. Xing, Q. Zhao, M. Wang, G. Wu, A new promising nucleating agent for polymer foaming: effects of hollow molecular-sieve particles on polypropylene supercritical CO_2 microcellular foaming, RSC Advances 8(36) (2018): 20061–20067. https://doi.org/10.1039/C8RA03071E

[24] H. Wen, M. Wang, S. Luo, Y. Zhou, and T. Liu, Mechanical properties of silicone rubber enhanced microcellular EPDM foams based on supercritical CO_2 foaming technology, Macromolecular Materials and Engineering 306(10) (2021) 2100310. https://doi.org/10.1002/mame.202100310

[25] P. Sun, T. Y. Qian, X. Y. Ji, C. Wu, Y. S. Yan, and R. R. Qi, HDPE/UHMWPE composite foams prepared by compression molding with optimized foaming capacity and mechanical properties, Journal of Applied Polymer Science 135, no. 46 (2018) 46768. https://doi.org/10.1002/app.46768

[26] R. Mishra, R. Varshney, N. Das, D. Sircar, P. Roy, Synthesis and characterization of gelatin-PVP polymer composite scaffold for potential application in bone tissue engineering. European Polymer Journal 119 (2019) 155–168 https://doi.org/10.1016/j.eurpolymj.2019.07.007

[27] J. Hou, J. Jiang, H. Guo, X. Guo, X. Wang, Y. Shen, Q. Li, Fabrication of fibrillated and interconnected porous poly (ε-caprolactone) vascular tissue engineering scaffolds by microcellular foaming and polymer leaching, RSC Advances 10(17) (2020) 10055–10066. https://doi.org/10.1039/D0RA00956C

[28] R. Xie, J. Hu, O. Hoffmann, Y. Zhang, F. Ng, T. Qin, and X. Guo, Self-fitting shape memory polymer foam inducing bone regeneration: A rabbit femoral defect study, Biochimica et Biophysica Acta (BBA)-General Subjects 1862(4) (2018) 936–945. https://doi.org/10.1016/j.bbagen.2018.01.013

[29] C. Li, D. Jiang, B. Huo, M. Ding, C. Huang, D. Jia, H. Li, C.-Y. Liu, and J. Liu, Scalable and robust bilayer polymer foams for highly efficient and stable solar desalination, Nano Energy 60 (2019) 841–849. https://doi.org/10.1016/j.nanoen.2019.03.087

[30] J. He, G. Zhao, P. Mu, H. Wei, Y. Su, H. Sun, Z. Zhu, W. Liang, and A. Li, Scalable fabrication of monolithic porous foam based on cross-linked aromatic polymers for efficient solar steam generation, Solar Energy Materials and Solar Cells 201 (2019) 110111 https://doi.org/10.1016/j.solmat.2019.110111

[31] X. Lü, T. Yu, F. Meng, W. Bao, Wide-range and high-ftability flexible conductive graphene/thermoplastic polyurethane foam for piezoresistive sensor applications, Advanced Materials Technologies. 6(10) (2021) 2100248. https://doi.org/10.1002/admt.202100248

[32] X. Huang, X. Yang, H. Liu, S. Shang, Z. Cai, and K. Wu, Bio-based thermosetting epoxy foams from epoxidized soybean oil and rosin with enhanced properties, Industrial Crops and Products 139 (2019) 111540. https://doi.org/10.1016/j.indcrop.2019.111540.

[33] J.A. Dicks, C. Woolard, Biodegradable polymeric foams based on modified castor oil, styrene, and isobornyl methacrylate. Polymers. 13(11) 2021 1872. https://doi.org/10.3390/polym13111872

[34] Y. Li, D. Yin, W. Liu, H. Zhou, Y. Zhang, and X. Wang, Fabrication of biodegradable poly (lactic acid)/carbon nanotube nanocomposite foams: significant improvement on rheological property and foamability, International Journal of Biological Macromolecules. 163 (2020) 1175–1186. https://doi.org/10.1016/j.ijbiomac.2020.07.094

[35] García-Casas, A. Montes, D. Valor, C. Pereyra, and E. J. Martínez de la Ossa, Foaming of polycaprolactone and its impregnation with quercetin using supercritical CO2, Polymers. 11(9) (2019) 1390. https://doi.org/10.3390/polym11091390

[36] B. Coppola, L. Courard, F. Michel, L. Incarnato, P. Scarfato, and L. Di Maio, Hygro-thermal and durability properties of a lightweight mortar made with foamed plastic waste aggregates, Construction and Building Materials. 170 (2018) 200–206. https://doi.org/10.1016/j.conbuildmat.2018.03.083.

[37] S Farhan, R Wang, H Jiang, and K Li, Use of waste rigid polyurethane for making carbon foam with fireproofing and anti-ablation properties, Materials & Design. 101 (2016) 332–339. https://doi.org/10.1016/j.matdes.2016.04.008.

[38] J.J. Lee, M. Yeon Cho, B.-H. Kim, and S. Lee, Development of eco-friendly polymer foam using overcoat technology of deodorant, Materials. 11(10) (2018) 1898. 10.3390/ma11101898.

[39] Q. Wu, Q. Zhang, L. Zhao, S.-N. Li, L.-B. Wu, J.-X. Jiang, and L.-C. Tang, A novel and facile strategy for highly flame retardant polymer foam composite materials: Transforming silicone resin coating into silica self-extinguishing layer, Journal of Hazardous Materials. 336 (2017) 222–231. https://doi.org/10.1016/j.jhazmat.2017.04.062.

[40] C. Wang, Y. Wu, Y. Li, Q. Shao, X. Yan, C. Han, Z. Wang, Z. Liu, and Z. Guo, Flame-retardant rigid polyurethane foam with a phosphorus-nitrogen single intumescent flame retardant, Polymers for Advanced Technologies. 29(1) (2018) 668–676. https://doi.org/10.1002/pat.4105

[41] F. Song, Z. Li, P. Jia, C. Bo, M. Zhang, L. Hu, and Y. Zhou, Phosphorus-containing tung oil-based siloxane toughened phenolic foam with good mechanical properties, fire performance and low thermal conductivity, Materials & Design. 192 (2020) 108668. https://doi.org/10.1016/j.matdes.2020.108668.

[42] Liu, L. Wang, W. Zhang, and Y. Han, Phenolic resin foam composites reinforced by acetylated poplar fiber with high mechanical properties, low pulverization ratio, and good thermal insulation and flame retardant performance, Materials. 13(1) (2019) 148. https://doi.org/10.3390/ma13010148

[43] Z. Wang, T. Zheng, C. Lu, X. Guo, H. Xiao, J. Jia, and D. Zhang, Preparation and properties of nano ZnO toughed phenol–urea-formaldehyde foam, Journal of Applied Polymer Science. 138(6) (2021) 49816. https://doi.org/10.1002/app.49816

[44] F. Song, P. Jia, C. Bo, X. Ren, L. Hu, and Y. Zhou, The mechanical and flame retardant characteristics of lignin-based phenolic foams reinforced with MWCNTs by in-situ polymerization, Journal of Dispersion Science and Technology. 42(7) (2021) 1042–1051. 10.1080/01932691.2020.1735410

[45] Z. Zhang, M. Xu, and B. Li, Preparation and characterization of polymethacrylimide/silicate foam, Polymers for Advanced Technologies. 29(12) (2018) 2982–2999. 10.1002/pat.4418

[46] T. Ge, K. Tang, X. Tang, Preparation and properties of acetoacetic ester-terminated polyether pre-synthesis modified phenolic foam. Materials. 12(3) (2019) 334. https://doi.org/10.3390/ma12030334

[47] Y. Yu, Y. Wang, P. Xu, J. Chang, Preparation and characterization of phenolic foam modified with bio-oil, Materials. 11(11) (2018) 2228. https://doi.org/10.3390/ma11112228

[48] C. Li, H. Ma, Z. Zhou, W. Xu, F. Ren, and X. Yang, Preparation and properties of melamine-formaldehyde rigid closed-cell foam toughened by ethylene glycol/carbon fiber, Cellular Polymers 40(2) (2021) 55–72. 1177/0262489320929232

21 Polymers for High-Performance Flame-Retardant Materials

Styliani Papatzani,[1,2] Dionysios E. Mouzakis,[2,3] and Panagiota Koralli[3]

[1] University of West Attica, 28 Agiou Spiridonos, 12243, Egaleo, Greece
[2] Hellenic Army Academy, Leoforos Eyelpidon (Varis – Koropiou) Avenue, 16673, Vari, Greece
[3] National Hellenic Research Foundation, 48 Vassileos Constantinou Avenue, 11635, Athens, Greece

1 INTRODUCTION

A number of fire hazards can be distinguished, with the most destructive for structures and most lethal for inhabitants including heat, smoke and toxic gases. In particular, the rate of heat release is considered to be the most important property by many scholars, as the higher it is, the faster the ignition and spread of flame and the greater the intensity of fire will be [1]. As a result, the more limited the time to evacuate structures under fire will also be [2]. Moreover, with respect to smoke, the darker and thicker it is, the greater the difficulty in detecting and rescuing victims trapped in buildings. The great importance of both parameters has led to the development of a number of standard test methods. Therefore, these can be evaluated using ASTM E1354 "Cone Calorimetry Testing" or ASTM 1354-09 "Standard Test Method for Heat and Visible Smoke Release Rates for Materials and Products Using an Oxygen Consumption Calorimeter". Finally, toxicity, mainly measured in terms of carbon monoxide, is responsible for thousands of deaths in Europe annually [2].

Apart from innumerable industrial applications, such as electronic and electrical materials, fabrics, furniture, aerospace and others, designing structural members and structures made of polymeric matrices is becoming increasingly popular due to the greater mechanical strength, ductility, lightness, robustness and, of course, due to the possibility of tailoring the material to specific shapes, needs and applications. At the same time, design standards and code specifications for designing with polymer-based composites have increased their applications and the confidence in their use [3]. One of the main issues with using such materials, their fire behaviour, still pertains. Fire safety has been improved over the decades with active or passive protection systems or even a combination of both [1]. Active systems comprise all advanced detection, alerting and fire-extinguishing mechanisms, while passive system encompass the modification of materials in order to minimize the probability of ignition or in the case of fire in prolonging flame retardancy in order to allow for the escape of those in the proximity and the possible extinguishing of the fire with the use of active systems. This chapter is dedicated to providing a presentation of all recent advances in the field of polymers that offer advanced flame retardancy.

The multi-stage process of combustion of polymers is thoroughly described in the literature, where two processes of interrupting and retarding have been distinguished: physical and chemical [4]. The former comprises cooling the polymer, creating a protective char layer or adding inert additives or fillers, diluting the fuel. The latter comprises adding flame retardants that directly terminate active H- and HO- radicals or allowing reactions between the flame retardants and the polymer.

DOI: 10.1201/9781003278269-21

Based on that, two central strategies can be distinguished:

1. Doping polymers with flame-retardant additives
   a. Using conventional/synthetic flame retardants
   b. Using bio-based flame retardants
2. Blending/compounding with flame-retardant polymers or synthesizing new polymers with increased flame-retardant properties
   a. Blending with intrinsically flame-retardant polymers
   b. Blending polymers with flame-retardant nanocomposites
   c. 3D printing for flame-retardant polymers

It is acknowledged that other fire-retarding technologies have evolved, such as encapsulating the polymer or protecting it from exposure, and the use of coatings and other intumescent systems, however the discussion of those areas is beyond the scope of this chapter.

2   DOPING POLYMERS WITH FLAME-RETARDANT ADDITIVES

One of the most well-established technologies in producing flame-retardant polymers is doping them with flame-retardant additives. In the next section a short discussion is provided on the main conventional additives.

2.1   USING CONVENTIONAL/SYNTHETIC FLAME RETARDANTS

Chemically, these can be classified into five main groups [4]:

- Halogenated compounds
- Metal hydroxides
- Phosphorus-containing compounds
- Phosphorus (P)- and nitrogen (N)-containing compounds
- Silica (Si)-containing compounds

Halogenated organic compounds are bromine- or chlorine-based additives. Of the two, the former is the most established due to its higher free radical trapping efficiency and lower decomposition temperature [5]. However, halogenated compounds pose a risk to the environment and have inherent increased toxicity risk. These health and environmental risks with respect to the use of brominated flame retardants have led to a ban and restriction on their use within the EU [5]. In fact, tetrabromobisphenol A has been upgraded to group 2A (probably carcinogenic to humans).

It has been argued that for using metal hydroxides such as aluminium trihydrate (ATH) or magnesium hydroxide, significant loading levels must be employed of over 50% wt., causing reductions in mechanical strength and bendability and difficulties for the extrusion process of polymers [2].

With respect to phosphorus-containing additives, one of the key representatives is red phosphorus, which however offers limited applications due to its colour, ease of migration over time, plasticization effect, and the release of toxic phosphine (PH_3). Another is the DOPO (9,10-dihydro-9-oxa-10-phosphaphenanth-rene-10-oxide) flame retardant, which, however, has limited toxicity and when incorporated directly into epoxy polymers lowers the mechanical properties and the glass transition temperature [6]. To overcome these shortcomings, surface modification methods or innovative synthesis procedures are often employed.

Phosphorus- and nitrogen-containing additives offer a more efficient solution as they exhibit low toxicity and low smoke emissions. However, their poor thermal stability, mechanical performance due to the difference in polarities between the additive and the polymer and leaching in the polymer matrix compromise their applications. In particular, the difference in polarities causes thermodynamic incompatibility and a weak interfacial interaction [7]. In a recent publication on the use of ammonium polyphosphate (APP) it was claimed that it is highly effective for epoxy resins, polyamide-6 and coatings [7]. To overcome these issues, microencapsulation or, in other words, wrapping of micro or nano quantities of the additive in a thin organic or inorganic membrane is necessary.

Silica-containing compounds are not as efficient as the previously described categories of fire retardants.

Creating flame-retardant polymers that are safe to the environment and to health, eco-friendly, with superior mechanical properties, while still being effective, poses a significant challenge to scientists, and bio-based flame additives offer a step forward in this regard.

2.2 Using Bio-Based Flame Retardants

The necessity for the development of new flame retardants arises not only from the pursuit of better performance but also from an environmentally friendly perspective in an effort to reduce the ecological footprint of materials and become more sustainable. The amelioration of the fire performance of polymers remains the most significant aim; however, ever-increasing environmental concerns drive the evolution of high-efficiency, natural, sustainable and renewable bio-based flame-retardant approaches, decreasing the use of traditional polymers that are generated from fossil fuel sources.

The bio-based flame retardants can be divided into two categories, as shown in Figure 21.1 [8]. One category comprises the flame retardants based on biomass, including cellulose, starch, tannins, phytic acid, proteins, lignin and oils, whilst the other category contains those with animal origins, such as deoxyribonucleic acid (DNA) and chitosan.

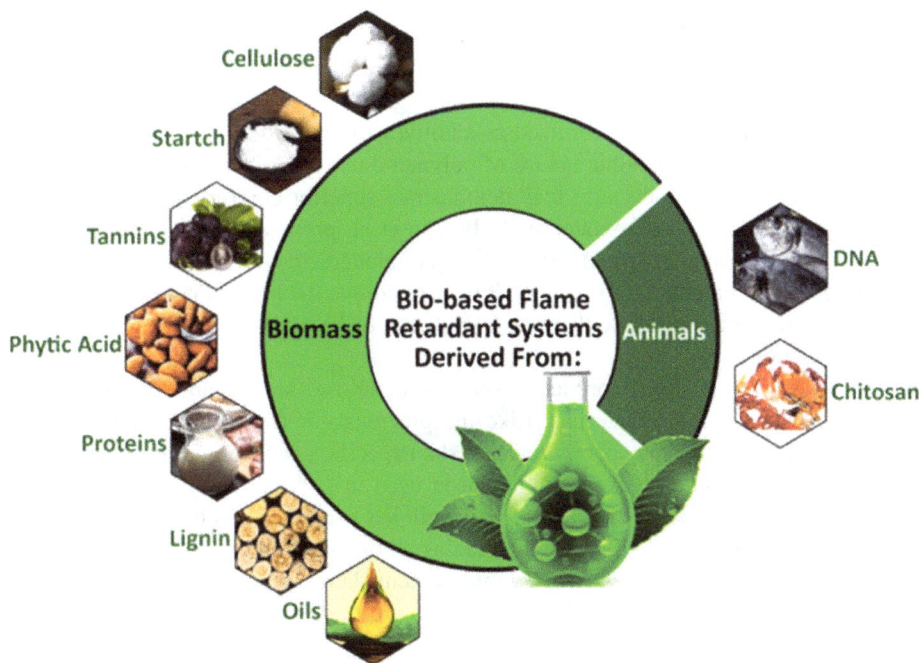

FIGURE 21.1 Schematic representation of the main bio-based flame retardants divided into two categories depending on their origin. Adapted with permission from Ref. [8]. Copyright 2021, Elsevier.

To break free from the halogenated organics that can pose a threat to the environment, the best candidate to replace them is phosphorous-rich materials, taking advantage of the ability of the phosphorus flame retardants to form compact char layers that act as an inhibitor to the spread of fire by effecting a self-extinguishing ability [9]. Phytic acid is a polyphosphate ester of inositol derived from plant tissues such as seeds and rice bran, and is a bio-based renewable compound. Since it contains six highly active phosphate groups, it can be combined with different organic groups to synthesize polymer-based flame retardants with high efficiency due to the production of polyphosphoric acid during combustion [10]. Several research groups have reported the use of phytic acid as a natural flame-retardant additive, utilizing it as a coating on cotton fabrics combined with chitosan or nitrogen- and silicon-based compounds [8]. Recently, Fang et al. developed a novel biomass flame retardant which exhibits both high-efficiency flame resistance and smoke suppression, achieving improved flame retardancy in the performance of epoxy resin [11]. In this work, the phenylphosphonate compound called EHPP and the phytic acid were used to fabricate EHPP@PA flame retardant, demonstrating different oxidation state phosphorus. It was found that the implantation of EHPP@PA into the epoxy resin provoked a reduction of the peak heat release rate of about 64%, whilst the total heat release decreased by 16%. Another approach refers to the synthesis of a self-assembled bio-derived microporous nanosheet using phytic acid [12]. In this design, the authors managed to produce 2D nanosheet intumescent flame retardants with microporous structure via self-assembly of hexakis (4-aminophenoxy) cyclotriphosphazene and phytic acid, as shown in Figure 21.2. It was observed that the incorporation of the resulting HACP-PA into PLA caused a decrease in the peak heat release rate and the total heat release of PLA at rates of 15.3% and 21.5%, respectively.

Other non-toxic alternatives for the development of flame retardants are bio-based aromatic compounds such as tannins and lignin, taking advantage of their ability for char formation during their pyrolytic decomposition due to their aromatic structure. Lignin is an abundant, highly cross-linked phenolic polymer that can be found in bark and wood. Moreover, it represents a plentiful industrial by-product, mainly from paper production. The flame-retardant performance of lignin has been widely investigated [13–15]. Very recently, the effects of lignin-based flame retardants in epoxy resin composites were reported with encouraging results [16]. Tannins represent the fourth most abundant material derived from plants, belonging to the class of natural polyphenols [17]. Even though tannins are not widely utilized as additives for synthetic polymers, probably due to incompatibility with the hydrophobic nature of polymers, they have become attractive candidates for bio-based flame retardants because they demonstrate superior charring ability in combination with antioxidant properties [18]. A later study by Lin et al. presented the combination of tannic acid with resveratrol, a natural trihydroxy stilbene compound, for the improvement of the flame retardancy of poly(vinyl alcohol) (PVA) [19]. It was found that the polyhydroxy structure of RETA stimulates char formation during the thermal degradation of PVA, restricting both its flammability and smoke production.

Furthermore, saccharide-based products, such as cellulose, starch, and chitosan have been investigated as flame-retardant materials. Like the previously noted aromatic compounds, these materials present good charring capacity under certain specific conditions, mainly in combination with phosphorus [20]. Additionally, DNA has been used in polymeric materials to impart them with flame-retarding properties. The group of Malucelli has been extensively involved in the flame-retarding properties of DNA [21–23] since it contains all the required elements (phosphate groups, nitrogenous base pairs and sugar) to act as intumescent flame retardant. Finally, there are also some studies claiming that bio-based flame retardants are more efficient than conventional flame retardants [24].

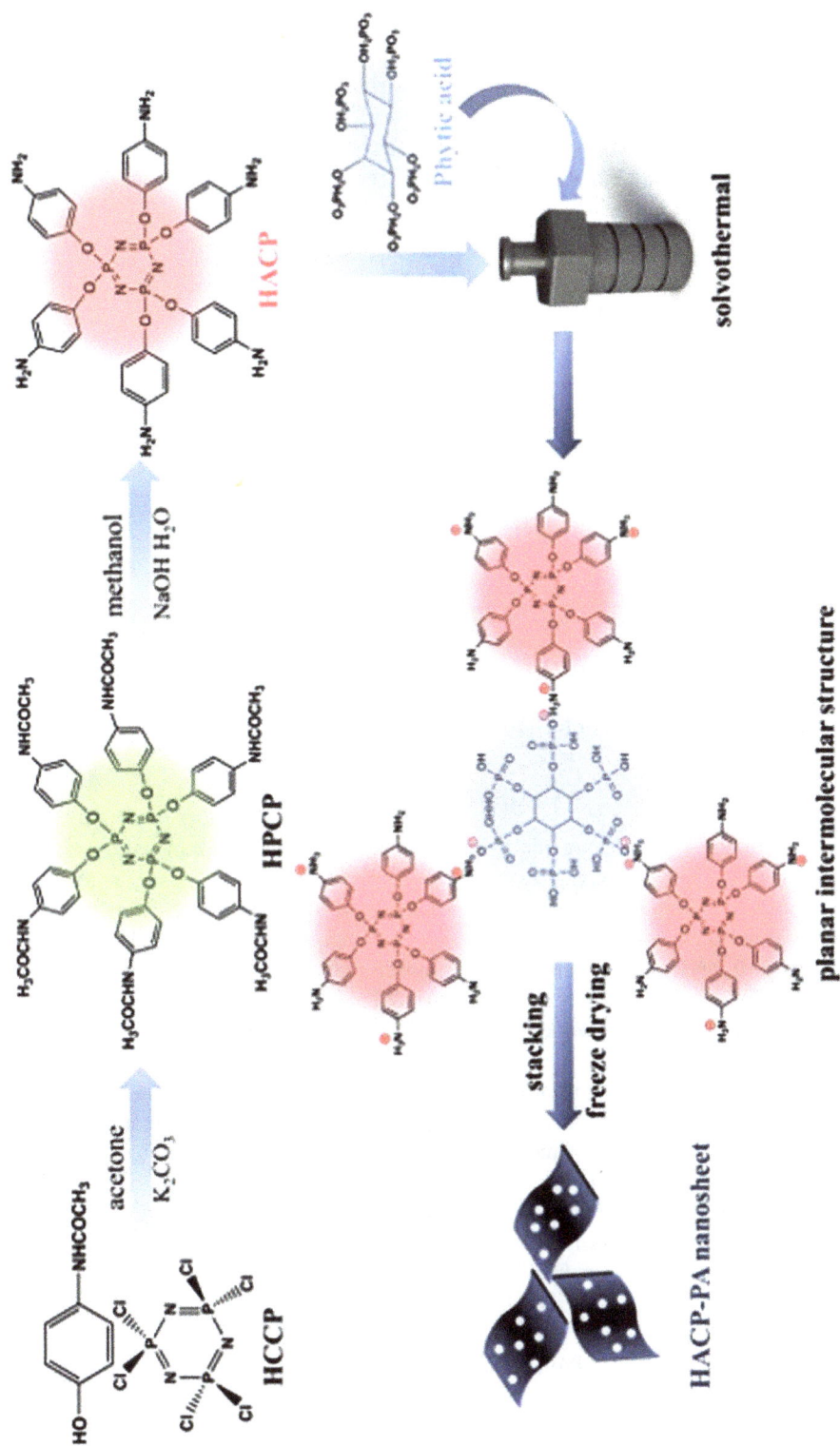

FIGURE 21.2   Synthesis of the HACP-PA microporous nanosheets. Adapted with permission from Ref. [12]. Copyright 2021, Elsevier.

3 BLENDING/COMPOUNDING WITH FLAME-RETARDANT POLYMERS OR SYNTHESIZING NEW POLYMERS WITH INCREASED FLAME-RETARDANT PROPERTIES

Blending with intrinsically flame-retardant polymers such as polyphosphazenes, polyvinyl chloride (PVC) or fluoropolymers offers another viable option for a variety of applications. With the advent of nanotechnology, new horizons in material science have been opened and polymer-based composite technology is one of the fields that have greatly benefited from it, as it offers the possibility of the nanomodification of polymers.

Most published research presents results on the surface modification or wrapping of a variety of polymers, such as:

- Polyphosphazenes
- Polyethylene terephthalate (PET)
- Polypropylene (PP)
- Poly(oxymethylene) (POM)
- Poly(tetrafluoroethylene) (PTFE)
- Poly(methyl methacrylate) (PMMA)
- Polyethylene (PE)
- Polystyrene (PS)
- Polyamide 6 (PA6)
- Polycarbonate (PC)
- Polyvinyl chloride (PVC)
- Poly(vinylidene fluoride) (PVDF)

In this subsection modification of the most widely used polymers is discussed.

3.1 BLENDING WITH INTRINSICALLY FLAME-RETARDANT POLYMERS

Polyphosphazene is an organic–inorganic polymer whose main chain exhibits single and double bonds of phosphorus and nitrogen [25], and therefore is inherently flame retardant. Linear polyphosphazenes offer high temperature resistance and improvement in mechanical performance. In a recent research study, epoxy resin was synthesized with a linear polyphosphazene-based additive. The phosphorus- and nitrogen-containing thermosets exhibited great mechanical performance, excellent flame resistance and good thermal stability [26].

Cyclocross-linked polyphosphazenes offer high thermal stability and are employed in applications as flame-retardant additives. The most significant advantage of the cyclocross-linked polyphosphazenes compared to the linear ones is that although they both maintain high thermal stability and heat and corrosion resistance, the former can be designed into nanotubes, nanospheres or nanosheets and both physical and chemical properties can be tailored by adjusting the number and nature of the functional groups, rendering this family of flame retardants the strongest candidate to replace halogen flame retardants. Furthermore, the addition levels have a mean of 5 wt.%, whereas for linear polyphosphazenes the addition levels range between 10–40.8 wt.%. The most researched nucleophilic substitution reagents include [25]:

- 4,4′-Diaminodiphenyl ether
- Amino compounds such as 4,4′-diaminodiphenyl sulfone
- Bisphenol A
- Melamine
- Phenols
- Trimethoxypyrimidine

In a recent review [25], 5% of polyphosphazene microspheres with various nucleophiles were synthesized. Thermal gravimetric results showed a pronounced reduction in the peak heat release rate ranging from 56.6% to 62.5%. Added to that, promotion of the char residue took place and the limiting oxygen index was also increased to 29.7%, while toxic gases were inhibited. Therefore, the flame retardancy and fire safety of the epoxy were significantly improved.

In another research work, epoxy resin was synthesized with 9.0 wt.% of a compound containing active amine groups of polyphosphazenes, that offered a reduction of 46.7% in the peak heat release rate, 29.3% in the total heat release and of 48% of the total smoke release of the composite [27].

Similarly, a thermoset synthesized with a simple one-pot reaction among 2-thenaldehyde, 2-aminobenzothiazole (ABZ) and DOPO with epoxy resin created a novel P/N/S-containing flame-retardant epoxy resin. The additive was added at 10 wt.% and offered a reduction in the peak heat release rate of 41.26%, a reduction in the total heat release of 35.7% and a reduction in the smoke release of 24.03% [28]. Again the composite exhibited excellent heat resistance and mechanical properties.

Ammonium polyphosphate (APP) was modified with two distinct synthesized polymers: phosphorus-containing organosilicon compound with (i) polyether segment with hydroxyl, PCOC–OH, creating MAPP-OH and (ii) polyether segment with amino group, PCOC–NH, creating MAPP-NH. The synthesized polymer was then added to low-density rigid polyurethane foam and was tested in terms of limited oxygen index (LOI) measurement, horizontal burning test and cone calorimeter test. [29]. The addition of MAPP-NH offered higher LOI measurements for the composite, together with a dramatical reduction in total hear release and smoke, owing to the free NH_3 radical, which in the gas phase enhanced the quenching effect and dilution effect.

Linear functional polymers were synthesized by polymerization with olefin metathesis. The resulting hyperbranched poly(phosphoester) is halogen-free and was added as a flame retardant to polyesters (ethyl 4-hydroxybenzoate) and epoxy resins (bisphenol A diglycidyl ether cured with isophorone diamine) and compared against a commercially available bisphenol A bis(diphenyl phosphate) [30]. The synthesized polymer exhibited superior fire protection for a greater variety of polymeric matrices which was attributed not only to its higher phosphorus content but also to the phosphorus's higher efficiency.

Two novel, halogen-free, DOPO-based flame retardants were synthesized and added in various combinations in three high-performance carbon fibre (70 wt.%) epoxy composites for aviation and three epoxy-only composites [6]. The phosphorus-based flame retardants were composed of DOPO units linked to the star-shaped aliphatic ground body tetra-[(acryloyloxy)ethyl] pentarythrit (DOPP), or heterocyclic tris-[(acryloyloxy)ethyl] isocyanurate (DOPI), respectively. Both flame retardants reduced the peak heat release rate by 31–49%, and the total heat release by 40–44%. The carbon fibres increased the residue and total heat release but inhibited the charring. Interestingly, the total heat release of the carbon fibre-reinforced epoxy doped with either of the two flame retardants was reduced by almost 25%.

The fire resistance of the organic poly(ethylene terephthalate) (PET) multifilaments was improved by adding flame-retardant agents based on a zinc phosphinate compound (Exolit OP950) via melt blending to produce textiles [31]. Specifically, polyhedral oligomeric silsesquioxane (POSS) nanoparticles were melt mixed with the Exolit OP950 (OMPOSS). Scanning electron microscopy imaging, thermal gravimetric analyses and cone calorimeter results showed that the POSS nanoparticles caused a minor drop in fire properties of the textile but suppressed smoke and lowered the combustion toxic fumes.

In another study 1 wt.% zinc borate addition was proven to effectively enhance the fire properties of polypropylene composites doped with ammonium polyphosphate (APP) and charring-foaming agent (CNCA-DA) in terms of LOI, cone calorimeter test (CCT), scanning electron microscopy (SEM), energy-dispersive spectrometry (EDS) and thermogravimetric analysis (TGA) [32]. In fact, it increased the LOI value from 27.1 to 30.7, and also the thermal stability was improved and the char residue was increased.

Halogen-free, silicon-nitrogen compounds have been proven effective flame retardants for polymers [33]. A branched benzoxazine-containing polysiloxane was obtained by polymerizing a benzoxazine-containing triethoxysilane compound with diphenylsilanediol. The resulting polymer is thermally reactive and exhibited high thermal stability and high flame retardancy since the limited oxygen index was over 45 at a low dosage of 5 wt.% simultaneously, the glass transition temperature was increased from 145°C to 205°C, proving the high efficiency of the flame retardant.

3.2 FLAME-RETARDANT NANOCOMPOSITES FOR POLYMERS

One of the greatest advantages of utilizing polymer nanocomposites is the enhanced properties and the fact that the filler/reinforcement content ranges between 2 and 10 wt.%. Since the 1990s, researchers in the US have studied the fire-retarding properties of organomodified clays dispersed in polymer matrices [34]. As early as 2000, the combination of nanosilicates with conventional fire-retarding additives was studied [4]. Since 2002, silicates, such as montmorillonite (MMT), saponite or hectorite, have been targeted for use in the synthesis of various nanocomposites [35], and polymer nanocomposites in particular [2]. Their structure is thoroughly described in the literature, together with the various options for surfactants (organic or inorganic) [35]. In short, silicates are mainly formed by stacks or platelets comprised of layers of silica or alumina. This structure is typically less than 1 nm thick, although the lateral dimension ranges from a few nanometres to a few hundreds of micrometres [36]. The platelets are held together by van der Waals or electrostatic forces, by hydrogen bonding or by interlayer cations creating a space between them called the intergallery space. The forces, which are moderately weak, can be overcome through the introduction of a modifier or a surfactant in the intergallery space, which will can have a chemical effect causing the exchange of cations leading to electrostatic repulsion of the platelets and consequent separation. If an organic modifier or surfactant is used in one of the above silicates, an "organoclay" is produced which is compatible with polymers to form polymer–clay nanocomposites. So far, three processes have been utilized in order to disperse a layered silicate in a polymer matrix, namely in situ polymerization, the solvent casting method and melt intercalation, details of which can be found in literature [2,34]. The resulting polymer–organoclay can either be microsized (Figure 21.3a) because the polymer did not manage to separate the organoclay or nanosized if the layers are intercalated, i.e. held in significant distances between them, however maintaining the orderly structure (Figure 21.3b) or completely dispersed in the polymer matrix, where the resulting nanocomposite is said to be "exfoliated" or "delaminated" (Figure 21.3c). This latter configuration is the most desirable as bonds holding the platelets together are completely broken and the entire surface area of each platelet is available for reactions. More often than not, intercalated particles coexist with exfoliated ones in polymer–organoclay nanocomposites [37].

The level of platelet separation can be confirmed by X-ray diffraction, transmission electron imaging and crystallographic calculations [38]. Particularly interesting for this chapter is the fact that polymer organoclays exhibit improved thermal stability and flame retardancy due to the low permeability char created by fire that can act as a barrier for mass and thermal transfer. Additionally, it has been proven that nanoparticles can modify the degradation pathway of polymers [34]. These properties can be studied using thermal gravimetry [38], which provides scholars with the decomposing masses over time. Therefore, delayed decomposition can be clearly distinguished and monitored. In fact, synergistic effects in terms of flame retardance between organoclays and phosphorus-based flame retardants and even with the use of thermoplastic matrices [39] or in polyurethane/organoclay nanocomposites [40] have been observed.

Apart from organoclays acting as nanofillers and nanoreinforcements in fire-retardant polymeric nanocomposites, carbon nanotubes offer another option. Cone calorimeter studies blended with

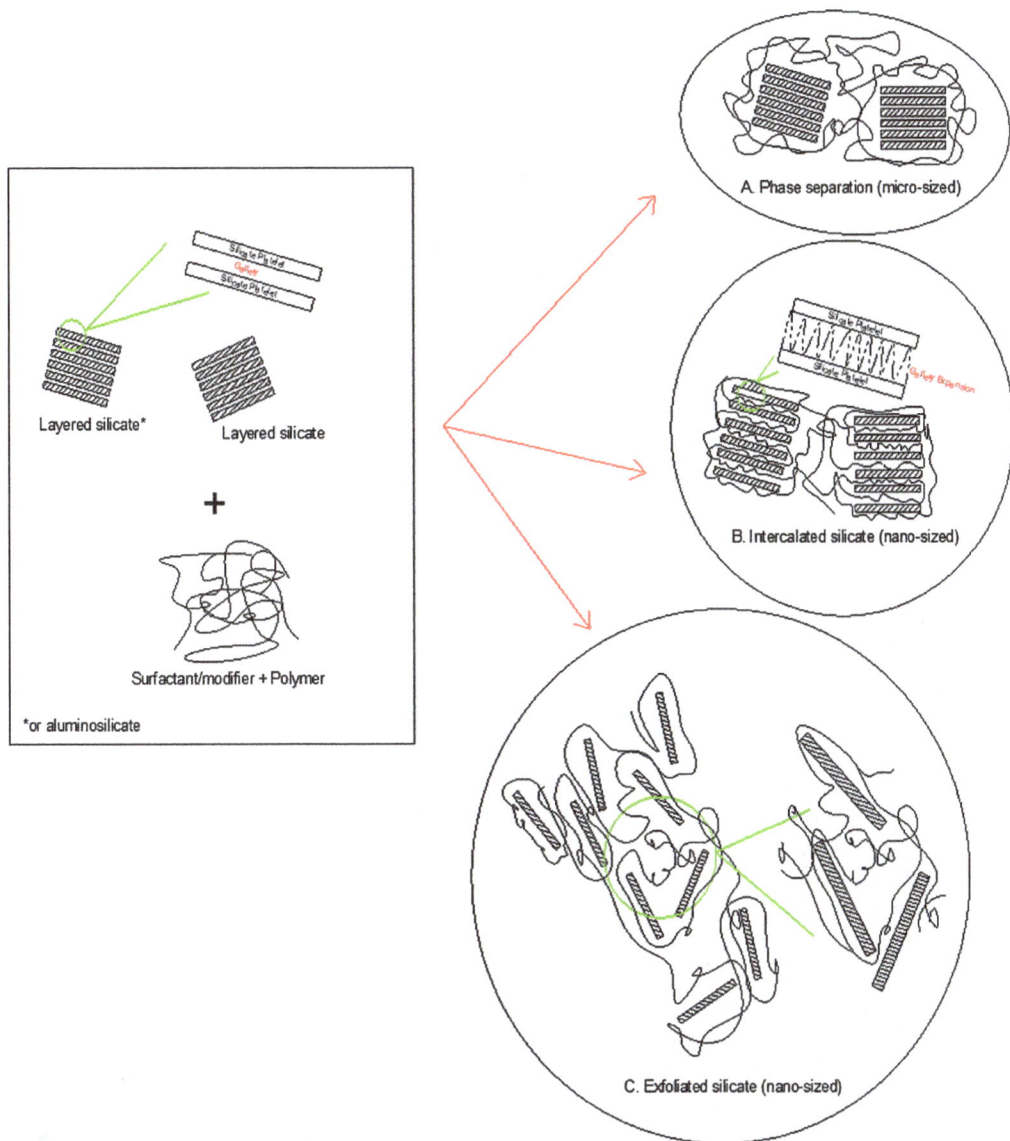

FIGURE 21.3 The three possible configurations of polymer–organoclay composites. Adapted with permission from Ref. [35]. Copyright 2016, Taylor & Francis.

melt-blending ethylene-vinyl acetate copolymer have shown a significant reduction of the peak heat release rates compared with the control sample [41]. The formulation containing carbon nanotubes and organoclays was the most successful in altering the structure of the char produced [41].

Melt-blending of bio-based polylactide with three different types of nanoparticles. nanosilica, halloysite and montmorillonite doped with aluminium diethylphosphinate flame retardant, was tested in terms of heat release rate, thermal gravimetry, SEM, TEM, XRD and tensile strength. Attenuated total reflectance Fourier transform infrared spectroscopy (ATR-FTIR) was used to analyse fire residues. It was concluded that the nanocomposite containing MMT exhibited a superior flame retardancy. This was attributed to the nanoparticle geometry, which, being plate-like, offered a greater surface area which was pivotal (i) in the delayed shar formation and (ii) in the formation of a thicker aluminium phosphate/MMT nanocomposite char that acted as a transport barrier [37]. The next best performance was obtained by the nanosilica spherical nanoparticles and the least successful were the rod-like halloysite nanoparticles.

The latest research is provided by Nazarenko et al. who studied an epoxy composite reinforced with 0.5 wt.% of multiwalled carbon nanotubes (MWCNTs) and 0 and 15 wt.% of boric acid and sodium bicarbonate separately, creating MWCNTs-boric acid and MWCNTs-sodium bicarbonate [42]. The sole addition of nanotubes did not enhance the thermal properties of the nanocomposite. Enhancements were offered only by the combination of MWCNTs and boric acid. In contrast, MWCNTs-sodium bicarbonate was proven unsuitable for the studied nanocomposite.

Further analysis on the various combinations of nanoparticles with conventional flame retardants is provided in literature [4].

3.3   3D Printing with Flame-Retardant Polymers

With respect to novel manufacturing technologies, the progress made and future prospects for creating precise structures using modern approaches, particularly 3D printing – as the state-of-the-art manufacturing methodology delivering innovative features in this field – and their flame-retardancy capabilities are of paramount significance.

Indeed, the techniques guiding the development of novel flame-retardant polymer materials and 3D printing technology are nearing a practical conclusion. It is vital for a variety of applications to ensure that 3D printed parts and components do not easily catch fire or spread flames. As a result, utilizing materials that can inhibit fire spread is prudent, and it is frequently required by regulatory agencies. In the aerospace and automobile industries, flame-retardant polymers are extremely significant due to the stringent safety regulations.

If a part or product must fulfil a self-extinguishing industry standard, there are various certified items on the market that meet international regulations, as discussed below. Many polymers, such as PEEK (poly-ether-ether-ketone) and ULTEM™ (plastic resins family of amorphous thermoplastic polyetherimide materials), are naturally flame-retardant and may not require special certification. Polymer makers may have conducted their own tests on the material to guarantee that it is flame-resistant, but they have not gone through the certification process. In the aerospace and automobile industries, flame-retardant polymers are extremely significant.

Small enterprises that print electronics housings or designer studios that 3D print lamps, for example, can gain from providing flame-resistant items. As an example, a burn test of a 3D printed part made from ABS versus a flame-retardant ABS from Filoalfa showed that regular ABS continues to burn and melt down while the flame-retardant material extinguishes itself. However, ULTEM combined with another polymer may no longer be flame retardant. The manufacturer may add a flame-retardant chemical to the material in this scenario. Without lab testing, it's impossible to identify the flame-retardancy of a 3D printing substance that is a blend of polymers. Not all ULTEM or PEEK-based filaments are flame-resistant. It is imperative to take a sharp look at the approved flame-retardant compounds in filament and polymer powders for selective laser sintering, as well as resins for SLA printing. Because they are more resistant to catching fire and staying on fire, flame-retardant materials inhibit the spread of fire, but they are not fire-proof. When in contact with a minor heat source, such as a small flame or an electrical fault, they are meant to reduce the chance of a fire developing.

Market research reveals the following results, shown in Table 21.1, on flame retardant polymers available as raw materials for 3D printing procedures.

There are several methods for making materials flame-resistant, as described below.

When heated, substances react endothermically. This means that the chemicals in the process absorb heat, which lowers temperatures and slows fire. When heated, it produces an inert gas that suffocates flames by cutting them off from oxygen. Through the charring process, having burning layers shields unburned layers.

TABLE 21.1
Flame-Retardant Materials for 3D Printing

Brand and Material	Polymer Type	Product Form	Certifications	Price
Formfutura ABSpro Flame Retardant	ABS	Filament	UL 94 V-0	$35/500 g
Markforged Onyx FR	carbon fibre-filled Nylon	Filament	UL 94 V-0	$260/800 cc
Essentium TPU 90A FR	TPU	Filament	UL 94 V-0 & FAR 25.853	$98/750 g
3DXTech Firewire Flame Retardant PC-ABS	ABS	Filament	UL 94 V-0	$78/1 kg
3DXTech Firewire ABS-FR	ABS	Filament	UL 94 V-0	$68/1 kg
BASF Ultrafuse PC/ABS FR	PC/ABS	Filament	UL 94 V-0	$52/750 g
DSM Novamid AM1030 FR	Nylon	Filament	UL 94 V-0	$212/1kg
Clairant PA6/66 GF20	Glass-filled Nylon	Filament	UL 94 V-0	$200/1 kg
Solvay Solef PVDF AM	PVDF	Filament	UL 94 V-0	$200/750 g
Kimya PEI 9085	PEI	Filament	UL 94 V-0, FAR 25.853 & EN45545	$336/1 kg
Kimya PEKK Carbon Fiber	PEKK carbon fibre	Filament	UL 94 V-0	$590/.5 kg
Sabic ULTEM 1010	PEI	Filament	UL 94 V-0 & FAR 25.853	$165/1 kg
Lehvoss Luvocom 3F Peek CF 9710 BK	PEEK carbon fibre	Filament	UL 94 V-0	$120/1 kg
EOS PA 2210 FR Nylon	Nylon	Polymer Powder	FAR 25.853	n.a.
EOS PA 2241 FR Nylon	Nylon	Polymer Powder	UL 94 V-0 & FAR 25.853	n.a.
3D Systems DuraForm FR 100	Polymer	Polymer Powder	UL 94 V-0 & FAR 25.853	n.a.
CRP Tech Windform FR2 & FR1	Composite	Polymer Powder	FAR 25.853	n.a.
Cubicure Evolution FR	resin	SLA	UL 94 V-0	n.a.
Henkel Loctite 3D 3955	Resin	SLA	UL 94 V-0 & FAR 25.853	n.a.
Carbon EPX 86FR	Resin	SLA (DLP)	UL 94 V-0 or FAR 25.853	n.a.

Any plastic component that comes into contact with an electrical current must be fire resistant or retardant. Components such as wire nuts, junction boxes, and internal cable supports are all at risk of catching fire if an electrical problem arises. Furthermore, flame-retardant components can be found in industrial machinery, DIY tools, and domestic appliances like microwaves, toasters, and dishwashers.

The interiors of vehicles, such as the inside of a plane or the cockpit of a racing car, must also resist the spread of fire to safeguard the human occupants in the automotive and aerospace industries.

There exists a variety of certification criteria that declare that a material is flame-retardant. The two most important ones are reviewed here.

3.3.1 Underwriters Laboratories UL 94

UL 94 [43] is the most widely used flame-retardant standard. This standard uses precise tests to determine how flame retardant a substance is. A part is set alight repeatedly using an external flame as part of the certification process.

The following attributes must be present in order for the part to receive the coveted 94 V-0 grade.

After the first and second burns, burning lasts for less than 10 seconds. This illustrates how quickly the material's sections would stop burning. The afterglow from the second flame has to be less than 30 seconds long. This is an excellent indicator of how rapidly a component cools after being ignited. The longer a part is hot enough to glow, the more probable it is to re-ignite a nearby component. After being set on fire 10 times, no substance that could ignite cotton batting was permitted to drip.

A component that drips substance that can ignite other materials is a hint that it could start other fires in its near vicinity. The sample could not be consumed entirely by burning. This is a useful metric for determining how difficult it is for a flame to move through a material. On a vertical specimen, 94-V.0 indicates that burning stops after 10 seconds, 94-V.1 and 94-V.2 indicate that burning stops within 30 seconds, and so on through 94-5VA and 94-5VB, which indicate that a specimen will flame but stop burning within 60 seconds.

3.3.2   FAR 25.853

The FAA (Federal Aviation Administration of the United States) developed a vertical Bunsen burner test to examine the flammability of materials used in airplane interiors, including both cabin and cargo compartments. This is known as FAR 25.853 [44], and some plastic products will have a "FAR 25.853 – authorized" stamp on them. Specimens are held vertically inside a cage and a Bunsen burner flame is administered from below for 60 seconds or 12 seconds to meet the standard. The material is examined once the flame application period has passed. There are records of flame time, ignition time, burn length, and material drip flame time.

One of the most pressing concerns for the AM industry in the near future will be the availability of more polymer types with relevant qualities, such as flame retardancy. As a result, flame-retardant adaptation to novel polymers employed in 3D printing technologies should be regarded as an essential research topic. PLA is currently one of the most widely utilized bio-based polymers in 3D printing technology. However, the flame-retardant solutions for PLA proposed by various authors and analysed in this chapter were not bio-based. In the future, bio-based flame retardants for PLA and other non-bio-based polymers should be developed. In general, the creation of bio-based FRs is a direct response to increased health consciousness.

4   EPILOGUE – OUTLOOK

Polymer flammability is a serious issue. FRs are critical to our personal safety. As a result, the science of flame retardancy and the development of flame-retardant polymer materials is always evolving and will continue to face challenges in the future as a result of fire and environmental regulations, as well as the widespread use of polymers in almost every aspect of our lives. For several industries, such as the automotive and construction industry, new fire standards are currently being developed. Meanwhile, new manufacturing technologies such as 3D printing are continually evolving, and flame-retardant systems must be flexible to these changes.

On the other hand, because of their remarkable flame retardancy and mechanical resilience, flame-retardant polymeric nanocomposites have received great attention in recent decades.

The combination effect of nanomaterials with varying dimensions and scales and/or traditional phosphorus-, nitrogen- and/or silicon-containing flame retardants has given them a lot of attention. The resultant polymeric nanocomposites have tuneable flame retardancy and mechanical properties thanks to the mixing of nanomaterials with varied geometries and traditional flame retardants via chemical or physical techniques. Given the ever-increasing abilities to master matter at the nanolevel and monitor the subsequent alterations, flame retardancy is expected to be exhaustively analysed in the near future and the new polymer nanocomposites, and those 3D printed in particular, are foreseen to be fully flame-retardant, protecting lives and property wherever utilized.

REFERENCES

[1] Marcelo M. Hirschler, Flame retardants and heat release: review of traditional studies on products and on groups of polymers, Fire Mater. 39 (2014) 207–231. doi: https://doi.org/10.1002/fam.2243.

[2] G. Beyer, Nanocomposites: A new class of flame retardants for polymers, Plast. Addit. Compd. 4 (2002) 22–28. doi:10.1016/S1464-391X(02)80151-9.

[3] V. Arumugaprabu, R.D.J. Johnson, M. Uthayakumar, P. Sivaranjana, eds., Polymer-Based Composites Design, Manufacturing, and Applications, 1st Editio, CRC Press, 2002.

[4] W. He, P. Song, B. Yu, Z. Fang, H. Wang, Flame retardant polymeric nanocomposites through the combination of nanomaterials and conventional flame retardants, Prog. Mater. Sci. 114 (2020) 100687. doi: https://doi.org/10.1016/j.pmatsci.2020.100687.

[5] K. Pivnenko, K. Granby, E. Eriksson, T.F. Astrup, Recycling of plastic waste: Screening for brominated flame retardants (BFRs), Waste Manag. 69 (2017) 101–109. doi:https://doi.org/10.1016/j.wasman.2017.08.038.

[6] B. Perret, B. Schartel, K. Stöß, M. Ciesielski, J. Diederichs, M. Döring, et al., Novel DOPO-based flame retardants in high-performance carbon fibre epoxy composites for aviation, Eur. Polym. J. 47 (2011) 1081–1089. doi:https://doi.org/10.1016/j.eurpolymj.2011.02.008.

[7] S. Qiu, C. Ma, X. Wang, X. Zhou, X. Feng, R.K.K. Yuen, et al., Melamine-containing polyphosphazene wrapped ammonium polyphosphate: A novel multifunctional organic-inorganic hybrid flame retardant, J. Hazard. Mater. 344 (2018) 839–848. doi:10.1016/j.jhazmat.2017.11.018.

[8] H. Vahabi, F. Laoutid, M. Mehrpouya, M.R. Saeb, P. Dubois, Flame retardant polymer materials: An update and the future for 3D printing developments, Mater. Sci. Eng. R Reports. 144 (2021) 100604. doi:10.1016/j.mser.2020.100604.

[9] K.P. Papaspyrides, C. D., Polymer green flame retardants, Elsevier B.V., 2014.

[10] L. Ahmed, B. Zhang, L.C. Hatanaka, M.S. Mannan, Application of polymer nanocomposites in the flame retardancy study, J. Loss Prev. Process Ind. 55 (2018) 381–391. doi:10.1016/j.jlp.2018.07.005.

[11] F. Fang, S. Huo, H. Shen, S. Ran, H. Wang, P. Song, et al., A bio-based ionic complex with different oxidation states of phosphorus for reducing flammability and smoke release of epoxy resins, Compos. Commun. 17 (2020) 104–108. doi:10.1016/j.coco.2019.11.011.

[12] W. Yang, H. Zhang, X. Hu, Y. Liu, S. Zhang, C. Xie, Self-assembled bio-derived microporous nanosheet from phytic acid as efficient intumescent flame retardant for polylactide, Polym. Degrad. Stab. 191 (2021) 109664. doi:10.1016/j.polymdegradstab.2021.109664.

[13] H. Yang, B. Shi, Y. Xue, Z. Ma, L. Liu, L. Liu, et al., Molecularly engineered lignin-derived additives enable fire-retardant, UV-shielding, and mechanically strong polylactide biocomposites, Biomacromolecules. 22 (2021) 1432–1444. doi:10.1021/acs.biomac.0c01656.

[14] L. Liu, M. Qian, P. Song, G. Huang, Y. Yu, S. Fu, Fabrication of green lignin-based flame retardants for enhancing the thermal and fire retardancy properties of polypropylene/wood composites, ACS Sustain. Chem. Eng. 4 (2016) 2422–2431. doi:10.1021/acssuschemeng.6b00112.

[15] H. Yang, B. Yu, X. Xu, S. Bourbigot, H. Wang, P. Song, Lignin-derived bio-based flame retardants toward high-performance sustainable polymeric materials, Green Chem. 22 (2020) 2129–2161. doi:10.1039/D0GC00449A.

[16] C. Li, B. Wang, L. Zhou, X. Hou, S. Su, Effects of lignin-based flame retardants on flame-retardancy and insulation performances of epoxy resin composites, Iran. Polym. J. (2022). doi:10.1007/s13726-022-01052-w.

[17] Z. Xia, W. Kiratitanavit, P. Facendola, S. Thota, S. Yu, J. Kumar, et al., Fire resistant polyphenols based on chemical modification of bio-derived tannic acid, Polym. Degrad. Stab. 153 (2018) 227–243. doi:10.1016/j.polymdegradstab.2018.04.020.

[18] C.E. Hobbs, Recent advances in bio-based flame retardant additives for synthetic polymeric materials, Polymers (Basel). 11 (2019) 949–962. doi:10.3390/polym11020224.

[19] Y. Lin, J. Chen, H. Li, Outstanding flame retardancy for poly(vinyl alcohol) achieved using a resveratrol/tannic acid complex, RSC Adv. 12 (2021) 285–296. doi:10.1039/d1ra08000h.

[20] L. Costes, F. Laoutid, S. Brohez, P. Dubois, Bio-based flame retardants: When nature meets fire protection, Mater. Sci. Eng. R Reports. 117 (2017) 1–25. doi:10.1016/j.mser.2017.04.001.

[21] J. Alongi, R.A. Carletto, A. Di Blasio, F. Carosio, F. Bosco, G. Malucelli, DNA: A novel, green, natural flame retardant and suppressant for cotton, J. Mater. Chem. A. 1 (2013) 4779–4785. doi:10.1039/c3ta00107e.

[22] J. Alongi, R.A. Carletto, A. Di Blasio, F. Cuttica, F. Carosio, F. Bosco, et al., Intrinsic intumescent-like flame retardant properties of DNA-treated cotton fabrics, Carbohydr. Polym. 96 (2013) 296–304. doi:10.1016/j.carbpol.2013.03.066.

[23] G. Alongi, Jenny; Di Blasio, Alessandro; Cuttica, Fabio; Carosio, Federico; Malucelli, Flame retardant properties of ethylene vinyl acetate copolymers melt-compounded with deoxyribonucleic acid in the presence of α-cellulose or β-cyclodextrins, Curr. Org. Chem. 18 (2014) 1651–1660.

[24] C. Réti, M. Casetta, S. Duquesne, S. Bourbigot, R. Delobel, Flammability properties of intumescent PLA including starch and lignin, Polym. Adv. Technol. 19 (2008) 628–635. doi:10.1002/pat.1130.

[25] X. Zhou, S. Qiu, X. Mu, M. Zhou, W. Cai, L. Song, et al., Polyphosphazenes-based flame retardants: A review, Compos. Part B Eng. 202 (2020) 108397. doi:https://doi.org/10.1016/j.composit esb.2020.108397.

[26] G. Yang, W.H. Wu, Y.H. Wang, Y.H. Jiao, L.Y. Lu, H.Q. Qu, et al., Synthesis of a novel phosphazene-based flame retardant with active amine groups and its application in reducing the fire hazard of epoxy resin, J. Hazard. Mater. 366 (2019) 78–87. doi:10.1016/j.jhazmat.2018.11.093.

[27] H. Liu, X. Wang, D. Wu, Synthesis of a novel linear polyphosphazene-based epoxy resin and its application in halogen-free flame-resistant thermosetting systems, Polym. Degrad. Stab. 118 (2015) 45–58. doi:10.1016/j.polymdegradstab.2015.04.009.

[28] J. Zou, H. Duan, Y. Chen, S. Ji, J. Cao, H. Ma, A P/N/S-containing high-efficiency flame retardant endowing epoxy resin with excellent flame retardance, mechanical properties and heat resistance, Compos. Part B Eng. 199 (2020) 108228. doi:10.1016/j.compositesb.2020.108228.

[29] Y. Chen, L. Li, X. Qi, L. Qian, The pyrolysis behaviors of phosphorus-containing organosilicon compound modified APP with different polyether segments and their flame retardant mechanism in polyurethane foam, Compos. Part B Eng. 173 (2019) 106784. doi:https://doi.org/10.1016/j.composit esb.2019.04.045.

[30] K. Täuber, F. Marsico, F.R. Wurm, B. Schartel, Hyperbranched poly(phosphoester)s as flame retardants for technical and high performance polymers, Polym. Chem. 5 (2014) 7042–7053. doi:10.1039/C4PY00830H.

[31] N. Didane, S. Giraud, E. Devaux, G. Lemort, G. Capon, Thermal and fire resistance of fibrous materials made by PET containing flame retardant agents, Polym. Degrad. Stab. 97 (2012) 2545–2551. doi:https://doi.org/10.1016/j.polymdegradstab.2012.07.006.

[32] C. Feng, Y. Zhang, D. Liang, S. Liu, Z. Chi, J. Xu, Influence of zinc borate on the flame retardancy and thermal stability of intumescent flame retardant polypropylene composites, J. Anal. Appl. Pyrolysis. 115 (2015) 224–232. doi:https://doi.org/10.1016/j.jaap.2015.07.019.

[33] C.-Y. Hsieh, W.-C. Su, C.-S. Wu, L.-K. Lin, K.-Y. Hsu, Y.-L. Liu, Benzoxazine-containing branched polysiloxanes: Highly efficient reactive-type flame retardants and property enhancement agents for polymers, Polymer (Guildf). 54 (2013) 2945–2951. doi:https://doi.org/10.1016/j.polymer.2013.03.060.

[34] J.M. Lopez-Cuesta, 16 – Flame-retardant polymer nanocomposites, in: F.B.T.-A. in P.N. Gao (Ed.), Woodhead Publ. Ser. Compos. Sci. Eng., Woodhead Publishing, 2012: pp. 540–566. doi:https://doi.org/10.1533/9780857096241.3.540.

[35] S. Papatzani, Effect of nanosilica and montmorillonite nanoclay particles on cement hydration and microstructure, Mater. Sci. Technol. 32 (2016) 138–153. doi:10.1179/1743284715Y.0000000067.

[36] A.A. Sapalidis, F.K. Katsaros, N.K. Kanellopoulos, PVA/Montmorillonite Nanocomposites: Development and Properties, in: J. Cuppoletti (Ed.), Nanocomposites Polym. with Anal. Methods, InTech, 2011. doi:10.5772/18217.

[37] N.A. Isitman, M. Dogan, E. Bayramli, C. Kaynak, The role of nanoparticle geometry in flame retardancy of polylactide nanocomposites containing aluminium phosphinate, Polym. Degrad. Stab. 97 (2012) 1285–1296. doi:https://doi.org/10.1016/j.polymdegradstab.2012.05.028.

[38] S. Papatzani, K. Paine, Inorganic and organomodified nano-montmorillonite dispersions for use as supplementary cementitious materials – A novel theory based on nanostructural studies, Nanocomposites. 3 (2017) 2–19. doi:10.1080/20550324.2017.1315210.

[39] N. Bitinis, M. Hernandez, R. Verdejo, J.M. Kenny, M.A. Lopez-Manchado, Recent advances in clay/polymer nanocomposites, Adv. Mater. 23 (2011) 5229–5236. doi:https://doi.org/10.1002/adma.201101948.

[40] L. Song, Y. Hu, Y. Tang, R. Zhang, Z. Chen, W. Fan, Study on the properties of flame retardant polyurethane/organoclay nanocomposite, Polym. Degrad. Stab. 87 (2005) 111–116. doi:https://doi.org/10.1016/j.polymdegradstab.2004.07.012.

[41] G. Beyer, Short communication: Carbon nanotubes as flame retardants for polymers, Fire Mater. 26 (2003) 291–293. doi:https://doi.org/10.1002/fam.805.

[42] O.B. Nazarenko, Y.A. Amelkovich, A.G. Bannov, I.S. Berdyugina, V.P. Maniyan, Thermal stability and flammability of epoxy composites filled with multi-walled carbon nanotubes, boric acid, and sodium bicarbonate, Polymers (Basel). 13 (2021). doi:10.3390/polym13040638.

[43] UL 94 Standard for Tests for Flammability of Plastic Materials for Parts in Devices and Appliances, Underwriters Laboratories, 2019.

[44] C. Jayakody, D. Myers, U. Sorathia, G.L. Nelson, Fire-retardant characteristics of water-blown molded flexible polyurethane foam materials, J. Fire Sci. 18 (2000) 430–455. doi:10.1106/4EGW-LH1C-XFBJ-AFWL.

22 Emerging Applications of Polymers in Biomedical and Pharmaceutical Fields

Melbha Starlin Chellathurai,[1] Syed Mahmood,[1] Zarif Mohamed Sofian,[1] and Najihah Binti Mohd Hashim[2]

[1] Department of Pharmaceutical Technology, Universiti Malaya, 50603 Kuala Lumpur, Malaysia
[2] Department of Pharmaceutical Chemistry, Universiti Malaya, 50603 Kuala Lumpur, Malaysia

1 INTRODUCTION

High-performance polymers include plastics, elastomers, fluids, and biomaterials that can retain their properties in harsh environment. They are not only used as an excipient in a pharmaceutical formulation but also as a processing aid and a packaging material. They are used to fabricate controlled-release, self-regulated, prolonged-release, targeted-release, and bio-responsive feedback-regulated drug-delivery systems. Polymers used in such advanced pharmaceutical preparations reduce drug fluctuations in the systemic circulation and increase the safety margin of high-potency drugs. Also, maximum drug utilization from these novel dosage forms reduces the total dose administered, leading to minimizing of the adverse effects. Modified polymers exhibit a range of benefits in short-acting solution dosage forms to long-acting implantable devices, for example, as a solubilizer (polyethylene glycol – PEG) in oral solutions containing the poorly aqueous soluble drug, binders (modified starch, hypromellose) in tablet manufacturing, enteric film forming (cellulose acetate phthalate, Eudragit) in capsules, rheology modifiers (copovidone) in emulsion and suspension, release controller (ethylene vinyl acetate copolymer) in intra-uterine implants, *in situ* gel formers (poly (lactic-co-glycolic acid – PLGA) in long-acting depot injections, etc.

2 PROPERTIES OF PERFORMANCE POLYMERS

The properties of a polymer are mainly based on its average molecular weight, chemical composition, and chain arrangement. Pure polymers and their blends lack unique characteristics such as antibacterial properties, osteoconductivity, low mechanical properties, uncontrollable degradation speed, lesser stability, and inability to control moisture uptake [1]. Specialty polymers possess exceptional heat resistance, chemical inertness, greater strength and toughness, less fatigue and wear resistance, resistance to corrosion, abrasion and weathering protection, water and stain repellence, electrical inertness, fire safety, biocompatibility, and transparency. Chemical cross-linking is effective for increasing the mechanical strength of polymers, improving their thermostability and healing capability, and decreasing their degradation rate [2].

The polymer's strength, hardness, and ductility depend on the nature and strength of atomic bonding forces. These forces determine the stability of the polymer under mechanical stress at different temperatures. The mechanical force required to separate atoms gives the proportionality constant, an elastic constant that is inversely proportional to the binding energy.

TABLE 22.1
Moduli of Some Polymers

Polymer	Young's Modulus E (GPa)
Nylon	2.8
Hard rubber	2.8
Rigid polyvinyl chloride	3.5
Polyethylene	0.096–2.2
Polypropylene	1.1–1.55
Acrylic resins	2.96–3.28
Teflon	0.5
Polyacetal	3.65
Polysulfone	2.52
Polycarbonate	2.45
Cellulose nanocrystals	100–200
Cellulose nanofibrils	100
Polyurethane	0.02
Silicone rubber	0.008
Poly ether ketone	8.3
Polyethylene terephthalate	2.85
Polymethylmethacrylate	2.55

$\sigma = E\epsilon$. In Hooke's equation, σ is the stress, force required per unit cross-sectional area (f/A), ϵ is the change in the length per original length (l/l_0), the strain, and E is the *elastic modulus* or *Young's modulus*.

The *elastic modulus* E of a hard and immobile polymer is due to the longer polymer chains. If the polymer chains are short and unequal, then they act as plasticizers, reducing the glass transition temperatures and rigidity. The small molecules facilitate chain movement and thus hinder the efficient packing of long-chain molecules. The Young's moduli of some polymers are given in Table 22.1.

Polymer properties could be changed by altering the chemical composition of the polymer backbone, substituting the main chain carbon, increasing the size of the side groups, cross-linking, and plasticizer addition. Also, polymers change properties as a function of the processing temperature. When the rate of cooling, heating, or loading is rapid, the polymer chains cannot move and behave as rigid chains, resulting in higher glass transition temperature (T_g). The T_g is the stiff glassy region, and the rubbery region is where the polymer tends to become flexible to flow. Polymers containing non-cross-linked branched, short chains and smaller-sized side groups have lower T_g values. Polymers with cross-linking, linear, long-chain, and bulky side groups have higher T_g values [3].

3   POLYMER TYPES AND THEIR APPLICATIONS

Polymer types and their biopharmaceutical applications are shown in Figure 22.1.

3.1   IONIC POLYMERS

These are organic or inorganic and their molecular structure has both covalent and ionic bonds. They are stimuli-responsive electroactive polymers (EAPs) due to their ionic groups. Ionic liquid polymers are used as solvents, plasticizers, coupling agents, and in fabricating biodegradable polymer composites [4].

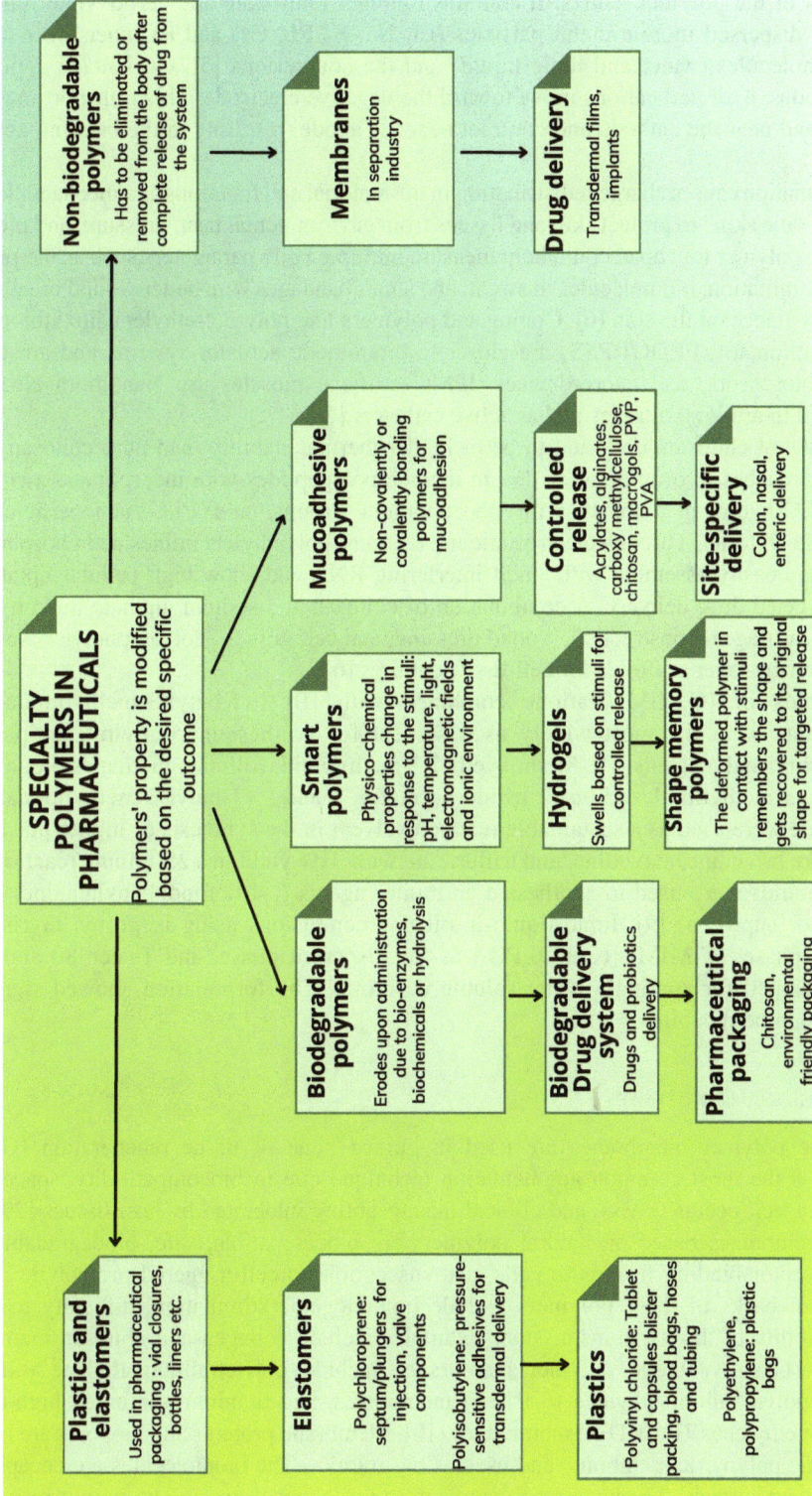

FIGURE 22.1 Types of specialty polymers.

The content of the figure is as follows:

SPECIALTY POLYMERS IN PHARMACEUTICALS
Polymers' property is modified based on the desired specific outcome

Non-biodegradable polymers
Has to be eliminated or removed from the body after complete release of drug from the system
- **Membranes** — In separation industry
- **Drug delivery** — Transdermal films, implants

Mucoadhesive polymers
Non-covalently or covalently bonding polymers for mucoadhesion
- **Controlled release** — Acrylates, alginates, carboxy methylcellulose, chitosan, macrogols, PVP, PVA
- **Site-specific delivery** — Colon, nasal, enteric delivery

Smart polymers
Physico-chemical properties change in response to the stimuli: pH, temperature, light, electromagnetic fields and ionic environment
- **Hydrogels** — Swells based on stimuli for controlled release
- **Shape memory polymers** — The deformed polymer in contact with stimuli remembers the shape and gets recovered to its original shape for targeted release

Biodegradable polymers
Erodes upon administration due to bio-enzymes, biochemicals or hydrolysis
- **Biodegradable Drug delivery system** — Drugs and probiotics delivery
- **Pharmaceutical packaging** — Chitosan, environmental friendly packaging

Plastics and elastomers
Used in pharmaceutical packaging : vial closures, bottles, liners etc.
- **Elastomers** — Polychloroprene: septum/plungers for injection, valve components; Polyisobutylene: pressure-sensitive adhesives for transdermal delivery
- **Plastics** — Polyvinyl chloride: Tablet and capsules blister packing, blood bags, hoses and tubing; Polyethylene, polypropylene: plastic bags

An IPMC has an outermost layer of metal electrodes (platinum and gold), and an intermediate layer of the polymer matrix. It contains ionomer (sulfonate and carboxylate functional groups) with dispersed mobile metal particles (Li, Na, K, Rb, Cs) and an inner layer that has the solvent molecules (water and ionic liquid) and the counterions [5]. On voltage application across electrodes, hydrated cations move toward the negative electrode. This transport makes the polymer expand near the cathode and contract near the anode, resulting in the bending actuation of IPMC.

This deformation causes charge redistribution in the material and functions as a mechanoelectrical sensor as of "ionic skin" to protect skin and tissues from environmental heat, pressure, and moisture. This skin-like polymer touchpad could help measure multiple body parameters such as the progress of muscle rehabilitation, biomolecules in sweat, and smart bandages with better wound breathability and moisture balance on the skin [6]. Conjugated polymers like poly(3,4-ethylenedioxythiophene)-poly(styrenesulfonate) (PEDOT:PSS) are closer to biomimetic actuator systems and are used in biorobotics, biosensors, and micro-devices. IPMC artificial muscles use Nanothorn electrodes. These are used in artificial muscles and as active catheters [7].

Ionic liquids of chitosan are found to be of higher thermal stability than pure chitosan matrix [8]. Other approaches to drug-delivery design include polypeptides with incorporated zwitterions sequence, PEG-hydrogel ionomer with phosphonium groups, and PEG nanoparticles with functionalizing chitosan. Through electrostatic interactions, polyethyleneimines and phosphonium-based ionomers easily assemble with small interfering RNA and show high cellular uptake efficiency for targeted drug delivery. A combination of chitosan and sodium alginate ionic hydrogel was used for cartilage reconstruction, wound dressing, and cell growth. Polyampholytes were used in making layer-by-layer scaffolds for cell-tissue cultures [6].

Ionic liquids of $[C_4MIM]$ cations combined with $[BF_4]$ 1-butyl-3-methylimidazolium tetrafluoroborate were the most widely used solvent for synthesizing antiviral, antiparasitic, and antileishmanial agents. 1-Methoxyethyl-3-methylimidazolium methanesulfonate, 1-methoxyethyl-3-methylimidazolium trifluoroacetate, and 1-butyl-3-methylimidazolium trifluoroacetate were used as a sustainable reaction solvent in the synthesis of higher-purity antiviral drugs like brivudine, stavudine, and trifluridine with 91% yield in a 25-minute reaction time. Also, ionic liquids were used to synthesize antitumor agents ($_L$-4-boronophenylalanine: $_L$-BPA, naproxen, and ibuprofen) [9]. Ionic liquid-in-oil microemulsions using isopropyl myristate as a continuous phase, $[C_1MIM]$ $[(CH_3O)_2PO_2]$ as the dispersed phase, and Tween-80 and Span-20 were developed for sparingly water-soluble acyclovir. The formulation showed significant increases in skin permeability.

3.2   BIODEGRADABLE POLYMERS

Biodegradable polymer membranes are used in guided bone or tissue regeneration (GBR or GTR). GBR is the most common augmentation technique due to biocompatibility, space maintenance ability, cell occlusiveness, and clinical manageability integrated by host tissues [29]. Bio-absorbable membranes based on natural polymers are biocompatible, safe, biodegradable, able to present receptor binding ligands to cells, and susceptible to cell-triggered proteolytic degradation. The drawbacks of these polymers include immunogenic stimulation, difficulty to purify, and the possibility of disease transmission. Examples include collagen and chitosan membranes [1]. Non-resorbable synthetic polymer membranes include polytetrafluoroethylene (e-PTFE), high-density polytetrafluoroethylene (d-PTFE) membranes, and titanium-reinforced high-density polytetrafluoroethylene (Ti-d-PTFE) membranes [10]. Membrane proteins and biopores are inserted in the synthetic polymeric membrane and used as bio-mimics. The biomolecules are encapsulated to the membrane–mimetic brushes or polymerosomes to deliver drugs *in situ* with a response to stimuli.

3.2.1 Liquid Crystal Polymers

Liquid-crystal polymers (LCPs) are aromatic polymers that are extremely inert and stable against hydrolysis. They are present in melted liquid or solid form. Biodegradable polyesters, co-polyamides, and polyester-co-amides are the most common liquid crystal polymers used in cosmetics and pharmaceutical-controlled drug-delivery devices.

Polymers like poly(lactic acid) (PLA), poly(ε-caprolactone) (PCL), PLGA, N-(2-hydroxypropyl)-methacrylamide copolymer (HPMA), poly(styrene-maleic anhydride) copolymer, gelatin, dextran, guar gum, chitosan, and collagen are the sources for preparing polymeric nanoparticles containing a drug. The drug is encapsulated in the nanoparticle by polymer matrix dispersion, micro-emulsification, emulsified nanoprecipitation, and conjugation. Controlled delivery is achieved through surface or bulk erosion, diffusion through the matrix, swelling followed by diffusion, or as a response to local stimuli (Figure 22.2). Antineoplastics like doxorubicin (PLA-PLGA conjugated with dextran and encapsulated in hydrogel), tamoxifen, and taxol (PLGA) loaded in nanoparticles showed reduced cytotoxicity and enhanced therapeutic efficacy [11] (Table 22.2).

3.2.2 Cross-Linked Collagen

Extracted collagen is cross-linked to improve its strength and enzymatic degradation resistance. Biomolecules like elastin, chitosan, and glycosaminoglycans were added during the cross-linking to enhance the cell differentiation, migration, strength, and proliferation capacity based on the intended use. Biomimetic collagen scaffolds that support cellular growth can be prepared by blending natural (fibroin, chitosan, and silk) and synthetic polymers (PLA and PEG) to improve the mechanical properties and structure of the scaffold, while collagen helps in tissue repair by cell signaling and providing binding sites.

Collagen sponges absorb wound exudates, maintaining moisture, and shield against physical trauma and microbial infection. They are used in burn treatment and diabetic ulcers. They are also used in sustained delivery of antibiotics such as vancomycin and gentamycin to treat sepsis and

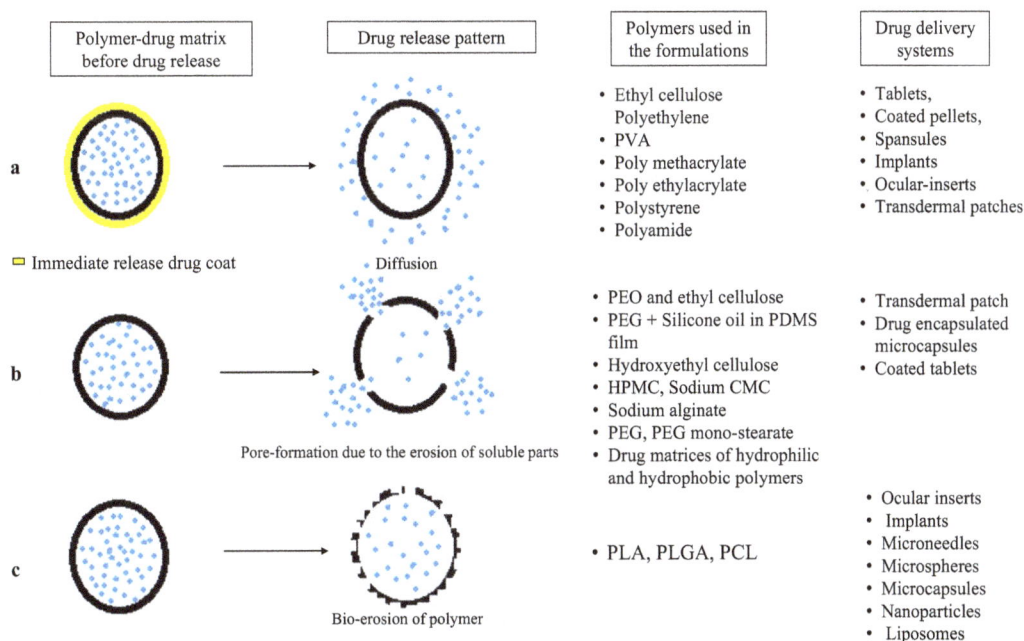

Polymer-drug matrix before drug release	Drug release pattern	Polymers used in the formulations	Drug delivery systems
a	Diffusion	• Ethyl cellulose Polyethylene • PVA • Poly methacrylate • Poly ethylacrylate • Polystyrene • Polyamide	• Tablets, • Coated pellets, • Spansules • Implants • Ocular-inserts • Transdermal patches
b	Pore-formation due to the erosion of soluble parts	• PEO and ethyl cellulose • PEG + Silicone oil in PDMS film • Hydroxyethyl cellulose • HPMC, Sodium CMC • Sodium alginate • PEG, PEG mono-stearate • Drug matrices of hydrophilic and hydrophobic polymers	• Transdermal patch • Drug encapsulated microcapsules • Coated tablets
c	Bio-erosion of polymer	• PLA, PLGA, PCL	• Ocular inserts • Implants • Microneedles • Microspheres • Microcapsules • Nanoparticles • Liposomes

□ Immediate release drug coat · Diffusion

FIGURE 22.2 Mechanism of drug release from polymer matrices [12,13].

TABLE 22.2

Liquid Crystal Polymer Nanoparticles and Their Characteristics [14–16]

Liquid Crystal Polymer Nanoparticles (LCP-NPs)	Application	Enhanced Characteristics/Benefits
Poly(lactic acid) (PLA)	Encapsulates hydrophobic drugs (DOX-Verampamil by MPEG-PLA nanoparticles) Barbaloin-loaded galactosamine conjugated with polydopamine-modified PLA-TPGS nanoparticles	Improve stability and solubility of drugs in blood. Reverses drug-resistant condition and better anti-tumor effects (ovarian cancer). Nanoparticle surface targets gastric cancer cells, maximum intracellular uptake, ROS generation, apoptosis, and induced autophagy
Poly(ε-caprolactone) (PCL)	Encapsulates hydrophobic drugs. Angiopep-2 (ANG)-ginsenoside Rg3 loaded PEG-PCL NPs for glioma treatment	Improve stability and solubility of drugs in blood. Dose-dependent, persistent drug release, crossed BBB for intracellular uptake
Poly(propylene fumarate) (PPF)	Injectable viscous liquids with initiators. 3D porous PPF nanoscaffolds with DOX-coated manganese oxide and iron oxide	Flexible drug loading, biodegradable, guided delivery systems formulation, constant drug release
Poly(lactic-co-glycolic acid) (PLGA)	Directed and regulated delivery of peptides, proteins, genes, growth factors, monoclonal antibodies, chemotherapy and radiotherapy drugs. Theranostics application when combined with imaging or contrast agents	Controlled delivery degrades more rapidly than other polyesters and increased cellular internalization. Chemotherapy or image-guided phototherapy or combination therapy. Molecular imaging
Polyhydroxyalkanoates (PHA)	Develop patches, sutures, stents, orthopedic pins, nerve guides, bone marrow scaffolds, and wound dressings	Compatible with bone, tissue, blood, cartilage, and cell lines
Poly(butylene succinate) (PBS)	Nanocarrier for camptothecin (CPT) loaded poly(ω-pentadecalactone-co-butylene-co-succinate) (PPBS) NPs	Thermoplastic, degradation product is non-toxic (CO_2 and water), higher intracellular uptake, longer circulation in blood
Poly(tert-butyl glycidyl ether-alt-phthalic anhydride) copolymer	Curcumin- and DOX-loaded poly(tBGE-alt-PA) NPs	Biodegradable, biocompatible, hemo-compatible, enhanced anti-tumor activity
Poly(cyclohexene oxide-alt-phthalic anhydride) copolymer	Poly(CHO-alt-PA) curcumin NPs	Sustained release of curcumin
Poly-glutamic acid	Paclitaxel/doxorubicin/daunorubicin and poly-L-glutamic acid conjugates and as a tablet coating agent. Used in skin, body and hair care products. Protein and peptide delivery (insulin microneedles) and pH-sensitive swelling composite microparticle, levofloxacin	Paclitaxel accumulation at tumor tissue due to enhanced permeation and retention. Also, decreased systemic exposure and increased the therapeutic index of the drug. Sustained delivery system
Poly(amino acid) polymers	PEGylated poly (amino acids) for micelles preparation for drug delivery	Degrade in the body to give bioactive amino acids. Micelles core entraps hydrophobic moiety and protects from precipitation and adsorption in the body

TABLE 22.2 (Continued)

Liquid Crystal Polymer Nanoparticles and Their Characteristics [14–16]

Liquid Crystal Polymer Nanoparticles (LCP-NPs)	Application	Enhanced Characteristics/Benefits
Tyrosine-derived polycarbonates (pseudo-poly amino acids)	Bone morphogenetic protein delivery (rhBMP-2)	Stimulates bone growth and healing
Poly(lysine)	Carrier for small and macromolecular drugs. It is conjugated with asialoorosomucoid to deliver genes to hepatocytes. Binds with targeting ligands like folate, transferrin, and galactose to target tumors	Easily degraded by cells. Cationic nature of polymer promotes internalization into tumor via endocytosis. Kills fungi, Gram-positive and Gram-negative bacteria, and used as food and pharmaceutical preservative
Poly(α-L-aspartic acid)	Carrier for drugs, design of dialysis membrane, hydrogels, orthopedic implants, self-propelled micro-device for doxorubicin and artificial skin	Biocompatible, biodegradable, non-antigenic. pH-dependent drug release to tumor site
Poly (amido amine) PAMAM dendrimer	Dendritic carrier for targeted delivery of genes, drugs, and in bio-imaging	Mimics biomolecules, ability to control size, biocompatibility, lack of immunogenicity, biodegradability, ability to adhere to cells and permeate the cell membrane via endocytosis
Poly(mannitol sebacate-co-mannitol citrate) (PMSeC) and poly(mannitol azelate-co-mannitol citrate) (PMAC)	Drug delivery, tissue engineering, and biomedical applications	Biodegradable, drug-loaded eco-friendly packing materials

to deliver retinoic acid intravaginally for cervical dysplasia treatment to avoid unwanted systemic effects. In addition, sponges loaded with growth factors (platelet-derived and fibroblast growth factors) promote capillary formation and wound healing.

Collagen hydrogels have hydrophilic functional groups on their polymeric backbone, which helps in water retention. They are amphoteric (adsorb to anions and cations) and form hydrophobic, dipole–dipole, electrostatic, and hydrogen bonding interactions to form a gel with aqueous systems. Greater microviscosity of the collagen hydrogel decreases the drug release 5–30-fold for a prolonged effect of local and central analgesics. They have a uniform surface area and are large when blended with synthetic polymers (PVA:PAA) and hyaluronic acid to deliver growth hormone in a controlled-release pattern. Collagen gel containing insulin or growth hormone liposomes released the drug in a controlled fashion by slow diffusion (3–5 days) due to the decrease in lipid peroxidation and permeability.

Collagen implants co-precipitated with chondroitin-6-sulfate or cross-linking by carbodiimide promoted axon growth and fibroblast proliferation and regulated the scarring process. Printed collagen–alginate hydrogels release the antibacterial drug sustainably, and collagen–alginate bio-inks are used for articular cartilage repair by accelerating the proliferation of human chondrocytes [17].

Therapeutic agents are loaded onto the films by covalent or hydrogen bonding, or by simple entrapment. The films are easily sterilized before wound dressing without compromising their strength and hydration properties. Collagen impregnated on silicon membrane knitted with a nylon membrane film is used in burn care due to its increased rate of epithelialization [18].

Collagen-coated polyurethane (PU) films (blended with atelocollagen) showed proliferation of fibroblasts, stimulating collagen synthesis and the formation of new connective tissue. These collagen membranes get resorbed via enzymatic degradation by collagenases/proteases and macrophages or polymorphonuclear leukocyte-derived enzymes and bacterial proteases [18]. Cross-linked collagen membrane gets resorbed after between 2–9 months, and a non-cross-linked collagen membrane has a resorption rate of 10 days to 16 weeks [19]. Cross-linked membranes display prolonged membrane integrity with surrounding tissues and blood vessels.

3.2.3   Cross-Linked Gelatin

Gelatin is a result of collagen denaturation and hydrolysis. Cross-linking is carried out using glutaraldehyde or natural less-toxic agents like caffeic acid, tannic acid, gallic acid, cocoa seed, genipin, and grape seed proanthocyanidin [20]. Gelatin microparticles are extensively used as drug carriers due to their ease of fabrication, stability, non-toxic nature, and compatibility with bioactives. Gelatin microparticles are prepared using solvent evaporation, spray drying, precipitation, and emulsification techniques. Small growth factors embedded in microparticles control the drug release rate *in vivo* and the larger microcarriers of embryonic stem cells with modified surfaces increase cell attachment and differentiation to help in bone regeneration [21].

3.2.4   Cellulose Nanocrystals (CNCs) and Cellulose Nanofibrils (CNFs)

Plant cellulose (wood, hemp, cotton, algae, and tunicin) is acid-hydrolyzed to remove amorphous cellulose regions and the crystalline structure is preserved for rigid CNCs. CNFs retain both the amorphous and crystalline cellulose regions and are flexible. Surface-modified nanocellulose supports conjugation with biomolecules like DNA, RNA, antimicrobial biomaterials, tissue bioscaffolds, active therapeutic agents, and medical implants, and is used in drug delivery and wound healing [22]. Cellulose nanocrystals can hold water-soluble drugs and have a slower release rate over 24 hours. Cetyltrimethylammonium bromide-modified nanocrystals are used in targeted drug delivery as they can bind and be taken up by KU-7 urothelial carcinoma cells [23]. CNFs are used in tablet coatings as they are more flexible and stronger than microcrystalline cellulose.

3.2.5   Chitosan

Chitosan can be fabricated as fibers, films, or 3D scaffold hydrogels and sponges. Their properties include low-cost, biocompatible, non-antigenic, appropriate degradation rate, flexibility in hydrated environments, hemostatic activity, and antimicrobial and wound-healing properties. Due to their cationic and mucoadhesive properties, hemostatic patches were fabricated containing an antifibrinolytic agent with an antibacterial agent (tranexamic acid with 1:1 PVA and chitosan). Drug release was by slow diffusion and this nanofiber membrane has good mechanical occlusion and had anti-biofilm formation properties [24]. Chitosan membrane of mitomycin C and 5-fluorouracil has been developed and compared to reduce intraocular pressure.

Electrospun chitosan with a PVA composite nanofiber wound-healing dressing stimulates fibroblast attachment and proliferation without any toxic effects on cells. Peptide drugs were adhered to and absorbed through the nasal epithelium when administered trans-nasally as chitosan microspheres. Chitosan nanoparticles protected the encapsulated insulin from degradation and oxidation within the gastrointestinal tract and increased their uptake by epithelial cells, thus increasing their stability and prolonged release [25].

3.2.6   Alginate Hydrogels

Cross-linking with divalent calcium cation, PEG amines, poly(aldehyde guluronate), and poly(acrylamide-co-hydrazide) forms alginate hydrogels with enhanced mechanical properties and lower degradation. Alginates lack biodegradation properties due to the incapability of enzymes in the body to break down their chains and reduced renal clearance due to their larger molecular weight. Alginate hydrogels are carriers for protein and peptide drugs, growth factors, antifungal agents, anti-inflammatory agents, and chemotherapeutics. Hydrophobic drugs are dissolved in the oil phase of the oil-in-water microemulsion and the Ca^{2+} cross-linked alginate gel acts as the outer delivery vehicle. Insulin–alginate microspheres completely release the active drug at the intestinal pH 6.8. Blending alginate with cellulose acetate phthalate enteric coating polymer followed by chitosan coating protects insulin at pH 1.2 and delivers it to the intestine. Proteins and cells of size smaller than 5 nm were quickly released from the porous alginate gel. Plasmid DNA of 100 nm diameter was released upon degradation, and larger molecules by gel dissociation [17].

4   PHARMACOLOGICAL ACTIONS OF SOME SPECIALTY POLYMERS

Blood substitute polymer regulates osmotic pressure and the viscosity of blood, and is used in heart failure, anaphylactic shock, burns, intoxication, diarrhea, embolic–thrombotic complications, and microcirculation impairment (Table 22.3).

TABLE 22.3
Pharmacological Actions of Modified Polymers [26,27]

Polymer	Action	Mechanism
Copolymer of divinyl ether–maleic anhydride	Antitumor and antiviral action	Stimulates glycoprotein production and suppresses viral RNA translocation in cells and division of cancer cells
Copolymer of ethylene and propylene glycols	Against constipation	Penetrates gut wall because of larger size
Linear polymer of uronic acid-alginic acid (mannuronic acid conjugated β-1,4 and L-guluronoic acid glycosidically conjugated α-1,4)	Anti-ulcer	Neutralizes hydrochloric acid. Detains water in the stomach reducing pain and irritation
Polyvinylpyrrolidone	Antidiarrheal, treats shock after burns	Normalizes pH in stomach and intestines through acids or bases adsorption
Synthetic protein hormone analogues of gonadoliberin	Prostate and breast cancer and endometriosis treatment	Unknown mechanisms
Somatoliberin	Treat children with growth hormone deficiency	
Somatostatin-octreotide		
Corticotrophins	Treat alimentary tract	
Oxitocin	Treat rheumatoid disease and severe asthma	
Vasopressin	Uterine contraction	
Ornipressin	Smooth muscle contraction	
	Added to anesthetics for vessels contraction	
Peptide antibiotics	Against Gram-positive, Gram-negative bacteria, fungi, and protozoa	Increased recognition of bacteria by host phagocytes

(continued)

TABLE 22.3 (Continued)
Pharmacological Actions of Modified Polymers [26,27]

Polymer	Action	Mechanism
Cyclosporin	Immunosuppressive drug in transplantology	Selectively inhibits lymphocytes' T function
Copolymer of divinyl benzene and styrene with quaternary trimethylammonium group and copolymer of diethyltriamine and epichlorohydrin	Anticholesterol	Inhibits absorption of cholesterol from the intestine, complexes with bile acids, and forms insoluble water polymers
Heparin	Treats arterial embolism and thrombosis, heart failure and before surgical operations	Shows effect on all blood-clotting phases
Dextran, modified starch, and gelatin	Blood substitutes	Plasma volume expanders

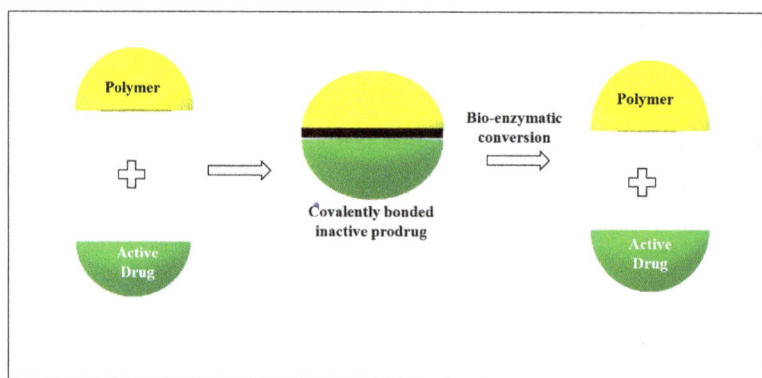

Polymers for prodrug	Prodrug applications

• Synthetic:
   • PEG, divinylethermaleic anhydride/acid copolymer,
• Polyethylenimine or polyaziridine and vinyl polymers:
   • N-(2-hydroxypropyl) methacrylamide, poly(styrene-co-maleic acid/anhydride).
• Poly(α-amino acids):
   Poly(L-glutamic acid), poly(L-lysine), poly(N-hydroxyalkyl) glutamines) and polyglycolic or polyglycolide acid
	Enhances solubility, absorption, distribution, site specific release and sustained drug action hence patient compliance.

FIGURE 22.3   Polymers for prodrugs.

5   SMART POLYMERS IN DRUG-DELIVERY SYSTEMS

Intelligent polymeric systems recognize and respond according to the physiological and pathological biochemical agents or processes in the body.

5.1   PRODRUG AND HYDROGEL-BASED DRUG-DELIVERY SYSTEM

Prodrugs are the inactive form of drugs with a polymer that, after enzymatic polymer biodegradation, yields the active drug (Figure 22.3). These prodrugs produce controlled and targeted delivery with less adverse effects and enhanced therapeutic action [28]. Controlled drug-delivery systems yield a predictable, reproducible, and constant amount of drug in the body for a longer duration (8–24 h). The drug is matrixed, coated, bonded on the ionites, or complexed for prolonged release.

Granules and hydrogel-based beads or pellets can be coated with different polymer coats to give prolonged and controlled release. Hydrogel-based beads absorb the physiological fluid from the administered site and swell. The swollen polymer matrix releases the drug either by diffusion or erosion of the polymer. Cross-linked hydrogel polymers used in drug-delivery systems include chitosan, hyaluronic acid, gelatin, sodium alginate, methylcellulose, collagen, carbopol with PEG, cellulose acetate phthalate (at 4.4 pH), hydroxyl methylcellulose, carboxymethyl chitosan, and poloxamer. Hydrogel-based drug-delivery systems [29] include tablets, beads or pellets, micro- and nanoparticles, *in situ* ocular microgels and nanogels, glucose-responsive biosensors, feedback-regulated drug-delivery systems, thixotropic injectable gels, transdermal patches, buccal film, and oral, vaginal, rectal, and nasal hydrogels [30].

Cellulose derivatives like ethyl cellulose (EC), hydroxyl ethyl cellulose (HEC), hydroxypropyl cellulose (HPC), hydroxyl propyl methylcellulose (HPMC), and carboxymethyl cellulose (CMC) are used in oral tablets, as film formers, and in transdermal patches. Imine-based liquid crystals have been used to deliver fragrances, flavor, bactericides, fungicides, insect attractants, repellants, and pharmaceuticals. The film releases the biomaterials by applying an electric field by diffusion. Silicone liquid crystals, silicone vesicles, and silicone gels with amines or epoxy groups are used to deliver personal care products to the skin, hair, and underarm by diffusion. Polyoxyethylene-glyceryl-trioleate, propylene glycol+isopropyl myristate, and the hyaluronic acid complex gel delivered estrogen and progestin in a controlled manner for transdermal hormone replacement. In the kollicoat + povidone oral delivery system, povidone is dissolved in water and small channels are formed. The drug dissolves and diffuses out through the pores leaving empty polymer. PLGA containing leuprolide acetate acid or alginic acid + collagen prevents air from leaking out of the damaged lung and is used as a collagen-based surgical adhesive [18].

5.2 SHAPE-MEMORY POLYMERS (SMPs) FOR DRUG DELIVERY

SMPs have high elastic deformation and low density and are biodegradable. They can recover their original shape after large deformation under external stimuli like temperature, magnetic field, moisture, water, pH, or light. Using SMP-PVA, gastro-retentive dosage forms and intravesical delivery systems have been prepared. Temperature-responsive polymers include polyurethanes and poly(styrene-block-butadiene). Poly(N-isopropyl acrylamide) was used to deliver insulin and calcitonin orally. By wet extrusion, casting, microfabrication, or injection molding (3D/4D printing) the polymer is deformed to its initial permanent shape, and then it is programmed to the temporary administrable shape. When the temporary shape comes in contact with the stimuli, the polymer recovers its original shape, resulting in targeted delivery and retention. Polymers with a shape-remembering property behave like a hard glass below the glass transition temperature and like rubber above the glass transition temperature. These SMP composites are used as stents and catheters. In one study, poly(ε-caprolactone) (PCL) blended with poly(vinyl chloride) (PVC) (7:3) showed excellent strain recovery once heated. In another study, focused ultrasound (FU) was used as the trigger to localize the heating effect noninvasively for initiating the controlled shape recovery process [31]. These SMPs are used in neuro-prosthetics, soft robotics, artificial skin, cardiac valves, and kidney dialysis. Due to the high specificity of the smart polymers, customized biomedical devices and personalized medicine for individual needs can be prepared.

5.3 SELF-HEALING POLYMERS FOR DRUG DELIVERY

Self-healing polymers can recover their original morphology and functionality after damage and are injectable through needles. Poly(glyceryl amine) has been used to deliver drugs dermally due to its high penetration efficiency. The rheological property of self-healing hydrogels can be tuned to give shear-thinning injectable polymers. They have found application in wound dressings, as cell/drug/protein carriers, and sensors in bioelectronics devices (polyaniline and phytic acid). Chitosan-based and hyaluronic acid-based hydrogels are used widely due to their biodegradability and biocompatibility [32].

6   DENDRIMERS IN DRUG DELIVERY

These macromolecular polymers (size: 1–15 nm diameter) are highly branched from a common center and there is no entanglement between each dendrimer molecule. As a result, their surfaces are easily modifiable. Their unique characteristics include nanoscale uniform size, high degree of branching, polyvalency, aqueous solubility, and internal cavities. The dendrimer can entrap molecules in the void spaces, hydrogen bonds at the branching points, or charge–charge interactions outside the surface groups.

Polyamidoamine (PAMAM) dendrimers are water-soluble and enhance hydrophobic drugs' solubility. Also, they prolong the circulation time of the drug in the body as the kidneys cannot filter particles larger than 5 nm. Weakly acidic indometacin showed the best solubility with a G4NH2 dendrimer. Oral bioavailability, stability, and dissolution of drugs were found to be improved after complexing with dendrimers. Multiple drug delivery is possible by entrapping two or more drugs into the dendrimer structure. Transdermal permeation has enhanced the drug complex in dendrimers. Tumor drugs can be passively targeted to the site via enhanced permeability and retention (EPR). Thiolated acyclovir-loaded dendrimers showed enhanced mucoadhesion and better drug entrapment with prolonged drug release. Self-assembled PAMAM dendrimers with anionic albumin nanoparticles loaded with paclitaxel showed a better cancer-targeting response than individual drug- and nanoparticle-delivery systems. Also, dendrimer–dendrimer nano-assembly showed a better release rate [33].

By combining dendrimers with PEG, novel controlled delivery formulations are prepared. These nanostructures are used in atherosclerosis, rheumatoid arthritis, and other associated inflammatory diseases, and have antimicrobial and antiviral properties. Bioinspired tryptophan-rich peptide dendrimers, docetaxel, oxaliplatin, and polyacylthiourea (PATU) dendrimers have anti-cancer properties. Cytotoxicity, limitations in incorporating drug efficiently in cavities, and inability to control the drug release rate hinder their application. To overcome the toxicity problem, biodegradable polyester dendrimers based on (R)-3-hydroxybutanoic acid and trimesic acid and polyester dendrimers based on 2,2-bis(hydroxymethyl) propanoic acid (bis-HMPA) monomers, polyacetal dendrimers with a β-cyclodextrin core and adamantane-terminated zwitterionic poly(sulfobetaine) have been synthesized [34].

In the biomedical field, dendrimers were used in the biomimetic regeneration of hydroxyapatite. PAMAM dendrimers capped with carboxylic acid, and alendronate conjugates (ALN-PAMAM-COOH) were absorbed by enamel to mimic the bio-mineralization process. They are also used in tissue and bone regeneration, cardiac assay diagnostics, transfection agents, anthrax-detecting agents, cell repair, and carriers for small antigens and vaccine development [35].

7   DESIGNED ENGINEERED PEPTIDES FOR DRUG DELIVERY

When triggered by external stimuli, the short peptides from the native protein get self-assembled entrapping cells or bioactive within the scaffolds and can bind and penetrate different biological barriers [36]. The side chains of the amino acids can be modified chemically with infinite sequence combinations. The primary structure of amino acids provides sites for attaching polymer substrates. These designed, engineered peptides are used in tissue engineering, drug delivery, vehicles for injections, biosensors, and immunotherapy. Some peptide hydrogels undergo a reversible transition from gel to sol when triggered by altering the ionic strength, mechanical forces, temperature, and pH of the medium [36].

8   MUCOADHESIVE POLYMER DRUG-DELIVERY SYSTEMS

Site-specific mucoadhesive polymers target particular body sites (colon, nasal, intestine, etc.) with respect to the stimuli for targeted drug release with increased residence time. Usually, high-molecular-weight anionic polyelectrolytes (PAA-based polymers), polymers with flexible long-chain containing hydroxyl, carboxyl, and amino groups in the concentration 1–2.5 wt.% produce effective bio-adhesion. Examples of mucoadhesive polymers include: poly(hydroxyethyl methyl acrylate), poloxamer (forms *in situ* gel), poly(ethylene oxide) (PEO), PVP, PVA, modified cellulose derivatives, and alginate-PEG-acrylate. Other multifunctional thiomers including chitosan–iminothiolane or thioglycolic acid or thioethylamidine, poly(acrylic acid)–cysteine or homocysteine, poly(methacrylic acid)–cysteine, alginate–cysteine, sodium carboxymethylcellulose–cysteine were used in matrix tablets [37].

9   POLYMER MICELLES FOR DRUG DELIVERY

Polymeric micelles are capable of entrapping hydrophobic drug substances within the core and hydrophilic agents (PEG) in the shell. They are nanostructures of 5–100 nm formed by amphiphilic block copolymers. Polymeric micelles coupled with targeting ligands enhance their uptake by specific cells, prolong circulation times, protect encapsulated drugs, and reduce adverse effect. In ocular delivery, polymer micelles form clear aqueous solutions that can penetrate well because of their small size and mucoadhesion helps in prolonged *in situ* delivery. Drug stability, constant release, simple and cost-effective fabrication, high drug-loading capacity, and ease of surface modification are the advantages of micelles. Smart polymeric micelles are stimuli-responsive and are used as nanocarriers to deliver diagnostic imaging agents and cancer drugs. Doxorubicin was incorporated into dual pH/redox-responsive amphiphilic diblock copolymers (PEG) methyl ether-b-poly(β-amino esters) (mPEG-b-PAE) and (PEG) methyl ether-grafted disulfide-poly(β-amino esters) (PAE-ss-mPEG) for controlled release. DOX was released at the tumor site due to the low pH and high glutathione concentration [29].

10   POLYMERS IN PHARMACEUTICAL PACKAGING

Hydrocolloids, lipids, chitosan, and cellulose derivatives are used as packaging materials. Poly ether, PEG, glycerol, and sorbitol are used as plasticizers to obtain a flexible film and to inhibit the growth of microbes. Cellulose acetate is used for laboratory and pharmaceutical packaging. Other packaging materials used are modified starches like starch-based thermoplastics, PVA, polybutylene succinate, and synthetic aliphatic polyester. Cross-linking and modification of protein side chains are used to make some pharmaceutical packaging components. PLA copolymers are used to manufacture blown biodegradable films and injection-molded objects. Polyhydroxyalkanoates and polyhydroxyl butyrates are thermoplastic polyesters that could be a substitute for plastics [38].

11   IMPLANTABLE POLYMERS

Polyether ether ketone (PEEK), a high-impact polymer material with polymethyl methacrylate (PMMA), is used in maxillary arch implants and prosthetic tooth rehabilitation. It is used in fabricating aesthetic orthodontic wires, spinal and orthopedic implants, removable dental prostheses, fixed dental prostheses, and resin-bonded retained tooth restorations, reducing the possibility of allergic reactions and corrosion [39,40]. Polymeric implantable biomaterials are capable of replacing living tissues and restoring their functions. They can be fabricated as viscous liquids, rods, fibers, textiles, and films. Silastic rubber, Teflon, Dacron, and Nylon are low-density resilient materials used as sutures for arteries, veins; in maxillofacial areas, such as the nose, ear, maxilla, mandible, teeth; cement; and as artificial tendons (Table 22.4).

TABLE 22.4
Implantable Polymers and Their Uses [41]

Implantable Polymer	Advantages and Disadvantages	Application
Heparin protein bonded on silicone and urethane rubbers		Prevention of blood clotting
Reconstituted natural collagen	Degradable polymer	Replacing arterial wall, heart valve, and skin
Poly-cyanoacrylates	Excellent adhesion	Tissue adhesives, drug delivery
Adhesive bio-polymers		Close wounds or lute orthopedic implants
Nylons (aramids): poly(p-phenylene terephthalate)	Hygroscopic and lose their strength at implanted site	
Polyolefins: polyethylene (LDPE, HDPE, UHDPE)	Readily crystallized	Orthopedic implant, tubing, shunts, catheters, plastic surgery implants and catheters
Polyolefins: polypropylene (PP)	High flexural fatigue life, excellent stress-resistance	Integrally molded hinges for finger joint prostheses, heart valve structure, oxygenator, plasmapheresis membranes
Polyacrylates	Excellent physical and coloring properties, easy to cast, mold, and machine, obtained as liquid, beads, sheets, rods	Hard and implantable contact lenses, bone cement, dentures, maxillofacial prostheses
Poly(hydroxyethyl methacrylate	Absorbs more than 30% of its weight the water	Soft lens
Silicone elastomers	High oxygen permeability and transparent	Lens material, wound dressing
Silicone (polydimethylsiloxane PDMS)	Easily moldable	Oxygenator membrane, tubing, shunts, prostheses, heart pacemaker leads, heart valve structure, burn dressing
Silicone gel	Biocompatible	Hypertrophic burn scars, recover soft tissue mass in breast, scrotum, chin, nose, cheek, and calf via reconstructive surgery
Polytetrafluoroethylene (PTFE): Teflon	Expands on a microscopic scale to microporous material, thermal insulator	Knits arterial grafts and does not leak blood
Rubber	Compatible with blood	Pharmaceutical closures, materials
Polyacetals		Implants
Polysulfones	Transparent	Implants
Polycarbonates	Transparent, excellent mechanical and thermal properties	Heart/lung assist devices, food packaging
Polyformaldehyde (polyoxymethylene)	Excellent mechanical properties, chemical resistance	
Poly-caprolactone	Biodegradable	Long-term drug delivery, orthopedic applications, staples, stents
Polydioxanones	Bioabsorbable	Sutures, wound clip

12   SUMMARY

Polymer-based drug targeting is an important research area in pharmaceutics where polymers of a specific property are utilized to enhance formulation parameters like solubility, entrapment, protection of drug, or to achieve the desired response of targeted release. There remains a great demand for discovering and developing polymers with new properties and applications. Existing polymers are also modified to yield better performance products. In the biomedical field, the uniqueness of polymer properties is utilized in designing diagnostics, wearable sensors, wound healing, and implantable devices. Advancements in polymer science will help to develop many useful biopolymers to obtain sustainable solutions for the current biopharmaceutical and biomedical challenges. Therefore, significant research should be carried out to develop innovative, target-specific carriers.

REFERENCES

[1] M. M. Islam, M. Shahruzzaman, S. Biswas, M. Nurus Sakib, and T. U. Rashid, Chitosan based bioactive materials in tissue engineering applications-A review, Bioact. Mater., vol. 5, no. 1, (2020) 164–183.

[2] Song, Pingan, and Hao Wang. High-performance polymeric materials through hydrogen-bond crosslinking. Advanced Materials 32, no.18 (2020) 1901244.

[3] P. J. Sinko, Pharmaceutical Polymers – Chapter 20, Martin's Physical Pharmacy and Pharmaceutical Sciences (2006) 647.

[4] A. A. Shamsuri, S. N. A. M. Jamil, and K. Abdan, A brief review on the influence of ionic liquids on the mechanical, thermal, and chemical properties of biodegradable polymer composites, Polymers (Basel)., vol. 13, no. 16, (2021) 1–14.

[5] Y. Wang and T. Sugino, Ionic Polymer Actuators: Principle, Fabrication and Applications, Actuators, United Kingdom: IntechOpen (2018) 55–86.

[6] J. E. Potaufeux, J. Odent, D. Notta-Cuvier, F. Lauro, and J. M. Raquez, A comprehensive review of the structures and properties of ionic polymeric materials, Polym. Chem., vol. 11, no. 37, (2020) 5914–5936.

[7] J. Kim, J. W. Kim, H. C. Kim, L. Zhai, H. U. Ko, and R. M. Muthoka, Review of soft actuator materials, Int. J. Precis. Eng. Manuf., vol. 20, no. 12, (2019) 2221–2241.

[8] M. Azmana, S. Mahmood, A.R. Hilles, A. Rahman, M.A.B. Arifin, and S. Ahmed, A review on chitosan and chitosan-based bionanocomposites: Promising material for combatting global issues and its applications. Int. J. Biol. Macromol., vol. 185, (2021) 832–848.

[9] I. M. Marrucho, L. C. Branco, and L. P. N. Rebelo, Ionic liquids in pharmaceutical applications, Annu. Rev. Chem. Biomol. Eng., vol. 5, no.7, (2014) 527–546.

[10] N. Y. Naung, E. Shehata, and J. E. Van Sickels, Resorbable versus nonresorbable membranes: When and why?, Dent. Clin. North Am., vol. 63, no. 3, (2019) 419–431.

[11] S. Senapati, A. K. Mahanta, S. Kumar, and P. Maiti, Controlled drug delivery vehicles for cancer treatment and their performance, Signal Transduct. Target. Ther., vol. 3, no. 1, (2018) 1–19.

[12] N. Kamaly, B. Yameen, J. Wu, and O. C. Farokhzad, Nanoparticles: Mechanisms of controlling drug release," Chem Rev., vol. 116, no. 4, (2016) 2602–2663.

[13] B. Mikolaszek, J. Kazlauske, A. Larsson, and M. Sznitowska, Controlled drug release by the pore structure in polydimethylsiloxane transdermal patches, Polymers (Basel)., vol. 12, no. 7, (2020) 1520.

[14] Gupta, P.K., Gahtori, R., Govarthanan, K., Sharma, V., Pappuru, S., Pandit, S., Mathuriya, A.S., Dholpuria, S. and Bishi, D.K, Recent trends in biodegradable polyester nanomaterials for cancer therapy, Mater. Sci. Eng. C, vol. 127, (2021) 112198.

[15] Boddu, S.H., Bhargav, P., Karla, P.K., Jacob, S., Adatiya, M.D., Dhameliya, T.M., Ranch, K.M. and Tiwari, A.K., Polyamide/poly (amino acid) polymers for drug delivery, Journal of Functional Biomaterials, 12(4) (2021) 58.

[16] A. Kesavan, T. Rajakumar, M. Karunanidhi, and A. Ravi, Synthesis and characterization of random copolymerization of aliphatic biodegradable reunite D-Mannitol, Mater. Today Proc., (in press). https://doi.org/10.1016/j.matpr.2021.01.522

[17] E. Troy, M. A. Tilbury, A. M. Power, and J. G. Wall, Nature-based biomaterials and their application in biomedicine, Polymers (Basel)., vol. 13, no. 19, (2021) 1–37.

[18] S. Chattopadhyay and R. T. Raines, Review collagen-based biomaterials for wound healing, Biopolymers, vol. 101, no. 8, (2014) 821–833.

[19] Wang, J., Wang, L., Zhou, Z., Lai, H., Xu, P., Liao, L. and Wei, J. Biodegradable polymer membranes applied in guided bone/tissue regeneration: A review, Polymers (Basel)., vol. 8, no. 4, (2016) 1–20.

[20] A. Ehrmann, Non-toxic crosslinking of electrospun gelatin nanofibers for tissue engineering and biomedicine—A Review, Polymers (Basel)., vol. 13, no. 12, (2021) 1973.

[21] A. B. Bello, D. Kim, D. Kim, H. Park, and S. H. Lee, Engineering and functionalization of gelatin biomaterials: From cell culture to medical applications, Tissue Eng. – Part B Rev., vol. 26, no. 2, (2020) 164–180.

[22] C. Agarwal, Chapter 6 – Surface-modified cellulose in biomedical engineering. In V. Grumezescu and A. M. Grumezescu (eds) Materials for Biomedical Engineering: Bioactive Materials, Properties, and Applications (2019) 2–3. Elsevier, Amsterdam.

[23] J. K. Jackson, K. Letchford, B. Z. Wasserman, L. Ye, W. Y. Hamad, and H. M. Burt, The use of nanocrystalline cellulose for the binding and controlled release of drugs., Int. J. Nanomedicine, vol. 6, (2011) 321–330.

[24] P. Sasmal and P. Datta, Tranexamic acid-loaded chitosan electrospun nanofibers as drug delivery system for hemorrhage control applications, J. Drug Deliv. Sci. Technol., vol. 52, (2019) 559–567.

[25] Jing, Z.W., Ma, Z.W., Li, C., Jia, Y.Y., Luo, M., Ma, X.X., Zhou, S.Y. and Zhang, B.L. Chitosan cross-linked with poly (ethylene glycol) dialdehyde via reductive amination as effective controlled release carriers for oral protein drug delivery, Bioorg. Med. Chem. Lett., vol. 27, no. 4, (2017) 1003–1006.

[26] E. Oledzka and M. Sobczak, Polymers in the pharmaceutical applications – Natural and bioactive initiators and catalysts in the synthesis of biodegradable and bioresorbable polyesters and polycarbonates, Innov. Biotechnol., (2012) 139–160.

[27] B. Karolewicz, A review of polymers as multifunctional excipients in drug dosage form technology, Saudi Pharm. J., vol. 24, no. 5, (2016) 525–536.

[29] P. M. Gandhi, A. R. Chabukswar, and S. C. Jagdale, Carriers for prodrug synthesis: A review, Indian J. Pharm. Sci., vol. 81, no. 3, (2019) 406–414.

[30] Luo, Y., Yin, X., Yin, X., Chen, A., Zhao, L., Zhang, G., Liao, W., Huang, X., Li, J. and Zhang, C.Y, Dual pH/redox-responsive mixed polymeric micelles for anticancer drug delivery and controlled release, Pharmaceutics, vol. 11, no. 4, (2019) 176.

[31] Zanna N, Focaroli S, Merlettini A, Gentilucci L, Teti G, Falconi M, Tomasini C., Thixotropic peptide-based physical hydrogels applied to three-dimensional cell culture, ACS Omega 2(5), (2017) 2374–2381.

[32] A. Maroni, A. Melocchi, L. Zema, A. Foppoli, and A. Gazzaniga, Retentive drug delivery systems based on shape memory materials, J. Appl. Polym. Sci., vol. 137, no. 25, (2020) 1–10.

[33] H. J. Huang, Y. L. Tsai, S. H. Lin, and S. H. Hsu, Smart polymers for cell therapy and precision medicine, J. Biomed. Sci., vol. 26, no. 1, (2019) 1–11.

[34] A. S. Chauhan, Dendrimers for drug delivery, Molecules, vol. 23, no. 4, (2018) 938.

[35] D. Huang and D. Wu, Biodegradable dendrimers for drug delivery, Mater. Sci. Eng. C, vol. 90, (2018) 713–727.

[36] Noriega-Luna, B., Godínez, L.A., Rodríguez, F.J., Rodríguez, A., Zaldívar-Lelo de Larrea, G., Sosa-Ferreyra, C.F., Mercado-Curiel, R.F., Manríquez, J. and Bustos, E., Applications of dendrimers in drug delivery agents, diagnosis, therapy, and detection, J. Nanomater., (2014) 507273.

[37] J. Chen and X. Zou, Self-assemble peptide biomaterials and their biomedical applications, Bioactive Materials, vol. 4. (2019) 120–131.

[38] G. Mythri, K. Kavitha, M. R. Kumar, and S. D. Jagadeesh Singh, Novel mucoadhesive polymers – A review, J. Appl. Pharm. Sci., vol. 1, no. 8, (2011) 37–42.

[39] S. Kumar and S. K. Gupta, Applications of biodegradable pharmaceutical packaging materials: A review, Middle East J. Sci. Res., vol. 12, no. 5, (2012) 699–706.

[40] P. Zoidis, The all-on-4 modified polyetheretherketone treatment approach: A clinical report, J. Prosthet. Dent., vol. 119, no. 4, (2018) 516–521.

[41] A. Haleem and M. Javaid, Polyether ether ketone (PEEK) and its 3D printed implants applications in medical field: An overview, Clin. Epidemiol. Glob. Heal., vol. 7, no. 4, (2019) 571–577.

[42] M. Zare, E. R. Ghomi, P. D. Venkatraman, and S. Ramakrishna, Silicone-based biomaterials for biomedical applications: Antimicrobial strategies and 3D printing technologies, J. Appl. Polym. Sci., (2021) 1–18.

23 Specialty Polymers for Biomedical Applications

Buwanila T. Punchihewa,[1] A.A.P.R. Perera,[2,3]
Felipe M. de Souza,[3] and Ram K. Gupta[2,3]

[1] Department of Chemistry, University of Missouri-Kansas City,
Missouri 64110, USA
[2] Department of Chemistry, Pittsburg State University, Pittsburg,
Kansas 66762, USA
[3] National Institute for Materials Advancement, Pittsburg State University,
Pittsburg, Kansas 66762, USA

1 INTRODUCTION

In medical applications, a variety of treatments, materials, and instruments are used to support, enhance, or replace damaged tissue or a biological function. Scientists have experimented with many materials including metals, ceramics, and polymers to develop technologies and instruments that allow physicians to make quality decisions to determine a person's health status. Various polymeric systems are thereby widely used in the field of biomedicine/biomedical engineering due to their special properties, including biocompatibility, non-toxicity, biodegradability, lightness, flexibility, and easy processing. However, the activity of polymers must be controlled to properly meet the requirements of biomedical applications such as biosensing, diagnostics, tissue engineering, regenerative medicine, and medical devices. Figure 23.1 shows some of the working principles of biomedical devices based on polymeric materials for bioelectronic devices that monitor the biological functions of the body and identify diseases to improve health. Figure 23.1a describes biomolecules that are absorbed to the surface of the polymer using physical absorption, covalent immunization, or electrochemical immunization. After being linked to specific analytics receptors, physicochemical and electrical signals can be generated. Figure 23.1b describes the applications of polymeric materials for the production of bioelectronic devices that provide a platform for medical treatments. Figure 23.1c describes the polymeric materials that can be applied in tissue engineering which can provide the required mechanical and structural needs to the target tissues and promote their healing by interacting with healing cells. Taking those working principles into consideration, polymers can be generally classified into two main groups: (a) bio-derived polymers and (b) synthetic polymers. Although natural polymers such as chitosan, alginate, and collagen were first used for biomedical applications, currently synthetic polymers have replaced them as the leading group. One of the reasons for this has been the development of conducting polymers (CPs) which can considerably enlarge the applications within the biomedical field such as the development of scaffolds for the growth of damaged tissue. CPs are organic polymers consisting of sp^2 hybrid carbon atoms with alternating single- and double-bonded backbone structures (conjugated) compared to non-conducting polymers (NCPs) (Figure 23.2).

CPs are electronically conductive due to the carbon atoms in the polymer backbone having unhybridized and parallel p orbitals in one direction, which may be a pathway for electron transport alongside the polymer chain. In another case, certain polymer structures can be doped to create positive or negative charges which can be performed through *p*-type doping (oxidized) or *n*-type doping (reduced), respectively. In this sense, the π-conjugated polymer backbone can be oxidized

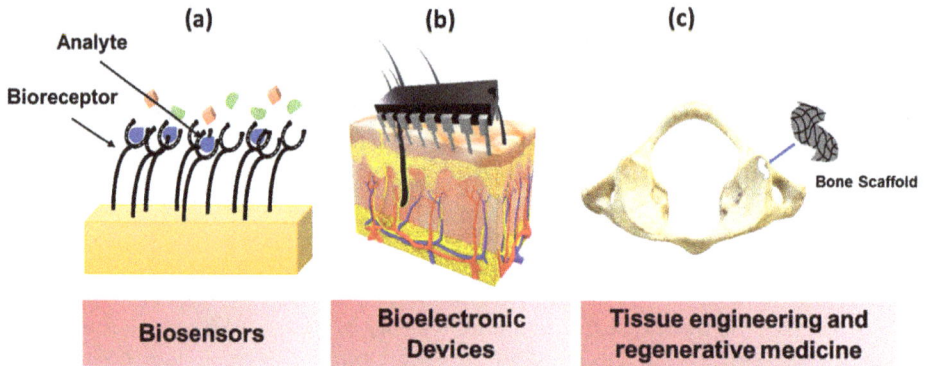

FIGURE 23.1 Schematic of the key features and functions of the biopolymers and their composites in biomedical applications separated in different fields such as (a) biosensors, (b) bioelectronic devices, and (c) tissue engineering and regenerative medicine.

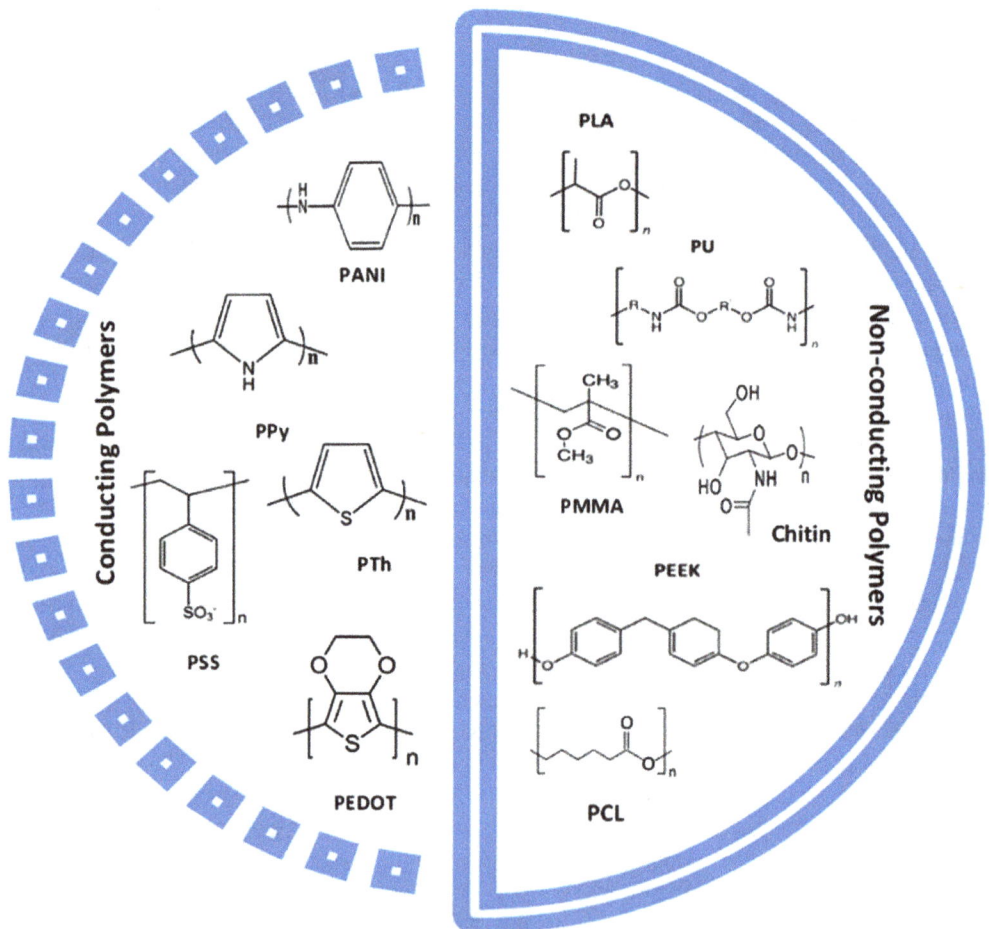

FIGURE 23.2 Chemical structures of common conducting and non-conducting polymers in biomedical applications, including polyaniline (PANI), polypyrrole (PPy), polythiophene (PTh), polystyrene sulfonate (PSS), poly(3,4-ethylenedioxythiophene) (PEDOT), polycaprolactone (PCL), poly (lactic acid) (PLA), polyether ether ketone (PEEK), polyurethanes (PU), and chitin.

or reduced to introduce positive or negative charge carriers. In addition, anionic or cationic dopants are incorporated into the polymer matrix to stabilize the net charge of the polymer backbone. These charge carriers are responsible for the movement of ions through the CP's backbone and dopants facilitate ionic transport across the CP. The ionic transporting ability makes CPs potential candidates for biosensors, bioelectronic, conductive tissue engineering applications, etc. [1].

NCPs are electrically insulating materials that are made up of synthetic monomers, whereas bio-derived polymers are derived from naturally occurring ones from plants or animals, for instance. Therefore, the properties of NCPs vary according to the type of monomer units that make up the polymer. The biocompatibility of bio-derived NCPs for internal and external medical treatments has been proven for decades. However, weak physical, mechanical, and chemical properties commonly hinder the wide application of bio-derived NCPs. Thus, a broad range of synthetic non-conductive biodegradable polymers has been investigated for biomedical applications over the years. Also, there have been studies that have focused on the development of biomaterial structure by combining synthetic and natural polymers. This approach has broadened the applications of polymers in the biomedical field [2–5].

Biosensors and bioelectronic devices play a significant and growing role in medical diagnostics and patient monitoring. These devices help to monitor specific metabolic actions in the body usually by being incorporated with a probe that can identify a specific species. The considerable versatility in this manner makes it highly desirable to fabricate new support platforms for biosensors and bioelectronic devices. Thus, as patients use biosensing and bioelectronic devices to support their health challenges, professionals are striving to develop high-sensitivity equipment to provide them with accurate information. Therefore, numerous attempts have been made to reduce the detection time by increasing the detection limit and specificity of biosensing and bioelectronic devices. Polymer-based hybrid materials have been developed with high selectivity and sensitivity to reduce some of the limitations found in biosensors and bioelectronic devices over the last decade in terms of performance. PANI, PPy, PTh conducting polymers, chitin, and chitosan can provide many of the above properties. Therefore, the scientific community in the biomedical field has been focusing on using these polymers in the manufacture of sensors and bioelectronic devices [6–10].

In advanced tissue engineering and regenerative medicine, researchers strive to regenerate and replace tissues and potentially entire organs lost through disease or trauma. The defects in body parts can be repaired by introducing engineered biomaterials. This biomaterial should be recognizable and compatible with body functions, cells, and signals. In tissue engineering and regenerative medicine, a support system is used to grow new cells into functional tissues or organs. The scaffold made of polymers supports the cellular interactions needed to form new tissues or organs. Scaffold materials can be produced from synthetic as well as bio-derived polymers. Different polymeric scaffold materials have been fabricated to treat damaged tissues or organs. These scaffoldings differ from each other in mechanical properties, including the persistence of the scaffolding shape and the adhesion sites for the growing cells.

Several examples of polymeric scaffold materials and fabrication techniques of scaffolds such as non-woven nets, sponges, and hydrogels are described herein. Natural polymers such as gelatin and collagen, and synthetic polymers such as PCL and polyurethanes are widely used as scaffolding biomaterials for tissue engineering due to their flexibility, biocompatibility, and biodegradability. Conductive biofilms, including carbon nanotubes and nanowires, graphene, and metal particles (non-metallic particles) have been used for tissue engineering due to their high conductivity and tensile strength. However, vulnerabilities, including long-term in vivo toxicity and nonbiodegradability, have limited the broader applications of biomaterials [11–13]. Further, advanced polymeric materials prepared in combination with nanotechnology have achieved tremendous progress in most biomedical applications over the past six decades. Nanomaterials (nanofibers, nanoparticles, and nanotubes) present unique and tunable properties, such as a greater surface area for cellular interaction, adequate absorption to facilitate binding sites for cell receptors, and active surface to be functionalized by probes, among others. In this chapter, recent information regarding biosensing, tissue engineering, regenerative medicine, and bioelectronic devices based on polymers is discussed in detail.

2   SYNTHESIS AND CHARACTERIZATION OF POLYMERS

The synthesis of polymers through their respective monomers can be performed through condensation polymerization, which includes most step-growth polymerization mechanisms, or addition polymerization which includes most chain-growth polymerization mechanisms. In the expansion of applications, new synthesis methods have been introduced for the production and modification of polymeric structures. Specifically, polymeric materials used in biomedical applications can be produced through several methods such as chemical oxidation polymerization (COP), electrochemical polymerization (EP), vapor phase synthesis, template-assisted, electrospinning, self-assembly, photochemical, inclusion, solid-state, and plasma polymerization [14]. Among these techniques, COP, EP, and electrospinning techniques are focused on further in the following sections.

2.1   CHEMICAL OXIDATIVE POLYMERIZATION

COP is one of the most promising polymerization methods for the synthesis of CPs widely used in biomedical applications such as PANI, PPy, PTh, among others. The monomers, i.e., aniline, pyrrole, thiophene, and others used in COP, are characterized by pronounced electron donor properties while also being prone to oxidation [15]. During this oxidation process, cations or radical sites are generated in the monomer to undergo a chain reaction. Commonly used oxidizing agents are $FeCl_3$, H_2O_2, ammonium persulfate, and potassium persulfate [16,17]. Most commonly, thin films of PANI nanofibers and composite PANI nanofibers are prepared by using in situ rapid-mixing COP methods. Within this line, a label-free electrochemical immunosensor was developed to detect the thyroxine hormone using Cu-MOF@PANI composite as a substrate material [18]. During this study, PANI was synthesized over the surface of Cu-MOF by an in situ COP method. The surface morphology was analyzed using field emission scanning electron microscopy (FESEM) (Figure 23.3) and energy-dispersive X-ray (EDX) analysis. The FESEM image of Cu-MOF in Figure 23.3a shows the crystalline nature of MOF with a particle size of around 1 μm. After the COP method, a cabbage-like structure was obtained due to the polymerization of PANI over the Cu-MOF surface (Figure 23.3b).

FIGURE 23.3   FESEM of Cu-MOF (a) and Cu-MOF@PANI (b) nanocomposite. Adapted with permission from Reference [18]. Copyright (2021) The Authors. Licensee Wiley-VCH GmbH. This is an open-access article under the terms of the Creative Commons Attribution License (https://creativecommons.org/licenses/by/4.0/).

This porous and non-smooth structure of MOF led to an enhancement in the immobilization of thyroxine. EDX analysis confirmed the presence of C, Cu, and O in the structure of synthesized Cu-MOFs and the presence of C, N, S, and Cl in the conductive polymer backbone. The EDX model of the Cu-MOF@PANI composite revealed the existence of C, N, O, Cu, and Cl, demonstrating that PANI completely covers the Cu-MOF surface.

2.2 ELECTROCHEMICAL POLYMERIZATION

The production of polymers and their composites using the EP method was introduced in the nineteenth century [19]. In EP technology, the production of a polymer begins with the oxidation of a monomer in an electrochemical cell and the final polymer is deposited directly on the surface (working electrode) to be coated. This polymerization can be performed under applied potentials based on the characteristics of the starting monomer and dopants. Potentiostat-based, galvanostatic, potential step, and potential sweep methods are frequently used electrochemical techniques for the EP of monomers and their derivatives [20]. These methods are usually performed in a three-electrode system that contains the working, counter, and reference electrodes under the presence of an electrolyte. The rate of polymer growth, the morphology, and the properties of the polymer films deposited on the electrode are strongly influenced by the nature of the supporting electrolyte, monomer concentration, pH, and doping [21,22]. In several studies, monomer polymerization of CPs such as aniline, thiophene, pyrrole, and their composites was performed by EP on a metal electrode in aqueous/organic electrolytes because of its simplicity and reproducibility.

In one study, a novel biosensor was developed for the detection of danazol by employing the PANI electrode [23]. The fabrication of PANI was carried out using the EP technique. A potentiostat connected to a conventional three-electrode cell was employed, including a modified carbon electrode as the working electrode. Polymerization of aniline took place on the surface of a modified carbon electrode in an acidic medium. Based on that, it has been observed that the mass and thickness of the electrochemical PANI films formed on the electrode were proportional to the anodic peak current (Ip), which can be used as an indicator for the rate of polymerization. Fourier transform infrared (FTIR) spectroscopic analysis was used to confirm PANI's functional groups. This analysis can be seen in Figure 23.4a, which shows that the optimum concentration of HCl to obtain the higher current was 1 M. In that sense, the higher concentration of acid may likely hinder the polymerization rate since aniline may not react with radical cations due to the higher load of Cl- in the media [24]. However, the FTIR spectrum displayed in Figure 23.4b shows the characteristics of peaks expected for the formation of PANI such as N–H stretching at 3422 cm^{-1}, C–N single bond at 1178 cm^{-1}, and benzoic and quinoid rings at 1483 and 1644 cm^{-1}, respectively. In addition, to detect the electrochemical response of danazol, PANI samples were deposited on a graphite electrode using cyclic voltammetry at a potential of –0.2 to 0.8 V [23].

The first step of polymerization is the formation of a cation radical. Then the cation radical alters its structure by localizing unpaired electrons, either the nitrogen atom or other carbon atoms, in the ring. The radicals are more likely to react with each other rather than react with a neutral monomer. Because of this, a dimer is created. The formed dimers can undergo a deprotonation process and form a deprotonated intermediate product. By repeating this oxidation–addition–deprotonation cycle, the monomers are added to the chain and, finally, the polymer is produced.

2.3 ELECTROSPINNING

Electrospinning is a versatile technique that has gained popularity for various innovative industrial applications. This approach has been used for the fabrication of nanofibers for biomedical devices, regeneration medicines, implants, wound healing, and prevention of dental caries. It consists of applied electrostatic forces that spin a polymer solution into whipped jets, resulting in a continuous,

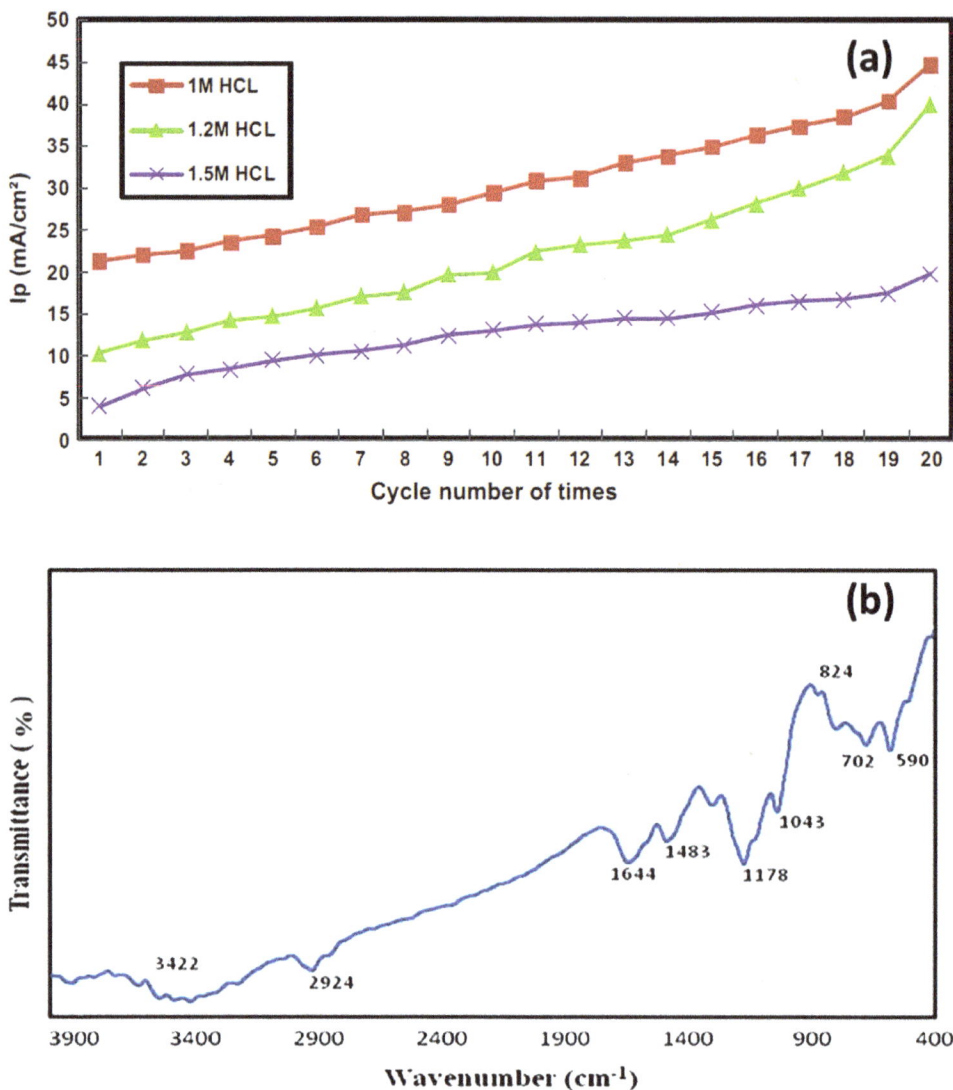

FIGURE 23.4 Characteristics of PANI films. (a) Ip vs. electrochemical cycles for polymerization of PANI formed from 0.88 M aniline and with different HCl concentrations. (b) FTIR spectrum of PANI. Adapted with permission from Reference [23]. Copyright (2017), Springer Nature.

ultrafine fiber fabrication with a solvent evaporation process under vacuum pressure. A typical electrospinning process is illustrated in Figure 23.5. Depending on the target application, composite electrospun nanofibers can be fabricated with different chemical, structural, and mechanical properties by spinning with various materials, incorporating bioactive molecules, etc. For instance, the placement of bioactive molecules (proteins or peptides) on the surface of electrospun nanofibers is essential for the development of functional nanofibers with more favorable biological properties for tissue engineering applications. Furthermore, the resulting electrospun nanofibrous materials can be obtained as random, aligned structures, or core–shell structures that closely mimic the behavior of the body's natural tissues. Yet, for this approach, two separate solutions have been employed simultaneously as one contained bioactive agents dissolved in water whereas the other contained a hydrophobic polymer. In these conditions, the hydrophobic polymer would expel the water solution

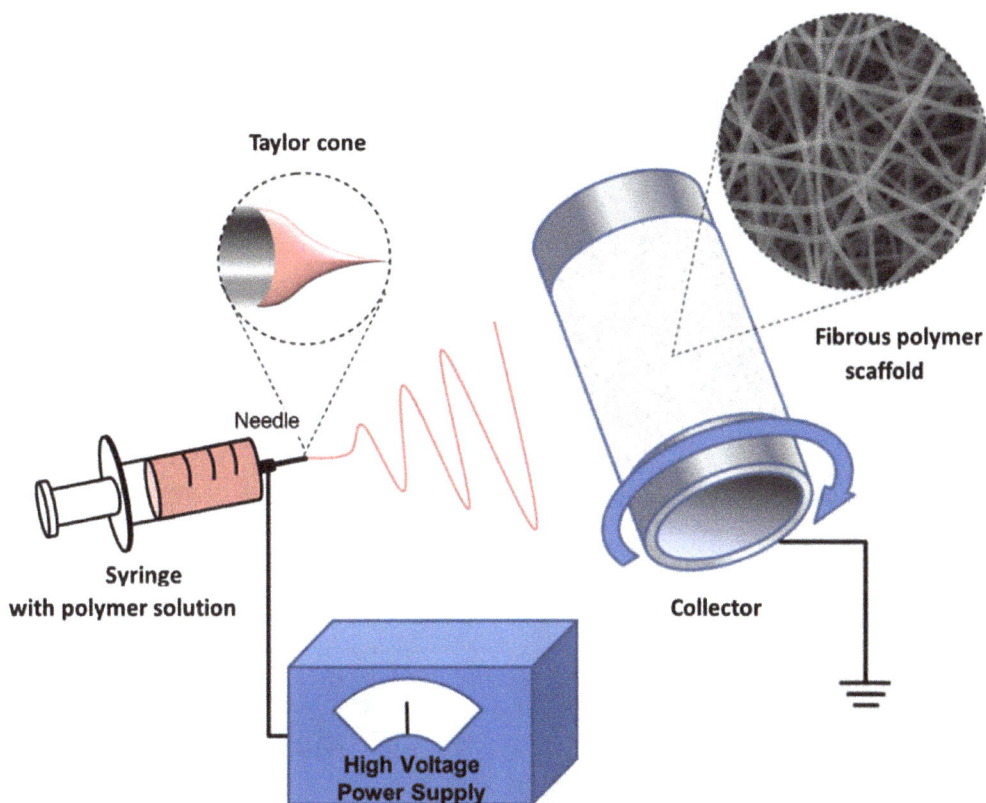

FIGURE 23.5 Scheme for a typical electrospinning process employed for the synthesis of polymers suitable for biomedical applications. Adapted with permission from Reference [25]. Copyright (2013), IOP Publishing, Ltd.

to the outer layer which would lead to the coating of bioactive material over its surface while also forming a 1D type of nanostructure. It is worth mentioning that several parameters can influence the fibers' formation such as concentration of both solutions, temperature, and supply voltage, among many others. The scheme for the electrospinning process is provided in Figure 23.6a. Also, this approach allowed proper control of the fibers' morphology, as both random (Figure 23.6b) and aligned (Figure 23.6c) arrangements were obtained, respectively.

This technique can deposit a 3D porous network of the nanofibrous mat with a large global pore volume on its surface. The pore size distribution can be tuned according to the application by changing the working parameters that include (i) molecular weight, molecular-weight distribution, and architecture (branched, linear, etc.) of the polymer, (ii) solution properties, (iii) electric potential, flow rate, and concentration, (iv) distance between the capillary and collection screen, (v) ambient parameters (temperature, humidity, and air velocity in the chamber), (vi) motion and size of target screen (collector), and (vii) needle gauge [25].

However, previous literature has pointed out that the typical electrospinning method poses several challenges in the production of biomaterials, such as toxicity of solvents, solubility, miscibility of polymers, instability of mass flow, and evaporation rate of solvents. Therefore, recently, scientists have developed a new solvent-free electrospinning technique for biomaterials fabrication called melt electrospinning (ME). This approach enables the production of filaments with higher surface quality as compared to solution electrospinning. Compared to the traditional setup, a heating system is integrated as an additional element, as can be seen in Figure 23.7a. In this sense, electrical

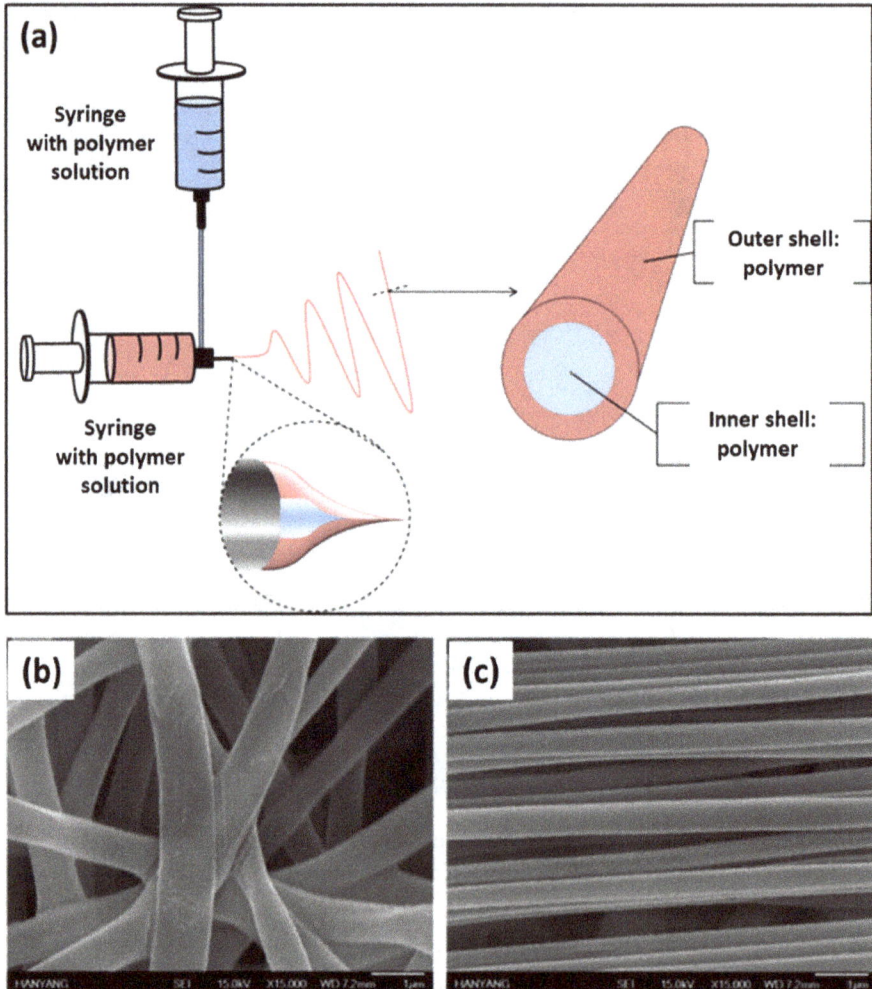

FIGURE 23.6 Various fibrous structures were fabricated via an electrospinning technique such as (a) core–shell, (b) random, and (c) align directional fiber structures. Adapted with permission from Reference [25]. Copyright (2013), IOP Publishing, Ltd.

potential, pressure, and surface tension are the forces exerted on a molten polymer jet. By increasing the voltage difference and reaching the threshold value, it will continue to protrude to the collector with a smaller diameter than the nozzle. In several studies, ME was used for 3D scaffold fabrication, including scaffolds for soft tissues such as skin, endosteum, nerves, and cardiac tissue engineering. Within this line, this method allowed the fabrication of multiphasic fibers that could form a highly organized network structure with defined diameters, as seen in Figure 23.7b, c. Furthermore, hydrogels were introduced between the micro-network structures as they can function as a matrix to receive biological material, which can be seen in Figure 23.7d. Also under this perspective, the most frequently reported materials to produce those scaffolds are based on PCL, PLA, PU, polymethyl methacrylate (PMMA), polypropylene, and their blends [26]. In one study, biomimetic soft network composites (SNCs) consisting of a water-swollen hydrogel matrix and a reinforcing fibrous network fabricated by melt electrospinning writing technology using medical grade poly(ε-caprolactone) (mPCL) were developed [27]. The printed fibrous networks were used to reinforce the hydrogel, and the resulting SNC was mechanically characterized. The fibrous network has demonstrated

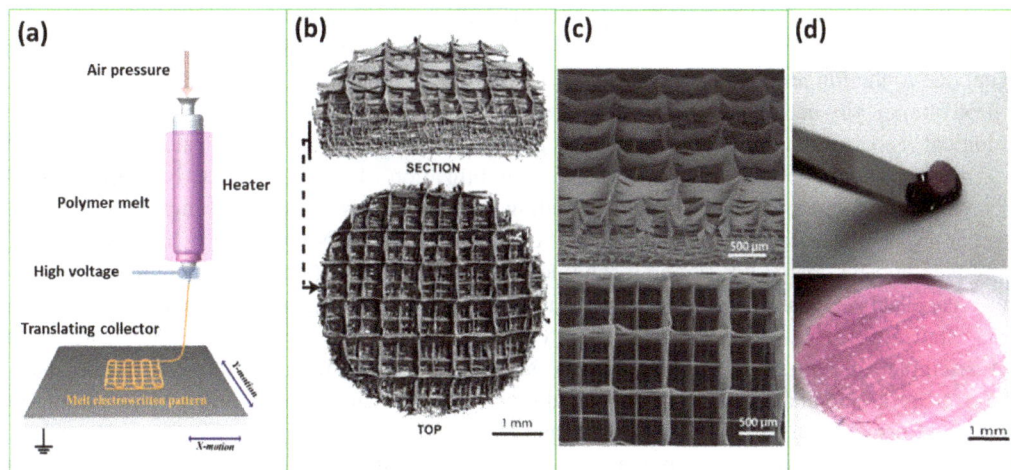

FIGURE 23.7 Development of the articular cartilage based on the natural design of the native tissue using ME. (a) Schematic representation of ME. Adapted with permission from Reference [26]. Copyright (2019) The Authors. Licensee MDPI, Basel, Switzerland. This is an open-access article under the terms of the Creative Commons Attribution License (https://creativecommons.org/licenses/by/4.0/). (b) 3D µCT reconstruction. (c) Scanning electron microscopy (SEM) images. (d) Images of a representative soft network composite. Adapted with permission from Reference [27]. Copyright (2018), Elsevier.

exceptional mechanical and biological properties, thus becoming a strong candidate for tissue engineering applications.

3 CONDUCTIVE POLYMERS FOR BIOMEDICAL APPLICATIONS

The first CPs were discovered in the 1970s. Due to their similar optical and electrical properties to inorganic and organic semiconductors, CPs are widely used in battery technology, biological field, and light-emitting diodes [28]. Among them, scientists use CPs for biomedical applications including biosensors, drug-delivery devices, tissue-engineering scaffolds, and bio-actuators, because they are compatible with many biological molecules. On top of that, CPs can controllably release biomolecules or drugs, and promote charge transfer steps in biochemical reactions. In addition, CPs are easy to synthesize, inexpensive, multifunctional, and can serve as a substrate with a high surface area compared to other electroplating materials, which has led to the application of CPs in the biomedical field [29].

3.1 CONDUCTING POLYMERS FOR BIOSENSORS

The biosensor consists of two main parts: a sensing element and a transducer. CPs are mainly used as transducers, whereas biomolecules are used as sensing elements. The sensing element first interacts with the target analyzer to generate a chemical signal, which is then transmitted to the converter and input into an electrical signal. The most common types of transducers applied in biosensing devices can be either amperometric or potentiometric. Based on that, the amperometric biosensors detect the current when the target product is oxidized or reduced at a constant potential. CPs are used as mediators for electron transfer between an enzyme and the final electrode. Redox mediators including Prussian blue and ferrocene are used to improve electron transfer from biochemical reactions to CPs. Those redox mediators can be inserted as a dopant or can chemically bind to monomers [30,31]. An ion-selective electrode is used as a transducer in a potentiometric biosensor, for instance, urea can be detected qualitatively and quantitatively by using urease enzyme via the production of NH_3 which interacts with PPy conductive polymer and produces an electrical signal due to the ion mobility in a polymer matrix with free ions in solution [32]. The immobilization of biologically active molecules on CPs promotes intimate contact between an electrical and biological recognition component in biosensors. Within this line, both non-covalent and covalent modification techniques can be used for the immobilization of biologically active molecules on CPs. Adsorption, entrapment, and affinity binding are used as non-covalent techniques and chemical conjugation is used as covalent conjugation [31]. A novel DNA biosensor has been created using the opposite-charge adsorption technique. This biosensor was fabricated using thin PPy films that were electrodeposited on a platinum surface. Then, an Au–Ag nanocomposite was bonded electrostatically with a positively charged PPy backbone. Lastly, a mercapto oligonucleotide was self-assembled on the Au–Ag nanocomposite surface. The DNA immobilization using opposite-charge adsorption on nanocomposite exhibited high selectivity and reproducibility. In addition, this method is versatile and suitable for direct reagentless analysis of DNA [6].

The modified electrodes were prepared using a three-electrode system. In this sense, PPy was electrochemically deposited over the Pt electrode. Then, the Pt-PPy was immersed in a suspension containing the Ag and Au nanoparticles in equal parts. Finally, the electrode composite was incubated in the presence of the bioactive probe, HS-ssDNA that could adhere to the surface of the nanoparticles, thus becoming functionalized [6]. In another approach, the layer-by-layer (LBL) self-assembled immobilization technique was used to create a novel glucose label-free biosensor [33]. The LBL self-assembled technique allowed precise control of composition along with good stability and a mild entrapment method for enzyme immobilization. Glucose oxidase (GOx) together with nitrogen-doped graphene (NG) was incorporated in the multilayer layer film on the Au electrode that was functionalized with electropolymerized PEDOT. The multilayer film contains assembled GOx, and NG dispersed biocompatible positively charged polymers chitosan and negatively charged PSS. PEDOT is used in this biosensor due to the presence of high electrochemical stability, good conductivity, low energy band gap, and biocompatibility. As a result, glucose label-free biosensors exhibited improved sensitivity that was attributed to the presence of PEDOT.

3.2 CONDUCTING POLYMERS FOR TISSUE ENGINEERING APPLICATIONS

Conducting polymer-based biomaterials including PANI, PPy, PTh, etc. are used in the treatment of electrical-sensitive tissues due to their better biocompatibility when compared to metals and inorganic biopolymers. Animal models confirmed the biocompatibility of CPs, and *in vivo* analysis showed that CPs do not have a significant long-term effect [34]. Moreover, CPs can promote cellular activities, for instance, cell migration, proliferation, adhesion, and protein secretion at the polymer tissue interface [35]. With this aim, pure CPs, conducting blends or composite films, conducting copolymer films, conducting nanofibers, and conducting hydrogels are used for tissue engineering [36]. Some of the conductive biomaterials along with their tissue engineering applications are demonstrated in Figure 23.8.

One of the challenges concerning the use of CPs is their inherent brittleness, which hinders the fabrication of biomedical devices based on neat materials. Therefore, CPs blended with different degradable polymers such as PLA, poly(lactic-co-glycolic acid) (PLGA), and PCL have become more popular. Within this scope, a network of interpenetrating PPy polymers was synthesized in a matrix of PCL and used as a facilitating platform for the formation of cardiac cell sheets. PCL membranes were treated with sodium hydroxide to enhance cell adhesion and improve hydrophilicity. Also, the PPy-PCL film produced by the chemical polymerization of PPy has similar resistance to native heart tissue (1.0 ± 0.4 kΩ cm) [37]. The selection of dopants for the synthesis of electroactive polymers with good biocompatibility plays an important role. CPs can be doped by a wide selection of anions, from small inorganic halogens (i.e., perchlorate, chloride) to large

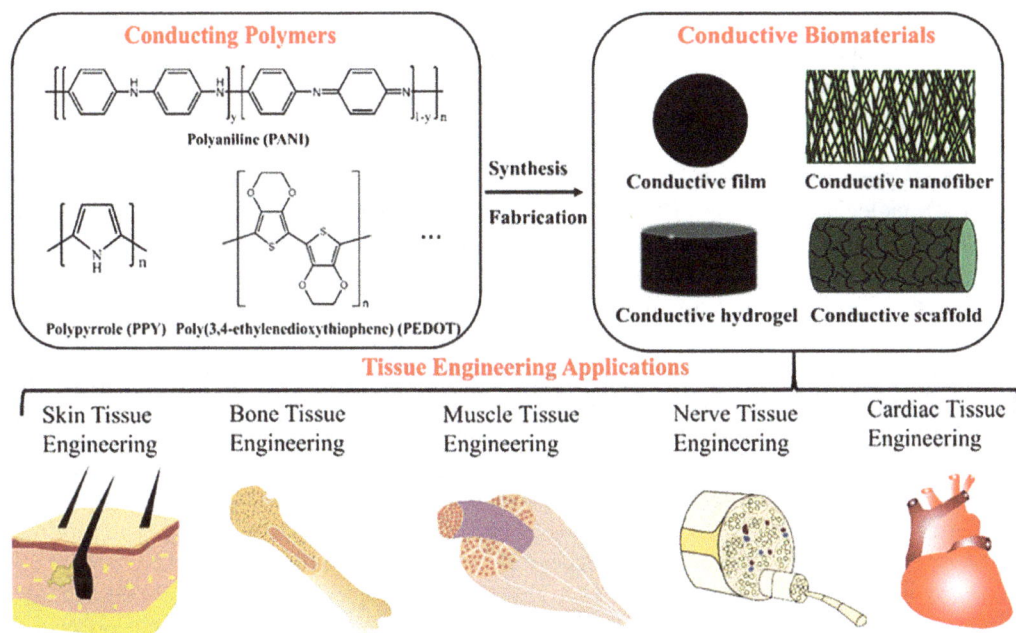

FIGURE 23.8 Conducting polymers used for the fabrication of conductive biomaterials that are suitable for tissue engineering applications. Adapted with permission from Reference [36]. Copyright (2018), American Chemical Society.

aromatic sulfonates (i.e., PSS). Dopant anions and a positively charged polymer backbone have charge interactions, and when CP is reduced or degraded, those charge interactions are lost and the doping ions can then move freely from the polymer backbone, which can cause harm to associated tissues [38]. Therefore, high-molecular-weight dopants have been widely used during the synthesis of biocompatible CPs, because larger dopants are not readily leached from polyelectrolyte complex. The biomedical characteristics of PPy films doped with tosylate (ToS), PSS, and Cl were tested using $PC1_2$ cells and Schwann cells. The PPy/ToS film exhibited higher electrical conductivity (53 S/cm) than PPy/Cl and PPy/PSS films (3 S/cm and 6 S/cm, respectively). PPy/PSS film exhibited superior chemical stability with phosphate buffer saline and water [39].

Biomaterials should be required to mimic the natural extracellular matrix (ECM) and its structure. Among the proteins contained in ECM, mainly collagen nanofibers regulate cell behavior and cell architecture. For that, scaffolds are designed to mimic natural collagen fibers and need to have high pores for efficient mass transport and cell ingrowth. In this sense, nanofibrous scaffolds have high surface areas, pores, and tunable diameters that help mimic the nanofiber collagen in ECM [12]. Polymeric nanofiber scaffolds can be synthesized using electrospinning, molecular assembly, and phase separation. Natural and synthetic biodegradable polymers have been blended with CPs to synthesize nanofibers for tissue engineering [36]. An electrospun PANI-gelatin nanocomposite doped with camphorsulfonic acid (CSA) was fabricated using N,N-dimethylformamide/water as solvents and stabilized using glutaraldehyde as cross-linker. Gelatin and gelatin + CSA exhibited very low electrical conductivity. Gelatin + CSA 5% + PANI 5% exhibited substantial electrical conductivity (4.2×10^{-3} s/cm) [40]. In addition, the *in situ* polymerized PPy uniform nanofibers were manufactured using an electrospun nanofiber template prepared by PCL and PLA. PPy tubes showed nanofibers with an outer diameter of ~ 320 ± 44 nm and a thickness of ~50 ± 3 nm, and dorsal root ganglia were found to adhere well to PPy core-sheath nanofiber. In addition, the PPy nanotubes were produced by soaking the PLA-PPy core-sheath nanofibers in dichloromethane to selectively remove the PLA cores. The polymerization was conducted with Fe^{3+} as an oxidant and Cl^- as a dopant. Therefore, PPy core-sheath nanofibers can be used as scaffolding in neural tissue engineering [41].

3.3 CONDUCTING POLYMERS FOR BIOELECTRONIC DEVICES

The nervous system generates electrical signals by moving ions through the membranes to control the body's biological functions. Therefore, manipulation and control of electrical signals in the body are important for tissue development and regeneration. In this sense, it can provide a platform for medical treatments using various implantable bioelectronic devices [42,43]. The latter uses an electrical signal to fix an anomaly in the body. In this sense, a pacemaker, spinal cord stimulator, and cochlear implant are examples of popular bioelectronic devices [1,42,43]. Most of the commercially available bioelectronic devices use metal electrodes including Pt, Au, and Pt-Ir alloy, which have the drawback of limited lifetimes from one to five years. The foreign body response (FBR) in the body acts as one of the major factors for bioelectronic devices' lifetime. Based on this, the FBR creates a barrier between biological tissue and the electrode by forming scar tissue that decreases the electrode's function and slowly leads to deactivation of the device. Because of that, surgery may be required to extract and replace the unactive device from the body with a new one [44]. Therefore, the manufacturing of biocompatible current devices is crucial for medical treatment.

Implantable materials that can be used for bioelectronic devices need to support cell adhesion because it facilitates close contact between the cellular environment and the device while preventing exposure to the FBR. In addition, implantable material must be conductive under biological conditions, and along with presenting satisfactory mechanical strength, biocompatibility, hydrophilicity, and biofunctionality as materials need to match with the targeted biological tissue [1]. However, the mechanical and electrical properties of most implantable devices used for medical treatment do not typically match with the targeted biological tissues [42]. Therefore, designing new materials for the development of fast-growing bioelectronic medicine is crucial. The inherent nature and characteristics of CPs allow them to be used in bioelectronic devices due to their similarities with biological tissue when compared to metallic-based implants. Some known examples of this are PANI, PPy, PTh, and their derivatives [7]. Part of introducing these materials into the body's system is to perform doping with either anionic or cationic materials to stabilize the CP's backbone net charge. In this line, small counterions, larger charge molecules, and polymers can be used as dopants for CPs. Attention toward polymeric dopants has increased because polymer dopants are unlikely to diffuse from the CP and therefore provide better stability in aqueous biological conditions. For instance, PSS is a polymer dopant that is used to stabilize positively charged oxidized PEDOT. Hence, PEDOT:PSS is used as *p*-doped CPs in many biological applications [10]. The *p*-doping consists of oxidation that takes place in the polymer chain, where electrons are removed from the π-system of the polymer, whereas the positively charged units remain in the conjugated polymer backbone. Electrons are introduced into the π-system during *n*-doping as the negatively charged units remain in the polymer backbone.

CPs can be employed for the fabrication of microelectrode arrays (MEAs) which are used for the *in vitro* study of electrophysiology of cell cultures. MEAs contain microfabricated electrodes made of Pt, Ir, TiN, and Au on a glass substrate, on top of which cells are cultured. In this sense, it allows studying the effect of various compounds on the neural activity of the hippocampal cells, for instance. In another case, CPs have been introduced as coatings to improve the electrical contact between electrodes and tissue. An expected effect of an increase in capacitance and decrease in impedance can occur which is due to both electronic and ionic conductivities. The MEA was fabricated using PEDOT:PSS, coated with polypeptide poly-L-lysine (PLL). Additionally, the introduction of nanorods, microneedles, and micropillars increased the CP surface area and further decreased the device impedance due to the tight electrode–tissue adhesion. For instance, lower impedance values (592 Ω vs. 7004 Ω) were obtained for galvanostatic polymerized PEDOT:PSS, doped using multiwalled carbon nanotubes (MWCNTs) [45,46].

Neural electrophysiology uses neural depth probes to record neural activities. However, the lack of stable recording, long-term usage, and the change of causing irreversible tissue damages are the main limitations of neural depth probes. PLGA and poly(vinyl alcohol) (PVA) were used to make flexible parylene probes that can penetrate brain cells. A polydimethylsiloxane (PDMS) molding was used to construct a PVA shuttle with a sharp tip that was required for piercing the brain tissue. Afterward, the PVA shuttle was dip-coated using PLGA to make a bi-layered shuttle. PLGA decreased the impedance of the PVA shuttles and parylene probe, which enabled it to record high-quality local field potentials after being implanted into mice [47]. The addition of the bioresorbable polymer into the neural device prevented the parylene probe from curling, which facilitated the handling and provided mechanical support as can be observed in Figure 23.9a. The fabrication steps for the neural device are summarized in Figure 23.9b. By the end of the process, a bi-layered composite of PVA/PLGA functionalized with a probe of 380 μm width was obtained (Figure 23.9c). The device's tip presented an average thickness of around 88 μm, as can be seen in Figure 23.9d.

Furthermore, CPs are used in the manufacture of electrochemical transistors that can be applied to biosensors and logic circuits due to their attractive thermoelectric properties, including stable electrical conductivity, excellent stability, and flexible mechanical properties. Based on that, PEDOT:PSS is usually used as a material to make organic electrochemical transistors (OECTs). OECTs have a planar monolayer structure composed of channels usually made of PEDOT:PSS. Therefore, the geometry of OECTs is simpler than organic field-effect transistors (OFETs). Electrons and ions in OECTs act as charge carriers, and polymer can undergo a reversible electrochemical redox process. That causes the electronic structure of the polymer to change during the reversible electrochemical oxidation and reduction. In this way, polymer-based electrochemical transistors have been used in flexography or ink-jet printing. Ink-jet printing is used for the fabrication of all PEDOT:PSS-based OECTs because ink-jet printing offers great flexibility in selecting substrates for printing and it is

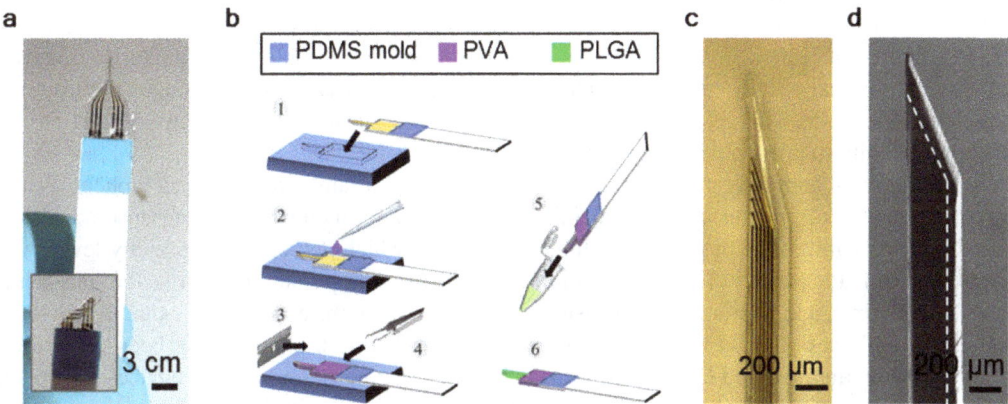

FIGURE 23.9   (a) Photocopy of the parylene-C probe which is supported by the PVA-PLGA device over PDMS with a flexible probe-containing device. (b) Scheme for the neural device fabrication process. Step 1: aligned probe attached to ZIF in PDMS mold. Step 2: Deposition of PVA solution. Step 3: PVA blade casting. Step 4: Probe removal from the mold after mild thermal treatment. Step 5: Dip coating of the PVA-supported probe in an acetone solution of PLGA. Step 6: Acetone evaporation to improve the device's rigidity. (c) Probe's contrasting image. (d) SEM for the PVA/PLGA-supported probe as the white dashed lines represent the boundaries of parylene inside the polymer shuttle. Adapted with permission from Reference [47]. Copyright (2018) The Authors. Licensee IOP Publishing Ltd. This is an open-access article under the terms of the Creative Commons Attribution License (https://creativecommons.org/licenses/by/3.0/).

suitable for low-cost disposable circuitry. The working range of OECTs can be controlled by changing the device's geometry, including the ratio between the gate and channel areas [47].

4 NON-CONDUCTIVE POLYMERS FOR BIOMEDICAL APPLICATIONS

Recently, the research field on the development of non-conductive polymers (NCPs) for biomedical applications has been gaining strength. The reasons for this include the excellent tunable properties of NCPs such as (i) biocompatibility, (ii) biodegradability, (iii) non-toxicity, (iv) immense diversity in the structure, (v) rich surface chemical alteration, (vi) modifiable mechanical properties, and (vii) high surface area with suitable porosity. NCPs can be categorized into synthetic and bioderived. As discussed previously, CPs are used for diverse medical applications such as biosensors, tissue engineering, regenerative medicine, and bioelectronic devices. In this section, synthetic, bio-derived, and composites based on NCPs are briefly discussed along with their applications in the mentioned fields.

4.1 SYNTHETIC NON-CONDUCTIVE POLYMERS

Several synthetic NCPs can be employed for biomedical applications. Among the most commonly used synthetic NCPs are polystyrene (PS), polymaleimidostyrene (PMS), Nafion, poly(glycolic acid) (PGA), PMMA, polyethylene terephthalate (PET), PCL, PU, poly(propylene fumarate) (PPF), etc.

4.1.1 Synthetic NCPs in Biosensors

Recently, researchers have been moved to use nonconductive polymers for biosensors to overcome the main challenges that are experienced with conductive polymer-based biosensors including decreased enzyme activity during electrodeposition in an organic solvent which also limits the amount of enzyme that can be loaded into the polymerized membrane. As already mentioned, modifying the surfaces of biosensors using polymeric materials can (i) smooth the interfacial surface and (ii) enhance the activation of biochemical receptor (enzyme) immobilization by introducing additional functional groups. Currently, various electrochemistry-driven biosensor methods using PS, PU, Nafion, and their composites have been introduced for simple and miniaturized analytical devices for analysis. In this sense, NCP nanoparticles can offer many advantages in biosensor applications, such as a large surface-to-volume ratio, high surface reaction activity, and strong adsorption ability to immobilize desired biomolecules [48,49].

Synthetic non-conductive polymer films such as poly(1,3-diaminobenzene), polyphenols, and poly(o-aminophenol), for instance, are mainly used as the outer membrane for biosensors because their films are synthesized from water-soluble monomers, which decreases the toxicity levels and improves enzymatic activity. The outer membrane in sensors is important to improve the sensor linearity and lifetime, and to control the diffusion of the substrate to the enzyme. An example is hydrophilic polyurethanes (HPUs) which are biocompatible polymeric films that have been applied on glucose oxidase immobilized electrodes for the fabrication of glucose-monitoring devices. However, some HPU-based devices have shown less reproducibility and a shorter lifetime due to the unstable membrane/enzyme layer interface. As a solution to this limitation, HPU was blended with polyvinyl alcohol/vinyl butyral copolymer (PVAB) at a 3:2 ratio (12:8 mg). The developed glucose biosensor exhibited continuous monitoring of glucose over 3 days in human serum and the outer membrane has high oxygen permeability and excellent control of glucose diffusion [50].

Another case is PS, which is a synthetic aromatic hydrocarbon polymer made from the polymerization of monomer styrene. The styrene monomer is the basic building block of plastic. The styrene monomer can be functionalized by a variety of active groups such as –COOH, –NH$_2$, –NO$_2$, and others through the aromatic benzene rings reactivity, giving it excellent surface activation of biomolecule immobilization for PS. Moreover, chemically modified styrene monomers were able to improve the electrical conductivity of the PSS composites. PS material is commonly used to make biosensor materials. However, developing a biosensor to meet the required sensitivity with high reproducibility remains a challenge. Therefore, various researchers have synthesized PS-based nanocomposites to enhance the covalent immobilization of enzymes, reusability, thermal, and pH stability of biosensors [49,51]. Recently, in one study, the immobilization of glucose oxidase enzyme on a new composite nanomaterial of PSS was investigated. The authors synthesized a modified PEDOT:PSS/titanium carbide, (Ti_3C_2)/graphene quantum dots screen-printed carbon electrode for glucose sensing. Their results confirmed that modified nanocomposite material showed higher performance compared to the PEDOT:PSS and PEDOT:PSS/Ti_3C_2 composites [49]. The detection results demonstrated that the fabricated biosensor had a linear voltammetry response in the glucose concentration range 0–500 μM with a relative sensitivity of 21.64 μAm/M.cm^2 and a detection limit of 65 μM (S/N = 3), with good stability and selectivity.

PS is widely used in combination with other organic molecules due to its advantageous interactions with certain enzymes. For instance, Shunichi Uchiyama et al. [48] attempted to prolong the stabilization of the GOx enzyme's immobilization in the PS membrane of glucose biosensors. In this way, the authors synthesized PMS by introducing maleimide groups on the PS's chain that was coated over a porous carbon paper. Hence, the presence of maleimide groups prompted the enzymes to covalently bond into the biosensor's surface, whereas the high surface area of carbon facilitated their adsorption. Therefore, enzymes could be easily immobilized through immersion in a PMS-coated porous material into an enzyme solution. The mechanism for that process is described in Figure 23.10. It is noted that a good linear correlation of the glucose and oxygen current response was obtained in the concentration range from 0.1 to 2 mM. The upper limit of the linear range was confirmed to be 3.0 mM.

Another widely explored polymeric material for the fabrication of biosensors is Nafion, which is a perfluoro sulfonated polymer made from copolymerization of tetrafluoroethylene and a derivative of a perfluoro (alkyl vinyl ether) monomer with sulfonyl acid fluoride. Nafion-based materials have been successfully applied to glucose biosensors and DNA biosensors [2,52]. These materials can improve the biosensor's sensitivity, detection limit, and storage stability due to their good chemical and thermal stability, excellent biocompatibility, and ability to resist interference from anions and biological macromolecules [3,52,53]. Past research works in the literature have discussed various applications of composite materials based on carbon nanotubes (CNTs) as a reliable biosensor material. Recent interest has focused on the design of modernized electrode surfaces using CNT/

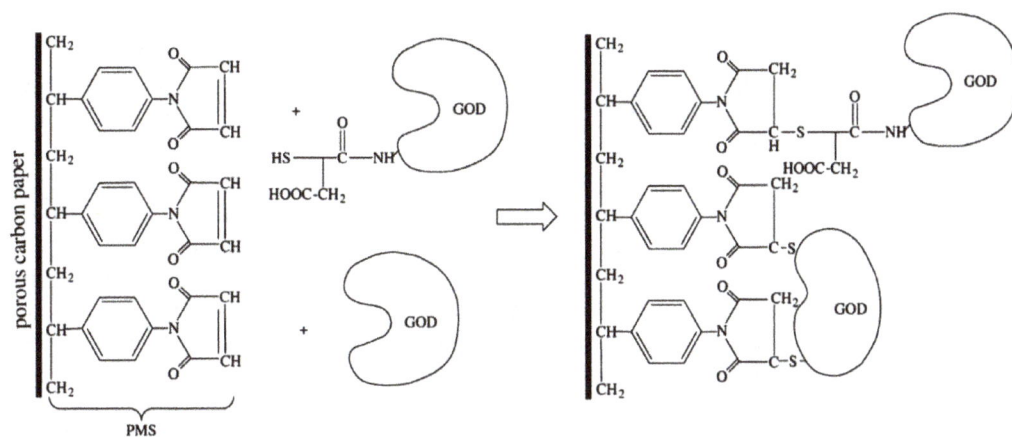

FIGURE 23.10 Illustration of immobilization types of glucose oxidase (GOD) enzyme using PMS. Reproduced with permission from Reference [48]. Copyright (2006), Elsevier.

Nafion composite films for biosensors. In 2002, CNT/Nafion-coated electrodes were fabricated by a research group for oxidase-based amperometric biosensors. Here, Nafion enhanced the redox activity of hydrogen peroxide, using its ability to disperse single-walled CNTs and MWCNTs. Thus, Nafion provides a feasible way to prepare CNT-based biosensors as a solubilizing agent for CNTs [53]. In addition, graphene-based composites have attracted considerable attention in recent years due to their high specific surface area and outstanding electrical conductivity for electrochemical sensors. Based on that, Qiaojuan Gong and coworkers developed an impedimetric DNA biosensor to determine the HIV-1 gene by employing a graphene–Nafion composite film modified glassy carbon electrode. Nafion has been used to effectively disperse graphene in an aqueous solution, which can improve the stability of the modified electrode [52].

4.1.2 Synthetic NCPs in Tissue Engineering

Biomaterials that are used for tissue engineering must provide the mechanical and structural requirements of the target tissue while promoting molecular interaction with tissue-healing cells. Hence, in the regeneration process, biomaterials provide sufficient space to develop new tissue and provide a suitable surface for direct cellular attachment, migration, and proliferation. In this sense, natural polymeric materials including proteins, polysaccharides, and synthetic polymeric materials are used for tissue engineering applications. Among those materials, synthetic polymers have gained increased attention due to their more flexible properties than natural materials. Biocompatibility, biodegradation into nontoxic products, and the ability to support cell growth and proliferation are other factors for the use of synthetic polymers for a wide range of TE applications [54]. Some examples of those used for tissue engineering applications are PS, poly(α-esters), PMMA, PPF, polyethylene (PE), and polyphosphoesters.

PMMA-based bone cement is widely used in tissue engineering to cover complex bone cavities and orthodontic applications to repair dental damages. In addition, bioactive inorganic reinforcement biomaterials (titania, bioactive glass-ceramics) have been introduced into the PMMA matrix to improve the clinical functionality of PMMA-based bone cement as well as hydroxyapatite (HA) which is one of the main components of teeth [55]. Through that, those biomaterials that were incorporated into PMMA-based bone cement promoted the enhancement of the mechanical properties and cytocompatibility. Also, other polymers such as PGA, PLA, PLG, polydioxanone, and PCL belong to the poly(α-ethers) which can be potentially used for this type of application. PGA, for instance, has high crystallinity which can lead to better mechanical support. This polymer is used to treat bone implants, fractures, and osteotomies, and to fabricate biodegradable sutures, for instance, DEXON which is a commercially available biodegradable synthetic suture. Based on that, compression molding, solvent casting, injection, and extrusion are the most common techniques used to synthesize polyglycolide-based biomaterials for biomedical applications. Poly(L-lactide) (PLLA) is a semicrystalline polymer, which has been used for load-bearing applications and high-strength fiber due to the presence of high modulus, high tensile strength, and low extension. Moreover, PLLA is used as a biomaterial for scaffold fabrication [11].

Polymer mixing with filler or polymer blending improves the mechanical properties and potential applications of polymers. PET is a synthetic thermoplastic biomaterial that can also be used as an implanted biomaterial for biomedical applications as it has good mechanical strength and biocompatibility. However, even though it has excellent mechanical properties, its non-biodegradable nature limits its application to the skin in the biomedical field. Biocompatibility and other bioactive properties of PET were enhanced by incorporating mineral polymer filler (HA) into the PET polymer. HA has great osteoconductive and osteoinductive capabilities, along with a strong affinity to host tissues due to its chemical similarity to the inorganic bone matrix component. Mixing of HA and PET can form nano-biocomposites with improved mechanical properties that can be applied to the skin. Nano-biocomposites with 98% of PET and 2% HA exhibited excellent properties (more flexible scaffolds and low values of modulus) for skin applications [56]. Also, PPF/polyethylene glycol-modified graphene oxide (PEG-GO) nanocomposite was used for bone tissue applications [57]. The hydrophilicity of PPF was enhanced after mixing with PEG-GO, which led to better cell attachment, spreading, and proliferation. Such a factor was attributed to the oxygenated groups from GO that were polarized and hydrophilic. Through that, the hydrophilicity, biodegradation rate, thermal stability, and other mechanical properties increased while increasing the concentration of GO in the PPF/PEG-GO composite.

4.2 BIO-DERIVED NON-CONDUCTING POLYMERS

Synthetic polymers have been popular for decades due to their interesting characteristics, which make them extremely useful for biomedical applications. However, the large use of synthetic polymers for biomedical applications may negatively affect people's health. A common feature of many bio-derived polymers is that they are naturally biodegradable. In this regard, scientists have attempted to incorporate bio-derived polymers (mostly non-conducting) into biomedical applications.

4.2.1 Bio-Derived Non-Conducting Polymers in Biosensors

The biosensor performance depends on its sensitivity, limit of detection, linear and dynamic ranges, reproducibility or response accuracy, selectivity, response time, ease of use, portability, and operational and storage stability parameters. Bio-derived NCPs are one class of materials that meet many of the prerequisites mentioned above and therefore find wide application in the development of biosensors. Unlike synthetic polymers that have a simple and random structure, bio-derived NCPs offer several advantages from the point of view of biosensor application, such as a well-defined three-dimensional structure which is crucial for their function, high monodispersed, and environmental friendliness. Moreover, bio-derived NCPs have aroused great interest among researchers due to their minimal or non-invasive label-free diagnostic process and cost-effectiveness. Typically, bio-derived NCPs are applied as a surface modifier of the electrodes/transducer surface in the biosensor. These modifiers can amplify the target signal by attenuating interfering signals originating from surface fouling (non-specific binding). Among the variety of bio-derived NCPs, cellulose and chitin/chitosan are the most intensively used polymers for biosensor development [58–61].

Cellulose is the most abundant biopolymer on Earth. The great interest in applying cellulose-based materials in biosensors is due to its beneficial characteristics, which include low thermal expansion, high Young's modulus, good optical properties, and large surface functional groups. In addition, cellulose is an excellent bonding agent and has readily available surface functional groups, which can immobilize the high-density biomolecules. Therefore, cellulose-based materials enhance the selectivity and sensitivity of biosensors compared to molecules that do not have functional groups [58,62]. Several cellulose matrices have been used in biological sensors such as cellulose nanocrystals, cellulose nanofibrils, bacterial cellulose, paper, gauze, and hydrogels. These materials are utilized in the production of biosensors that can be used to detect urea, glucose, cells, amino acids, proteins, lactates, hydroquinones, genes, and cholesterol, among others [63–67]. Intensive and rapid signal generation using high-surface-area electrode-substrate materials is currently emerging for biosensors. Examples of such innovations are cellulose nanoparticles, which combine with inorganic nanoparticles such as metal ion oxides [63–65,68], carbon-based materials (nanotubes, graphene, graphene oxide) [66,68,69], and conductive polymers [69]. Chitin is the second most abundant natural biopolymer after cellulose. However, the application of chitin in the biomedical field is challenging due to its low solubility. Chitosan is a natural biomacromolecule produced by N-deacetylation of chitin, which is a natural polysaccharide. The popularity of these polysaccharides in the biosensor is mostly based on their desired properties, such as biocompatibility, biodegradability, and non-toxicity, as well as hypoallergenic properties. Chitin–chitosan-based compounds have unique structural and functional properties. For instance, chitin–chitosan compounds can form uniform films and hydrogels and a high surface-area-to-volume ratio. Therefore, they increase the number of binding sites available for biomolecule immobilization, resulting in faster mass transfer rates, lower detection limits, and faster detection rates. Another important characteristic of chitin–chitosan is the chemical structure, which has intrinsic oxygen- and nitrogen-based functional groups that can be chemically modified. In addition, chitin–chitosan materials can be obtained cheaply from a variety of commercial sources. In numerous studies, the use of chitosan hydrogel-based metal-organic frameworks (CS-MOFs) as immobilization matrices for biologically active materials (enzyme/antibody/nucleic acid) in biosensors has been reported. This type of biosensor exhibited a rapid response and a comparatively broad linear calibration range with a low detection limit for lactose, glucose, and H_2O_2 detection [59,70,71]. Similarly, various researchers have developed CP/chitin–chitosan-based hydrogels using PANI, PPy, and PTh. The results of their findings confirmed polymer/chitosan-based hydrogels also as a promising candidate for biosensor applications [4,5,61,72].

Moreover, dextran- and collagen-based compounds are also used to fabricate biosensors. Dextran is a heterogeneous polysaccharide made from the polymerization of the α-d-glucopyranosyl moiety of sucrose. The common feature of dextran-based compounds is a preponderance of (1 → 6)-linked α-d-glucopyranosyl units. These compounds are often used to modify the sensor chip surface of biological sensors in the form of hydrogel layers. This is due to the high water content, swelling, and flexible structure of their hydrogel matrix, which can offer superior anti-fouling properties. Within another area, the surface plasmon resonance (SPR) biosensors are also highly sensitive and are generally used in monitoring biomolecular interactions, including proteins and nucleic acids. Recently, scientists have introduced a thick dextran matrix on the probes of SPR biosensors to enhance the signal generated by low-molecular-weight targets. Additionally, dextran-based nanocomposites could be used as a reductive and protective agent in biosensors to detect *Escherichia coli* and glucose [73,74].

Collagen is the primary structural component of connective tissues. Because of its inherent biocompatibility, high adhesion ability, and high dispersity, collagen-based compounds are widely used in biosensor applications. Collagen-based composites consist of biosensors that are commonly used to detect hydrogen peroxide. For example, novel hydrogen peroxide biosensors were fabricated by different research groups, based on a hemoglobin (Hb)–collagen composite. They synthesized Hb immobilized collagen nanofibers using the electrospinning method to increase the performance of the biosensor [60]. Some researchers incorporated CNT, zirconia nanoparticles, etc., into Hb–collagen composite to improve the sensitivity of the biosensor [60,75,76].

4.2.2 Bio-Derived Non-Conducting Polymers in Tissue Engineering and Regenerative Medicine

The non-toxic and low inflammatory reactions, metabolic compatibility, and ability to communicate with the biological system are the main factors that have attracted more attention to biologically derived polymers for tissue engineering and regenerative medicine applications. However, temperature sensitivity and the complex nature of the biological polymers can be considered disadvantages of these biologically derived polymers [77]. Polysaccharides, for instance, chitin, chitosan alginate or proteins, collagen, and gelatin are used as biologically derived polymers for tissue engineering and regenerative medicine applications. High strength, non-toxicity, biodegradability, and bio-adhesion, allow a broad application of chitin–chitosan in tissue engineering and regenerative medicine. In this respect, there are different types of methods that can be applied for their synthesis including freeze-drying, layer-by-layer assembly, photo-initiated grafting, alkaline hydrolysis followed by grafting-coating, etc. Based on these approaches, the freeze-drying process has been employed for the synthesis of biodegradable and highly porous scaffolds used for the proper accommodation of mammalian cells and their growth [78]. In this sense, chitosan nanofibers have been introduced using freeze-drying into parallel-aligned microfibers of poly(propylene carbonate) (PCC) that were prepared through electrospinning and oxygen plasma treatment, to achieve superior cell attachment, cell–scaffold interactions, elevated hydrophilicity, and cell proliferation.

As the range of potential tissue engineering systems expands, there is an ongoing search for materials that have specific tissue properties that can be widely applied in several systems [79]. Chitin blended polymers, for instance, can be used for the regeneration of bone defects. Within this area, chitin-poly(ε-caprolactone) (PCL) has facile handling, lack of physical harm to the adjacent tissues, non-toxic degradation product formation, and the ability to fit a defective shape without forming any voids. PCL-based nanohydroxyapatite (nHAp) incorporated with injectable microgels was developed for the regeneration of irregularly shaped and complex bone defects [80].

Another factor for the proper application of polymeric materials for tissue engineering is that they should present antimicrobial properties while preventing an allergic reaction. In this sense, the antimicrobial resistance of microorganisms can cause a delay in wound healing which prompts the need for novel antibacterial drugs that can aid in the treatment of wounds infected with such microorganisms. Based on that, Ag nanoparticles incorporated in chitin composite scaffolds have been used for wound-healing treatments because Ag is considered a potent antibacterial agent. α-Chitin/nano-Ag composite scaffolds and β-chitin/nano-Ag composite scaffolds have been used for wound-healing applications. Both nano-Ag composite scaffolds exhibited excellent antibacterial activity for *S. aureus* and *E. coli* and exhibited good blood clotting ability. The cell adhesion ability of β-chitin/nano-Ag composite scaffolds was determined using epithelial cells. The adhesion of epithelial cells confirmed that β-chitin/nano-Ag composite scaffolds have excellent adhesion properties in addition to their antibacterial nature [13].

Another important biological material within the biomedical field is collagen, which is a protein made of three polypeptide chains that can also be found in animals' extracellular matrix (ECM). Collagen protein plays a vital role in maintaining the structural as well as the biological integrity of the ECM and provides physical support to tissues. Its advantages include biocompatibility, being easily degradable by enzymatic action, strong cell binding, and low stimulation of the immune system. Through that, collagen-based materials can be applied for wound healing, delivery systems, and tissue engineering [81]. Lack of mechanical strength and lack of structural stability upon hydration limit the applications of collagen scaffolds in certain tissues. However, intermolecular cross-linking and blending it with other materials, including inorganic materials and natural or synthetic polymers, can improve the mechanical strength of collagen-based scaffolds without inducing cell toxicity.

Electrospinning and lyophilization are useful methods for pure collagen scaffolds fabrication, and those scaffolds exhibited improvements in cell growth and penetration capacity. Treatments including ultraviolet (UV) irradiation, hydrothermal treatment, γ radiation, or chemical agents such as glutaraldehyde can be used to form intermolecular cross-linking of collagen. Based on that, γ radiation-induced cross-linking in non-fibrillar collagen can improve the osteogenic differentiation and elongation of mesenchymal stem cells. In addition, collagen type I plays an important role in neuronal pathfinding and provides a permissive environment for the regeneration of injured axons. Electrospun collagen nanofibers have been used to treat spinal cord injury (SCI) and could be evaluated in both *in vitro* and *in vivo* cases. Collagen nanofibers prevented astrocyte proliferation and glial fibrillary acidic protein and facilitated dorsal root ganglia outgrowth *in vitro* [82]. Natural polymers (chitosan, alginate, silk fibroin, etc.) can be blended with collagen to improve the mechanical and biological properties of scaffolds. Also, freeze-drying synthesized composites of collagen–chitosan scaffold have been used for the transplantation of bone marrow mesenchymal stem cells into the ischemic area in animal models [13,81,82].

4.2.3 Bio-Derived Non-Conducting Polymers in Bioelectronic Devices

Natural biomaterials can be used as structural and/or functional components, including within the field of bioelectronic devices. In this sense, tunable mechanical strength, flexibility, biological activity, and biocompatibility are direct factors for the use of natural biomaterials in biodevices. Protein-based polymeric materials including silk, keratin, collagen, and gelatin, and polysaccharide-based polymeric materials including cellulose, chitin, and agarose can be used for bioelectronic devices. Silk is naturally synthesized by domesticated *Bombyx mori* silkworms and cannibalistic spiders. Silk fibers are relatively strong materials found naturally. Yet, silk fibroin has been used as a dielectric layer in organic field-effect transistors (OFET) that has applications in e-skins, biosensors, e-papers, etc. In this sense, silk fibroin can be mixed with PVA to decrease the dielectric surface roughness to allow its use as an OFET [8]. Another study demonstrated a similar effect for the decrease of surface roughness and charge trap density at the dielectric/semiconductor interface with an operating voltage reported as –1 V [83]. Under this scope, silk fibroins can be used in actuators, electrodes, wound healing, patterning, etc. Silk fibroin can be processed into films, hydrogels, and woven and non-woven matrices that are used as wound dressing materials, and those silk fibroins can be mixed with antibiotics and biocides. Antibiotics can be easily loaded into the silk fibroin matrices through simple immersion techniques or blended with an antibiotic mixture with silk fibroin solution before preparing different matrices [84].

Keratin is a structural protein present in feathers, horns, claws, and hair. It has a coiled and twisted structure. In that sense, disulfide and hydrogen bonding between fibers improve its strength as well as increase its hydrophilicity. However, even though the biodegradability of keratin is slow by common proteolytic enzymes, its water insolubility and long-term stability are advantageous for the fabrication of multilayered devices. Based on that, keratin protein extracted from chicken feathers is used as a gate dielectric in organic thin-film transistors (OTFTs). Such a factor is desirable in comparison to other polymeric gate dielectric materials such as PS and PMMA, which are used in OTFTs. They can be dissolved during the deposition of organic semiconductor materials using spin and dip coating. The insolubility in organic solvents, ability to improve the electrical performance of solution-processable OTFTs, and environmentally friendly characteristics have directed the use of keratin as the gate dielectric in OTFTs [85]. Additionally, keratin-based composites are used to develop artificial organs and skins. Also, keratin-elastin has been used to produce keratin-based composites with flexibility, surface roughness, and rheological behavior similar to that of natural human skin. In another case, keratin extracted from waste human hair was mixed with flexible thermally processable thermoplastic polyurethane (TPU) to form keratin polyurethane composite (PUK). TPU can mimic the elastic part of the keratin-elastin composition of the skin and PUK exhibited rheological and tribological behaviors like the human skin [86].

In another line, extensive cross-linking in collagen decreases its solubility in a wide range of solvents and improves elasticity and tensile strength. Characteristic properties of collagen and its derivatives lead to their use as suitable materials for flexible and biodegradable electronics. Additionally, scientists have produced different forms including sheets, films, and aerogels to facilitate the use of collagen and its derivatives. Electrically conducting collagen–PPy hybrid aerogels which can be applied in biosensors, biomedical implants, and TEs have been synthesized using *in situ* oxidative polymerization combined with freeze-drying. Based on that, the combination of collagen with other active materials is important to get extra strength and stability due to its cross-linking and self-aggregation. In addition, collagen promotes cell proliferation, attachment, and regeneration of new cells [87].

5 CONCLUSION

Biocompatibility, biodegradability, non-toxicity, easy processing, and lightness have led to the application of synthetic and naturally derived polymers for various sectors in the field of biomedicine. Synthetic and naturally derived polymers can be divided into conductive and non-conductive depending on their conductivity. CPs are a special type of polymer used in the biomedical field and the conductivity of those polymers can be improved using doping or charge carriers. When CPs are mixed with biological molecules, the bioavailability of CPs improves, which affects the morphophonology of CPs. The electrical conductivity of CPs helps to communicate with the body and provides a better platform for medical treatment. Therefore, CPs are widely used for biosynthesis, tissue engineering, and bioelectronic devices. Also, synthetic, and bio-derived non-conducting polymers are used for biosensing, tissue engineering, and bioelectronic devices. These polymeric biomaterials provide excellent surfaces to immobilize biomolecules. Moreover, polymeric materials are used to produce scaffolds in tissue engineering, provide required mechanical and structural requirements to the target tissues, and promote tissue healing. The low cost, flexibility, and biocompatibility of non-conducting polymers have led to their use in bioelectronic devices. Non-conducting polymers can be applied as pure polymers when they have certain mechanical and structural properties to provide the required function. Thus, throughout this chapter, it is notable that several types of polymers available can greatly contribute to the development of biomedical devices.

REFERENCES

[1] C. Baker, K. Wagner, P. Wagner, D.L. Officer, D. Mawad, Biofunctional conducting polymers: synthetic advances, challenges, and perspectives towards their use in implantable bioelectronic devices, Adv. Phys. X. 6 (2021) 1.

[2] L. Yang, X. Ren, F. Tang, L. Zhang, A practical glucose biosensor based on Fe_3O_4 nanoparticles and chitosan/nafion composite film, Biosens. Bioelectron. 25 (2009) 889–895.

[3] F. Davis, S.P.J. Higson, Polymers in biosensors, Woodhead Publishing Limited, 2007.

[4] A. Sassolas, L.J. Blum, B.D. Leca-Bouvier, Immobilization strategies to develop enzymatic biosensors, Biotechnol. Adv. 30 (2012) 489–511.

[5] H. Huang, J. Wu, X. Lin, L. Li, S. Shang, M.C. Yuen, G. Yan, Self-assembly of polypyrrole/chitosan composite hydrogels, Carbohydr. Polym. 95 (2013) 72–76.

[6] Y. Fu, R. Yuan, Y. Chai, L. Zhou, Y. Zhang, Coupling of a reagentless electrochemical DNA biosensor with conducting polymer film and nanocomposite as matrices for the detection of the HIV DNA sequences, Anal. Lett. 39 (2006) 467–482.

[7] Y. Chen, Y.S. Kim, B.W. Tillman, W.H. Yeo, Y. Chun, Advances in materials for recent low-profile implantable bioelectronics, Materials (Basel). 11 (2018) 1–24.

[8] L.-D. Koh, J. Yeo, Y.Y. Lee, Q. Ong, M. Han, B.C.-K. Tee, Advancing the frontiers of silk fibroin protein-based materials for futuristic electronics and clinical wound-healing (invited review), Mater. Sci. Eng. C. 86 (2018) 151–172.

[10] S. Pradhan, A.K. Brooks, V.K. Yadavalli, Nature-derived materials for the fabrication of functional biodevices, Mater. Today Bio. 7 (2020) 100065.

[11] N. Sultana, H.C. Chang, S. Jefferson, D.E. Daniels, Application of conductive poly(3,4-ethylenedioxythiophene):poly(styrenesulfonate) (PEDOT:PSS) polymers in potential biomedical engineering, J. Pharm. Investig. 50 (2020) 437–444.

[12] X. Tang, S.K. Thankappan, P. Lee, S.E. Fard, M.D. Harmon, K. Tran, X. Yu, Polymeric Biomaterials in Tissue Engineering and Regenerative Medicine, Elsevier, 2014.

[13] J.M. Holzwarth, P.X. Ma, Biomimetic nanofibrous scaffolds for bone tissue engineering, Biomaterials. 32 (2011) 9622–9629.

[14] R. Jayakumar, D. Menon, K. Manzoor, S. V Nair, H. Tamura, Biomedical applications of chitin and chitosan based nanomaterials—A short review, Carbohydr. Polym. 82 (2010) 227–232.

[15] K. Namsheer, C.S. Rout, Conducting polymers: a comprehensive review on recent advances in synthesis, properties and applications, RSC Adv. 11 (2021) 5659–5697.

[16] N.Y. Abu-Thabit, Chemical oxidative polymerization of polyaniline: a practical approach for preparation of smart conductive textiles, J. Chem. Educ. 93 (2016) 1606–1611.

[17] B. Senthilkumar, P. Thenamirtham, R. Kalai Selvan, Structural and electrochemical properties of polythiophene, Appl. Surf. Sci. 257 (2011) 9063–9067.

[18] E.N. Zare, P. Makvandi, B. Ashtari, F. Rossi, A. Motahari, G. Perale, Progress in conductive polyaniline-based nanocomposites for biomedical applications: A Review, J. Med. Chem. 63 (2020) 1–22.

[19] Mradula, R. Raj, S. Mishra, Voltammetric immunosensor for selective thyroxine detection using Cu-MOF@PANI composite, Electrochem. Sci. Adv. n/a (2021) e2100051.

[20] A.F. Diaz, K.K. Kanazawa, G.P. Gardini, Electrochemical polymerization of pyrrole, J. Chem. Soc. Chem. Commun. (1979) 635–636.

[21] M.M. Gvozdenović, B. Jugović, J. Stevanović, B.N. Grgur, Electrochemical synthesis of electroconducting polymers, Hem. Ind. 68 (2014) 673–684.

[22] X. Li, Y. Li, Electrochemical preparation of polythiophene in acetonitrile solution with boron fluoride–ethyl ether as the electrolyte, J. Appl. Polym. Sci. 90 (2003) 940–946.

[23] G. Odian, Principles of polymerization, John Wiley & Sons, 2004.

[24] H.H. Hamid, M.E. Harb, A.M. Elshaer, S. Erahim, M.M. Soliman, Electrochemical preparation and electrical characterization of polyaniline as a sensitive biosensor, Microsyst. Technol. 24 (2018) 1775–1781.

[25] J. Stejskal, R.G. Gilbert, Polyaniline. Preparation of a conducting polymer(IUPAC Technical Report), Pure Appl. Chem. 74 (2002) 857–867.

[26] N.G. Rim, C.S. Shin, H. Shin, Current approaches to electrospun nanofibers for tissue engineering, Biomed. Mater. 8 (2013) 14102.

[27] F. Afghah, C. Dikyol, M. Altunbek, B. Koc, Biomimicry in bio-manufacturing: developments in melt electrospinning writing technology towards hybrid biomanufacturing, Appl. Sci. 9 (2019) 3540.

[28] O. Bas, S. Lucarotti, D.D. Angella, N.J. Castro, C. Meinert, F.M. Wunner, E. Rank, G. Vozzi, T.J. Klein, I. Catelas, Rational design and fabrication of multiphasic soft network composites for tissue engineering articular cartilage: a numerical model-based approach, Chem. Eng. J. 340 (2018) 15–23.

[29] A.J. Heeger, Semiconducting and metallic polymers: the fourth generation of polymeric materials, J. Phys. Chem. B. 105 (2001) 8475–8491.

[30] X. Cui, J.F. Hetke, J.A. Wiler, D.J. Anderson, D.C. Martin, Electrochemical deposition and characterization of conducting polymer polypyrrole/PSS on multichannel neural probes, Sensors Actuators, A Phys. 93 (2001) 8–18.

[31] M. Gereadr, A. Choubey, B.. Malhotra, Review: Application of conducting polymer to biosensors, Biosens. Bioelectron. 17 (2001) 345–359.

[32] N.K. Guimard, N. Gomez, C.E. Schmidt, Conducting polymers in biomedical engineering, Prog. Polym. Sci. 32 (2007) 876–921.

[33] A. Gambhir, M. Gerard, A.K. Mulchandani, B.D. Malhotra, Coimmobilization of urease and glutamate dehydrogenase in electrochemically prepared polypyrrole-polyvinyl sulfonate films, Appl. Biochem. Biotechnol. – Part A Enzym. Eng. Biotechnol. 96 (2001) 249–257.

[34] M. David, M.M. Barsan, C.M.A. Brett, M. Florescu, Improved glucose label-free biosensor with layer-by-layer architecture and conducting polymer poly(3,4-ethylenedioxythiophene), Sensors Actuators, B Chem. 255 (2018) 3227–3234.

[35] P. Humpolicek, V. Kasparkova, P. Saha, J. Stejskal, Biocompatibility of polyaniline, Synth. Met. 162 (2012) 722–727.

[35] G. Kaur, R. Adhikari, P. Cass, M. Bown, P. Gunatillake, Electrically conductive polymers and composites for biomedical applications, RSC Adv. 5 (2015) 37553–37567.

[36] B. Guo, P.X. Ma, Conducting polymers for tissue engineering, Biomacromolecules. 19 (2018) 1764–1782.

[37] B.S. Spearman, A.J. Hodge, J.L. Porter, J.G. Hardy, Z.D. Davis, T. Xu, X. Zhang, C.E. Schmidt, M.C. Hamilton, E.A. Lipke, Conductive interpenetrating networks of polypyrrole and polycaprolactone encourage electrophysiological development of cardiac cells, Acta Biomater. 28 (2015) 109–120.

[38] S. Baek, R.A. Green, L.A. Poole-Warren, Effects of dopants on the biomechanical properties of conducting polymer films on platinum electrodes, J. Biomed. Mater. Res. – Part A. 102 (2014) 2743–2754.

[39] J.M. Fonner, L. Forciniti, H. Nguyen, J.D. Byrne, Y.F. Kou, J. Syeda-Nawaz, C.E. Schmidt, Biocompatibility implications of polypyrrole synthesis techniques, Biomed. Mater. 3 (2008).

[40] S. Ostrovidov, M. Ebrahimi, H. Bae, H.K. Nguyen, S. Salehi, S.B. Kim, A. Kumatani, T. Matsue, X. Shi, K. Nakajima, S. Hidema, M. Osanai, A. Khademhosseini, Gelatin-polyaniline composite nanofibers enhanced excitation-contraction coupling system maturation in myotubes, ACS Appl. Mater. Interfaces. 9 (2017) 42444–42458.

[41] J. Xie, M.R. MacEwcm, S.M. Willerth, X. Li, D.W. Moran, S.E. Sakiyama-Elbert, Y. Xia, Conductive core-sheath nanofibers and their potential application in neural tissue engineering, Adv. Funct. Mater. 19 (2009) 2312–2318.

[42] D. Fitzpatrick, Implantable electronic medical devices, Elsevier, 2014.

[43] E. Cingolani, J.I. Goldhaber, E. Marbán, Next-generation pacemakers: From small devices to biological pacemakers, Nat. Rev. Cardiol. 15 (2018) 139–150.

[44] M. Kastellorizios, N. Tipnis, D.J. Burgess, Foreign body reaction to subcutaneous implants, Adv. Exp. Med. Biol. 865 (2015) 93–108.

[45] D.A. Koutsouras, A. Hama, J. Pas, P. Gkoupidenis, B. Hivert, C. Faivre-Sarrailh, E. Di Pasquale, R.M. Owens, G.G. Malliaras, PEDOT:PSS microelectrode arrays for hippocampal cell culture electrophysiological recordings, MRS Commun. 7 (2017) 259–265.

[46] Z. Aqrawe, J. Montgomery, J. Travas-Sejdic, D. Svirskis, Conducting polymers for neuronal microelectrode array recording and stimulation, Sensors Actuators, B Chem. 257 (2018) 753–765.

[47] J. Pas, A.L. Rutz, P.P. Quilichini, A. Slézia, A. Ghestem, A. Kaszas, M.J. Donahue, V.F. Curto, R.P. O'Connor, C. Bernard, A. Williamson, G.G. Malliaras, A bilayered PVA/PLGA-bioresorbable shuttle to improve the implantation of flexible neural probes, J. Neural Eng. 15 (2018).

[48] S. Uchiyama, R. Tomita, N. Sekioka, E. Imaizumi, H. Hamana, T. Hagiwara, Application of polymaleimidostyrene as a convenient immobilization reagent of enzyme in biosensor, Bioelectrochemistry. 68 (2006) 119–125.

[49] S.N.A. Nashruddin, J. Abdullah, M.A.S. Mohammad Haniff, M.H. Mat Zaid, O.P. Choon, M.F. Mohd Razip Wee, Label free glucose electrochemical biosensor based on poly(3,4-ethylenedioxy thiophene):Polystyrene sulfonate/titanium carbide/graphene quantum dots, Biosensors. 11 (2021) 267.

[50] J.H. Han, J.D. Taylor, D.S. Kim, Y.S. Kim, Y.T. Kim, G.S. Cha, H. Nam, Glucose biosensor with a hydrophilic polyurethane (HPU) blended with polyvinyl alcohol/vinyl butyral copolymer (PVAB) outer membrane, Sensors Actuators, B Chem. 123 (2007) 384–390.

[51] N. Kumar, L.S.B. Upadhyay, Enzyme immobilization over polystyrene surface using cysteine functionalized copper nanoparticle as a linker molecule, Appl. Biochem. Biotechnol. 191 (2020) 1247–1257.

[52] Q. Gong, Y. Wang, H. Yang, A sensitive impedimetric DNA biosensor for the determination of the HIV gene based on graphene-Nafion composite film, Biosens. Bioelectron. 89 (2017) 565–569.

[53] J. Wang, M. Musameh, Y. Lin, Solubilization of carbon nanotubes by Nafion toward the preparation of amperometric biosensors, J. Am. Chem. Soc. 125 (2003) 2408–2409.

[54] E.S. Place, J.H. George, C.K. Williams, M.M. Stevens, Synthetic polymer scaffolds for tissue engineering, Chem. Soc. Rev. 38 (2009) 1139–1151.

[55] S. Soleymani Eil Bakhtiari, H.R. Bakhsheshi-Rad, S. Karbasi, M. Tavakoli, S.A. Hassanzadeh Tabrizi, A.F. Ismail, A. Seifalian, S. RamaKrishna, F. Berto, Poly(methyl methacrylate) bone cement, its rise, growth, downfall and future, Polymer International 70(9) (2021) 1182–1201.

[56] S.A.P. Sughanthy, M.N.M. Ansari, A. Atiqah, Dynamic mechanical analysis of polyethylene terephthalate/hydroxyapatite biocomposites for tissue engineering applications, J. Mater. Res. Technol. 9 (2020) 2350–2356.

[57] A.M. Díez-Pascual, A.L. Díez-Vicente, Poly(propylene fumarate)/polyethylene glycol-modified graphene oxide nanocomposites for tissue engineering, ACS Appl. Mater. Interfaces. 8 (2016) 17902–17914.

[58] S. Kamel, T. A. Khattab, Recent advances in cellulose-based biosensors for medical diagnosis, Biosens. 10 (2020).

[59] S. Kempahanumakkagari, V. Kumar, P. Samaddar, P. Kumar, T. Ramakrishnappa, K.-H. Kim, Biomolecule-embedded metal-organic frameworks as an innovative sensing platform, Biotechnol. Adv. 36 (2018) 467–481.

[60] F. Guo, X.X. Xu, Z.Z. Sun, J.X. Zhang, Z.X. Meng, W. Zheng, H.M. Zhou, B.L. Wang, Y.F. Zheng, A novel amperometric hydrogen peroxide biosensor based on electrospun Hb–collagen composite, Colloids Surfaces B Biointerfaces. 86 (2011) 140–145.

[61] C. Ulutürk, N. Alemdar, Electroconductive 3D polymeric network production by using polyaniline/chitosan-based hydrogel, Carbohydr. Polym. 193 (2018) 307–315.

[62] K.B.R. Teodoro, R.C. Sanfelice, F.L. Migliorini, A. Pavinatto, M.H.M. Facure, D.S. Correa, A review on the role and performance of cellulose nanomaterials in sensors, ACS Sensors. 6 (2021) 2473–2496.

[63] S.K. Mahadeva, J. Kim, Conductometric glucose biosensor made with cellulose and tin oxide hybrid nanocomposite, Sensors Actuators B Chem. 157 (2011) 177–182.

[64] S. Mun, M. Maniruzzaman, H.-U. Ko, A. Kafy, J. Kim, Preparation and characterisation of cellulose ZnO hybrid film by blending method and its glucose biosensor application, Mater. Technol. 30 (2015) B150–B154.

[65] M. Maniruzzaman, S.-D. Jang, J. Kim, Titanium dioxide–cellulose hybrid nanocomposite and its glucose biosensor application, Mater. Sci. Eng. B. 177 (2012) 844–848.

[66] S. Palanisamy, V. Velusamy, S. Balu, S. Velmurugan, T.C.K. Yang, S.-W. Chen, Sonochemical synthesis and anchoring of zinc oxide on hemin-mediated multiwalled carbon nanotubes-cellulose nanocomposite for ultra-sensitive biosensing of H2O2, Ultrason. Sonochem. 63 (2020) 104917.

[67] F.G. Torres, O.P. Troncoso, K.N. Gonzales, R.M. Sari, S. Gea, Bacterial cellulose-based biosensors, Med. Devices Sensors. 3 (2020) 1–13.

[68] P. Lv, H. Zhou, A. Mensah, Q. Feng, D. Wang, X. Hu, Y. Cai, L. Amerigo Lucia, D. Li, Q. Wei, A highly flexible self-powered biosensor for glucose detection by epitaxial deposition of gold nanoparticles on conductive bacterial cellulose, Chem. Eng. J. 351 (2018) 177–188.

[69] A. Jasim, M.W. Ullah, Z. Shi, X. Lin, G. Yang, Fabrication of bacterial cellulose/polyaniline/single-walled carbon nanotubes membrane for potential application as biosensor, Carbohydr. Polym. 163 (2017) 62–69.

[70] L. Wang, H. Yang, J. He, Y. Zhang, J. Yu, Y. Song, Cu-hemin metal-organic-frameworks/chitosan-reduced graphene oxide nanocomposites with peroxidase-like bioactivity for electrochemical sensing, Electrochim. Acta. 213 (2016) 691–697.

[71] H.S. Choi, X. Yang, G. Liu, D.S. Kim, J.H. Yang, J.H. Lee, S.O. Han, J. Lee, S.W. Kim, Development of Co-hemin MOF/chitosan composite based biosensor for rapid detection of lactose, J. Taiwan Inst. Chem. Eng. 113 (2020) 1–7.

[72] T.A.P. Hai, R. Sugimoto, Surface modification of chitin and chitosan with poly(3-hexylthiophene) via oxidative polymerization, Appl. Surf. Sci. 434 (2018) 188–197.

[73] R.J. Russell, M. V Pishko, C.C. Gefrides, M.J. McShane, G.L. Coté, A fluorescence-based glucose biosensor using concanavalin A and dextran encapsulated in a poly(ethylene glycol) hydrogel, Anal. Chem. 71 (1999) 3126–3132.

[74] Y. Ma, N. Li, C. Yang, X. Yang, One-step synthesis of amino-dextran-protected gold and silver nanoparticles and its application in biosensors, Anal. Bioanal. Chem. 382 (2005) 1044–1048.

[75] J. Li, H. Mei, W. Zheng, P. Pan, X.J. Sun, F. Li, F. Guo, H.M. Zhou, J.Y. Ma, X.X. Xu, Y.F. Zheng, A novel hydrogen peroxide biosensor based on hemoglobin-collagen-CNTs composite nanofibers, Colloids Surfaces B Biointerfaces. 118 (2014) 77–82.

[76] S. Zong, Y. Cao, Y. Zhou, H. Ju, Hydrogen peroxide biosensor based on hemoglobin modified zirconia nanoparticles-grafted collagen matrix, Anal. Chim. Acta. 582 (2007) 361–366.

[77] T.A. Sonia, C.P. Sharma, An overview of natural polymers for oral insulin delivery, Drug Discov. Today. 17 (2012) 784–792.

[78] X. Jing, H.-Y. Mi, J. Peng, X.-F. Peng, L.-S. Turng, Electrospun aligned poly(propylene carbonate) microfibers with chitosan nanofibers as tissue engineering scaffolds, Carbohydr. Polym. 117 (2015) 941–949.

[79] S. V Madihally, H.W.T. Matthew, Porous chitosan scaffolds for tissue engineering, Biomaterials. 20 (1999) 1133–1142.

[80] R. Arun Kumar, A. Sivashanmugam, S. Deepthi, S. Iseki, K.P. Chennazhi, S. V Nair, R. Jayakumar, Injectable chitin-poly(ε-caprolactone)/nanohydroxyapatite composite microgels prepared by simple regeneration technique for bone tissue engineering, ACS Appl. Mater. Interfaces. 7 (2015) 9399–9409.

[81] F. Asghari, M. Samiei, K. Adibkia, A. Akbarzadeh, S. Davaran, Biodegradable and biocompatible polymers for tissue engineering application: a review, Artif. Cells, Nanomedicine, Biotechnol. 45 (2017) 185–192.

[82] T. Liu, J.D. Houle, J. Xu, B.P. Chan, S.Y. Chew, Nanofibrous collagen nerve conduits for spinal cord repair, Tissue Eng. – Part A. 18 (2012) 1057–1066.

[83] X. Zhuang, W. Huang, X. Yang, S. Han, L. Li, J. Yu, Biocompatible/degradable silk fibroin:poly(vinyl alcohol)-blended dielectric layer towards high-performance organic field-effect transistor, Nanoscale Res. Lett. 11 (2016) 439.

[84] A.J. Choudhury, D. Gogoi, J. Chutia, R. Kandimalla, S. Kalita, J. Kotoky, Y.B. Chaudhari, M.R. Khan, K. Kalita, Controlled antibiotic-releasing Antheraea assama silk fibroin suture for infection prevention and fast wound healing, Surgery. 159 (2016) 539–547.

[85] R. Singh, Y.-T. Lin, W.-L. Chuang, F.-H. Ko, A new biodegradable gate dielectric material based on keratin protein for organic thin film transistors, Org. Electron. 44 (2017) 198–209.

[86] H. Li, T.K. Sinha, J. Lee, J.S. Oh, Y. Ahn, J.K. Kim, Melt-compounded keratin-TPU self-assembled composite film as bioinspired e-skin, Adv. Mater. Interfaces. 5 (2018) 1800635.

[87] B.T. Mekonnen, M. Ragothaman, C. Kalirajan, T. Palanisamy, Conducting collagen-polypyrrole hybrid aerogels made from animal skin waste, RSC Adv. 6 (2016) 63071–63077.

24 Advanced Polymers for Biomedical Applications

Sinem Özlem Enginler,[1] Nazlı Albayrak,[2] and Selcan Karakuş[3]

[1] Department of Obstetrics and Gynecology, Faculty of Veterinary Medicine, Istanbul University-Cerrahpasa, Avcılar, 34320, Istanbul, Turkey
[2] School of Medicine, Acibadem M. A. Aydınlar University, Istanbul, 34752, Turkey
[3] Department of Chemistry, Faculty of Engineering, Istanbul University-Cerrahpasa, Avcılar, 34320, Istanbul, Turkey

1 THE FABRICATION AND CHARACTERIZATION OF NEW POLYMER-BASED FORMULATIONS

With the new developments and achievements in science and technology, polymer-based systems with excellent surface and structural properties have started to be in demand to obtain advanced materials such as polymer-based drug-delivery systems, smart energy-storage devices, skin-based wearable devices, printed electronics, and renewable and flexible batteries. In recent years, a wide range of polymeric structure-based materials has been used for different purposes. As an example of these, electrically conductive polymers have recently attracted attention for use in polymeric systems for smart textile and sensor applications. Biodegradable and biocompatible polymers are proposed for innovative and smart polymer-based systems for monitoring and clinical aspects of pharmaceutical applications. In particular, recent studies have been paying increased attention to obtaining water-soluble polymers for food and pharmaceutical applications. In the literature, natural and synthetic polymers have been utilized to develop smart materials in wound dressings, tissue-engineered scaffolds, drug-delivery systems, theranostic systems, contact lenses, soft electronics, and biosensors in nanobiotechnology applications. Previous studies have reported that natural polymer-based systems have excellent biological properties such as being easy to metabolize, non-toxic, biodegradable, and biocompatible with tissues and biological fluids. The majority of U.S. Food and Drug Administration (FDA)-approved doxorubicin-loaded polyethylene glycol (PEG)/liposome (Doxil: PEGylated Liposomal doxorubicin) and paclitaxel-loaded albumin (Abraxane: Albumin-particle bound paclitaxel) are used as anticancer agents.Among natural polymers, chitin, chitosan, heparin, starch, silk fibroin protein, cellulose, alginate, xanthan gum, pectin, guar gum, konjac gum, kappa carrageenan, dextran, agarose, collagen, locust bean gum, hyaluronic acid, pullulan, and gelatin are widely used biopolymers in biomedical applications. On the other hand, the fabrication and characterization of polycarbonates (PC), poly(lactide-co-glycolide) (PLGA), poly(lactide) (PLA), polyanhydrides (PAN), poly(ε-caprolactone) (PCL), PEG, poly (alkyl cyanoacrylates), polyhydroxyalkanoates (PHA), polyacetals (PA), poly(propylene fumarate) (PPF), polyvinyl alcohol (PVA), poly(glycolide) (PGA), polyphosphoester (PPE), polyorthoesters (POEs), poly(diolcitrates), polyurethane (PUR), polyphosphazenes (PPZ), poly(amino acids) (PAAs), poly(methyl methacrylate) (PMMA), poly(trimethylene carbonate) (PTMC), polyether, and their blends have been reported for designing smart materials over the past 20 years.

DOI: 10.1201/9781003278269-24

Many previous studies have shown that advanced polymer-based nanostructures have been prepared using different methods such as controlled radical polymerization, emulsion polymerization, micro-emulsion polymerization, interfacial polymerization, ring-opening polymerization, sonication, microwave, and ball milling for biomedical applications. For polymer-based bio-engineered materials, Lv et al. developed a novel phospholipid biomimetic PC as a hemocompatible drug-stent coating material in biodegradable cardiovascular stent systems [1]. In this study, the ring-opening polymerization method was used to prepare PC using a cyclic carbonate monomer. Experimental results showed that the prepared phospholipid biomimetic system had low fibrinogen adhesion, good anticoagulant property, and an effective feature in the inhibition of smooth muscle cells (SMCs) proliferation. In another study, Duan et al. prepared a novel integrated bilayered porous PLGA-based polymeric scaffold with a small pore size in the range of 100–450 micrometers in the tissue layer [2]. They assessed the *in vitro* prolonged therapeutic efficacy of osteochondral defect repair of biocompatible PLGA-based scaffolds implanted in rabbits. From the experimental results, they proved that bone marrow stromal cells (BMSCs) and PLGA bilayered scaffolds had a high repair potential on the critical size defects of the rabbit knee in high load-bearing sites (at 6 months and 24 weeks).

To date, various common preparation methods have been developed to obtain advanced polymers, including: solution mixing, supercritical fluid technology (SCF), sonochemical, melt mixing, nanoprecipitation, sol–gel process, mechanical milling method, salting-out, spray pyrolysis, dialysis, emulsion, solvent-evaporation, suspension, precipitation, bulk, and dispersion polymerization methods. Previous reports have proved that surfactants, initiators, molecular weight, viscosity, co-stabilizer, and surface tension were significant factors in the architecture, shape, and size distributions of nanostructures. Different surfactants such as tween 80, tween 20, betaines, amino oxides, poloxamer 188, vitamin E TPGS, sodium dodecyl sulfate (SDS), and polysorbate 20 have commonly been used for the formation of polymer-based core–shell nanoparticles, nanocapsules, nanospheres, and nanotubes. As is well known, different experimental factors such as concentration, solvent, sonication factors, cross-linking agent, time, injection rate, polymer blend ratio, ultrasonic amplitude, agitation rate, pH, extraction time, and temperature affect the morphology and biological properties of micron and sub-micron polymer-based structures. Homogeneous size and shape distributions of structures in the polymeric matrix are significant problems encountered in the fabrication of advanced polymers. Moreover, it is a very critical issue to select the most suitable approach for the preparation method of advanced polymers with reduced problems of safety hazards and low manufacturing costs. In addition, recent trends in digital manufacturing technology have led to a proliferation of studies on constructing three-dimensional (3D) nanostructures in the Industry Revolution (IR) 4.0. The rapid advances in 3D printing systems have focused on the development of innovative nanostructures to improve the biocompatibility, functionality, and biological safety of existing conventional drugs or devices. Targonska et al. fabricated a novel filler and polymeric blend-based bone implant using a 3D-printed technique with bioresorbability and biodegradability properties. The composite including poly(L-lactide-co-D,L-lactide) (PLDLLA), and small hydroxyapatite particles was prepared using the melt-mixing method at 210°C [3].

2   SENSOR APPLICATIONS OF NEW POLYMER-BASED FORMULATIONS

The advanced polymer/biopolymer-based nanosystems are used in different biomedical fields such as antimicrobial systems, regenerative medicine, tissue engineering, gene-delivery systems, polymer-based biosensor, cell imaging systems, nano-carriers in cancer therapy, and targeting cancer stem cells with nano-therapy. In Figure 24.1, a schematic diagram of advanced polymer-coated biosensors is provided with different analytes, bioreceptors, polymer-coated transducers, and polymer-based coating materials.

With the nanotechnological approach, cancer, which is a deadly disease, is increasingly recognized as a serious public health problem worldwide. Cancer, unlike infectious and environmental diseases,

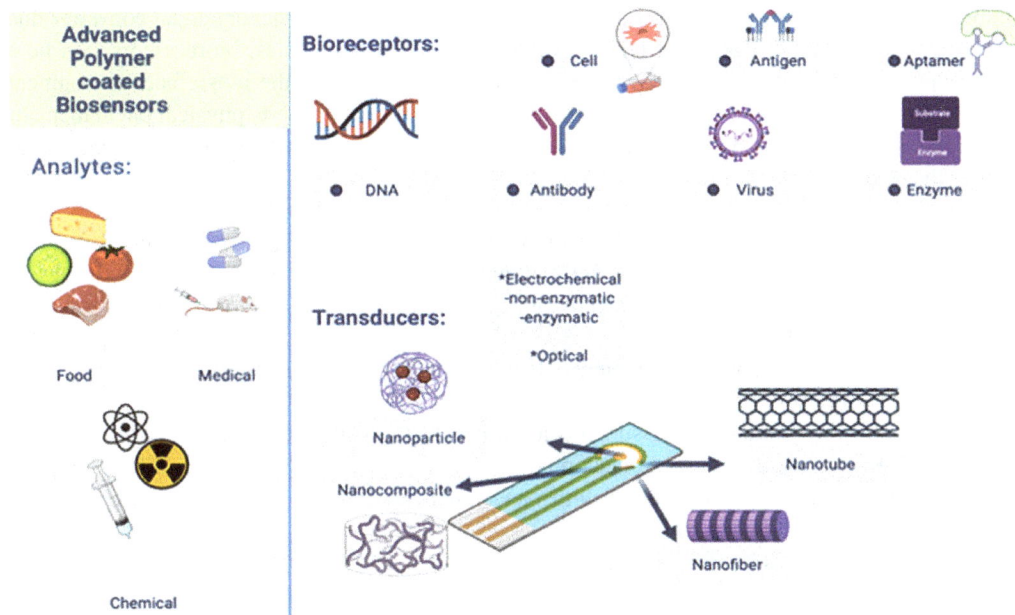

FIGURE 24.1 Advanced polymer-based biosensors.

is not caused by an organism; the tumor cells are part of the body creating the disease. Each year, over 11 million people are diagnosed with cancer, and this number is increasing. A tumor is an abnormal cell growth without any functional purpose. Since cancer is life threatening and one of the major causes of death worldwide, detection and treatment strategies are important in the research fields. Besides traditional diagnosis and treatment strategies, biosensor technology is a rising technology as a feature of nanotechnology in this field. The use of biosensors in the diagnosis of cancer is helps to monitor or detect more than one tumor marker and creates the opportunity to reveal more precise results to identify tumor cells earlier. In addition, the main achievement of this nanotechnology will be to be to understand the pathophysiology and disease progression. Cancer can spread to healthy body parts, therefore the early diagnosis and detection of the progression of cancer according to the treatment outcome can reduce the mortality rate from this disease. Today, there are many diagnostic methods for tumor diagnosis that are time-consuming, require experienced technicians, and can be very expensive, such as X-ray, computerized tomography, positron emission tomography, ultrasonography, endoscopy, thermography, and cytology. Tumor markers (biomarkers) are the products of normal cell metabolism and their production increases due to malignancy. They can be detected in blood, serum, urine, cerebrospinal fluid, or tumoral cells. Molecular tools which are sensitive to specific cancer biomarkers such as enzyme-linked immunosorbent assay (ELISA), polymer chain reaction (PCR), immunohistochemistry (IHC), radioimmunoassay (RIA), and flow cytometry require time for the results, an instrumented laboratory environment, and expensive kits to ensure accurate results. New developed biosensors in recent years, by enabling the detection of tumor biomarkers even at very low concentrations such as altered genomic, circulating tumor DNA, microRNA, cancer antigens, cytokines, matrix metalloproteins, exosomes, and circulating tumor cells, help in the early diagnosis of cancers and increase the efficacy of treatment for patients.

In the literature, various electrochemical measurement methods, such as impedimetric techniques, amperometry, field-effect transistors, capacitive sensing systems, conductometric sensing systems, potentiometry, voltammetry, chronopotentiometry, bio-surface, and receptors used in this field, have been described. Polymer-based biosensors are used as smart analytical devices that are the combination of target analytes of origin, a signal transducer, and a physicochemical converter due to the measurable signal. These include biological recognition molecules, bioreceptors, can be a receptor protein, antigen, antibody, aptamer, nucleic acid, or enzyme; the biosurface environment provides proper functioning of the bioreceptor; a transducer which converts physical phenomena or chemical answers into readable signals; signal amplifier, signal processor, visualization, and evaluation of the data.

1. Receptor recognition element: Activation or inactivation of receptor molecules induces various signaling events within the cell that may be beneficial for biosensors. The efficacy of cancer therapeutics can be examined by these receptors on the cell surfaces. The receptor can be composed of these units.
2. Antigen/antibody recognition elements: These elements act like a "lock-key" binding mechanism in biosensors.
3. Enzymes: Mostly, the catalytic subunits are transducers, while the regulatory subunit has the role of recognition in enzymatic reactions.
4. DNA: Gene mutations associated with cancer disease can be detected very precisely by biosensors that recognize DNA.
5. Aptamers: Highly selective oligonucleotides (aptamers) are very useful for the detection of cancer biomarkers.
6. Lectins: These are proteins that are found usually in plants and are very beneficial as a recognition element of glycoprotein-based tumor markers for biosensors.
7. Peptides: These are proteins that are composed of 10–20 amino acids with high binding affinity to biosensors.

The transmission mode of biosensors consists of a variety of approaches, including (1) electrochemical (amperometric and potentiometric), (2) optical [(a) colorimetric, (b) fluorescent, (c) luminescent, (d) interferometric], (3) mass measurement [piezoelectric and acoustic wave], and (4) calorimetric (temperature-based). Various biosensors in the literature use polymer-based enzyme biosensors, polymer-based immunosensors, polymer-based DNA biosensors, polymer-based whole-cell biosensors, and biosensors based on molecularly imprinted polymers. There has been a study with electrochemical DNA biosensors in which single-stranded DNA molecules attached to it for BRCA1 sensing were applied successfully to human plasma samples for early detection of breast cancer in women[4].

Recent developments in the detection of canine mammary tumors have promoted using advanced sensor systems. Canine mammary tumors are among the most common tumor types, with about half of them diagnosed as malignant. In dogs, it is possible to diagnose different types of mammary tumors in different mammary glands on the same animal. There are several risk factors for the formation of canine mammary tumors, such as hormonal, nutritional, and genetic factors. Especially, early spaying has an effect that prevents the formation of mammary tumors in dogs due to hormonal effects. Usually, tumors in the mammary glands are noticed by their owners when they increase in size. However, the possibility of metastasis to other organs increases due to tumor growth. The prognosis of canine mammary tumors depends on the size of the mammary tumor, distant metastasis, clinical stage, and lymph node status. Early diagnosis of tumors is very important for the patient's prognosis. Determination of biomarkers in dogs is very important for early diagnosis of tumor formation, the evaluation of tumor progression, and the animal's response to chemotherapy. It has been proven that human breast cancer is immunohistochemically similar to that of canine mammary tumors. Therefore, biomarkers used in humans can also be used in the diagnosis of canine mammary tumors. Anemnesis, physical examination, and palpation of the mammary gland, and hematologic and serum chemistry profiles are the first step in the diagnosis. Thoracic radiography and ultrasonography can be used to detect possible lung and internal organ metastases.

Fine-needle aspirates can be performed from the mammary gland and suspicious lymph nodes. The most important diagnosis technique for mammary tumors which remains the most preferred is the histopathological examination of excised specimens after mastectomy. Target biomarkers in canine mammary cancer zones have been preferred for their effects on cancer grading, early diagnosis of cancer, the prognosis of cancer cells, and monitoring of the response to treatment. Various target biomarkers have been reported on apoptosis, proliferation, metastasis, angiogenesis, inflammation, cancer stem cells, hormone receptors, mutagenic genes, MiRNA, and tumoral cells in circulation for the detection of canine mammary cancer treatment. Kaszak et al. reported that tumor markers can be used to identify the tumor in the early stages [5]. However, to obtain a result from these biomarkers immunohistochemical labeling and/or ELISA technique are needed. These techniques require a laboratory environment, and are very costly. Recently, biosensors, which provide ease of use in terms of cheap, easy, fast, and reliable results, have been produced for use in the early diagnosis of human cancers. Due to the limited number of studies on this subject in the literature, these biosensors also could be a good choice for early canine mammary tumor diagnosis by using these tumor biomarkers. There has been a report about a surface plasmon resonance (SPR) immunosensor to detect the baculoviral inhibitor of apoptosis repeat containing-5 (BIRC5) protein as a prognostic biomarker in the sera of dogs with mammary tumors. According to the results of this study, detection of BIRC5 in dogs' serum samples with SPR immunosensor could be a good tool for diagnosing mammary tumors in dogs. As this sensor is automated, it could be used for scanning a large number of samples, and they have suggested that this SPR biosensor system could be transformed into a portable biosensor for use with patients in a clinic in the near future.

There are several prognostic factors used to detect the prognosis of canine mammary tumors; these are tumor size and staging, lymph node involvement, distant metastasis, histological type, lymphoid cellular activity around the tumor, steroid receptors, expression of oncogenes, tumor suppressors, and adhesion molecules, proliferation markers, microvessel density, and circulating tumor cells. There have been no studies on prognostic factors related to tumor progression and patient survival. The detection of circulating tumor cells includes mostly cytometric and cost-effective nucleic-acid-based techniques, such as reverse transcription-polymerase chain reaction (RT-PCR). To date, several tumor markers have been investigated to detect the progression of canine mammary tumors. However, these studies have shown that different results can be obtained in studies with the same marker. In biosensor applications, it is one of the most recognized tumor biomarkers which is used for screening prostate cancer. Generally, the most widely preferred nanocarriers for the detection of the tumor biomarker for cancer treatment are polymeric nanocarriers (polymeric nanogels, polymeric micelles, dendrimers, polymeric nanocapsules, and polymersomes), which allow covalent binding of specific antibodies. It has been reported that above 4.0 ng/ml of this biomarker level is related to prostate cancer, however higher PSA levels can be found in other diseases such as prostatitis, benign prostatic hyperplasia, and small tumoral formations that are not deadly [6]. The detection of Cancer Antigen 125 (CA 125), known as cancer antigen, has been gaining momentum using advanced biopolymer/polymer-based biosensors in recent years. A high level of this biomarker can be related to lung, uterus, cervix, pancreas, liver, colon, breast, and digestive tract cancer, and the progression of ovarian cancer, while its level can be increased in pregnancy and menstruation [7]. Alfa-fetoprotein (AFP), human chorionic gonadotropin, and lactate dehydrogenase are other biomarkers used to detect ovarian cancer in women. AFP, a biomarker that is a protein produced in the liver of the developing fetus, is of interest in biosensor studies. It is an important biomarker of hepatocellular carcinoma of the liver, ovary, and testicular cancers [8]. Other biomarker studies used for early detection in the field of breast cancer include estrogen receptor (ER)/progesterone receptor (PR), which are the biomarkers of breast tumor. Estrogen can promote tumorigenesis and is very important in detecting the results of breast tumor treatment [9].

Endometriosis is a common gynecologic disease that is seen mostly during the patient's reproductive years. The pathophysiology behind this disease is having the endometrial tissue growing outside the uterus where it behaves just like the endometrium of the uterine cavity. Kalyani et al. recently studied carbohydrate antigen 19-9 (CA19-9) as a potential marker of endometriosis for its diagnosis [10]. Biosensor technology in this specific disease senses CA19-9 in the physiological range and suggests application for early-stage diagnostics, monitorization, and optimization of treatment [10].

Human papilloma virus (HPV) is the most common sexually transmitted disease causing cervical cancer in women. Since cervical cancer is the third most common female cancer in the world, the detection and treatment of this particular virus have gained increased attention. There is a screening program for the detection of cervical cancer in the early onset with cytological methods. With the use of nanotechnology, Espinosa et al. recently suggested a rapid electrochemical resistive and low-cost DNA biosensor [11]. Since there are many subtypes of the HPV virus, this technology suggests a DNA probe specific to one subtype for each. The simple electrochemical resistive biosensor is mentioned as a useful option for the diagnosis of HPV at low cost and in a short time.

Uterine leiomyomas (fibroids) have affected women's life quality since antiquity, causing heavy uterine bleeding, abdominal pain, and sometimes also bowel problems due to compression on the intestines related to the location and size of the leiomyoma. During pregnancy, leiomyomas may grow, and degenerate, becoming calcified and resulting in great pain for the pregnant women. which may result in them going to hospital due to confusing the pain with a delivery signal. Since this particular disease is an important complaint of many of women, studies are on-going for its treatment without surgery. There is one option of medical treatment rather than the surgery which involves surpassing the progesterone and estrogen hormones via gonadotropin-releasing hormone agonists. Although there is no specific study in the literature related to the role of biosensor technology in leiomyoma treatment, focusing on the gonadotropin-releasing hormone for this technology might suggest further investigations [12].

There are both benign and malignant ovarian masses described in the literature. While malignant ones may be fatal, benign ovarian or adnexal masses usually need follow-ups to prevent them from becoming malignant. Besides ultrasonography, tumor markers also are used frequently for adnexal masses. However, the main problem with this is that the tumor markers can be increased nonspecifically both in malignant and benign masses. Nanomaterial biosensor technology might suggest differentiating the nature of the mass through these markers. For the treatment, surgery is the first option for the adnexal masses both for pathological diagnosis and also for debulking. However, sometimes surgery may not be enough for the residues of the mass, so chemotherapeutic medication is used to ensure no malignant cells remain. Biosensor technology also has an emerging important role in chemotherapeutics in this field.

Bacteria are an important reservoir of biosensing proteins. Chamas et al. in 2017 created three *Arxula adeninivorans* yeast strains where each strain expressed a different recombinant human hormone receptor and reporter protein [13]. The use of different reporter proteins may allow instant incubation of a sample with strains producing various receptor proteins to detect estrogen, progestogens, and androgens instantly.

3　DRUG-DELIVERY SYSTEMS OF NEW POLYMER-BASED FORMULATIONS

Recent studies have focused on the development of novel polymer/biopolymer-based drug-delivery systems with high drug performance, low side effects, and low patient compliance. Smart drugs have been fabricated using different polymer-carriers such as polymer blends, biopolymer hydrogels, polymer clay composites, polymer nanofibers, polymer nanocomposites, polymer nanogels, liposome–polymer complexes, polymeric nanoparticles, additive–polymer hybrid systems, polymeric films, polymer-based core–shell nanocarriers, and polymeric microneedles for transdermal drug delivery for targeted delivery of chemotherapeutic drugs using polymer-based magnetic nanoparticles for micro- and nanoscale drug-delivery systems. Recent studies have shown that polymer networks in drug-delivery systems have excellent properties with high therapeutic effects, such as high drug efficacy, controlling the drug rate, high solubility, high permeability, use of low dosages of drugs, and high biocompatibility and biodegradability. Moreover, these biodegradable polymeric nanocarrier systems have also overcome the problem of low drug loading and solubility of hydrophobic drugs. Key developments and advances in these drug-delivery systems have allowed the fabrication of novel and more efficient therapeutics that are used in different treatments. From cancer to inflammatory diseases, many new strategies using different smart therapeutic nanoagents, such as nanotherapeutic drugs, antibody-conjugated nanoparticles, and nano-gene delivery systems, have been reported. In this regard, the fabrication of promising polymeric chemotherapy drugs using green strategies has been gaining attention in biomedical applications. Generally, the biopolymer-based matrix is widely used in the development of therapeutic nanoagents. Nevertheless, scientific studies are currently underway to improve the compatibility between hydrophobic drugs and biopolymer carriers, and to determine the mechanisms of pharmacokinetic interactions and profiles.

In the literature, Khodarahmi developed a novel liposome/alginate/chitosan complex-based carrier for targeted 5-fluorouracil delivery systems with high biocompatibility, high biodegradability, and low toxicity properties [14]. Interestingly, the characterization results exhibited that these nanosystems had spherical shapes and small sizes (<200 nm). The unique morphology of the prepared nanosystems showed that they had a high endocytosis rate and easily penetrate cancer zones with high permeability and retention (EPR) properties. In another study, Allafchian et al. fabricated a novel PVA–carboxymethyl cellulose (CMC) nanofiber-based nanocarrier for a flufenamic acid drug-delivery system using the electrospinning method [15]. The surface characterization results showed that the prepared nanofibers had a small size in nanofiber diameters, ranging from 176 nm to 285 nm. These results demonstrated that the polymer matrix had a role in the uniform distribution of the nanocarrier and the increased drug release system. In 2021, Yan et al. reported that pH-sensitive PVA/polycaprolactone (PCL) blend-based core–shell nanofibers are promising doxorubicin (DOX)-delivery systems for cervical cancer therapy [16]. According to the characterization results, it was observed that these nanosystems have a core–shell structure with small particle sizes at approximately 200 nm and with a shell thicknesses ranging from 51 nm to 85 nm. *In vitro* drug release results showed that the morphology of the polymer-based nanocarrier was a major factor in controlling the cumulative DOX release systems and induction of apoptosis in cancer cells. With successive increases in performance of the polymer-based nanocarriers, Abouzeid et al. fabricated green PEG- phosphatidylethanolamine (PE)–vitamin E-based micelles for paclitaxel- and curcumin-delivery systems in ovarian cancer treatments [17]. In this study, it was assumed that the use of polymeric blend systems with the combination of paclitaxel and curcumin therapy would be an effective solution for the multi-drug resistance with cytotoxic effects in cancer cell lines. According to the drug release results, it was observed that the polymeric matrix had a role in the stability of the paclitaxel- and curcumin-loaded carriers and the slow-release profile of the system. Ehsanimehr et al. prepared a novel pH-sensitive polyacrylamide grafted nanocrystalline cellulose-based nanocarrier for targeted DOX-delivery systems on MCF7 cancer cells for breast cancer therapy [18]. In summary, the release profile of the prepared DOX-loaded polyacrylamide grafted nanocrystalline cellulose-based nanocarrier was examined in different pH media, such as pH = 5 and pH = 7.4. Experimental results showed that the novel nanocarriers had good biocompatibility, high drug penetration, and a rapid DOX release profile at pH = 5. Recently, Shalabalija and co-workers [19] developed a multifunctional nanomaterial using bioinspired rosemary extract-loaded PEG-based nanostructure for Alzheimer's disease therapy. According to the experimental results it was observed that the multifunctional nanostructure had a small particle size of ~120 nm with uniform distribution, negative zeta potential values ranging from −18.50 to −48.3 mV, a high drug encapsulation efficiency (~90%) at 24 h, and high antioxidant capacity. The most interesting finding was that the proposed polymeric matrix had a significant function on the *in vitro* stability experiments [19].

In tissue applications, novel nano-ZnO and nano-bioactive glass-based chitosan/PVA bionanocomposites as an extracellular matrix for bone tissue engineering have been reported by Christy et al [20]. These bionanocomposites were prepared using the sol–gel method and the anti-bacterial activity of the prepared nanostructure was examined against different pathogens such as *Salmonella typhi* and *Enterococcus faecalis*. The results showed that the prepared bionanocomposites had an antibacterial inhibition activity for *Enterococcus faecalis* and high blood compatibility with a hemolytic ratio of less than 2%. In another study, Hu et al. fabricated a novel chitosan/gelatin blend-based nano-hydroxyapatite-bone as a multilayer scaffold to repair cartilage regions in osteochondral tissue engineering [21].

In 2021, Luo et al. highlighted the high potential performance of functionalized polymer scaffolds for the repair of spinal cord injuries with advanced clinical therapy strategies [22]. Also in this review article, they presented that 3D printing supported the preparation of novel scaffolds using different polymers and combinations of polymer blends for damaged spinal cords. These experimental results exhibited that the novel biopolymer blend-based material has a role in the improvement of cell regulation for osteochondral regeneration. In summary, according to previous *in vivo* and *in vitro* studies, biopolymer blend-based materials have been shown to be green promising nanomaterials with excellent biological, chemical, and mechanical properties for future clinical applications.

4   CELL CULTURE APPLICATIONS OF NEW POLYMER-BASED FORMULATIONS

The term "tissue culture" is defined as cell culture and organ culture. Tissue culture allows the possibility of controlling the cell's environment such as pH, temperature, osmotic pressure, oxygen, and carbon dioxide. After a few passages, the cell lines show an exemplary structure and as the cells randomly mix during each passage, the most dominant of the cells begin to be grown in the cell culture medium. One of the innovations brought about by cell culture to science is that it allows many studies to be conducted in this prepared environment rather than in animal experiments. Thus, it does not require the permission of an ethics committee and allows to obtain results from studies conducted cheaply and quickly. In tissue culture, there may be two types of cells that grow as a suspension and adherent cell culture. After cell proliferation occurs, cells become a cluster. A subculture is the cell strain that remains after cells are selected from a culture. These strains can be used to provide a more in-depth description. Various cell types can be cultured from various cell lineages such as mammary gland duct cells, cells from the gastrointestinal tract, macrophages, and lymphoid cells for a hemopoietic system, etc. Cell sources can be obtained from cell banks or the cell stocks of previous workers in this field, or they can be produced from the original tissue. Cell cultures are subjected to many tests in terms of viability, contamination, viral, drug ,and cancer susceptibility, and tumorigenicity, and good-quality cell cultures are produced by many successful companies working in this field. Laminar flow cabinets, incubators, inverted microscopes, counting chambers, disposable plastic vessels, spinner vessels, culture bottles with filters, water purifiers, etc. may be needed for a laboratory to carry out a cell culture. Since mammary tumors are very common in dogs and chemotherapy drugs are not as effective in their treatment as in humans, cell culture provides an important model to study this disease. Cell lines allow *in vitro* investigation of human breast cancer and canine mammary tumors to understand processes such as proliferation, apoptosis, and migration in cancer. This is a model that allows the identification of cell signaling pathways, effective genes in cancer such as oncogenes and tumor suppressors, evaluation of the action mechanisms of drugs thought to be effective in cancer, and the discovery of new drugs. According to Uyama et al.'s study using four novel canine mammary tumor cancer cell lines derived from primary and metastatic lesions, reduced E-cadherin function is related to the invasion and metastasis potential of canine mammary gland tumor cells [23]. In 2021, Li et al. focused on the *in vitro* imaging of a therapeutic drug for a new canine mammary cancer cell line (B-CMT) in canine mammary gland tumors [24]. According to the experimental results, it was found that B-CMT cell lines were neoplastic in mice but they did not observe the possibility of detectable metastasis and they also found that rapamycin and imatinib had a significant effect on B-CMT cell cycle arrest. As shown by the compiled results of a number of studies, cell culture studies in canine mammary tumors in recent years are promising for the future. However, there is an opinion that studies on this subject should be increased. A cancer stem cell (CSC) can be described as an undifferentiated cell that can renew itself and be differentiated into multiple lineages. They can start tumors that mimic the parent tumor. As tumors are heterogeneous, they have very few cancer stem cells. However, the origin of cancer stem cells is unclear; there are two proposals about their origin, i.e., that they can be derived from adult stem cells or mutations in a progenitor. As is well known, cancer stem cells have many roles such as the initiation of tumors, angiogenesis, maintenance, and metastasis. They are less affected by radiation and chemotherapy treatments than the main tumor population. Another important report was that by Gilbert and Ross who mentioned that they could use their extensive knowledge of the signaling pathways required for stem cells during cell development to overcome the resistance to evidence-based medical treatments, and additionally added that there have been advances in cancer stem cell cultures, although the heterogeneity of the tumor cells complicates the purification of CSCs [25]. Cancer stem cells vary significantly from tumor to tumor. Although canine mammary tumors are among the most frequently encountered tumor types, their formation mechanisms have not been fully elucidated until recently. Studies conducted in recent years suggest that cancer stem cells may also be effective in the formation of canine mammary tumors. Understanding CSCs is crucial for targeting new therapeutic strategies for early detection of this disease and improving diagnostic biomarkers. Michishita mentioned in 2020 that the genetic changes in the canine mammary gland, especially those in cancer stem cells, are not fully understood. It has been reported that cancer stem cells isolated from tumor specimens can be evaluated after serial transplantation by serial dilution xenotransplantation, *in vitro* replication tests of spheres, *in vitro* susceptibility tests with anticancer drugs, and radiation. CSCs have been investigated in *in vitro* studies through the expression of different biomarkers, such as CD44 and CD24, for canine mammary adenocarcinoma treatment [26]. Cancer studies conducted in recent years have shown that cancer stem cells, which are rare, have an important role in the initiation, development, and progression of the tumor and even in its recurrence. These results indicate how important canine mammary tumor studies are in understanding human mammary tumors.

Li et al. developed stem cells labeled with supported gold nanoparticles/polyethylenimine @ PEG nanotracers as an *in vivo* imaging strategy to determine the behavior of human transplanted mesenchymal stem cells for pulmonary fibrosis therapy in stem cell applications [27]. The experimental results showed that the biocompatible polyethylenimine @ PEG polymer blend had a role in the good biocompatibility, uniform morphology, small hydrodynamic diameter of 58.33 nm, and cell labeling efficiency. Sladkova et al. reported that green PEG particle-based macroporous calcium phosphate cements bone scaffolds for human induced pluripotent stem cell-based mesenchymal progenitors [28]. Surprisingly, they demonstrated that polymeric nanoparticles affected the porosity, pore size, and mechanical properties of the material. In another study, Ghandforoushan et al. fabricated biodegradable gelatin/PLGA – PEG hydrogels embedded with transforming growth factor- β1 (TGF-β1) for cartilage injury therapy in human stem cell applications [29]. They observed that the proposed biodegradable and thermosensitive polymer ternary blend was a promising biomimetic nanocarrier for cartilage tissue applications. In 2022, Ozbay et al. reported mitochondria-targeted PLGA–PEG–triphenylphosphonium nanoparticles with CoQ10 nephropathies [30]. The main advantage of the prepared nanocarrier was that the structure showed a small size of 150 nm and a positive surface charge of + 20 mV as a green and promising strategy in CoQ10-related nephropathy treatments. Consequently, from macro- to nano-systems, these advanced polymer-based systems deserve increased investigation to enable an advanced strategy and design materials in biomedical applications.

5 CONCLUSIONS AND FUTURE PERSPECTIVES

Over the past decades, advanced polymers such as polymer engineering, polymer matrix composites, polymer nanocomposites, carbon fiber, engineering plastic, nanofibers, and glass fiber have been fabricated to develop advanced diagnostic strategies and sensor approaches. Regarding the comprehensive literature searches, it has been revealed that many reports have focused on the biomedical applications of polymer-based nanostructures to improve the therapeutic efficiency, reduce toxicity, and have side effects of therapeutic agents. From all the above-presented biocompatible polymers, various green polymeric nanostructures have been developed using the top-down and bottom-up approaches with rapid metabolism, high encapsulation efficiency, and high drug loading of the polymer-based nanoformulations. In this chapter, we have reviewed the novel advanced polymeric structures fabricated for biomedical applications. The use of polymeric materials has many attractive features, such as: (a) biological properties; (b) unique surface properties; and (c) high biocompatibility and biodegradation for novel pharmaceutical materials and electronic devices. Various polymer- or polymer blend-based structures have been developed for biosensors and drug-delivery systems to improve the efficiency of agents in cumulative drug release, sensing, and imaging systems in biomedical applications. In addition they have been used as nanocarriers for therapeutic polymer-based nanoagents in the treatment of cancer with synergistic anticancer effects. Some earlier studies focused on the use of biosensors for the detection of target molecules, toxic substances, pathogens, early diagnosis, and treatment with a small particle size under different media conditions, such as temperature, pH, redox conditions, and ionic strength. This review has provided a detailed overview of the fabrication and characterization of new polymer-based formulations with different applications such as sensors, drug-delivery systems, and cell culture applications. In this context, new polymeric nanocarriers present performance efficiency and safety considerations for future clinical applications with green strategies.

REFERENCES

[1] D. Lv, P. Li, L. Zhou, R. Wang, H. Chen, X. Li, Y. Zhao, J. Wang, N. Huang, Synthesis, evaluation of phospholipid biomimetic polycarbonate for potential cardiovascular stents coating, Reactive and Functional Polymers. 163 (2021) 104897. https://doi.org/10.1016/J.REACTFUNCTPO LYM.2021.104897.

[2] P. Duan, Z. Pan, L. Cao, J. Gao, H. Yao, X. Liu, R. Guo, X. Liang, J. Dong, J. Ding, Restoration of osteochondral defects by implanting bilayered poly(lactide-co-glycolide) porous scaffolds in rabbit joints for 12 and 24 weeks, Journal of Orthopaedic Translation. 19 (2019) 68–80. https://doi.org/10.1016/J.JOT.2019.04.006.

[3] S. Targonska, M. Dobrzynska-Mizera, M. Wujczyk, J. Rewak-Soroczynska, M. Knitter, K. Dopierala, J. Andrzejewski, R.J. Wiglusz, New way to obtain the poly(L-lactide-co-D,L-lactide) blend filled with nanohydroxyapatite as biomaterial for 3D-printed bone-reconstruction implants, European Polymer Journal. 165 (2022) 110997. https://doi.org/10.1016/J.EURPOLYMJ.2022.110997.

[4] S. Shahrokhian, R. Salimian, Ultrasensitive detection of cancer biomarkers using conducting polymer/electrochemically reduced graphene oxide-based biosensor: Application toward BRCA1 sensing, Sensors and Actuators B: Chemical. 266 (2018) 160–169. https://doi.org/10.1016/J.SNB.2018.03.120.

[5] I. Kaszak, A. Ruszczak, S. Kanafa, K. Kacprzak, M. Król, P. Jurka, Current biomarkers of canine mammary tumors, Acta Veterinaria Scandinavica 2018 60:1. 60 (2018) 1–13. https://doi.org/10.1186/S13 028-018-0417-1.

[6] M.R. Safarinejad, Population-based screening for prostate cancer by measuring free and total serum prostate-specific antigen in Iran, Annals of Oncology. 17 (2006) 1166–1171. https://doi.org/10.1093/ANNONC/MDL087.

[7] P. Samadi Pakchin, M. Fathi, H. Ghanbari, R. Saber, Y. Omidi, A novel electrochemical immunosensor for ultrasensitive detection of CA125 in ovarian cancer, Biosensors and Bioelectronics. 153 (2020) 112029. https://doi.org/10.1016/J.BIOS.2020.112029.

[8] G.J. Mizejewski, Biological role of α-fetoprotein in cancer: prospects for anticancer therapy, http://Dx.Doi.Org/10.1586/14737140.2.6.709. 2 (2014) 709–735. https://doi.org/10.1586/14737140.2.6.709.

[9] H. Hua, H. Zhang, Q. Kong, Y. Jiang, Mechanisms for estrogen receptor expression in human cancer, Experimental Hematology & Oncology 2018 7:1. 7 (2018) 1–11. https://doi.org/10.1186/S40164-018-0116-7.

[10] T. Kalyani, A. Sangili, A. Nanda, S. Prakash, A. Kaushik, S. Kumar Jana, Bio-nanocomposite based highly sensitive and label-free electrochemical immunosensor for endometriosis diagnostics application, Bioelectrochemistry. 139 (2021) 107740. https://doi.org/10.1016/J.BIOELECHEM.2021.107740.

[11] J.R. Espinosa, M. Galván, A.S. Quiñones, J.L. Ayala, V. Ávila, S.M. Durón, Electrochemical resistive DNA biosensor for the detection of HPV type 16, Molecules 2021, Vol. 26, Page 3436. 26 (2021) 3436. https://doi.org/10.3390/MOLECULES26113436.

[12] J.H. Segars, A. Al-Hendy, Uterine leiomyoma: new perspectives on an old disease, Seminars in Reproductive Medicine. 35 (2017) 471–472. https://doi.org/10.1055/S-0037-1606569/BIB.

[13] A. Chamas, H.T.M. Pham, K. Jähne, K. Hettwer, S. Uhlig, K. Simon, A. Einspanier, K. Baronian, G. Kunze, Simultaneous detection of three sex steroid hormone classes using a novel yeast-based biosensor, Biotechnology and Bioengineering. 114 (2017) 1539–1549. https://doi.org/10.1002/BIT.26249.

[14] M. Khodarahmi, H. Abbasi, H. Kouchak, M. Mahdavinia, S. Handali, N. Rahbar, Nanoencapsulation of aptamer-functionalized 5-Fluorouracil liposomes using alginate/chitosan complex as a novel targeting strategy for colon-specific drug delivery, Journal of Drug Delivery Science and Technology. 71 (2022) 103299. https://doi.org/10.1016/J.JDDST.2022.103299.

[15] A. Allafchian, H. Hosseini, S.M. Ghoreishi, Electrospinning of PVA-carboxymethyl cellulose nanofibers for flufenamic acid drug delivery, International Journal of Biological Macromolecules. 163 (2020) 1780–1786. https://doi.org/10.1016/J.IJBIOMAC.2020.09.129.

[16] E. Yan, J. Jiang, X. Yang, L. Fan, Y. Wang, Q. An, Z. Zhang, B. Lu, D. Wang, D. Zhang, pH-sensitive core-shell electrospun nanofibers based on polyvinyl alcohol/polycaprolactone as a potential drug delivery system for the chemotherapy against cervical cancer, Journal of Drug Delivery Science and Technology. 55 (2020) 101455. https://doi.org/10.1016/J.JDDST.2019.101455.

[17] A.H. Abouzeid, N.R. Patel, V.P. Torchilin, Polyethylene glycol-phosphatidylethanolamine (PEG-PE)/vitamin E micelles for co-delivery of paclitaxel and curcumin to overcome multi-drug resistance in ovarian cancer, International Journal of Pharmaceutics. 464 (2014) 178–184. https://doi.org/10.1016/J.IJPHARM.2014.01.009.

[18] S. Ehsanimehr, P. Najafi Moghadam, W. Dehaen, V. Shafiei-Irannejad, Synthesis of pH-sensitive nanocarriers based on polyacrylamide grafted nanocrystalline cellulose for targeted drug delivery to folate receptor in breast cancer cells, European Polymer Journal. 150 (2021) 110398. https://doi.org/10.1016/J.EURPOLYMJ.2021.110398.

[19] D. Shalabalija, L. Mihailova, M.S. Crcarevska, I.C. Karanfilova, V. Ivanovski, A.K. Nestorovska, G. Novotni, M.G. Dodov, Formulation and optimization of bioinspired rosemary extract loaded PEGylated nanoliposomes for potential treatment of Alzheimer's disease using design of experiments, Journal of Drug Delivery Science and Technology. 63 (2021) 102434. https://doi.org/10.1016/J.JDDST.2021.102434.

[20] P.N. Christy, S.K. Basha, V.S. Kumari, Nano zinc oxide and nano bioactive glass reinforced chitosan/poly(vinyl alcohol) scaffolds for bone tissue engineering application, Materials Today Communications. 31 (2022) 103429. https://doi.org/10.1016/J.MTCOMM.2022.103429.

[21] X. Hu, S. Zheng, R. Zhang, Y. Wang, Z. Jiao, W. Li, Y. Nie, T. Liu, K. Song, Dynamic process enhancement on chitosan/gelatin/nano-hydroxyapatite-bone derived multilayer scaffold for osteochondral tissue repair, Materials Science and Engineering: C. (2022) 112662. https://doi.org/10.1016/J.MSEC.2022.112662.

[22] Y. Luo, F. Xue, K. Liu, B. Li, C. Fu, J. Ding, Physical and biological engineering of polymer scaffolds to potentiate repair of spinal cord injury, Materials & Design. 201 (2021) 109484. https://doi.org/10.1016/J.MATDES.2021.109484.

[23] R. Uyama, T. Nakagawa, S.-H. Hong, M. Mochizuki, R. Nishimura, N. Sasaki, Establishment of four pairs of canine mammary tumour cell lines derived from primary and metastatic origin and their E-cadherin expression, Veterinary and Comparative Oncology. 4 (2006) 104–113. https://doi.org/10.1111/J.1476-5810.2006.00098.X.

[24] R. Li, H. Wu, Y. Sun, J. Zhu, J. Tang, Y. Kuang, G. Li, A novel canine mammary cancer cell line: preliminary identification and utilization for drug screening studies, Frontiers in Veterinary Science. 8 (2021) 493. https://doi.org/10.3389/FVETS.2021.665906/BIBTEX.

[25] C.A. Gilbert, A.H. Ross, Cancer stem cells: Cell culture, markers, and targets for new therapies, Journal of Cellular Biochemistry. 108 (2009) 1031–1038. https://doi.org/10.1002/JCB.22350.

[26] M. Michishita, Understanding of tumourigenesis in canine mammary tumours based on cancer stem cell research, The Veterinary Journal. 265 (2020) 105560. https://doi.org/10.1016/J.TVJL.2020.105560.

[27] X. Li, C. Yu, H. Bao, Z. Chen, X. Liu, J. Huang, Z. Zhang, CT/bioluminescence dual-modal imaging tracking of stem cells labeled with Au@PEI@PEG nanotracers and RfLuc in nintedanib-assisted pulmonary fibrosis therapy, Nanomedicine: Nanotechnology, Biology and Medicine. 41 (2022) 102517. https://doi.org/10.1016/J.NANO.2022.102517.

[28] M. Sladkova, M. Palmer, C. Öhman, R.J. Alhaddad, A. Esmael, H. Engqvist, G.M. de Peppo, Fabrication of macroporous cement scaffolds using PEG particles: In vitro evaluation with induced pluripotent stem cell-derived mesenchymal progenitors, Materials Science and Engineering: C. 69 (2016) 640–652. https://doi.org/10.1016/J.MSEC.2016.06.075.

[29] P. Ghandforoushan, J. Hanaee, Z. Aghazadeh, M. Samiei, A.M. Navali, A. Khatibi, S. Davaran, Novel nanocomposite scaffold based on gelatin/PLGA-PEG-PLGA hydrogels embedded with TGF-β1 for chondrogenic differentiation of human dental pulp stem cells in vitro, International Journal of Biological Macromolecules. 201 (2022) 270–287. https://doi.org/10.1016/J.IJBIOMAC.2021.12.097.

[30] H. Sena Ozbay, S. Yabanoglu-Ciftci, I. Baysal, M. Gultekinoglu, C. Can Eylem, K. Ulubayram, E. Nemutlu, R. Topaloglu, F. Ozaltin, Mitochondria-targeted CoQ10 loaded PLGA-b-PEG-TPP nanoparticles: Their effects on mitochondrial functions of COQ8B-/- HK-2 cells, European Journal of Pharmaceutics and Biopharmaceutics. 173 (2022) 22–33. https://doi.org/10.1016/J.EJPB.2022.02.018.

25 Recent Advancements in Polymers for Biomedical Applications

Mattia Bartoli[1,2] and Alberto Tagliaferro[2,3,4]

[1] Center for Sustainable Future, Italian Institute of Technology, Via Livorno 60, 10144 Turin, Italy
[2] Consorzio Interuniversitario Nazionale per la Scienza e Tecnologia dei Materiali (INSTM), Via G. Giusti 9, 50121 Florence, Italy
[3] Department of Applied Science and Technology, Politecnico di Torino, C.so Duca degli Abruzzi 24, 10129 Turin, Italy
[4] Faculty of Science, Ontario Tech University, 2000 Simcoe Street North, Oshawa, Canada

1 INTRODUCTION

Advanced solutions for next-generation therapies and diagnostic tools represent a fast-growing field of medicinal chemistry. Nowadays, there are many challenges, such as biocompatibility, selectivity, stability, and administration route, that must be duly tackled to successfully produce materials for biomedical applications. The great complexity of this topic requires the enforcement of flexible solutions that could be tuned based on the input provided by medical specialists and pharma companies. This exciting research field is exploiting several different approaches. The most solid and ready to be quickly developed is the one based on polymers. Polymers for biomedical applications is a branch of materials science where competencies from different scientific sectors are merging to accomplish challenging tasks. Polymeric matrices provide unique opportunities for the synthesis of the biomaterials used in tissue engineering, drug delivery, and theragnostic applications. These applications can be developed using different strategies based on polymeric nano- and microparticles, fiber mats, or hydrogels. Hydrogels are among the most interesting polymer derivatives as they are cross-linked water-insoluble hydrophilic formulations based on different polymeric species.

In this chapter, we present a general overview of the polymers with biomedical uses focusing on the specific applications of tissue engineering, drug delivery, and theragnostics. We report also on the most advanced results achieved using several classes of bio-derived or synthetic polymers.

2 AN OVERVIEW OF POLYMERS FOR THE BIOMEDICAL FIELDS

2.1 TISSUE ENGINEERING

Tissue engineering is an interdisciplinary field that targets the repair, maintenance, and improvement of the functionalities of tissues affected by native or pathological defects through their replacement or reconstruction using bio- and synthetic polymers, as summarized in Figure 25.1.

The field of tissue engineering has been greatly developed in the last few decades due to the wide range of applications aimed at repairing and/or regenerating human tissues and organs. The use of tissue engineering represents also a very promising approach for the treatment of several diseases when chemotherapy treatments have failed. Accordingly, the efforts of the scientific community

DOI: 10.1201/9781003278269-25

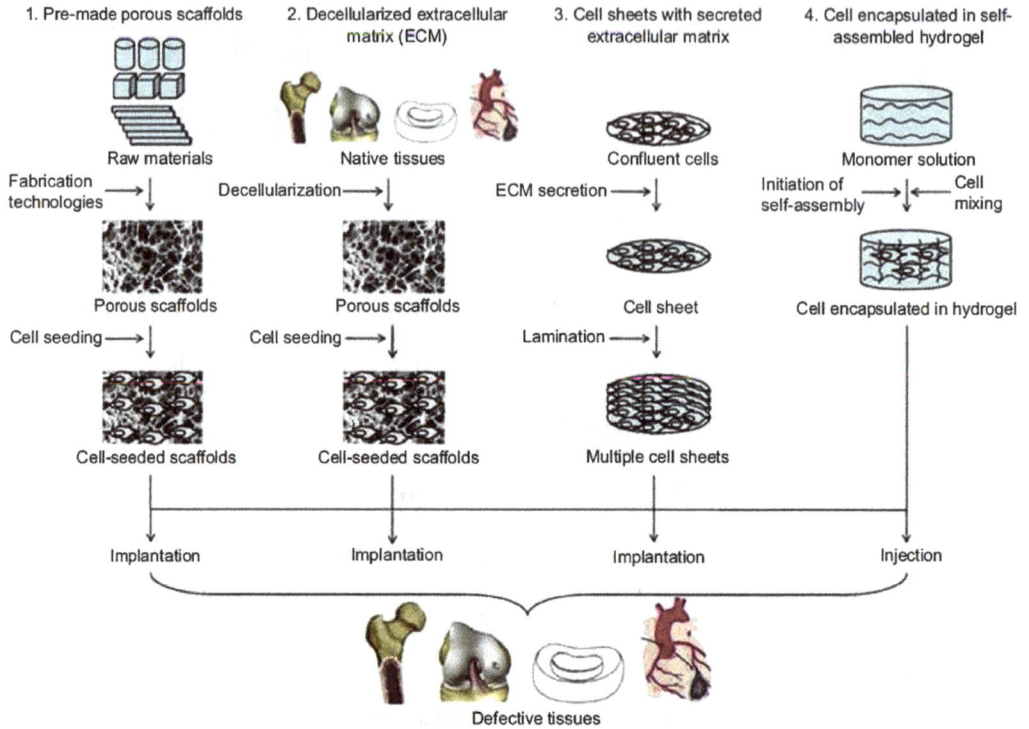

FIGURE 25.1   Summary of different approaches to tissue engineering. Adapted with permission from Reference [1], Copyright (2019), Elsevier.

have been devoted to the development of new solutions that are able to quickly achieve a high technological readiness level. Nevertheless, this ambitious goal is difficult to achieve due to the great complexity of biological systems. The most used routes enforced are based on the implantation of cells or cell replacements, the insertion of artificial tissue replacements, or the delivery of growth factors to tissues. All these routes require the design of an appropriate formulation involving or based on polymeric materials for the stabilization of the cell lines and/or the production of designed scaffolds.

The design of polymeric scaffolds playing a great and critical role due to their compulsory use for all tissue engineering applications. Broadly speaking, a polymeric scaffold is a three-dimensional solid with an intrinsic porosity that is able to promote the interaction between the cell/tissues and a biomaterial allowing at the same time the correct transport of nutrients, chemical messengers, and gases, and controlled cell proliferation and survival. A polymeric scaffold must be non-toxic, biocompatible, and biodegradable with approximately the same rate of tissue regeneration, allowing its replacement with biological tissues without producing harmful compounds. The tunability of polymeric scaffolds allows the production of devices for biological applications with tailored properties of strength, porosity, micro- and nanostructure, and degradability, while maintaining high reproducibility. However, only a few polymeric scaffolds, known as regenerative polymeric scaffolds, can act as regenerative materials that induce the spontaneous regeneration of tissues and organs while simultaneously degrading. Such scaffolds were reported for the first time in 1975 by Yannas and co-workers [2] who discovered them using collagen modified with glycosaminoglucans.

A few years later, the first systematic study on the relevance of the degradative rate of biological scaffolds, focusing on skin repairs, was published by the same group [3]. Many different tissues (i.e., nerves, derma, and conjunctiva) have been targeted in further works [4]. The number of new polymeric scaffolds and biomaterials for tissue engineering has grown year by year and today many of them are unrecognizable by the immune human system. They are used for many applications such as sutures, bone and joint replacements, and dental, vascular, and ocular implants.

2.2 Drug Delivery

Drug delivery can be defined as a method for administering a chemical with pharmaceutical activity to achieve a therapeutic effect in a living organism by using a suitable vector, as summarized in Figure 25.2. In the first place, the choice of the biological route for the administration is critical for the preparation of an effective formulation. The most common way to administer a chemical is the intravenous route, but nasal, ocular, and pulmonary routes are also used. Depending on the administration procedure, a different formulation can be used, such as liposomes, microspheres, gels, prodrugs, and cyclodextrins. Each solution must be carefully engineered using the appropriate combination of polymers to improve the biocompatibility and simultaneously enhance the affinity for specific receptors on cell membranes and tissues.

The complexity of human physiology prevents the selection of a unique drug-delivery system. The necessity of many solutions has been the driving force for the development of an immense number of drug-delivery systems based on polymers. It is an overwhelming challenge to report a comprehensive view of the plethora of different polymers used for drug delivery but we can categorize them into two main groups: (i) directly bonded or (ii) non-directly bonded to the active molecule being carried. The wider class of pro-drugs is that of directly bonded polymers. Pro-drugs are chemical species containing a silenced active molecule covalently bonded to other species and able to deliver it untouched to the biological target. This drug-delivery system is appealing due to its intrinsic stability, although its applicability is hindered by the unavoidable challenge of drug release. The chemical lysis of the bond between the active species and the polymeric carrier is a very delicate and difficult-to-tune chemical property. Several solutions have been proposed but the final word is far from being had on this topic. The other class of polymer-based drug-delivery systems contains all non-covalent systems such as micellar and host–guest ones. These systems

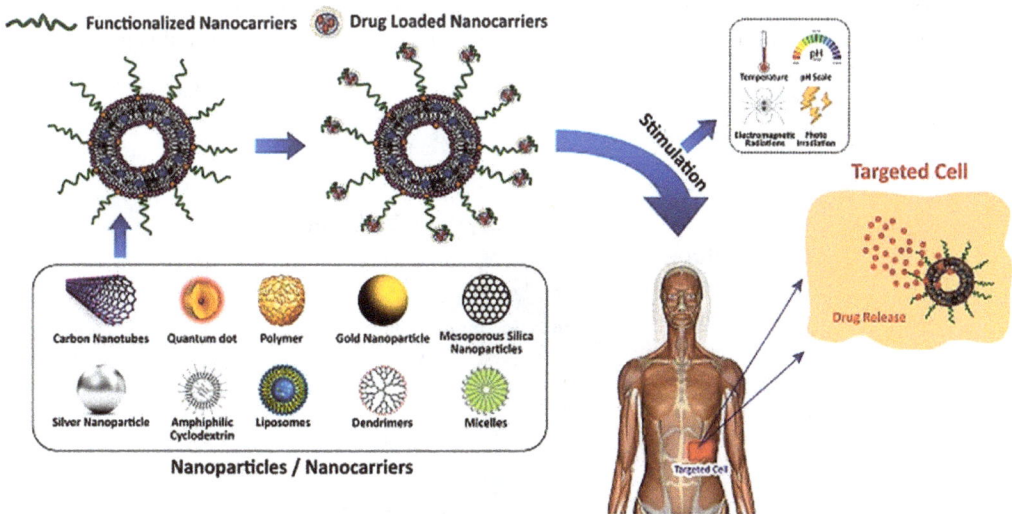

FIGURE 25.2 Overview of different drug-delivery platforms. Adapted with permission from Reference [5], Copyright (2021), Elsevier.

are very sensitive to environmental conditions and the achievement of controlled release is a challenging task. Nevertheless, the tailoring of polymers might tackle the instability of these systems, establishing a complex network of weak interactions able to stabilize the drug-delivery system sufficiently to be used in a biological environment.

Another issue related to polymeric drug-delivery systems is related to the chemical species carried. In the case of organic molecules, the polymeric delivery systems must ensure not only the release but also the preservation of the molecule until the targeted biological structure has been reached. Similarly, a metal-based drug should be delivered and expelled without any leaching or metal bioaccumulation.

2.3 THERAGNOSTICS

Theragnostics is defined as treatments that combine the diagnostic and therapeutic processes, as illustrated in Figure 25.3.

The simultaneous exploitation of both diagnostic and therapeutic effects represents a great improvement in the development of highly specific medical treatments able to monitor drug release, interaction, and excretion using a single chemical formulation. Nonetheless, theragnostics is not a panacea, but a challenging field due to the immense complexity related to the production of suitable multifunctional materials. Advancements in theragnostics require combined efforts from several research fields such as proteomics, genomics, molecular biology, biochemistry, and physiology. The difficulties in developing reliable theragnostic platforms are, however, counterbalanced by their potentially widespread use. The immediate advantage provided by the theragnostic approach is related to the possibility of monitoring in real time the effect of drug release, and improving the knowledge of pharmacokinetics. Furthermore, theragnostics could be exploited by coupling drug release with a simple imaging process providing a fast, reliable, and easy-to-use tool for the medical operator to assess the effect of the drug. Even if the cost of a working theragnostic agent is currently far too expensive to be competitive, their development could slowly replace the traditional approaches in routine healthcare procedures. A long-term economic breakthrough for theragnostics will surely be a drop in the price of drugs. By using a theragnostic approach, the experimental step of drug testing and pharmacokinetic evaluation could run simultaneously, with beneficial effects.

Polymers represent a perfect class of materials for the development of theragnostic platforms and the implementation of polymeric carbon dots could be the game-changer event in this field.

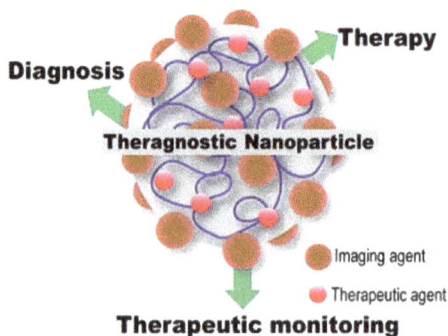

FIGURE 25.3 Theragnostic approach. Adapted with permission from Reference [6], Copyright (2012), Elsevier.

3 POLYMERS FOR BIOMEDICAL APPLICATIONS: AN OVERVIEW

3.1 BIO-DERIVED POLYMERS

3.1.1 Collagen

Collagen is a hydrophilic fibrous protein insoluble in organic solvents that is the main component of any connective tissue. It represents up to 30 wt.% of the protein amount in animals. Furthermore, collagens inhibit the tissue damage related to stretching, preserving the structural integrity of blood vessels, skin, bones, and cartilage. Collagen is the main component of the extracellular matrix in animals. The history of collagen as a material for biomedical applications started at the end of the 19th century with the first attempt to produce biodegradable intestinal sutures [7]. Collagen is one of the most consolidated polymers for the replacement of corneal tissue due to its astonishing bio-adhesivity, reduced immunogenicity, high biocompatibility, biodegradability, and good mechanical properties. Furthermore, collagen could induce adhesion, growth, and differentiation of several cellular lines, enhancing tissue regeneration [8].

The high biocompatibility of collagen is a great driving force for its use in biomedical applications, but its complex mechanisms of degradation are a pending issue even if cross-linking modifications could improve its stability by reducing the hydrolysis rate. Collagen and its derivatives show also a remarkable ability to induce the growth of epithelial cells. It has been proved that collagen and collagen-derivative fibrils can pass across the corneal epithelial tissue and orientate themselves by stacking onto circumferential limbal native collagen [9]. Currently, the use of native collagen for biomedical applications is slowed down by its high hydrophilicity, which could lead to denaturation and swelling *in vivo* [10].

3.1.2 Keratin

Keratin is a cysteine-rich protein (up to 13 wt.%) that represents the major constituent of horn and feathers. Keratin represents another promising polymer source for many biomedical applications due to its high biocompatibility. The use of keratin is not as straightforward as in the case of collagen as it requires several chemical processes to be used.

As a first step, keratin is extracted in harsh chemical conditions, along thermochemical or microbial routes. After this mandatory step, keratin could be converted into films with very remarkable transparency, biostability, biocompatibility, and promoting cell adhesion in living organisms such as rabbits [11]. Nonetheless, the cytocompatibility of keratin remains a matter of debate and only a few studies have focused on its use for tissue regeneration, although the role of keratin in the hemostasis processes remains unclear.

Human hair keratin-based hydrogels have been found to be an effective hemostatic agent in rabbits with lethal liver injury [12]. Aboushwareb et al. have shown the improved performances of keratin-based materials compared with commercial hemostat formulations such as QuickClot® and HemCon®. The effect of keratin hemostatic hydrogel was attributed to the formation of a physical seal on the wound site acting like a porous scaffold promoting cellular infiltration with enhanced granulose tissue formation. Another interesting field of application for keratin materials is the use of keratin hydrogel for the regeneration of peripheral nerves. Keratin-based hydrogel was used for inducing the regeneration of nerves in a mouse model [13]. Sierpinski et al. used human hair keratin to show the improved proliferation of Schwann cells upregulating the specific gene expression relevant to neuronal functionality. They also reported the effect of keratin gel in the stimulation process of axon regeneration in an injured mouse tibial nerve. Apel et al. [14] reported the accelerated regeneration of nerves with improvements in electrophysiological recovery and axon density. They reported also good connectivity between regenerated nerves and muscles with appreciable axon myelination after six months. However, keratin and keratin derivatives could induce inflammation and neovascularization, reducing the adhesion of the film on some soft tissues [15].

3.1.3   Hyaluronic Acid

Hyaluronic acid is a polysaccharide composed of disaccharide units based on glucuronic acid and N-acetyl-D-glucosamine which is present in almost all organic tissues. Hyaluronic acid has generally a high molecular mass (up to millions of Daltons) and remarkable viscoelastic properties. Hyaluronic acid has found many uses ranging from simple cosmetic to advanced biomedical applications [16]. Hyaluronic acid is used in so-called viscosurgery for the protection of tissue, allowing manipulation in ophthalmological interventions [17]. Nevertheless, the main ophthalmological use of hyaluronic acid is as a replacement for humor vitreous during cataract surgery or lens implantations.

Another interesting use of hyaluronic acid is the protection of healthy or wounded tissue surfaces from dryness or external agents, thus actively promoting the healing process. Edmonds et al. [18] reported the efficiency of hyaluronic acid to treat ulcers and wound infections. Hyaluronic acid-based formulations were also tested in the separation of the connective tissue surfaces regeneration after the trauma of a surgical procedure or common injuries avoiding massive scar formation. Hyaluronic acid is also of common use as a replacement for synovial fluids and in cosmetic surgery for filling up and augmenting the tissue space in skin muscles [19] and pharyngeal tissues [20]. Hyaluronic acid also showed interesting features as an active molecule able to stimulate chondrocyte proliferation and inhibit cartilage degradation.

In addition, several studies have assessed the healing effect of hyaluronic acid administration as an effective anti-inflammatory remediation. This is due to modulation of the cytokine expression reducing the concentration of reactive oxygen species content in the tissue with a consequent analgesic effect.

Hyaluronic acid could be also used as a drug-delivery platform due to the ability of its carboxylic residue to form a quite stable hydrogel. Hyaluronic-based hydrogels have been used to shape microcapsules with high biocompatibility and the ability to drag and drop chemicals, antibodies, and plasmid DNA [21].

3.1.4   Cellulose

Cellulose is the most abundant natural polymer on Earth and is composed of units of D-glucose bonded together through a β-1-4 glycosidic bond. Cellulose is the main constituent of plants but it is also present in the cell walls of bacteria, algae, and tunicates. An interesting feature of cellulose is the wide range of morphologies in which it can shape, and form microfibrils up to nanocrystals. The main use of cellulose is as feedstock for the pulp and mill industry, but its use in the preparation of biomedical formulations is a promising field. Cellulose represents a very easy-to-tune material as it is possible to functionalize its residual groups without compromising its mechanical properties. This has led to widespread use of cellulose in tissue engineering as an additive or even as a primary scaffold material, allowing mechanical properties to be achieved that match those of real tissues [22]. This is due to the strong network of hydrogen bonds established between cellulose chains allowing the formation of nanostructures with suitable mechanical properties.

The main drawback of the direct use of cellulose as a scaffold is related to cellular adhesion and proliferation. Mammalian cells are unable to attach to unmodified cellulosic media due to their high hydrophilicity. This issue can be tackled by tailoring the cellulose surface chemistry or by inserting a specific ligand such as collagen into the cellulose matrix [23]. The use of cellulose-based materials mixed with other polymers is also a choice. Mao and co-workers [24] combined cellulose with (poly)lactic acid and hydroxyapatite, achieving remarkable results in bone repair.

Surface modification of cellulose is a vast area, but it is currently mainly focused on the introduction of unipolar residues on hydroxyl functions, such as ethyl, methyl, or acetate groups. As an example, hydroxyethylcellulose derivative is a non-ionic water-soluble material able by itself to stimulate cell viability, proliferation, and growth [22]. Cellulose materials are also used in wound dressings, promoting water retention and avoiding bacterial adhesion. Nanostructured cellulose has shown great potential for wound-healing applications due to its moisture absorption and water retention ability, contributing also to decreasing the inflammatory responses and promoting fibroblast proliferation. Nanofibrils of cellulose are particularly promising for wound healing due to their large surface/volume ratio, porosity, and water-holding ability. Cellulose nanofibrils are also able to match the extracellular architecture, controlling excessive wound contraction as verified both *in vitro* and *in vivo* by Nuutila et al. [25]. Similar results have been obtained also by using cellulose nanocrystals and oxidized cellulose nanofibers. These effects were due to the regulation of angiogenesis and connective tissue formation promoted by the cellulose and surface-modified cellulose materials.

Cellulose materials were also extensively used as drug carrier platforms due to their hydrosolubility and the easy tuning of both chemical and morphological features. The first use of cellulose as a drug carrier was as a coating of a tablet for buccal administration, preventing rapid degradation in the stomach. So far, cellulose powder and nanostructured cellulose have been shown to be the ideal combination of stability and adsorbing properties for use as excipients for all tablet preparations. More advanced techniques are based on the encapsulation of chemicals into cellulose nanoparticles of pristine or modified cellulose [26]. Modified cellulose can exploit amphiphilic properties and represents one of the preferred choices for the production of drug carrier systems. As an example, the Federal and Drug Administration allowed the use of hydroxyl propyl methyl cellulose mixed with chitosan/glycerophosphate for the preparation of hydrosoluble vancomycin-containing drugs. Neat cellulose nanofibers also have been tested for the transdermal administration of poorly soluble drugs, while cellulose acetate phthalate nanofibers were tested as anti-HIV drug carriers and in the preparation of HIV infection prevention devices for risky sexual intercourse [27].

3.1.5 Chitin and Chitosan

Chitin is a polysaccharide composed of repetitive units of β-(1,4)-N-acetyl-*d*-glucosamine and is the second most abundant biopolymer after cellulose. Chitin is biosynthesized in nature as an ordered highly crystalline microfibril forming the structural part of the arthropod exoskeleton cell walls of both yeast and fungi. The most important chitin derivative is chitosan, obtained by partial deacetylation of chitin under alkaline conditions or by controlled enzymatic hydrolysis using chitin deacetylase producing an irregular distribution of acetyl group alongside the chitosan chains. Chitin is used in tissue engineering mainly in the form of electrospun-produced fibers. Shalumon et al. [28] used carboxymethyl chitin blended with (poly)vinyl alcohol to realize a fibrous mat reticulated using glutaraldehyde vapors. The chitin fibers were bioactive and biocompatible, promoting both cell adhesion/attachment and proliferation with negligible cytotoxicity. However, both chitin and chitosan show very poor mechanical properties and generally require the addition of a filler such as hydroxyapatite to strengthen them. Peter et al. [29] reported the production of highly porous chitosan–gelatin composites containing hydroxyapatite. This material produced a positive biological response from MG-63 cells, thereby improving cell attachment and proliferation.

A lot of research work has been devoted to improving the properties of chitin and chitosan-filled composites producing the first chitin and chitosan bioactive glass–ceramic composite. First developed by Hench in 1991 [30], bioactive glass–ceramic are nowadays widely used for bone repair in orthopedics and dentistry. These ceramics promote the formation of carbonated apatite and are able to influence osteoblast and bone marrow stromal cell proliferation and differentiation [31]. Composite scaffolds of chitosan with nano-bioactive glasses were first prepared by Peter et al. [32] using gelatin as a compatibilizer. These authors reported the production of materials with pores having sizes ranging from 150 to 300 μm. The study of biomineralization showed a great number of mineral deposits on the nanocomposite that allowed the composites to be used for alveolar bone regeneration.

Another widely used filler is silica. As reported by Karlsson et al. [33], silica induces the formation of apatite through a repolymerization process induced by pH that allows the deposition of phosphates and calcium salts on the hydroxyl-rich surface of silica. The addition of silica to chitin-based hydrogel scaffolds improved their bioactivity and swelling ability and reduced their cytotoxicity, as reported by Madhumathi et al. [34]. These authors also tested the chitin/silica scaffold *in vivo* using the MG63 cell line and reporting good biocompatibility. Chitosan and chitin play major roles in the development of innovative wound dressings by mixing them with active particles like silver ones. This allows exploitation of the activity of silver nanoparticles entering into the bacteria cell interacting with sulfur-rich proteins and inducing apoptosis. Chitin and chitosan also have been considered for the production of drug carrier platforms. The procedure is the same as that above described for cellulose based on the formation of nano- and microcapsules filled with active chemicals. Many literature studies have proved the viability of chitosan for the delivery of antibiotics, antivirals, and chemotherapy drugs [35].

3.2 SYNTHETIC POLYMERS

3.2.1 Polyesters

Polyesters are widely used as drug carriers and also for tissue engineering due to their mechanical properties. A first-class polyester widely used for biomedical purposes is represented by polyhydroxyalkanoates. These polyesters are produced by polymerization of cyclic lactones, known as glycolides, mainly through ring-opening polymerization using alcohol as an initiator of the polymerization. Among them, poly(caprolactone) (PCL) is one of the most commonly used due to its biocompatibility and biodegradability, and it is semicrystalline at human body temperatures. This rubbery behavior allows efficient permeability of the body's metabolites through PCL membranes.

PCL displays a lot of interesting features such as adjustable biodegradability, blending properties with a wide range of different polymers, and the possibility of incorporation in block co-polymeric structures leading to full tuneability of its hydrophobicity. Electrospun nanofibers of PCL, either alone or in combination with other polyhydroxyalkanoates such as poly(lactic acid) (PLA), have found interesting applications in surgical sutures [36]. PCL-based surgical suture biodegradability is easily tuned by balancing the PCL molecular weight and the amount of non-PCL blocks. This property also allows the use of PCL for the preparation of very effective wound-dressing formulations. The more effective PCL-based wound dressing is achieved by combining the polymers with silver nanoparticles [37], however the combination with polysaccharides and lipopolysaccharides is also very useful [38]. PCL films and mats have been widely investigated in soft tissue engineering for the treatment of corneal disease [39]. In this field, PCL-based scaffolds showed improved mechanical properties compared with collagen and collagen derivatives, and Hashem et al. [40] reported its successful use in a clinical trial as patch grafts in ophthalmology. PCL films were also used for cartilage and bone tissue engineering [41]. PCL nano- and microparticles also have been used as drug-delivery systems for many chemicals.

In contrast to PCL, PLA displays improved mechanical properties even if the cell adhesion is limited due to the high hydrophobicity of PLA materials. Nevertheless, the combination of PLA with hydroxyapatite improved the cell adhesion, prompting a boost in PLA-based exploits for tissue engineering. Furthermore, isotactic PLA could be used for the formation of highly stable structures called stereocomplexes with superior mechanical properties and hydrolytic resistance. PLA stereocomplexes are suitable candidates for bone repairing. Bandelli et al. proposed materials based on poly(mandelic acid) stereocomplexes for further improvement of the mechanical properties [42]. PLA and PLA stereocomplexes are also used for the preparation of microspheres with remarkable performances in drug delivery and high biocompatibility using three-dimensional PLA fiber networks[43]. Polycarbonates are another great family of polyesters used for the preparation of biological scaffolds. The main use of polycarbonates is in the field of bone repair due to their superior mechanical strength and hydrolytic resistance [44].

3.2.2   Polyolefins

Polyolefins are the most diffuse plastic material on Earth, with the production of up to several Gton/year. For biomedical applications the most used are poly(vinyl alcohol) (PVA), poly(ethylene) (PE), and poly(styrene) (PS). PVA is a linear polyolefin produced through the full or partial hydrolysis of poly(vinyl acetate) to remove the acetate residues. The hydrolysis degree of PVA defines its chemical and physical properties and this is the reason behind PVA water solubility. The water solubility of PVA represents an issue for the preparation of hydrogel, requiring the addition of a cross-linking agent that is able to guarantee the structural integrity of the biological fluids. PVA is among the most used polymers for biomedical applications, finding uses in many fields including tissue engineering and drug delivery [45]. PVA is very promising for the preparation of cartilaginous tissue replacements and for soft tissue repair.

PE is a linear polyolefin produced from the polymerization of ethylene in the presence of a catalyst. The properties of PE are strongly related to its molecular weight, inducing different densities leading to PE classification into low-density PE (LDPE), high-density PE (HDPE), and ultra-high-density PE (UHDPE). HDPE has been used for biomedical applications since the middle of the 20th century as a lightweight replacement for metals and reaching a considerable market share. PE implants showed a remarkable low inflammatory response due to PE having an inert behavior with consequently limited cytotoxicity and high molding customizability. Kumar et al. [46] developed a very interesting approach coupling UHDPE with chitosan and loading the scaffold with gentamicin. The *in vivo* results showed a drastic reduction in bacterial proliferation with a drop in infections. PS is less used compared to PE and PVA, but it has recently found applications as a cell culture medium and for some specific applications such as studying cellular interactions. Fluorinated polyolefins, such as Teflon® or poly(divinyliden fluoride) (PVDF), are also used to induce the proliferation of neuronal cells or simply as a coating to improve the durability and interphase properties of implants.

3.2.3   Carbon Quantum Dots

Carbon quantum dots (CQDs) are carbonaceous nanoparticles whose sizes are in the nanometers range and are characterized by a great fluorescence yield. The CQD family is wide and can be split into graphene quantum dots (CGQDs), carbon nitride dots (CNDs), and carbon polymeric dots (CPDs), as summarized in Figure 25.4.

CGQDs were the first to be discovered in 2004 as a byproduct of the purification of the residue of carbon nanotube oxidation [48]. CGQDs are composed of fragments of graphene sheets, with an average size lower than 20 nm decorated on the edges with functional residues. CQDs are usually quasi-spherical and show fluorescence emission due to the quantum confinement effect related to the small size of conjugated $\pi-\pi$ domains. Similarly, CNDs are composed of graphene layers doped with nitrogen. CNDs' fluorescence mainly arises from defects introduced by nitrogen atoms included in the graphene-like domains. CPDs are the most tunable species among CQDs. They are formed by a hybrid structure in which polymer chains surround a carbon-rich core that can resemble those of pure sp^2 carbon or carbon nitride dots. CPDs are characterized by a larger outer shell compared to CGDs and CNDs, and the polymer crown accounts for their fluorescence.

CPDs have found many theragnostic applications due to their ability to cross the blood–brain barrier and to interact with many different molecular targets. One of the most consolidated routes for the production of CPDs is the reaction of organic polyacids with aliphatic and aromatic [49] diamines or glycols [50] for the synthesis of randomly organized nanostructures. CPDs have been used for cutting-edge nanomedicine applications combining cancer treatment, gene therapy, tissue regeneration, and drug carriers with diagnostic effects due to their fluorescence emission. CPD-based theragnostic platforms have shown high biocompatibility and shelf stability. Furthermore, CPDs have been excreted through the kidneys without great stress on the organism's metabolism.

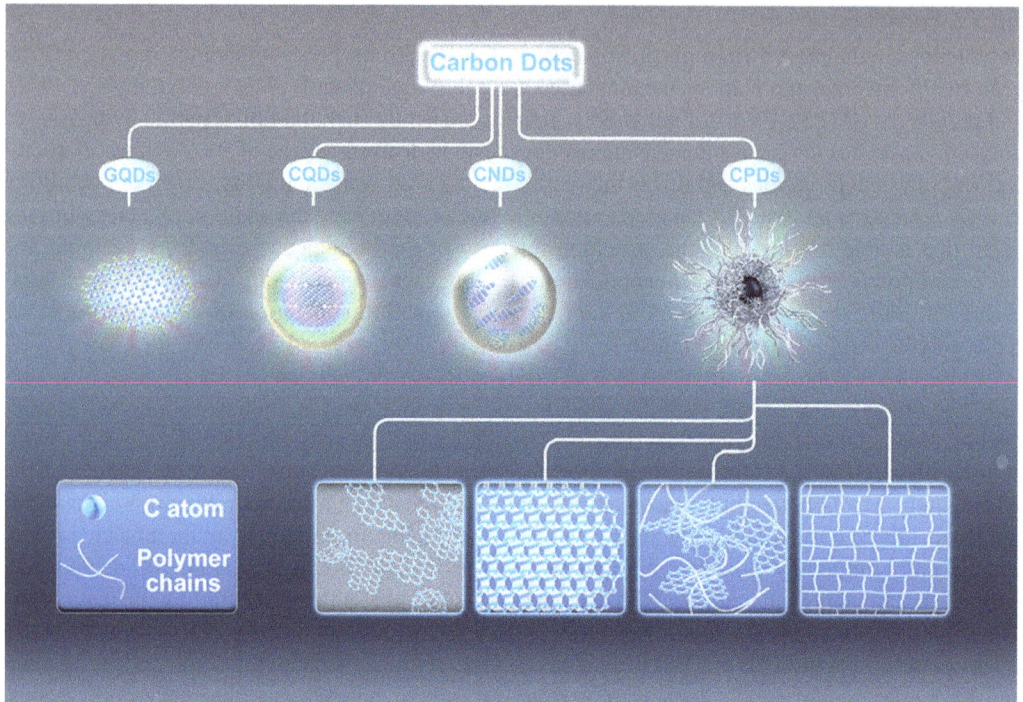

FIGURE 25.4   Classifications of CQDs. Adapted from Reference [47]. Copyright (2019). Distributed under a Creative Commons Attribution License 4.0 (CC BY) https://creativecommons.org/licenses/by/4.0/.

4   CONCLUSIONS AND FUTURE PERSPECTIVES

The recent advancements in engineered polymer derivatives have allowed the production of materials that are able to combine superior properties and attractive biological features. Tissue engineering is one of the frontiers of medicine aiming to overcome the classical transplant procedure through the production of hybrid materials able to perform the same activity as the replaced original tissues. Polymeric materials have shown very attractive results in the production of composites and hydrogels with great mechanical performances together with high biocompatibility for several applications ranging from bone to soft tissue repair. Outstanding results have been achieved in drug delivery by using chemicals entrapped in micro- and nanopolymeric structures able to achieve great selectivity and improved pharmacokinetics. Polymers are facing several competitors such as peptide and inorganic-based solutions in both tissue engineering and drug carrying. Nevertheless, polymers are the most promising tools for the development of game-changing theragnostic platforms that are able to couple diagnosis, chemical treatment, and pharmacokinetic monitoring. In this field, polymer materials such as CPDs are the most advanced formulations able to combine biological and diagnostic effects. The endless research into new polymeric species will lead to the development of new materials paving the way for the next generation of medical solutions.

REFERENCES

[1] F. Sefat, M. Mozafari, A. Atala, 1 – Introduction to tissue engineering scaffolds, in: M. Mozafari, F. Sefat, A. Atala (Eds.) Handbook of Tissue Engineering Scaffolds: Volume One, Woodhead Publishing, 2019, pp. 3–22.

[2] I. Yannas, Suppression of in vivo degradability and of immunogenicity of collagen by reaction with glycosaminoglycans, Polym Prepr Am Chem Soc, 16 (1975) 209–214.

[3] I. Yannas, J.F. Burke, Design of an artificial skin. I. Basic design principles, Journal of Biomedical Materials Research, 14 (1980) 65–81.

[4] A. Tathe, M. Ghodke, A.P. Nikalje, A brief review: biomaterials and their application, International Journal of Pharmacy and Pharmaceutical Sciences, 2 (2010) 19–23.

[5] A. Shah, S. Aftab, J. Nisar, M.N. Ashiq, F.J. Iftikhar, Nanocarriers for targeted drug delivery, Journal of Drug Delivery Science and Technology, 62 (2021) 102426.

[6] J.H. Ryu, H. Koo, I.-C. Sun, S.H. Yuk, K. Choi, K. Kim, I.C. Kwon, Tumor-targeting multi-functional nanoparticles for theragnosis: New paradigm for cancer therapy, Advanced Drug Delivery Reviews, 64 (2012) 1447–1458.

[7] S. Chattopadhyay, R.T. Raines, Collagen-based biomaterials for wound healing, Biopolymers, 101 (2014) 821–833.

[8] A.M. Ferreira, P. Gentile, V. Chiono, G. Ciardelli, Collagen for bone tissue regeneration, Acta biomaterialia, 8 (2012) 3191–3200.

[9] K.M. Meek, C. Boote, The organization of collagen in the corneal stroma, Experimental Eye Research, 78 (2004) 503–512.

[10] G. Laurent, Dynamic state of collagen: pathways of collagen degradation in vivo and their possible role in regulation of collagen mass, American Journal of Physiology-Cell Physiology, 252 (1987) C1–C9.

[11] M. Borrelli, N. Joepen, S. Reichl, D. Finis, M. Schoppe, G. Geerling, S. Schrader, Keratin films for ocular surface reconstruction: Evaluation of biocompatibility in an in-vivo model, Biomaterials, 42 (2015) 112–120.

[12] T. Aboushwareb, D. Eberli, C. Ward, C. Broda, J. Holcomb, A. Atala, M. Van Dyke, A keratin biomaterial gel hemostat derived from human hair: evaluation in a rabbit model of lethal liver injury, Journal of Biomedical Materials Research Part B: Applied Biomaterials, 90 (2009) 45–54.

[13] P. Sierpinski, J. Garrett, J. Ma, P. Apel, D. Klorig, T. Smith, L.A. Koman, A. Atala, M. Van Dyke, The use of keratin biomaterials derived from human hair for the promotion of rapid regeneration of peripheral nerves, Biomaterials, 29 (2008) 118–128.

[14] P.J. Apel, J.P. Garrett, P. Sierpinski, J. Ma, A. Atala, T.L. Smith, L.A. Koman, M.E. Van Dyke, Peripheral nerve regeneration using a keratin-based scaffold: long-term functional and histological outcomes in a mouse model, The Journal of hand surgery, 33 (2008) 1541–1547.

[15] F. Loschke, K. Seltmann, J.-E. Bouameur, T.M. Magin, Regulation of keratin network organization, Current opinion in cell biology, 32 (2015) 56–64.

[16] M. Dovedytis, Z.J. Liu, S. Bartlett, Hyaluronic acid and its biomedical applications: A review, Engineered Regeneration, 1 (2020) 102–113.

[17] F.T. Kretz, I.-J. Limberger, G.U. Auffarth, Corneal endothelial cell coating during phacoemulsification using a new dispersive hyaluronic acid ophthalmic viscosurgical device, Journal of Cataract & Refractive Surgery, 40 (2014) 1879–1884.

[18] M. Edmonds, M. Bates, M. Doxford, A. Gough, A. Foster, New treatments in ulcer healing and wound infection, Diabetes/Metabolism Research and Reviews, 16 (2000) S51–S54.

[19] G.D. Monheit, K.M. Coleman, Hyaluronic acid fillers, Dermatologic Therapy, 19 (2006) 141–150.

[20] D.K. Chhetri, A.H. Mendelsohn, Hyaluronic acid for the treatment of vocal fold scars, Current Opinion in Otolaryngology & Head and Neck Surgery, 18 (2010) 498–502.

[21] G. Huang, H. Huang, Application of hyaluronic acid as carriers in drug delivery, Drug Delivery, 25 (2018) 766–772.

[22] H. Seddiqi, E. Oliaei, H. Honarkar, J. Jin, L.C. Geonzon, R.G. Bacabac, J. Klein-Nulend, Cellulose and its derivatives: Towards biomedical applications, Cellulose, 28 (2021) 1893–1931.

[23] S. Kalia, S. Boufi, A. Celli, S. Kango, Nanofibrillated cellulose: surface modification and potential applications, Colloid and Polymer Science, 292 (2014) 5–31.

[24] D. Mao, Q. Li, N. Bai, H. Dong, D. Li, Porous stable poly(lactic acid)/ethyl cellulose/hydroxyapatite composite scaffolds prepared by a combined method for bone regeneration, Carbohydrate Polymers, 180 (2018) 104–111.

[25] K. Nuutila, A. Laukkanen, A. Lindford, S. Juteau, M. Nuopponen, J. Vuola, E. Kankuri, Inhibition of skin wound contraction by nanofibrillar cellulose hydrogel, Plastic and Reconstructive Surgery, 141 (2018) 357e–366e.

[26] D. Klemm, F. Kramer, S. Moritz, T. Lindström, M. Ankerfors, D. Gray, A. Dorris, Nanocelluloses: a new family of nature-based materials, Angewandte Chemie International Edition, 50 (2011) 5438–5466.

[27] L.S.C. Wan, W.K. Chui, Deviation of the ratio of drugs in a two-component mixture encapsulated in cellulose phthalate microspheres, Journal of Microencapsulation, 12 (1995) 417–423.

[28] K. Shalumon, N. Binulal, N. Selvamurugan, S. Nair, D. Menon, T. Furuike, H. Tamura, R. Jayakumar, Electrospinning of carboxymethyl chitin/poly (vinyl alcohol) nanofibrous scaffolds for tissue engineering applications, Carbohydrate Polymers, 77 (2009) 863–869.

[29] M. Peter, N. Ganesh, N. Selvamurugan, S. Nair, T. Furuike, H. Tamura, R. Jayakumar, Preparation and characterization of chitosan–gelatin/nanohydroxyapatite composite scaffolds for tissue engineering applications, Carbohydrate Polymers, 80 (2010) 687–694.

[30] L. Hench, H. Stanley, A. Clark, M. Hall, J. Wilson, Dental applications of Bioglass® implants, in: Bioceramics, Elsevier, 1991, pp. 231–238.

[31] M. Bosetti, M. Cannas, The effect of bioactive glasses on bone marrow stromal cells differentiation, Biomaterials, 26 (2005) 3873–3879.

[32] M. Peter, P.T.S. Kumar, N.S. Binulal, S.V. Nair, H. Tamura, R. Jayakumar, Development of novel α-chitin/nanobioactive glass ceramic composite scaffolds for tissue engineering applications, Carbohydrate Polymers, 78 (2009) 926–931.

[33] K.H. Karlsson, K. Fröberg, T. Ringbom, A structural approach to bone adhering of bioactive glasses, Journal of Non-Crystalline Solids, 112 (1989) 69–72.

[34] K. Madhumathi, P.S. Kumar, K. Kavya, T. Furuike, H. Tamura, S. Nair, R. Jayakumar, Novel chitin/nanosilica composite scaffolds for bone tissue engineering applications, International Journal of Biological Macromolecules, 45 (2009) 289–292.

[35] A. Bernkop-Schnürch, S. Dünnhaupt, Chitosan-based drug delivery systems, European Journal of Pharmaceutics and Biopharmaceutics, 81 (2012) 463–469.

[36] A. Vieira, R. Medeiros, R.M. Guedes, A. Marques, V. Tita, Visco-elastic-plastic properties of suture fibers made of PLA-PCL, in: Materials Science Forum, Trans Tech Publ, 2013, pp. 56–61.

[37] R. Augustine, N. Kalarikkal, S. Thomas, Electrospun PCL membranes incorporated with biosynthesized silver nanoparticles as antibacterial wound dressings, Applied Nanoscience, 6 (2016) 337–344.

[38] F. Croisier, G. Atanasova, Y. Poumay, C. Jérôme, Polysaccharide-coated PCL nanofibers for wound dressing applications, Advanced Healthcare Materials, 3 (2014) 2032–2039.

[39] A.O. Mahmoud Salehi, S. Heidari Keshel, F. Sefat, L. Tayebi, Use of polycaprolactone in corneal tissue engineering: A review, Materials Today Communications, 27 (2021) 102402.

[40] H. Hashemi, S. Asgari, S. Shahhoseini, M. Mahbod, F. Atyabi, H. Bakhshandeh, A.H. Beheshtnejad, Application of polycaprolactone nanofibers as patch graft in ophthalmology, Indian Journal of Ophthalmology, 66 (2018) 225.

[41] P. Yilgor, R.A. Sousa, R.L. Reis, N. Hasirci, V. Hasirci, 3D plotted PCL scaffolds for stem cell based bone tissue engineering, in: Macromolecular Symposia, Wiley Online Library, 2008, pp. 92–99.

[42] D. Bandelli, J. Alex, C. Weber, U.S. Schubert, Polyester stereocomplexes beyond PLA: could synthetic opportunities revolutionize established material blending?, Macromolecular Rapid Communications, 41 (2020) 1900560.

[43] M. Singhvi, S. Zinjarde, D. Gokhale, Polylactic acid: synthesis and biomedical applications, Journal of Applied microbiology, 127 (2019) 1612–1626.

[44] R.P. Brannigan, A.P. Dove, Synthesis, properties and biomedical applications of hydrolytically degradable materials based on aliphatic polyesters and polycarbonates, Biomaterials Science, 5 (2017) 9–21.

[45] A. Kumar, S.S. Han, PVA-based hydrogels for tissue engineering: A review, International Journal of Polymeric Materials and Polymeric Biomaterials, 66 (2017) 159–182.

[46] R.M. Kumar, P. Gupta, S.K. Sharma, A. Mittal, M. Shekhar, V. Kumar, B.M. Kumar, P. Roy, D. Lahiri, Sustained drug release from surface modified UHMWPE for acetabular cup lining in total hip implant, Materials Science and Engineering: C, 77 (2017) 649–661.

[47] C. Xia, S. Zhu, T. Feng, M. Yang, B. Yang, Evolution and Synthesis of Carbon Dots: From Carbon Dots to Carbonized Polymer Dots, Advanced Science, 6 (2019) 1901316.

[48] X. Xu, R. Ray, Y. Gu, H.J. Ploehn, L. Gearheart, K. Raker, W.A. Scrivens, Electrophoretic Analysis and Purification of Fluorescent Single-Walled Carbon Nanotube Fragments, Journal of the American Chemical Society, 126 (2004) 12736–12737.

[49] K.J. Mintz, M. Bartoli, M. Rovere, Y. Zhou, S.D. Hettiarachchi, S. Paudyal, J. Chen, J.B. Domena, P.Y. Liyanage, R. Sampson, D. Khadka, R.R. Pandey, S. Huang, C.C. Chusuei, A. Tagliaferro, R.M. Leblanc, A deep investigation into the structure of carbon dots, Carbon, 173 (2021) 433–447.

[50] Z. Peng, C. Ji, Y. Zhou, T. Zhao, R.M. Leblanc, Polyethylene glycol (PEG) derived carbon dots: Preparation and applications, Applied Materials Today, 20 (2020) 100677.

26 Advanced Polymers for Craniomaxillofacial Reconstruction

Dinesh Rokaya,[1] Ashutosh K. Singh,[2] Sasiwimol Sanohkan,[3] and Suresh Nayar[4, 5]

[1] Department of Clinical Dentistry, Walailak University International College of Dentistry, Bangkok 10400, Thailand
[2] Department of Oral and Maxillofacial Surgery, Tribhuvan University Teaching Hospital, Institute of Medicine, Kathmandu 44600, Nepal
[3] Department of Prosthetic Dentistry, Faculty of Dentistry, Prince of Songkla University, Hat Yai, Songkhla 90110, Thailand
[4] Institute for Reconstructive Sciences in Medicine, 16940, 87 Avenue, Edmonton, Alberta, Canada
[5] Department of Surgery, Faculty of Medicine and Dentistry, University of Alberta, Canada

1 INTRODUCTION

Reconstruction of the maxillofacial and craniofacial defects can be done by autologous bone grafts, which are the gold standard as they have no immunogenicity and biologically integrate with the native bone. However, they have been undermined by their associated complications, namely, donor site morbidity and risk of failure to uptake. There has been renewed interest in the search for alternative techniques that could potentially assuage this problem with reliance on autologous tissue. Various alloplastic biomaterials have been used for this purpose, specifically titanium (Ti) alloys in combination with polymers.

Maxillofacial defects result from congenital, trauma, and cancer surgery which affect the quality of the life. Various prostheses are used for craniomaxillofacial reconstruction which improves the aesthetics and functions to improve the quality of life of patients. Maxillofacial fixation and reconstruction are done following maxillofacial surgery such as trauma surgery, skeletal reconstruction, and orthognathic surgery using various biomaterials.

Maxillofacial defects can be classified as extraoral (involving the eye, orbit, face, nose, ear), intraoral (involving the maxilla and mandible), and combination or complex (involving extraoral and intraoral) [1]. The primary objective of extraoral reconstruction is the aesthetics and the objective of intraoral reconstruction is function. A small defect involves a small area, whereas a large defect involves multiple areas causing difficult reconstruction. Brown and Shaw classified the maxillary and midface defects into the horizontal extent defect (a–d), and the vertical extent defect (classes I–VI) [2] (Figure 26.1a). Vertical extent defects indicate extension into the maxilla and cranium, whereas the horizontal extent defects determine an anterior–posterior extension. Similarly, Cantor and Curtis classified the mandibular defects for guiding surgical and prosthetic rehabilitation (classes I–VI) [3] (Figure 26.1b). This classification provides the extent of the involvement from simple to complex, and guides the rehabilitation. The condylar involvement defects are difficult to resolve during reconstruction as they involve the opening and closing of the mandible.

DOI: 10.1201/9781003278269-26

A. Maxillary Defect   **B. Mandibular Defect**

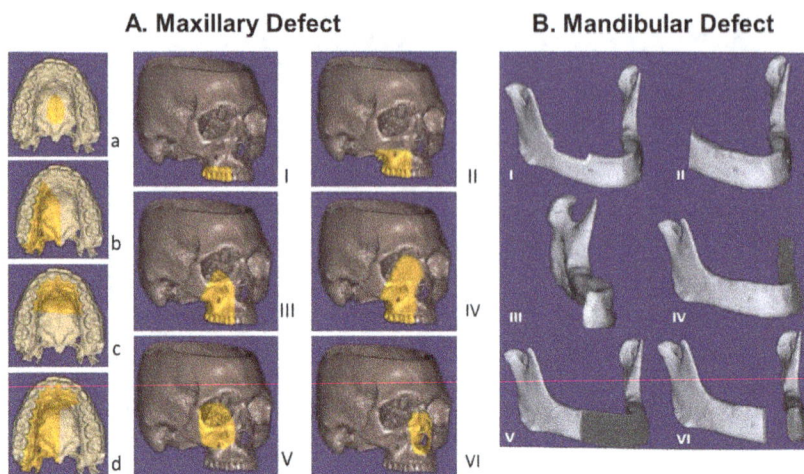

FIGURE 26.1 Various classifications of the maxillary and mandibular defects. Classification of maxillary defects by Brown and Shaw as horizontal extent defect (a–d), and vertical extent defect (classes I–VI) was used [2]. Classification of mandibular defects by Cantor and Curtis (classes I–VI) for surgical and prosthetic rehabilitation [3]. Adapted with permission from Reference [1]. Copyright (2021) The Authors, some rights reserved; exclusive licensee [MDPI]. Distributed under a Creative Commons Attribution License 4.0 (CC BY).

The use of various reconstruction biomaterials has been studied for many decades. The need for safe and effective materials is still increasing due to increasing longevity and highly functional and aesthetic demands (Figure 26.2) [4]. Potential biomaterials should be biocompatible and not cause adverse effects such as inflammation, allergy, carcinogenicity, or toxicity. In addition, the biomaterial should have adequate mechanical properties; adequate mechanical strength to withstand the forces to which they are subjected under varying loading conditions, and high wear resistance in a highly corrosive body environment. In addition, a biomaterial should remain intact for the required period without failure. For long-term implanted biomaterials, the service period in patients is over 15–20 years [5].

Broadly, biomaterials that are used in the human body are classified as (1) metals, (2) ceramics, and (3) polymers. Stainless steel (SS) is not recommended due to the likelihood of corrosion, hypersensitivity, toxicity, and stress protection, as a long-term or permanent implant in maxillofacial fixation [7]. Ti alloy biomaterials are used widely and most often in the fabrication of screws and plates for bone fixation as Ti alloys are biocompatible and show suitable mechanical properties such as sufficient hardness, strength, and Young's modulus for rigid fixation, and Ti is also bone-friendly and can be retained asymptomatically. However, some limitations with Ti alloy biomaterials include that they may result in growth disturbances in children, radiological imaging interference, hypersensitivity reactions, and stress shielding effects. Although ceramics have strengths and are biocompatible, they are brittle. Considering this, various polymers have been tried and utilized for reconstruction of the midface, orbital, temporomandibular joint, and cranial defects with varying success. Figure 26.3 shows the common polymeric biomaterials used for maxillofacial reconstruction.

The interest in absorbable plates and screws has resulted in the development of biodegradable osteofixation devices, which include magnesium (Mg) alloys, bioceramics, bioglasses, and polymers [8]. These biomaterials have unique physical properties, chemical properties, and clinical performance. Magnesium (Mg)-based materials have attracted attention for a few decades as a potential metallic biomaterial alternative to Ti and have shown success in vascular applications [9].

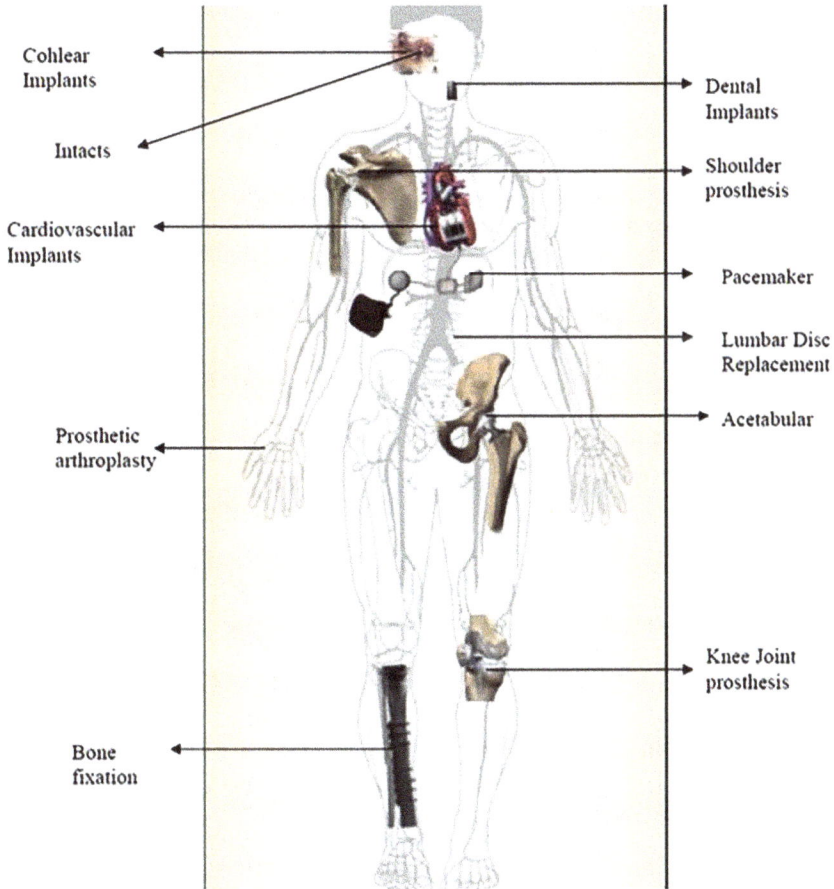

FIGURE 26.2 Various implanted prosthetic biomaterial applications in the human body. Adapted with permission from Reference [6]. Copyright (2009) The Authors, some rights reserved; exclusive licensee [Bentham Open]. Distributed under a Creative Commons Attribution License Non-commercial 3.0 (CC BY NC).

However, Mg-based implants have shown a high corrosion rate which can lead to degradation of the metal with structural failure and cytotoxicity. To prevent such complications, alloying Mg with other elements and surface treatment is on-going. Further long-term studies are needed for their clinical application for osteofixation, especially in the maxillofacial region.

2   POLYMERIC BIOMATERIALS FOR CRANIOMAXILLOFACIAL RECONSTRUCTION

Polymers are chains of macromolecules consisting of repeating structural units of the corresponding monomers [10]. They are produced by chemical reactions by sequential joining of chains of monomers and they can contain similar kinds of monomers or more than one different kind of monomer. Polymers are gaining popularity in various fields of medical science. They are widely used in surgery, prosthetic systems, and drug delivery. At present, special polymers called shape-memory polymers have been used increasingly. Polymers can be natural, semisynthetic, or synthetic [11]. Commonly used polymers for craniofacial reconstruction are silicon, polymethyl methacrylate (PMMA), polyether ether ketone (PEEK), polyether ketone ketone (PEKK), etc.

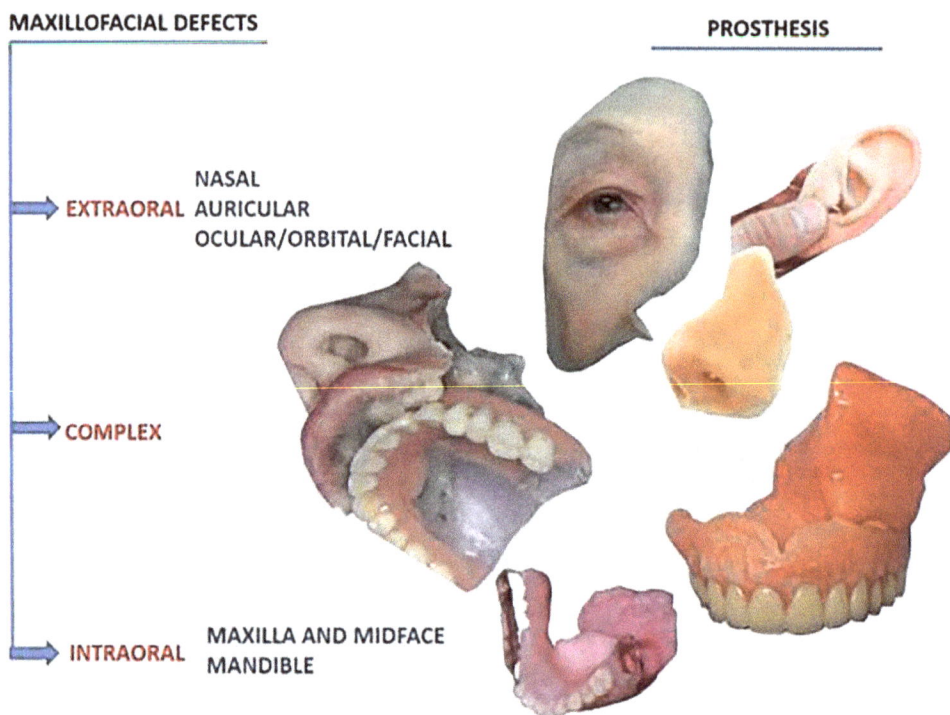

FIGURE 26.3 Use of polymeric materials for maxillofacial prostheses. Adapted with permission from Reference [1]. Copyright (2021) The Authors, some rights reserved; exclusive licensee [MDPI]. Distributed under a Creative Commons Attribution License 4.0 (CC BY).

2.1 SILICONE

Silicone polymers have a long history and silicone prostheses are the most popular material for oropharyngeal reconstruction and are suitable in terms of biocompatibility, morbidity, and functionality. Prostheses made from silicone are highly aesthetic and show a lifelike appearance and present rehabilitative and psychological advantages [12]. The retention of silicone prostheses can be anatomic, or through implants or adhesives.

2.2 PMMA

PMMA is commonly used for the manufacture of various dental prostheses and craniofacial reconstruction such as artificial teeth, temporary or provisional crowns, dentures, and obturators, and is also used for the repair of artificial crowns and dentures. PMMA is aesthetic, low cost, low density, and has ease of manipulation and handling, which make it a popular material for various dental prosthetic applications [13]. To improve further the physical and mechanical properties of PMMA such as wear resistance, impact strength, flexural strength, solubility, and water sorption, different chemical modifications and mechanical reinforcements are being carried out using various nanoparticles, fibers, and nanotubes [13].

2.3 PEEK AND PEKK

PEEK and PEKK are thermoplastic polymers that are chemically inert, biocompatible, and have nearly identical elastic modulus to human bone (7–30 GPa), and present good integration with adjacent bone tissues upon implantation [14]. They have been increasingly used for the fabrication of various prostheses and maxillofacial rehabilitation [14,15]. The PEEK surface can be modified using nanotechnology to improve its bioactivity and osteoconductive properties.

2.4 OTHER POLYMERS

Other polymers used for craniofacial reconstructions include polydimethylsiloxane (PDMA), polydioxanone (PDS), castor oil-derived biopolymer (CODB), polylactic acid (PLA), polyethylene (PE), polydioxanone (PDS), polyglactin (PG), etc. Thus, patient-specific implants with 3D printable polymers are going to be the future for midface and orbital reconstruction as they provide customized patient solutions without any additional morbidity [16].

3 POLYMER MODIFICATIONS

For improvement of the physical, mechanical, and biological properties of polymeric materials, they can be modified. The addition of silver nanoparticles (AgNPs) into polymeric materials has improved their antimicrobial properties by reducing the colonization by microorganisms such as bacteria and fungi from the biomaterials [17,18]. However, the addition of AgNPs into the polymeric materials results in changes in the mechanical properties of the final biomaterials. The addition of AgNPs into silicone has resulted in a reduction of the fatigue strength, although it has increased flexibility and prevents the development of microbial growth [17]. Mechanical testing, scanning electron microscopy (SEM), and Raman spectrometry provide useful information regarding the functional parameters of silicone prostheses with AgNPs. For example, in silicone prostheses, such blending techniques can help to study the polymer network cross-linkage on $-Si-O-$ on $-CH_2-CH_2-$, the vinyl group which increases the elasticity of silicone [17].

Furthermore, a successful coating includes biocompatible, very thin, good adhesion, good flexibility, hardness, and strength. Both synthetic and natural polymers are being extensively studied as polymeric biomaterials. In addition, various polymers and polymer composites that have been attempted to coat NiTi alloys [19] include pyrrole, polypyrrole/HA nanocomposite, polyurethane (PU), graphite–polyurethane (G-PU), polyamide (PA), polyetheretherketone (PEEK), polytetrafluoroethylene (PTFE), parylene, inorganic fullerene like-tungsten disulfide (IF-WS2), hexamethyldisilazane (HMDS), and N-isopropylacrylamide/N-tert-butylacrylamide (NIPAAm/NTBAm) copolymer. It is always difficult to produce successful polymer coatings and the drawbacks of the coatings include roughness, porosity, thickness, non-uniformity, toxicity, and detachment of the coatings [20].

Within the limitations of each research, the polymer to some extent helps to improve the surface properties. Research is focusing on the development of a biocompatible new composite coating with better and long-lasting performance. At present, shape-memory polymers are being used increasingly for biomedical applications.

4 CRANIOFACIAL RECONSTRUCTIONS

Various polymers have been utilized for the reconstruction of the midface, orbit, temporomandibular joint, and cranial defects with varying success.

4.1 MIDFACE AND ORBIT

Midface and orbital wall defects have been conventionally reconstructed with autografts, but with the advent of advanced biocompatible polymers, and 3D printing, the trend has shifted toward alloplastic reconstruction. Alloplastic materials for the orbit and midface include titanium, castor oil-derived biopolymer (CODB), polydioxanone (PDS), L-lactic acid and DL-lactic acid ([L/DL] LA) copolymer, polypropylene (PP), polyethylene (PE), and polydioxanone and polyglactin (PDS/PG) [21]. 3D-printed polymer models can be used to shape the Ti mesh for orbital roof reconstruction [22]. In Figure 26.4, the PMMA model was used to pre-shape the Ti mesh before insertion in a patient with an orbital defect caused by traumatic injury. Finally, the orbital defect was reconstructed using Ti mesh.

PDS is resorbable, thus avoiding displacement of the graft, and there is no need for removal, however the degradation products are inflammatory and incite foreign body reactions. Resorption also alters their mechanical properties, and the material may not support a globe with a large defect area. Thus, resorbable biomaterials [PDS and PL(DL/LA)] are limited to defects smaller than 2 mm, whereas non-resorbable (PP and Ti) biomaterials should be used in the case of larger defects [23]. PP is versatile enough for reconstruction of the ear, nasal midface defects, and malar reconstruction or augmentation [24]. PE implants are porous, thus allowing native tissue ingrowth, which increases implant stability. However, a major limitation is the infection rate, implant exposure or extrusion,

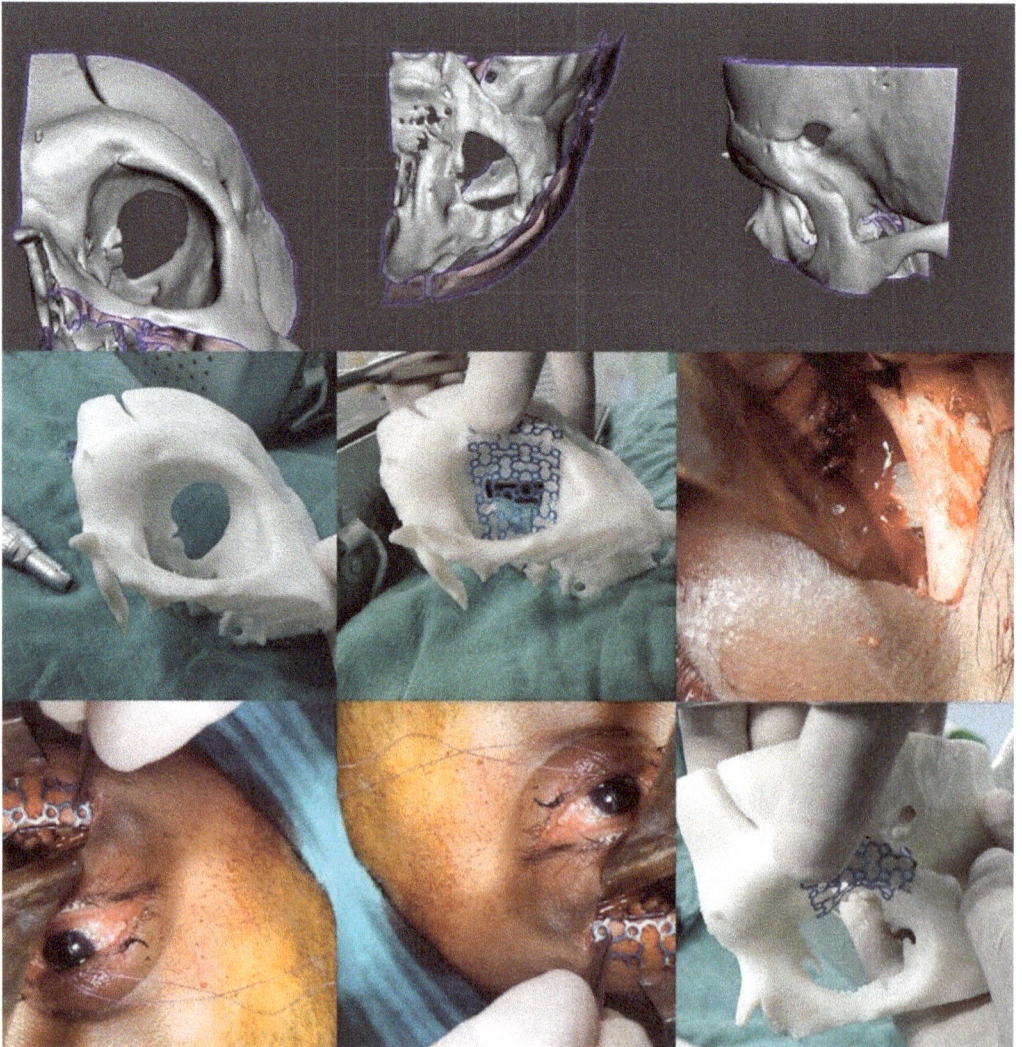

FIGURE 26.4 3D-printed models used to pre-shape the titanium mesh for orbital roof reconstruction.

and less than desired aesthetic appearance [25]. In a recent study [26], various maxillofacial defect reconstruction was done with computer-aided design and computer-aided manufacturing (CAD/CAM). The patients were satisfied functionally and cosmetically with the PEEK implants and no complications were seen in these patients in successive follow-ups.

4.2 MANDIBLE AND TMJ

For mandibular reconstruction, the gold standard is vascular autografts which have effectively been used with a high success rate [27]. Nevertheless, autografts can sometimes be limited by the graft size that can be harvested, dissimilarity in morphology to the native recipient tissue, and donor-site morbidity [28]. These limitations may be overcome by an alloplastic mandible that may satisfy some or all of the prerequisites of an ideal mandible reconstruction material. However, there may be other limitations, which will become apparent with outcome data. Patient-specific implants (PSIs) are customized mandibular implants with dental fixtures that can be the least demanding treatment option for the patient, as no autograft is harvested [29,30]. These implants are manufactured with various titanium alloys, with or without additional polymers, by using CAD/CAM technology [31]. The use of polymers in mandible reconstruction with PSIs is used in fabricating the models and cutting guides. PE polymeric models and cutting guides, and Ti alloy custom PSIs can be used for mandibular reconstruction (Figure 26.5).

Temporomandibular joint replacement is a rapidly growing field of maxillofacial surgery as the treatment of choice for a multitude of temporomandibular disorders, and the material of choice is customized alloplastic joint implants. The TMJ is reconstructed with two components; one rigid metal alloy component for the mandibular segment, and the other is a polymer-based glenoid fossa component. The devices are classified into two types: metal on metal (MoM) and metal on polymer (MoP) [32]. The polymer used for glenoid fossa is dense ultra-high-molecular-weight PE along with

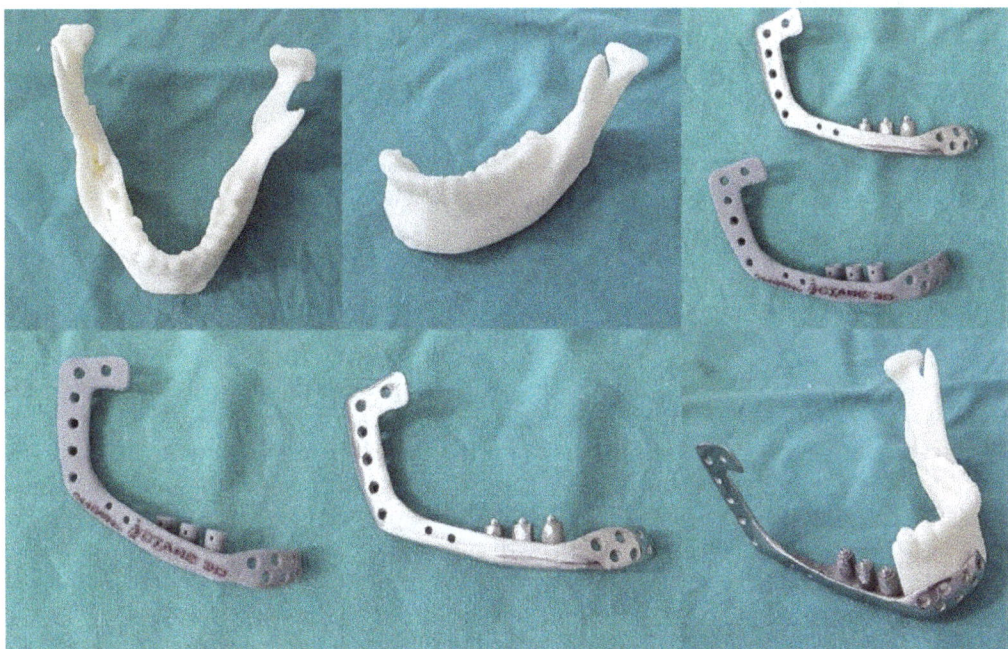

FIGURE 26.5 Polyethylene (PE) polymeric models and cutting guides, and titanium alloy customized patient-specific implants (PSI) for mandibular reconstruction.

titanium alloy, whereas the mandibular component is usually MoM titanium alloy [33]. Long-term outcomes data still need to be considered before this modality becomes mainstream.

4.3　CRANIUM

Craniotomy defects include a loss of cranial bone tissues caused by head injuries, cerebral tumors, cerebrovascular accidents, cerebral infections, and an abscess that requires delayed reconstruction later. Cranioplasty is the procedure for the repair of cranial defects using a diverse group of materials, including but not limited to autologous bone and alloplastic materials. The ideal material for cranioplasty should protect the brain, provide structural integrity, and regain the aesthetic appearance of the head while being free of complications. Autologous bone can be used for a small defect, but a larger defect requires large bone coverage, which is not possible with autografts. Additionally, autografts require a donor site and may cause local site infections and other morbidities. Thus, alloplastic materials are preferred by neurosurgeons for their ease of use and availability. The most commonly used materials are Ti mesh and PMMA. Ti mesh is used in the form of a mesh that is cut and shaped to the defect, whereas the PMMA is available as "bone cement"; a term used for its dough-like form which is then molded into the craniotomy defect by hand [34].

The conventional technique did not satisfy some desired properties for cranioplasties such as regaining the aesthetics, and was marred by complications such as mesh exposure in the case of titanium, and potential foreign body reactions with freehand-shaped PMMA. These problems led to the development of custom 3D-printed cranioplasty implants. These are manufactured by an additive manufacturing process, utilizing a variety of conventional and novel polymers. PMMA is the most widely used polymer for 3D-printed cranioplasty. Figure 26.6 shows the reconstruction of an orbital defect with a 3D-printed PMMA implant. The advantages are that it is customizable, hard, and biocompatible. The impact behavior of the printed PMMA plays a vital role in the design process of craniofacial implants [35]. However, PMMA implants do not allow infiltration by new bone tissue due to the lack of porosity; additionally, they interfere with osteoconduction and neovascularization, thus they are susceptible to infection [36].

Recently, PEEK has also been widely studied and has shown better properties over PMMA in biocompatibility, strength, and elastic modulus [37]. In addition, various studies have found an advantage of PEEK over Ti, though superiority over PMMA could not be established [38]. The 3D-printable PEEK potentially could be the desired cranioplasty biomaterial in the near future. In addition, the integration of bioactive compounds and surface modifications in PEEK may be the way forward for cranioplasty.

5　POLYMERIC MATERIALS AND SCAFFOLDS FOR TISSUE ENGINEERING IN DENTISTRY

With the development of the tissue bioengineering and regenerative medicine discipline, various clinical advancements in the reconstruction of maxillofacial defects are evolving. There have been widespread studies on the development and progress of strategies to restore and replace the biological tissues and functions [39]. Recent developments include the effective regeneration of damaged or necrotic tissue by replenishing cells and regeneration of the cells/tissues [40].

Tissue engineering is facilitated by scaffolds which are thin temporary frameworks to provide a 3D environment for cell attachment, differentiation, and proliferation, and to produce the desired tissue. They help the inflow of oxygen for the metabolism of cells for tissue regeneration and permit the delivery of growth factors. Scaffolds are made from polymers, ceramics, or composite films. Polymer scaffold possesses favorable properties such as adequate surface roughness and porosity, degradation of the scaffold, suitable rheological behavior, and mechanical properties. Various forms

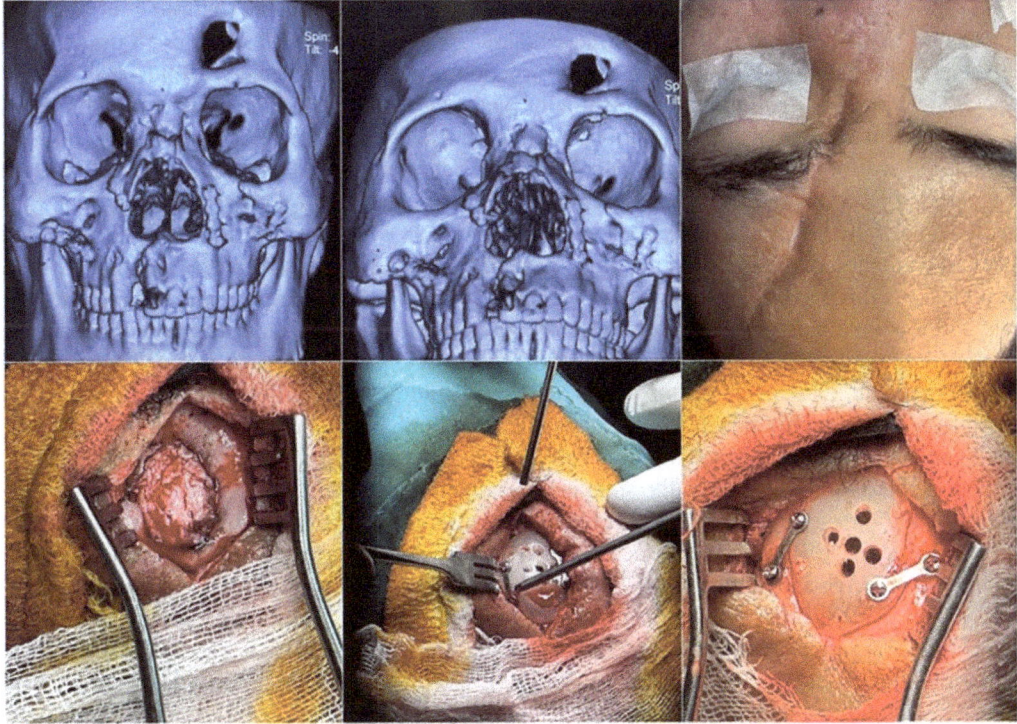

FIGURE 26.6 3D-printed PMMA implant used in reconstruction of the frontal defect.

of polylactic glycolic acid (PLGA)-based scaffolds and matrices can be used for oral and craniofacial tissue engineering and regeneration [41].

Additionally, 3D-printed scaffolds of various polymers are being used for tissue engineering efforts toward a novel treatment strategy for mandibular defects. Novel scaffolds which are used are made from composite materials which include beta-tricalcium phosphate (β-TCP), hydroxyapatite (HAP), polylactic-co-glycolic acid (PLGA), customized borosilicate glass, ceramics, or alloys containing Ti. Scaffolds for small mandibular defects are sponges, hydrogels, and other malleable biomaterials, whereas for larger defects, large scaffolds and larger solid constructs made from load-bearing materials ceramics and alloys are used [42].

For the alveolar ridge reconstruction in prosthetic and implant dentistry, guided bone regeneration is a promising method. In this technique, a barrier membrane is placed between the gingival tissue and the bone-defect site, and this material prevents the invasion of soft-tissue cells but helps the osteoblasts and osteoprogenitor cells from the surrounding bone. Thus, this membrane helps in guiding new bone formation in the preferred shape [43]. The GBR membrane should have sufficient mechanical strength and be biocompatible and flexible. Non-resorbable membranes consist of metal and synthetic polymers, whereas resorbable membranes consist of synthetic polymers and collagen. Both non-resorbable and resorbable membranes can be used as GBR membranes. Polytetrafluoroethylene (PTFE) expanded polytetrafluoroethylene (ePTFE) or Ti reinforced polytetrafluoroethylene (TR-PTFE) have been commonly used for convenience [44]. PTFE is a fluorinated linear thermoplastic polymer and ePTFE is its porous form produced by emulsion polymerization. ePTFE films are hydrophobic and have high toughness and non-adhesiveness, and wide biomaterial applications. Their nanofibrillar structure is favorable in cell cultures, which can provide the biomimetic environment for reproducible cell phenotypes. The microporous structure of ePTFE is obtained by rapid stretching of the extruded tube at high temperatures.

For the scaffold to be effective, porosity and pore interconnectivity in the scaffold with its mechanical integrity are important for the successful transport of mechanical stimulus to the scaffold cells. Over the traditional autografting, reconstruction using bioengineered scaffolds is an advantage as it helps in preventing the provocation of pain and reduces the site morbidity which creates an environment to stimulate cell migration, proliferation, and reattachment of cells [40]. In addition, ideally, the scaffold should reabsorb after completing its function (forming a template for tissue regeneration). Various materials have been attempted for use as scaffold film. McBane et al. [45] studied a polar, hydrophobic, ionic polyurethane film in vivo which was degradable, and they found that human monocytes seeded into the films were differentiated into macrophages and that the cells could migrate through the 3D matrix. They concluded that a high wound-healing phenotype monocyte and low inflammatory can control the foreign body reaction and order the cells into suitable phenotypes for effective tissue engineering.

Stem cells are clonogenic cells that can self-renew and generate differentiated progenies. It is possible to isolate highly proliferative clonogenic dental pulp and characterize other kinds of cells like bone marrow stem cells and multiple cell lineages, such as neurons, chondrocytes, adipocytes, and odontoblasts [46]. Similarly, the stem cells obtained from deciduous teeth can undergo endothelial, chondrogenic, adipogenic, odontoblastic, and osteogenic differentiation. The combination of polymeric biomaterials and growth factors has led to the development of new treatment opportunities in biomedicine and dentistry. In dentistry, these may be important in periodontal tissue regeneration, jaw bone defects, peri-implantitis, soft tissue defects, and restorative procedures for the regeneration of pulp, dentin, and enamel. Wang et al. [47] investigated the differentiation of human dental pulp tissue cells on nanofibrous (NF)-poly(L-lactic acid) (PLLA) scaffolds in vitro and in vivo. They found that the combination of BMP-7 and dexamethasone can induce odontogenic differentiation more efficiently than dexamethasone alone. The NF-PLLA scaffold with odontogenic inductive factors resulted in a suitable environment for dental pulp stem cells to regenerate dentin and dental pulp.

Periodontal diseases destroy the gingival tissue and alveolar bone. Tissue engineering can be an alternative to the common treatments of periodontal diseases such as scaling for the regeneration of gingival tissue. Gingival tissues are highly vascularized for enabling the transportation of nutrients and metabolites. Consequently, it would be expected that tissue-engineered structures could be produced under perfusion to develop more quickly and hold a more physiologic phenotype. Similarly, Cheung et al. [48] studied human gingival fibroblasts that were cultured on degradable polar hydrophobic ionic polyurethane scaffolds in a dynamic culture in tissue-engineered scaffolds. They found that the growth of gingival fibroblasts was continuous over 4 weeks (3-fold increase) and was reduced after 2 weeks of culture with no flow condition. This shows that gingival fibroblasts in the diffused scaffolds show various cell phenotypes which favor tissue regeneration.

Recently, graphene-based scaffolds have found interest in tissue engineering and regeneration. Zhang et al. [49] studied graphene oxide nanosheets coated and aligned on PLLA nanofibrous scaffolds. The graphene oxide-coated and PLLA nanofibrous scaffolds with nerve growth factors considerably supported the proliferation of various nerve cells (Schwann cells pheochromocytoma 12 cells) and neurite growth on the nanofibrous alignment. Hence, graphene oxide-based scaffolds with nanofibrous surface topography may be used in nerve regeneration in medicine and dentistry.

Recently, a 3D printing method has been implemented in tissue engineering which can accurately control scaffold micron-scale architecture [50]. Wang et al. [50] mentioned that cold atmospheric plasma is a quick and economic method to modify the composition and roughness of 3D-printed scaffold films. The cold atmospheric plasma has wide applications, such as sterilization, surface modification, antitumor, blood coagulation, and low temperature. Studies have shown that both nanoscale roughness and hydrophilicity changes to these scaffolds play a vital role in helping attachment of the bone and mesenchymal stem cells [50].

6   LIMITATIONS AND FUTURE DEVELOPMENTS

Alternatives to Ti plates and screws for bone fixation surgery are polymeric materials. Such bioresorbable materials use is increasing, to avoid secondary surgery to remove plates and screws, and they have adequate satisfactory and physical properties that are comparable with those of Ti. There are some limitations to the use of polymeric materials in clinical practice such as questionable strength and cumbersome molding procedures. Still, there is increasing interest in such resorbable metal materials for osteofixation. The 3D-printed polymeric scaffolds have some limitations such as less mimicking of the nanoscale structure of the tissues. Furthermore, the introduction of nanoparticles as bacteriostatic agents or drugs for local therapy of different agents in polymeric materials is a real possibility, but long-term studies are needed.

7   CONCLUSION

Various prostheses are used for craniomaxillofacial reconstruction, which improves the aesthetics and functions to improve the quality of life of patients. Various polymeric thin films have a wide range of applications in dentistry. Broadly, biomaterials that are used in the human body are classified as metals, ceramics, and polymers. The polymeric biomaterials have provided alternatives to treat patients in the clinics and have also provided a therapeutic strategy for maxillofacial surgery and prosthetic applications. The patient-specific implants with 3D-printable polymers are going to be the future of midface and orbital reconstruction as they provide customized patient solutions without any donor site morbidity. In addition, advanced polymers are being used in tissue engineering in dentistry as a tissue growth scaffold, with both cell and extracellular matrix support. Polymeric materials can be modified to improve their mechanical and biological properties. However, outcome data associated with their use will need to be evaluated before these materials can be considered to be mainstream.

REFERENCES

1. C.M. Cristache, I. Tudor, L. Moraru, G. Cristache, A. Lanza, M. Burlibasa, Digital workflow in maxillofacial prosthodontics–An update on defect data acquisition, editing and design using Open-Source and commercial available software. Appl. Sci. 11 (2021) 973.
2. J.S. Brown, R.J. Shaw, Reconstruction of the maxilla and midface: introducing a new classification. Lancet Oncol. 11 (2010) 1001–1008.
3. (a) Cantor, R. and T.A. Curtis, Prosthetic management of edentulous mandibulectomy patients. Part I. Anatomic, physiologic, and psychologic considerations. J. Prosthet. Dent. 25 (1971) 446–457.
   (b) A. Nouri, C. Wen, Introduction to surface coating and modification for metallic biomaterials, in Surface Coating and Modification of Metallic Biomaterials, C. Wen, Editor. Elsevier Science & Technology. Woodhead Publishing Series in Biomaterials: Sydney. (2015) 3–60.
4. Y. Sato, N. Kitagawa, A. Isobe, Implant treatment in ultra-aged society. Jpn. Dent. Sci. Rev. 54 (2018) 45–51.
5. G. Manivasagam, D. Dhinasekaran, A. Rajamanickam, Biomedical implants. Corrosion and its prevention – a review. Recent Pat. Mater. Sci. 2 (2010) 40–54.
6. M.C. Devendrappa, M.D. Kulkarni, N. Haidry, P. Kulkarni, F. Verma, D.A. Pawar, Evaluation of surface changes of stainless steel miniplates and screws following retrieval from maxillofacial trauma and orthognathic surgery patients: A comparative study. Natl. J. Maxillofac. Surg. 12 (2021) 357–360.
7. H. Zhou, B. Liang, H. Jiang, Z. Deng, K. Yu, Magnesium-based biomaterials as emerging agents for bone repair and regeneration: from mechanism to application. J. Magnes. Alloy. 9 (2021) 779–804.
8. S. Amukarimi, M. Mozafari, Biodegradable magnesium-based biomaterials: An overview of challenges and opportunities. MedComm 2 (2021) 123–144.
9. F. Schellhammer, M. Walter, A. Berlis, H.-G. Bloss, E. Wellens, M. Schumacher, Polyethylene terephthalate and polyurethane coatings for endovascular stents: Preliminary results in canine experimental arteriovenous fistulas. Radiology 221 (1999) 169–175.

10. C. Carolina Gutierrez, B. Veerle, M. Arn, Synthetic, natural, and semisynthetic polymer carriers for controlled nitric oxide release in dermal applications: A review. Polymers 13 (2021) 760.

11. P. Amornvit, D. Rokaya, K. Keawcharoen, S. Raucharernporn, N. Thongpulsawasdi, One- vs two stage surgery technique for implant placement in finger prosthesis. J. Clin. Diagnostic Res. 7 (2013) 1956–1968.

12. M.S. Zafar, Prosthodontic applications of polymethyl methacrylate (PMMA): An update. Polymers 12 (2020) 2299.

13. H. Alqurashi, Z. Khurshid, A.U.Y. Syed, S. Rashid Habib, D. Rokaya, M.S. Zafar, Polyetherketoneketone (PEKK): An emerging biomaterial for oral implants and dental prostheses. J. Adv. Res. 28 (2021) 87–95.

14. P. Amornvit, D. Rokaya, S. Bajracharya, K. Keawcharoen, W. Supavanich, Management of obstructive sleep apnea with implant retained mandibular advancement device. World J. Dent. 5 (2014) 184–189.

15. P. Honigmann, N. Sharma, B. Okolo, U. Popp, B. Msallem, F.M. Thieringer, Patient-specific surgical implants made of 3D printed PEEK: Material, technology, and scope of surgical application. BioMed Res. Int. 2018 (2018) 4520636.

16. R. Grigore, B. Popescu, V.G. Berteşteanu Ş, C. Nichita, I.D. Oaşă, G.S. Munteanu, A. Nicolaescu, P.L. Bejenaru, C.B. Simion-Antonie, D. Ene, R. Ene, The role of biomaterials in upper digestive tract transoral reconstruction. Materials 14 (2021) 1436.

17. D. Rokaya, V. Srimaneepong, J. Qin, K. Siraleartmukul, V. Siriwongrungson, Graphene oxide/silver nanoparticle coating produced by electrophoretic deposition improved the mechanical and tribological properties of NiTi alloy for biomedical applications. J. Nanosci. Nanotechnol. 19 (2019) 3804–3810.

18. S.K. Patel, B. Behera, B. Swain, R. Roshan, D. Sahoo, A. Behera, A review on NiTi alloys for biomedical applications and their biocompatibility. Mater. Today: Proc. 33 (2020) 5548–5551.

19. M.A. Raza, Z.U. Rehman, F.A. Ghauri, A. Ahmad, R. Ahmad, M. Raffi, Corrosion study of electrophoretically deposited graphene oxide coatings on copper metal. Thin Solid Films 620 (2016) 150–159.

20. A.K. Saha, S. Samaddar, A. Kumar, A. Chakraborty, B. Deb, A comparative study of orbital blow out fracture repair, using autogenous bone graft and alloplastic materials. Indian J. Otolaryngol. Head Neck Surg. 71 (2019) 542–549.

21. (a) S. Raisian, H.R. Fallahi, K.S. Khiabani, M. Heidarizadeh, S. Azdoo, Customized titanium mesh based on the 3D printed model vs. manual intraoperative bending of titanium mMesh for feconstructing of orbital bone fracture: A randomized clinical trial. Rev. Recent Clin. Trials 12 (2017) 154–158.
    (b) A. Hudecki, W. Wolany, W. Likus, J. Markowski, R. Wilk, A. Kolano-Burian, K. Łuczak, M. Zorychta, M. Kawecki, M.J. Łos, Orbital reconstruction – applied materials, therapeutic agents and clinical problems of restoration of defects. Eur. J. Pharmacol. 892 (2021) 173766.

22. Ridwan-Pramana, J. Wolff, A. Raziei, C.E. Ashton-James, T. Forouzanfar, Porous polyethylene implants in facial reconstruction: Outcome and complications. J. Craniomaxillofac. Surg. 43 (2015) 1330–1334.

23. M. Khorasani, P. Janbaz, F. Rayati, Maxillofacial reconstruction with Medpor porous polyethylene implant: a case series study. J. Korean Assoc. Oral Maxillofac. Surg. 44 (2018) 128–135.

24. D. Tantawi, S. Eberlin, J. Calvert, Midface implants: surgical and nonsurgical alternatives. Clin. Plast. Surg. 42 (2015) 123–127.

25. E.M. Genden, Reconstruction of the mandible and the maxilla: the evolution of surgical technique. Arch. Facial Plast. Surg. 12 (2010) 87–90.

26. M. Bak, A.S. Jacobson, D. Buchbinder, M.L. Urken, Contemporary reconstruction of the mandible. Oral Oncol. 46 (2010) 71–76.

27. K.T. Vakharia, N.B. Natoli, T.S. Johnson, Stereolithography-aided reconstruction of the mandible. Plast. Reconstr. Surg. 129 (2012) 194e–195e.

28. J.P. Levine, A. Patel, P.B. Saadeh, D.L. Hirsch, Computer-aided design and manufacturing in craniomaxillofacial surgery: the new state of the art. J. Craniofac. Surg. 23 (2012) 288–293.

29. K. Darwich, M.B. Ismail, M.Y.A. Al-Mozaiek, A. Alhelwani, Reconstruction of mandible using a computer-designed 3D-printed patient-specific titanium implant: a case report. Oral Maxillofac. Surg. 25 (2021) 103–111.

30. A.J. Sidebottom, Alloplastic or autogenous reconstruction of the TMJ. J. Oral Biol. Craniofac. Res. 3 (2013) 135–139.

31. S.K.R. Chowdhury, V. Saxena, K. Rajkumar, R.A. Shadamarshan, Evaluation of total alloplastic temporomandibular joint replacement in TMJ ankylosis. J. Oral Maxillofac. Surg. 18 (2019) 293–298.

32. Morselli, I. Zaed, M.P. Tropeano, G. Cataletti, C. Iaccarino, Z. Rossini, F. Servadei, Comparison between the different types of heterologous materials used in cranioplasty: a systematic review of the literature. J. Neurosurg. Sci. 63 (2019) 723–736.

33. S. Petersmann, M. Spoerk, P. Huber, M. Lang, G. Pinter, F. Arbeiter, Impact Optimization of 3D-printed poly(methyl methacrylate) for cranial implants. Macromol. Mater. Eng. 304 (2019) 1900263.

34. M.P. Nikolova, M.S. Chavali, Recent advances in biomaterials for 3D scaffolds: A review. Bioact. Mater. 4 (2019) 271–292.

35. S. Najeeb, M.S. Zafar, Z. Khurshid, F. Siddiqui, Applications of polyetheretherketone (PEEK) in oral implantology and prosthodontics. J. Prosthodont. Res., 2016. 60(1): p. 12–19.

36. Z.Y. Ng, I. Nawaz, Computer-designed PEEK implants: a peek into the future of cranioplasty? J. Craniofac. Surg. 25 (2014) e55–e58.

37. V. Rosa, A.D. Bona, B.N. Cavalcanti, J.E. Nör, Tissue engineering: from research to dental clinics. Dent. Mater. 28 (2012) 341–348.

38. H.E. Jazayeri, M.D. Fahmy, M. Razavi, B.E. Stein, A. Nowman, R.M. Masri, L. Tayebi, Dental applications of natural-origin polymers in hard and soft tissue engineering. J. Prosthodont. 25 (2016) 510–517.

39. E.K. Moioli, P.A. Clark, X. Xin, S. Lal, J.J. Mao, Matrices and scaffolds for drug delivery in dental, oral and craniofacial tissue engineering. Adv. Drug Deliv. Rev., 2007. 59: p. 308–324.

40. J.M Latimer, S. Maekawa, Y. Yao, D.T. Wu, M. Chen, W.V. Giannobile, Regenerative medicine technologies to treat dental, oral, and craniofacial defects. Front. Bioeng. Biotechnol. 9 (2021) 704048.

41. Y. Kinoshita, H. Maeda, Recent developments of functional scaffolds for craniomaxillofacial bone tissue engineering applications. ScientificWorldJournal 2013 (2013) 863157.

42. A.I. Cassady, N.M. Hidzir, L. Grøndahl, Enhancing expanded poly(tetrafluoroethylene) (ePTFE) for biomaterials applications. J. Appl. Polym. Sci. 131 (2014) 40533.

43. J.E. McBane, D. Ebadi, S. Sharifpoor, R.S. Labow, J.P. Santerre, Differentiation of monocytes on a degradable, polar, hydrophobic, ionic polyurethane: Two-dimensional films vs. three-dimensional scaffolds. Acta Biomater. 7 (2011) 115–122.

44. N. Koyama, Y. Okubo, K. Nakao, K. Bessho, Evaluation of pluripotency in human dental pulp cells. J. Oral Maxillofac. Surg. 67 (2009) 501–506.

45. J. Wang, X. Liu, X. Jin, H. Ma, J. Hu, L. Ni, P.X. Ma, The odontogenic differentiation of human dental pulp stem cells on nanofibrous poly(L-lactic acid) scaffolds in vitro and in vivo. Acta Biomater. 6 (2010) 3856–3863.

46. Cheung, J.W.C., E.E. Rose, J.P. Santerre, Perfused culture of gingival fibroblasts in a degradable/polar/hydrophobic/ionic polyurethane (D-PHI) scaffold leads to enhanced proliferation and metabolic activity. Acta Biomater. 9 (2013) 6867–6875.

47. K. Zhanga, H. Zheng, S. Liang, C. Gao, Aligned PLLA nanofibrous scaffolds coated with graphene oxide for promoting neural cell growth. Acta Biomater. 37 (2016) 131–142.

48. M. Wang, P. Favi, X. Cheng, N.H. Golshan, K.S. Ziemer, M. Keidar, T.J. Webster, Cold atmospheric plasma (CAP) surface nanomodified 3D printed pol ylactic acid (PLA) scaffolds for bone regeneration. Acta Biomater. 46 (2016) 256–265.

27 Emerging Applications of Polymers for Supercapacitors

Larissa Bach-Toledo,[1] Bruna M. Hryniewicz,[1] Gabriela De Alvarenga,[1] Tatiana L. Valerio,[1] Rafael J. Silva,[1] Jean Gustavo de A. Ruthes,[1] Andrei Deller,[1] Vanessa Klobukoski,[1] Franciele Wolfart,[2] and Marcio Vidotti[1]

[1] Grupo de Pesquisas em Macromoléculas e Interfaces (GPMIn), Universidade Federal do Paraná, CP 19032, CEP 81531-980 Curitiba, PR, Brazil
[2] Instituto Federal de Educação, Ciência e Tecnologia Farroupilha – Campus São Borja, Rua Otaviano Castilho Mendes, 355, Betim, CEP 97670-000, São Borja – RS, Brazil

1 INTRODUCTION

The global demand for new and renewable energy sources is still growing, but more than just for generating energy, one of the major challenges is the lack of a stable mechanism to store the generated energy [1]. Several technologies have attempted to solve this problem, constructing high-performance batteries, capacitors, supercapacitors, and even the combination of different systems, such as the supercapacitor/battery hybrid [1].

In energy-storage devices, power and energy density are the key parameters, and the relation between those is shown in the Ragone plot, represented in Figure 27.1 [2]. The energy density (Wh/kg) in the vertical axis represents the amount of available energy, while the power density (W/kg) in the horizontal axis shows how quickly that energy can be delivered. Both axes are on a logarithmic scale, which allows the inclusion of different types of devices.

The ratio between energy and power density results in the discharge time, represented in the graph by cross lines, which indicate the time required for charge/discharge of the device. Supercapacitors are in the middle of the graph, having high power density and rapid charge/discharge times [2].

The high capacitance of supercapacitors is a direct consequence of the chosen design and material, which are summarized into two major forms of storage – the electric double-layer capacitance (EDLC) and pseudocapacitance – and the combination of both forms is also explored and studied [1,3].

EDLC supercapacitors store charge electrostatically, similarly to dielectric capacitors, where the charge accumulation happens at the electrode–electrolyte interface, relying on the physical adsorption of electrolyte ions at the electrode surface. Carbon-based materials are the most used in EDLC electrodes, due to their large surface area, which maximizes the electric double-layer effect [4]. Pseudocapacitors rely on the faradaic reactions occurring inside the electroactive material. These redox reactions are fast and reversible electronic transfers [3,4]. Electroactive transition-metal oxides (e.g. MnO_2) and conducting polymers (CPs) are two of the most used materials for these devices [3,5,6].

DOI: 10.1201/9781003278269-27

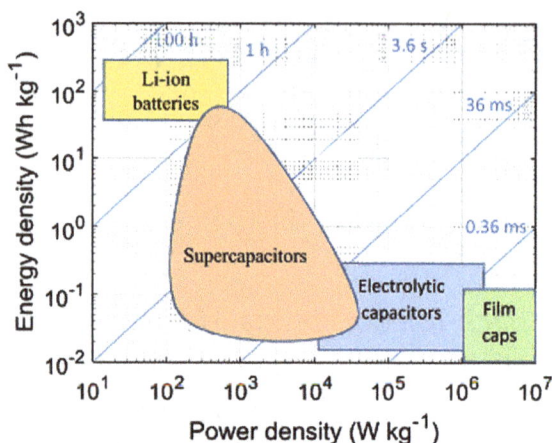

FIGURE 27.1 Ragone plot of energy storage systems. Lines in blue correspond to characteristic times for device charge/discharge. Adapted with permission from Reference [2]. Copyright 2019. Copyright Berrueta et al., some rights reserved; exclusive licensee IEEE. Distributed under a Creative Commons Attribution License 4.0 (CC BY) https://creativecommons.org/licenses/by/4.0/.

The interest in CPs for supercapacitors comes from their efficient electrical and optical properties, high carrier mobility, and stimuli-responsiveness [7,8]. CPs are largely used for the construction of pseudocapacitors, where the charging of the electrical double layer and fast reversible redox reaction is key to the device's operation [9].

There are many CPs and their derivatives, but polyaniline (PANI), polypyrrole (PPy), polythiophene (PTh), and poly(3,4-ethylenedioxythiophene) (PEDOT) are the most used. The structure of CPs has conjugated double bonds, where the π-electrons can be added or removed, forming a polymeric ion. This process is called doping, and is the key point for the CP's variable conductivity, where the polymer oscillates between its conductive and non-conductive forms depending on its redox state. The charge that the polymer acquires during this process is counterbalanced by the presence of counter-ions called dopants [7,8,10].

The use of CPs also allows the production of flexible devices capable of maintaining high performance. Also, they offer a significant cost reduction when compared to other materials. The combination of CPs with other materials, such as metals and oxides, further optimizes their properties. The combination of CPs with metal nanoparticles, for example, can improve both thermal and electrical conductivity, processability, mechanical and cycling stability, and increase surface area. The overall enhancement in the polymer-chain stability and electrode–electrolyte interfacial contact is capable of improving greatly the resulting supercapacitor performance [7,11].

The electrolyte – a solvent combined with a salt – also plays a vital role in the electrochemical properties, stability, and lifespan of a supercapacitor [12]. The nature of the electrolyte depends on the ion type, size and concentration, and also on the solvent concentration [13]. To achieve high energy and power densities, the electrolytes must have: high ionic conductivity, minimizing the internal resistance, especially during large current discharges; and wide electrochemical stability and operating temperature window, ensuring a long life for the device. For practical applications, these electrolytes must also be chemically inert, environmentally friendly, and relatively inexpensive [14].

The several existent types of electrolytes are divided into aqueous, organic, ionic liquid, and solid or quasi-solid-state [12–14]. Aqueous electrolytes are the most widespread, but their main disadvantages are high volatility and narrow electrochemical potential window [14]. The electrochemical potential window can be increased by organic electrolytes, but they have low ionic conductivity and safety issues due to their flammable nature [13]. Ionic liquids (ILs) – liquids consisting of separate ions without any solvent – are also used as electrolytes, having low vapor pressure, wide and stable electrochemical potential windows, and high thermal and electrochemical stability. However, ILs are sensitive to moisture, which requires sophisticated purification and ultra-dry manufacturing procedures [13,14].

Solid-state and quasi-solid-state electrolytes can be classified as polymer-based and inorganic. They have high specific energy, a wide electrochemical potential window, and no leakage, aiding in the development of ultrathin, flexible, and miniaturized supercapacitors [14]. The polymeric solid-state electrolytes stand out, usually being solvent-free, lightweight, flexible, thin, and transparent, while maintaining high ionic conductivity and a wide electrochemical window [15].

Polymeric electrolytes (PEs) are composed of a host polymer and inorganic salts, where the polymer acts as a matrix, expanded by incorporating the solution species [15,16]. There are two types of PE based on their sources and origins: (i) synthetic and (ii) natural [15,17]. Several synthetic polymers have been used, such as poly(ethylene oxide) (PEO), polyethylene glycol (PEG), polypropylene oxide (PPO), poly(vinyl alcohol) (PVA), and also copolymers.

Among the synthetic polymers, PVA is one of the most widely investigated [13], due to its inherent characteristics such as good tensile and mechanical strength, non-toxicity, cost-effectiveness, good optical properties, high-temperature resistance, biocompatibility, ease of preparation, and high flexibility. PVA also has a greater extent of polar groups (hydroxyl groups) and high chain flexibility, which promotes salt-solvation [13,15]. Polysaccharide-based polymers have been used to replace synthetic polymers, such as chitosan [18], cellulose [19], and alginate [20], mainly because there are biodegradable, non-toxic, reusable, and recyclable, minimizing waste [15,17].

Herein, this chapter discusses conducting polymer-based electrodes and polymeric electrolytes used in the development of supercapacitors. It covers the synthetic strategies of conducting polymers and their hybrids, and of the polymeric electrolytes, wrapping up with the mechanisms and characteristics of the devices themselves.

2 CONDUCTING POLYMERS

Intrinsically conducting polymers are an outstanding group of organic materials that can be easily synthesized to form a wide variety of micro- and nano-architectures, attaining customized chemical and physical behaviors. CPs can be synthesized by different methods, and the electrochemical and chemical approaches are the most widespread methodologies [21].

The best known characteristic of CPs is their tunable conductivity, which arises from their extended π-conjugated molecular backbone. The oxidation of a CP creates localized positive charges, usually known as polarons, that are responsible for increasing the mobility of π-electrons, also increasing the polymer conductivity. To maintain electroneutrality in the polymer backbone, negatively charged counterions, known as dopants, interact with the polymeric matrix, and can increase their stability and conductivity through stabilization of the positive charges. Hence, the conductive behavior of these polymers can be fine-tuned from the insulative range to the metal range via electrochemical or chemical doping/de-doping [22].

2.1 CPs-Based Supercapacitors

Two charge-storage mechanisms are involved in the operation of supercapacitors. The EDLC supercapacitor stores charge at the electrode–electrolyte interface, through the electric double-layer capacitance, which responds rapidly to the potential changes and does not interfere with the properties of the materials, resulting in a longer cycle life. It is usually formed by carbonaceous materials with a high surface area, such as powders, felts, and activated-carbon electrodes [4]. The pseudocapacitor stores charge through fast and reversible redox reactions (faradaic processes) of the electroactive material, in addition to the double-layer capacitance. The faradaic processes are potential-dependent and lead to a higher capacitance and energy density than EDLC, but the multiple redox reactions can induce changes in material morphology and electroactivity, leading to a shorter cycle life [23].

Conducting polymers have been extensively explored to make pseudocapacitors [24]. For CPs, the doping/de-doping of the polymeric matrix is responsible for the pseudocapacitance [25]. In this way, CPs use both the electrode–electrolyte interface (formation of an electric double layer) and the inner surface area (through redox reactions) for the charge storage [24]. There are many ways to evaluate the supercapacitor metrics, but the most used instruments are cyclic voltammetry (CV), galvanostatic charge and discharge cycles (GCDC), cycling stability, and Ragone plot.

CV is performed by a linear change in the potential in the working electrode with respect to the reference electrode in a three-electrode configuration, or between the positive and negative electrodes in a two-electrode device, in a direct and reverse scan between two predefined potentials [23]. For a three-electrode configuration, the potential window is defined by the electrolyte and materials stability, while for a two-electrode system, the CV potential window is in the positive range (from 0.0 V) to the maximum operating voltage [26]. CV generates a graph of current vs. the applied potential, where both capacitive and faradaic current will be shown overlapped. EDLC materials present a quasi-rectangular CV profile (Figure 27.2a), while pseudocapacitor materials display the redox peaks from the faradaic processes (Figure 27.2b). Pseudocapacitor materials are often confused with battery-type materials. These materials show well-defined redox peaks in the CV (Figure 27.2c) [27], while for pseudocapacitors a more rectangular shape is observed. Furthermore, pseudocapacitor materials should respond linearly to the scanning rate, have reversible charge transfer processes, and should not present phase changes during cycling [26].

The GCDC technique uses a constant current and measures the potential overtime in charging and discharging processes. The potential range is chosen for the CV. The two supercapacitor sub-categories show distinct GCDC profiles: while EDLC materials present a linear charge–discharge profile with a triangular shape (Figure 27.2d), the pseudocapacitor materials show a nonlinear profile due to the redox reactions (Figure 27.2e) [28]. In the GCDC of batteries, the potential remains

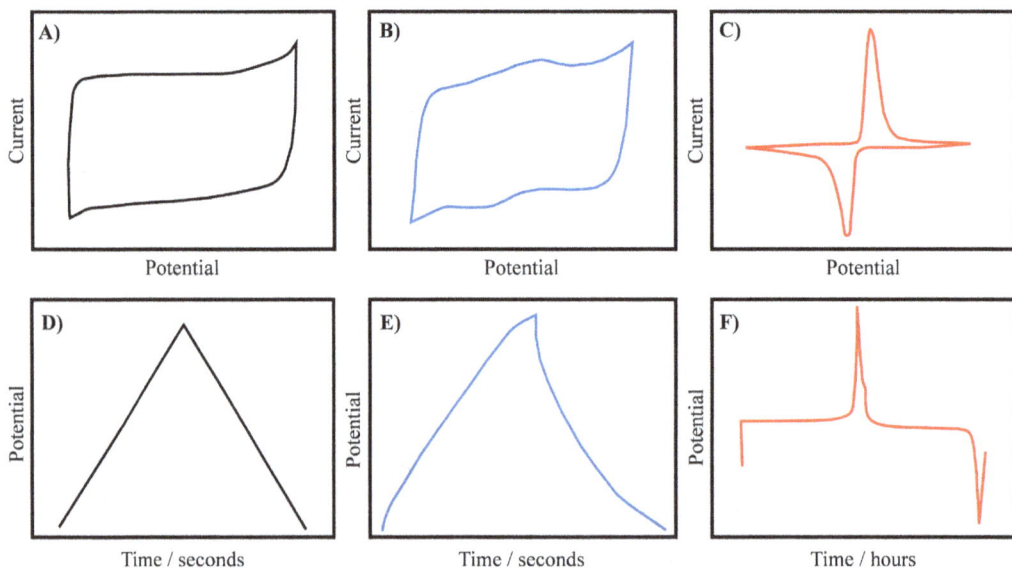

FIGURE 27.2 CV and GCDC of different energy-storage mechanisms: (A, D) electric double-layer capacitor; (B, E) pseudocapacitor, and (C, F) battery electrode.

constant at specific values in both charge and discharge cycles due to the redox reactions. [26] These processes increase the charge–discharge time and a complete GCDC cycle takes hours to be completed (Figure 27.2f).

GCDC is the most used technique to calculate the specific capacitance. For CPs and other pseudocapacitor materials, the specific capacitance can be calculated by equation 1 [23]:

$$C_s = \frac{I \cdot \Delta t}{\Delta V \cdot s} \tag{1}$$

where I is the applied current, Δt is the discharge time, ΔV is the potential window of the discharge process, and s is the normalization factor, which can be the mass of the electroactive material, the area of the electrode, or the volume of the device.

The cycle life is an important parameter for the supercapacitor performance evaluation since the high cycling stability is one of the greatest advantages of this type of device when compared to batteries. GCDC is the most used method for determining the cycling stability of the material (in a three-electrode system) or the device (in a two-electrode configuration). Usually, EDLC materials can withstand more than 10,000 cycles, while pseudocapacitors are stable until 5,000–10,000 cycles.

From GCDC cycles, the energy density (ED) of the supercapacitor device can be calculated by equation 2 [26]:

$$ED = \frac{1}{m} \int_{0}^{t_d} VI \, dt \tag{2}$$

where V is the potential window, t_d is the discharge time, I is the current, and m is the total mass of the electroactive material in both electrodes. Furthermore, the power density (PD) can be calculated by equation 3 [26]:

$$PD = \frac{ED}{td} \tag{3}$$

ED and PD values calculated from GCDC cycles in different current densities are expressed in a Ragone plot, allowing the comparison of energy storage in terms of how much energy is stored and how quickly this energy is delivered, as described in Figure 27.1.

2.2 NANOSTRUCTURED CPs

Energy harvesting and storage can have major benefits with the advent of nanostructured CPs. The electrodes prepared with bulk CPs are compact structures, and usually hinder the penetration of electrolyte ions, harming the charge carriers' mobility and effective charge transfer processes. The nanostructured arrays provide a huge increase in surface area, associated with an improved diffusion path for the electrolyte ions, improving the charge transfer processes [29].

The most common approach to forming nanostructured CPs is the bottom-up technique. In this methodology, the materials are prepared from monomers, which offers several advantages, such as control of reaction kinetics, and allows the formation of complex morphologies through self-assembly techniques [30]. In general, the synthesis of CPs consists of three steps: oxidation, coupling, and propagation. The oxidation of the monomer starts the polymerization reaction and is usually triggered by chemical oxidants (chemical approach), or electrochemical potential (electrochemical approach). To control the growth of the polymer and keep the final structure in the nanoscale, two main approaches have been developed: template-based approaches, including hard and soft templates; and template-free approaches, including interfacial methods, seeding methods, and electrospinning methods [31].

In the template-based approaches, the size and shape of the final nanostructure are determined by the size and shape of the template used. Hard-template methodologies usually employ colloidal particles or nanostructured channels which allow complete control over the size and shape of the nanostructures and are compatible with both the chemical and electrochemical polymerization. However, post-processing for template removal can be difficult and influence the final shape of hollow nanostructures. Soft-template methodologies employ soft matter, such as micelles or emulsion droplets, to orient the growth of the polymer and determine the shape of the nanostructure. Usually, this approach is very simple and inexpensive when compared with the hard template, but it may be more difficult to control the size of the final nanostructure [30]. Both template-based approaches can be used either with chemical or electrochemical polymerization.

The template-free approaches are mostly self-assembly methodologies and are usually easier and more economic than the template-based technique, as it requires no template and no post-treatment. However, the main limitation of this technique is that it is limited to some precursors and must be done in most cases via chemical polymerization [31]. The interfacial approach is often used to form nanofibers of CPs. The seeding approach is a relatively new methodology used to synthesize bulk quantities of nanowires and nanofibers of CPs [31]. Finally, the electrospinning technique is one of the most effective methods to obtain 1D nanostructures (nanofibers, nanorods, etc.), and is based on the pulling and stretching of a jet of material, bent by an electrical force. However, electrospinning can only be used for soluble and thermoplastic CPs [30,31].

2.3 CPs AND METALLIC NANOPARTICLES

Another type of nanocomposite with very interesting properties arises from the combination of CPs and metal nanoparticles. Noble metal nanoparticles have remarkable electron conductivity and thermal stability, while transition metal oxides (Co_3O_4, NiO, RuO_2, SnO_2, and MnO_2) have short conduction lengths and electrochemical pseudocapacitive properties, which can result in interesting combinations with CPs [31]. Functional properties obtained by embedding metallic nanoparticles in a dielectric matrix may include electronic conductivity ranging from single-electron hopping and tunneling to percolation; and particle surface plasmon resonances giving rise to optical features, magnetic properties governed by ferromagnetic single domain behavior, or superparamagnetism [32].

Nanohybrids of metal nanoparticles can improve the electrical and thermal conductivity and also increase the stability of polymeric films. The low stability of CP films due to swelling and shrinking is one of the main problems in energy-storage applications [29], and an increase in the stability of polymer films was observed by the incorporation of silver or gold nanoparticles in PPy and PEDOT materials [33,34]. Finally, the use of silica nanoparticles and carbon materials, such as carbon nanotubes, graphene, and graphene oxide (GO), with a CP films chain can also be explored, resulting in high surface area materials and larger specific capacitances [31].

3 POLYMERIC ELECTROLYTES

Polymeric electrolytes (PEs) have interesting features, such as mechanical resistance, non-flammable character, good tolerance to elevated temperatures, easy processing, no leakage problems, design flexibility, and low manufacturing cost. PEs can be classified into two classes, according to their physical type and composition: solid polymer electrolytes (SPE) and gel polymer electrolytes (GPE). Moreover, they can also be classified according to their source, synthetic or natural, which is better described in this section.

3.1 SYNTHETIC POLYMER ELECTROLYTES

Synthetic polymer electrolytes are the most used electrolytes in supercapacitors, due to their better mechanical properties. PEO, PEG, PPO, and PVA are examples of synthetic polymers, and PVA is by far the most used for PEs, and so the discussion will be focused on this polymer [13].

PVA was discovered by Haehnel and Herrmann in 1924 and is a non-toxic, chemically stable, biodegradable, and cheap polymer [35]. Its synthesis is different from that of most other vinyl polymers as it is not made from monomer polymerization but by hydrolysis of poly(vinyl acetate) under alkaline conditions [13]. The resulting PVA is a linear polymer with high availability of hydroxyl groups, which gives great hydrophilicity. Several electrolytes can be employed to synthesize PVA, such as H_2SO_4, H_3PO_4, KOH, and LiCl [5]. Also, inorganic materials such as metal oxides (e.g., TiO_2, SiO_2) and graphene oxide have been incorporated into the matrix to improve the electrical conductivity and mechanical properties [10].

PVA hydrogels are widely used in biological and medical applications, from food packaging to 3D bioprinting. More recently, the potential of PVA to be used as GPE for supercapacitors has been investigated and highlighted in several works. One of the advantages of using synthetic PEs is that the device can be assembled in diverse ways, such as coin-shaped or wire-shaped [36]. Although PVA is the most used polymer because of its improved mechanical properties, it is not the only synthetic polymer employed to fabricate flexible and stretchable SC devices. PEO is also often used as a substitute, which can also be combined with different materials, such as TiO_2 nanopowders, to increase conductivity [37].

Another strategy for the synthesis of synthetic PEs is to combine different polymers. PVA and PEG combined have shown improved ionic conductivity and mechanical resistance when compared with bare PVA. Also, a PVA/PEG gel was used to act as a matrix for the copolymerization of aniline (An) and pyrrole (Py), fabricating a gel electrode based on poly(Py-co-An) in the PVA/PEG gel matrix [38].

3.2 NATURAL POLYMER ELECTROLYTES

Polymers from natural and renewable sources have become a replacement for non-biodegradable polymers, due to their non-toxicity and biocompatibility. They offer the opportunity for the development of sustainable, cheap, and scalable supercapacitors because of their properties and low cost. Many natural polymers such as lignin, cellulose, alginate, pectin, and agar have been used in the construction of these devices.

Polysaccharides are the main type of natural polymer used in solid or gel electrolytes. One of the most abundant natural polymers is cellulose. This linear polysaccharide, which is an important structural component of the primary cell wall of green plants, is derived from the condensation of D-glucose units through β(1→4) glycosidic bonds. Cellulose and modified cellulose (e.g., cellulose acetate, methyl cellulose, hydroxyethyl cellulose, and hydroxyl-propyl cellulose) are promising candidates for electrolytes due to their high mechanical strength, good ionic conductivity, and wide electrochemical window [19].

Chitosan is a polysaccharide composed of D-glucosamine and N-acetyl-D-glucosamine residues, derived from the chitin presented in crustaceans. Chitosan-based hydrogels have been reported as electrolytes for supercapacitors due to their biocompatibility, biodegradability, non-toxicity, antimicrobial activity, and film-forming property. However, chitosan usually dissolves only in acids, which can limit the use of this polymer. To overcome this issue, chitosan-based hydrogel film has been prepared by a simple process of cross-linking with hydrochloric acid [18].

Another polysaccharide of interest is alginate, which is a linear anionic copolymer consisting of two saccharides, β-D-mannuronic acid, and α-L-guluronic acid, which are joined together through 1→4 glycosidic covalent bonds. It is a structural component of the cell wall of brown algae, of the *Phaeophyceae* class. The gel formation occurs in a very simple way, through the complexation of carboxylate groups with di- or trivalent cations. Alginate is one of the most widely applied supercapacitor gel electrolytes, and was first introduced by Ishikawa [39], who demonstrated that alginate has a good affinity toward carbon electrode materials, resulting in better electrochemical performance than the liquid electrolyte.

The second most abundant natural polymer is lignin, which comes from the plant's cell wall and is one of the principal byproducts of the pulp/paper industry. Lignin is primarily composed of three different phenylpropane units, namely, p-coumaryl, coniferyl, and sinapyl alcohols. Similarly to other biopolymers, lignin offers many advantages and also has been reported to be resistant to most biological attacks [16], while also exhibiting a significant improvement in mechanical strength and high ionic conductivity [40].

Natural polymers are interesting alternatives for the construction of supercapacitors for all the characteristics mentioned, but the gels formed with these polymers have little mechanical resistance. This problem can be overcome through the formation of blends between natural and synthetic polymers, as well as the formation of composites with oxides or nanomaterials.

3.3 COMPARISON BETWEEN POLYMER ELECTROLYTES AND LIQUID ELECTROLYTES

The main problem concerning liquid electrolytes is the risk of leakage and even combustion of organic electrolytes that require high standard encapsulation of materials to prevent the leakage. This issue is surpassed when PEs are employed instead of liquids. The main advantage in using PEs is the safety of the system, because they reduce or even eliminate the risk of leakage. They are also non-flammable, avoiding the risk of combustion and explosion. Due to their flexible nature, they allow novel cell configurations with minimal packaging – sandwiched, interdigitated, and fiber-shaped. In addition, PE has shown the capability to fulfill the dual roles of electrolyte and separator, which makes it possible to decrease the volume of devices and facilitate assembly.

Nowadays, the major issue with PEs is the limited contact with the electrodes due to the poor accessibility of the micropores. Thus, ion transport does not happen at the same rate as in liquid electrolytes. This disadvantage may cause low specific capacitance and, consequently, low-density energy. A limited contact area between the electrolyte and electrode also increases the equivalent-series resistance, resulting in a decrease in the supercapacitor performance.

Another problem related to SPEs is the low conductivity at room temperature to the high crystallization degree, which is a severe restriction. Therefore, the development and improvement of PEs rely on different synthetic methods to provide high ionic conductivity, reliable contact with electrodes, low resistance, and high energy/power densities. Figure 27.3 summarizes the mentioned advantages and disadvantages of PEs.

4 ENGINEERING OF SUPERCAPACITORS

Nowadays, high energy density storage and reliable cycle life are the major focus of the assembly of new apparatuses to supply the energy industry demands. Supercapacitors offer relatively higher energy and power densities when compared to typical electrochemical capacitors and batteries. The structure of these apparatuses enhances the amount of energy stored by the device compared to common electrochemical capacitors, allowing them to reach great values of capacitance, ranging from 50 F to over 7000 F [41]. Table 27.1 summarizes the information about some commercially available supercapacitors [42].

Supercapacitors are also subdivided into two main classes according to the forms of energy storage, the EDLC, and the pseudocapacitor, as previously explained. The EDLC still draws the major part of the industry's attention, as this technology allows high cyclability and the capacitance is directly linked to the surface area of contact between the electrode and electrolyte [2].

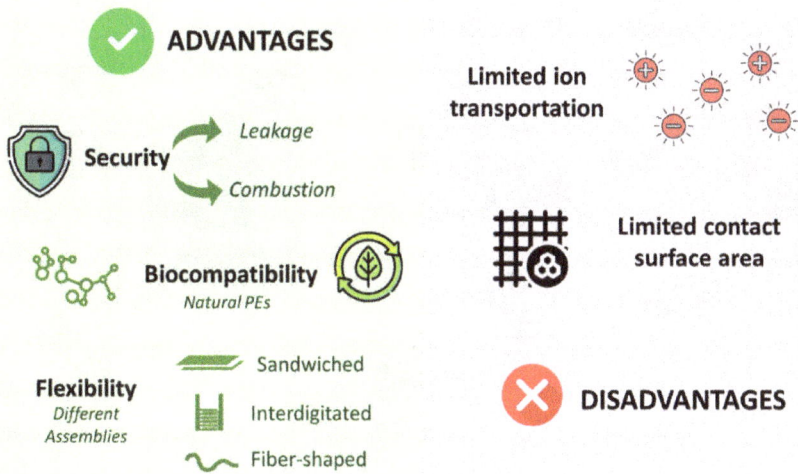

FIGURE 27.3 Principal advantages and disadvantages of PEs as compared with aqueous and organic electrolytes.

TABLE 27.1

Examples of Commercially Available Supercapacitor Devices

Manufacturer	Voltage (V)	Capacitance (F)
APowerCap	2.70	55
Asahi Glass	2.70	1375
BatScap	2.70	2680
Fuji	3.80	1800
Ioxus	2.70	3000
JSR Micro	3.80	2300
Maxwell	3.0	3400
Yunasko	2.70	7200

Source: Adapted with permission from Reference [42]. Copyright 2017. Copyright Yassine and Drazen, some rights reserved; exclusive licensee MDPI. Distributed under a Creative Commons Attribution License 4.0 (CC BY) https://creativecommons.org/licenses/by/4.0/.

Usually, the electrodes for commercial devices tend to be foils of conductive material modified with carbon materials, separated by a thin sheet of semi-insulator denominated separator (e.g. porous cloth) soaked with an electrolyte solution with the major objective to avoid short circuits among the electrodes. Both the electrodes and separator are immersed in the electrolyte solution and hermetically sealed in a case to prevent electrolyte leakage and contamination. Figure 27.4 presents a commercially available device, its structure, and a scheme of how it works [2].

According to the world's leading manufacturer, Maxwell Technologies® [43], their method of fabrication consists of modifying the surfaces of aluminum foils with electrochemically active material, and later inserting the separator and rolling them into a cylindrical form. After rolling, the system is impregnated with a solution containing electrolytes and inserted into a case.

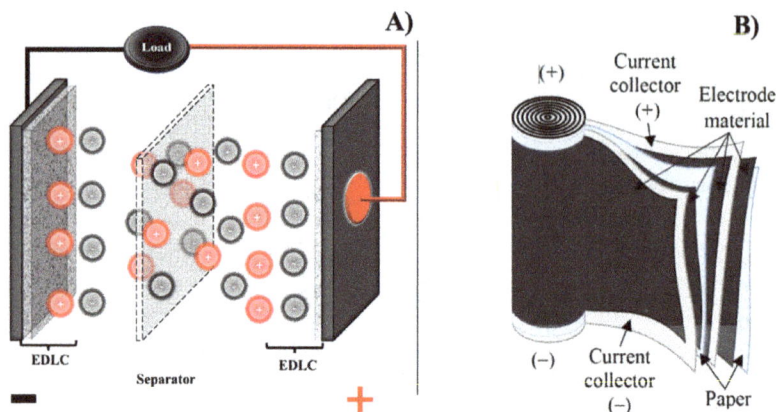

FIGURE 27.4 Brief presentation of the (A) schematic of the double-layer formation and (B) structure model of the commercial device's assembly. Adapted with permission from Reference [2]. Copyright 2019. Copyright Berrueta et al., some rights reserved; exclusive licensee IEEE. Distributed under a Creative Commons Attribution License 4.0 (CC BY) https://creativecommons.org/licenses/by/4.0/.

TABLE 27.2
Characteristics of the SCs Offered by the Main Manufacturers

Company	Country	Voltage (V)	Capacitance (F)	Temperature of Operation (°C)	Applications
Maxwell	USA	2.3–2.85	1–3400	–40/60	Transport, energy generation
Ioxus Inc.	USA	2.7	1250–3150	–40/85	Transport, renewable energy, backup generators
SPS Cap	China	2.5–2.7	1–5000	–40/60	Transport, winds turbines, micro-grids
Yunasko	Ukraine	2.7	400–3000	–40/60	Transport, industry, electronics
Vinatech	Rep. Korea	2.5–3	1–3000	–40/70	Transport, UPS, wind turbines

Source: Adapted with permission from Reference [2]. Copyright 2019. Copyright Berrueta et al., some rights reserved; exclusive licensee IEEE. Distributed under a Creative Commons Attribution License 4.0 (CC BY) https://creative commons.org/licenses/by/4.0/.

However, the performance of supercapacitors directly depends on the temperature that they are submitted to during the system operation. If the temperature increases, the cylindrical devices tend to suffer power loss proportionally. The company Ioxus® introduces a structure to minimize the effect of temperature, by stacking current collectors, electrodes, and separators. In this configuration, the electrodes would be connected in parallel, and the structure sealed into a plastic bag to prevent electrolyte leakage, being denominated Pouch Supercapacitors. Nevertheless, this geometry makes the devices more expensive, and with comparatively lower values of power densities. Table 27.2 presents some applications of commercially available supercapacitors.

4.1 SUPERCAPACITOR DEVICES APPLYING CPs

Conducting polymers allow the structuring of devices with unique advantages such as good conductivity, flexibility, ease to handle, and low prices. Figure 27.5 shows CPs and CP-based composites used in the structuration of supercapacitors [44]. The major application of CPs for the development

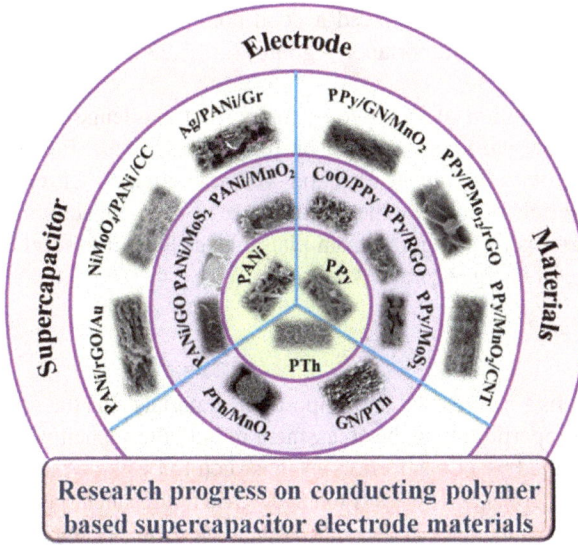

FIGURE 27.5 CP and CP-based composites used in the elaboration of SCs. Adapted with permission from Reference [44]. Copyright 2017, Elsevier.

TABLE 27.3
Advantages and Disadvantages of PANi, PPy, and PTh

Conducting Polymer	Advantages	Disadvantages
PANi	Flexibility, large specific capacitance range, ease of synthesis, ease of doping, high theoretical specific capacitance, controllable conductivity	Specific capacitance relies on synthesis conditions, poor cycling stability
PPy	Flexibility, ease of synthesis, relatively high specific capacitance, high cycling stability	Difficulty of doping
PTh	Flexibility, ease of synthesis, favorable cycling stability, and environment stability	Poor conductivity and poor specific capacitance

Source: Adapted with permission from Reference [**44**]. Copyright 2017, Elsevier.

of new pseudocapacitors relies on electrode modification. The most widely evaluated CPs are PPy-, PANI-, and PTh-based materials [41].

Several factors can influence the properties of CP-based electrodes, especially the preparation method and substrate. Adjusting and optimizing these conditions allow the improvement of the CP's electrochemical properties [44]. Problems in the cycling stability of PANi, for instance, can be addressed by associating the polymer with carbon nanotubes and/or metal oxides to improve the performance of the device [45]. Table 27.3 summarizes the advantages and disadvantages of some of the most commonly used CPs.

As an example that shows the prospects of what these supercapacitors can become, one can refer to the all-solid-state highly stretchable supercapacitor prepared by Kim et al. [23], using a symmetric arrangement of two electrodes of PEDOT/carbon nanotube sheets on an elastomeric substrate, with a LiCl-PVA-based solid electrolyte, and absorbed nylon as a stretchable separator, sealed by Sil-Poxy glue. The device showed a good combination of energy density and stretchability – a feature that has gained importance – going from 7.28 to 6.87 Wh/kg at a strain of 0% to biaxial 600%.

Due to the constant evolution of technology, the current challenge is to structure devices that are closer to the solid state, no longer relying on liquid electrolytes. Flexibility has also become important and attractive, with the recent focus on wearable devices. CP-modified electrodes have proven to be useful materials to enhance the performance of supercapacitors due to their versatility, allowing a variety of arrangements, going from plate-plane to flexible solid electrodes, giving one more step in the science of energy-storage devices.

5   CONCLUSION

Conducting polymers has proven to be important materials in the development of better supercapacitors. Their properties allow the assessment of both the capacitive and faradaic current, in the construction of so-called pseudocapacitors. Nanostructured CPs go one step further, increasing the surface area of the materials, which leads to an increase in the electrode–electrolyte interface for ion storage, which enhances the material-specific capacitance, further improving the devices. In the metal nanoparticle/CP hybrids, there are double advantages, with the CP acting as a matrix to stabilize the growth of the nanoparticles and prevent aggregation, and the nanoparticles improving the polymer properties, especially the charge-transfer processes. Metal nanoparticles can also improve the stability of CP films, increasing the devices' useful life.

The electrolyte plays a vital role in the performance of the supercapacitor since it determines the type of charge interaction. An electrolyte with a high ionic conductivity minimizes the internal resistance of supercapacitors, resulting in high energy and power densities. Moreover, chemically inert, environmentally friendly, and relatively inexpensive electrolytes are more attractive. Although liquid electrolytes are the most used, they present several limitations, such as high volatility, narrow electrochemical potential window, and the risk of leakage. To overcome these drawbacks, the use of polymeric electrolytes in supercapacitors is increasing, either synthetic, natural, or a combination of both.

The research into supercapacitors based on conducting polymers is still growing quickly, with new synthesis strategies dedicated to developing composite nanomaterials for both electrodes and electrolytes, associated with new device arrangements and geometries, offering promising results to achieve high-performance energy devices in the close future.

REFERENCES

[1] N.N. Loganathan, V. Perumal, B.R. Pandian, R. Atchudan, T.N.J.I. Edison, M. Ovinis, Recent studies on polymeric materials for supercapacitor development, Journal of Energy Storage. 49 (2022) 104149. https://doi.org/10.1016/j.est.2022.104149.

[2] A. Berrueta, A. Ursua, I.S. Martin, A. Eftekhari, P. Sanchis, Supercapacitors: electrical characteristics, modeling, applications, and future trends, IEEE Access. 7 (2019) 50869–50896. https://doi.org/10.1109/ACCESS.2019.2908558.

[3] M.Z. Iqbal, U. Aziz, Supercapattery: Merging of battery-supercapacitor electrodes for hybrid energy storage devices, Journal of Energy Storage. 46 (2022) 103823. https://doi.org/10.1016/j.est.2021.103823.

[4] A.G. Olabi, Q. Abbas, A. Al Makky, M.A. Abdelkareem, Supercapacitors as next generation energy storage devices: Properties and applications, Energy. 248 (2022) 123617. https://doi.org/10.1016/j.energy.2022.123617.

[5] C. Zhong, Y. Deng, W. Hu, J. Qiao, L. Zhang, J. Zhang, A review of electrolyte materials and compositions for electrochemical supercapacitors, Chemical Society Reviews. 44 (2015) 7484–7539. https://doi.org/10.1039/C5CS00303B.

[6] M.M. Pérez-Madrigal, F. Estrany, E. Armelin, D.D. Díaz, C. Alemán, Towards sustainable solid-state supercapacitors: electroactive conducting polymers combined with biohydrogels, Journal of Materials Chemistry A. 4 (2016) 1792–1805. https://doi.org/10.1039/C5TA08680A.

[7] G. de Alvarenga, B.M. Hryniewicz, I. Jasper, R.J. Silva, V. Klobukoski, F.S. Costa, T.N.M. Cervantes, C.D.B. Amaral, J.T. Schneider, L. Bach-Toledo, P. Peralta-Zamora, T.L. Valerio, F. Soares, B.J.G. Silva, M. Vidotti, Recent trends of micro and nanostructured conducting polymers in health and environmental applications, Journal of Electroanalytical Chemistry. 879 (2020) 114754. https://doi.org/10.1016/j.jelechem.2020.114754.

[8] Y. Han, L. Dai, Conducting polymers for flexible supercapacitors, Macromolecular Chemistry and Physics. 220 (2019) 1800355. https://doi.org/10.1002/macp.201800355.

[9] Q. Meng, K. Cai, Y. Chen, L. Chen, Research progress on conducting polymer based supercapacitor electrode materials, Nano Energy. 36 (2017) 268–285. https://doi.org/10.1016/j.nanoen.2017.04.040.

[10] R. Balint, N.J. Cassidy, S.H. Cartmell, Conductive polymers: Towards a smart biomaterial for tissue engineering, Acta Biomaterialia. 10 (2014) 2341–2353. https://doi.org/10.1016/j.actbio.2014.02.015.

[11] F. Hu, B. Yan, G. Sun, J. Xu, Y. Gu, S. Lin, S. Zhang, B. Liu, S. Chen, Conductive polymer nanotubes for electrochromic applications, ACS Applied Nano Materials. 2 (2019) 3154–3160. https://doi.org/10.1021/acsanm.9b00472.

[12] S. Sardana, A. Gupta, K. Singh, A.S. Maan, A. Ohlan, Conducting polymer hydrogel based electrode materials for supercapacitor applications, Journal of Energy Storage. 45 (2022) 103510. https://doi.org/10.1016/j.est.2021.103510.

[13] S. Alipoori, S. Mazinani, S.H. Aboutalebi, F. Sharif, Review of PVA-based gel polymer electrolytes in flexible solid-state supercapacitors: Opportunities and challenges, Journal of Energy Storage. 27 (2020) 101072. https://doi.org/10.1016/j.est.2019.101072.

[14] W. Ye, H. Wang, J. Ning, Y. Zhong, Y. Hu, New types of hybrid electrolytes for supercapacitors, Journal of Energy Chemistry. 57 (2021) 219–232. https://doi.org/10.1016/j.jechem.2020.09.016.

[15] K.S. Ngai, S. Ramesh, K. Ramesh, J.C. Juan, A review of polymer electrolytes: fundamental, approaches and applications, Ionics (Kiel). 22 (2016) 1259–1279. https://doi.org/10.1007/s11581-016-1756-4.

[16] V.K. Thakur, M.K. Thakur, Recent advances in green hydrogels from lignin: a review, International Journal of Biological Macromolecules. 72 (2015) 834–847. https://doi.org/10.1016/j.ijbiomac.2014.09.044.

[17] F.G. Torres, G.E. De-la-Torre, Algal-based polysaccharides as polymer electrolytes in modern electrochemical energy conversion and storage systems: A review, Carbohydrate Polymer Technologies and Applications. 2 (2021) 100023. https://doi.org/10.1016/j.carpta.2020.100023.

[18] H. Yang, Y. Liu, L. Kong, L. Kang, F. Ran, Biopolymer-based carboxylated chitosan hydrogel film crosslinked by HCl as gel polymer electrolyte for all-solid-sate supercapacitors, Journal of Power Sources. 426 (2019) 47–54. https://doi.org/10.1016/j.jpowsour.2019.04.023.

[19] M.M. Pérez-Madrigal, M.G. Edo, M.G. Saborío, F. Estrany, C. Alemán, Pastes and hydrogels from carboxymethyl cellulose sodium salt as supporting electrolyte of solid electrochemical supercapacitors, Carbohydrate Polymers. 200 (2018) 456–467. https://doi.org/10.1016/j.carbpol.2018.08.009.

[20] Z. Zhai, B. Ren, Y. Xu, S. Wang, L. Zhang, Z. Liu, Green and facile fabrication of Cu-doped carbon aerogels from sodium alginate for supercapacitors, Organic Electronics. 70 (2019) 246–251. https://doi.org/10.1016/j.orgel.2019.04.028.

[21] G. de Alvarenga, B.M. Hryniewicz, I. Jasper, R.J. Silva, V. Klobukoski, F.S. Costa, T.N.M. Cervantes, C.D.B. Amaral, J.T. Schneider, L. Bach-Toledo, P. Peralta-Zamora, T.L. Valerio, F. Soares, B.J.G. Silva, M. Vidotti, Recent trends of micro and nanostructured conducting polymers in health and environmental applications, Journal of Electroanalytical Chemistry. 879 (2020). https://doi.org/10.1016/j.jelechem.2020.114754.

[22] G. Wang, A. Morrin, M. Li, N. Liu, X. Luo, Nanomaterial-doped conducting polymers for electrochemical sensors and biosensors, Journal of Materials Chemistry B. 6 (2018) 4173–4190. https://doi.org/10.1039/c8tb00817e.

[23] B.K. Kim, S. Sy, A. Yu, J. Zhang, Electrochemical supercapacitors for energy storage and conversion, in: Handbook of Clean Energy Systems, 2015. https://doi.org/10.1002/9781118991978.hces112.

[24] D. Sarmah, A. Kumar, Conducting Polymer-Based Ternary Composites for Supercapacitor Applications, in: Conducting Polymer-Based Energy Storage Materials, CRC Press, 2019: pp. 301–332. https://doi.org/10.1201/9780429202261-19.

[25] C.I. Idumah, Novel trends in conductive polymeric nanocomposites, and bionanocomposites, Synthetic Metals. 273 (2021). https://doi.org/10.1016/j.synthmet.2020.116674.

[26] T.S. Mathis, N. Kurra, X. Wang, D. Pinto, P. Simon, Y. Gogotsi, Energy storage data reporting in perspective—Guidelines for interpreting the performance of electrochemical energy storage systems, Advanced Energy Materials. 9 (2019). https://doi.org/10.1002/aenm.201902007.

[27] A. Balducci, D. Belanger, T. Brousse, J.W. Long, W. Sugimoto, Perspective—A guideline for reporting performance metrics with electrochemical capacitors: From electrode materials to full devices, Journal of The Electrochemical Society. 164 (2017) A1487–A1488. https://doi.org/10.1149/2.0851707jes.

[28] N.R. Chodankar, H.D. Pham, A.K. Nanjundan, J.F.S. Fernando, K. Jayaramulu, D. Golberg, Y. Han, D.P. Dubal, True meaning of pseudocapacitors and their performance metrics: Asymmetric versus hybrid supercapacitors, Small. 16 (2020) 2002806. https://doi.org/10.1002/smll.202002806.

[29] Y. Shi, L. Peng, G. Yu, Nanostructured conducting polymer hydrogels for energy storage applications, Nanoscale. 7 (2015) 12796–12806. https://doi.org/10.1039/c5nr03403e.

[30] Y. Xue, S. Chen, J. Yu, B.R. Bunes, Z. Xue, J. Xu, B. Lu, L. Zang, Nanostructured conducting polymers and their composites: Synthesis methodologies, morphologies and applications, Journal of Materials Chemistry C. 8 (2020) 10136–10159. https://doi.org/10.1039/d0tc02152k.

[31] L. Zhang, W. Du, A. Nautiyal, Z. Liu, X. Zhang, Recent progress on nanostructured conducting polymers and composites: synthesis, application and future aspects, Science China Materials. 61 (2018) 303–352. https://doi.org/10.1007/s40843-017-9206-4.

[32] F. Faupel, V. Zaporojtchenko, T. Strunskus, M. Elbahri, Metal-polymer nanocomposites for functional applications, in: Advanced Engineering Materials, 2010: pp. 1177–1190. https://doi.org/10.1002/adem.201000231.

[33] A.L. Soares, B.M. Hryniewicz, A.E. Deller, J. Volpe, L.F. Marchesi, D.E.P. Souto, M. Vidotti, Electrodes based on PEDOT nanotubes decorated with gold nanoparticles for biosensing and energy storage, ACS Applied Nano Materials. 4 (2021) 9945–9956. https://doi.org/10.1021/acsanm.1c02677.

[34] B.M. Hryniewicz, I. C. Gil, M. Vidotti, Enhancement of polypyrrole nanotubes stability by gold nanoparticles for the construction of flexible solid-state supercapacitors, Journal of Electroanalytical Chemistry. 911 (2022). https://doi.org/10.1016/j.jelechem.2022.116212.

[35] N.A. Choudhury, S. Sampath, A.K. Shukla, Hydrogel-polymer electrolytes for electrochemical capacitors: an overview, Energy Environ. Sci. 2 (2009) 55–67. https://doi.org/10.1039/B811217G.

[36] R. Zhang, C. Chen, H. Yu, S. Cai, Y. Xu, Y. Yang, H. Chang, All-solid-state wire-shaped asymmetric supercapacitor based on binder-free CuO nanowires on copper wire and PPy on carbon fiber electrodes, Journal of Electroanalytical Chemistry. 893 (2021) 115323. https://doi.org/10.1016/j.jelechem.2021.115323.

[37] T.G. Yun, B. il Hwang, D. Kim, S. Hyun, S.M. Han, Polypyrrole–MnO_2-coated textile-based flexible-stretchable supercapacitor with high electrochemical and mechanical reliability, ACS Applied Materials & Interfaces. 7 (2015) 9228–9234. https://doi.org/10.1021/acsami.5b01745.

[38] L. Guo, W.-B. Ma, Y. Wang, X.-Z. Song, J. Ma, X.-D. Han, X.-Y. Tao, L.-T. Guo, H.-L. Fan, Z.-S. Liu, Y.-B. Zhu, X.-Y. Wei, A chemically crosslinked hydrogel electrolyte based all-in-one flexible supercapacitor with superior performance, Journal of Alloys and Compounds. 843 (2020) 155895. https://doi.org/10.1016/j.jallcom.2020.155895.

[39] K. Soeda, M. Yamagata, M. Ishikawa, Outstanding features of alginate-based gel electrolyte with ionic liquid for electric double layer capacitors, Journal of Power Sources. 280 (2015) 565–572. https://doi.org/10.1016/j.jpowsour.2015.01.144.

[40] T. Liu, X. Ren, J. Zhang, J. Liu, R. Ou, C. Guo, X. Yu, Q. Wang, Z. Liu, Highly compressible lignin hydrogel electrolytes via double-crosslinked strategy for superior foldable supercapacitors, Journal of Power Sources. 449 (2020) 227532. https://doi.org/10.1016/j.jpowsour.2019.227532.

[41] S. Suriyakumar, P. Bhardwaj, A.N. Grace, A.M. Stephan, Role of polymers in enhancing the performance of electrochemical supercapacitors: A review, Batteries & Supercaps. 4 (2021) 571–584. https://doi.org/10.1002/batt.202000272.

[42] M. Yassine, D. Fabris, Performance of commercially available supercapacitors, Energies (Basel). 10 (2017) 1340. https://doi.org/10.3390/en10091340.

[43] Maxwell Technologies ®, Datasheet – 3.0V 3400F ULTRACAPACITOR CELL BCAP3400 P300 K04/05 (n.d.).

[44] Q. Meng, K. Cai, Y. Chen, L. Chen, Research progress on conducting polymer based supercapacitor electrode materials, Nano Energy. 36 (2017) 268–285. https://doi.org/10.1016/j.nan oen.2017.04.040.

[45] J. Xu, K. Wang, S.-Z. Zu, B.-H. Han, Z. Wei, Hierarchical nanocomposites of polyaniline nanowire arrays on graphene oxide sheets with synergistic effect for energy storage, ACS Nano. 4 (2010) 5019–5026. https://doi.org/10.1021/nn1006539.

28 A Facile Approach to Recycling Used Facemasks for High-Performance Energy-Storage Devices

Anjali Gupta,[1,2] Cassia A. Allison,[1,2] Mahesh Chaudhari,[1,3] Priyesh Zalavadiya,[1] Felipe M. de Souza,[1] Ram K. Gupta,[1,3] and Tim Dawsey[1]

[1] National Institute for Materials Advancement, Pittsburg State University, Pittsburg, KS 66762, USA
[2] Pittsburg High School, Pittsburg, KS 66762, USA
[3] Department of Chemistry, Pittsburg State University, Pittsburg, KS 66762, USA

1 INTRODUCTION

The introduction of polymeric materials on a commercial scale has changed the living standards of society, provided more comfort, and contributed significantly to the advancement of science and technology. Polymeric materials are extremely versatile, and find applications in consumer and industrial sectors such as in packaging, automobiles, construction, biomedicals, energy, and many more. Polyethylene (PE), polypropylene (PP), polyethylene terephthalate (PET), polyvinyl chloride (PVC), polyvinylidene fluoride (PVDF), polytetrafluoroethylene (PTFE), polystyrene, polyacrylonitrile butadiene styrene, polyamides, polycarbonates, and polyurethanes are some of the most commercially important polymers. Although they are being used in many applications, their utilization in biomedicals is an emerging, and substantial, area. For example, polyurethanes with a structure containing delocalized positive and negative charges (zwitterions) can be used to produce tubes for body fluid transportation to perform dialysis, blood, or plasma transfusions [1]. In addition, blends of polybutadiene and polyacrylonitrile are used in surgical gloves. In addition, a broad range of polymers is being used to manufacture personal protective equipment (PPE).

Recently, PPE became a significant preventive measure against the worldwide coronavirus disease 19 (COVID-19). There continues to be massive consumption of disposable medical masks to impede the further spread of diseases [2–4]. Polypropylene is among many polymers employed for the manufacture of disposable masks with the classifications of N95, FFP2, and FFP3 along with surgical-grade masks (Figure 28.1). However, during the escalation of the pandemic, the demand for PP grew dramatically, necessitating the use of other polymers such as PET, PVDF, and PTFE. These polymers can withstand relatively aggressive pressurized environments such as autoclaves and peroxide treatments for sterilization so they can be safely used in hospitals and other biomedical-related applications [2,5,6].

Disposable masks have led to an environmental concern since some of them are being discarded improperly, allowing them to reach the ocean or be scattered in the environment. The former case leads to the release of microplastic fibers that can be consumed by marine life. Since these microplastics are non-digestible they can cause the death of aquatic creatures, leading to instability in the ecosystem. Also, due to bioaccumulation, the residues of microplastics can eventually be

DOI: 10.1201/9781003278269-28

Part	Chemical Compositions	Weight Percentage (wt.%)
Out Layer	PP	41.83
Middle Layer	PP	17.52
Inner Layer	PP	17.5
Nose wire	PP	2.52
Ear strap	PET	16.46
Nose frame	Metals	4.17 (Fe: 4.125, Zn: 0.026, Mn: 0.012, Mg: 0.0007)

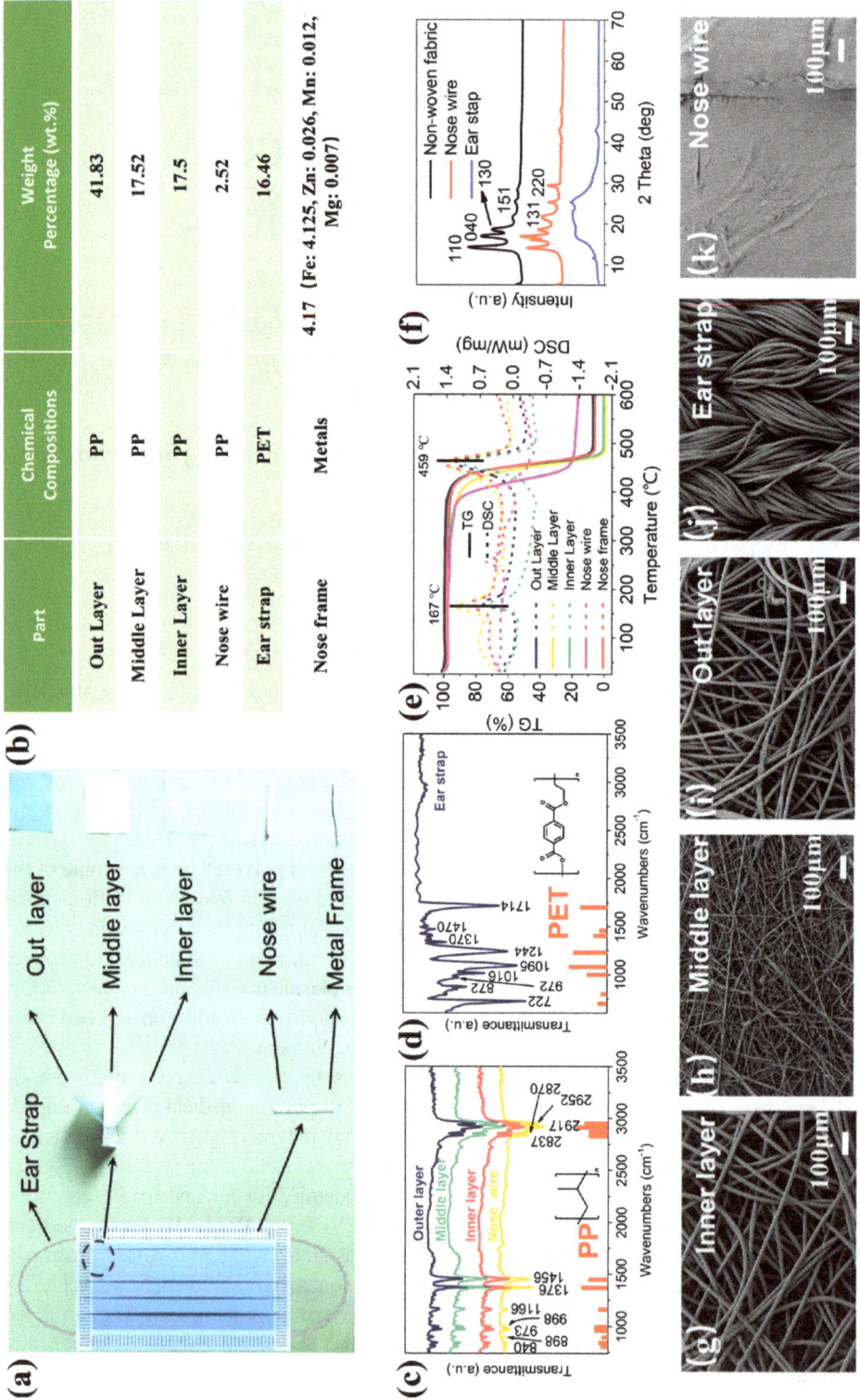

FIGURE 28.1 (a) Components and (b) chemical composition of a disposable mask. Adapted with permission from Reference [2]. Copyright (2022), Elsevier.

consumed by humans since they are not digestible by the organisms [7]. Hence, finding a sustainable end for used disposable masks is a follow-up matter of the pandemic that must be addressed. Up to now, they have been either incinerated or tossed into a landfill, neither of which is a feasible long-term solution [8,9]. It is necessary to address this situation more sustainably. Since these masks contain a high amount of carbon (~85.7% of their total composition), they can undergo chemical treatment to prepare high-performance carbon for many applications such as energy-storage devices, purifications, catalysis, etc., thus creating a sustainable circle of consumption and reintegration of C into the mainstream [10,11].

One example of using such sustainable conversion techniques to generate material suitable for use in energy-storage devices was demonstrated by Hu *et al.* [12] who synthesized porous carbon-based materials, derived from disposable masks, through a hydrothermal method followed by activation with KOH at high temperatures. Such processes promote the etching of the carbonaceous structure to increase its porosity and surface area considerably. Through that, 10.4 Wh/kg of energy density along with 81.1% of capacity retention after 3000 cycles were obtained. In another approach, a composite based on carbon nanotubes (CNTs) with Ni was prepared through a carbonization process of disposable masks, performed at 700°C in the presence of nickel hydroxide along with around 50 wt.% of catalysts, for applications in microwave absorption [13]. In a separate interesting work, carried out by Yuwen *et al.* [2], a sulfonation and oxidation process of disposable masks was performed through a microwave-assisted method, along with highly acidic media in the presence of concentrated sulfuric acid, in under 8 min. Such a process activates the carbon-based polymers found in disposable masks, enabling their use as electrodes for energy-storage devices. On top of that, the co-doping process with S and O can introduce defects in the structure that aid in the ionic adsorption, increasing overall capacitance.

From the perspective of reincorporating the used disposable masks and providing a value-added application, this chapter provides a study for their conversion into carbon-based materials for energy-storage devices. A facile procedure consists of performing acid digestions of the polymers over time followed by thermal treatment to obtain a material with a high surface area. Along with that, different types of doping or activation with alkaline media can be performed as these are commonly utilized strategies to increase the surface area of electroactive materials.

2 TYPES OF ELECTROCHEMICAL DEVICES

Finding a sustainable cycle for the generation and consumption of energy is one of the main concerns of every nation as the depletion of non-renewable resources becomes a greater concern. It is crucial to optimize the efficiency of the currently used energy-storage devices, such as batteries and supercapacitors, as these can provide a feasible alternative to the excessive use of non-renewable resources. This section addresses the concepts and main types of electrochemical devices, focusing on supercapacitors. Supercapacitors are devices made of two electrodes that are isolated through a separator immersed in an electrolyte. A current collector is used to harvest the electrons and transport them through an external circuit. The energy storage takes place at the interface between the electrode and electrolyte. In principle, an electrode should present a high surface area to allow proper permeation of the ionic species from the electrolyte to perform the energy-storage process.

The types of supercapacitors and their energy-storage mechanisms are as follows. The electrochemical double-layer capacitor (EDLC) consists of the formation of an electric double layer of ions at the interface between the electrolyte and electrode which creates a potential. This process is based on the polarization of the electrode's surface and the adsorption of ions. Hence, the main strategies to improve the number of ions that can be adsorbed on the electrode's surface usually consist of increasing the surface area and introducing defects in the electroactive material's crystal lattice. Therefore, it does not carry out any reaction during the charging or discharging process since it is mostly based on an electrostatic process. As previously mentioned, the surface properties of the material used in the electrodes play a major role in its properties toward EDLC. Because of that, it is necessary to functionalize or optimize the surface area of the materials used to make the electrodes. In this sense, carbon-based materials have been widely used for that purpose from the earlier stages of EDLC development to nowadays. Some of the reasons for that are attributed to the low cost of carbon, high surface area, chemical stability, and relatively high structural versatility which include carbon aerogels, carbon nanotubes (CNTs), foams, graphene, and carbon black, among many other materials derived from carbon.

The second type of supercapacitor is referred to as a pseudocapacitor and is based on the electrochemical reactions that occur between the electrode and electrolyte interface. A redox process occurs at the interface which leads to the release of electrons that contribute to the overall capacitance, making this type of energy storage mechanism somewhat similar to that of a battery [14]. The pseudocapacitance arises when an electrical potential is applied in the system which induces a faradaic current. This phenomenon is observed in several classes of materials such as transition metal-based oxides, sulfides, bimetallics, as well as conducting polymers which would include polyaniline (PANI) and polythiophene (PPy) [14–16]. Usually, pseudocapacitance contributes greater energy density relative to an EDLC, however, the electroactive materials that perform the energy-storage process through pseudocapacitance may suffer from lower conductivity, cycling stability, and relatively lower power density [17].

There are several factors to be considered to improve the overall capacitance when employing materials that perform energy storage through the pseudocapacitance mechanism. One of these aspects is related to the ionic diffusion to access a greater surface area of the redox process [18]. Also, the affinity of the electrode toward the ion adsorbed may be influenced by the pore dimensions, structure, surface-related features, and chemical nature, which can greatly influence the pseudocapacitive properties [19]. Based on these aspects, pseudocapacitance can arise from both redox reactions as well as chemisorption processes [20,21]. One of the main focuses of research within this field lies in enhancing the number of active sites over the electrode's surface and adopting synthetic routes that can yield an optimal structure for proper interaction with ions from the electrolyte, to achieve high capacitance, rate capability, and stability.

The third type of supercapacitor is classified as a hybrid, which consists of a combination of EDLC which includes carbon-based materials such as graphite, graphene, CNTs, graphene aerogel, as well as pseudocapacitive materials which include conducting polymers, transition metal-based oxides or sulfides, and their derivatives. One of the goals in incorporating materials with different energy-storage mechanisms is to increase the working potential, which leads to higher capacitance. Along with that, higher energy densities are more likely to be achieved [22]. The hybrid supercapacitors can be symmetric or asymmetric. The former consists of a system in which the same materials are employed in both electrodes. The latter is the opposite, in the sense that the cell is built using different materials in each of the electrodes. Usually, hybrid supercapacitors that present electrodes composed of different materials tend to display a more satisfactory electrochemical behavior as compared to their symmetric counterparts.

As an example of a hybrid supercapacitor, Amatucci *et al.* [23] utilized activated carbon that performed the energy storage through EDLC along with $Li_4Ti_5O_{12}$ which functioned through pseudocapacitance. An organic electrolyte was employed as it has a higher working voltage compared to an aqueous electrolyte. It should be noted that aqueous electrolytes are cheaper and eco-friendly, however, they start degrading at potentials above 1.23 V due to water splitting. The overall performance of a device can be enhanced by improving the conductivity, optimizing the surface structure, as well as the presence of active sites of the electrode materials. Supercapacitors can be designed in several ways to improve their performance. Figure 28.2 displays a scheme of the classifications of capacitors along with some examples of materials that can be implemented in their design.

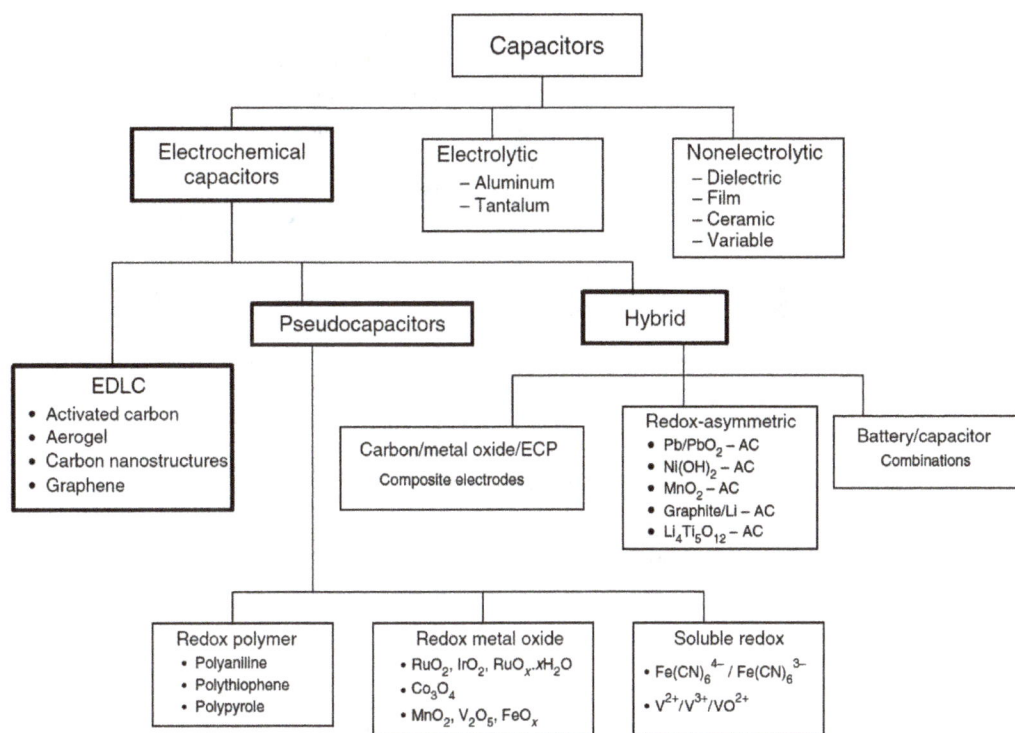

FIGURE 28.2 Scheme of the classifications of capacitors. Adapted with permission from Reference [24]. Copyright (2013), John Wiley and Sons.

One of the latest goals in the field of energy storage is to obtain a system that can provide high power as well as high energy densities. Supercapacitors provide high power density due to their fast charge discharging process through an EDLC mechanism. The energy density of a supercapacitor can be enhanced by introducing materials that are capable of redox reactions (pseudocapacitance) as this is a slower process. Batteries, on the other hand, display a different behavior in which higher energy densities are provided usually with lower power densities. Such effects are observed because in a battery the energy is stored through an electrochemical reaction that is relatively slower since it often involves the intercalation of ions within the crystal lattice of the electroactive materials that compose the electrode. Hence, based on these differences there is a current push to further optimize the performance of energy-storage devices by combining the advantageous properties of supercapacitors, based on higher power density, along with positive aspects of batteries, such as higher energy densities. Such combinations have been incorporated into the definition of novel energy-storage devices that consist of a hybrid between supercapacitors and batteries, which are further elucidated in Figure 28.3.

3   RECYCLED FACEMASKS FOR SUPERCAPACITORS

3.1   SYNTHESIS OF CARBON USING RECYCLED FACEMASKS

The facemasks used in this study were made of PP. The used facemask was chemically treated without any pretreatments. For chemical treatment, the used facemask was cut into small pieces and soaked in concentrated sulfuric acid (5 g of a facemask in 50 ml sulfuric acid) for varying times, such as 2, 4, 6, and 12 h at 120°C. The treated facemasks were washed with distilled water many

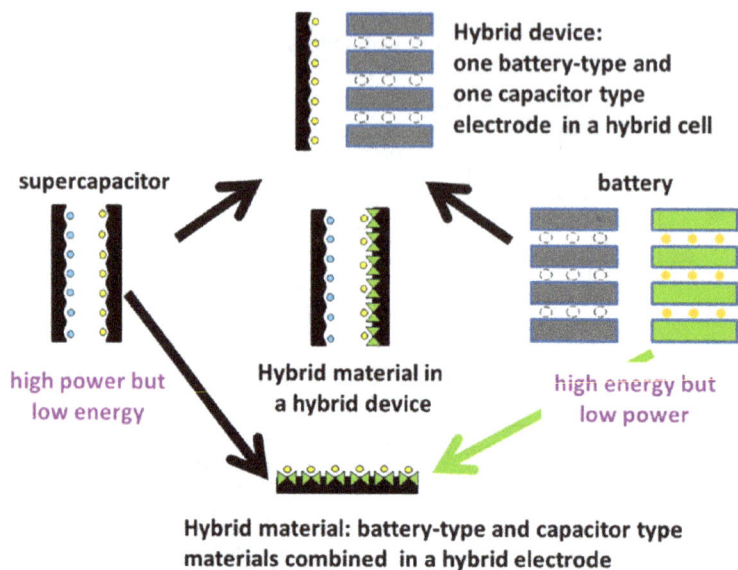

FIGURE 28.3 Scheme of the hybrid designs between supercapacitors and batteries. Adapted with permission from Reference [25]. Copyright (2015), Royal Society of Chemistry.

times and kept for drying in a conventional oven overnight at 90°C. The sulfuric acid-treated and dry facemasks were carbonized in a tube furnace at 1000°C for 2 h with a ramp rate of 5°C/min under a nitrogen atmosphere. The carbonized facemasks were washed with distilled water several times and dried in a conventional oven for 12 h. The untreated facemask was called FM. Facemasks treated with sulfuric acid for 2, 4, 6, and 12 h followed by carbonization were named as 2 Hr, 4 Hr, 6 Hr, and 12 Hr. The proposed reaction for the process is provided in Figure 28.4.

3.2 STRUCTURAL CHARACTERISTICS

Structural characterizations of facemasks and treated facemasks were performed using scanning electron microscopy (SEM), Fourier-transform infrared spectroscopy (FTIR), X-ray powder diffraction (XRD), thermogravimetric analysis (TGA), and Brunauer-Emmett-Teller (BET) surface area analysis. SEM was performed on the untreated FM to elucidate its microstructure. It was observed that both the inner and outer layers (Figure 28.5a and c) presented a fibrous structure of PP, whereas the middle layer (Figure 28.5b) presented a network-like structure with fibers of a relatively long length. The difference in microstructure is likely due to the processing of the layers. A previous report showed that the middle layer of an N95 FM is processed through the melt-blown method, which grants it the stacked fibrous morphology forming a 3D structure [27,28]. A cross-section of an N95 FM displaying the differences in microstructure of the inner and outer layers in comparison with the middle layer is provided in Figure 28.6a. It is worth noting that the intrinsic efficiency of the FM in terms of the flow of air can be considerably low since there is a relatively large void space between the small-diameter fibers. Such conditions can make its filtration ineffective. To address that, the microfibers can be exposed to a corona discharge process which leads to the formation of quasi-permanent dipoles that are called electrets that can hinder the permeation of harmful particles while maintaining the initial air permeability. A schematic for this process is presented in Figure 28.6b.

To further characterize the neat FM, FT-IR was performed as it can provide valuable structural information. Since the spectra of the inner, middle, and outer layers were indistinguishable from

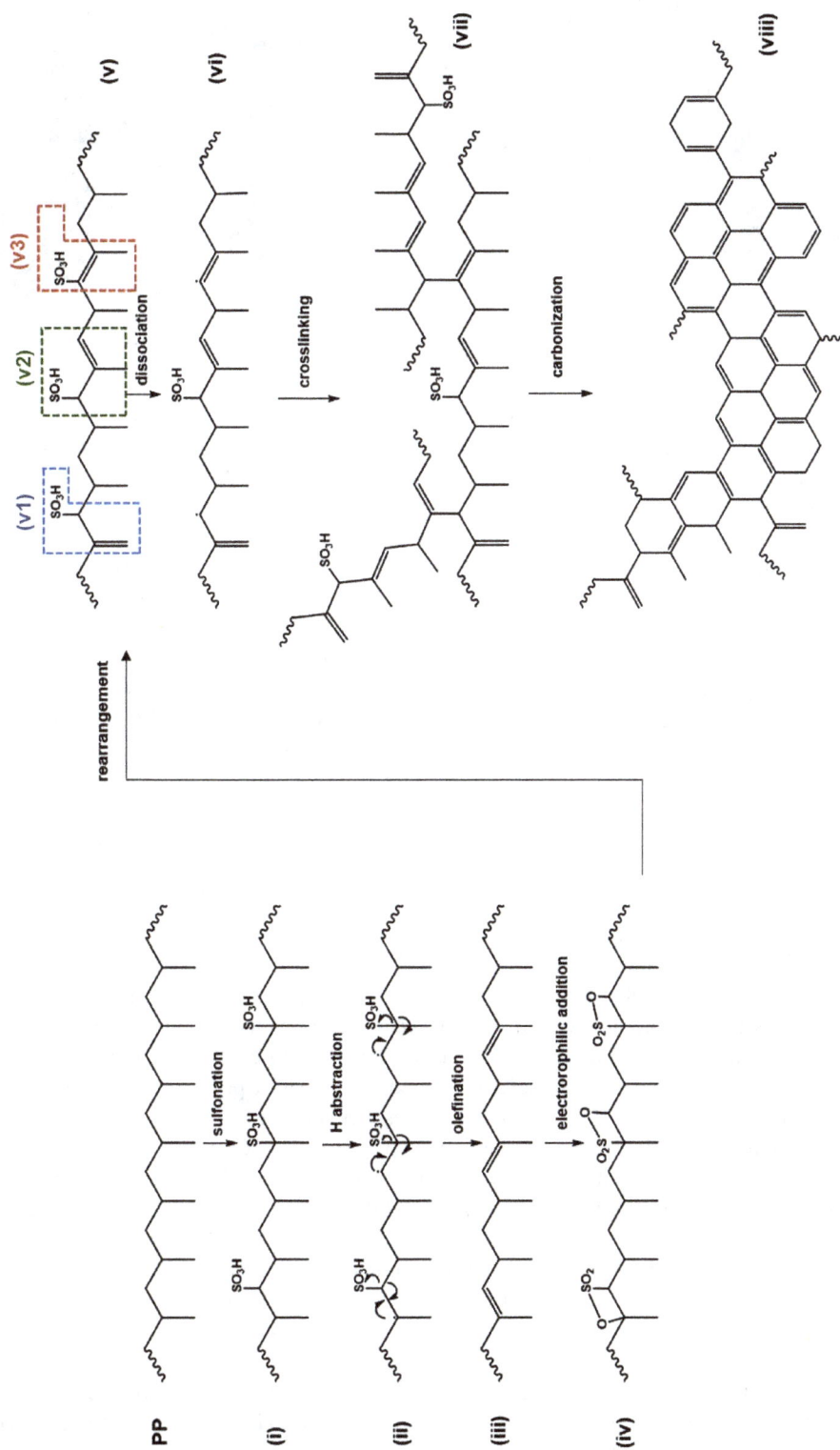

FIGURE 28.4 The proposed stabilization mechanism of PP during its sulfonation. Adapted with permission from Reference [26]. Copyright (2022), Elsevier.

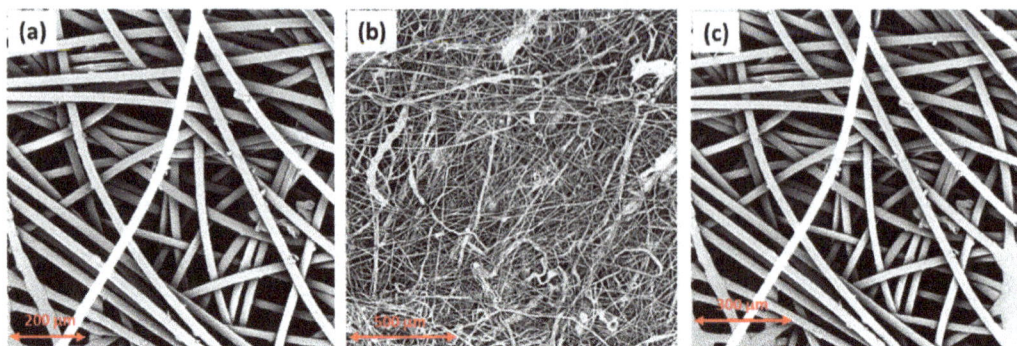

SEM images of the facemask (a) inner layer, (b) middle layer, and (c) outer layer.

each other, Figure 28.7a presents a spectrum of the middle layer. The FT-IR spectra of the FM presented two peaks at around 991 and 1166 cm^{-1} which were ascribed to the $-CH_3$ rocking vibration [26]. Also, the strong peaks at around 1370 and 2970 cm^{-1} were attributed to the $-CH_3$ symmetric bending vibration and asymmetric stretching vibrations, respectively. In addition, the peaks around 2915, 2850, and 1456 cm^{-1} were assigned to asymmetric stretch, symmetric stretch, and symmetric bending of $-CH_2$, respectively [29]. These peaks were characteristic of pristine PP [2,30]. XRD is another important characterization to define the material's morphology in terms of crystalline or amorphous nature. Figure 28.7b displays the XRD analysis of the FM which presented clear peaks at around 14.0, 17.0, 18.6, and 21.3°, which are correlated to the (110), (040), (130), and (111) planes of the monoclinic α-crystal phase [2,31–33].

The treatment with concentrated sulfuric acid of the pristine facemask with increasing time was noted to provide a considerable difference in its properties and morphology, which was likely due to the insertion of sulfur in the structure along with an increase in the material's surface area. Figure 28.8a shows the effect of acid treatment time followed by carbonization on the thermal stability of the carbons. It was observed that the thermal stability of the samples increased with increasing sulfuric acid treatment time. Hence, the 12 Hr sample presented the highest amount of residue along with the lowest weight loss at 600°C. The initial weight loss for the acid-treated samples started at around 400°C, whereas the pristine FM presented an abrupt decomposition at around 325°C. Based on this thermal analysis it is notable that the acid-treated facemasks presented remarkable stability even at higher temperatures. In addition, aside from the thermal analysis of the FM-based samples, it is important to elucidate the effect observed in its morphology. The XRDs for the FM samples treated with acid under different times followed by carbonization are presented in Figure 28.8b. It was observed that the samples presented two signature peaks at around 24.5 and 43° which were related to (002) and (100) crystal planes, respectively [26].

Chemical treatment of PP-based FM was performed by Lee *et al.* [26]. They treated the PP masks in concentrated sulfuric acid solution at 2, 4, and 6 h. Then, a thermal treatment was performed at significantly higher temperatures (in the range of 2000–2400°C) under an argon atmosphere to improve the porosity and surface area. During the thermal testing, it was noted that the samples treated with acid presented higher thermal stability compared to the neat FM as they presented a larger amount of residue. The TGA plots for the samples are presented in Figure 28.9. It should be noted that our process was performed at 800°C, which is higher energy efficiency than the process performed by Lee *et al.* [26].

Improving the surface area of carbon-based materials is a core aspect of the improvement in capacitance, as it allows a larger number of ions from the electrolyte to be absorbed on the electrode's surface. To understand the effect of acid treatment time on the surface characteristics, Brunauer-Emmett-Teller (BET) surface area measurement was performed. The nitrogen adsorption–desorption

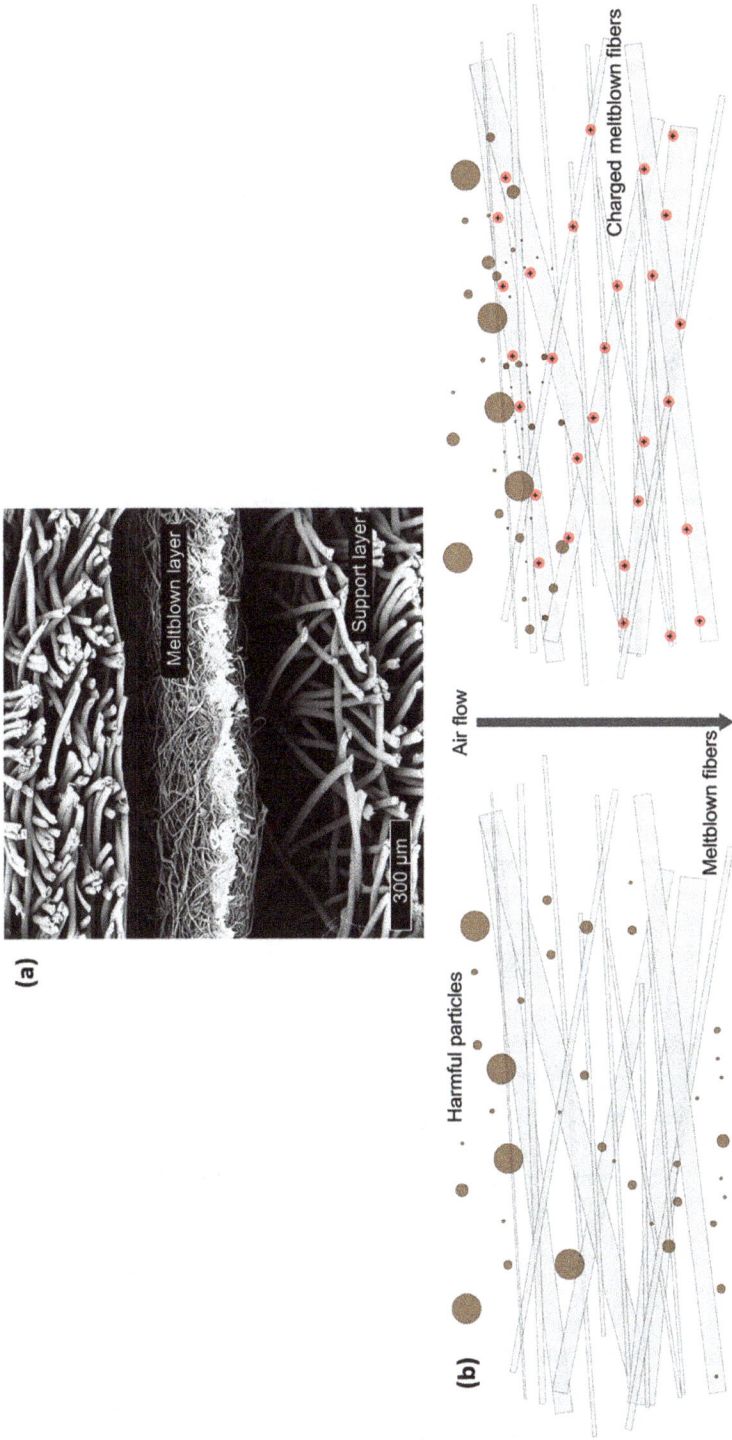

FIGURE 28.6 (a) SEM image of a cross-section of an N95 FM displaying a middle layer of PP obtained through the melt blow method which is thinner than the support layers around it. (b) Scheme of the fibers obtained through the melt-blown approach (left) without and (right) with the charged structure induced through corona discharge method. Adapted from Reference [27]. Copyright (2020), American Chemical Society.

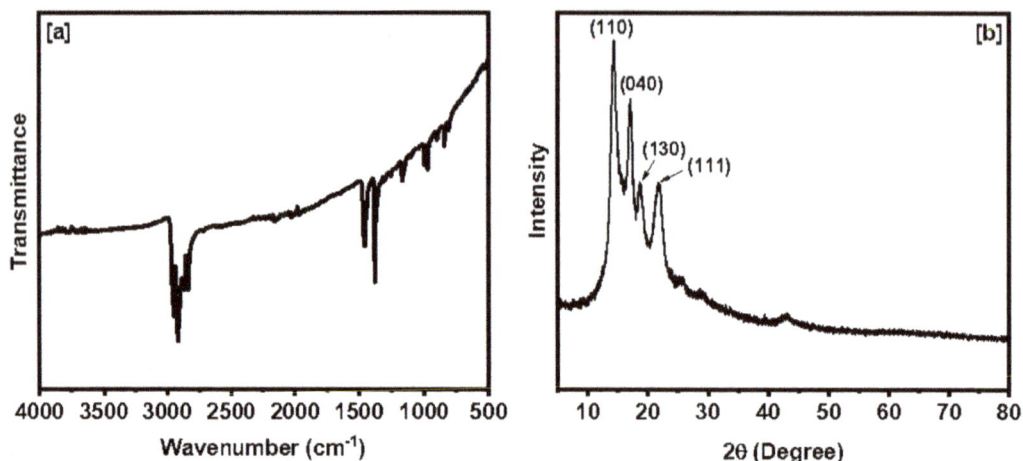

FIGURE 28.7   (a) FT-IR spectra and (b) XRD patterns of the facemask.

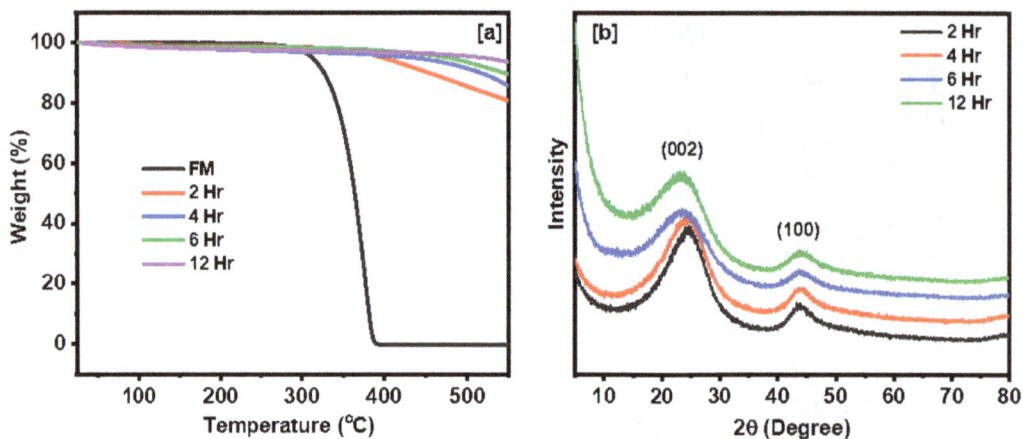

FIGURE 28.8   (a) TGA plots of FM and treated FM, and (b) XRD patterns of various treated facemask samples.

isotherms are provided in Figure 28.10a, and the BET surface area and t-plot micropore area are given in Figure 28.10b. Based on the results obtained, it is noted that the highest surface area of 275 m^2/g was obtained for the 6 Hr sample and further exposure to acid beyond 6 h led to a decrease in the surface area. For the t-plot micropore area, it was observed that the 12 Hr sample presented the highest area value followed by the 6 Hr. In the work reported by Lee *et al.* [26], the chemically treated FM presented the highest BET surface area of 27.4 m^2/g for the PP-based FM soaked in sulfuric acid for 16 h.

3.3   ELECTROCHEMICAL PERFORMANCE

The electrochemical performance of the recycled and treated facemasks was studied using a multi-channel potentiostat. The carbon from the recycled and treated facemasks was used as an active material for the fabrication of coin-cell type supercapacitors. A symmetrical coin cell type device was fabricated in a 6M KOH electrolyte. For this, first, the carbon from the recycled and treated

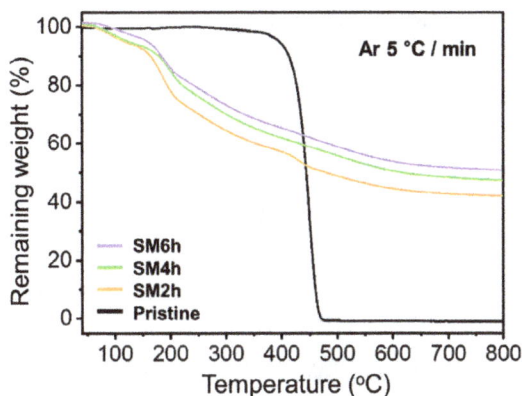

FIGURE 28.9 TGA for sulfonated and thermally treated PP-based FM. Adapted with permission from Reference [26]. Copyright (2021), Elsevier.

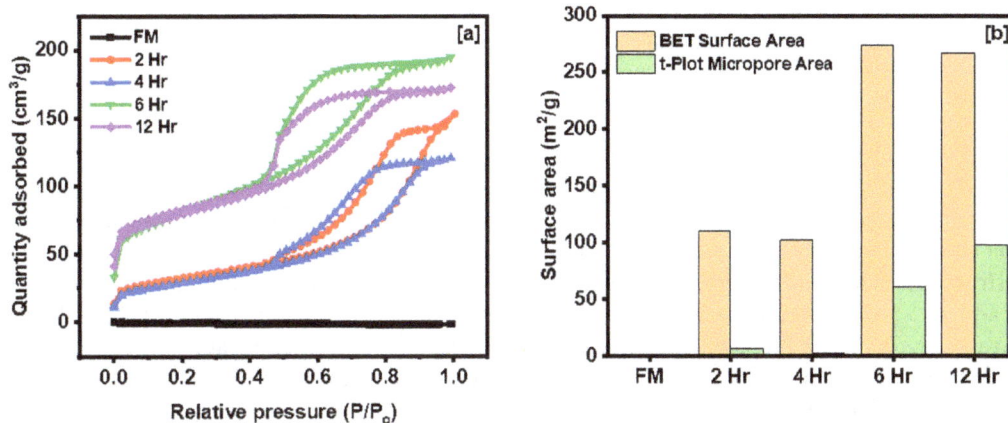

FIGURE 28.10 (a) Nitrogen adsorption–desorption isotherms and (b) BET surface area and t-plot microporous surface area of various samples.

facemask (80%) was mixed with acetylene black (10%) and polyvinylidene fluoride (PVDF, 10%) in n-methyl pyrrolidone (NMP). After thorough mixing, the slurry was coated on a circular conducting carbon cloth and dried in a vacuum oven for 24 h at 90°C. A schematic of the device is shown in Figure 28.11. Cyclic voltammetry (CV) at various scan rates, galvanostatic charge–discharge (GCD) at various current densities, and electrochemical impedance spectroscopy (EIS) were used to study the electrochemical properties of the coin cell type supercapacitors. The effect of sulfurization time on the electrochemical properties was studied.

The electrochemical properties of the carbonized FM were studied in detail. Figure 28.12a shows the specific capacitance values (F/g) for various samples at different current densities. It was noticeable that the 6 Hr sample presented the highest specific capacitance of around 27 F/g at 1 A/g, which gradually decreased to around 23 F/g at 10 A/g. Based on these results it is likely to infer that the chemical treatment with sulfuric acid for 6 Hr led to an optimized structure to adsorb the ions from the electrolyte. It is worth noting that the 6 Hr presented a surface area that was relatively similar to the 12 Hr, however, it possesses a relatively smaller micropore area. One possibility is that the smaller micropores of the 6 Hr were more likely to adsorb the ions than the 12 Hr, which could explain the larger specific capacitance for the 6 Hr. The galvanostatic charge–discharge (GCD) of all

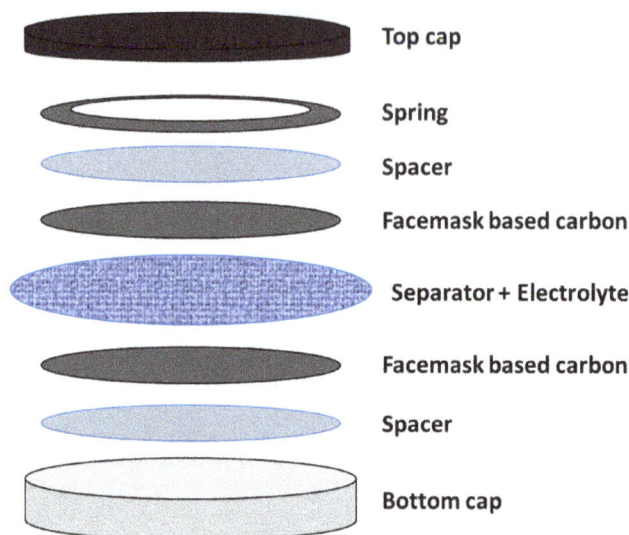

FIGURE 28.11   Schematics of the coin-cell type supercapacitor.

the samples was also performed. The GCD of the 6 Hr sample at various current densities is shown in Figure 28.12b. The graph presents a defined triangular shape which suggests the predominance of an EDLC type of energy-storage mechanism. Based on this type of profile, it could be inferred that the treatment with sulfuric acid led to the formation of defects in the structure to enable the adsorption of ions rather than doping the carbonaceous surface with sulfur. Following the same trend, the 6 Hr sample delivered the highest energy density of around 2.5 Wh/kg, as well as a power density of around 5.5 kW/kg both at a current density of 10 A/g, which are presented in Figures 28.12c and d, respectively. The 6 Hr sample outperformed the other tested samples regardless of the current density. Based on that, it is logical to infer that performing the treatment with acid for 6 h promoted the optimal surface area, which allowed better adsorption of the ions from the electrolyte.

Yuwen and his team converted used facemasks into active material for Li-S batteries by treating the disposable PP-based facemasks with sulfuric acid in a microwave-assisted reactor for 6, 8, and 10 min [2]. Through that, co-doping with S and O was performed to obtain the functionalized PP-based disposable medical mask (SO@DMM). Then, self-activation pyrolysis was performed by placing the SO@DMM in a tube furnace that was heated up to 900°C at a low Ar flow rate, and a ramp rate of 5°C/min for 120 min leading to the pyrolyzed SO@DMM (P-SO@DMM). The technique employed for the synthesis of P-SO@DMM led to an appreciable increase in some properties such as a surface area that went from 2 to 830.9 m²/g. Hence, co-doping with S and O followed by the considerable increase in the surface area created a synergy effect that improved the overall electrochemical performance. Following that, sulfur cathodes were prepared by mixing the P-SO@DMMs and S with 30 and 70 wt.%, respectively. The mixture was placed in a PTFE reactor and heated up to 155°C for 12h. Then the composite functionalized with S was obtained and named P-SO@DMM-S. The P-SO@DMM-s obtained at 6, 8, and 10 min of microwave reaction were analyzed through the CV. Figures 28.13a–c show the CV profile for those samples. It could be observed that all the CVs displayed two reduction peaks that are related to the polysulfide conversions of S_8 to Li_2S_n where $4 \leq n \leq 8$, and the conversion of Li_2S_n into Li_2S/Li_2S_2 [34]. In addition to that, the reverse conversion of Li_2S/Li_2S_2 into larger lithium polysulfides and sulfur led to the oxidation peak presented at 2.41 V. Another observation extracted from this plot was noted at the peaks near 2.31 V seen in the first cycles which either shifted to 2.38 V or disappeared in the following cycles. This behavior is attributed to the slow diffusion of electrolytes within the anode along with the formation of a solid

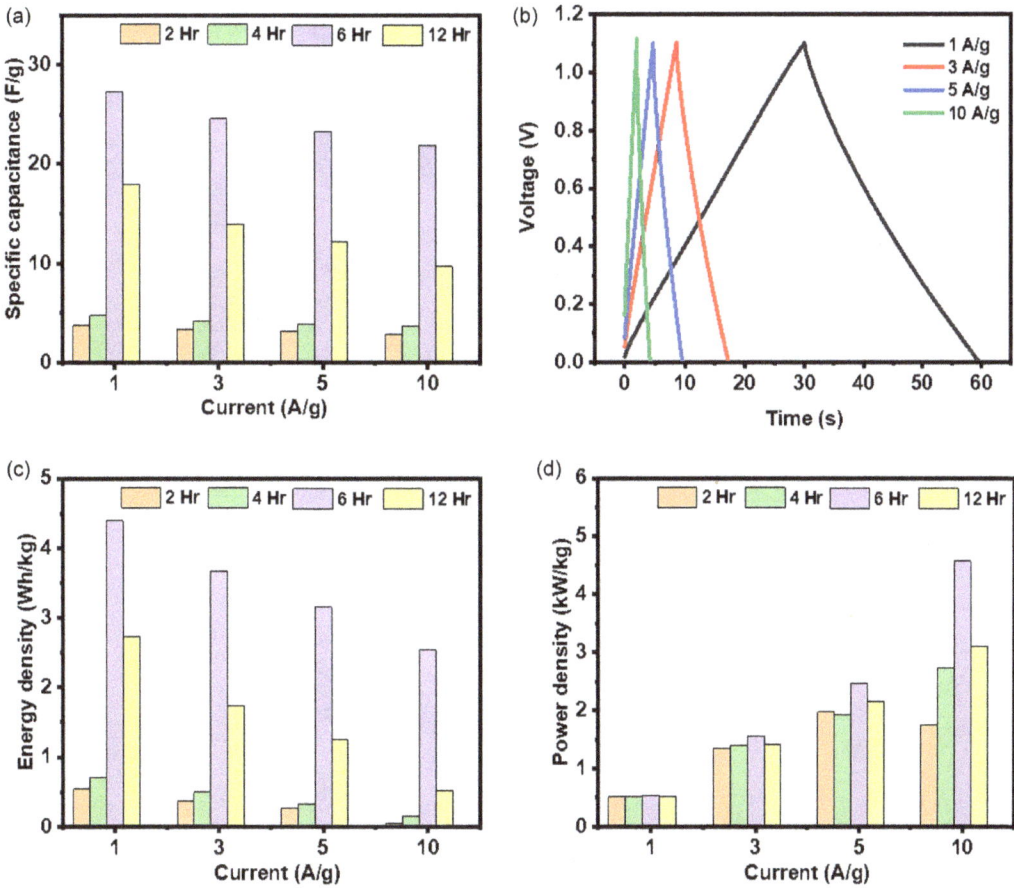

FIGURE 28.12 (a) Specific capacitance at different current densities for variously treated and carbonized samples, (b) GCD characteristics of 6 Hr sample at various applied currents, (c) energy density, and (d) power density at different current densities for sulfuric acid-treated and carbonized samples.

electrolyte interfacial film [35]. Alongside these results, it was observed that the sample irradiated through 10 min of microwave presented the highest specific capacitance of 1313.6 mAh/g at 0.1°C, however, the Li-S battery showed a considerable capacity loss of around 61.9% after 400 cycles.

The CV plot from the work by Yuwen *et al.* [2] for P-SO@DMM synthesized through microwave-assisted and self-activating pyrolysis presented a clear redox-dominated behavior due to the presence of S atoms. On the other hand, the CV results of the FM 6 Hr presented a signature EDLC (rectangular-like shape) type of behavior (absence of redox peaks). Such behavior is attributed to the permeation of ionic species within the porous structure of the FM 6 Hr without the occurrence of a chemical reaction in the process. The electrochemical behavior of the FM 6 Hr can be seen in Figure 28.14a. In addition to EDLC behavior, the FM 6 Hr presented outstanding electrochemical stability for over 10,000 galvanostatic charge–discharge cycles (Figure 28.14b). The high electrochemical stability was reflected in its cycling behavior which was maintained at nearly constant due to the non-faradaic process of ions from the electrolyte being attracted to the carbonaceous material's surface through electrostatic forces. Alongside the CV and electrochemical stability studies, electrochemical impedance spectroscopy was also used to understand their electrochemical behavior. The Bode and Nyquist plots of 6 Hr samples are presented in Figures 28.14c and d, respectively. In the Bode plot (Figure 28.14c), it is observed that the impedance decreased

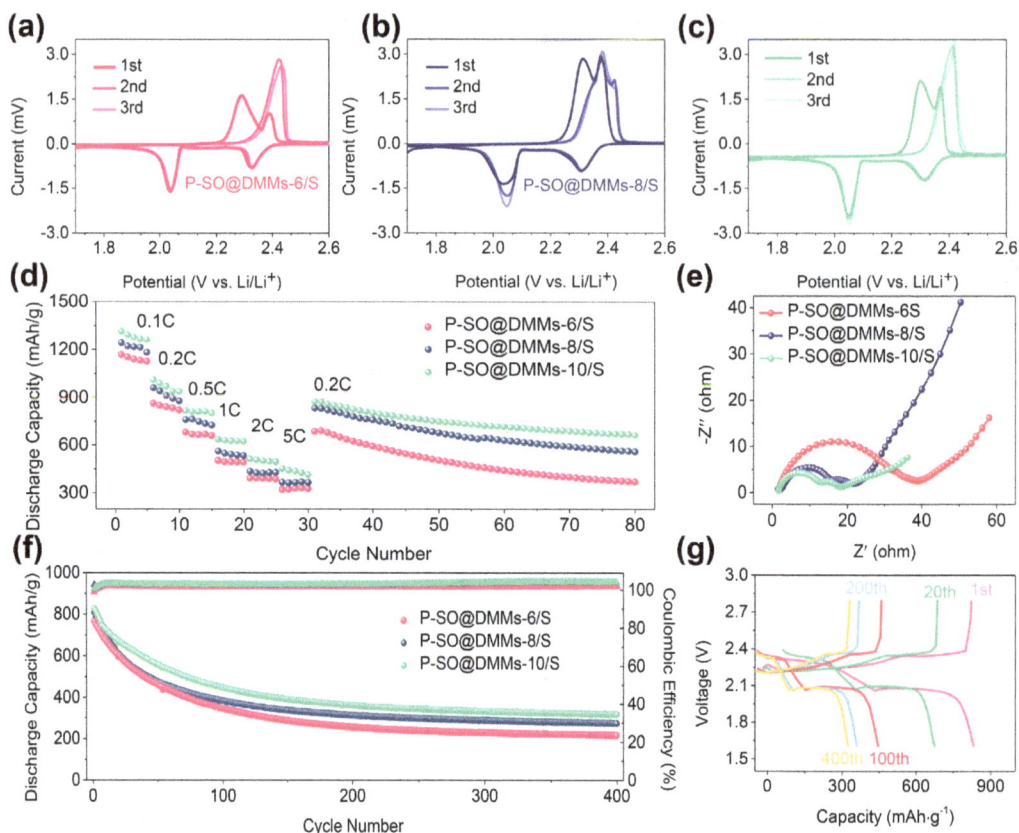

FIGURE 28.13 CV profiles for the P-SO@DMMs at (a) 6 min, (b) 8 min, and (c) 10 min of microwave reaction. Adapted with permission from Reference [2]. Copyright (2022), Elsevier.

exponentially with the increase in frequency, which is the expected behavior for a supercapacitor. In addition to that, the Bode's plot profile remained nearly the same after 10,000 cycles of GCD studies. Since the electrochemical behavior of the FM 6 Hr occurs predominantly through EDLC it could be inferred that ions from the electrolyte can easily permeate through the electrode's structure. Since no chemical reactions are occurring, the adhesion of ions on the electrode's surface can take place with relative ease, which explains the decrease in impedance of the system at higher frequencies. Such behavior suggests that ions can easily permeate through the pores of the FM 6 Hr without degrading the electrode's surface structure which can be seen through the high electrochemical stability. Also, through the analysis of the Nyquist plot (Figure 28.14d), it can be noted that the impedance decreases after 10,000 GCD cycles, meaning that ions are able to permeate through the electrode's structure more easily after several cycles. Such an effect may be attributed, perhaps, to the large porosity and constant flux of ions which may have led to more defined paths within the carbonaceous-based electrode culminating in a decrease in impedance.

4 CONCLUSIONS AND FUTURE REMARKS

Recent events related to the COVID-19 pandemic have led to the massive production of facemasks to avoid the spread of the virus. Since the masks were meant to be discarded after use there continues to be a high amount of solid waste generated. Even though these products served their initial purpose, this chapter demonstrates examples studied by the authors of this chapter, as well as from the

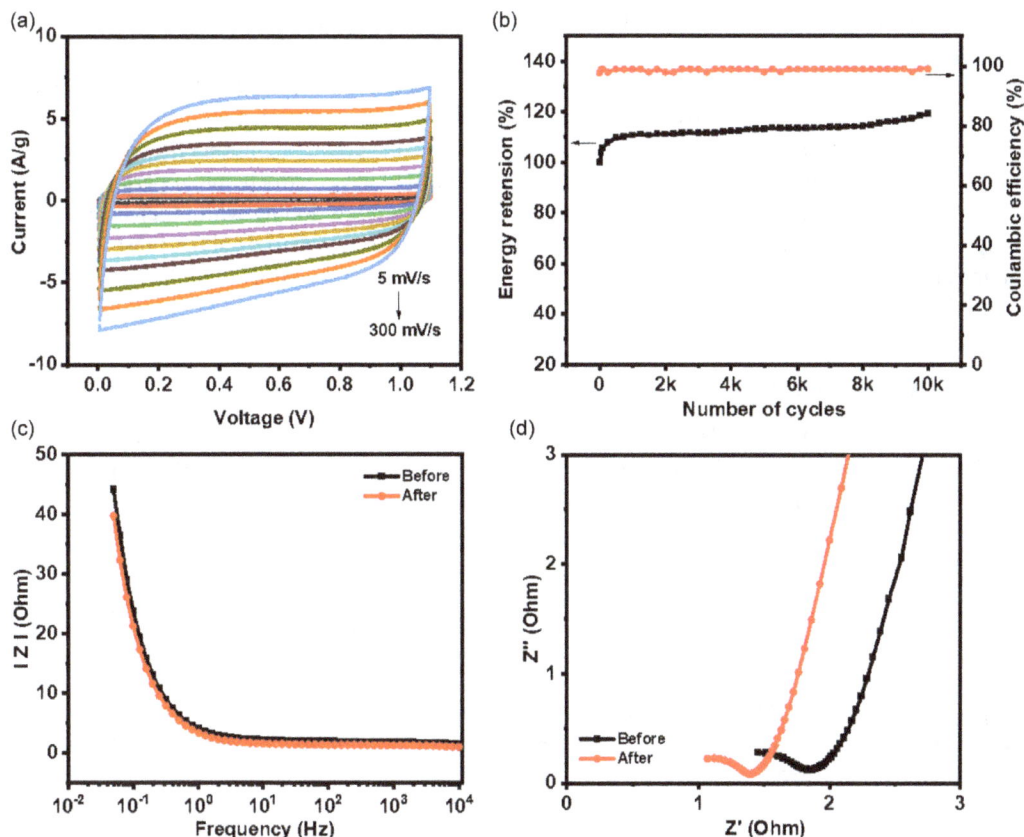

FIGURE 28.14 (a) CV at various scan rates, (b) energy retention and Coulombic efficiency up to 10,000 cycles of a charge–discharge study of a coin cell device fabricated using 6 Hr sample, (c) impedance versus frequency, and (d) variation in Z' and Z" of a coin cell device fabricated using 6 Hr sample before and after stability test.

literature, that PP-based masks can be converted into carbon-based materials that are suitable as an electrode for supercapacitors or batteries. In this sense, simple processes that consist of the digestion of facemasks in concentrated acid such as H_2SO_4, followed by thermal treatment at high temperature, can lead to the formation of a highly porous carbon-based material. This type of carbonaceous matrix can then serve as an electrode for supercapacitors or batteries since the techniques employed can greatly increase the material's surface area, which allows a facile permeation and flux of ions within its structure. A well-defined EDLC type of electrochemical behavior was observed. The inherent chemical stability of carbon provided excellent performance stability during 10,000 cycles of charge–discharge studies, which suggests that the carbonaceous matrix can properly accommodate ions from the electrolyte in its structure without considerable capacitance loss.

Other studies have explored the versatility of carbon by utilizing a facile and fast microwave-assisted approach that allowed the co-doping of O and S. After that, the co-doped carbon-based material was thermally treated and mixed with S. Through that, an electrode for Li-S batteries was obtained. Despite the relatively low electrochemical stability due to the shuttling effect of S, the electrode material displayed satisfactory capacitance values. Interestingly, since there was a relatively high S content, the CV behavior of the co-doped carbon material with O and S presented a pseudocapacitive profile. Based on these examples, it is notable that facemasks used during the COVID-19 pandemic can be converted into value-added electrode materials for supercapacitors and batteries. Such a process can be performed through relatively facile methods which may include acid digestion or microwave-assisted reactions. It is worth noting that carbon-based materials are already widely used in energy-storage and energy-generation devices. Thus, finding innovative ways to convert solid waste such as used facemasks into valued-added electroactive materials can provide an interesting route for recycling and reusing materials that could be improperly discarded and harm the environment. There is still a need to further improve the electrochemical performance of these carbon-based materials which likely can be achieved through doping with other materials or compositing with conducting polymers or transition metal-based materials since these materials can introduce pseudocapacitive properties that can create a synergistic effect with the carbon-based material, hence improving their overall electrochemical performance.

ACKNOWLEDGMENTS

The authors are grateful to the National Institute of Standards and Technology (NIST award number 70NANB20D146) and the U.S. Economic Development Administration (US-EDA award number 05-79-06038) for providing research infrastructure funding.

REFERENCES

[1] S.P. Nikam, P. Chen, K. Nettleton, Y.-H. Hsu, M.L. Becker, Zwitterion surface-functionalized thermoplastic polyurethane for antifouling catheter applications, Biomacromolecules. 21 (2020) 2714–2725.

[2] C. Yuwen, B. Liu, Q. Rong, L. Zhang, S. Guo, Porous carbon materials derived from discarded COVID-19 masks via microwave solvothermal method for lithium-sulfur batteries, Sci. Total Environ. 817 (2022) 152995.

[3] A.L. García-Basteiro, C. Chaccour, C. Guinovart, A. Llupià, J. Brew, A. Trilla, A. Plasencia, Monitoring the COVID-19 epidemic in the context of widespread local transmission, Lancet Respir. Med. 8 (2020) 440–442.

[4] S. Feng, C. Shen, N. Xia, W. Song, M. Fan, B.J. Cowling, Rational use of face masks in the COVID-19 pandemic, Lancet Respir. Med. 8 (2020) 434–436.

[5] S. Dharmaraj, V. Ashokkumar, S. Hariharan, A. Manibharathi, P.L. Show, C.T. Chong, C. Ngamcharussrivichai, The COVID-19 pandemic face mask waste: A blooming threat to the marine environment, Chemosphere. 272 (2021) 129601.

[6] I. Anastopoulos, I. Pashalidis, Single-use surgical face masks, as a potential source of microplastics: Do they act as pollutant carriers?, J. Mol. Liq. 326 (2021) 115247.

[7] K. Selvaranjan, S. Navaratnam, P. Rajeev, N. Ravintherakumaran, Environmental challenges induced by extensive use of face masks during COVID-19: A review and potential solutions, Environ. Challenges. 3 (2021) 100039.

[8] S.-L. Wu, J.-H. Kuo, M.-Y. Wey, Thermal degradation of waste plastics in a two-stage pyrolysis-catalysis reactor over core-shell type catalyst, J. Anal. Appl. Pyrolysis. 142 (2019) 104641.

[9] V.G. Pol, Upcycling: Converting waste plastics into paramagnetic, conducting, solid, pure carbon microspheres, Environ. Sci. Technol. 44 (2010) 4753–4759.

[10] J. Gong, J. Liu, Z. Jiang, J. Feng, X. Chen, L. Wang, E. Mijowska, X. Wen, T. Tang, Striking influence of chain structure of polyethylene on the formation of cup-stacked carbon nanotubes/carbon nanofibers under the combined catalysis of CuBr and NiO, Appl. Catal. B Environ. 147 (2014) 592–601.

[11] J. Gong, J. Liu, Z. Jiang, X. Chen, X. Wen, E. Mijowska, T. Tang, Converting mixed plastics into mesoporous hollow carbon spheres with controllable diameter, Appl. Catal. B Environ. 152–153 (2014) 289–299.

[12] X. Hu, Z. Lin, Transforming waste polypropylene face masks into S-doped porous carbon as the cathode electrode for supercapacitors, Ionics (Kiel). 27 (2021) 2169–2179.

[13] R. Yu, X. Wen, J. Liu, Y. Wang, X. Chen, K. Wenelska, E. Mijowska, T. Tang, A green and high-yield route to recycle waste masks into CNTs/Ni hybrids via catalytic carbonization and their application for superior microwave absorption, Appl. Catal. B Environ. 298 (2021) 120544.

[14] B. Xu, F. Wu, S. Chen, C. Zhang, G. Cao, Y. Yang, Activated carbon fiber cloths as electrodes for high performance electric double layer capacitors, Electrochim. Acta. 52 (2007) 4595–4598.

[15] C.-C. Hu, K.-H. Chang, M.-C. Lin, Y.-T. Wu, Design and tailoring of the nanotubular arrayed architecture of hydrous RuO_2 for next generation supercapacitors, Nano Lett. 6 (2006) 2690–2695.

[16] T. Cottineau, M. Toupin, T. Delahaye, T. Brousse, D. Bélanger, Nanostructured transition metal oxides for aqueous hybrid electrochemical supercapacitors, Appl. Phys. A. 82 (2006) 599–606.

[17] M.J. Bleda-Martínez, J.M. Pérez, A. Linares-Solano, E. Morallón, D. Cazorla-Amorós, Effect of surface chemistry on electrochemical storage of hydrogen in porous carbon materials, Carbon N. Y. 46 (2008) 1053–1059.

[18] S. Roldán, M. Granda, R. Menéndez, R. Santamaría, C. Blanco, Mechanisms of energy storage in carbon-based supercapacitors modified with a quinoid redox-active eElectrolyte, J. Phys. Chem. C. 115 (2011) 17606–17611.

[19] G. Lota, E. Frackowiak, Striking capacitance of carbon/iodide interface, Electrochem. Commun. 11 (2009) 87–90.

[20] D.N. Futaba, K. Hata, T. Yamada, T. Hiraoka, Y. Hayamizu, Y. Kakudate, O. Tanaike, H. Hatori, M. Yumura, S. Iijima, Shape-engineerable and highly densely packed single-walled carbon nanotubes and their application as super-capacitor electrodes, Nat. Mater. 5 (2006) 987–994.

[21] P.-L. Taberna, G. Chevallier, P. Simon, D. Plée, T. Aubert, Activated carbon–carbon nanotube composite porous film for supercapacitor applications, Mater. Res. Bull. 41 (2006) 478–484.

[22] A. Burke, R&D considerations for the performance and application of electrochemical capacitors, Electrochim. Acta. 53 (2007) 1083–1091.

[23] G.G. Amatucci, F. Badway, A. Du Pasquier, T. Zheng, An asymmetric hybrid nonaqueous energy storage cell, J. Electrochem. Soc. 148 (2001) A930.

[24] T. Pandolfo, V. Ruiz, S. Sivakkumar, J. Nerkar, General properties of electrochemical capacitors, Supercapacitors. (2013) 69–109.

[25] D.P. Dubal, O. Ayyad, V. Ruiz, P. Gómez-Romero, Hybrid energy storage: the merging of battery and supercapacitor chemistries, Chem. Soc. Rev. 44 (2015) 1777–1790.

[26] G. Lee, M. Eui Lee, S.-S. Kim, H.-I. Joh, S. Lee, Efficient upcycling of polypropylene-based waste disposable masks into hard carbons for anodes in sodium ion batteries, J. Ind. Eng. Chem. 105 (2022) 268–277.

[27] L. Liao, W. Xiao, M. Zhao, X. Yu, H. Wang, Q. Wang, S. Chu, Y. Cui, Can N95 respirators be reused after disinfection? How many times?, ACS Nano. 14 (2020) 6348–6356.

[28] A. Ghosal, S. Sinha-Ray, A.L. Yarin, B. Pourdeyhimi, Numerical prediction of the effect of uptake velocity on three-dimensional structure, porosity and permeability of meltblown nonwoven laydown, Polymer (Guildf). 85 (2016) 19–27.

[29] X. Zhu, D. Yan, H. Yao, P. Zhu, In situ FTIR spectroscopic study of the regularity bands and partial-order melts of isotactic poly(propylene), Macromol. Rapid Commun. 21 (2000) 354–357.

[30] M.R. Jung, F.D. Horgen, S. V Orski, V. Rodriguez C., K.L. Beers, G.H. Balazs, T.T. Jones, T.M. Work, K.C. Brignac, S.-J. Royer, K.D. Hyrenbach, B.A. Jensen, J.M. Lynch, Validation of ATR FT-IR to identify polymers of plastic marine debris, including those ingested by marine organisms, Mar. Pollut. Bull. 127 (2018) 704–716.

[31] L. Huang, Q. Wu, S. Li, R. Ou, Q. Wang, Toughness and crystallization enhancement in wood fiber-reinforced polypropylene composite through controlling matrix nucleation, J. Mater. Sci. 53 (2018) 6542–6551.

[32] D.G. Papageorgiou, D.N. Bikiaris, K. Chrissafis, Effect of crystalline structure of polypropylene random copolymers on mechanical properties and thermal degradation kinetics, Thermochim. Acta. 543 (2012) 288–294.

[33] Y. Liu, C.H.L. Kennard, R.W. Truss, N.J. Calos, Characterization of stress-whitening of tensile yielded isotactic polypropylene, Polymer (Guildf). 38 (1997) 2797–2805.

[34] Y. Tian, G. Li, Y. Zhang, D. Luo, X. Wang, Y. Zhao, H. Liu, P. Ji, X. Du, J. Li, Z. Chen, Low-bandgap Se-deficient antimony selenide as a multifunctional polysulfide barrier toward high-performance lithium–sulfur batteries, Adv. Mater. 32 (2020) 1904876.

[35] Y. Gao, C. Wang, J. Zhang, Q. Jing, B. Ma, Y. Chen, W. Zhang, Graphite recycling from the spent lithium-ion batteries by sulfuric acid curing–leaching combined with high-temperature calcination, ACS Sustain. Chem. Eng. 8 (2020) 9447–9455.

29 Polymers in Display Devices

Neena George,[1] Ajalesh B. Nair,[2] Simi Pushpan K.,[2] and Rani Joseph[3]

[1] Post Graduate and Research Department of Chemistry, Maharaja's College, Eranakulam-682 011, Kerala, India
[2] Post Graduate and Research Department of Chemistry, Union Christian College, Aluva-683 102, Kerala, India
[3] Department of Polymer Science and Rubber Technology, Cochin University of Science and Technology, Cochin-22, Kerala, India

1 INTRODUCTION

The present digital era demands instantaneous audio-visual information, even from remote global locations. People favor compact, light, low-power displays when they're on the go, while huge and flat/thin panels are preferred at home. In the current scenario, we prefer high-resolution colored photographs with a lot of brightness and contrast that can be seen from any angle. In man–machine communication, the display is crucial. New display concepts are popping up. Polymer-based LED displays, electro-chromic displays, and electro-phoretic displays based on nanoparticles are all recent examples. The polymer-driven market is a highly competent area for transparent, robust, light-weight, and flexible displays which are comparatively less expensive materials and suitable for bulk production.

Polymer-based flexible electronics allow border-line technology and are used to offer novel methods to interact with the wider world and unlock doors in the direction of innovative applications such as wearable display devices [1], e-skin (electronic skins) [2], flexible/bendable display devices [3], and electronic papers. The main compensation of flexible electronics includes inter-facial conformation changes, and their flexibility, stretchability, and lower weight.

Polymeric materials are integrated into the construction and performance of liquid crystal displays (LCDs). Polymeric or organic LEDs are semiconductor display devices that can be used for inkjet and screen-printing technologies because of their conductivity, flexibility, and transparency, and also their structural properties. Electrochromic polymers are widely used in transparent semiconductor technology. In addition, thin-film transistors (TFTs) and electronic papers are a key component in flexible thin-film electronics.

The aim of this chapter is to provide an understand of the use of conjugated/conducting polymers in LEDs, LCDs, TFTs, and other display devices such as electrochromic and electrophoretic displays. The use of transparent polymer- and nanomaterial-based conductive electrodes in electro-optic devices is also discussed. Additionally, research gaps/future perspectives of polymers in display devices are also highlighted in this chapter.

2 POLYMER-BASED LCDS

Liquid crystals (LCs) are rod-like complex organic molecules with an organized meso-phase change between the melting temperature of crystal form and its isotropic liquid. In solid-state liquid crystal, there exist parallel arrangements of molecules. In the cloudy liquid state at slightly higher temperatures, the molecules have more freedom than in the solid state but they are inclined to line up in the same direction which results in reflecting light that creates cloudiness, while in the higher temperature range, the disconcerted molecules tend to create a clear liquid.

DOI: 10.1201/9781003278269-29

Four common constituents are found in almost all LCDs. The first component is a set of two glass plates (usually 0.7 mm thick) separated by roughly 5 μm. Second, between the substrates is an orientated LC layer. Later, the transparent electrodes apply potential across the LC layer on the inner surfaces of the substrates. Finally, two polarizing filters are inserted between the glass substrates.

Liquid crystal displays (LCDs) are used to demonstrate alpha-numeric information in various machines including laptop screens, scientific instruments, clocks, HDTVs, and other modern-day gadgets. Voltage transforms a segment of liquid crystal from the transparent phase to the cloudy phase, which is a part of a numeral or character or dots arranged in rows and columns. Individual switching to block or to permit polarized light to pass through the LCD creates dark or a bright spot, respectively, on the reflecting screen. LCDs are generally classified into passive matrix the active matrix LCDs depending on the application of electricity on unit pixels. The passive matrix involves uniform distribution of voltage to all display pixels, while in active matrix LCDs, thin-film transistors control the image quality of each pixel resulting in better images, but this is a highly expensive technology. Various components of a working LCD consist of a power supply, display glass, mechanical package, drive, and control electronics. LCD makers also make use of thin sheets like polarizer material which enhances the contrast effect.

The worldwide consumer trend has been in such a manner that travellers prefer smaller, portable, and lighter devices which utilize low power displays, while large, thin, flat screens are preferred for home use. Current LCDs are manufactured from glass substrates, indium tin oxide (ITO) electrodes, active matrix TFT,s and a thin film based on organic polymers. In the cell manufacturing process, lithographic polymers are used to structure ITO and to produce TFTs. The cell thickness is controlled by polymer spheres, and adhesives are utilized to join parallel aligned glass plates with a liquid crystal-filled display cell in between. UV-curing adhesive seals the filling opening and liquid crystal orientation is regulated by an extremely thin polyimide layer aligned to circumvent bonding surface areas. Black-pigmented lithographic acrylate is used to avoid diminishing contrast, in the vacant areas of primary color pixels filled black matrix. To reduce the thickness of LCDs, cold cathode fluorescent tubes or LEDs are used as the light source. These are positioned at the planar waveguide of poly(methyl methacrylate) or polycarbonate origin containing scattering units that permit the light to reach the observer. Special light-collimating films with an acrylate coating on a polyester base like cholesteric networking with a chiral-nematic acrylate monomer coating or poly(ethylene-2,6-naphtylenedicarboxylate) (PEN) are used to diffuse the reflection so as to improve the brightness and contrast. The polarizer film of the backlight system is a composite of PVA [poly(vinyl alcohol)] interspersed in cellulose triacetate films. PVA affiliated with iodine crystals takes care of the polarization-dependent absorption of light [4].

Hence, it is important to note that in the present scenario, LCD production utilizes various polymeric products in the manufacturing stage, display creation, alignment of liquid crystals, and in the operational optical display. Quintessential applications of digital displays and the possibility of improving the quality of imaging on flexible devices and bending capability has been a desirable topic of interest for a long time. Polymer-dispersed liquid crystals (PDLCs), as shown in Figure 29.1, combine the mechanical aspects of conducting polymers and electro-optical properties of liquid crystals [4]. PDLC was distinguished as an electrically switchable material that has incited scientific interest for the past four decades and is prepared using phase separation and emulsion methods. PDLC composite films are comprised of nanosized LC droplets aligned in a haphazard fashion which demonstrate a milky-white scattered state implanted in a polymer medium/matrix. On voltage application this film turns transparent due to the orientation of LC droplets along the electric field direction, as shown in Figure 29.2 [5].

The phase-separation methodology which was induced by the action of heat, solvent evaporation, and polymerization for the preparation of PDLCs is the preferred technique owing to its morphological control and final film characteristics. LC droplets prepared by different methodologies vary according to the induction method adopted for the formation. In thermally induced

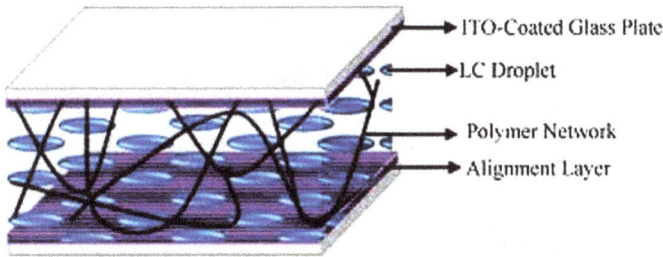

FIGURE 29.1 Polymer stabilized liquid crystal. Adapted with permission from Reference [4], under an open access license.

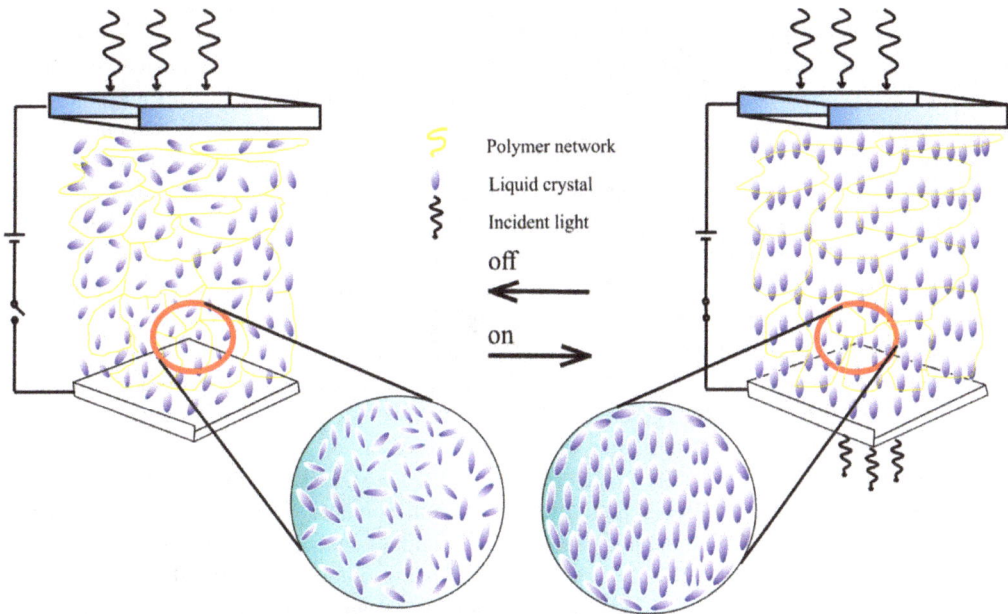

FIGURE 29.2 Working of a PDLC device: (a) off state and (b) on state. Adapted with permission from Reference [5], under an open access license, MDPI.

phase separation (TIPS), polymer and LC are mixed at elevated temperature conditions and subsequent cooling at a controlled rate creates the phase separation which reinforces the LC domains. In the case of solvent-induced phase separation (SIPS) rate-controlled evaporation of the solvent results in phase separation, while in polymerization-induced phase separation (PIPS), a prepolymer or monomer solution is combined with LC to form a homogeneous mixture. The most commonly used monomer is (meth)acrylate and the structural and E-O properties are modulated in the literature by incorporating hydroxy, epoxy, straight/branched/cyclic methylene and methyl groups, phenyl and bisphenol moieties [6]. Optimization of electro-optical properties of PDLC composite films are mainly done by modulating viscosity, refractive index, and hydrogen bond interactions existing in the networking polymer matrix. Additionally, a nanoparticle (NP) doping technique has been adopted to enhance the optical, mechanical, and thermal properties of the polymer matrix as it impacts the refractive index, stabilizing forces and dielectric constant of the crossing point of LC and polymer matrix. ITO NPs impart thermal insulation, thereby contributing to energy efficiency, while Nanjing University scientists achieved a decrease in operating voltages by lowering the anchoring force at the LC–polymer interface by doping 1.5 wt.% submicron ITO powders in the mixture (E7 + NOA65). Zhang et al. utilized ITO NPs along with 3-methacryloxypropyltrimethoxysilan which resulted in an increase in the driving voltage, CR lowering, and enhancement of NIR absorption [7]. Inorganic NPs like ZnO, MgO, CuO, $BaTiO_3$, Fe_3O_4, and TiO_2 were also doped into PDLCs to decrease the driving voltage, while ferroelectric dopant NPs resulted in improved E-O properties for the PDLC composites [8].

3 POLYMERS IN LEDS (OLEDS)

OLEDs are used in devices such as smartphones, personal digital assistants (PDAs), and television screens, and are a prominent research field for solid-state lighting applications. An OLED is a semi-conductor light source or light-emitting diode wherein an organic electroluminescent film responds to an electric current by getting illuminated. An organic layer is placed between two electrodes, of which at least one is transparent. There are two categories of OLED, namely, small-molecule and polymer-based. When the OLED is modified by the addition of mobile ions it creates a light-emitting electrochemical cell (LEC). OLED displays are divided broadly into two categories, passive matrix (PMOLED) and active matrix (AMOLED). Each row in the PMOLED display is sequentially con-trolled and in AMOLED it is a thin-film transistor backplane which admits and controls each indi-vidual pixel, permitting enhanced display and resolution. Unlike LCD, the OLED display is devoid of backlight, thus decreasing operating voltage and exhibiting deep black levels and higher contrast ratio even in dim light conditions, and it has the added advantage of slimmer and lighter versions. OLED is a solid-state-based thin-film device which can be easily spread to form flexible displays owing to the simple fabrication process, improved light-emitting efficiency, and reduced geometric distortion, and they have replaced CRTs and LCDs in recent times.

The challenge in fabricating a full-color OLED display lies in patterning of pixels and in attaining a continually emitting light. Red, green, and blue pixels have good optical performance; however, the short life time of blue-emitting materials and difficulties of deposition of individual pixels is a major concern. Princeton University invented stacked OLED (SOLED) having red, blue, and green emitters in each pixel which are separated by transparent contacts. They successfully fabricated OLED with improved resolution but at the expense of brightness which was considered as a demerit of the system. A variety of plastic substrates/foils are being pursued with the desirable mechan-ical and optical properties for lightweight and bendable organic light-emitting devices. Highly transparent and flexile thin films/foils show a low degradation temperature compared to the rigid substrates. Foils that exhibit high thermal stability lack transparency, while polycarbonate (PC)-based transparent substrates possess moderate solvent resistance [9].

Polyethylene terephthalate (PET) and polyethylene naphthalate (PEN) are polymers with excel-lent mechano-optical properties with modest thermal stability, low moisture and oxygen absorp-tion and solvent resistance are successfully utilized in flexible electronics. Thermal stabilization of plastic materials results in a loss of bendability [10], which limits the rolling of the device. Disadvantages of low-cost plastic substrates are associated with the fabrication processes which require high temperatures leading to degradation of the material [11]. Even though foils remain unstable at extreme conditions due to high voltage and maximum brightness, they have been found to be chemically stable and resistant to cleaning solutions [12].

Good OLEDs should have the desired combination of transparency, sheet resistance, and flex-ible fabrication. A lack of highly flexible transparent electrodes is identified as a major drawback of OLED technology. A variety of transparent flexible electrodes, such as conducting polymers [13], graphene/MWCNT [14], and SWCNT films [15], have been researched in an attempt to realize consistent and exceedingly flexible OLED displays. Analogous approaches, wherein there is the incorporation of an intermediate metal film with desirable refractive index between two polymeric films, have been introduced to enhance the electrical conductivity [16]. In flexible OLEDs, serious implications related to the mechanical stability were not identified with the color organic films although there were some limitations in patterning for pixel formation [17].

4 POLYMERIC THIN-FILM TRANSISTORS (PTFTS)

PTFTs, also known as organic thin-film transistors (OTFTs), are among the futuristic devices meant to be integral to the development of a wide range of affordable and large-area electronics applications, including active-matrix displays and modifiable microelectronics. Organic transistors have great mobility, speed, and enhanced current characteristics, making them ideal for flexible display-driving elements [18].

OTFTs are transistors that use an organic semiconductor as the active layer to regulate the potential. Applications of OTFTs include removable storage devices, smartphones, wearable cloths, electronic displays, automobiles, flexible integrated circuits, sensors, etc. OTFTs have two major benefits over inorganic semiconductor TFTs: they can be synthesized at lower temperatures and they are inexpensive. Inorganic semiconductors require a very high fabrication temperature. Organic semiconductors, on the other hand, are soft and should be handled mostly at ambient temperature, as elevated temperatures can damage the organic components. Organic transistors have the potential to operate at low voltages while still delivering good performance.

PTFT is made up of a source, drain, polymeric thin semiconducting layer, gate, and gate dielectric. The first three components are in the same layer (Figure 29.3). Organic materials such as poly(3-octylthiophene) (P3OT), pentacene, poly(3-hexylthiophene) (P3HT), and poly(3-alkylthiophene) (P3AT) have been reported to be used as semiconductors in TFTs, and are currently being explored to substitute the dielectric layer with suitable organic insulating materials. Remarkably, the transistor can operate at even lower voltages when we combine an organic semiconductor layer with an organic insulator layer.

The performance of OTFTs is determined by the organic semiconductors and insulator materials used in their construction. The most often utilized organic compounds for semiconducting layers are pentacene, poly(3-octylthiophene) (P3OT), poly(3-alkylthiophene) (P3AT), and poly(3-hexylthiophene) (P3HT). P3HT is highly soluble in a wide range of organic solvents. Polymers including polyethylene naphthalate (PEN), polyimide, and polyethylene are employed as substrates. As electrodes, poly-(3, 4- ethylenedioxythiophene):(styrene sulfonic acid) (PEDOT:PSS) and polyaniline (PANI) doped with camphorsulfonic acid (PANI-CSA) deposited by spin coating can be used.

The manner in which a polymeric TFT (PTFT) is fabricated has a significant impact on its performance. PTFTs can be made in a variety of ways, using diverse inorganic or organic constituents as the components, namely substrates (n-doped silicon or a plastic foil), semiconducting layers, electrodes, and insulator gates [19]. A thermally generated oxide is employed as a gate insulator for devices on Si wafers.

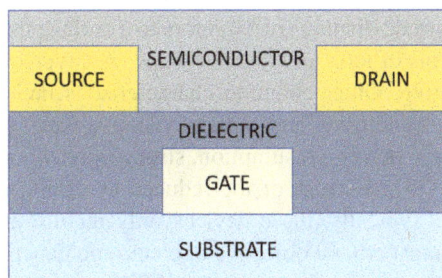

FIGURE 29.3 Configuration of PTFT.

The charge mobility in polymers and crystals is very different. In the conduction or valence band of crystals, carriers move at the thermal velocity, and scattering regulates carrier mobility. Unlike charge transport in crystals, localized charges are observed in amorphous polymer films. This involves a conduction mechanism, wherein there is charge hopping between energetically and geographically dispersed states in polymer films. PTFT's performance is critically dependent on the material of the source and drain connections.

4.1 Applications of TFTs

Flexible memories and display devices use OTFTs extensively, as they enable appealing qualities and advantages, such as being budget friendly, low temperature processability, lower manufacturing cost, and good strength. These devices usually contain (a) metal nanoparticles (NPs) embedded in gate dielectrics, (b) ferroelectric polymer dielectric materials, and (c) an organic semiconductor layer incorporated with metal NPs [20]. To create efficient memory devices, nano-sized particles are employed in polymer matrices which render energy efficiency, smaller device dimensions, excellent SLIC immunity, and large data density, coupled with extended data retention [21]. Organic electronic memory, which can be made at low cost, can be used to provide memory chip capabilities to goods that require data storage [22].

OTFTS can also be used in the construction of radio frequency identification (RFID) tags for consumer item tracking. An OTFT-based DNA sensor requires a total analysis time of less than 40 minutes, compared to roughly 24 hours for traditional approaches [23]. As a result, OTFTs can be used to implement quick and low-cost DNA detection. Low-cost and flexible OTFTs built on an insulating substrate provide a number of advantages over silicon-based devices, particularly in disposable applications.

Over the last decade, the polymeric thin-film transistors have improved steadily and consistently to be a worthy competitor for inorganic TFTs. OTFTs deposited with active layers of pentacene have been the outcome of previous research efforts. To make OTFTs even more cost effective, better process techniques are used to deposit organic molecules which ultimately have an impact on its performance. In addition, chemical modification of organic semiconducting materials augments chemical stability. The construction of OTFTs can be modulated by choosing the right materials for its components, as well as the thicknesses of the layers and the device topology. The next crucial step is the advent of novel devices that utilize organic semiconductor features and prepare multifunctional systems that are currently impossible to fabricate with inorganic semiconductors. To reduce the drive voltage, all solutions require improvements in charge mobilities. Despite the remarkable progress made in recent years, numerous critical concerns with PTFTs/OTFTs must be addressed, including stability, device repeatability, dependability, better carrier mobility, longer life, and cost-effective large-scale manufacture.

5 ELECTRONIC PAPER (E-PAPER)

Electronic paper or electrophoretic display (EPD) refers to a surface that is electrically charged and resembles ink on paper in terms of look and touch. e-Paper is a type of reflective display that does not require any internal lighting. Other common characteristics include lightweight, low energy usage, insensitivity to daylight conditions, and angle of vision. e-Paper technology has demonstrated increased flexibility, nearly zero power consumption, strong optical contrast in bright sunlight, and even solar panel integration [24]. Because of the reduced eyestrain, many people prefer e-paper devices for portable reading. Typically, these devices only require a small battery, allowing for innovative designs [25]. The most crucial objective of electronic paper is paper-like readability.

e-Paper display technology was developed in the 1970s, and commercialization began in the 1990s. e-Paper uses thin, flexible, foldable, rollable, reflective displays which need very low power.

They find application as wearable computer screens, newspapers, magazines, smart identity cards, labels, stickers, wallpapers, and smart windows [26]. Some of the techniques being used in e-paper production are cholesteric, ferroelectric, PDLC, electrochromic, electrophoretic, electrodeposition, and OLED.

5.1 Electrophoretic Displays

Electrophoretic displays use the flow of charged particles under an electric field to switch images. It usually contains a clear fluid comprised of millions of microcapsules carrying positively charged white and negatively charged black pigments. When a negative electric field is supplied to the "ink," the white particles travel to the top of the capsule, causing the surface to look white at that particular location. The black particles ascend to the top when the technique is reversed, causing the surface to appear dark. Generation of bicolored letters and graphics is possible by dividing the display into many independently controlled pixels. A colorful image is generated by adding a filter matrix [27].

Unlike traditional LCD flat panel screens with backlighting, these screens are bi-stable. LCD displays need a continuous power supply, whereas e-paper displays show images even without electricity. Electricity is used once the content of an e-paper changes, such as when a supermarket's e-paper shelf label is updated with a new price. The display will merely show the content you want it to show the rest of the time, drawing no power until the next update. Another advantage is the reflective property, which allows light from the environment to be reflected off the surface of the e-paper display and into the user's eyes, much like light is reflected from conventional paper. It may be read in direct sunlight because of the wide viewing angle available.

Unlike LCDs and OLEDs, electrophoretic displays can be made in large sizes. The biggest drawback is the slow switching speed, and hence these are inappropriate for video or even quick scrolling. e-Paper and e-books are common uses for electrophoretics because of their bistability.

Xerox (USA) invented the first electronic paper, a twisting ball panel display called Gyricon, in 1976 [28]. A thin sheet of elastomer containing bi-colored spherical particles is the basic display structure. Each sphere has its own silicon oil-filled chamber which allows it to rotate. One hemisphere is white and the other is a different color, and each hemisphere has a distinct permanent charge. The balls drift to the cavity on one side and maintain a firm grip on the wall. When an electric field is applied to the elastomer sheet, the ball is released from the cavity's side, enabling it to rotate in order to line up with the field. The ball returns to the cavity wall after rotation and remains strongly attached until a reverse electric current is supplied.

Sipix Imaging Inc. developed Microcup® electrophoretic displays where a suspension of particles is injected into 80–160 mm diameter chambers engraved into epoxy substrates. The Microcups hold and top-seal an electrophoretic fluid made up of microparticles carrying charged pigment (TiO_2) scattered in a colored dielectric solvent. Color rendition can be achieved by using a color filter or progressively filling and closing red (R), green (G), and blue (B) electrophoretic fluids in the Microcups [29].

When most people think of e-paper nowadays, they're thinking about items that use E Ink's EPD technology. This technology, combined with features added to the Amazon Kindle and other comparable items, has made e-books a competitive product, with e-paper being the preferred immersion reading technology. Technology advances at a rapid pace, improving in terms of performance, reliability, and affordability.

However, success fosters competition, and the e-paper type of goods is anticipated to evolve rapidly in the next few years. With growing volume, the costs of these devices will continue to fall.

Hanvon announced the first commercialized colored EPD product, with a diagonal of 9.68 inches, that was released in China in November 2010. Still, grayscale video quality is a key challenge for vertical EPDs. If there is a strong demand for color and video in e-paper products, and vertical EPD technology cannot meet it, vertical EPD devices' existing supremacy may be threatened.

Vertical switching electrophoretic displays are now common. Because vertical switching has a color constraint, horizontal switching remains a viable option. Horizontal displays could find a product niche where print-like color is needed, such as electronic signs and/or electronic skins on the exteriors of electronic gadgets. This technique allows for personal-use devices like e-readers by balancing the color and image quality of the final product.

Flexible electrophoretic displays were developed by Plastic Logic Germany, where they use glass-free backplanes, where the transistor matrix is arranged on plastic. Rather than using typical silicon transistors, they use organic thin-film transistors (OTFTs) constructed of PET polymer in their active-matrix backplane. To build a fully flexible display with endless possibilities, a flexible backplane is combined with a flexible display material, such as flexible OLED.

5.2 Electrochromic Polymers for Display Applications

An electrochromic material is one that undergoes a reversible color change as a result of electron reduction (gain) or oxidation (loss) when an electrical current passes through it after a suitable electrode voltage is applied.

Metal oxide films [30], conjugated polymers, and molecular dyes [31] (mainly viologen-based organic Type I) are the most popular electrochromic materials utilized in electrochromic devices. Due to the simplicity of manipulating characteristics through structural alterations, ease of preparation, flexibility, and low cost, conjugated polymers have greater attracted interest than others. Conjugated polymers provide excellent color variety, huge optical contrasts, rapid response times, and low energy usage when it comes to electrochromic (EC) performance.

Conjugated polymers, viz. poly(thiophene) (PTh), poly-(pyrrole) (PPy), and poly(aniline) (PANI) derivatives, exhibit electrochromism. Alkoxy-substituted poly(thiophene) derivatives, such as PEDOT, have been investigated due to their simplicity of manufacturing, high optical contrast values between redox states, and high chemical stabilities in the oxidatively doped state [32]. In conjugated polymers electrochromism is caused by changes in the electronic character of the conjugated polymer, as well as reversible insertion and extraction of ions through the polymer film due to electrochemical oxidation and reduction. These polymers exhibit semiconducting behavior in their neutral (insulating) states, with an energy gap (Eg) between the valence band (HOMO) and the conduction band (LUMO). The neutral polymer's band structure is altered by electrochemical or chemical doping ("p-doping" for oxidation and "n-doping" for reduction). This causes higher conductivity and optical modulation due to lower energy intraband transitions and the creation of charged carriers (polarons and bipolarons).

The growing interest in industrial applications has fueled demand for electrochromic devices based on conjugated polymers. Smart windows, sunglasses, optical shutters in airplanes, rear-view mirrors, and reflecting displays are among these applications. Numerous researches have been done on the smart window design as a polymer electrochromic technology application, which uses a seven-layer design (Figure 29.4) [33]. Smart windows are often made using chemical vapor deposition (CVD) and sputtering methods. Various constraints in producing electrochromic devices still exist due to the required high temperature and vacuum (ECDs).

6 TRANSPARENT ELECTRODES

Electrically conductive transparent electrodes are required for electronic display devices like LCDs, OLEDs, TFTs, and electronic papers. Indium-tin-oxide (ITO) coated on a glass substrate is the most common electrode material in electro-optic devices. However, its disadvantages include lack of flexibility due to brittleness, high processing temperatures, and high cost. Further options include transparent conducting oxides (TCOs), such as fluorine-doped tin oxide and aluminum-doped zinc oxide. When compared to ITO, these TCOs have low electrical conductivities.

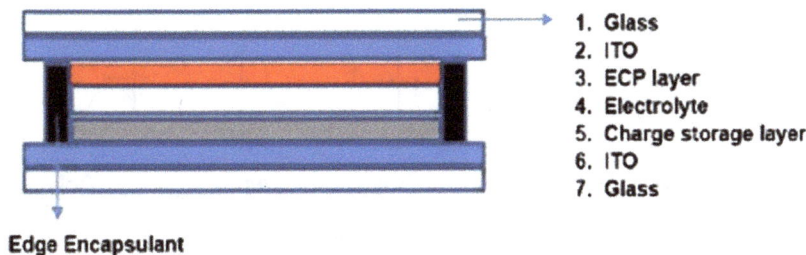

1. Glass
2. ITO
3. ECP layer
4. Electrolyte
5. Charge storage layer
6. ITO
7. Glass

Edge Encapsulant

FIGURE 29.4 Typical polymer EC device architecture for glass/ITO substrate. Adapted with permission from Reference [33], Copyright (2022), Elsevier.

Despite recent advancements, TCOs still have constraints, such as the cost of deposition processes and the brittleness of TCO films, which makes them unsuitable for flexible electronic applications. Features such as flexibility and bendability are necessary for applications such as paper-like displays and wearable devices, although they are suitable for windows and mirrors. Flexible electrodes, such as conducting polymers and nanomaterials, have been used as alternatives to TCOs for these reasons.

6.1 CONDUCTING POLYMERS

Polymers had long been thought to be electrical insulators until Heeger et al. (MacDiarmid, Shirakawa, and Heeger) discovered polyacetylene, a much better conductive polymer, in 1976 [34]. Polyacetylene, polyphenylene, polythiophene, polyaniline, polypyrrole, and poly(p-phenylene vinylene) are examples of low-cost, stable conducting polymers, as shown in Figure 29.5. PANI-CSA, or polyaniline doped with camphor sulfonic acid (CSA), exhibits metallic conductivity ($>10^3$ S/cm) [35]. These have been extensively used as alternatives for ITO in flexible displays due to their good strength and electro-optic stability during bending [36]. The best choice for transparent conducting polymer anodes is polythiophene, specifically poly(3,4-ethylenedioxythiophene) (PEDOT). It has a very good conductivity of about 300 S/cm and is transparent in thin, oxidized films, and has excellent oxidation stability [32]. Using poly(styrene sulfonic acid) (PSS) as the charge-balancing dopant during PEDOT polymerization, a water-soluble polyelectrolyte system PEDOT/PSS can be developed. With a conductivity of 10 S/cm, exceptional stability, and high visible light transmission, this system possesses good film-forming capabilities. PEDOT/PSS films may be heated in air at 100°C for over 1000 hours with just minor conductivity changes

6.2 NANOMATERIALS

Graphene and carbon nanotubes (CNTs) are a promising electrode alternative for ECDs due to properties such as transparency, electrical conductivity, chemical stability, and good electrochromic features such as fast color-switching speed, good cyclic stability, and high coloration efficiency. Flexible ECDs have been made utilizing graphene-coated PET [37] in one study and multilayer graphene in another study [38]. Single-walled carbon nanotubes on glass have also been used as electrode substrates in polymer-based ECD [39].

CNTs are among the most promising materials for flexible electronics applications. There are many reports claiming the use of CNTs in electronic displays. Martel et al. [40] used transparent SWCNT sheets, whereas Zhou et al. [41] used arc-discharge nanotubes to create hole injection electrodes for OLEDs. Marks et al. [42] used SWCNT films on flexible PET substrates to make polymer-based OLEDs. Rowell et al. [43] created flexible transparent electrodes for flexible solar cells by printing SWCNT networks on plastic films. CNT networks of various densities can be

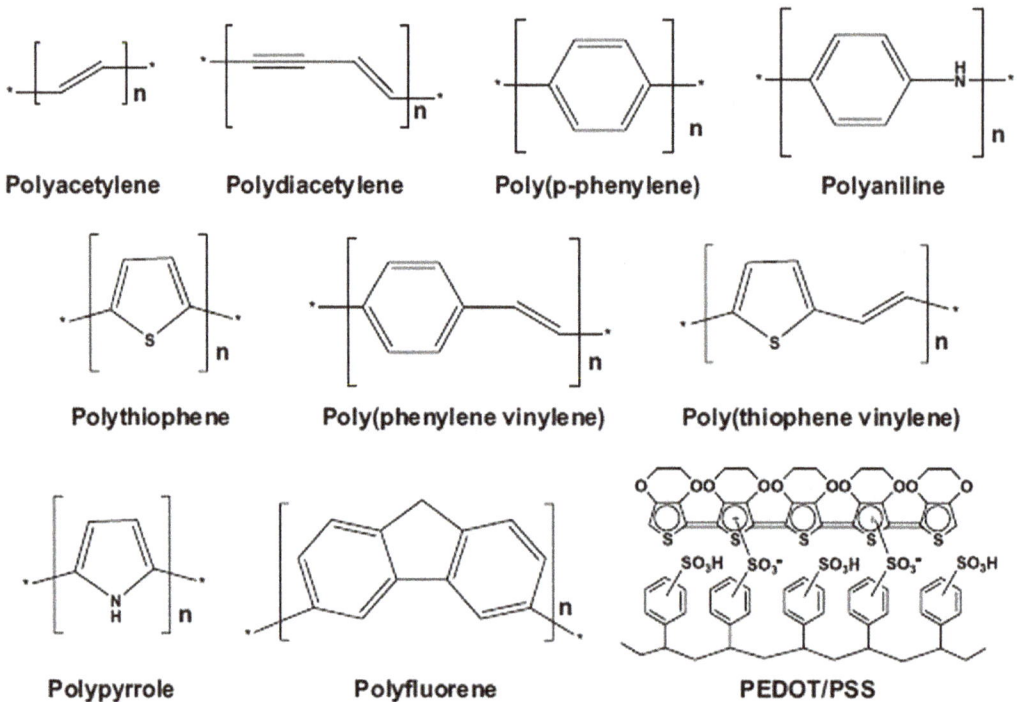

FIGURE 29.5   Examples of conducting polymers. Adapted with permission from Reference [35]. Copyright (2022) Elsevier.

utilized in making both bottom gate and conducting channels of transparent and flexible transistors [44]. CNT-based electrode patterning techniques include spin-coat, the Langmuir–Blodgett method, vacuum filtering, line-patterning, transfer with poly(dimethysiloxane) (PDMS) stamps, and regulating droplets with a gas flow.

Devices can be flexible and bendable with the introduction of flexible electrodes and appropriate substrates. The use of nanocellulose-based material as the substrate was described by Kang et al. Even after being folded, the paper-like prototype still showed promising EC behavior [45]. Yan et al. created a stretchable device that can survive twisting, folding, and crumpling utilizing polydimethylsiloxane (PDMS) as the substrate [46]. Spin coating, blade coating, inkjet printing, and spray coating are some of the solution-based processing techniques used in the fabrication of ECDs [47].

7   FUTURE PERSPECTIVES AND CHALLENGES

Recently, polymer electronic (polytronic) research has shown the potential to transform the entire world of electronics by providing a major platform for electronic circuit improvements, particularly printed circuit boards (PCBs). This will open up new dimensions in flexible electronic/semiconductor technologies in the future. The development of sustainable green products in eco-marketing and environmentally friendly materials is predicted to grow rapidly in the next years.

The development of flexible LCDs, which have excellent designability and storability, enhanced portability and stability, and are affordable, conformable, low cost, and large-area differently shaped panels, has broadened the display opportunities and found applications in everyday life, and so is another important step in human interfaces. The bending tolerance of flexible LCDs limits curved

LC devices, resulting in image quality reduction. A solution to this problem is being pursued as a future path in this area.

Because of their portability and low cost of production, flexible OLEDs/field transistors will be the preferred option in the near future. A dry printing technique, which needs the formation of dry transferable substances, should be employed to replace the solvents used in processing with conducting/conjugated polymers with harmless materials in the future.

In recent years, the properties of EC polymers and their use in a wide range of device designs have evolved substantially. By modifying the composition of copolymers, composites, and blends, the multiple color states of EC polymers can be varied over the visible spectrum and applied to wavelengths outside of the visible range. The fact that an EC polymer has at least two distinct color states can be continuously manipulated as a function of applied voltage, and in some cases it can show numerous distinct color states that provide a lot of versatility for display and window-type devices. Proper electrode designs can produce rapidly switching polymers with subsecond response times. Surface functionalization, for example, is being used to make electrode materials more compatible with the EC polymer structure. Because the redox states are set by the applied potential and physically separated from one another, polymer-based ECDs can have a significant of electrochromic memory.

8 CONCLUSION

This chapter provides a brief account of polymer-based electronic displays. Nowadays, polymers are widely used in making LCDs, OLEDs, transparent electrodes, TFTs, and electrophoretic and electrochromic display devices. Polymers are preferred for use in all types of electronic display devices that may develop in the near future due to their better characteristics, low temperature processing, low cost, flexibility, and bendability. However, there are numerous obstacles to overcome. Polymer materials need to increase their thermal stability, thermal expansion coefficients, chemical and solvent resistance, and air and water vapor permeability to match glass substrates.

REFERENCES

[1] Nathan, A.; Chalamala, B. Special issue on flexible electronics technology, Part II: Materials and devices. Proc. IEEE 2005, 93, 1391–1393.

[2] Rogers, J.A.; Someya, T.; Huang, Y. Materials and mechanics for stretchable electronics. Science 2010, 327, 1603–1607.

[3] Lee, J.K.; Lim, Y.S.; Park, C.H.; Park, Y.I.; Kim, C.D.; Hwang, Y.K. a-Si:H thin-film transistor-driven flexible color e-paper display on flexible substrates. IEEE Electron Device Lett. 2010, 31, 833–835.

[4] Akins, R.; Displays for hand-held portable lectronic products. SID International Symposium: Digest of Technical Papers XXXI, 2000, 510

[5] Mohsin Hassan Saeed, Shuaifeng Zhang, Yaping Cao, Le Zhou, Junmei Hu, Imran Muhammad, Jiumei Xiao, Lanying Zhang, and Huai Yang, Recent advances in the polymer dispersed liquid crystal composite and its applications. Molecules 2020, 25, 5510; doi:10.3390/molecules25235510

[6] Chang, W.; Fourth-generation TFT-LCD production line; SID International Symposium: Digest of Technical Papers XXXI, 2000, 64

[7] Chen, M.; Liang, X.; Hu, W.; Zhang, L.; Zhang, C.; Yang, H. A polymer microsphere-filled cholesteric-liquid crystal film with bistable electro-optical characteristics. Mater. Des. 2018, 157, 151–158

[8] Zhang, C.; Wang, D.; Cao, H.; Song, P.; Yang, C.; Yang, H.; Hu, G.-H. Preparation and electro-optical properties of polymer dispersed liquid crystal films with relatively low liquid crystal content. Polym. Adv. Technol. 2013, 24, 453–459

[9] W. S. Wong and A. Salleo, Flexible Electronics: Materials and Applications, Springer Science & Business Media, Berlin, Germany, 2009.

[10] W. A. MacDonald, M. K. Looney, D. Mackerron et al., Latest advances in substrates for flexible electronics, Journal of the Society for Information Display, vol. 15, no. 12, pp. 1075–1083, 2007

[11] Y. Leterrier, L. Médico, F. Demarco et al., Mechanical integrity of transparent conductive oxide films for flexible polymer-based displays, Thin Solid Films, vol. 460, no. 1–2, pp. 156–166, 2004.

[12] V. Zardetto, T. M. Brown, A. Reale, and A. DiCarlo, Substrates for flexible electronics: a practical investigation on the electrical, film flexibility, optical, temperature, and solvent resistance properties, Journal of Polymer Science, Part B: Polymer Physics, vol. 49, no. 9, pp. 638–648, 2011

[13] G.-X. Ni, Y. Zheng, S. Bae et al., Graphene-ferroelectric hybrid structure for flexible transparent electrodes, ACS Nano, vol. 6, no. 5, pp. 3935–3942, 2012.

[14] B.-J. Kim, S.-H. Han, and J.-S. Park, Properties of CNTs coated by PEDOT:PSS films via spin-coating and electrophoretic deposition methods for flexible transparent electrodes, Surface and Coatings Technology, vol. 271, pp. 22–26, 2015.

[15] S. Yadav, V. Kumar, S. Arora, S. Singh, D. Bhatnagar, and I. Kaur, Fabrication of ultrathin, free-standing, transparent and conductive graphene/multiwalled carbon nanotube film with superior optoelectronic properties, Thin Solid Films, vol. 595, pp. 193–199, 2015.

[16] J. Sun and R. Wang, Carbon nanotube transparent electrode, in Syntheses and Applications of Carbon Nanotubes and Their Composites, S. Suzuki, Ed., chapter 14, pp. 313–335, InTech, Rijeka, Croatia, 2013.

[17] J. Jensen, H. F. Dam, J. R. Reynolds, A.L. Dyer, and F.C. Krebs, Manufacture and demonstration of organic photovoltaicpowered electrochromic displays using roll coating methods and printable electrolytes, Journal of Polymer Science, Part B:Polymer Physics, vol. 50, no. 8, pp. 536–545, 2012

[18] Kumar, B., Kaushik, B. K., Negi, Y. S., & Mittal, P. (2011, March). Characteristics and applications of polymeric thin film transistor: Prospects and challenges. In 2011 International Conference on Emerging Trends in Electrical and Computer Technology (pp. 702–707). IEEE.

[19] C. J. Bettinger and Z. Bao, Organic thin-film transistors fabricated on resorbable biomaterial substrates, Adv. Mater. vol. 22, pp. 651–655, 2010.

[20] W. L. Leong, P. S. Lee, A. Lohani, Y. M. Lam, T. Chen, S. Zhang, A. Dodabalapur, and S. G. Mhaisalkar, Non-volatile organic memory applications enabled by in situ synthesis of gold nanoparticles in a selfassembled block copolymer, Adv. Mater., vol. 20, pp. 2325, 2008.

[21] Shashi Paul, Realization of nonvolatile memory devices using small organic molecules and polymer, IEEE Trans. Nano Techology, vol. 6, no. 2, pp. 191–195, 2007.

[22] D. Prime and S. Paul, Overview of organic memory devices, Phil.Trans. R. Soc. A, vol. 367, pp. 4141–4157, 2009.

[23] Q. T. Zhang and V. Subramanian, DNA hybridization detection with organic thin film transistors: toward fast and disposable DNA microarray chips, Biosens. Bioelectron., vol. 22 no. 12, pp. 3182–3187, 2007.

[24] Gelinck, G. H., Huitema, H. E. A., Van Mil, M., Van Veenendaal, E., Van Lieshout, P. J. G., Touwslager, F., ... & McCreary, M. D. (2006). A rollable, organic electrophoretic QVGA display with field-shielded pixel architecture. Journal of the Society for Information Display, 14(2), 113–118.

[25] Heikenfeld, J., Drzaic, P., Yeo, J. S., & Koch, T. (2011). A critical review of the present and future prospects for electronic paper. Journal of the Society for Information Display, 19(2), 129–156.

[26] Henzen, A., & van de Kamer, J. (2006). The present and future of electronic paper. Journal of the Society for Information Display, 14(5), 437–442.

[27] Chen, Y., Au, J., Kazlas, P., Ritenour, A., Gates, H., & McCreary, M. (2003). Flexible active-matrix electronic ink display. Nature, 423(6936), 136–136.

[28] Sprague, R. A. (2006, April). A distributed system of wireless signs using Gyricon electronic paper displays. In Photonics in Multimedia (Vol. 6196, pp. 179–190). SPIE.

[29] Liang, R. C., Hou, J., Zang, H., Chung, J., & Tseng, S. (2003). Microcup® displays: Electronic paper by roll-to-roll manufacturing processes. Journal of the Society for Information Display, 11(4), 621–628.

[30] Cossari, P., Cannavale, A., Gambino, S., & Gigli, G. (2016). Room temperature processing for solid-state electrochromic devices on single substrate: From glass to flexible plastic. Solar Energy Materials and Solar Cells, 155, 411–420.

[31] Palenzuela, J., Vinuales, A., Odriozola, I., Cabanero, G., Grande, H. J., & Ruiz, V. (2014). Flexible viologen electrochromic devices with low operational voltages using reduced graphene oxide electrodes. ACS Applied materials & interfaces, 6(16), 14562–14567.

[32] Groenendaal, L., Jonas, F., Freitag, D., Pielartzik, H., & Reynolds, J. R. (2000). Poly (3,4-ethylenedioxythiophene) and its derivatives: past, present, and future. Advanced Materials, 12(7), 481–494.

[33] Wang, H., Barrett, M., Duane, B., Gu, J., & Zenhausern, F. (2018). Materials and processing of polymer-based electrochromic devices. Materials Science and Engineering: B, 228, 167–174.

[34] Shirakawa, H. (2001). The discovery of polyacetylene film–the dawning of an era of conducting polymers. Current Applied Physics, 1(4–5), 281–286.

[35] Choi, M. C., Kim, Y., & Ha, C. S. (2008). Polymers for flexible displays: From material selection to device applications. Progress in Polymer Science, 33(6), 581–630.

[36] Heuer, H. W., Wehrmann, R., & Kirchmeyer, S. (2002). Electrochromic window based on conducting poly (3, 4-ethylenedioxythiophene)–poly (styrene sulfonate). Advanced Functional Materials, 12(2), 89–94.

[37] Polat, E. O., Balcı, O., & Kocabas, C. (2014). Graphene based flexible electrochromic devices. Scientific Reports, 4(1), 1–9.

[38] Choi, D. S., Han, S. H., Kim, H., Kang, S. H., Kim, Y., Yang, C. M., ... & Yang, W. S. (2014). Flexible electrochromic films based on CVD-graphene electrodes. Nanotechnology, 25(39), 395702.

[39] Vasilyeva, S. V., Unur, E., Walczak, R. M., Donoghue, E. P., Rinzler, A. G., & Reynolds, J. R. (2009). Color purity in polymer electrochromic window devices on indium–tin oxide and single-walled carbon nanotube electrodes. ACS Applied Materials & Interfaces, 1(10), 2288–2297.

[40] Aguirre, C. M., Auvray, S., Pigeon, S., Izquierdo, R., Desjardins, P., & Martel, R. (2006). Carbon nanotube sheets as electrodes in organic light-emitting diodes. Applied Physics Letters, 88(18), 183104.

[41] Zhang, D., Ryu, K., Liu, X., Polikarpov, E., Ly, J., Tompson, M. E., & Zhou, C. (2006). Transparent, conductive, and flexible carbon nanotube films and their application in organic light-emitting diodes. Nano letters, 6(9), 1880–1886.

[42] Li, J., Hu, L., Wang, L., Zhou, Y., Grüner, G., & Marks, T. J. (2006). Organic light-emitting diodes having carbon nanotube anodes. Nano letters, 6(11), 2472–2477.

[43] Rowell, M. W., Topinka, M. A., McGehee, M. D., Prall, H. J., Dennler, G., Sariciftci, N. S., ... & Gruner, G. (2006). Organic solar cells with carbon nanotube network electrodes. Applied Physics Letters, 88(23), 233506.

[44] Artukovic, E., Kaempgen, M., Hecht, D. S., Roth, S., & Grüner, G. (2005). Transparent and flexible carbon nanotube transistors. Nano letters, 5(4), 757–760.

[45] Kang, W., Yan, C., Foo, C. Y., & Lee, P. S. (2015). Foldable electrochromics enabled by nanopaper transfer method. Advanced Functional Materials, 25(27), 4203–4210.

[46] Yan, C., Kang, W., Wang, J., Cui, M., Wang, X., Foo, C. Y., ... & Lee, P. S. (2014). Stretchable and wearable electrochromic devices. ACS nano, 8(1), 316–322.

[47] Jensen, J., Hösel, M., Dyer, A. L., & Krebs, F. C. (2015). Development and manufacture of polymer-based electrochromic devices. Advanced Functional Materials, 25(14), 2073–2090.

30 Smart Polymers in Flexible Devices

Khairunnisa Amreen and Sanket Goel

MEMS, Microfluidics and Nanoelectronics Lab, Department of Electrical and Electronics Engineering, Birla Institute of Technology and Science Pilani, Hyderabad Campus, Hyderabad 500078, India

1 INTRODUCTION

The growth in the fabrication of flexible and wearable electronic devices has increased life expectancy tremendously recently. This, in turn, has led to a rise in demand for novel materials and approaches for the preparation of these devices on a larger scale. In this context, polymers fit quite well, because they are versatile, have different forms, with desired properties. The physiological and cell organelle poses complex polymers like carbohydrates, starch, proteins, cellulose, etc. The significant property of these biopolymers is that they are responsive to external stimuli. Synthetic and structurally engineered polymers are being developed to mimic this behavior, and these are designated as smart polymers. The last couple of decades has seen enormous growth and demand for smart material-based polymers. These are also referred to as "intelligent polymers" or "stimuli-responsive" or "environmentally sensitive" polymers [1][2]. The distinctive salient property of these materials which makes them intelligent or smart is their capability to react to changes in the external environmental parameters. Figure 30.1 shows a schematic overall representation of the stimulation factors that affect the properties of these polymers.

These polymers show instant structural changes that can be clearly seen through macroscopic characterization. Transition in terms of shape, surface morphology, molecular model, solubility, gel-to-sol, and sol-to-gel formation, etc. are a few significant advantages of these. Further, to stabilize the physical changes, these polymers can either collapse, swell, or expand based on their chemical properties. The hydrophilic behavior of these polymer surfaces can also be altered. These changes are reversible and can be restored by fluctuating the environmental parameters [3]. Parameters like pH change, temperature, ionic concentration [4], metabolites and chemicals [5], charge, ionic complex formation [6], magnetic [7], electric [8], and photo or light properties [9] affect the properties of the smart polymer. Table 30.1 gives examples of common smart polymers with the stimuli they react to. Due to this nature, the applications of smart polymers have increased significantly in the manufacturing of flexible, wearable devices such as low invasive injectable devices, drug delivery, tissue and cell culture, etc. This chapter discusses types of smart polymers, and their synthesis and applications in flexible devices.

2 CLASSIFICATION OF SMART POLYMERS

Figure 30.2 gives an overall diagrammatic representation of the types of smart polymers. Based on their stimuli behavior, smart polymers are classified as follows.

DOI: 10.1201/9781003278269-30

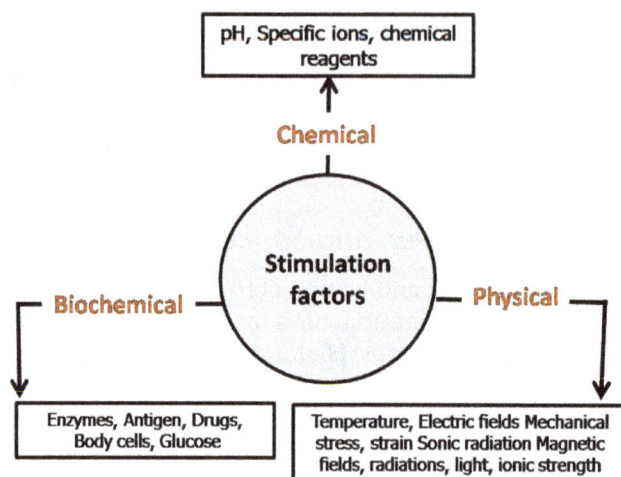

FIGURE 30.1 Schematic representation of all the stimulation factors that change the properties of smart polymers.

TABLE 30.1
Examples of Common Smart Polymers with Their Corresponding Stimuli

Stimulus	Example
Temperature	Chitosan, cellulose, poly(N-alkylacrylamide)s, poly(N-vinylcaprolactam)s, xyloglucan
pH	Poly(methacrylic acid)s, poly(vinylimidazole)s, poly(vinylpyridine)s
Ultrasound	Ethylenevinylacetate
Light	Poly(acrylamide)s
Electric field	Poly(ethyloxazoline), sulfonated polystyrenes, poly(thiophene)s

2.1 TEMPERATURE-SENSITIVE POLYMERS

These kinds of smart polymers change their soluble nature and can transform sol to gel and vice versa in response to an alteration in temperature. Moieties like ethyl, propyl, and methyl are commonly present in these. Thermosensitive polymers demonstrate two behaviors: (1) lower critical solution temperature (LCST) and (2) upper critical solution temperature (UCST). LCST undergoes changes from monophasic to biphasic upon temperature elevation. Beyond LCST, the polymers become insoluble and show hydrophobicity, whereas the same polymers under LCST are hydrophilic and soluble. UCST is quite rare and gets converted from biphasic to monophasic. LCST is most commonly used over UCST as it needs high temperatures. Examples of LCST are poly(vinyl amide), poly(oligoethylene glycol (meth)acrylate), poly(N-substituted acrylamide), poly(2-oxazoline)s, poly(vinyl ether)s, poly(N,N-diethylacrylamide), poly(N-vinylalkylamide), poly(N-vinylcaprolactam), phosphazene derivatives, chitosan, polysaccharide derivatives, tetronics, pluronics, etc. These are applied for drug delivery as they are resistant to sharp drops in temperatures.

There are three major types of temperature-resistive polymers: liquid-crystalline, shape-memory, and responsive polymer solutions. The main benefit of these polymeric materials is that they diminish the side effects and provide target-specific drug delivery, consistent, sustained release of drug [10]. Table 30.2 summarizes some of the successful applications of smart polymers for drug delivery using various target-specific drugs.

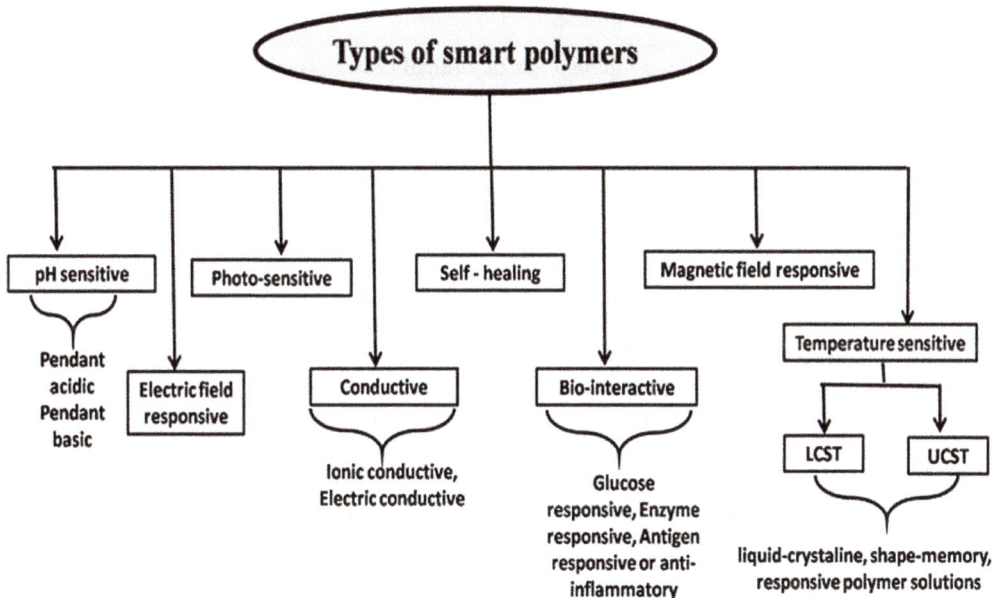

FIGURE 30.2 Schematic overall representation of types of polymers.

TABLE 30.2
Applications of Smart Polymers for Target-Specific Drug Delivery

Polymer	Drug	Disease	References
Pluronic and linoleic acid conjugated	Docetaxel	Gastric cancer	[13]
Polybenzofulvene	Leuprolide	Tumor	[14]
Chitosan-beta, glycerophosphate	Desferroxamine	Limbic ischemia	[12]
Glycerol, chitosan with lycerophosphate disodium salt	Ethosuximide	Depot therapy	[15]
Poly(dl-lactide-co-glycolide-b-ethylene glycol-b-dl-lactide-co-glycolide) (PLGA-PEG-PLGA)	Exenatide	Diabetes	[16]

2.2 pH-SENSITIVE POLYMERS

These contain a pendant acidic or a pendant basic group, which is capable of accepting and releasing protons when stimulated by the surrounding pH. Certain polymers have ionizable groups called polyelectrolytes, which are either weak polyacids or weak polybases. These weak polyacids are proton acceptors at low pH, whereas, at high and neutral pH, they release protons. Poly(methacrylic acid) (PMAAc) and poly(acrylic acid) (PAAc) are common pH-sensitive smart polymers. Sulfonamide-containing polyacid polymers operate within a narrow pH range with better sensitivity than carboxylic acid polymer. Chitosan is another common polycationic pH-sensitive biopolymer. Similarly, amino group-containing polymers are polybasic. Poly(N,N-diethylaminoethylmethacrylate) (PDEAEMA) and poly(N,N-dimethylaminoethylmethacrylate) (PDMAEMA) are common examples. Some of the specific smart polymers like poly(vinyl-immidazole) (PVI), poly(4- or 2- vinylpyrrolidine) (PVP), and poly(propyleneimine) have also been used for drug delivery. However, for drug-delivery systems, thermos-sensitive is preferred to pH-sensitive. Apart from drug delivery, biomedical, water remediation, and industrial processes are some other applications [15–17].

2.3   LIGHT- OR PHOTOSENSITIVE POLYMERS

These types of polymers can change their polarity, optical chiral property, conformations, charge, conjugation, and amphiphilicity due to stimuli from light. These changes are both reversible and irreversible, based on their functional groups. These groups are called chromophores. Coumarin, azobenzenes, diaryl ethane, and spiropyran are reversible chromophores. Nitrobenzylphotolabile is an irreversible chromophore. These types of polymers are often used as hydrogels and photodegradable materials, and in drug delivery, etc. [18].

2.4   CONDUCTIVE OR ELECTROACTIVE POLYMERS

These are broadly of two types: (1) ionic conductive polymer: exhibits conductivity due to ionic functional groups in their structure as well as electrolyte medium. (2) Electric conductive polymer: exhibits conductivity due to electron movement in their chemical bonds. These are further subdivided into intrinsic electric and percolated electric polymers. Figure 30.3a is a reproduction of a schematic representation of these various conductive mechanisms [19].

2.4.1   Intrinsic Conductive Polymers

Intrinsic conductive polymers contain conjugated highly localized carbon–carbon σ single bonds and low localized carbon–carbon π double bonds. The overlap of the p-orbitals in the π bond offers high mobility of electrons between the atoms. Further, the conductivity is enhanced by doping. The polarity of the molecules allows the movement of dopants in or out of the polymer [20]. Figures 30.3b and c are reproductions of the conductivity range of these polymers and a diagrammatic representation of the conjugated chain [19]. Polyaniline, polypyrrole, and PEDOT are such types of intrinsic conductive polymers used commonly. However, they have limitations in terms of mechanical properties, thus, doping using dopants like hyaluronic acid is useful.

2.4.2   Percolated Composites

The conductive electric particles are percolated into the insulator polymers as fillers which form composites with enhanced conductivity. This filler volume and percolation determines the conductivity of the polymer composites. Carbon nanotubes (CNTs) are widely used as fillers. Some of the other fillers are metal nanoparticles of gold, silver, copper, nickel, palladium, and carbon-based fillers like carbon fibers and graphite. The major advantage of these is that they improve flexibility.

2.4.3   Polyelectrolyte Conductive Hydrogels

3D polymer networks with hydrophilic nature and extensive water absorption properties are hydrogels. They are soft and porous. Hence, they are ideal for mimicking physiological tissue. The hydrogen bonds and ionic bonds can be altered to manipulate this property. Their conductivity depends on the ions, water content, polarity of the polymer, and mobility of ions.

2.5   BIORESPONSIVE OR BIOINTERACTIVE POLYMERS

The polymers that respond to the physiological species, i.e., synthetic polymers conjugated to natural bio-components, are called biointeractive polymers. These are of three kinds. (1) Glucose-responsive polymers: these are sensitive to sugar and can biomimic the insulin secretion process. These polymers get stimulated by glucose. These are applied in drug delivery for the release of insulin in a controlled manner. Further, they are also used for sensing glucose. However, biocompatibility is a major limitation here. (2) Enzyme-responsive polymers: these interact with specific enzyme catalysis. A synthetic polymer can be integrated with biological enzymes. The biological process can be monitored with the manipulation of enzyme activity and co-factors. These are applied

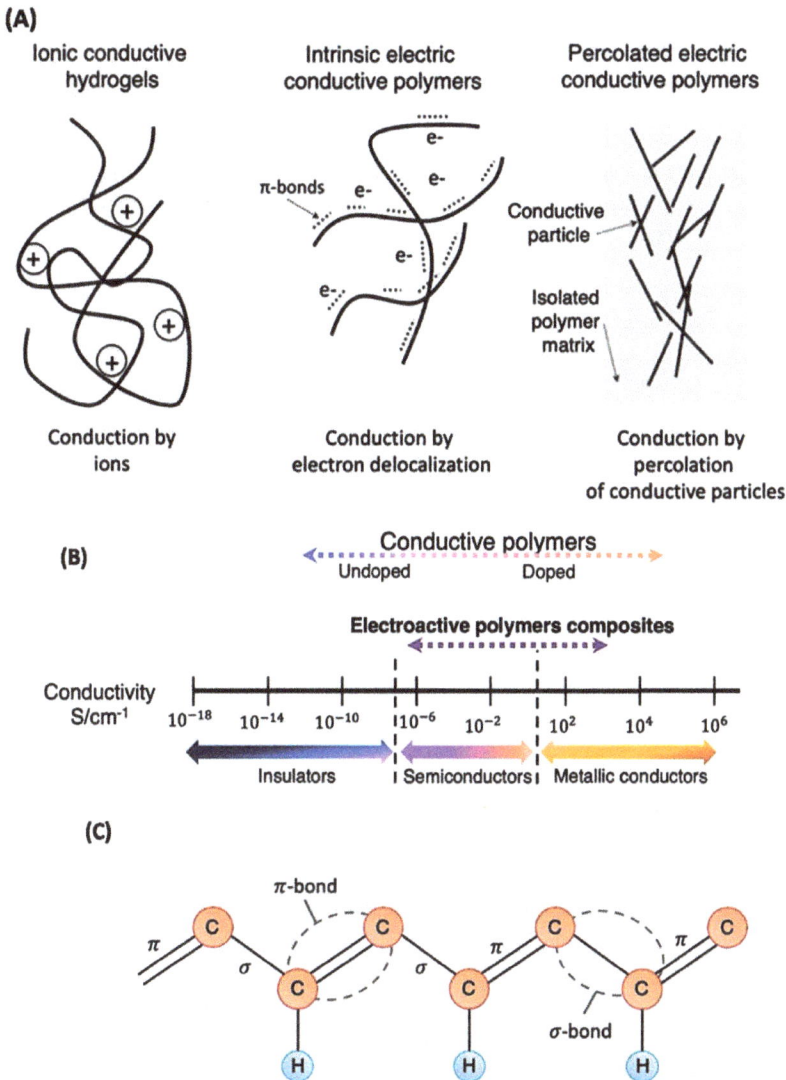

FIGURE 30.3 (A) Reproduction of the schematic representation of various conductivity mechanism of different polymers. (B) Reproduction of the conductivity range of various intrinsic polymers. (C) Reproduction of the diagrammatic representation of the conjugated bonds of polymers. Adapted from Reference [19]. Copyright (2019) The Authors, some rights reserved; exclusive licensee [MDPI]. Distributed under a Creative Commons Attribution License 4.0 (CC BY) https://creativecommons.org/licenses/by/4.0/.

in drug delivery, scaffolds, etc. (3) Antigen- or inflammation-responsive polymers: these are reactive to the presence of an antigen and can be used in the treatment of diseases. Some are responsive to phagocytic cells, hence giving an effective treatment for inflammation [21,22].

2.6 Self-Healing Polymers

The self-healing polymers are those which can restore lost functional properties with the absence of any kind of external aid. The healing phenomenon can be extrinsic, wherein the compound is segregated from the polymer as capsules and fibers, or intrinsic, wherein the polymer structural chains increase the flow of ions to the damaged surface. This, in turn, determines the restoration, structure integrity, conductivity, and hydrophobic and hydrophilic nature [23].

2.7 MAGNETIC FIELD-RESPONSIVE POLYMERS

The magnetic nanoparticles are embedded in the polymer matrix. Hence, upon exposure to the magnetic field, the polymer matrix can be deformed without the release of heat and noise. Therefore, these are applied in micromachines, transducer energy devices, sensors, etc. Coupling of the magnetic nanoparticles with polymer gel is a limitation here which can be resolved by cross-linking the nodes of the hydrogel [24].

2.8 ELECTRIC FIELD-RESPONSIVE POLYMERS

These polymers can alter their properties with stimulation from an electric current. They possess several ionizable groups and can convert electrical energy to mechanical energy. The change in the electric field changes the pH, breaking the hydrogen bonds in the polymeric chain and causing bending of the matrix. Hence, they are used in drug delivery, muscle actuators, energy transducers, etc. [25].

3 GENERAL SYNTHESIS OF SMART POLYMER NANOCOMPOSITES

Broadly, there are four synthetic routes for obtaining smart polymer-based nanocomposite materials with desired characteristics, morphology, and properties. These are:

1. *Solution-based method*: in this approach, the polymer with specified properties is taken as a solution. Nanoparticles as fillers are dispersed homogeneously in a solvent. To embed these nanoparticles in the polymer matrix, both solutions are mixed and a homogeneous solution is obtained. Finally, the solvent is either evaporated or the nanocomposite is precipitated to obtain pure smart polymer nanocomposite.
2. *Melt-mixing method*: Herein, a polymer and the nanoparticle fillers are mixed and melted directly together for incorporation in the polymer matrix.
3. *In situ polymerization*: In this approach, a monomer of the desired polymer is taken in a liquid form. Nanofillers are mixed with liquid monomer and polymerization is done which automatically incorporates the nanoparticles in the matrix.
4. *Template synthesis*: In this method, a precursor solution and a reagent are taken as a solution with a template polymer. Nanoparticles are synthesized from this, which directly embeds the nanoparticles in the polymer [26].

4 FABRICATION OF SMART POLYMER-BASED FLEXIBLE MICRODEVICES

The common methods employed for the preparation of flexible microdevices with complete integration and portability are discussed here with a brief discussion of the procedures and the base materials used.

1. *Direct laser writing*: This is a relatively newer approach that is being adapted widely now for fabricating flexible devices over surfaces like polymer sheets, glass, plastic, paper, etc. The exposure of laser over these substrates or any other flexible carbon-containing substrate will form a layer of laser-induced graphene (LIG). LIG is transferrable to any other flexible substrate via a simple adhesive tape leaving the cavity as a microchannel pattern. Hence, this approach can be used to draw patterns as well. Generally used lasers are CO_2, ultraviolet,

diode, etc. The intensity, power, and speed of the laser determine the functionality of the devices. The smart polymer sheets can either be directly ablated with a laser for the design or they can be integrated as coatings in the microchannels or substrates.

2. *3D printing*: This is also referred to as an additive manufacturing technique in which conductive filaments are given a 3D structure by layer-by-layer deposition. Herein, a computer-aided design (CAD) software is used for designing the microdevice. A specific type of 3D printer, i.e., stereolithography (SLA) or 3D resin printer and fused deposition modeling (FDM) can only be employed for flexible substrate microdevices fabrication. Smart polymers can be melted and deposited as flexible 3D microdevices. Nadgorny et al. gave a detailed review of the functional polymeric materials and nanoparticles for 3D smart devices. Figure 30.4 is a reprint from Reference [27] showing a schematic of various approaches for 3D printing.

3. *Lamination*: This is a simple approach in which individual cut layers of the flexible substrates like polymer sheets are stacked together. Basically, three layers are present, a bottom support layer, a middle or intermediate layer which has the design of microchannels, and a top layer to seal. Adhesives are used for bonding these layers, whereas a cutter plotter, laser, or lamination machine can be used for assembling the device. In this approach, smart polymer sheets also can be directly used or they can be liquefied and modified on another flexible substrate.

4. *Screen printing method*: A conductive ink of choice, liquid smart polymer etc. is used for printing over the flexible substrate materials like polymer sheets, cloth, paper, fiber, etc. using a mask with specified design or pattern.

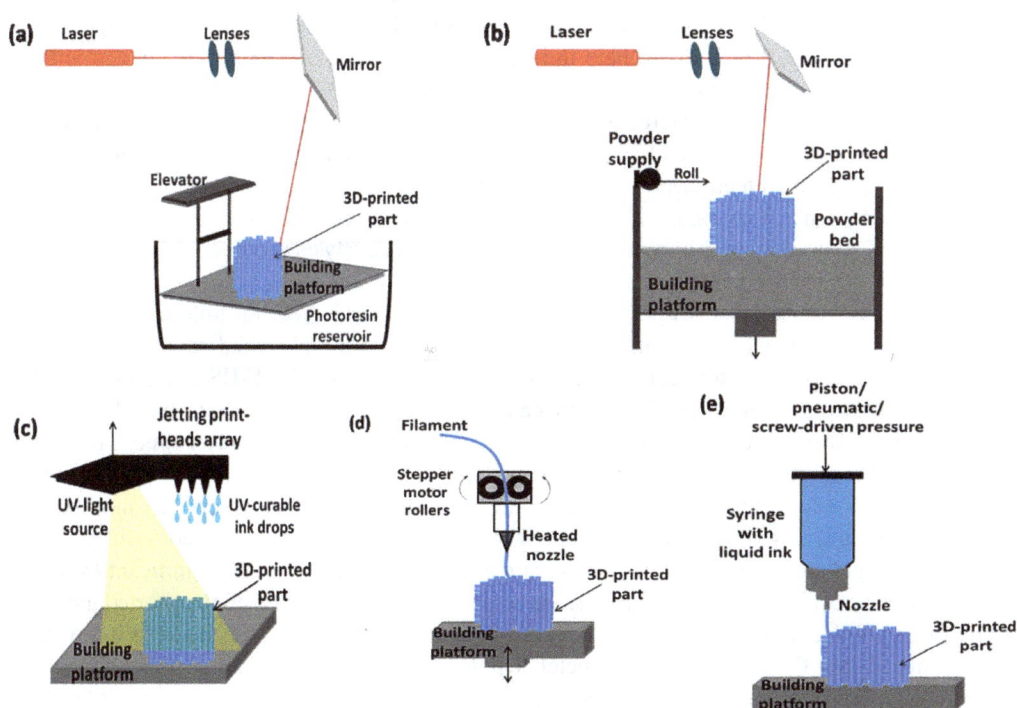

FIGURE 30.4   Diagrammatic representation of (A) laser-assisted stereolithography (SLA), (B) laser-assisted selective laser sintering (SLS), (C) inkjet printing, (D) fused deposition modeling (FDM)€nd (E) liquid deposition modeling (LDM). Adapted with permission from Reference [27]. Copyright (2018) American Chemical Society.

5. *Ink-jet printing*: This approach is more automated than screen printing, and an ink-jet printer is used. The printer is fed with the micro device design through software. A conductive ink, liquid polymer etc. of a specific viscosity can be filled in the nozzle of the printer. Flexible substrates like polymer, paper, fiber, cloth, etc. are affixed on the platform of the printer and the nozzle sprays the ink as the pattern [28,29].

5 APPLICATION IN FLEXIBLE DEVICES

5.1 BIOMEDICAL DEVICES

Several biomedical devices in angiology, endocrinology, cardiology, nephrology, orthodontics, neurology, etc. for diagnosis, treatment and therapy been developed using smart polymers. Shape-memory smart polymers (SMPs) are commonly applied in clinical flexible devices owing to the fact that they can recover their shape upon stimulation. SMPs which are light responsive, chemical responsive, and temperature responsive are applied in these devices. They can easily restore their original structure and morphology with simple body heat. Similarly, water-sensitive SMPs are also used as they regain their original shape after being immersed in water. Heat-responsive SMPs are employed in cardiac valve rings. The ring is inserted as a deformed shape which appears the same as the physiological shape of a natural heart valve. Later, magnetic nanoparticles or electric resistance are applied to generate heat so the SMP regains its actual shape. This mechanism helps to avoid reversal of blood flow. Certain smart polymers have been explored for blood vessel reconnections instead of the conventional method of stitching the distorted ends surgically. Bilayered tubes of polypyrrole have been investigated for this application. Such polymers have also been used in implants for governing nervous system function as neuroprosthetic devices in diseases like Parkinson's disease. Another significant use of these is as biodegradable, self-inflatable implants and as intragastric balloons. These would help in aiding metabolism upon coming in contact with the heart and not restricting the food flow through the gastrointestinal tract. This feature could help in chronic obesity issues. Another interesting application of these polymers is their use in orthodontic braces for aligning teeth. Materials like polynorborene and polurethane thin wires are used. These exert high force and align teeth over the course of time. Further, the physiological temperatures do not deform them, hence a continuous force is applied. In the treatment of kidney ailments through dialysis, certain smart polymers have been used as adapters for dialysis needles. These are temperature dependent, tube shaped, and help to prevent hemodynamic stress [30,31].

Over a decade, the use of smart polymers for fabricating microdevices for minimal invasive surgery (MMIS) has emerged significantly. In the case of implanting medical aid devices in a physiological system, complicated surgical procedures are generally required. MMIS has been proven to reduce risky surgical approaches and therefore cause less damage to organs and tissues. In addition, this minimal invasive procedure has enabled the recovery of patients in a shorter time. The size of MMIS has to be optimized to small diameters as these can be inserted into the body via minute incisions through endovascular or laparoscopic methods. Since smart polymers can modify their shapes upon stimuli, they are useful as implants. They can be inserted in compressed forms and upon reaching the target area inside the body can regain their original shape. Some significant examples of MMIS with smart polymers are mentioned here. For instance, Maitland et al. reported a polyurethane smart polymer device for the treatment of blood clots. This device was made as a rod shape and introduced into the artery with a catheter. Upon photothermal stimulation, this changes the shape to an umbrella or coil form which can trap the clot in the blood vessel. Therefore, it has been proven as an effective treatment for strokes. Another cardiac ailment is bulging or ballooning of arteries, called "aneurysms," which cause strokes and heart attacks. As a treatment, platinum coils are introduced into the bulged area to clot the blood, preventing rupture of the vessel. A remarkable alternative to this is a laser-integrated smart polymer device which releases coils in the blood vessel. Herein, a photothermal-responsive smart polymer, polyurethane foam, is used. The inbuilt laser increases the temperature causing the polymer to expand and release the coil. When the temperature is reduced automatically, the polymer contracts. This release can be monitored in a timely fashion as a continuous treatment [32,33].

The application of smart polymer-based stent medical devices is also being explored. Stents are basically narrow tubes that are inserted into blood vessels or renal arteries to prevent narrowing of the vessels. These stents ease the normal blood flow by keeping the vessel open. Presently, metal-based stents are clinically used. However, side effects like thrombosis, wherein blood clots block the vessels, or restenosis, i.e., narrowing of the operated vessel, are common with these metal stents. Smart polymer stents can resolve this to a certain extent as drugs to limit the restenosis, thrombosis, and rejection of the stent post implant can be incorporated into the polymer. Further, these polymers are smaller and are stimulated by body heat. Since they are flexible, unlike metallic stents, they can be introduced into complex networks of vessels. In terms of device economics, in comparison to metallic stents, they are 50% cheaper. Poly(3-caprolactone) and poly(ϵ-3- hydroxybutyrate-ϵ(R)-3-hydroxyvalerate) acrylate are examples of smart polymers that can be used to make stents. Polyurethane has also been studied for fabricating stents with a drug elution property. It enabled continued and steady release of drug. These smart polymer-based stents are responsive to the body temperatures and can expand slowly, avoiding creating sudden pressure on arteries. The smart polymer-based stents are used in nephrology for releasing antibiotics in the urinary tract. In addition to these, smart polymers can also be used to suture the incisions without any damage to surrounding tissues. Oligo(p-dioxanone)diol and oligo(ε-caprolactone)diol fibers are used for this application. A wound closure on the stomach of a rat was demonstrated wherein the suture was closed by expansion of the polymer upon reaching body temperature [34,35].

The application of the smart polymer microdevices for cancer diagnosis and treatment has also been investigated. Early diagnosis and target-specific treatment are being developed. In a target-driven treatment, destruction of malignant tumor via thermal or heat ablation therapy has been emerging as an alternative to radiation/chemotherapy, especially in those body parts where surgical procedures are risky. Magnetic nanoparticles are being used for this therapy, wherein an external magnetic field is applied to heat these particles once they are at the tumor site. Destruction of nearby healthy tissues is a risk with these. To overcome this, nanoparticles can be coated with smart polymers which are bioresponsive. Antigen interactive smart polymers are used wherein antibodies specific to target tumor antigen are conjugated which interact with the tumor cells. In addition, a thermo-sensitive or pH-sensitive smart polymer can also be used for drug delivery at the tumor site. For instance, iron oxide magnetic nanoparticles coated with poly(N-isopropylacrylamide) (temperature sensitive) polymer was demonstrated for this application. Other such smart polymers are poly((2-dimethylamino)ethylmethacrylate) and poly(d,l-lactide-co-glycolide) [36,37]. Such a type of magnetic nanoparticles coated with smart polymers is also being used for magnetic resonance imaging (MRI) for the diagnosis of tumors. In this regard, bioactive coatings with iron oxide nanoparticles have been exploited. A remarkable example was reported by Hu et al., wherein polyethylene glycol (PEG)-coated magnetic nanoparticles conjugated with monoclonal rch24 antibody which could capture carcinoantigen were demonstrated [38,39].

Smart polymer-based flexible microfluidic medical devices for diagnosis, drug delivery, and treatment have been emerging recently. These are advantageous as they require minimal sample volume, with precise flow control for good sensitivity. Hydrogels which are responsive to moisture when placed in microfluidic devices act as microvalves and open or close the route for flow by shrinking or swelling. Hence, additional power sources for valve operation can be avoided [40]. For example, Geiger et al. demonstrated a thermosensitive hydrogel microvalve-based device in which the polymer shrinks below the LCST, allowing flow of fluid [41]. Similarly, smart polymers are employed as micropumps for controlling the flow rate. Agarwal et al. demonstrated a thermosensitive hydrogel, poly(HEMA-co-DMAEMA), which pumps fluid at an elevated temperature and so the flow rate can be controlled by lowering the temperature [42].

5.2   BIOSENSOR DEVICES

Smart polymers can biomimic physiological receptors. Hence, they can be used for the preparation of devices for the detection of disease biomarkers, antigens, enzymes, and other metabolites like glucose, creatinine, etc. Glucose biosensors have been explored widely in this context as there are smart polymers that are selective to glucose interaction. The general mechanism for glucose detection in the case of smart polymers is enzymatic. Herein, glucose oxidase (GOx) enzyme catalyzes the reaction of glucose and oxygen to form gluconic acid as a product and hydrogen peroxide as a by-product. Hence, the smart polymers sensitive toward gluconic acid, glucose, and hydrogen peroxide can be used. Certain pH-sensitive polymers react to the presence of gluconic acid as it lowers the overall pH. Chitosan, glycopolymer–lectin complexes, poly(dimethyldiallyl ammonium chloride), poly(acrylamideran-3-acrylamidophenylboronic acid), etc. are examples of smart polymers used in glucose biosensing. Wearable glucose biosensors which continuously monitor the glucose levels and alert patients, record and store the data for analysis at regular intervals are present in the market [43].One of the extraordinary attempts to fabricate a microfluidic device was done by Huang et al. They reported an automated glucose sensor for real-time sample analysis and insulin injection. A PDMS device with embedded microchannels, micropumps, and microvalves was made. An insulin reservoir, glucose biosensor, and a flow sensor were the other components. Polypyrrole smart polymer entrapped with glucose oxidase enzyme is the biosensor material [44]. In another remarkable work, Hoffman et al. prepared a microfluidic device for biomarker detection. Herein, magnetic nanoparticles were coated with a thermoresponsive poly(N-isopropylacrylamide) polymer-antibody. Upon increasing the temperature, the polymer undergoes structural conformational change, and the nanoparticles aggregate and are separated by a magnetic field and the antigen then can be released by lowering the temperature again [45].

An interesting work wherein a pH-sensitive polymer-based hydrogel sensor was developed for detection of carbon dioxide gas in the stomach was reported by Herber et al. This was actually designed for the diagnosis of gastrointestinal ischemia. There is a limited flow of blood to the gastro-intestinal tract, which hampers the oxygen supply and increases the carbon dioxide levels in the tract. Poly(dimethylaminoethyl methacrylate) (PDMAEMA)-based hydrogel was prepared, which is a highly pH-sensitive smart polymer. This hydrogel was mixed with a bicarbonate solution and kept in a porous cover which was facilitated with a pressure sensor. Upon excess carbon dioxide generation, a reaction occurs with bicarbonate solution resulting in lowering of the pH. This reduction in pH causes the hydrogel to swell but the porous cover limits the space, which in turn generates pressure. The pressure sensor records and compares it with the carbon dioxide partial pressure, giving the analysis [46]. Likewise, Tagit et al. developed nano-thermometers based on poly(N-isopropylacrylamide) smart polymer. Luminescent quantum dots were affixed with this polymer and assembled over a gold substrate. This polymer was a thermo-sensitive polymer, therefore, when the temperature was elevated above the LCST, the luminescence of the quantum dots was quenched and then reversed upon decreasing the temperature [47].

5.3   FLEXIBLE ELECTRONIC DEVICES

Owing to the fact that flexible electronic devices are portable, wearable, and can be user friendly, there has been a significant demand for their fabrication and applications. Electronic devices ranging from capacitive touch screens to smart electronic skins, energy harvesting (fuel cells) to energy storage (supercapacitors), bioelectronic devices for implants, etc. have been fabricated. The major advantage of these types of devices is that they can perform with the same precision even after stretching, folding, and bending. Basically, the base material used plays the key role in enhancing the performance and functionality. Soft, moldable, smart polymeric materials with special structural and geometrical configuration are used. Some of the commonly used polymers are polyimide (PI), poly(ethylene terephthalate) (PET), polypyrrole (PPy), polyaniline (PANI), polydimethylsiloxane

(PDMS), polythiophene (PT), etc. Conductive and conjugated polymers showcase tunable electronic properties, which makes them an ideal material for device fabrication. Recently, Gong et al. gave a detailed review of polymer nanocomposites-based flexible electronic devices. Herein, they discussed the fabrication, several broad-spectrum applications, and the challenges for real-time use of these devices [48].

Conductive hydrogels have also proven to be a significant substrate material for flexible devices due to their excellent mechanical bendability, electronic conductivity, and simple preparation. Further, they are also biocompatible in nature and have salient features such as antimicrobial, self-healing, and self-adhesion properties. Hence, they have been employed as wearable health-monitoring devices. Conductive polymers, metals, ions, and carbon-based hydrogels are commonly prepared. Some of the applications include the following:

1 *Flexible touch pads*: for instance, silver nanowire and PEDOT:PSS conductive polymer nanocomposite was reported as an excellent touch panel. Similarly, polyacrylamide hydrogel embedded with lithium salt composite was also explored in this context. Mechanocromic hydrogels are another category of smart material demonstrated for capacitive touch pad application.
2. *Energy storage*: energy storage as batteries and supercapacitors is due to the functionality of the electrodes in them. Therefore, several materials for developing flexible electrodes have been investigated. For example, paper-based carbon nanotube electrodes, Li ion paper battery, Si nanoparticles embedded in polyaniline 3D porous hydrogel, metal nanoparticles like TiO_2, Si, or Fe_3O_4 in polypyrrole hydrogel have been used for battery fabrication. Similarly, graphene hydrogels, hybrid hydrogel, i.e., the combination of carbon and polymer hydrogel, supercapacitors have been reported.
3. *Sensors*: net-shaped, coiled, and ripple-shaped circuit-based sensors were designed for measuring several physical parameters in fields such as robotics, wearable medical devices, mechanical and electronics using conductive hydrogels. Conductive hydrogels with fillers were generally used, for instance, chitosan, polyvinyl alcohol, organo-hydrogels, ethylene glycols, glycerol with carbon nanotubes, etc. [49].

6 CONCLUSIONS AND FUTURE OUTLOOK

In recent years, the development of flexible smart polymer materials and their widespread applications has witnessed tremendous progress. The advancement of interdisciplinary research has led to the fabrication of highly efficient flexible devices for biomedical diagnosis, bioelectronics, electronic skin, flexible biosensors, energy devices, etc. Owing to these emerging applications, there has been a remarkable impact on daily life. Smart polymers, polymer composites, and polymer hydrogels that exhibit a flexible nature, and viable and changeable chemical and physical properties are promising materials for the fabrication of these devices. Smart polymers research, being sensitive toward various stimuli, is being focused to manipulate their properties as per the need of the application. Further, polymeric materials are being embedded with various kinds of fillers, including metallic and conductive materials, to increase their real-time applicability. Diverse approaches for the preparation of these polymers have been adapted for integration into flexible electronic devices. In this chapter, the classification of smart polymers based on chemical, physical, and biochemical properties is described with some common examples. General wet chemical preparation methods are also highlighted. Fabrication and incorporation of these polymers into microdevices via methods like printing, lithography, etc. are also discussed. Some of the remarkable reported flexible devices in the biomedical, biosensor, and energy sectors have been mentioned briefly here. Despite their significant growth, the clinical applications have limitations as biocompatibility is a major issue. Medical devices requiring implantations in the physiological system need to be biodegradable and non-toxic.

In future, natural polymer-based devices could resolve this issue to a certain extent. Likewise, the implantable biosensors need to be improved in durability as presently these devices cannot be used for a long time. Strategies for selective multiple analyte detection and stability have to be explored in future. Micro-bioactuator devices also require a detailed study about their mechanism. Further, miniaturized and microfluidic platforms with real-time multiplexed analyte quantification and wearability of these devices remains challenging. The other challenges to be worked on in future include cost-effectiveness, less complexity, and ease of operation.

REFERENCES

[1] B. Jeong, A. Gutowska, Lessons from nature: Stimuli-responsive polymers and their biomedical applications, Trends Biotechnol. 20 (2002) 305–311.

[2] A.S. Hoffman, P.S. Stayton, V. Bulmus, G. Chen, J. Chen, C. Cheung, A. Chilkoti, Z. Ding, L. Dong, R. Fong, C.A. Lackey, C.J. Long, M. Miura, J.E. Morris, N. Murthy, Y. Nabeshima, T.G. Park, O.W. Press, T. Shimoboji, S. Shoemaker, H.J. Yang, N. Monji, R.C. Nowinski, C.A. Cole, J.H. Priest, J.M. Harris, K. Nakamae, T. Nishino, T. Miyata, Really smart bioconjugates of smart polymers and receptor proteins, J. Biomed. Mater. Res. 52 (2000) 577–586.

[3] A. Kikuchi, T. Okano, Intelligent thermoresponsive polymeric stationary phases for aqueous chromatography of biological compounds, Prog. Polym. Sci. 27 (2002) 1165–1193.

[4] B.R. Twaites, C. De Las Heras Alarcón, D. Cunliffe, M. Lavigne, S. Pennadam, J.R. Smith, D.C. Górecki, C. Alexander, Thermo and pH responsive polymers as gene delivery vectors: Effect of polymer architecture on DNA complexation in vitro, J. Control. Release. 97 (2004) 551–566.

[5] N. Lomadze, H.J. Schneider, Ternary complex formation inducing large expansions of chemomechanical polymers by metal chelators, aminoacids and peptides as effectors, Tetrahedron Lett. 46 (2005) 751–754.

[6] L. Leclercq, M. Boustta, M. Vert, A physico-chemical approach of polyanion – Polycation interactions aimed at better understanding the in vivo behaviour of polyelectrolyte-based drug delivery and gene transfection, J. Drug Target. 11 (2003) 129–138.

[7] M. Zrinyi, Intelligent polymer gels controlled by magnetic fields, Colloid Polym. Sci. 278 (2000) 98–103.

[8] G. Filipcsei, J. Fehér, M. Zrínyi, Electric field sensitive neutral polymer gels, J. Mol. Struct. 554 (2000) 109–117.

[9] S. Juodkazis, N. Mukai, R. Wakaki, A. Yamaguchi, S. Matsuo, H. Misawa, Reversible phase transitions in polymer gels induced by radiation forces, Nature. 408 (2000) 178–181.

[10] C.L. Hastings, H.M. Kelly, M.J. Murphy, F.P. Barry, F.J. O'Brien, G.P. Duffy, Development of a thermoresponsive chitosan gel combined with human mesenchymal stem cells and desferrioxamine as a multimodal pro-angiogenic therapeutic for the treatment of critical limb ischaemia, J. Control. Release. 161 (2012) 73–80.

[11] W.K. Bae, M.S. Park, J.H. Lee, J.E. Hwang, H.J. Shim, S.H. Cho, D.E. Kim, H.M. Ko, C.S. Cho, I.K. Park, I.J. Chung, Docetaxel-loaded thermoresponsive conjugated linoleic acid-incorporated poloxamer hydrogel for the suppression of peritoneal metastasis of gastric cancer, Biomaterials. 34 (2013) 1433–1441.

[12] M. Licciardi, G. Amato, A. Cappelli, M. Paolino, G. Giuliani, B. Belmonte, C. Guarnotta, G. Pitarresi, G. Giammona, Evaluation of thermoresponsive properties and biocompatibility of polybenzofulvene aggregates for leuprolide delivery, Int. J. Pharm. 438 (2012) 279–286.

[13] M.H. Hsiao, M. Larsson, A. Larsson, H. Evenbratt, Y.Y. Chen, Y.Y. Chen, D.M. Liu, Design and characterization of a novel amphiphilic chitosan nanocapsule-based thermo-gelling biogel with sustained in vivo release of the hydrophilic anti-epilepsy drug ethosuximide, J. Control. Release. 161 (2012) 942–948.

[14] K. Li, L. Yu, X. Liu, C. Chen, Q. Chen, J. Ding, A long-acting formulation of a polypeptide drug exenatide in treatment of diabetes using an injectable block copolymer hydrogel, Biomaterials. 34 (2013) 2834–2842.

[15] K.H.M. Kan, J. Li, K. Wijesekera, E.D. Cranston, Polymer-grafted cellulose nanocrystals as pH-responsive reversible flocculants, Biomacromolecules. 14 (2013) 3130–3139.

[16] A. Chan, R.P. Orme, R.A. Fricker, P. Roach, Remote and local control of stimuli responsive materials for therapeutic applications, Adv. Drug Deliv. Rev. 65 (2013) 497–514.

[17] C.L. Peng, L.Y. Yang, T.Y. Luo, P.S. Lai, S.J. Yang, W.J. Lin, M.J. Shieh, Development of pH sensitive 2-(diisopropylamino)ethyl methacrylate based nanoparticles for photodynamic therapy, Nanotechnology. 21 (2010).

[18] C. Ohm, M. Brehmer, R. Zentel, Liquid crystalline elastomers as actuators and sensors, Adv. Mater. 22 (2010) 3366–3387.

[19] H. Palza, P.A. Zapata, C. Angulo-Pineda, Electroactive smart polymers for biomedical applications, Materials (Basel). 12 (2019).

[20] R. Balint, N.J. Cassidy, S.H. Cartmell, Conductive polymers: Towards a smart biomaterial for tissue engineering, Acta Biomater. 10 (2014) 2341–2353.

[21] V. Ravaine, C. Ancla, B. Catargi, Chemically controlled closed-loop insulin delivery, J. Control. Release. 132 (2008) 2–11.

[22] J. Hu, G. Zhang, S. Liu, Enzyme-responsive polymeric assemblies, nanoparticles and hydrogels, Chem. Soc. Rev. 41 (2012) 5933–5949.

[23] B. Aïssa, D. Therriault, E. Haddad, W. Jamroz, Self-healing materials systems: Overview of major approaches and recent developed technologies, Adv. Mater. Sci. Eng. 2012 (2012).

[24] Y. Li, G. Huang, X. Zhang, B. Li, Y. Chen, T. Lu, T.J. Lu, F. Xu, Magnetic hydrogels and their potential biomedical applications, Adv. Funct. Mater. 23 (2013) 660–672.

[25] S.U.-N. Toyoichi Tanaka, Izumi Nishio, Shao-Tang Sun, Repors, Dep. Phys. Cent. Mater. Sci. Eng. Massachusetts Inst. Technol. Cambridge 02139. 218 (1982) 4–6.

[26] L. Peponi, A. Tercjak, L. Martin, I. Mondragon, J.M. Kenny, Morphology-properties relationship on nanocomposite films based on poly(styrene-block-diene-block-styrene) copolymers and silver nanoparticles, Express Polym. Lett. 5 (2011) 104–118.

[27] M. Nadgorny, A. Ameli, Functional polymers and nanocomposites for 3D printing of smart structures and devices, ACS Appl. Mater. Interfaces. 10 (2018) 17489–17507.

[28] A. Plecis, Y. Chen, Fabrication of microfluidic devices based on glass-PDMS-glass technology, Microelectron. Eng. 84 (2007) 1265–1269.

[29] B.K. Gale, A.R. Jafek, C.J. Lambert, B.L. Goenner, H. Moghimifam, U.C. Nze, S.K. Kamarapu, A review of current methods in microfluidic device fabrication and future commercialization prospects, Inventions. 3 (2018).

[30] W. Small IV, P. Singhal, T.S. Wilson, D.J. Maitland, Biomedical applications of thermally activated shape memory polymers, J. Mater. Chem. 20 (2010) 3356–3366.

[31] E. Smela, Conjugated polymer actuators for biomedical applications, Adv. Mater. 15 (2003) 481–494.

[32] D.J. Maitland, M.F. Metzger, D. Schumann, A. Lee, T.S. Wilson, Photothermal properties of shape memory polymer micro-actuators for treating stroke, Lasers Surg. Med. 30 (2002) 1–11.

[33] W. Sokolowski, Shape memory polymer foams for biomedical devices, Open Med. Devices J. 2 (2010) 20–23.

[34] K. Gall, C.M. Yakacki, Y. Liu, R. Shandas, N. Willett, K.S. Anseth, Thermomechanics of the shape memory effect in polymers for biomedical applications, J. Biomed. Mater. Res. – Part A. 73 (2005) 339–348.

[35] H.M. Wache, D.J. Tartakowska, A. Hentrich, M.H. Wagner, Development of a polymer stent with shape memory effect as a drug delivery system, J. Mater. Sci. Mater. Med. 14 (2003) 109–112.

[36] S.H. Huang, R.S. Juang, Biochemical and biomedical applications of multifunctional magnetic nanoparticles: A review, J. Nanoparticle Res. 13 (2011) 4411–4430.

[37] S. Purushotham, R. V. Ramanujan, Thermoresponsive magnetic composite nanomaterials for multimodal cancer therapy, Acta Biomater. 6 (2010) 502–510.

[38] H. Lee, E. Lee, D. Kim, N. Jang, Y. Jeong, S. Jon, Antibiofouling polymer-coated superparamagnetic iron oxide nanoparticles as potential magnetic …, J. Am. Chem. Soc. (2006) 7383–7389.

[39] F. Hu, L. Wei, Z. Zhou, Y. Ran, Z. Li, M. Gao, Preparation of biocompatible magnetite nanocrystals for in vivo magnetic resonance detection of cancer, Adv. Mater. 18 (2006) 2553–2556.

[40] I. Tokarev, S. Minko, Stimuli-responsive hydrogel thin films, Soft Matter. 5 (2009) 511–524.

[41] E.J. Geiger, A.P. Pisano, F. Svec, A polymer-based microfluidic platform featuring on-chip actuated hydrogel valves for disposable applications, J. Microelectromechanical Syst. 19 (2010) 944–950.

[42] A.K. Agarwal, S.S. Sridharamurthy, D.J. Beebe, H. Jiang, Programmable autonomous micromixers and micropumps, J. Microelectromechanical Syst. 14 (2005) 1409–1421.

[43] G.S. Wilson, R. Gifford, Biosensors for real-time in vivo measurements, Biosens. Bioelectron. 20 (2005) 2388–2403.

[44] C.J. Huang, Y.H. Chen, C.H. Wang, T.C. Chou, G. Bin Lee, Integrated microfluidic systems for automatic glucose sensing and insulin injection, Sensors Actuators, B Chem. 122 (2007) 461–468.

[45] A.S. Hoffman, P.S. Stayton, Conjugates of stimuli-responsive polymers and proteins, Prog. Polym. Sci. 32 (2007) 922–932.

[46] S. Herber, W. Olthuis, P. Bergveld, A. Van Den Berg, Exploitation of a pH-sensitive hydrogel for CO_2 detection, Sensors and Actuators B: Chemical 103 (2004) 284–289.

[47] O. Tagit, N. Tomczak, E.M. Benetti, Y. Cesa, C. Blum, V. Subramaniam, J.L. Herek, G. Julius Vancso, Temperature-modulated quenching of quantum dots covalently coupled to chain ends of poly(N-isopropyl acrylamide) brushes on gold, Nanotechnology. 20 (2009).

[48] M. Gong, L. Zhang, P. Wan, Polymer nanocomposite meshes for flexible electronic devices, Prog. Polym. Sci. 107 (2020) 101279.

[49] Q. Rong, W. Lei, M. Liu, Conductive hydrogels as smart materials for flexible electronic devices, Chem. – A Eur. J. 24 (2018) 16930–16943.

Index

For Product Safety Concerns and Information please contact our EU
representative GPSR@taylorandfrancis.com
Taylor & Francis Verlag GmbH, Kaufingerstraße 24, 80331 München, Germany

www.ingramcontent.com/pod-product-compliance
Lightning Source LLC
Chambersburg PA
CBHW080124220326
41598CB00032B/4945